APPLIED MATHEMATICS–II

FOR POLYTECHNIC STUDENTS

Dr. Hari Arora

APPLIED MATHEMATICS–II

Dr. Hari Arora
M.Sc. (Gold Medalist), M.B.A. (Marketing)
M.Phil (Operations Research), Ph.D. (Applied Mathematics)
Professor
Department of Applied Mathematics
Amity Institute of Applied Sciences,
Amity University Uttar Pradesh, Noida

Dr. Anjali Naithani
Department of Applied Mathematics
Amity Institute of Applied Sciences,
Amity University Uttar Pradesh, Noida

S.K. KATARIA & SONS®

Publisher of Engineering Books
4885/109, Prakash Mahal, Dr. Subhash Bhargav Lane,
Opposite Delhi Medical Association, Daryaganj, New Delhi-110002 (India)
Tel.: +91-11-43551243, Mobile: +91-9871775858, 7678425209
Email: katariabook@gmail.com, katariabooks@yahoo.com
Website: www.skkatariaandsons.com
(e-books, subscriptions and bulk buying also available on our website)

Published By :

4885/109, Prakash Mahal, Dr. Subhash Bhargav Lane,
Opposite Delhi Medical Association, Daryaganj, New Delhi-110002 (India)
Tel.: +91-11-43551243, Mobile: +91-9871775858, 7678425209
Email: katariabook@gmail.com, katariabooks@yahoo.com
Website: www.skkatariaandsons.com

First Edition: 2008, Fourth Edition: 2022, Reprint: 2024

Typesetting at: Goswami Associates

Printed at: AL-Aziz Book Binding

Word to the Readers

It gives me great pleasure in presenting the fourth edition of this book. In this edition modifications have been done for covering various syllabuses of Indian universities offering Mathematics course for polytechnic students for the second semester. In this edition great emphasis has been placed on the fundamental concepts. A large number of illustrative examples have been provided on each concept. Exercises and their answers have been organized very carefully.

First and Foremost, I wish to acknowledge my profound and deep regards to **Dr. Ashok K. Chauhan,** Founder President, Amity University for his constant encouragement and inspiration. I am also indebted to **Prof. (Dr.) Balvinder Shukla,** Vice Chancellor, Amity University, Noida for her help and suggestions to develop confidence in me and also leading me during writing me this book. I am also obliged to **Prof. (Dr.) Sunita Rattan,** Director, Amity Institute of Applied Sciences, Amity University, Noida for providing me necessary counselling and support during writing this book. It has been a pleasure working with the staff of **M/s S.K. Kataria & Sons** publisher who has shown their usual care, concern, untiring sincere efforts, cooperation and enthusiasm in bringing out this book of quality in time.

I am deeply indebted to my wife **Dr. Anita Arora**, daughter **Sadhika** and son **Tanishq** for encroachment upon their time because the book was mostly written during the moments which were exclusively for them.

Last but not the least, I would like to thank my parents **Dr. G.R. Arora** and **Mrs. Tara Arora** for their sound counselling and cheerful support. They have always inspired me and kept my spirit up. Without their blessings, this task of writing a book could never been completed.

The coauthor of the book **Dr. Anjali Naithani** takes this opportunity to record the heart felt gratitude and affection to her family members specially her husband **Mr. Gaurav Arora**, her daughter **Yashvini** and son **Garveet** for their endurance and moral support which greatly encouraged her in the effective completion of the book.

I have honestly and sincerely tried in the book to serve the purpose of a complete text which could be useful for the teachers as well as the students. I have also tried to remove all errors and misprints which escaped from my notice in the last edition. Suggestions for further improvements of the book will be gratefully received and acknowledged.

Dr. Hari Arora

Dedicated to

My Reverend Parents *Dr. G.R. Arora* and *Mrs. Tara Arora* whose sense of responsibility is a permanent source of inspiration to me

&

My Beloved wife *Dr. Anita Arora*, Darling Daughter *Sadhika* and Son *Tanishq* for their appreciative understanding

Contents

UNIT - I

ALGEBRA

1. Complex Numbers
2. Matrices and Determinants
3. Vector Algebra

CHAPTER 1

Complex Numbers

INSIDE THIS CHAPTER

1.1. INTRODUCTION

We know that the square of a real number is non-negative, thus no real number can satisfy an equation of the type $x^2 + k = 0$, $k > 0$. This shows the inadequancy of the real number system R.

Euler was the first mathematician to introduce the symbol i (iota) for the square root of – 1 with the property of $i^2 = -1$. He also called this symbol as the imaginary unit. We have, for example,

$$\sqrt{-9} = \sqrt{-1}\cdot\sqrt{9} = i \cdot 3 = 3i$$

$$\sqrt{-49} = \sqrt{-1}\cdot\sqrt{49} = i \cdot 7 = 7i$$

1.2. INTEGRAL POWERS OF *i*

Since $\sqrt{-1} = i$, we have $i^2 = (\sqrt{-1})^2 = -1$

Further, $$i^3 = i^2 \times i = (-1) \times i = -i$$

and $$i^4 = (i^2)^2 = (-1)^2 = 1$$

Let n be a positive integer greater than 4. Then n can be expressed as

$$n = 4m + r$$

where m is the quotient obtained on dividing n by 4, and r is the remainder. Obviously $0 \le r < 4$.

This give us $$i^n = i^{4m+r} = (i^{4m}) \cdot (i^r) = (1)^m \cdot i^r = i^r$$

Thus, any higher power of i can be expressed in terms of one of the four values viz. i, 1, $-i$ and 1.

1.3. COMPLEX NUMBERS

A number of the form $a + ib$, where a and b are real number and $i = \sqrt{-1}$, is called a complex number. The complex number $a + ib$ is usually denoted by z. The set of complex numbers is denoted by C.

If $z = a + ib$ is a complex number, than a is called the real part of z and we write $\text{Re}(z) = a \cdot b$ is called the imaginary part of z and we write $\text{Im}(z) = b$.

If $b = 0$, then $z = a + i \cdot 0 = a$ is a real number

If $a = 0$ and $b = 0$, then $z = 0 + i0 = 0$ called the zero complex number.

We observe that a complex number $a + ib$, can be uniquely determined by an ordered pair (a, b) of real number in which the first member is the imaginary part of the complex number.

1.4. EQUALITY OF COMPLEX NUMBERS

Two complex number $z_1 = x_1 + iy_1$ and $z_2 = x_2 + iy_2$ are equal if their real and imaginary parts are equal that is $x_1 = x_2$ and $y_1 = y_2$

For example, if $z_1 = -2 + iy$ and $z_2 = -x - 3i$ are equal, then $x = 2$ and $y = -3$.

1.5. CONJUGATE OF A COMPLEX NUMBERS

If $z = x + iy$, $x, y \in R$, is complex number, then the complex number $x - iy$ is called the conjugate of the complex number $x + iy$. The conjugate of complex number z is denoted by $\bar{z}$.

Thus, we have $\overline{x+iy} = x - iy$

If z_1 is conjugate of z_2, then z_2 is conjugate of z_1. The conjugate of complex numbers differ only in sign of their imaginary parts.

For example, the complex number $3 - 5i$ and $3 + 5i$ are conjugate of each other, we may write

$$\overline{3-5i} = 3 + 5i \quad \text{and} \quad \overline{3+5i} = 3 - 5i$$

1.6. ALGEBRA OF COMPLEX NUMBERS

We shall now discuss the operations of addition, subtraction multiplication and division of complex numbers.

1.6.1. Addition of Complex Numbers

If $z_1 = a_1 + ib_1$ and $z_2 = a_2 + ib_2$ be two complex numbers, then their sum $z_1 + z_2$ is the number $(a_1 + a_2) + i(b_1 + b_2)$

It is, therefore, observed that while adding two complex numbers the real and imaginary parts of the sum is obtained by adding the real and imaginary parts of the summands.

1.6.2. Subtraction of Complex Numbers

If $z_1 = a_1 + ib_1$ and $z_2 = a_2 + ib_2$ be two complex numbers, then their difference $z_1 - z_2$ is the number $z_1 + (-z_2)$

Symbolically

$$\begin{aligned} z_1 - z_2 &= z_1 + (-z_2) \\ &= (a_1 + ib_1) + (-a_2 - ib_2) = (a_1 - a_2) + i(b_1 - b_2) \end{aligned}$$

1.6.3. Multiplication of Complex Numbers

The product of two complex number $z_1 = a_1 + ib_1$ and $z_2 = a_2 + ib_2$ is defined as a complex number obtained by the multiplication of these two numbers as binomials governed by the rules of algebra and substituting -1 for i^2 we have

$$\begin{aligned} z_1z_2 &= (a_1 + ib_1)(a_2 + ib_2) = a_1a_2 + ia_1b_2 + ib_1a_2 + i^2b_1b_2 \\ &= (a_1a_2 - b_1b_2) + i(a_1b_2 + a_2b_2) \end{aligned}$$

1.6.4. Division of a Complex Numbers

If $z_1 = a + ib$ and $z_2 = c + id$ be any two complex number then the quotient $\frac{z_1}{z_2}$ is defined as $z_1 z_2^{-1}$, where z_2^{-1} is the multiplicative inverse of z_2. In other words,

$$\frac{z_1}{z_2} = \frac{a+ib}{c+id} = (a+ib)\left(\frac{c}{c^2+d^2} - \frac{id}{c^2+d^2}\right)$$

$$= \left(\frac{ac}{c^2+d^2} + \frac{bd}{c^2+d^2}\right) + i\left(\frac{bc}{c^2+d^2} - \frac{ad}{c^2+d^2}\right)$$

$$\frac{z_1}{z_2} = \left(\frac{ac+bd}{c^2+d^2}\right) + i\left(\frac{bc-ad}{c^2+d^2}\right)$$

1.7. GRAPHICAL REPRESENTATION OF A COMPLEX NUMBER

Consider a system of rectangular coordinate axes X′OX and Y′OY. Let $z = x + iy$ be a complex number. It can be uniquely determined by the ordered pair (x, y) or real number. Since their exists one-one correspondence between the ordered pairs of real numbers and the points on a plane, every complex number $z = x + iy$ can be presented by a unique point P(x, y) with regard to two rectangular axes X′OX and Y′OY.

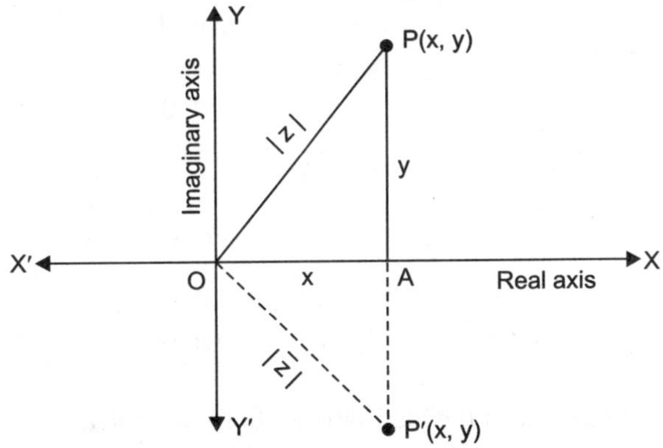

The plane representing complex number as ordered pairs of real numbers is called the complex plane or *Argand plane* or *Gausian plane*. The representation of a complex number is due to Argand. The diagram drawn above to represent a complex number is known as *Argand Diagram*.

The point $(x, 0)$ on the x-axis represent complex number $x + i0$, *i.e.*, the real number x and every real number is represented as a point on the axis. As such x-axis is called the real axis. In fact, positive number are represented by points on the positive x-axis while the negative real numbers by points on the negative x-axis and the real number 0 by the origin. Similarly, the point $(0, y)$ on the y-axis represents complex numbers $0 + iy$, *i.e.*, the imaginary number iy. Therefore y-axis is called the imaginary axis. All imaginary numbers are represented as points on this axis.

The distance from the origin to the point P(x, y) is defined as the modulus (or absolute value) of the complex number $z = x + iy$ and is denoted by $|z|$, *i.e.*

$$|z| = \sqrt{x^2 + y^2}$$

The conjugate $\bar{z}$ of the complex number z is represented by the point P′ which is symmetric to P with respect to x-axis *i.e.*, P′ is the mirror image of P on the x-axis as shown.

1.8. POLAR FORM OF A COMPLEX NUMBER

Every complex number $x + iy$ can be written in the form $r(\cos\theta + i\sin\theta)$. r is called the modulus and θ is called the amplitude or argument and it is written as

$$\theta = \text{amp } z \quad \text{or} \quad \arg z$$

The complex number $x + iy$ is called cartesian form and $r(\cos\theta + i\sin\theta)$ is called the polar form.

Let $$x + iy = r(\cos\theta + i\sin\theta) = r\cos\theta + i\cdot r\sin\theta$$

Equating the real parts and imaginary parts

$$r\cos\theta = x \quad ...(i)$$

$$r\sin\theta = y \quad ...(ii)$$

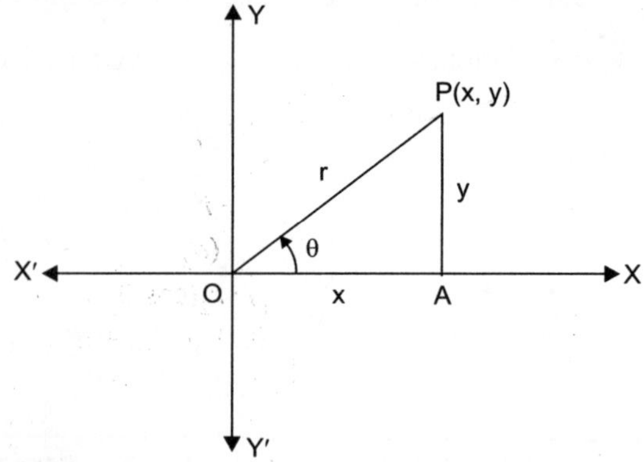

i.e., $$r^2 = x^2 + y^2$$

$\therefore$ $$r = \sqrt{x^2 + y^2}$$

Thus the modulus of the complex number is

$$r = \sqrt{x^2 + y^2}$$

It is denoted by $|z|$ or $|x + iy|$

$\therefore$ $$r = |z|$$

$$r = |x + iy| = \sqrt{x^2 + y^2}$$

Also $$\tan\theta = \frac{y}{x} \quad \text{[Dividing (ii) by (i)]}$$

$\Rightarrow$ $$\theta = \arg z = \arg(x + iy) = \tan^{-1}\frac{y}{x}$$

Also from (*i*) and (*ii*), we get

$$\cos\theta = \frac{x}{r} = \frac{x}{\sqrt{x^2 + y^2}} \quad ...(iii)$$

$$\sin\theta = \frac{y}{r} = \frac{y}{\sqrt{x^2 + y^2}} \quad ...(iv)$$

The value of θ which satisfies both (*iii*) and (*iv*) and which is such that $-\pi < \theta \le \pi$, is called the principle value of the amplitue of $x + iy$ and it is written as amp $(x + iy)$ or amp (z). 'θ' is also called argument of $(x + iy)$ and is written as arg $(x + iy)$ or arg (z). If we use the general value of the argument θ, then the polar form of z is

$$z = r[\cos(2n\pi + \theta) + i\sin(2n\pi + \theta)],\ n \in z.$$

1.9. EULERIAN FORM OF A COMPLEX NUMBER

Let $z = x + iy$ be a complex number with

$$|z| = r \quad \text{and} \quad \arg(z) = \theta. \text{ Then}$$

$$z = r(\cos\theta + i\sin\theta) \quad ...(i)$$

Using the Euler's formula $\cos\theta + i\sin\theta = e^{i\theta}$, (*i*) becomes

$$\boxed{z = re^{i\theta} \quad \text{or} \quad z = |z|e^{i\arg(z)}}$$

This form is called the Eulerian form or Exponential form of z

Note: Since $\cos\theta + i\sin\theta = e^{i\theta}$. Therefore

$$e^{-i\theta} = \cos(-\theta) + i\sin(-\theta) = \cos\theta - i\sin\theta$$

SOLVED EXAMPLES

Example 1.1. Express the following in the form $a + ib$:

(*i*) $\dfrac{(1+i)(1-2i)}{1+3i}$ (*ii*) $\dfrac{2-\sqrt{3}i}{1+i}+\dfrac{3+i}{(1-i)^2}$

Solution. (*i*) We have

$$\frac{(1+i)(1-2i)}{1+3i} = \frac{1+i-2i-2i^2}{1+3i},$$

$$= \frac{1-i+2i}{1+3i} = \frac{3-i}{1+3i} = \frac{3-i}{1+3i}\times\frac{1-3i}{1-3i} = \frac{3-i-9i+3i^2}{1-9i^2}$$

$$= \frac{3-i-9i+3i^2}{1+9} = \frac{1}{10}(-10i) = -i = 0-1i$$

(*ii*) We have

$$\frac{2-\sqrt{3}i}{1+i}+\frac{3+i}{(1-i)^2} = \frac{2-\sqrt{3}i}{1+i}+\frac{3+i}{1-2i+i^2}$$

$$= \frac{(2-\sqrt{3}i)(1-i)}{(1+i)(1-i)}-\frac{(3+i)}{2i} = \frac{(2-\sqrt{3})-i(\sqrt{3}+2)}{2}-\frac{(3+i)i}{2i^2}$$

$$= \frac{(2-\sqrt{3})-i(\sqrt{3}+2)}{2}+\left(\frac{3i+i^2}{2i}\right) = \frac{(2-\sqrt{3})-i(\sqrt{3}+2)}{2}+\left(\frac{3i-1}{2}\right)$$

$$= \left(\frac{1-\sqrt{3}}{2}\right)+i\left(\frac{1-\sqrt{3}}{2}\right)$$

Example 1.2. Find the modulus and amplitude of $\dfrac{(1+i)(1+\sqrt{3}i)}{1-i}$.

Solution. Let
$$z = \frac{(1+i)(1+\sqrt{3}i)}{1-i}$$

Then
$$z = \frac{(1+i)(1+\sqrt{3}i)}{1-i}\times\frac{1+i}{1+i}$$

$$= \frac{(1+i)^2(1+\sqrt{3}i)}{1^2-i^2} = \frac{(1+i^2+2i)(1+\sqrt{3}i)}{1+1}$$

$$= \frac{(1-1+2i)(1+\sqrt{3}i)}{2} = i(1+\sqrt{3}i) = i+\sqrt{3}i^2 = -\sqrt{3}+i$$

$\therefore$
$$|z| = \sqrt{(-\sqrt{3})^2+(1)^2} = \sqrt{3+1} = \sqrt{4} = 2$$

and
$$\text{amp } z = \tan^{-1}\left(\frac{1}{\sqrt{3}}\right) = \pi+\frac{\pi}{6} = \frac{7\pi}{6}$$

Example 1.3. Express the complex number in the polar form:

(*i*) $\dfrac{2+6\sqrt{3}i}{6+\sqrt{3}i}$ (*ii*) $\dfrac{(2+5i)(-3+i)}{(1-2i)^2}$

Solution. (*i*) Let $z = \dfrac{2+6\sqrt{3}i}{5+\sqrt{3}i}$, and let r (cos θ + *i* sin θ) be the polar form of z. Then $r = |z|$ and θ = arg (z)

Now, $$z = \frac{2+6\sqrt{3}i}{5+\sqrt{3}i}$$

⇒ $$z = \frac{2+6\sqrt{3}i}{5+\sqrt{3}i}\cdot\frac{(5-\sqrt{3}i)}{(5+\sqrt{3}i)} = \frac{28+28\sqrt{3}i}{28} = 1+i\sqrt{3}$$

∴ $$|z| = r = \sqrt{1+3} = 2$$

As $$r\cos\theta + ir\sin 1+i\sqrt{3} = 1+i\sqrt{3}$$

∴ $$r\cos\theta = 1 \quad\text{and}\quad r\sin\theta = \sqrt{3}$$

⇒ $$r\cos\theta = \frac{1}{2} \quad\text{and}\quad \sin\theta = \frac{\sqrt{3}}{2}$$

⇒ $$\theta = \frac{\pi}{3}$$

Hence, $1+i\sqrt{3} = 2\left(\cos\dfrac{\pi}{3} + i\sin\dfrac{\pi}{3}\right)$

(*ii*) Let $z = \dfrac{(2+5i)(-3+i)}{(1-2i)^2}$ and let r (cos θ + *i* sin θ) be the polar form of z. Then $r = |z|$ and θ = arg (z)

Now, $$z = \frac{(2+5i)(-3+i)}{(1-2i)^2} = \frac{(2+5i)(-3+i)}{1-4i+4i^2}$$

$$= \frac{-11-13i}{1-4i-4} = \frac{-11-13i}{-3-4i} = \frac{11+3i}{3+4i} = \frac{11+3i}{3+4i}\times\frac{3-4i}{3-4i}$$

$$= \frac{(33+52)+i(39-44)}{9+16} = \frac{85-5i}{25} = \frac{17}{5}-\frac{1}{5}i$$

$$|z| = r = \sqrt{\left(\frac{17}{5}\right)^2+\left(\frac{-1}{5}\right)^2} = \sqrt{\frac{290}{5}}$$

As $$r\cos + ir\sin\theta = \frac{17}{5}-\frac{1}{5}i$$

⇒ $$r\cos\theta = \frac{17}{5} \quad\text{and}\quad r\sin\theta = -\frac{1}{5}$$

⇒ $$\cos\theta = \frac{17}{\sqrt{290}} \quad\text{and}\quad \sin\theta = -\frac{1}{\sqrt{290}}$$

⇒ $$\tan\theta = -\frac{1}{17}$$

As the point $\left(\frac{17}{5}, -\frac{1}{5}\right)$ lies in the fourth quadrant.

$$\therefore \qquad \arg(z) = -\theta = -\tan^{-1}\left(-\frac{1}{17}\right)$$

$$\text{Hence,} \qquad \frac{17}{5} - \frac{1}{5}i = \frac{\sqrt{290}}{5}(\cos\theta + i\sin\theta) \qquad \text{where } \theta = -\tan^{-1}\left(-\frac{1}{17}\right)$$

Example 1.4. Find the smallest positive integer n for which $\left(\frac{1+i}{1-i}\right)^n = 1$.

Solution. We have
$$\frac{1+i}{1-i} = \frac{(1+i)(1+i)}{(1-i)(1-i)} = \frac{1+2i+i^2}{1-i^2} = \frac{1+2i-1}{1+1} = i$$

$$\therefore \qquad \left(\frac{1+i}{1-i}\right)^n = 1 \quad \Rightarrow \quad i^m = 1$$

$\Rightarrow$ n is a multiple of 4

So, the smallest positive integral value of n is 4.

Example 1.5. If $(a_1 + ib_1)(a_2 + ib_2)(a_3 + ib_3) \ldots\ldots (a_n + ib_n) = A + iB$, prove that

(*a*) $(a_1^2 + b_1^2)(a_2^2 + b_2^2)(a_3^2 + b_3^2) \ldots\ldots (a_n^2 + b_n^2) = A^2 + B^2$ and

(*b*) $\tan^{-1}\left(\frac{b_1}{a_1}\right) + \tan^{-1}\left(\frac{b_2}{a_2}\right) + \ldots\ldots + \tan^{-1}\left(\frac{b_n}{a_n}\right) = \tan^{-1}\left(\frac{B}{A}\right)$

Solution. Let $\qquad a_k = r_k \cos\theta_k,\ b_k = r_k \sin q_k;\ k = 1, 2, \ldots\ldots, n$

Then $\qquad r_k^2 = a_k^2 + b_k^2,\ \theta_k = \tan^{-1}\left(\frac{b_k}{a_k}\right)$

$$a_k + ib_k = r_k(\cos\theta_k + i\sin\theta_k);\ k = 1, 2 \ldots\ldots, n$$

As, $\quad (a_1 + ib_1)(a_2 + ib_2)(a_3 + ib_3) \ldots. (a_n + ib_n) = A + iB$

$\Rightarrow \quad r_1(\cos\theta_1 + i\sin\theta_1) \cdot r_2(\cos\theta_2 + i\sin\theta_2) \cdot r_3(\cos\theta_3 + i\sin\theta_3) \ldots\ldots r_n(\cos\theta_n + i\sin\theta_n) = A + iB$

$\Rightarrow \quad r_1 r_2 r_3 \ldots.. r_n[\cos(\theta_1 + \theta_2 \ldots.. + \theta_n) + i\sin(\theta_1 + \theta_2 \ldots.. + \theta_n)] = A + iB$

$\Rightarrow \quad r_1 r_2 r_3 \ldots.. r_n \cos(\theta_1 + \theta_2 \ldots.. + \theta_n) = A \qquad \ldots(i)$

and $\quad r_1 r_2 r_3 \ldots.. r_n \sin(\theta_1 + \theta_2 \ldots.. + \theta_n) = B \quad$ [Equating real and imaginary parts] $\qquad \ldots(ii)$

(*a*) Squaring and adding (*i*) and (*ii*) we obtain

$$r_1^2 r_2^2 r_3^2 \ldots. r_n^2 = A^2 + B^2$$

$$\Rightarrow \quad (a_1^2 + b_1^2)(a_2^2 + b_2^2) \ldots. (a_n^2 + b_n^2) = A^2 + B^2 \qquad (\because \ r_k^2 = a_k^2 + b_k^2)$$

(*b*) Dividing (*ii*) by (*i*), we have

$$\frac{\sin(\theta_1 + \theta_2 + \ldots.. + \theta_n)}{\cos(\theta_1 + \theta_2 + \ldots.. + \theta_n)} = \frac{B}{A}$$

or
$$\tan(\theta_1 + \theta_2 + \ldots.. + \theta_n) = \frac{B}{A}$$

or
$$\theta_1 + \theta_2 + \ldots.. + \theta_n = \tan^{-1}\left(\frac{B}{A}\right)$$

$$\Rightarrow \quad \tan^{-1}\left(\frac{b_1}{a_1}\right) + \tan^{-1}\left(\frac{b_2}{a_2}\right) + \ldots. + \tan^{-1}\left(\frac{b_n}{a_n}\right) = \tan^{-1}\left(\frac{B}{A}\right) \qquad \left(\because \ \theta_k = \tan^{-1}\left(\frac{b_k}{a_k}\right)\right)$$

EXERCISE 1.1

1. Find the value of $x^3 - 4x^2 - 9x + 97$ if $x = 4+\sqrt{7}i$ **[Ans.** 5]

2. If $(x + iy)^3 = 4 + iv$, then show that $\frac{u}{v}+\frac{v}{y}=4(x^2-y^2)$.

3. Find a complex number which when multiplied by $2 + 5i$ gives $3 - 7i$. **[Ans.** $-1-i$]

4. Find the values of x and y, if $\frac{(1+i)x-2i}{3+i}+\frac{(2-3i)y+i}{3-i}=i$. **[Ans.** $x = 3; y = -1$]

5. If $\frac{a+ib}{c+id} = x + iy$, prove that $x^2 + y^2 = \frac{a^2+b^2}{c^2+d^2}$.

6. Find the conjugates of the following:

   (*i*) $\frac{1+i}{1-i}-\frac{1-i}{1+i}$ (*ii*) $\frac{(3-2i)(2+3i)}{(1+2i)(2-i)}$ $\left[\textbf{Ans. } (i)\ -2i;\ (ii)\ \frac{63}{25}+\frac{16}{25}i\right]$

7. Find the modulus of the following complex numbers:

   (*i*) $\frac{(1+i)(2+i)}{3+i}$ (*ii*) $\frac{(3+4i)(4+5i)}{(4+3i)(6+7i)}$ $\left[\textbf{Ans. } (i)\ 1;\ (ii)\ \sqrt{\frac{41}{85}}\right]$

8. Find the modulus and aptitude of $(1 + 2i)^2 (1 - i)$. **[Ans.** $5\sqrt{2}, \tan^{-1}(7)$]

9. If z_1, z_2, z_3 be the vertices of an isosceles triangle, right angled at z_2, prove that $z_1^2 + z_3^2 + 2z_2^2 = 2z_2(z_1 + z_3)$.

10. Show that $|\exp(z^2)| \le \exp(|z|^2)$.

11. What are the loci given by $|z - 3| = k|z + 1|$ for $k = 1$ and 2.

12. What domain of the z-plane is represented by $|z + 1| + |z - 1| < 4$.

    [Ans. For $k = 1$, right bisector & for $k = 2$, circle]

13. Find the modulus and principal value of the argument of

    (*i*) $1+\sqrt{2}+i$

    (*ii*) $\frac{(3-\sqrt{2}i)^2}{1+2i}$ $\left[\textbf{Ans. } (i)\ \frac{11}{\sqrt{5}}\tan^{-1}\left(\frac{6\sqrt{2}+14}{12\sqrt{2}-7}\right);\ (ii)\ \sqrt{4+2\sqrt{2}};\ \tan^{-1}\left(\frac{1}{\sqrt{2}+1}\right)\right]$

14. Express the following complex number in the polar form

    (*i*) $\frac{1+i}{1-i}$

    (*ii*) $\frac{-35+5i}{4\sqrt{2}+3\sqrt{2}i}$ $\left[\textbf{Ans. } (i)\ 1\left(\cos\frac{\pi}{2}+i\sin\frac{\pi}{2}\right);\ (ii)\ 5\left(\cos\frac{3\pi}{4}+i\sin\frac{3\pi}{4}\right)\right]$

15. If $a^2 + b^2 = 1$ prove that $\left(\frac{1+b+ia}{1+b-ia}\right) = b + ia$.

16. If z_1 and z_2 are two complex numbers, then

    (*i*) $|z_1+z_2| \le |z_1|+|z_2|$ (*ii*) $|z_1-z_2| \le |z_1|+|z_2|$

    (*iii*) $|z_1+z_2| \ge |z_1|-|z_2|$ (*iv*) $|z_1-z_2| \ge |z_1|-|z_2|$

17. Draw and Argand's diagram to represent $\frac{1}{(2+i)^2}-\frac{1}{(2-i)^2}$.

18. The centre of a regular hexagon is at origin and one vertex is given by $(1 + i)$ on the Argand's diagram. Find the remaining vertices.

[**Ans.** Vertices are at a distance $\sqrt{2}$, the origin amplitud are 45°, 105°, 165°, 225°, 265° and 345°]

19. Evaluate: $\dfrac{i^{4m+3} \cdot (-i)^{8m+3}}{(-)^{4m+1} \cdot (-i)^{4m-1}}$.

20. Show that the origin and the complex numbers represented by the roots of the equation $z^2 + az + b = 0$ form an equilateral triangle if $a^2 = 3b$.

21. If p is real and the complex numbers $z = \dfrac{1+i}{2+ip} + \dfrac{2+3i}{3+i}$ is represented by the Argand's diagram by a point on the line $y = z$; show that $p = -5 + \sqrt{21}$.

22. Express in the form $a + ib$:

(*i*) $\dfrac{(2-8i)(7+8i)}{1+i}$ (*ii*) $\dfrac{1}{(2+i)^2} - \dfrac{1}{(2-i)^2}$ $\left[\textbf{Ans. } (i)\, 19 - 59i;\ (ii) -\dfrac{8}{25}i\right]$

23. Put the following in the form $r(\cos\theta + i\sin\theta)$, where r is a + ve real number and $-\pi < \theta < \pi$: $\dfrac{1+7i}{(2-i)^2}$.

$\left[\textbf{Ans. } \sqrt{2}\left(\cos\dfrac{3\pi}{4} + i\sin\dfrac{3\pi}{4}\right)\right]$

24. If $x = -5 + 2\sqrt{-4}$, find the value of $x^4 + 9x^3 + 35x^2 - x + 4$ [**Ans.** – 160]

25. Prove that the points $x + iy$ and $\dfrac{1}{-x+iy}$ lie on a straight line through the origin.

1.10. DE MOIVRE'S THEOREM

Statement:

(i) If $n \in z$ (the set of integers), then

$$(\cos\theta + i\sin\theta)^n = \cos n\theta + i\sin n\theta$$

(ii) If $n \in Q$ (the set of rational numbers), then $\cos n\theta + i\sin n\theta$ is one of the values of $(\cos\theta + i\sin\theta)^n$

Proof. (*i*) The theorem is trivially true for $n = 0$. If $n \neq 0$, then we consider the following different cases:

Case I. When $n \in z$:

We shall prove the theorem by induction on the index n. The theorem is obviously true for $n = 1$. Let us assume that the theorem is true for $n = m$. Then

$$(\cos\theta + i\sin\theta)^m = \cos m\theta + i\sin m\theta \quad \ldots(i)$$

Now we shall show that the theorem is true for $n = m + 1$.

We have
$$\begin{aligned}(\cos\theta + i\sin\theta)^{m+1} &= (\cos\theta + i\sin\theta)^m(\cos\theta + i\sin\theta)\\ &= (\cos m\theta + i\sin\theta)(\cos\theta + i\sin\theta) \quad [\text{using } (i)]\\ &= (\cos m\theta\cos\theta - \sin m\theta\sin\theta) + i(\sin m\theta\cos\theta + \cos m\theta\sin\theta)\\ &= \cos(m+1)\theta + i\sin(m+1)\theta\end{aligned}$$

So, the theorem hold good for $n = m + 1$

Hence by induction the theorem holds good for all $n \in N$.

Case II. When n is a negative integer.

Let $n = -m$ where $m \in N$

Now, $(\cos\theta + i\sin\theta)^n = (\cos\theta + i\sin\theta)^{-m} = \dfrac{1}{(\cos\theta + i\sin\theta)^m}$

$$= \frac{1}{(\cos m\theta + i\sin m\theta)\,(\cos m\theta - i\sin n\theta)}$$

$$= \frac{\cos m\theta - i\sin m\theta}{\cos^2 m\theta + \sin^2 m\theta} = \cos m\theta - i\sin m\theta$$

$$= \cos(-m)\theta - i\sin(-m)\theta = \cos n\theta + i\sin n\theta \quad (\because \quad n = -m)$$

$\therefore$ $(\cos\theta + i\sin\theta)^n = \cos n\theta + i\sin n\theta$, if n is a negative integer

(*ii*) Let $n = \dfrac{p}{q}$ where $q \in N$ and $p \in z$

We know, $\left(\cos\dfrac{\theta}{p} + i\sin\dfrac{\theta}{q}\right)^q = \cos\theta + i\sin\theta$ [by case I and case II]

Taking power $\left(\dfrac{p}{q}\right)$ on both sides, we get

$$\left[\left(\cos\frac{\theta}{q} + i\sin\frac{\theta}{q}\right)^q\right]^{p/q} = (\cos\theta + i\sin\theta)^{p/q}$$

or $$\left(\cos\frac{\theta}{q} + i\sin\frac{\theta}{q}\right)^p = (\cos\theta + i\sin\theta)^n \qquad \left[\because\ n = \frac{p}{q}\right]$$

or $$(\cos\theta + i\sin\theta)^n = \left(\cos\frac{p}{q}\theta + i\sin\frac{p}{q}\theta\right)$$

or $$(\cos\theta + i\sin\theta)^n = \cos n\theta + i\sin n\theta \qquad \left[\because\ n = \frac{p}{q}\right]$$

is the required result.

Remark: The theorem is also true for $\cos\theta - i\sin\theta$ *i.e.*,

$(\cos\theta - i\sin\theta)^n = \cos n\theta - i\sin n\theta$, because $(\cos\theta - i\sin\theta)^n$

$= (\cos n\theta - i\sin\theta)^{-n} = \cos(-n\theta) + i\sin(-n\theta) = \cos n\theta - \sin n\theta$

Note 1. $(\cos\theta + i\sin\theta)^{-1} = \cos(-\theta) + i\sin(-\theta) = \cos\theta - i\sin\theta$

Note 2. $(\sin\theta \pm i\cos\theta)^n \neq \sin n\theta + i\cos n\theta$

Note 3. $(\cos\theta \pm i\sin\phi)^n \neq \sin n\theta + i\cos n\theta$

Note 4. $(\sin\theta + i\cos\theta)^n = \left[\cos\left(\dfrac{\pi}{2} - \theta\right) + i\sin\left(\dfrac{\pi}{2} - \theta\right)\right]^n$

$$= \cos\left(\frac{n\pi}{2} - n\theta\right) + i\sin\left(\frac{n\pi}{2} - n\theta\right)$$

Note 5. $(\cos\theta_1 + i\sin\theta_1)(\cos\theta_2 + i\sin\theta_2)\ \ldots.\ (\cos\theta_n + i\sin\theta_n)$

$$= \cos(\theta_1 + \theta_2 + \ldots. + \theta_n) + i\sin(\theta_1 + \theta_2 + \ldots. + \theta_n)$$

SOLVED EXAMPLES

Example 1.6. Prove that $\dfrac{(\cos 5\theta - i\sin 5\theta)^2 (\cos 7\theta + i\sin 7\theta)^{-3}}{(\cos 4\theta - i\sin 4\theta)^9 (\cos\theta + i\sin\theta)^5} = 1.$

Solution. We have

$$(\cos 5\theta - i\sin 5\theta)^2 = (\cos 10\theta - i\sin 10\theta) = (\cos\theta + i\sin\theta)^{-10}$$
$$(\cos 7\theta + i\sin 7\theta)^{-3} = (\cos 21\theta - i\sin 21\theta) = (\cos\theta + i\sin\theta)^{-21}$$
$$(\cos 4\theta - i\sin 4\theta)^9 = (\cos 36\theta - i\sin 36\theta) = (\cos\theta + i\sin\theta)^{-36}$$
$$(\cos\theta + i\sin\theta)^5 = (\cos 5\theta + i\sin 5\theta) = (\cos\theta + i\sin\theta)^5$$

$$\text{L.H.S.} = \frac{(\cos 5\theta - i\sin 5\theta)^2 (\cos 7\theta + i\sin 7\theta)^{-3}}{(\cos 4\theta - i\sin 4\theta)^9 (\cos\theta + i\sin\theta)^5}$$
$$= \frac{(\cos\theta + i\sin\theta)^{-10} (\cos\theta + i\sin\theta)^{-21}}{(\cos\theta + i\sin\theta)^{-36} (\cos\theta + i\sin\theta)^5} = \frac{(\cos\theta + i\sin\theta)^{-31}}{(\cos\theta + i\sin\theta)^{-31}}$$
$$= 1 = \text{R.H.S.} \quad \textbf{Proved}$$

Example 1.7. Simplify $\dfrac{(\cos 5\theta - i\sin 5\theta)^{2/5}\left(\cos\frac{2}{7}\theta + i\sin\frac{2}{7}\theta\right)^7}{(\cos 4\theta + i\sin 4\theta)^{1/4}\left(\cos\frac{2}{3}\theta - i\sin\frac{2}{3}\theta\right)^3}.$

Solution. We have

$$(\cos 5\theta - i\sin 5\theta)^{2/5} = [(\cos\theta + i\sin\theta)^{-5}]^{2/5} = (\cos\theta + i\sin\theta)^{-2}$$
$$\left(\cos\frac{2}{7}\theta + i\sin\frac{2}{7}\theta\right)^7 = [(\cos\theta + i\sin\theta)^{2/7}]^7 = (\cos\theta + i\sin\theta)^2$$
$$(\cos 4\theta + i\sin 4\theta)^{1/4} = [(\cos\theta + i\sin\theta)^4]^{1/4} = (\cos\theta + i\sin\theta)$$
$$\left(\cos\frac{2}{3}\theta - i\sin\frac{2}{3}\theta\right)^3 = [(\cos\theta + i\sin\theta)^{-2/3}]^3 = (\cos\theta + i\sin\theta)^{-2}$$

$$\therefore \quad \frac{(\cos 5\theta - i\sin 5\theta)^{2/5}\left(\cos\frac{2}{7}\theta + i\sin\frac{2}{7}\theta\right)^7}{(\cos 4\theta + i\sin 4\theta)^{1/4}\left(\cos\frac{2}{3}\theta - i\sin\frac{2}{3}\theta\right)^3} = \frac{(\cos\theta + i\sin\theta)^{-2} (\cos\theta + i\sin\theta)^2}{(\cos\theta + i\sin\theta)(\cos\theta + i\sin\theta)^{-2}}$$
$$= \cos\theta + i\sin\theta$$

Example 1.8. If $x = \cos\alpha + i\sin\alpha$, $y = \cos\beta + i\sin\beta$, **prove that** $\dfrac{x-y}{x+y} = i\tan\left(\dfrac{\alpha-\beta}{2}\right)$

Solution. We have $\quad x = \cos\alpha + i\sin\alpha \quad$ and $\quad y = \cos\beta + i\sin\beta$

$$\text{L.H.S.} = \frac{x-y}{x+y} = \frac{(\cos\alpha + i\sin\alpha) - (\cos\beta + i\sin\beta)}{(\cos\alpha + i\sin\alpha) + (\cos\beta + i\sin\beta)}$$
$$= \frac{(\cos\alpha - \cos\beta) + i(\sin\alpha - \sin\beta)}{(\cos\alpha + \cos\beta) + i(\sin\alpha + \sin\beta)}$$

$$= \frac{-2\sin\left(\frac{\alpha+\beta}{2}\right)\sin\left(\frac{\alpha-\beta}{2}\right)+i\,2\cos\left(\frac{\alpha+\beta}{2}\right)\cdot\sin\left(\frac{\alpha-\beta}{2}\right)}{2\cos\left(\frac{\alpha+\beta}{2}\right)\cos\left(\frac{\alpha-\beta}{2}\right)+i\,2\sin\left(\frac{\alpha+\beta}{2}\right)\cdot\cos\left(\frac{\alpha-\beta}{2}\right)}$$

$$= \frac{i^2\sin\left(\frac{\alpha+\beta}{2}\right)\cdot\sin\left(\frac{\alpha-\beta}{2}\right)+i\cos\left(\frac{\alpha+\beta}{2}\right)\cdot\sin\left(\frac{\alpha-\beta}{2}\right)}{\cos\left(\frac{\alpha+\beta}{2}\right)\cdot\cos\left(\frac{\alpha-\beta}{2}\right)+i\sin\left(\frac{\alpha+\beta}{2}\right)\cdot\cos\left(\frac{\alpha-\beta}{2}\right)}$$

$$= \frac{i\sin\left(\frac{\alpha+\beta}{2}\right)\left[i\sin\left(\frac{\alpha+\beta}{2}\right)+\cos\left(\frac{\alpha+\beta}{2}\right)\right]}{\cos\left(\frac{\alpha+\beta}{2}\right)\left[\cos\left(\frac{\alpha+\beta}{2}\right)+i\sin\left(\frac{\alpha+\beta}{2}\right)\right]} = i\tan\left(\frac{\alpha-\beta}{2}\right) \qquad \textbf{Proved}$$

Example 1.9. Prove that $\dfrac{(1+\sin\phi+i\cos\phi)^n}{(1+\sin\phi-i\cos\phi)^n}=\cos\left(\dfrac{n\pi}{2}-n\phi\right)+i\sin\left(\dfrac{n\pi}{2}-n\phi\right)$.

Solution. $$\frac{(1+\sin\phi+i\cos\phi)^n}{(1+\sin\phi-i\cos\phi)^n} = \frac{\left[1+\cos\left(\frac{\pi}{2}-\phi\right)+i\sin\left(\frac{\pi}{2}-\phi\right)\right]^n}{\left[1+\cos\left(\frac{\pi}{2}-\phi\right)-i\sin\left(\frac{\pi}{2}-\phi\right)\right]^n}$$

$$= \frac{\left[2\cos^2\left(\frac{\pi}{4}-\frac{\phi}{2}\right)+2i\sin\left(\frac{\pi}{4}-\frac{\phi}{2}\right)\cos\left(\frac{\pi}{4}-\frac{\phi}{2}\right)\right]^n}{\left[2\cos^2\left(\frac{\pi}{4}-\frac{\phi}{2}\right)-2i\sin\left(\frac{\pi}{4}-\frac{\phi}{2}\right)\cos\left(\frac{\pi}{4}-\frac{\phi}{2}\right)\right]^n}$$

$$= \frac{\left[\cos\left(\frac{\pi}{4}-\frac{\phi}{2}\right)+i\sin\left(\frac{\pi}{4}-\frac{\phi}{2}\right)\right]^n}{\left[\cos\left(\frac{\pi}{4}-\frac{\phi}{2}\right)-i\sin\left(\frac{\pi}{4}-\frac{\phi}{2}\right)\right]^n}$$

$$= \frac{\left[\cos\left(\frac{\pi}{4}-\frac{\phi}{2}\right)+i\sin\left(\frac{\pi}{4}-\frac{\phi}{2}\right)\right]^n}{\left[\left\{\cos\left(\frac{\pi}{4}-\frac{\phi}{2}\right)+i\sin\left(\frac{\pi}{4}-\frac{\phi}{2}\right)\right\}^{-1}\right]^n}$$

$$= \frac{\left[\cos\left(\frac{\pi}{4}-\frac{\phi}{2}\right)+i\sin\left(\frac{\pi}{4}-\frac{\phi}{2}\right)\right]^n}{\left[\cos\left(\frac{\pi}{4}-\frac{\phi}{2}\right)+i\sin\left(\frac{\pi}{4}-\frac{\phi}{2}\right)\right]^{-n}}$$

$$= \left[\cos\left(\frac{\pi}{4}-\frac{\phi}{2}\right)+i\sin\left(\frac{\pi}{4}-\frac{\phi}{2}\right)\right]^{2n}$$

$$= \cos\left(\frac{n\pi}{2}-n\phi\right)+i\sin\left(\frac{n\pi}{2}-n\phi\right) \qquad \textbf{Proved}$$

Example 1.10. **If $2\cos\theta = x + \frac{1}{x}$ and $2\cos\phi = y + \frac{1}{y}$, show that one of the values of**

(i) $x^m y^n + \frac{1}{x^m y^n}$ is $2\cos(m\theta + n\phi)$ **(ii) $\frac{x^m}{y^n} + \frac{y^n}{x^m}$ is $2\cos(m\theta - n\phi)$**

Solution. We have

$$x + \frac{1}{x} = 2\cos\theta,\ y + \frac{1}{y} = 2\cos\phi$$

Now $$x + \frac{1}{x} = 2\cos\theta \Rightarrow x^2 - 2x\cos\theta + 1 = 0$$

$\Rightarrow$ $$x = \frac{2\cos\theta \pm \sqrt{4 + \cos^2\theta - 4}}{2}$$

$\Rightarrow$ $$x = \cos\theta \pm i\sin\theta$$

$\Rightarrow$ $$x = \cos\theta + i\sin\theta, \text{ taking + ve sign only}$$

Similarly one of the value of y is $\cos\phi + i\sin\phi$.

(*i*) One of the values of $x^m y^n = (\cos\theta + i\sin\theta)^m (\cos\phi + i\sin\phi)^n$

$$= (\cos m\theta + i\sin m\theta)(\cos n\phi + i\sin n\phi)$$

$$= \cos(m\theta + n\phi) + i\sin(m\phi + n\phi)$$

One of the values of $\frac{1}{x^m y^n}$, *i.e.*, $(x^m y^n)^{-1}$

$$= [\cos(m\theta + n\phi) + i\sin(m\theta + n\phi)]^{-1}$$

$$= \cos(m\theta + n\phi) - i\sin(m\theta + n\phi)$$

Hence one of the values of $x^m y^n + \frac{i}{x^m y^n}$ is $2\cos(m\theta + n\phi)$

(*ii*) One of the values of

$$\frac{x^m}{y^n} = \frac{(\cos\theta + i\sin\theta)^m}{(\cos\phi + i\sin\phi)^n} = \frac{\cos m\theta + i\sin m\theta}{\cos n\phi + i\sin n\phi}$$

$$= \cos(m\theta - n\phi) + i\sin(m\theta - n\phi)$$

One of the values of $\frac{y^n}{x^m}$, *i.e.*, $\left[\frac{x^m}{y^n}\right]^{-1}$

$$= [\cos(m\theta - n\phi) + i\sin(m\theta - n\phi)]^{-1}$$

$$= \cos(m\theta - n\phi) + i\sin(m\theta - n\phi)$$

Hence one of the values of $\frac{x^m}{y^n} + \frac{y^n}{x^m}$ is $2\cos(m\theta - n\phi)$.

EXERCISE 1.2

1. Simplify $\dfrac{(\cos 3\theta + i \sin 3\theta)^4 (\cos 4\theta - i \sin 4\theta)^5}{(\cos 4\theta + i \sin 4\theta)^3 (\cos 5\theta + i \sin 5\theta)^{-4}}$. **[Ans.** 1]

2. Simplify $\dfrac{(\cos\theta + i \sin\theta)^5}{(\sin\theta + i\cos\theta)^7}$. **[Ans.** – sin 12θ + *i* cos 12θ]

3. Prove $(1 + \cos\theta + i \sin\theta)^n + (1 + \cos\theta - i \sin\theta)^n = 2^{n+1} \cos^n\left(\dfrac{\theta}{2}\right)\cdot\cos\left(\dfrac{\theta}{2}\right)$.

4. If $a = \cos 2\alpha + i \sin 2\alpha$, $b = \cos 2\beta + i \sin 2\beta$, $c = \cos 2\gamma + i \sin 2\gamma$ and $d = \cos 2\delta + i \sin 2\delta$ prove that

   (*i*) $\sqrt{\dfrac{ab}{c}} + \sqrt{\dfrac{c}{ab}} = 2\cos(\alpha+\beta-\gamma)$ (*ii*) $\sqrt{\dfrac{ab}{c}} + \sqrt{\dfrac{c}{ab}} = 2\cos(\alpha+\beta-\gamma-\delta)$

   (*iii*) $\sqrt{abcd} + \dfrac{1}{\sqrt{abcd}} = 2\cos(\alpha+\beta-\gamma+\delta)$

5. If $2\cos\theta = x + \dfrac{1}{x}$, prove that

   (*i*) $2\cos r\theta = x^r + \dfrac{1}{x^r}$ (*ii*) $\dfrac{x^{2n}+1}{x^{2n-1}+x} = \dfrac{\cos n\theta}{\cos(n-1)\theta}$

6. Prove that $(a + ib)^{m/n} + (a - ib)^{m/n} = 2(a^2+b^2)m/2n\cos\left(\dfrac{m}{n}\tan^{-1}\dfrac{b}{a}\right)$.

7. If n is a positive integer, prove that $(\sqrt{3}+i)^n + (\sqrt{3}-i)^n = 2^{n+1}\cos\dfrac{n\pi}{6}$.

8. Prove that the general value of q which satisfies the equation $(\cos\theta + i\sin\theta)\cdot(\cos 2\theta + i \sin 2\theta)$ ....

   $(\cos n\theta + i \sin n\theta) = 1$ is $\dfrac{4m\pi}{n(n+1)}$, where m is any integer

9. If $(a_1 + ib_1)(a_2 + ib_2) \ldots (a_n + ib_n) = A + iB$, prove that

   (*i*) $(a_1^2 + b_1^2)(a_2^2 + b_2^2) \ldots (a_n^2 + b_n^2) = A^2 + B^2$

   (*ii*) $\tan^{-1}\dfrac{b_1}{a_1} + \tan^{-1}\dfrac{b_2}{a_2} + \ldots\ldots + \tan^{-1}\dfrac{b_n}{a_n} = \tan^{-1}\left(\dfrac{B}{A}\right)$.

10. State and prove De Moivre's theorem. If $a = \cos\alpha + i\sin\alpha$, $b = \cos\beta + i \sin\beta$, $c = \cos\gamma + i\sin\gamma$, then prove that

$$\frac{(b+c)+(c+a)(a+b)}{abc} = 8\cos\frac{\beta-\gamma}{2}\cos\frac{\gamma-\alpha}{2}\cos\frac{\alpha-\beta}{2}$$

11. If α, β be the roots of $x^2 - 2x + 4 = 0$, prove that

$$\alpha^n + \beta^n = 2^{n+1}\cos\frac{n\pi}{3}$$

12. Prove that

$$[(\cos\theta - \cos\phi) + i(\sin\theta - \sin\phi)]^n + [(\cos\theta - \cos\phi) - i(\sin\theta - \sin\phi)]^n = 2^{n+1}\sin^n\left(\frac{\theta-\phi}{2}\right)\cos^n\frac{(\pi+\theta+\phi)}{2}$$

13. If $p = \cos\theta$ and $q = \cos\theta$, show that

$$\frac{(p+q)(pq-1)}{(p-q)(pq+1)} = \frac{\sin\theta + \sin\phi}{\sin\theta - \sin\phi}$$

1.11. EXPONENTIAL AND CIRCULAR FUNCTION OF A COMPLEX VARIABLE

For any real value of x, we have

$$e^x = 1 + x + \frac{x^2}{2!} + \frac{x^3}{3!} + \frac{x^4}{4!} + \dots \infty \qquad \dots(i)$$

$$\cos x = 1 - \frac{x^2}{2!} + \frac{x^4}{4!} - \frac{x^6}{6!} + \dots \infty \qquad \dots(ii)$$

$$\sin x = x - \frac{x^3}{3!} + \frac{x^5}{5!} - \frac{x^7}{7!} + \dots \infty \qquad \dots(iii)$$

Let us extend these definitions to the case when x is a complex number and say that e^x, $\cos x$, $\sin x$ are only short forms of denoting the above series (*i*), (*ii*) and (*iii*) respectively. Hence we can say that the above three relations hold good for all values x, real or complex.

Replacing x by ix in (*i*), we obtain

$$e^{ix} = 1 + ix + \frac{(ix)^2}{2!} + \frac{(ix)^3}{3!} + \frac{(ix)^4}{4!} + \frac{(ix)^5}{5!} + \dots\dots.$$

$$e^{ix} = 1 + ix + \frac{i^2x^2}{2!} + \frac{i^3x^3}{3!} + \frac{i^4x^4}{4!} + \frac{i^5x^5}{5!} + \dots\dots.$$

$$= 1 + ix - \frac{x^2}{2!} - i\frac{x^3}{3!} + \frac{x^4}{4!} + i\frac{x^5}{5!} + \dots\dots.$$

$$= \left(1 - \frac{x^2}{2!} + \frac{x^4}{4!} - \dots\dots\right) + i\left(x - \frac{x^3}{3!} + \frac{x^5}{5!} - \dots\dots\right)$$

$$= \cos x + i \sin x \qquad \text{[Using (ii) and (iii)]}$$

$$\therefore \quad e^{ix} = \cos x + i \sin x \qquad \dots(iv)$$

Again replacing x by $-x$ in the above result, we get

$$e^{-ix} = \cos x - i \sin x \qquad \dots(v)$$

Adding (*iii*) and (*iv*), we get

$$e^{ix} + e^{-ix} = 2 \cos x$$

$$\Rightarrow \quad \boxed{\cos x = \frac{e^{ix} + e^{-ix}}{2}} \qquad \dots(vi)$$

Subtracting (*v*) from (*iv*), we get

$$e^{ix} - e^{-ix} = 2i \sin x$$

$$\boxed{\sin x = \frac{e^{ix} - e^{-ix}}{2i}} \qquad \dots(vii)$$

The formula (*vi*) and (*vii*) we know as **Euler's formulae** for circular functions. If z is a complex variable, then we define

$$\sin z = \frac{e^{iz} - e^{-iz}}{2i} \quad \text{and} \quad \cos z = \frac{e^{iz} + e^{-iz}}{2}$$

Obviously, $\cos z + i \sin z = e^{iz}$

The value of sin z and cos z in terms of exponential functions are also called **Euler's formulae.**

Using (*iv*) z can be written as $z = re^{i\theta}$, where $r = |z|$ and $\theta = \arg(z)$. This called **exponential form** of a complex number.

SOLVED EXAMPLES

Example 1.11. Express $6e^{\frac{5\pi i}{6}}$ in the form $a + ib$.

Solution. We have $6e^{\frac{5\pi i}{6}} = 6\left(\cos\frac{5\pi}{6} + i\sin\frac{5\pi}{6}\right) = 6\left(-\frac{\sqrt{3}}{2} + \frac{i}{2}\right) = -3\sqrt{3} + 3i.$

Example 1.12. Using Euler's exponential forms, prove that

(*i*) $\sin^2 x + \cos^2 x = 1.$ **(*ii*) $\sin 2x = 2 \sin x \cos x.$**

(*iii*) $\cos 2x = \cos^2 x - \sin^2 x = 2\sin^2 x - 1 = 1 - \sin^2 x.$

(*iv*) $\tan 2x = \dfrac{2\tan x}{1 - \tan^2 x}.$ **(*v*) $\sin 3x = 3 \sin x - 4 \sin^2 x.$**

(*vi*) $\sin x + \sin y = 2\sin\left(\dfrac{x+y}{2}\right)\cos\left(\dfrac{x-y}{2}\right).$

Solution. (*i*) We have $\sin x = \dfrac{e^{ix} - e^{-ix}}{2i}$ and $\cos x = \dfrac{e^{ix} + e^{-ix}}{2}$

$$\text{L.H.S.} = \sin^2 x + \cos^2 x$$

$$= \left(\frac{e^{ix} - e^{-ix}}{2i}\right)^2 + \left(\frac{e^{ix} + e^{-ix}}{2}\right)^2 = \frac{e^{2ix} + e^{-2ix} - 2}{4i^2} + \frac{e^{2ix} + e^{-2ix} + 2}{4}$$

$$= \frac{e^{2ix} + e^{-2ix} - 2}{-4} + \frac{e^{2ix} + e^{-2ix} + 2}{4}$$

$$= \frac{1}{4}[-e^{2ix} - e^{-2ix} + 2 + e^{2ix} - e^{-2ix} + 2] = 1. \quad \text{R.H.S.}$$

(*ii*)

$$\text{R.H.S.} = 2 \sin x \cos x$$

$$= 2\left(\frac{e^{ix} - e^{-ix}}{2i}\right)\left(\frac{e^{ix} + e^{-ix}}{2}\right) = \frac{e^{2ix} - e^{-2ix}}{2i} = \sin 2x = \text{L.H.S.}$$

(*iii*) We have $\cos^2 x - \sin^2 x = \left(\dfrac{e^{ix} + e^{-ix}}{2}\right)^2 - \left(\dfrac{e^{ix} - e^{-ix}}{2i}\right)^2$

$$= \frac{1}{4}[(e^{2ix} + e^{-2ix} + 2) + (e^{2ix} + e^{-2ix} - 2)]$$

$$= \frac{e^{-2ix} + e^{-2ix}}{2} = \cos 2x = \text{L.H.S.}$$

$$2\cos^2 x - 1 = 2\left(\frac{e^{ix} + e^{-ix}}{2}\right)^2 - 1$$

$$= \frac{1}{2}(e^{2ix} + e^{-2ix} + 2) - 1 = \frac{e^{2ix} + e^{-2ix}}{2} = \cos 2x = \text{L.H.S.}$$

$$1 - 2\sin^2 x = 1 - 2\left(\frac{e^{ix} + e^{-ix}}{2i}\right)^2$$

$$= 1 + \frac{1}{2}(e^{2ix} + e^{-2ix} - 2) = \frac{e^{2ix} + e^{-2ix}}{2} = \cos 2x = \text{L.H.S.}$$

(*iv*) $$\text{R.H.S.} = \frac{2\tan x}{1 - \tan^2 x} = \frac{2.\dfrac{e^{ix} - e^{-ix}}{i(e^{ix} + e^{-ix})}}{1 - \left[\dfrac{e^{ix} - e^{-ix}}{i(e^{ix} + e^{-ix})}\right]}$$

$$= \frac{2(e^{ix} - e^{-ix})(e^{ix} + e^{-ix})}{i\left[(e^{ix} + e^{-ix})^2 + (e^{ix} - e^{-ix})^2\right]} = \frac{2(e^{2ix} - e^{-2ix})}{i\,.\,2(e^{2ix} + e^{-2ix})}$$

$$= \frac{e^{2ix} - e^{-2ix}}{i(e^{2ix} + e^{-2ix})} = \tan 2x = \text{L.H.S.}$$

(*v*) We have $$\sin 3x = \frac{e^{3ix} - e^{-3ix}}{2i} = \frac{a^3 - b^3}{2i}, \text{ where } a = e^{ix}, b = e^{-ix}$$

$$= \frac{(a-b)^3 + 3ab(a-b)}{2i}$$

$$= \frac{1}{2i}[(e^{ix} - e^{-ix})^3 + 3e^{ix} \cdot e^{-ix}(e^{ix} - e^{-ix})]$$

$$= \frac{1}{2i}[(2i\sin x)^3 + 2\,(2i\sin x)]$$

$$= \frac{1}{2i}[-8i\sin^3 x + 6i\sin x]$$

$$= 3\sin x - 4\sin^3 x = \text{R.H.S.}$$

(*vi*) We have $$\sin x + \sin y = \frac{e^{ix} - e^{-ix}}{2i} + \frac{e^{iy} - e^{-iy}}{2i} = \frac{1}{2i}[e^{ix} + e^{iy} - e^{-ix} - e^{-iy}] \quad ...(i)$$

and $$2\sin\left(\frac{x+y}{2}\right)\cos\left(\frac{x-y}{2}\right) = 2\left[\frac{e^{i\left(\frac{x+y}{2}\right)} - e^{-i\left(\frac{x+y}{2}\right)}}{2i} \cdot \frac{e^{i\left(\frac{x-y}{2}\right)} + e^{-i\left(\frac{x-y}{2}\right)}}{2}\right]$$

$$= \frac{1}{2i}[e^{ix} + e^{iy} - e^{-iy} - e^{-ix}] \quad ...(ii)$$

From (*i*) and (*ii*), we get

$$\sin x + \sin y = 2\sin\left(\frac{x+y}{2}\right)\cos\left(\frac{x-y}{2}\right).$$

EXERCISE 1.3

1. (*i*) $\sin(-x) = -\sin x$ (*ii*) $\cos(-x) = \cos x$

2. (*i*) $\cos x = 1 - 2\sin^2 \dfrac{x}{2}$. (*ii*) $\cos x = 2\cos^2 \dfrac{x}{2} - 1$.

3. (*i*) $\cos 3x = 4\cos^3 x - 3\cos x$. (*ii*) $\tan 3x = \dfrac{3\tan x - \tan^3 x}{1 - 3\tan^2 x}$.

4. (*i*) $\sin x - \sin y = 2\cos\left(\dfrac{x+y}{2}\right)\sin\left(\dfrac{x-y}{2}\right)$. (*ii*) $\cos x + \cos y = 2\cos\left(\dfrac{x+y}{2}\right)\cos\left(\dfrac{x-y}{2}\right)$.

5. (*i*) $\sin x = 2\sin\dfrac{x}{2}\cos\dfrac{x}{2}$. (*ii*) $\sin 4x = 8\cos^3 x \sin x - 4\sin x$.

6. (*i*) $\sin(x+y) = \sin x\cos y + \cos x\sin y$. (*ii*) $\cos(x+y) = \cos x\cos y - \sin x\sin y$.

1.12. HYPERBOLIC FUNCTIONS

Definition: For any x (real or complex) the quantities $\dfrac{e^x - e^{-x}}{2}$ and $\dfrac{e^x + e^{-x}}{2}$ are respectively defined as the hyperbolic sine of x and hyperbolic cosine of x and are written as $\sinh x$ and $\cosh x$ respectively.

Thus, $$\sinh x = \frac{e^x - e^{-x}}{2} \quad \text{and} \quad \cosh x = \frac{e^x + e^{-x}}{2}$$

The other hyperbolic functions are defined in terms of hyperbolic sine and cosine as follows:

$$\tanh x = \frac{\sinh x}{\cosh x} = \frac{e^x - e^{-x}}{e^x + e^{-x}}$$

$$\coth x = \frac{\cosh x}{\sinh x} = \frac{e^x + e^{-x}}{e^x - e^{-x}}$$

$$\text{sech}\, x = \frac{1}{\cosh x} = \frac{2}{e^x + e^{-x}}$$

$$\text{cosech}\, x = \frac{1}{\sinh x} = \frac{2}{e^x - e^{-x}}$$

Remarks.

1. $\cosh x$ and $\sinh x$ are named as hyperbolic cosine and sine of x because $(a\cosh x, b\sinh x)$ is any point on the hyperbola.
2. It is very evident from the definition that

$$\sinh(-x) = -\sinh x, \quad \cosh(-x) = \cosh x$$
$$\tanh(-x) = -\tanh x, \; \coth(-x) = -\cot hx$$
$$\text{sech}(-x) = \text{sech}\, x \quad \text{and} \quad \text{cosech}(-x) = -\text{cosech}\, x.$$

Also $$\sinh 0 = \frac{e^0 - e^0}{2} = \frac{1-1}{2} = 0$$

$$\cosh 0 = \frac{e^0 + e^0}{2} = \frac{1+1}{2} = 1 \quad \text{and} \quad \tanh 0 = \frac{e^0 - e^0}{e^0 + e^0} = \frac{1-1}{1+1} = 0.$$

1.12.1. Relation Between Hyperbolic and Circular Functions

We have $e^{ix} = \cos x + i\sin x$

Similarly, $$e^{-ix} = \cos x - i\sin x$$

$\therefore$ $$\sin x = \frac{1}{2i}(e^{ix} - e^{-ix}) \qquad \ldots(i)$$

and $$\cos x = \frac{1}{2}(e^{ix} + e^{-ix}) \qquad ...(ii)$$

Now by definition $$\sinh x = \frac{e^x - e^{-x}}{2} \qquad ...(iii)$$

and $$\cosh x = \frac{e^x + e^{-x}}{2} \qquad ...(iv)$$

(1) $$\sin(ix) = \frac{1}{2i}\left[e^{i(ix)} - e^{-i(ix)}\right] \qquad \text{[Replacing } x \text{ by } ix \text{ in } (i)]$$

$$= \frac{1}{2i}\left(e^{i^2x} - e^{-i^2x}\right) = \frac{1}{2i}(e^{-x} - e^{x}) = \frac{-1}{i}\left(\frac{e^{-x} - e^x}{2}\right)$$

$$= i\left(\frac{e^{-x} - e^x}{2}\right) = i \sinh x$$

$\therefore$ $$\boxed{\sin(ix) = i \sinh x.}$$

(2) $$\cos(ix) = \frac{1}{2}[e^{i(ix)} + e^{-i(ix)}] = \frac{1}{2}\left(e^{i^2x} + e^{-i^2x}\right)$$

$$= \frac{1}{2}(e^{-x} + e^{x}) = \frac{1}{2}(e^x + e^{-x}) = \cosh x \qquad [\because \text{ by } (iv)]$$

$\therefore$ $$\boxed{\cos(ix) = \cosh x.}$$

(3) Replacing x by ix in (iii), we have

$$\sinh(x) = \frac{e^x - e^{-x}}{2} = i\left(\frac{e^{ix} - e^{-ix}}{2i}\right)$$

$$= i \sin x \qquad [\because \text{ by } (i)]$$

$\therefore$ $$\boxed{\sinh(ix) = i \sin x}$$

(4) Replacing x by ix in (iv), we have

$$\cosh(ix) = \frac{e^{ix} - e^{-ix}}{2} = \cos x \qquad [\because \text{ by } (ii)]$$

$\therefore$ $$\boxed{\cosh(ix) = \cos x}$$

(5) It can shown that

$$\tanh(ix) = \frac{\sin(ix)}{\cos(ix)} = \frac{i \sinh x}{\cosh x} = i \tan x$$

$$\tanh(ix) = i \tan x$$

So, we have the following relations between circular and hyperbolic functions:

$\sin(ix) = i \sinh x$	$\sinh(ix) = i \sin x$
$\cos(ix) = \cosh x$	$\cosh(ix) = \cos x$
$\tan(ix) = i \tanh x$	$\tanh(ix) = i \tan x$
$\cot(ix) = i \coth x$	$\coth(ix) = i \cot x$
$\sec(ix) = \text{sech } x$	$\text{sech}(ix) = \sec x$
$\text{cosec}(ix) = i \text{ cosech } x$	$\text{cosech}(ix) = i \text{ cosec } x$

1.12.2. Fundamental Relations between Hyperbolic Functions

We shall now establish now some fundamental relations between hyperbolic functions:

(I) ***(i)*** $\cosh^2 x - \sinh^2 x = 1$ ***(ii)*** $1 - \tanh^2 x = \text{sech}^2 x$

(iii) $\coth^2 x - 1 = \text{cosech}^2 x$.

Proof. (*i*) We have

$$\cosh^2 x - \sinh^2 x = \left(\frac{e^x + e^{-x}}{2}\right)^2 \left(\frac{e^x - e^{-x}}{2}\right)^2$$

$$= \frac{1}{4}[(e^{2x} + e^{-2x} + 2) - (e^{2x} + e^{-2x} - 2)] = 1$$

(*ii*) We know that

$$\cosh^2 x - \sinh^2 x = 1$$

Dividing both sides by $\cosh^2 x$, we get

$$1 - \tanh^2 x = \text{sech}^2 x$$

(*iii*) We know that

$$\cosh^2 x - \sinh^2 x = 1$$

Dividing both sides by $\sinh^2 x$, we get

$$\coth^2 x - 1 = \text{cosech}^2 x$$

(II) ***(i)*** $\sinh 2x = 2 \sinh x \cosh x = \dfrac{2 \tanh x}{1 - \tanh^2 x}$

(ii) $\cosh 2x = \cosh^2 x + \sinh^2 x = 2 \cosh^2 x - 1$

$$= 1 + 2 \sinh^2 x = \frac{1 + \tanh^2 x}{1 - \tanh^2 x}$$

(iii) $\tanh 2x = \dfrac{2 \tanh x}{1 + \tanh^2 x}$.

Proof. (*i*) We know that

$$\sin 2x = 2 \sin x \cos x$$

Putting $x = ix$, we get

$$\sin (2ix) = 2 \sin (ix) \cos (ix)$$

or $$i \sinh 2x = 2i \sinh x \cosh x$$

or $$\sinh 2x = 2 \sinh x \cosh x$$

Also $$\sin 2x = \frac{2 \tanh x}{1 + \tanh^2 x}$$

Putting $x = ix$, we get

$$\sin (2ix) = \frac{2 \tanh ix}{1 + \tanh^2 ix} = \frac{2 . i \tanh x}{1 + (i \tanh x)^2}$$

or $$i \sinh 2x = \frac{2i \tanh x}{1 - \tanh^2 x}$$

or $$\sinh 2x = \frac{2 \tanh x}{1 - \tanh^2 x}$$

The other two parts are left as in exercise for the reader.

(III) **(*i*) $\sinh 3x = 3\sinh x + 4\sinh^3 x$** **(*ii*) $\cosh 3x = 4\cosh^3 x - 3\cosh x$**

(*iii*) $\tanh x = \dfrac{3\tanh x + \tanh^3 x}{1 + 3\tanh^2 x}$

Proof. (*i*) We know that $\sin 3x = 3\sin x - 4\sin^3 x$

Putting $x = ix$, we get

$$\sin(3ix) = 3\sin(ix) - 4\sin^3(ix)$$

or $$i\sinh 3x = 3i\sinh(x) - 4(i\sinh x)^3$$

or $$i\sinh 3x = 3i\sinh x + 4i\sinh^3 x$$

or $$\sinh 3x = 3\sinh x + 4\sinh^3 x$$

The other two parts are left as an exercise for the reader.

(IV) **(*i*) $\sinh(x + y) = \sinh x\cosh y - \cosh x\sinh y$** **(*ii*) $\sinh(x - y) = \sinh x\cosh y - \cosh x\sinh y$**

(*iii*) $\cosh(x + y) = \cosh x\cosh y + \sinh x\sinh y$ **(*iv*) $\cosh(x - y) = \cosh x\cosh y - \sinh x\sinh y$**

(*v*) $\tanh(x + y) = \dfrac{\tanh x + \tanh y}{1 + \tanh x\tanh y}$ **(*vi*) $\tanh(x - y) = \dfrac{\tanh x - \tanh y}{1 - \tanh x\tanh y}$**

Proof. (*i*) We know that $\sin(x + y) = \sin x\cos y + \cos x\sin y$

Putting $x = ix$; $y = iy$; we get

$$\sin(ix + iy) = \sin(ix)\cos(iy) + \cos(ix)\sin(iy)$$

$$\sin i(x + y) = i\sin x\cos y + i\cos x\sin y$$

or $$i\sinh(x + y) = i\sin x\cos y + i\cos x\sin y$$

or $$\sinh(x + y) = \sin x\cos y + \cos x\sin y$$

Similarly we can prove other formulae.

(V) **(*i*) $\sinh A + \sinh B = 2\sinh\left(\dfrac{A+B}{2}\right)\cosh\left(\dfrac{A-B}{2}\right)$**

(*ii*) $\sinh A - \sinh B = 2\cosh\left(\dfrac{A+B}{2}\right)\sinh\left(\dfrac{A-B}{2}\right)$

(*iii*) $\cosh A + \cosh B = 2\cosh\left(\dfrac{A+B}{2}\right)\cosh\left(\dfrac{A-B}{2}\right)$

(*iv*) $\cosh A - \cosh B = 2\sinh\left(\dfrac{A+B}{2}\right)\sinh\left(\dfrac{A-B}{2}\right)$

Proof. (*i*) We know that

$$\sin x + \sin y = 2\sin\left(\frac{x+y}{2}\right)\cos\left(\frac{x-y}{2}\right)$$

Putting $x = iA$; $y = iB$, we get

$$\sin(iA) + \sin(iB) = 2\sin\left[\frac{iA+iB}{2}\right]\cos\left[\frac{iA-iB}{2}\right]$$

or $$i\sinh A + i\sinh B = 2i\sinh\left(\frac{A+B}{2}\right)\cosh\left(\frac{A-B}{2}\right)$$

or $$\sinh A + \sinh B = 2\sinh\left(\frac{A+B}{2}\right)\cosh\left(\frac{A-B}{2}\right).$$

The other three part are left as an exercise for the reader.

(VI) (*i*) $2 \sinh A \cosh B = \sinh (A + B) + \sinh (A - B)$

(*ii*) $2 \cosh A \sinh B = \sinh (A + B) - \sinh (A - B)$

(*iii*) $2 \cosh A \cosh B = \cosh (A + B) + \cosh (A - B)$

(*iv*) $2 \sinh A \sinh B = \cosh (A + B) - \cosh (A - B)$.

Proof. (*i*) We know that

$$2 \sinh x \cosh y = \sinh (x + y) + \sinh (x - y)$$

Putting $x = iA$; $y = iB$; we get

$$2 \sinh (iA) \cosh (iB) = \sinh [iA + iB] + \sinh [iA - iB]$$

or $$2i \sinh A \cosh B = \sinh i(A + B) + \sinh i(A - B)$$

or $$2i \sinh A \cosh B = i \sinh (A + B) + i \sinh (A - B)$$

or $$2 \sinh A \cosh B = \sinh (A + B) + \sinh (A - B)$$

The other three parts are left as an exercise for the reader.

SOLVED EXAMPLES

Example 1.13. Separate into real and imaginary parts:

(*i*) $\sin (x + iy)$ **(*ii*) $\cos (x + iy)$**

(*iii*) $\sec (x + iy)$ **(*iv*) $\sinh (x + iy)$**

(*v*) $\tanh (x + iy)$ **(*vi*) $\log \sin (x + iy)$.**

Solution. (*i*) $\sin (x + iy) = \sin x \cos (iy) + \cos x \sin (iy)$

$$= \sin x \cos y + \cos x \sin y \qquad [\because \cos (iy) = \cosh y, \sin (iy) = i \sinh y]$$

$\therefore$ Real part of $\sin (x + iy) = \sin x \cosh y$

Imaginary part of $\sin (x + iy) = \cos x \sinh y$

(*ii*)
$$\cos (x + iy) = \cos x \cos (iy) - \sin x \sin (iy)$$
$$= \cos x \cosh y - \sin x \,.\, i \sinh y$$
$$= \cos x \cosh y - i \sin x \sinh y$$

$\therefore$ Real part of $\cos (x + iy) = \cos x \cosh y$

Imaginary part of $\cos (x + iy) = - \sin x \sinh y$

(*iii*)
$$\sec (x + iy) = \frac{1}{\cos (x + iy)} = \frac{2 \cos (x - iy)}{2 \cos (x + iy) \cos (x - iy)}$$

$$= \frac{2 (\cos x \cos (iy) + \sin x \sin (iy))}{\cos 2x + \cos 2iy} = \frac{2 (\cos x \cosh y + \sin x \,.\, i \sinh y)}{\cos 2x + \cosh 2y}$$

$$= \frac{2 \cos x \cosh y}{\cos 2x + \cosh 2y} + i \frac{2 \sin x \,.\, \sinh y}{\cos 2x + \cosh 2y}$$

$\therefore$ Real part of $\sec (x + iy) = \dfrac{2 \cos x \cosh y}{\cos 2x + \cosh 2y}$

Imaginary part of $\sec (x + iy) = \dfrac{2 \sin x \,.\, \sinh y}{\cos 2x + \cosh 2y}$

(iv) $\sinh(x+iy) = \frac{1}{i}\sin i(x+iy)$ $\qquad [\because \; i\sinh x = \sin ix]$

$$= \frac{1}{i^2}\sin(ix-y) = -i(\sin ix\cos y - \cos ix\sin y)$$

$$= -i(i\sinh x\cos y - \cosh x\sin y) = \sinh x\cos y + i\cosh x\sin y$$

$\therefore$ Real part of $\sinh(x+iy) = \sinh x\cos y$

Imaginary part of $\sinh(x+iy) = \cosh x\sin y$

(v) $\tanh(x+iy) = \frac{1}{i}\tan i(x+iy)$ $\qquad [\because \; i\tanh x = \tan ix]$

$$= \frac{1}{i^2}\tan(ix-y) = -i\frac{2\sin(ix-y)}{\cos(ix-y)}$$

$$= -i.\frac{2\sin(ix-y)\cos(ix+y)}{2\cos(ix-y)\cos(ix+y)} = -i.\frac{\sin 2ix - \sin 2y}{\cos 2ix + \cos 2y}$$

$$= -i.\frac{i\sinh 2x - \sin 2y}{\cosh 2x + \cos 2y} = \frac{\sinh 2x}{\cosh 2x + \cos 2y} + i.\frac{\sin 2y}{\cosh 2x + \cos 2y}$$

$\therefore$ Real part of $\tanh(x+iy) = \frac{\sinh 2x}{\cosh 2x + \cos 2y}$

Imaginary part of $\tanh(x+iy) = \frac{\sin 2y}{\cosh 2x + \cos 2y}$

(vi) $\log\sin(x+iy) = \log(\sin x\cos iy + \cos x\sin iy)$

$$= \log(\sin x\cosh y + i\cos x\sinh y)$$

$$= \log(A+iB) \quad \text{where } A = \sin x\cosh y,\ B = \cos x\sinh y$$

$$= \frac{1}{2}\log(A^2+B^2) + i\tan^{-1}\frac{B}{A}$$

$$= \frac{1}{2}\log\left[\frac{1-\cos 2x}{2}.\frac{\cosh 2y+1}{2} + \frac{1+\cos 2x}{2}.\frac{\cosh 2y - 1}{2}\right] + i\tan^{-1}(\cot x\tanh y)$$

$$= \frac{1}{2}\log\left[\frac{1}{4}(2\cosh 2y - 2\cos 2x)\right] + i\tan^{-1}(\cot x\tanh y)$$

$$= \frac{1}{2}\log\left[\frac{1}{2}(\cosh 2y - \cos 2x)\right] + i\tan^{-1}(\cot x\tanh y).$$

Example 1.14. If $x + iy = \tan(A+iB)$, prove that

(i) $x^2 + y^2 + 2x\cot 2A = 1$. (ii) $x^2 + y^2 - 2y\coth 2B + 1 = 0$.

Solution. $x + iy = \tan(A+iB)$

$\therefore$ $x - iy = \tan(A+iB)$

Now (i) $\tan 2A = \tan[(A+iB)+(A-iB)]$

$$= \frac{\tan(A+iB) + \tan(A+iB)}{1 - \tan(A+iB)\tan(A-iB)} = \frac{(x+iy)+(x-iy)}{1-(x+iy)(x-iy)} = \frac{2x}{1-(x^2+y^2)}$$

or $$\cot 2A = \frac{1-(x^2+y^2)}{2x}$$

or $$2x \cot 2A = 1-(x^2+y^2)$$

or $$x^2+y^2+2x\cot 2A = 1$$

Again $$2iB = (A+iB)-(A-iB)$$

$$\tan(2iB) = \tan[(A+iB)-(A-iB)]$$

$$= \frac{\tan(A+iB)+\tan(A+iB)}{1-\tan(A+iB)\tan(A-iB)} = \frac{(x+iy)-(x-iy)}{1+(x+iy)-(x-iy)} = \frac{2iy}{1+x^2+y^2}$$

or $$i\tan 2B = \frac{2iy}{1+x^2+y^2} \quad \text{or} \quad \frac{1}{\coth 2B} = \frac{2iy}{1+x^2+y^2}$$

or $$1+x^2+y^2 = 2y\coth 2B \quad \text{or} \quad x^2+y^2-2y\coth 2B = 1.$$

Example 1.15. If $x+iy = \sin(A+iB)$, prove that

$$(i)\ \frac{x^2}{\cosh^2 B}+\frac{y^2}{\sinh^2 B} = 1 \qquad (ii)\ \frac{x^2}{\sin^2 A}+\frac{y^2}{\cos^2 A} = 1.$$

Solution. We have $x+iy = \sin(A+iB)$

$$= \sin A\cos(iB)+\cos A\sin(iB)$$

$$= \sin A\cosh B + i\cos A\sinh B$$

Equating real and imaginary parts from both sides, we get

$$x = \sin A\cosh B \text{ and } y = \cos A\sinh B$$

$$\Rightarrow \quad \sin A = \frac{x}{\cosh B},\ \cos A = \frac{y}{\sinh B}$$

Squaring and adding, we have

$$\sin^2 A+\cos^2 A = \frac{x^2}{\cosh^2 B}+\frac{y^2}{\sinh^2 B} \quad \text{or} \quad \frac{x^2}{\cosh^2 B}+\frac{y^2}{\sinh^2 B} = 1$$

Again $x = \sin A\cosh B$ and $y = \cos A\sinh B$

$$\Rightarrow \quad \cosh B = \frac{x}{\sin A} \quad \text{and} \quad \sinh B = \frac{y}{\cos A}$$

Squaring and subtracting, we have

$$\cosh^2 B-\sinh^2 B = \frac{x^2}{\sin^2 A}-\frac{y^2}{\cos^2 A} \quad \text{or} \quad \frac{x^2}{\sin^2 A}-\frac{y^2}{\cos^2 A} = 1.$$

EXERCISE 1.4

Resolve the following into real and imaginary parts:

1. $\tan(x+iy)$. $\left[\textbf{Ans. } \text{R. P.} = \frac{\sin 2x}{\sin^2 y+\cosh 2y}, \text{I.P.} = \frac{\sinh 2y}{\cos 2x+\cosh 2y}\right]$

2. $\cot(x+iy)$. $\left[\textbf{Ans. } \text{R. P.} = \frac{\sin 2x}{\cosh 2y-\cosh 2x}, \text{I.P.} = \frac{-\sinh 2y}{\cosh 2y-\cosh 2x}\right]$

3. sec $(x + iy)$. $\left[\textbf{Ans. } \text{R. P.} = \dfrac{2\cos x\cosh y}{\cos 2x = \cosh 2y}, \text{I.P.} = \dfrac{2\sin x, \sinh y}{\cos 2y + \cosh 2y}\right]$

4. cosec $(x + iy)$. $\left[\textbf{Ans. } \text{R. P.} = \dfrac{2\sin x\cosh y}{\cosh 2y - \cos 2x}, \text{I.P.} = \dfrac{-2\cos x, \sinh y}{\cosh 2y - \cos 2x}\right]$

5. cosh $(x + iy)$. [**Ans.** R.P. = $\cosh x \cos y$, I.P. = $\sinh x \sin y$]

6. coth $(x + iy)$. $\left[\textbf{Ans. } \text{R. P.} = \dfrac{\sinh 2y}{\cosh 2x - \cos 2y}, \text{I.P.} = \dfrac{-\sin 2y}{\cosh 2x - \cos 2y}\right]$

7. sech $(x + iy)$. $\left[\textbf{Ans. } \text{R. P.} = \dfrac{2\cosh x\cos y}{\cosh 2x + \cos 2y}, \text{I.P.} = \dfrac{-2\sinh x, \sin y}{\cosh 2x + \cos 2y}\right]$

8. If $\cosh(A + iB) = x + iy$, then prove that

(*a*) $\dfrac{x^2}{\cosh^2 A} + \dfrac{y^2}{\sinh^2 A} = 1$ (*b*) $\dfrac{x^2}{\cos^2 B} - \dfrac{y^2}{\sin^2 B} = 1.$

9. If $\cos(A + iB) = x + iy$, then prove that

(*a*) $\dfrac{x^2}{\cos^2 A} - \dfrac{y^2}{\sin^2 A} = 1$ (*b*) $\dfrac{x^2}{\cosh^2 B} + \dfrac{y^2}{\sinh^2 B} = 1.$

10. If $\sin(\alpha + \beta) = r(\cos\theta + i\sin\theta)$, prove that

$2r^2 = \cos 2\theta - \cos 2\alpha$ and $\tan\theta = \cot\alpha \tanh\beta$.

11. If $u = \log\left[\tan\left(\dfrac{\pi}{4} + \dfrac{\theta}{2}\right)\right]$, prove that

(*i*) $\sinh(u) = \tan(\theta)$ (*ii*) $\tanh(u) = \sin(\theta)$

12. If $\tan(\theta + i\phi) = \cos\alpha + i\sin\alpha = e^{i\alpha}$, prove that $\alpha = \dfrac{n\pi}{2} + \dfrac{\pi}{4}$ and $\phi = \dfrac{1}{2}\log\tan\left(\dfrac{\pi}{4} + \dfrac{\alpha}{2}\right)$.

13. If $\tan(x + iy) = \sin(u + iv)$, prove that $\dfrac{\sin 2x}{\sinh 2y} = \dfrac{\tan u}{\tanh v}$.

14. If $\sin[\log(A + iB)] = x + iy$, show that $\dfrac{x^2}{\sin^2 u} - \dfrac{y^2}{\cos^2 u} = 1$, where $A^2 + B^2 = e^{2u}$.

15. If $u = \log\tan\left(\dfrac{\pi}{4} + \dfrac{\theta}{2}\right)$, prove that $\theta = -i\log\tan\left(\dfrac{\pi}{4} + \dfrac{iu}{2}\right)$.

1.13. APPLICATIONS OF COMPLEX NUMBERS

Complex numbers are an important mathematical tool that is widely use in many physics domains, including electronics. The concept might seem odd, but their manipulation is simple and their efficiency notable. In this, we will see how phasors are a convenient, graphic way to represent sinusoidal voltages and currents in terms of their magnitude and phase angle. Later, we will see how phasors can also represents other ac circuit quantities.

Phasor diagrams are an abstract method of representing quantities that have both magnitude and direction. In the case of sinusoidal voltages and currents, the magnitude is the amplitude of the sine wave and the direction is its phase angle. Phasors provide a way to diagram sine waves and their phase relationships with other sine waves. The complex number system provides a way to mathematically express a phasor quantity and allows phasor quantities to be added, subtracted, multiplied or divided.

1.14. PHASOR

Phasors provide a graphic means for representing quantities that have both magnitude and direction (angular position). Phasors are especially useful for representing sine waves in terms of their magnitude and phase angle and also for analysis for reactive circuits.

When a phasor rotates through 360 degrees or 2π radians, one complete cycle is traced out. Therefore, the time required for the phasor to go through 2π radians is the period of the sine wave. Because the phasor rotates through 2π radians in a time equal to the period T, the angular velocity can be expressed as

$$\omega = \frac{2\omega}{T}$$

In electronics, a phasor is similar to a vector but generally refers to quantities that vary with time such as sine waves. Examples of phasors are shown in below figure. The length of the phasor "arrow" represents the magnitude of a quantity. The angle, 0 (relative to 0°), represents the angular position, as shown in part (*a*) for a positive angle. The specific phasor example in part (*b*) has a magnitude of 2 and a phase angle of 45°. The phasor in part (*c*) has magnitude of 3 and a phase angle of 180°. The phasor in part (*d*) ha a magnitude of 1 and a phase angle of 45° (or + 315°). Notice that positive angles are measured counter clockwise (CCW) from the reference (0°) and negative angles are measured clockwise (CW) from the reference.

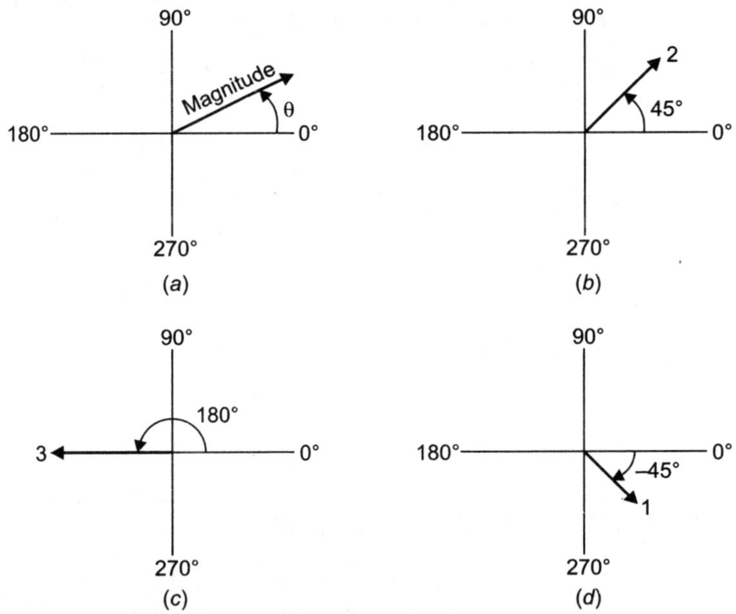

Examples of phasors

1.15. PHASOR DIAGRAMS

1. **Phasor diagram of $f(t) = A \sin \omega t$:** Consider a sinusodial waveform given by the function $f(t) = A \sin \omega t$. Clearly, the amplitude of the waveform is A and period $2\pi/\omega$. The quantity ω is termed as the radian frequency of the waveform. The graph of the waveform is shown in figure. This waveform can be generated by a uniformly rotating arrow of length A as discussed below.

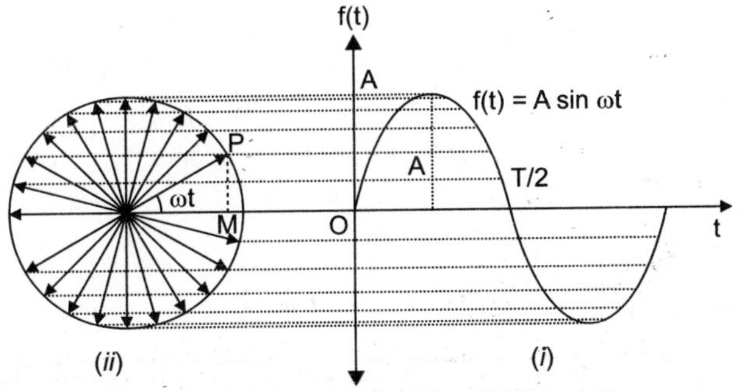

2. **Phasor diagram of $f(t) = A \cos \omega t$:** The equation of the wave form is

$$f(t) = A \cos \omega t \quad \text{or} \quad f(t) = A \sin (\omega t + \pi/2)$$

So, the phase angle of the waveform given by $f(t) = A \cos \omega t$ is $\pi/2$ and hence to represent a cosinusoidal waveform $f(t) = A \cos \omega t$ we draw a phasor of length A which makes the angle of 90° with the horizontal axis as shown in below figure.

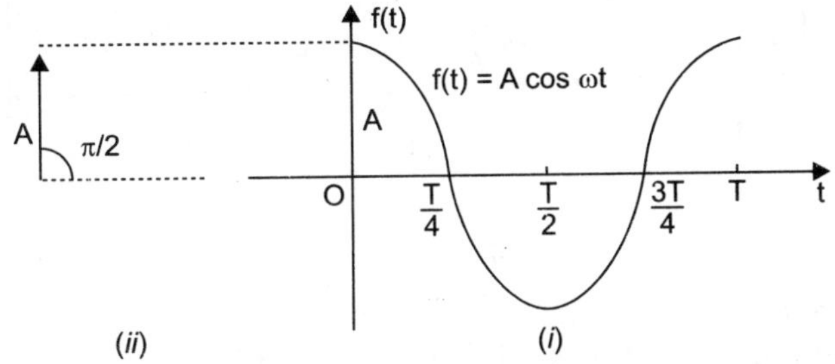

3. **Phasor diagram of $f(t) = -A \sin \omega t$:** The equation of the waveform is

$$f(t) = -A \sin \omega t \quad \text{or} \quad f(t) = A \sin (\omega t + \pi)$$

So, phase angle of the waveform given by $f(t) = -A \sin \omega t$ is π and hence to represent a sinusoidal waveform $f(t) = -A \sin \omega t$ we draw a phasor of length A which makes an angle of 180° with the horizontal axis as shown in below figure.

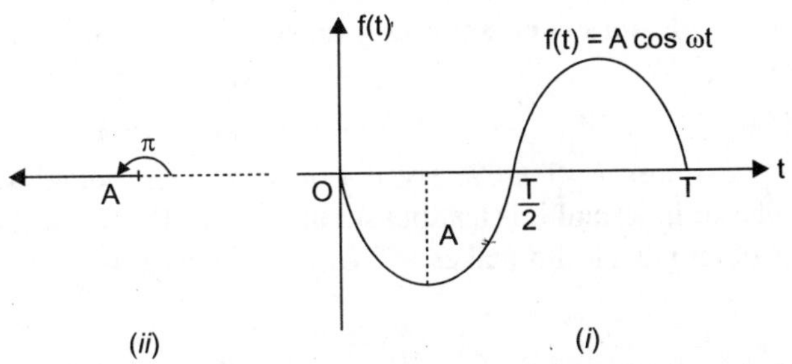

4. **Phasor diagram of $f(t) = -A \sin \omega t$:** The equation of the waveform is

$$f(t) = -A \cos \omega t \quad \text{or} \quad f(t) = A \sin (\omega t + 3\pi/2)$$

So, the phase angle of the waveform given by $f(t) = -A \cos \omega t$ is $3\pi/2$ and hence to represent a cosinusoidal waveform given by $f(t) = -A \cos \omega t$ we draw a phasor of length A which makes an angle of 270° with the horizontal axis as shown in below figure, because its projection on vertical axis determines the position of waveform at $t = 0$.

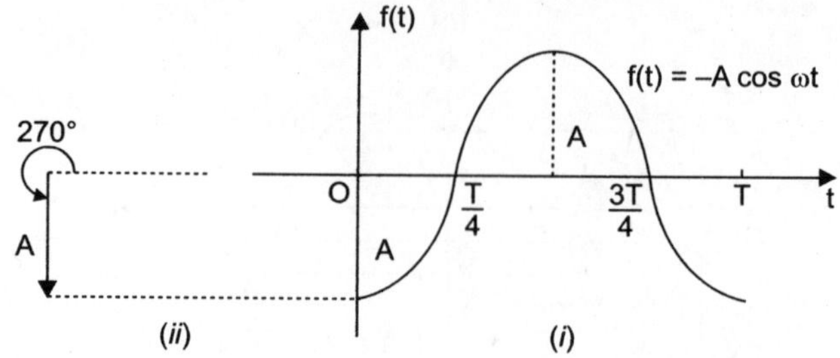

SOLVED EXAMPLES

Example 1.16. For the phasor in each part of figure shown below, determine the instantaneous sine wave value. Also express each positive angle shown as an equivalent negative angle. The length of each phasor represents the peak value of the sine wave.

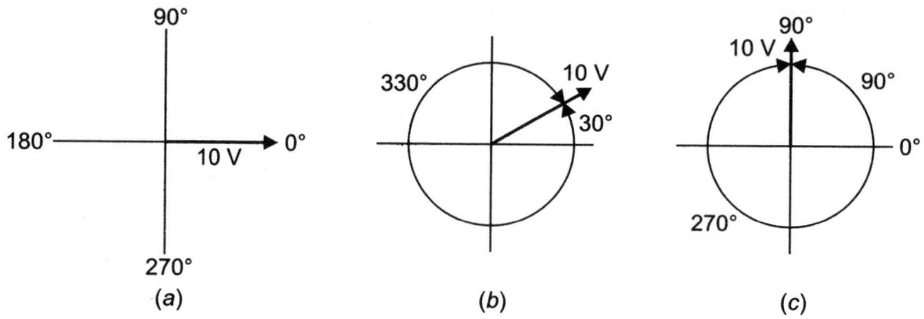

Solution. (*a*) $V = 10 \sin 0° = 10(0) = 0V$

$\theta = 0°$

(*b*) $V = 10 \sin 30° = 10(0.5) = 5V$

$\theta = 30° = \theta - 360° = 30 - 360° = -330°$

(*c*) $V = 10 \sin 90° = 10(1) = 10V$

$\theta = 90° = -270°$

Example 1.17. Draw the phasors corresponding to each of the following waveforms:

(*i*) $12 \sin(\omega t + 45°)$ **(*ii*) $6 \cos(\omega t + 30°)$**

(*iii*) $-5 \cos(\omega t - 30°)$ **(*iv*) $-6 \cos(\omega t - 45°)$**

Solution. (*i*) Let $f(t) = 12 \sin(\omega t + 45°)$. Clearly, it represents a sinusoidal waveform of amplitude 12. Threfore, the length of the phasor is 12 and it is leading $\sin \omega t$ axis by 45°. Hence, the phasor corresponding to $f(t) = 12 \sin(\omega t + 45°)$ is of length 12 and makes 45° with $\sin \omega t$ axis as shown in below figure.

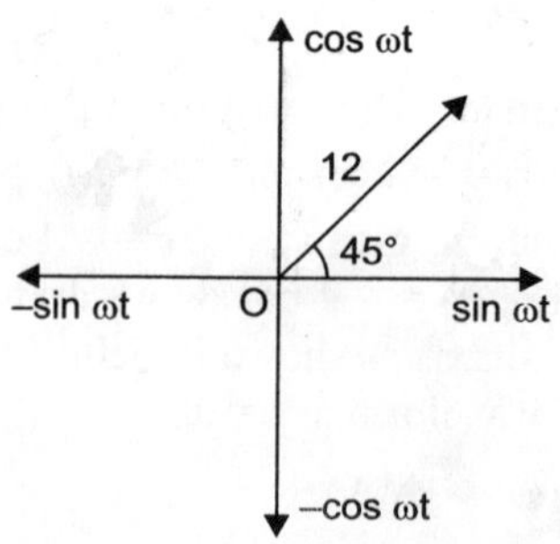

Phasor of f(t) = 12 sin (ωt + 45°)

(*ii*) Let $f(t) = 6 \cos(\omega t + 30°)$. Clearly, $f(t)$ represents a cosinusoidal waveform of amplitude 6. Therefore, the length of the phasor representing it is 6 and it is leading $\cos \omega t$ axis by 30°.

Hence, the phasor corresponding to $f(t) = 6 \cos(\omega t + 30°)$ is as shown in below figure.

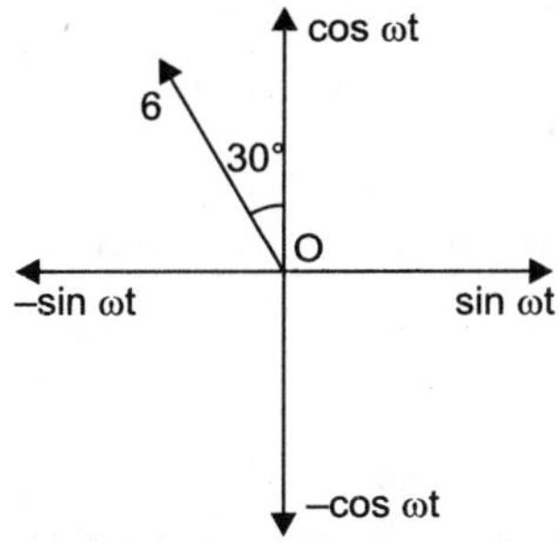

Phasor of V(t) = 6 cos (ωt + 30°)

(*iii*) Let $f(t) = -5 \cos(\omega t - 30°)$. Clearly, $f(t)$ represents a cosinusoidal waveform of amplitude 5. Therefore, the length of the phasor representing $f(t)$ is 5 and it is lagging the $-\cos \omega t$ axis by 30°. Hence, the phasor generating $f(t) = -5 \cos(\omega t - 30°)$ is as shown in Figure.

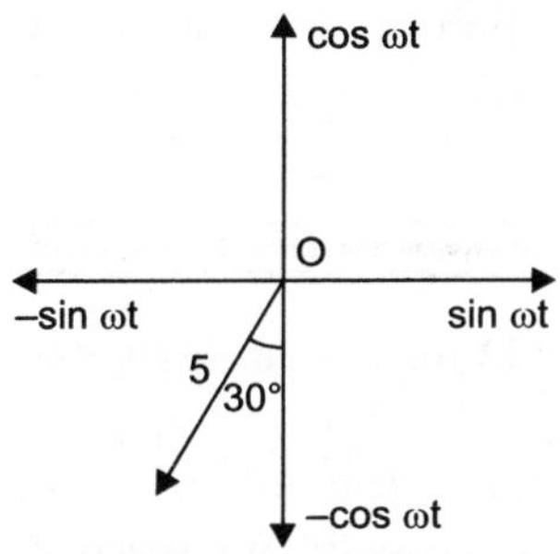

Phasor of f(t) = –5 cos (ωt – 30°)

EXERCISE 1.5

1. What is a phasor.
2. What is the angular velocity of a phasor representing a sine wave with a frequency of 1500 Hz? [**Ans.** 9425 rad/s]
3. A certain phasor has an angular velocity of 628 rad/s. To what frequency does this correspond? [**Ans.** 100 Hz]

Draw the phasor diagrams corresponding to the following functions:

4. $3 \sin(\omega t - 60°)$
5. $-4 \cos(\omega t - 50°)$
6. $-5 \sin(\omega t - 45°)$
7. $5 \cos(\omega t - 60°)$

1.16. PHASOR REPRESENTATION OF THE SUM OF TWO SINUSOIDAL FUNCTIONS

The addition of the two real sinusoidal functions can be carried out using phasor approach, if both functions are treated as the real (or imaginary) part of a rotating vectors in the complex plane. Their arithmetic operations (addition, multiplication, etc.) can be more conveniently carried out based on *phasor* representations of the sinusoidal variables.

Specifically, the sum $z(t) = x(t) + y(t)$ of the two sinusoidal functions once represented in phasor form in complex plane can be found as the real part of the vector sum in the following three steps:

- Represent both sinusoids by their corresponding phasors

$$x(t) = A\cos(\omega t + \phi) = Re[Ae^{j\phi}e^{j\omega t}] = Re[\dot{X}e^{j\omega t}]$$

$$y(t) = B\cos(\omega t + \psi) = Re[Be^{j\psi}e^{j\omega t}] = Re[\dot{Y}e^{j\omega t}]$$

where $\dot{X} = Ae^{j\phi} = A\angle\phi, \quad \dot{Y} = Be^{j\psi} = B\angle\psi$

are the phases represented as vectors in the complex plane in terms of their magnitudes and initial phases.

- Find the phasor $\dot{Z}$ of the sum $z(t) = x(t) + y(t)$ as the vector sum of the two phasors (carry out operation in frequency domain)

$$\begin{aligned}\dot{Z} &= \dot{X} + \dot{Y} = Ae^{j\phi} + Be^{j\psi} = (A\cos\phi + jA\sin\phi) + (B\cos\psi + jB\sin\psi)\\ &= (A\cos\phi + B\cos\psi) + j(A\sin\phi + B\sin\psi)\\ &= C_r + jC_i = Ce^{i\theta}\end{aligned}$$

where $\begin{cases} C_r = A\cos\phi + B\cos\psi \\ C_i = A\sin\phi + B\sin\psi \end{cases}, \quad \begin{cases} C = \sqrt{C_r^2 + C_i^2} \\ \theta = \tan^{-1}(C_i/C_r) \end{cases}$

- Get the time function $z(t)$ by multiplying its phasor $\dot{Z}$ by $e^{j\omega t}$ and taking its real part

$$z(t) = Re(\dot{Z}e^{j\omega t}) = Re(Ce^{j\phi}e^{j\omega t}) = C\cos(\omega t + \theta)$$

SOLVED EXAMPLES

Example 1.18. Express 5 cos ωt + 12 sin ωt in a single sinusoidal quantity. Also, draw the phasor diagram of the function.

Solution. Let $f_1(t) = 12\sin\omega t$ and $f_2(t) = 5\cos\omega t$

Clearly, phasor OP and OQ represent $f_1(t)$ and $f_2(t)$ respectively.

The sum of these two functions is generated by the phasor represented by the diagram OR of the parallelogram $OPRQ$ as shown in below figure.

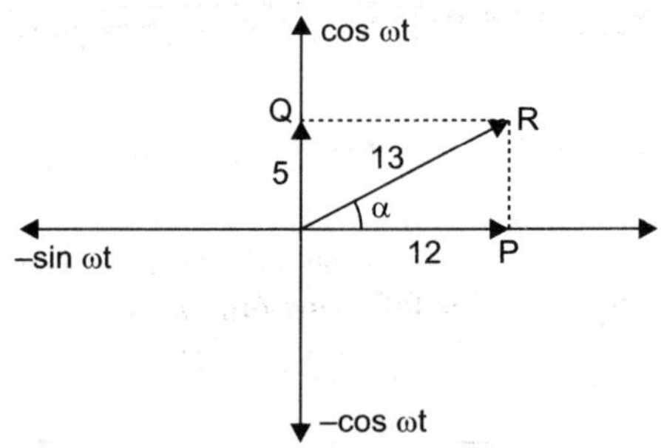

From figure, we obtain

$$OR = \sqrt{OP^2 + OQ^2} = \sqrt{12^2 + 5^2} = 13$$

Let α be the angle which OR makes with the horizontal axis. Then

$$\tan\alpha = \frac{5}{12} \Rightarrow \alpha = \tan^{-1}\frac{5}{12}$$

We find that *OR* leads the sin ω*t* axis by $\tan^{-1}\frac{5}{12}$. Therefore, equation of the sum function is given by $13\sin\left(\omega t+\tan^{-1}\frac{5}{12}\right)$.

Example 1.19. Determine the following sum by phasor method, expressing the answer as a single sinusoidal quantity 10 sin (ω*t* + 30°) + 8 sin (ω*t* + 60°).

Solution. Clearly, 10 sin (ω*t* + 30°) and 8 sin (ω*t* + 60°) represent sinusoidal waveforms of amplitudes 10 and 8 respectively. Therefore magnitude of the phasors generating them are 10 and 8 respectively as shown in below figure. Complete the parallelogram *OPRQ*. The phasor given by *OR* generates the sum functions.

Now, horizontal component of $OR = 10\cos 30° + 8\cos 60° = 5\sqrt{3}+4 = 12.66$

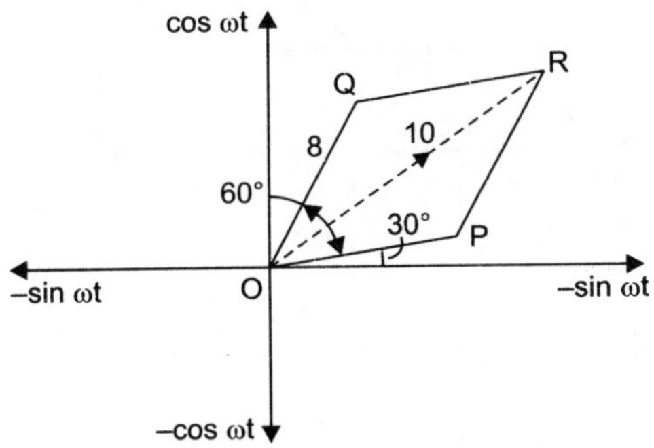

Vertical component of $OR = 10\sin 30° + 8\sin 60° = 5+4\sqrt{3} = 11.93$

$$\therefore \qquad OR = \sqrt{(12.66)^2+(11.93)^2} = 17.40$$

Clearly, *OR* leads the sin ω*t* axis by an angle α given by

$$\tan\alpha = \frac{\text{Vertical component of } OR}{\text{Horizontal component of } OR} = \frac{11.93}{12.66}$$

$$\Rightarrow \qquad \alpha = \tan^{-1}\left(\frac{11.93}{12.66}\right) = 43°30'$$

So, the equation of the sinusoidal waveform generated by *OR* is 17.40 sin (ω*t* + α).

Hence, 10 sin (ω*t* + 30°) + 8 sin (ω*t* + 60°) = 17.40 sin (ω*t* + α).

EXERCISE 1.6

Determine the following sums by phasor method, expressing answer as a single sinusoidal quantity:

1. 3 cos ω*t* + 4 sin ω*t*. $\left[\textbf{Ans.}\ 5\sin\left\{\omega t+\tan^{-1}\frac{3}{4}\right\}\right]$
2. $\sqrt{3}\sin\omega t+\cos\omega t$ $\left[\textbf{Ans.}\ 2\sin\left\{\omega t+\frac{\pi}{6}\right\}\right]$
3. 10 sin ω*t* – 5 cos ω*t* $\left[\textbf{Ans.}\ 5\sqrt{5}\sin(\omega t-26°34')\right]$
4. 6 cos ω*t* + 8 sin ω*t* [**Ans.** – 108.47 sin ω*t* – 11°9′]
5. 10 cos (ω*t* + 40°) – 15 cos (ω*t* – 60°) [**Ans.** 16 sin (ω*t* + 41°17′)]
6. – 6 sin (ω*t* + 30°) + 3 cos ω*t* $\left[\textbf{Ans.}\ -3\sqrt{3}\sin\omega t\right]$
7. 10 sin (ω*t* + 60°) + 8 sin (ω*t* + 30°) [**Ans.** 16 sin (ω*t* + 41°17′)]

1.17. PHASOR DIAGRAMS OF R-L, R-C AND R-L-C CIRCUITS

1.17.1. Phasor Diagrams of R-L Circuit

A circuit that contains a pure resistance R ohms connected in series with a coil having a pure inductance of L (Henry) is known as RL Series Circuit. When an AC supply voltage V is applied, the current, I flows in the circuit.

So, I_R and I_L will be the current flowing in the resistor and inductor respectively, but the amount of current flowing through both the elements will be same as they connected in series with each other. The circuit diagram of RL Series Circuit is shown below:

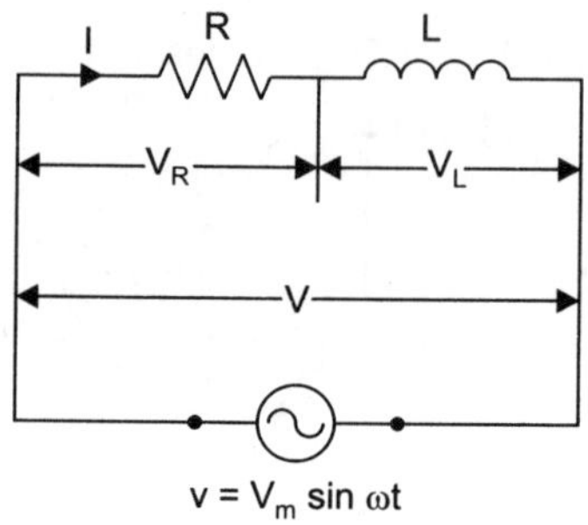

where, V_R = Voltage across the resistor R

V_L = Voltage across the inductor L

V = Total voltage of the circuit

The phasor diagram of the RL Series circuit is shown below:

Steps to draw the Phasor Diagram of RL Series Circuit

The following steps are given below which are followed to draw the phasor diagram step by step:

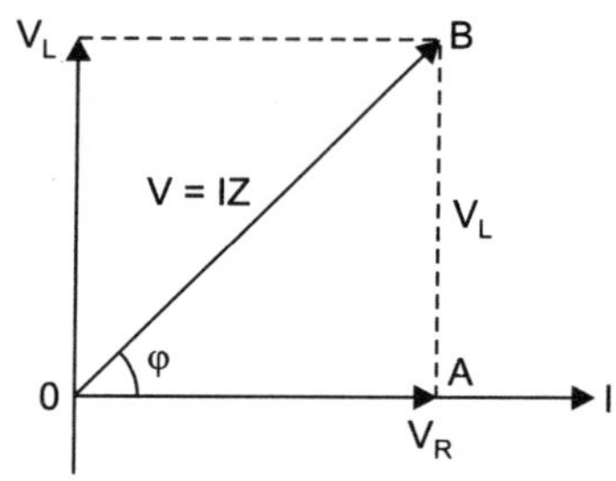

Step 1: Current I is taken as a reference.

Step 2: The voltage drop across the resistance $V_R = I_R$ is drawn in phase with the current I.

Step 3: The voltage drop across the inductive resistance $V_L = IX_L$ is drawn ahead of the current I. As the current lags voltage by an angle of 90 degrees in the pure inductive circuit.

Step 4: The vector sum of the two voltages drops, V_R and V_L is equal to the applied voltage V.

Now,

In right-angle triangle OAB

$$V_R = I_R \quad \text{and} \quad V_L = IX_L \quad \text{where } X_L = 2nfl$$

$$V = \sqrt{(V_R)^2 + (V_L)^2} = \sqrt{(IR)^2 + (IX_L)^2}$$

$$V = I\sqrt{R^2 + X_L^2} \qquad \text{or} \qquad I = L = \frac{V}{Z}$$

where, $$Z = \sqrt{R^2 + X_L^2}$$

Z is the total opposition offered to the flow of alternating current by an RL series circuit and is called impedance of the circut. It is measured in ohms (Ω).

In *RL* series circuit the current lags the voltage by 90 degrees angle known as *phase angle*. It is given by the equation:

$$\tan\varphi = \frac{X_L}{V_R} = \frac{IX_L}{IR} = \frac{X_L}{R} \qquad \text{or} \qquad \varphi = \tan^{-1}\frac{X_L}{R}$$

SOLVED EXAMPLES

Example 1.20. For the circuit shown in figure (*a*), draw the phasor circuit, impedance diagram and voltage phasor diagram.

Solution. $V = 100\angle 0$, so the phasor circuit is shown in figure (*b*)

$$Z_T = Z_R + Z_L = 3\ \Omega + j4\ \Omega = 5\angle 53.13°$$

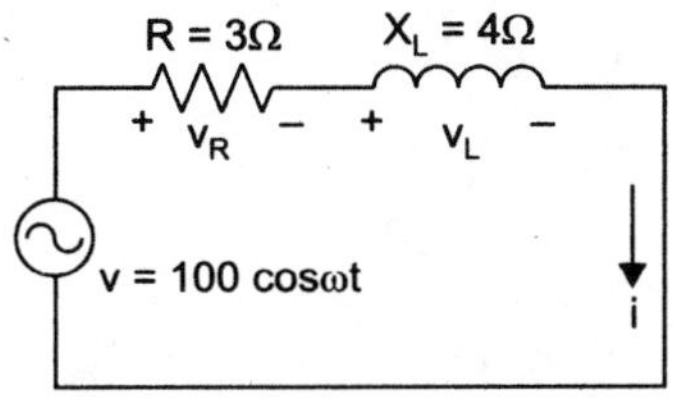

(*a*)

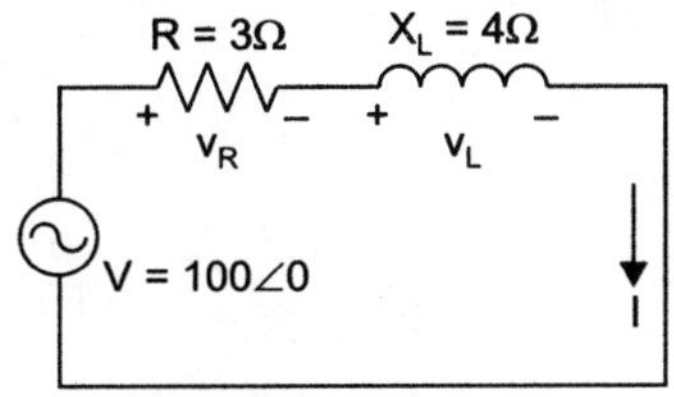

(*b*)

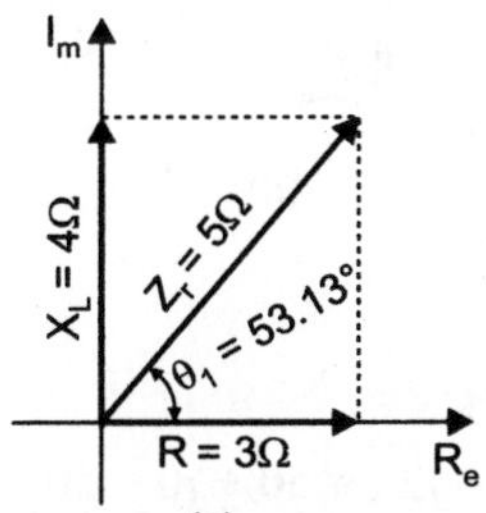

(*c*)

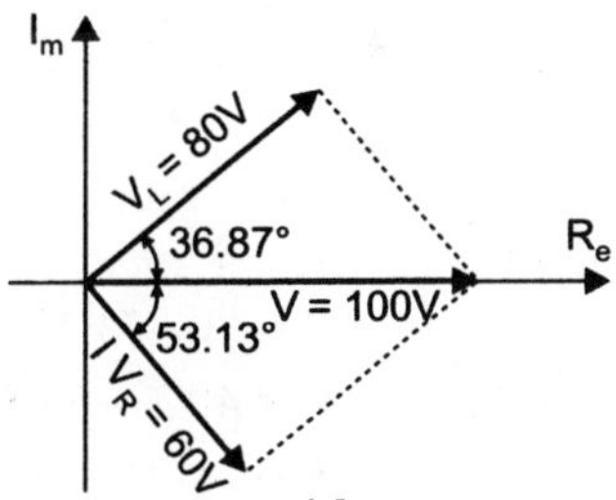

(*d*)

Impedance diagram is shown in figure (*c*)

$$I = \frac{v}{Z_T} = \frac{100\angle 0°}{5\angle 53.13°} = 20\angle -53.13°$$

$$V_R = IZ_R = (20\angle -53.13°\ A) = (3\angle 0\Omega) = 60\angle -53.13°\ \text{V}$$
$$V_L = IZ_L = (20\angle -53.13°\ A) = (4\angle 90\Omega) = 80\angle 36.87°\ \text{V}$$

Phasor diagram is shown in figure (*d*)

In rectangular form

$$V_R = 60\angle -53.13° = 36 - j48\ V$$
$$V_L = 80\angle 39.87° = 64 + j48V$$
$$V = V_R + V_L = 36 - j48 + 64 + j48$$
$$= 100 + j0V = 100\angle 0V$$

Example 1.21. For the circuit shown in figure (*a*), draw the phasor circuit, impedance diagram and voltages phasor diagram.

Solution. $I = 5\angle 0$, so the phasor circuit is shown in figure (*b*)

$$Z_T = Z_R + Z_C = 6\ \Omega - j8\ \Omega = 10\angle -53.13°$$

Impedance diagram is shown in figure (*c*)

$$V = IZ_T = (5\angle 53.13°)\ (10\angle - 53.13°) = 50\angle 0\ V$$
$$V_R = IZ_R = (5\angle 53.13°)\ (6\angle 0°) = 30\angle 53.13°0\ V$$
$$V_C = IZ_C = (5\angle 53.13°)\ A(8\angle - 90\Omega) = 40\angle - 36.87°\ V$$

Phasor diagram is shown in figure (*d*).

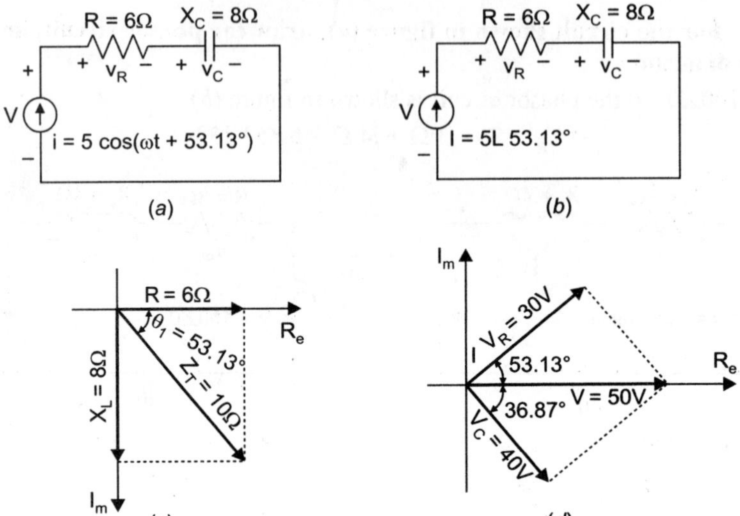

(*a*) (*b*) (*c*) (*d*)

In rectangular form

$$V_R = 30\angle 53.13° = 18 + j24V$$
$$V_C = 40\angle - 36.87° = 32 - j24V$$
$$V = V_R + V_C = 18 + j24 + 32 - j24 = 50 + j0 = 50\angle 0V$$

1.17.2. Phasor Diagrams of R-C Circuit

Consider an A.C. circuit containing a resistance of R ohms and a capacitor of C farads connected in series as shown in below figure. Let I be the current flowing through R and C. Then

Voltage drop across the resistance = $V_R = IR$

and Voltage drop across the capacitor $C = V_C = IX_C$, where X_C is the capacitive reactance.

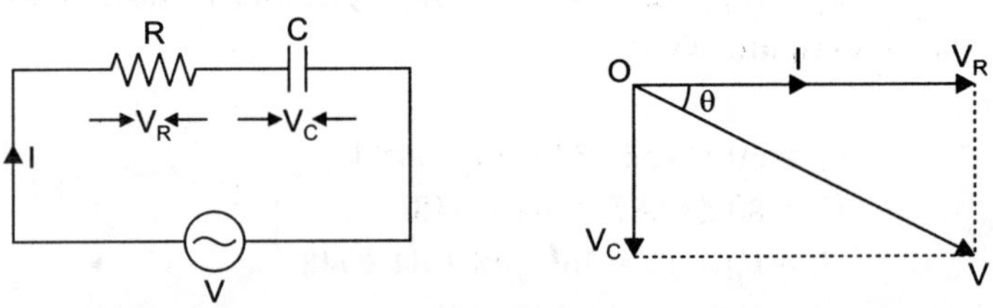

(*a*) (*b*)

Since I is common to both V_R and V_C. Therefore to draw the phasor diagram of V_R and V_C, I is taken as the reference phasor. We shall find that V_R is in phase with I and V_C lags I by 90°.

Let V be the voltage of the source, so voltage phasor V is represented by the resultant of the phasor V_C and V_R [see above figure (*b*)]

$$\therefore \quad V^2 = V_R^2 + V_C^2$$

$$\Rightarrow \quad V = \sqrt{V_R^2 + V_C^2} = \sqrt{I^2R^2 + I^2X_C^2} = I\sqrt{R^2 + X_C^2}$$

The applied voltage is lagging the current in the circuit by an angle θ given by

$$\tan\theta = \frac{V_C}{V_R} \Rightarrow \theta = \tan^{-1}\frac{V_C}{V_R} = \tan^{-1}\frac{X_C}{R}$$

The phasor diagram shown above suggests that, if we construct a right angled triangle having perpendicular sides V_R, V_C and diagonal V as shown in below, then a right angled triangle having perpendicular sides $\frac{V_R}{I}, \frac{V_C}{I}$ and diagonal $\frac{V}{I}$ will be similar to the triangle in figure shown. This new triangle is known as impedance triangle, because its each side represents $\frac{\text{Voltage}}{\text{Ampere}}$ ratio, and defines one of three: resistance, reactance or impedance.

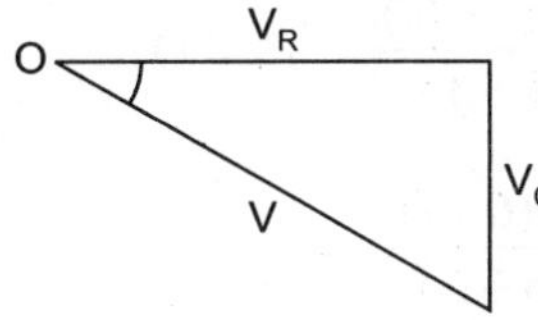

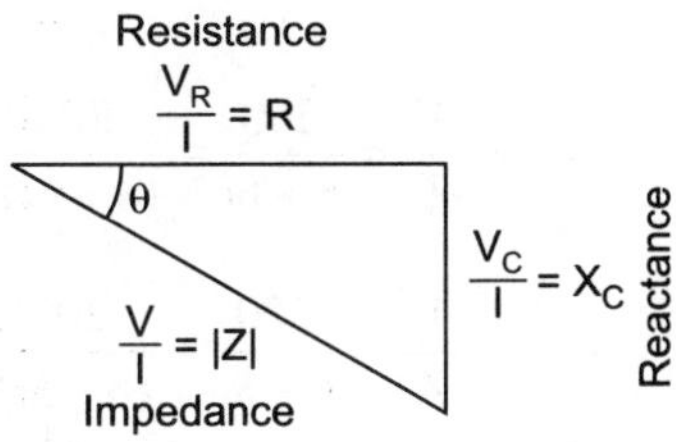

From the impedance triangle, we find that

$$|Z| = \sqrt{R^2 + X_C^2} \quad \text{and} \quad \tan\theta = \frac{X_C}{R}$$

If the phasor diagram in the above figure is superimposed on the Argand plane such that V_R is taken along the real axis and V_C along imaginary axis as shown in adjoining figure, then V can be written as a complex number given by

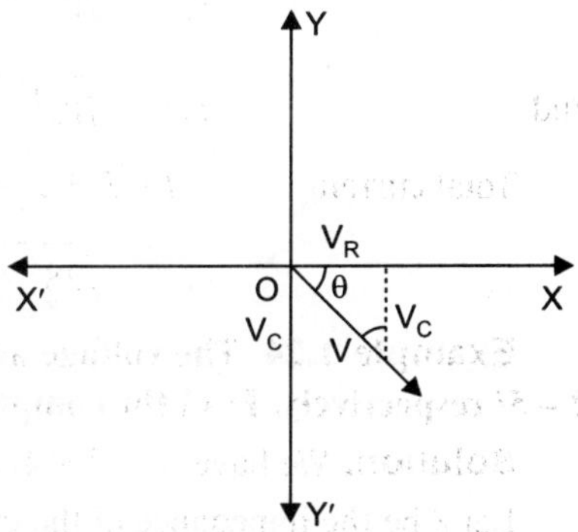

$$V = V_R - iV_C$$

$$\Rightarrow \quad \frac{V}{I} = \frac{V_R}{|I|} - i\frac{V_R}{|I|} \Rightarrow Z = R - iX_C$$

Hence, impedance is a complex number Z such that $|Z| = \sqrt{R^2 + X_C^2}$ and, $\tan\theta = \frac{X_C}{R}$

The power consumed and power factor are given by the following formulae:

$$\text{Power} = I^2 R \text{ watts, Power factor} = \frac{R}{|Z|} = \frac{\text{Resistance}}{\text{Impedance}}$$

SOLVED EXAMPLES

Example 1.22. If two impedances $Z_1 = 50\angle -40°$ and $Z_2 = 70\angle 30°$ are connected in series, find the total impedance in polar form.

Solution. We have

$$Z_1 = 50\angle -40° = 50\{\cos(-40°) + i\sin(-40°)\}$$

$$\Rightarrow \quad Z_1 = 50(\cos 40° - i\sin 40°) = 50[(0.7660) - i(0.6428)] = 38.3 - 32.14i$$

and

$$Z_2 = 70\angle 30° = 70(\cos 30° + i\sin 30°) = 70(0.8660 + 0.5\,i) = 60.62 + 35i$$

It is given that Z_1 and Z_2 are connected in series

$\therefore$ Total impedance $Z = Z_1 + Z_2 = (38.3 + 32.14i) + (60.62 + 35i) = 98.92 + 2.86i$

$$\Rightarrow \quad |Z| = \sqrt{(98.92)^2 + (2.86)^2} = 98.94$$

The argument θ of Z is given by

$$\tan\theta = \frac{2.86}{98.92} \quad \Rightarrow \quad \theta = 1.656°$$

Hence, the polar form of Z is $98.94 \angle 1.656°$ or $98.94 (\cos\theta + \sin\theta)$, where $\theta = 1.656°$.

Example 1.23. Two impedances $Z_1 = 8 + 6i$ ohms and $Z_2 = 6 - 8i$ ohms are connected in parallel across 200 volts, 50 cycles per sec in A.C. mains. Find the magnitude of the current in each branch and the total current in the circuit by symbolic method.

Solution. Let I_1 and I_2 be the current in amperes in the two branches containing impedance Z_1 and Z_2 respectively. Then

$$I_1 = \frac{V}{Z_1} = \frac{200}{8+6i} = \frac{200(8-6i)}{(8+6i)(8+6i)} = \frac{200(8-6i)}{64+36} = 16 - 12i$$

and
$$I_2 = \frac{V}{Z_2} = \frac{200}{6-8i} = \frac{200(6+8i)}{(6-8i)(6-8i)} = \frac{200(6+8i)}{36+64} = 12 + 16i$$

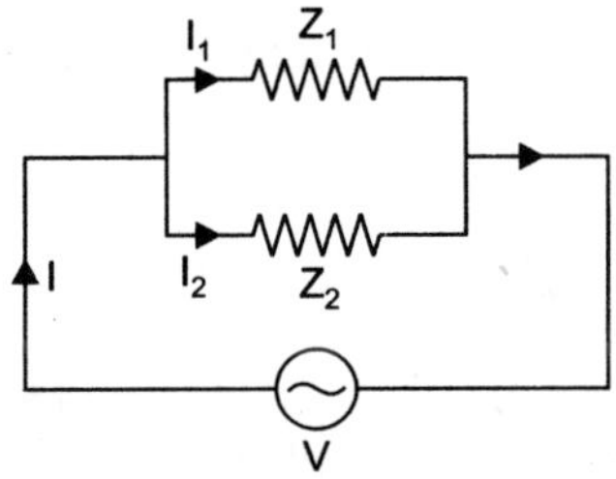

$$\therefore \quad |I_1| = \sqrt{16^2 + (-12)^2} = 20 \text{ ampere}$$

and
$$|I_2| = \sqrt{12^2 + 16^2} = 20 \text{ ampere}$$

Total current $\quad I = I_1 + I_2 = 16 - 12i + 12 + 16i = 28 + 4i$

$$\therefore \quad |I| = \sqrt{28^2 + 4^2} = 20\sqrt{2} \text{ amperes}$$

Example 1.24. The voltage and current of a circuit are given by the complex numbers $3 + 4i$ and $2 - 5i$ respectively. Find the complex number of the impedance of the circuit.

Solution. We have, $V = 3 + 4i$ and $I = 2 - 5i$

Let Z be the impedance of the circuit. Then,

$$Z = \frac{V}{I} = Z = \frac{3+4i}{2-5i} = \left(\frac{3+4i}{2-5i}\right)\frac{(2+5i)}{(2+5i)} = \frac{-14+23i}{29} = \frac{-14}{29} + \frac{23}{29}i$$

1.17.3 Phasor Diagrams of R-L-C Circuit

The **RLC circuit** is shown below:

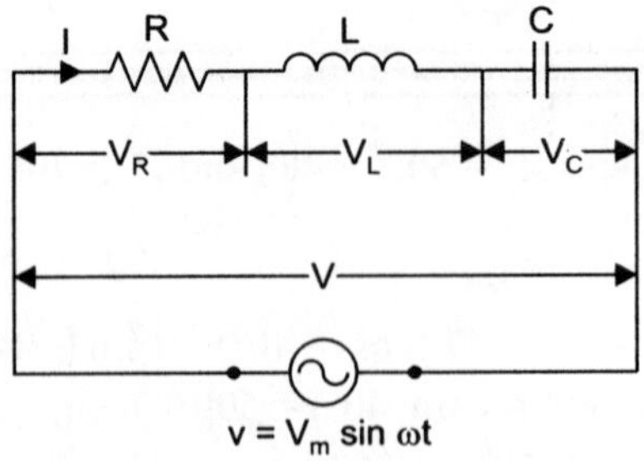

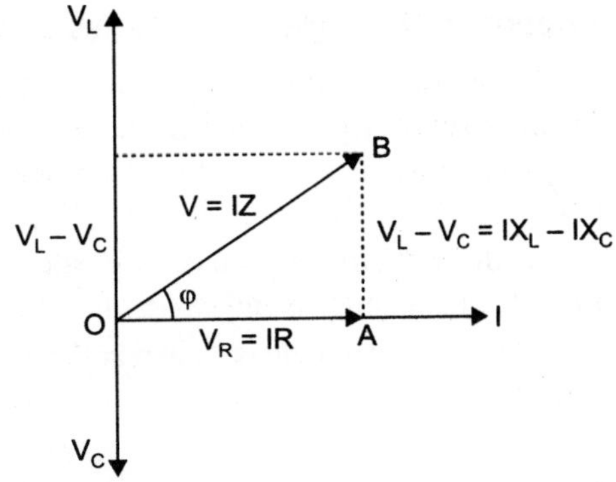

When the AC voltage is applied through the *RLC* series circuit the resulting current *I* flows through the circuit, and thus the voltage across each element will be:

- $V_R = IR$ that is the voltage across the resistance *R* and is in phase with the current *I*.
- $V_L = IX_L$ that is the voltage across the inductance *L* and it leads the current *I* by an angle of 90 degrees.
- $V_C = IX_C$ that is the voltage across the capacitor *C* and it lags the current *I* by an angle of 90 degrees.

The phasor diagram of the *RLC* series circuit when the circuit is acting as an inductive circuit that means ($V_L > V_C$) is shown below and if ($V_L < V_C$) the circuit will behave as a capacitive circuit.

Steps to draw the Phasor Diagram of the RLC Series Circuit

Step 1: Take current *I* as the reference as shown in the figure above.

Step 2: The voltage across the inductor *L* that is V_L is drawn leads the current *I* by a 90° angle.

Step 3: The voltage across the capacitor *c* that is V_C is drawn lagging the current *I* by a 90° angle because in capacitive load the current leads the voltage by an angle of 90 degrees.

Step 4: The two vector V_L and V_C are opposite to each other.

$$V = \sqrt{(V_R)^2 + (V_L - V_C)^2} = \sqrt{(IR)^2 + (IX_L - IX_C)^2}$$

or

$$V = I\sqrt{R^2 + (X_L - X_C)^2}$$

or

$$I = \frac{V}{\sqrt{R^2 + (X_L - X_C)^2}} = \frac{V}{Z}$$

where, $Z = \sqrt{R^2 + (X_L - X_C)^2}$

It is the total opposition offered to the flow of current by an *RLC* circuit and is known as **Impedance** of the circuit.

From the phasor diagram, the value of phase angle will be

$$\tan\varphi = \frac{V_L - V_C}{V_R} = \frac{X_L - X_C}{R} \quad \text{or} \quad \varphi = \tan^{-1}\frac{X_L - X_C}{R}$$

Power in RLC Series Circuit

The product of voltage and current is defined as power

$$P = VI\cos\varphi = I^2R$$

where cos ϕ is the power factor of the circuit and is expressed as

$$\cos\varphi = \frac{V_R}{V} = \frac{R}{Z}$$

The Three Cases of RLC Series Circuit

- When $X_L > X_C$, the phase angle ϕ is positive. The circuit behaves as *RL* series circuit in which the current lags behind the applied voltage and the power force is lagging.
- When $X_L > X_C$, the phase angle ϕ is negative, and the circuit acts as a series *RC* circuit in which the current leads the voltage by 90 degrees.
- When $X_L = X_C$, the phase angle ϕ is zero, as a result, the circuit behaves like a purely resistive circuit. In this type of circuit, the current and voltage are in phase with each other. The value of the power factor is **unity**.

Impedance Triangle of RLC Series Circuit

When the quantities of the phasor diagram are divided by the common factor I then the right angle triangle is obtained known as impedance triangle. The impedance triangle of the RL series circuit, when $(X_L > X_C)$ is shown below:

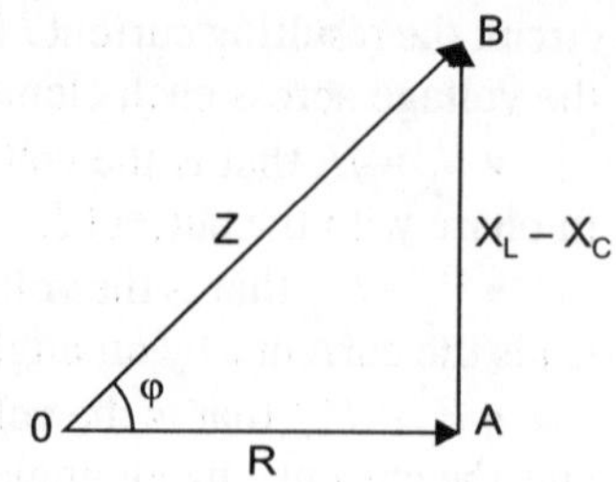

If the inductive reactance is greater than the capacitive reactance than the circuit reactance is inductive giving a lagging phase angle.

Impedance triangle is shown below when the circuit acts as an RC series circuit $(X_L < X_C)$.

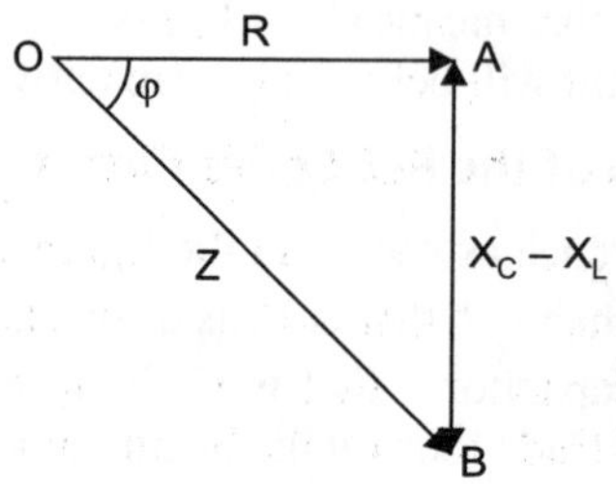

When the capacitive reactance is greater than the inductive reactance the overall circuit reactance acts as capacitive and the phase angle will be leading.

SOLVED EXAMPLES

Example 1.25. A sinusoidal voltage of *R.M.S.* of value 100 volts and frequency of 50 cycles per second is applied to the circuit which has two resistances of 3 ohms and 4 ohms, two inductances of 0.0159 Henry and 0.0477 Henry and a capacitor of 318 micro-farad. Calculate the current in the circuit and find its phase relative to the applied voltage.

Solution. We have

$$f = 50 \text{ cycles/sec}, R = 3 + 4 = 7 \text{ ohms}, L = (0.0159 + 0.0477) = 0.0636 \text{ Henry}$$

and $$C = 318 \times 10^{-6} \text{ farad}$$

$\therefore$ $$\omega = 2\pi f = 100\,\pi \text{ radians/sec}, X_L = L\omega = 0.0636 \times 100\pi = 19.981 \text{ ohms}$$

and $$X_C = \frac{1}{\omega C} = \frac{1}{100\pi \times 318 \times 10^{-6}} = 10.0097 \text{ ohms}$$

Let Z be the impedance of the circuit. Then

$$Z = R + i(X_L - X_C) \Rightarrow Z = 7 + i(19.981 - 10.0097) = 7 + 9.9713i$$

Let I be the current in the circuit. Then

$$I = \frac{V}{Z}$$

$\Rightarrow$ $$I = \frac{100}{7 + 9.9713i} = \frac{100(7 - 9.9713)i}{(7 + 9.9713i)(7 - 9.9713i)}$$

$$= \frac{700 - 997.13i}{49 + (9.9713)^2} = \frac{700 - 997.13i}{148.4268} = 4.716 - 6.718i$$

$\Rightarrow$ $$|I| = \sqrt{(4.716)^2 + (-6.718)^2} = 8.21$$

Let θ be the arrangement of I. Then

$$\tan\theta = \left(\frac{-6.718}{4.716}\right) \Rightarrow \theta = \tan^{-1}\left(\frac{-6.718}{4.716}\right) = \tan^{-1}(-1.4245) = -54°56'$$

Example 1.26. A resistance of 20 ohms, an inductance of 0.2 Henry and a capacitance of 100 micro-farad are connected in series across 220 volts, 50 cycles/sec. mains. Determine the following: (*i*) impedance, (*ii*) current, (*iii*) voltage across *L*, *R* and *C*.

Solution. We have

$$R = 20 \text{ ohms } L = 0.2 \text{ Henry}, C = 100 \times 10^{-6} \text{ farad}, V = 200 \text{ volts}, \omega = 2\pi \times 50 \text{ radians/sec}$$

$$\therefore \quad Y_L = L\omega = 0.2 \times 100\pi = 62.8319 \text{ ohms}$$

and
$$X_C = \frac{1}{C\omega} = \frac{1}{100\times10^{-6}\times100\pi} = 31.8319 \text{ ohms}$$

(*i*) Let Z be the impedance of the circuit. Then

$$Z = R + i(X_L - X_C)$$

$$\Rightarrow \quad Z = 20 + i(62.8319 - 31.8319) = 20 + 31i$$

$$\therefore \quad |Z| = \sqrt{(20)^2 + (31)^2} = 36.89 \text{ ohms}$$

(*ii*) Let I be the current in the circuit. Then

$$I = \frac{V}{|Z|} \Rightarrow I = \frac{200}{36.89} = 5.42 \text{ amperes}$$

(*iii*) Let V_L be the voltage across L. Then

$$L = V_L = 1\,X_L = 5.42(62.8319) = 340.55 \text{ volts}$$

Voltage across $R = V_R = IR = 5.42 \times 20 = 108.4$ volts

Voltage across $C = V_C = IX_C = 5.42 \times 31.8319 = 172.52$ volts

EXERCISE 1.7

1. The current in a circuit is given by $4.5 + 12i$ when the applied voltage is $100 + 150i$. Determine the complex expression for the impedance stating whether it is inductive or capacitive. [**Ans.** $13.7 - 3.2i$; Capacitive]
2. If the potential difference across a circuit is represented by $40 + 25i$ volts, and the circuit consists of a resistance of 20 ohms in series with an inductance of 0.06 Henry and the frequency is 79.5 cycles per second. Find the complex number representing the current in amperes. [**Ans.** $1.19 - 0.54i$ amperes]
3. If the impedances $Z_1 = 50\angle -40°$ and $Z_2 = 70\angle 30°$ are connected in series, find the total impedance in polar form. [**Ans.** $\theta = 1.656°\ 98.94\ (\cos\theta + \sin\theta)$]
4. The voltage and current of a circuit are given by the complex numbers $3 + 4i$ and $2 - 5i$ respectively. Find the complex number of the impedance of the circuit. $\left[\textbf{Ans. } z = \frac{-14}{29} + \frac{23}{29}i\right]$
5. A coil of resistance 10 ohms and inductance 0.1 Henry is connected in series with a condensor of 150 micro-farad across a 200 volt, 50 cycles per second supply. Find the impedance of the circuit. [**Ans.** $10 + 10.21i$; $\theta = 45°6'$]
6. A resistance of 200 ohms, a capacitor of 50 micro-farad and an inductance of 100 mH are connected in series. Find an equivalent parallel circuit, containing two elements, at a frequency of 50 Hz. [**Ans.** $R = 205.2$ ohms, $C = 2.5\ \mu$F]

❑❑❑

CHAPTER 2

Matrices and Determinants

INSIDE THIS CHAPTER

2.1. MATRICES

Suppose we wish to express that Anita has 7 books. We may express it as [7] with the understanding that the number inside [ ] denotes the number of books that Anita has. Next, suppose we want to express that Anita has 7 books, 4 pencils. We may express it as [7, 4] with the understanding that the first entry inside [ ] denotes the number of books while the second entry, the numbers of pencils possessed by Anita. Let us now suppose that we have the three persons—Amar, Akbar and Anthony.

Suppose Amar has 7 books and 4 pencils, Akbar has 6 books and 8 pencils and Anthony has 10 books and 3 pencils. This information can be represented in the tabular form as follows:

	Books	*Pencils*
Amar	7	4
Akbar	6	8
Anthony	10	3

We can briefly write these as follows:

$$\begin{array}{r c} & \begin{array}{cc}\text{First column} & \text{Second column}\\ \downarrow & \downarrow\end{array}\\ \begin{array}{r}\text{First row}\rightarrow\\ \text{Second row}\rightarrow\\ \text{Third row}\rightarrow\end{array} & \begin{bmatrix} 7 & 4\\ 6 & 8\\ 10 & 3\end{bmatrix}\end{array}$$

The entries in the first, second and third row represent the number of books and pencils possessed by Amar, Akbar and Anthony, respectively, whereas the entries in the first and second column represent in the number of books and the number of pencils, respectively.

Definition: A set of mn number (real or complex) arranged in the form of rectangular array having m rows and n columns is called a matrix of order $m \times n$ or simply $m \times n$ or simply $m \times n$ matrix (to be real 'm' by 'n' matrix)

An $m \times n$ matrix is usually written as:

$$A = \begin{bmatrix} a_{11} & a_{12} & \cdots & a_{1n} \\ a_{21} & a_{22} & \cdots & a_{2n} \\ a_{31} & a_{32} & \cdots & a_{3n} \\ \vdots & \vdots & \vdots & \vdots \\ a_{m1} & a_{m2} & \cdots & a_{mn} \end{bmatrix}$$

In a compact form the above matrix is represented by $A = [a_{ij}]$, $i = 1, 2, \ldots m$, $j = 1, 2, \ldots n$ or simply by $[a_{ij}]_{m \times n}$, we write the general element of the matrix and enclose it in brackets of the type [ ] or ( ).

The numbers $a_{11}, a_{12}, \ldots$ etc. of this rectangular array are called the **element** or **entries** of the matrix. The element a_{ij} belongs to the ith row and jth column and is sometimes called the (i, j)th element of the matrix.

Some examples of matrices are:

(*i*) $\begin{bmatrix} 1 & 2 \\ 3 & 4 \end{bmatrix}$ is a 2×2 matrix, whose elements are 1, 2, 3, 4.

(*ii*) $\begin{bmatrix} -1 & 2 & -3 \\ -4 & 0 & 1 \end{bmatrix}$ is a matrix of order 2×3, whose elements are – 1, 2, – 3, – 4, 0, 1.

(*iii*) $\begin{bmatrix} 1-i & 3 & 4 \\ 1 & -1+i & 3+i \\ 5 & -1 & 1+i \end{bmatrix}$ is a 3×3 matrix.

Notations: A matrix is denoted by a capital letter such as A, B, C, etc.

Note: A matrix is essentially an arrangement of numbers and has no value.

2.2. TYPES OF MATRICES

Certain types of matrices, which play important roles in matrix theory are now considered.

(a) **Row Matrix.** If a matrix has only one row any number of columns, it is called a row matrix. For example, the matrix

$$A = [2 \quad 7 \quad 9 \quad 3] \text{ is row matrix of order } 1 \times 4.$$

(b) **Column Matrix.** A matrix consisting of a single column any number of rows, is called a column matrix. For example, the matrix

$$B = \begin{bmatrix} 4 \\ 5 \\ 6 \end{bmatrix} \text{ is a column matrix of order } 3 \times 1.$$

(c) **Zero or Null Matrix.** A matrix is said to be zero matrix or null matrix if all its elements are zero. For example,

$$O = \begin{bmatrix} 0 & 0 & 0 \\ 0 & 0 & 0 \end{bmatrix} \text{ is a } 2 \times 3 \text{ zero matrix.}$$

(*d*) **Square Matrix.** A matrix in which the number of rows is equal to the number of columns, say n, is called a square matrix of order $n \times n$ or simply or n. For example,

$$\begin{bmatrix} 1 & 2 \\ 3 & 4 \end{bmatrix} \text{ and } \begin{bmatrix} 5 & 4 & 6 \\ -3 & 1 & 0 \\ 6 & 3 & -1 \end{bmatrix} \text{ are square matrices of order 2 and 3.}$$

(*e*) **Diagonal Matrix.** A square matrix is said to be diagonal if each of its entries not falling on the main diagonal is zero. Thus, a square matrix $A = (a_{ij})$ is diagonal if $a_{ij} = 0$ for $i ^1 j$. For example,

$$\begin{bmatrix} 1 & 0 & 0 \\ 0 & 2 & 0 \\ 0 & 0 & 3 \end{bmatrix} \text{ is a diagonal matrix.}$$

A diagonal matrix of order n with diagonal elements $(d_1, d_2, ..., d_n)$ is denoted as diagonal $(d_1, d_2, ..., d_n)$.

(*f*) **Scalar Matrix.** A square matrix whose all non-diagonal elements are zero and diagonal elements are equal is called a scalar matrix. Thus, if $A = (a_{ij})$ is a square matrix and

$$a_{ij} = \begin{cases} K, & \text{if } i = j \\ 0, & \text{if } i \neq j \end{cases}, \text{ then A is a scalar matrix.}$$

For example, $\begin{bmatrix} 2 & 0 & 0 \\ 0 & 2 & 0 \\ 0 & 0 & 2 \end{bmatrix}$ is a scalar matrix of order 3.

(*g*) **Identify Matrix or Unit Matrix.** A square matrix is said to be identity matrix or unit matrix if all its main diagonal entries are 1's and all other entries are 0's. Thus, the square matrix $A = (a_{ij})$ is an identity matrix if

$$a_{ij} = \begin{cases} 1, & \text{if } i = j \\ 0, & \text{if } i \neq j \end{cases}$$

An identity of order n is denoted by I_n or more simply by I. For example, [1], $\begin{bmatrix} 1 & 0 \\ 0 & 1 \end{bmatrix}$, $\begin{bmatrix} 1 & 0 & 0 \\ 0 & 1 & 0 \\ 0 & 0 & 1 \end{bmatrix}$ are identity matrices of order 1, 2 and 3 respectively.

(*h*) **Upper Triangular Matrix.** A square matrix $A = [a_{ij}]$ is called upper triangular matrix if $a_{ij} = 0$ for all $i > j$.

$$A = \begin{bmatrix} 2 & 4 & 3 & 1 \\ 0 & 5 & 2 & 3 \\ 0 & 0 & 3 & 4 \\ 0 & 0 & 0 & 5 \end{bmatrix} \text{ is an upper triangular matrix.}$$

(*i*) **Lower Triangular Matrix.** A square matrix $A = [a_{ij}]$ is called a lower triangular matrix if $a_{ij} = 0$ for all $i < j$.

$$A = \begin{bmatrix} 5 & 0 & 0 & 0 \\ 3 & 4 & 0 & 0 \\ 5 & 2 & 3 & 0 \\ 1 & 2 & 3 & 4 \end{bmatrix} \text{ is a lower triangular matrix.}$$

2.3. OPERATIONS ON MATRICES

2.3.1. Equality of Matrices

Two matrices A and B are said to be equal if they are of same order and if all the corresponding entries are equal. It is written as $A = B$. For example, if

$$\begin{bmatrix} a & b \\ c & d \end{bmatrix} = \begin{bmatrix} 1 & 2 \\ 3 & 4 \end{bmatrix}$$

then, $a = 1, b = 2, c = 3, d = 4$.

2.3.2. Addition of Matrices

Let A and B be two matrices each of order $m \times n$. Then the sum matrix $A + B$ is a matrix of order $m \times n$ and is obtained by adding the corresponding elements of A and B.

Thus, if $A = [a_{ij}]_{m \times n}$ and $B = [b_{ij}]_{m \times n}$ are two matrices of same order then the sum $A + B$ is defined to be the matrix of order $m \times n$ such that $(A + B)_{ij} = a_{ij} + b_{ij}$, $i = 1, 2, ..., m$ and $j = 1, 2, ..., n$ *i.e.*, (i, j)th element of $A + B = a_{ij} + b_{ij}$.

For example, if $A = \begin{bmatrix} 3 & 4 & 5 \\ 2 & 0 & 4 \end{bmatrix}$ and $B = \begin{bmatrix} -1 & 2 & -2 \\ 1 & -1 & 3 \end{bmatrix}$

then $A + B = \begin{bmatrix} 3+(-1) & 4+2 & 5+(-2) \\ 2+1 & 0+(-1) & 4+3 \end{bmatrix} = \begin{bmatrix} 2 & 6 & 3 \\ 3 & -1 & 7 \end{bmatrix}.$

If A is any matrix, the negative of A, denoted by $-A$, is the matrix obtained by replacing each entry in A by its negative. For example, if

$$A = \begin{bmatrix} 2 & -1 \\ 5 & 4 \\ -6 & 0 \end{bmatrix}, \text{ then } -A = \begin{bmatrix} -2 & 1 \\ -5 & -4 \\ 6 & 0 \end{bmatrix}.$$

Properties of Addition of Matrices

1. *Matrix addition is commutative.* That is, if A and B be two matrices of the same order, then $A + B = B + A$.
2. *Matrix addition is associative.* That is, if A, B, and C be three matrices of the same order, then $(A + B) + C = A + (B + C)$.
3. *Existance of additive identity.* That is, if O is the zero matrix of the same order as that of the matrix A, then $A + O = A = O + A$.
4. *Existance of additive inverse.* That is, if A be any matrix, then $A + (-A) = O = (-A) + A$.

2.3.3. Subtraction of Matrices

Let A and B be two matrices of the same order. Then by $A - B$, we mean $A + (-B)$. In other words, to find $A - B$ we subtract each entry of B from the corresponding entry of A. For example,

Let $A = \begin{bmatrix} 5 & -2 \\ 3 & 4 \\ 6 & -8 \end{bmatrix}$ and $B = \begin{bmatrix} 0 & 2 \\ 5 & -2 \\ 6 & 1 \end{bmatrix}$

Then, $A - B = \begin{bmatrix} 5-0 & -2-2 \\ 3-5 & 4-(-2) \\ 6-6 & -8-1 \end{bmatrix} = \begin{bmatrix} 5 & -4 \\ -2 & 6 \\ 0 & -9 \end{bmatrix}$

2.3.4. Multiplication of a Matrix by a Scalar

Let $A = [a_{ij}]$ be an $m \times n$ matrix and k be any scalar. Then the matrix obtained by multiplying each element of A. by k is called the scalar multiple of A by k and is denoted by kA. Thus, if

$$A = [a_{ij}]_{m \times n} \text{ then } kA = [kA]_{m \times n}$$

For example, if $A = \begin{bmatrix} 5 & 6 & 4 \\ 2 & -3 & -1 \end{bmatrix}$ then

$$4A = 4\begin{bmatrix} 5 & 6 & 4 \\ 2 & -3 & -1 \end{bmatrix} = \begin{bmatrix} 20 & 24 & 16 \\ 8 & -12 & -4 \end{bmatrix}$$

Properties:

1. If A and B are two matrices of the same order and if k be a scalar, then
$$k(A + B) = kA + kB$$
2. If k_1 and k_2 are two scalars and if A is a matrix, then
$$(k_1 + k_2) A = k_1A + k_2A.$$
3. If k_1 and k_2 are two scalars and if A is a matrix, then
$$(k_1k_2)A = k_1(k_2A) = k_2(k_1A)$$
4. If A is any matrix, then
$$1 \cdot A = A$$

2.3.5. Multiplication of Matrices

The product AB of two matrices A and B is defined when the number of columns in A is equal to the number of rows in B.

Thus, if A is matrix of order $m \times n$ and B is matrix of order $n \times p$, then A and B are said to be conformal for the product AB. In that case the product AB gives rise to a matrix of order $m \times p$ and is given by

$$4B = [C_{ik}] = \left[\sum_{j=1}^{n} a_{ij}\, b_{jk}\right]; \; i = 1, 2, ..., m; \qquad k = 1, 2, ..., p$$

where C_{ik} = Element of ith row, kth column

$$= a_{i1}\, b_{1k} + a_{i2}\, b_{2k} + ... + a_{in}\, b_{nk} = [a_{i1} \quad a_{i2} \quad \cdots \quad a_{in}]\begin{bmatrix} b_{1k} \\ b_{2k} \\ \vdots \\ b_{nk} \end{bmatrix}$$

For example, if $A = \begin{bmatrix} a_1 & a_2 & a_3 \\ b_1 & b_2 & b_3 \\ c_1 & c_2 & c_3 \end{bmatrix}$ and $B = \begin{bmatrix} \alpha_1 & \beta_1 \\ \alpha_2 & \beta_2 \\ \alpha_3 & \beta_3 \end{bmatrix}$

then, $AB = \begin{bmatrix} a_1\alpha_1 + a_2\alpha_2 + a_3\alpha_3 & a_1\beta_1 + a_2\beta_2 + a_3\beta_3 \\ b_1\alpha_1 + b_2\alpha_2 + b_3\alpha_3 & b_1\beta_1 + b_2\beta_2 + b_3\beta_3 \\ c_1\alpha_1 + c_2\alpha_2 + c_3\alpha_3 & c_1\beta_1 + c_2\beta_2 + c_3\beta_3 \end{bmatrix}$

In the product AB, A is called the and B is called the **post-factor.**

Note:

(*i*) Product AB is different from product BA.

(*ii*) If A is of order $m \times n$ and both AB and BA are defined then B has to be of order $n \times m$.

(*iii*) $AB = AC$ does not necessaily imply $B = C$. However, if $AB = AC$ and $BA = CA$, then it necessarily implies $B = C$.

Properties:

1. *Matrix multiplication is associative.* That is, if A, B and C are matrices of order $m \times n$, $n \times p$ and $p \times q$ respectively, then
$$(AB)\,C = A(BC)$$
2. *Matrix multiplication distributes over matrix addition.* That is if A is matrix of order $m \times n$ and if B and C are matrices of order $n \times p$, then
$$A(B + C) = AB + BC.$$
3. *Multiplication identity.* If A is an $m \times n$ matrix and if I_n denotes the identity matrix of order $n \times n$ and I_m denotes the identity matrix of order $m \times m$, then
$$I_m A = A \quad \text{and} \quad AI_n = A$$
In particulars if A a square matrix of order n, then
$$AI_n = I_n A = A$$
4. *If A is a square matirx.* then
$$A^m A^n = A^{m+n}\ \forall m, n \in N \quad \text{and} \quad (A^m)^n = A^{mn}\ \forall\ m, n \in N.$$

2.3.6. Transpose of a Matrix

Let $A = [a_{ij}]$ be an $m \times n$ matrix. Then, the transpose of A, denoted by A' or A^T, A^t, is an $n \times m$ matrix such that
$$(A')_{ij} = a_{ij} \text{ for all } i = 1, 2, ..., m; j = 1, 2, ..., n$$
Thus, A' is obtained from A by changing its rows into column and its columns into rows. For example, if
$$A = \begin{bmatrix} 1 & 3 \\ 5 & 7 \\ 2 & -3 \end{bmatrix}, \quad \text{then} \quad A' = \begin{bmatrix} 1 & 5 & 2 \\ 3 & 7 & -3 \end{bmatrix}$$
The first row of A' is the first column of A. The second row of A' is the second column of A and so on.

Properties:

Let A and B be two matrices. Then

1. $(A')' = A$
2. $(A + B)' = A' + B'$
3. $(kA)' = kA'$, where k is any scalar.
4. $(AB)' = B'A'$, where A and B are conformal for the product AB. **[Reversal law]**

Symmetric Matrix

A is said to be symmetric if $A' = A$,

or $A = [a_{ij}]_{m \times n}$ is said tobe symmetric if $a_{ij} = a_{ij}$ for all i and j.

For example,
$$A = \begin{bmatrix} a & f & e \\ f & b & d \\ e & d & c \end{bmatrix} \quad \text{and} \quad A' = \begin{bmatrix} a & f & e \\ f & b & d \\ e & d & c \end{bmatrix}$$
are equal and hence A is symmetric.

Skew-Symmetric Matrix

A square matrix A is said to be skew-symmetric if $A' = -A$. For example,

$$A = \begin{bmatrix} 0 & a & b \\ -a & 0 & c \\ -b & -c & 0 \end{bmatrix} \quad \text{and} \quad A' = \begin{bmatrix} 0 & -a & -b \\ a & 0 & -c \\ b & c & 0 \end{bmatrix}$$

are such that $A = -A'$ and hence A is skew-symmetric.

Note: If A is a skew-symmetric matrix, its diagonal elements must be zero.

Related Properties

1. If A is a any square matrix, then $A + A'$ is a symmetric matrix and $A - A'$ is a skew-symmetric matrix.
2. Every square matrix can be expressed as the sum of a symmetric matrix and a skew-symmetric matrix as shown below:

$$A = \frac{1}{2}(A + A') + \frac{1}{2}(A - A)'$$

SOLVED EXAMPLES

Example 2.1. Find x, y, z, w when $\begin{bmatrix} x+y & 2x-5 \\ 2x+y & 3z-w \end{bmatrix} = \begin{bmatrix} 1 & 5 \\ 0 & 13 \end{bmatrix}$.

Solution. Since the corresponding elements of two equal matrices are equal, therefore

$$x + y = 1, \quad 2x - z = 5$$
$$2x + y = 0, \quad 3z - w = 13$$

On solving these equations, we get $x - 1, y = 2, z = -7, w = -34$.

Example 2.2. If $A = \begin{bmatrix} 3 & 2 \\ 4 & 1 \end{bmatrix}$ and $B = \begin{bmatrix} a & b \\ 3 & 5 \end{bmatrix}$, find a and b such that $AB = BA$. Compute $3A + 5B$.

Solution. We have $\quad A = \begin{bmatrix} 3 & 2 \\ 4 & 1 \end{bmatrix}$ and $B = \begin{bmatrix} 3 & 2 \\ 4 & 1 \end{bmatrix}$

$$\therefore \quad AB = \begin{bmatrix} 3 & 2 \\ 4 & 1 \end{bmatrix}\begin{bmatrix} a & b \\ 3 & 5 \end{bmatrix} = \begin{bmatrix} 3a+6 & 3b+10 \\ 4a+3 & 4b+5 \end{bmatrix}$$

and

$$BA = \begin{bmatrix} a & b \\ 3 & 5 \end{bmatrix}\begin{bmatrix} 3 & 2 \\ 4 & 1 \end{bmatrix} = \begin{bmatrix} 3a+4b & 2a+b \\ 29 & 11 \end{bmatrix}$$

$$AB = BA \Rightarrow 4a + 3 = 29 \text{ and } 4b + 5 = 11$$

$$\therefore \quad a = \frac{13}{2} \text{ and } b = \frac{3}{2}$$

Now,

$$3A + 5B = \begin{bmatrix} 9 & 6 \\ 12 & 3 \end{bmatrix} + \begin{bmatrix} \frac{65}{2} & \frac{15}{2} \\ 15 & 25 \end{bmatrix} = \begin{bmatrix} \frac{83}{2} & \frac{27}{2} \\ 27 & 28 \end{bmatrix}.$$

Example 2.3. Find the value of x such that $[1 \quad 1 \quad x]\begin{bmatrix}1 & 0 & 2\\0 & 2 & 1\\2 & 1 & 0\end{bmatrix}\begin{bmatrix}1\\1\\1\end{bmatrix} = 0.$

Solution. We have $[1 \quad 1 \quad x]\begin{bmatrix}1 & 0 & 2\\0 & 2 & 1\\2 & 1 & 0\end{bmatrix}\begin{bmatrix}1\\1\\1\end{bmatrix} = 0$

$$\Rightarrow \quad [1+0+2x \quad 0+2+x \quad 2+1+0]\begin{bmatrix}1\\1\\1\end{bmatrix} = 0$$

$$\Rightarrow \quad [1+2x \quad 2+x \quad 3]\begin{bmatrix}1\\1\\1\end{bmatrix} = 0$$

$$\Rightarrow \quad 1 + 2x + 2 + x + 3 = 0$$

$$\Rightarrow \quad 3x + 6 = 0 \quad \Rightarrow \quad x = -\frac{6}{3} = -2.$$

Example 2.4. If $A = \begin{bmatrix}1 & 2 & 3\\3 & -2 & 1\\4 & 2 & 1\end{bmatrix}$, **show that** $A^3 - 23A - 40I = 0$.

Solution.

$$A^2 = A.A = \begin{bmatrix}1 & 2 & 3\\3 & -2 & 1\\4 & 2 & 1\end{bmatrix}\begin{bmatrix}1 & 2 & 3\\3 & -2 & 1\\4 & 2 & 1\end{bmatrix}$$

$$= \begin{bmatrix}1+6+12 & 2-4+6 & 3+2+3\\3-6+4 & 6+4+2 & 9-2+1\\4+6+4 & 8-4+2 & 12+2+1\end{bmatrix} = \begin{bmatrix}19 & 4 & 8\\1 & 12 & 8\\14 & 6 & 15\end{bmatrix}$$

$$A^3 = A.A^2 = \begin{bmatrix}1 & 2 & 3\\3 & -2 & 1\\4 & 2 & 1\end{bmatrix}\begin{bmatrix}19 & 4 & 8\\1 & 12 & 8\\14 & 6 & 15\end{bmatrix}$$

$$= \begin{bmatrix}19+2+42 & 4+24+18 & 8+16+45\\57-2+14 & 12-24+6 & 24-16+15\\76+2+14 & 16+24+6 & 32+16+15\end{bmatrix} = \begin{bmatrix}63 & 46 & 69\\69 & -6 & 23\\92 & 46 & 63\end{bmatrix}.$$

$$23A = 23\begin{bmatrix}1 & 2 & 3\\3 & -2 & 1\\4 & 2 & 1\end{bmatrix} = \begin{bmatrix}23 & 46 & 69\\69 & -46 & 23\\92 & 46 & 23\end{bmatrix}$$

and

$$40I = 40\begin{bmatrix}1 & 0 & 0\\0 & 1 & 0\\0 & 0 & 1\end{bmatrix} = \begin{bmatrix}40 & 0 & 0\\0 & 40 & 0\\0 & 0 & 40\end{bmatrix}$$

$$\therefore \quad A^3 - 23A - 40I = \begin{bmatrix} 63 & 46 & 69 \\ 69 & -6 & 23 \\ 92 & 46 & 63 \end{bmatrix} - \begin{bmatrix} 23 & 46 & 69 \\ 69 & -46 & 23 \\ 92 & 46 & 23 \end{bmatrix} - \begin{bmatrix} 40 & 0 & 0 \\ 0 & 40 & 0 \\ 0 & 0 & 40 \end{bmatrix}$$

$$= \begin{bmatrix} 63-23-40 & 46-46-0 & 69-69-0 \\ 69-69-0 & -6-(-46)-40 & 23-23-0 \\ 92-92-0 & 46-46-0 & 63-23-40 \end{bmatrix} = \begin{bmatrix} 0 & 0 & 0 \\ 0 & 0 & 0 \\ 0 & 0 & 0 \end{bmatrix} = 0.$$

Example 2.5. If A and B be symmetric matrices of the same order, then show that

(i) $A + A'$ is a symmetric matrix.

(ii) $A - A'$ is a skew symmetric matrix.

(iii) AA' and $A'A$ are symmetric matrices.

Solution. (*i*) Let $P' = A + A'$. Then

$$P' = (A + A)' = A' + (A')' \qquad [\because (A+B)' = A' + B']$$

$$= A' + A \qquad [\because (A') = A]$$

$$= A + A' = P$$

$\therefore$ P is a symmetric matrix.

(*ii*) Let $Q = A - A'$. Then

$$Q' = (A - A')' = A' - (A')' \qquad [\because (A+B)' = A' + B']$$

$$= A' - A \qquad [\because (A')' = A]$$

$$= -(A - A') = -Q.$$

$\therefore$ Q is a skew symmetric matrix.

(*iii*) We have

$$(AA')' = [(A')' A'] \qquad \text{[By Reversal Law]}$$

$$= AA' \qquad [\because (A')' = A]$$

$\therefore$ AA' is symmetric.

Example 2.6. For the matrices $A = \begin{bmatrix} 1 & -4 \\ 0 & 5 \\ 6 & 7 \end{bmatrix}, B = \begin{bmatrix} 2 & 3 & -1 \\ 1 & 0 & -7 \end{bmatrix}$ verify that $(AB)' = B'A'$.

Solution.

$$AB = \begin{bmatrix} 1 & -4 \\ 0 & 5 \\ 6 & 7 \end{bmatrix}, B = \begin{bmatrix} 2 & 3 & -1 \\ 1 & 0 & -7 \end{bmatrix}$$

$$= \begin{bmatrix} 2-4 & 3+0 & -1+28 \\ 0+5 & 0+0 & 0-35 \\ 12+7 & 18+0 & -6-49 \end{bmatrix} = \begin{bmatrix} -2 & 3 & 27 \\ 5 & 0 & -35 \\ 19 & 18 & -55 \end{bmatrix}$$

$$\therefore \quad (AB)' = \begin{bmatrix} -2 & 5 & 19 \\ 3 & 0 & 18 \\ 27 & -35 & -55 \end{bmatrix}$$

Again

$$B' = \begin{bmatrix} 2 & 1 \\ 3 & 0 \\ -1 & -7 \end{bmatrix}, A' = \begin{bmatrix} 1 & 0 & 6 \\ -4 & 5 & 7 \end{bmatrix}$$

$$\therefore \qquad B'A' = \begin{bmatrix} 2 & 1 \\ 3 & 0 \\ -1 & -7 \end{bmatrix} \begin{bmatrix} 1 & 0 & 6 \\ -4 & 5 & 7 \end{bmatrix}$$

$$= \begin{bmatrix} 2-4 & 0+5 & 12+7 \\ 3+0 & 0+0 & 18+0 \\ -1+28 & 0-35 & -6-49 \end{bmatrix} = \begin{bmatrix} -2 & 5 & 19 \\ 3 & 0 & 18 \\ 27 & -35 & -55 \end{bmatrix}$$

Hence $(AB)' = B'A'$.

Example 2.7. Express the matrix $\begin{bmatrix} -1 & 7 & 1 \\ 2 & 3 & 4 \\ 5 & 0 & 5 \end{bmatrix}$ as the sum of a symmetric and a skew-symmetric matrix.

Solution. Given matrix is $A = \begin{bmatrix} -1 & 7 & 1 \\ 2 & 3 & 4 \\ 5 & 0 & 5 \end{bmatrix}$

$$\therefore \qquad A' = \begin{bmatrix} -1 & 2 & 5 \\ 7 & 0 & 3 \\ 1 & 4 & 5 \end{bmatrix}$$

$$\therefore \qquad A + A' = \begin{bmatrix} -1-1 & 7+2 & 1+5 \\ 2+7 & 3+3 & 4+0 \\ 5+1 & 0+4 & 5+5 \end{bmatrix} = \begin{bmatrix} -2 & 9 & 6 \\ 9 & 6 & 4 \\ 6 & 4 & 10 \end{bmatrix}$$

$$\therefore \qquad B = \frac{1}{2}(A + A') = \begin{bmatrix} -1 & \frac{9}{2} & 3 \\ \frac{9}{2} & 3 & 2 \\ 3 & 2 & 5 \end{bmatrix}$$

$$A - A' = \begin{bmatrix} -1+1 & 7-2 & 1-5 \\ 2-7 & 3-3 & 4-0 \\ 5-1 & 0-4 & 5-5 \end{bmatrix} = \begin{bmatrix} 0 & 5 & -4 \\ -5 & 0 & 4 \\ 4 & -4 & 0 \end{bmatrix}$$

$$\therefore \qquad C = \frac{1}{2}(A - A') = \begin{bmatrix} 0 & \frac{5}{2} & -2 \\ -\frac{5}{2} & 0 & 2 \\ 2 & -2 & 0 \end{bmatrix}$$

$$A = B + C = \begin{bmatrix} -1 & \frac{9}{2} & 3 \\ \frac{9}{2} & 3 & 2 \\ 3 & 2 & 5 \end{bmatrix} + \begin{bmatrix} 0 & \frac{5}{2} & -2 \\ -\frac{5}{2} & 0 & 2 \\ 2 & -2 & 0 \end{bmatrix}$$

where B is a symmetric and C is a skew-symmetric matrix.

EXERCISE 2.1

1. Construct a matrix $A = [a_{ij}]_{3 \times 4}$ whose elements are given by $a_{ij} = \dfrac{i-j}{i+j}$.

$$\left[\textbf{Ans.} \begin{bmatrix} 0 & -\frac{1}{3} & -\frac{1}{2} & -\frac{3}{5} \\ \frac{1}{3} & 0 & -\frac{1}{5} & -\frac{1}{3} \\ \frac{1}{2} & \frac{1}{5} & 0 & -\frac{1}{7} \end{bmatrix}\right]$$

2. Find the values of x and y if $X = Y$, where

$$X = \begin{bmatrix} x+10 & y^2+2y \\ 0 & -4 \end{bmatrix} \text{ and } Y = \begin{bmatrix} 3x+4 & 3 \\ 0 & y^2-5y \end{bmatrix}.$$

[**Ans.** $x = 3$; $y = 1$]

3. (*a*) If $A = \begin{bmatrix} 1 & 3 & 4 \\ 2 & 0 & 1 \\ -3 & 2 & 3 \end{bmatrix}, B = \begin{bmatrix} 0 & 2 & -1 \\ 5 & 7 & 2 \\ -1 & 0 & 3 \end{bmatrix}$ and $C = \begin{bmatrix} 2 & -1 & 3 \\ 6 & 8 & 5 \\ 0 & 1 & 4 \end{bmatrix}$ find $4A - 2B + 3C$.

$$\left[\textbf{Ans.} \begin{bmatrix} 10 & 5 & 27 \\ 16 & 10 & 15 \\ -10 & 11 & 18 \end{bmatrix}\right]$$

(*b*) If $A = \text{diag}\,[1, -2, 3]$; $B = \text{diag}\,[3, 4, -6]$ and $C = \text{diag}\,[0, 1, 2]$, find $A - 2B + 3C$.

[**Ans.** diag $[-5, -7, 21]$]

4. If $A = \begin{bmatrix} 4 & -1 & -4 \\ 4 & 0 & -4 \\ 3 & -1 & -3 \end{bmatrix}$, show that $A^2 = I_3$, where I_3 is the identity matrix of order 3.

5. If $A = \begin{bmatrix} 1 & 2 \\ 3 & 4 \end{bmatrix}, B = \begin{bmatrix} 1 & 0 \\ 2 & -3 \end{bmatrix}$ and $C = \begin{bmatrix} 1 & -1 \\ 0 & 1 \end{bmatrix}$, show that $A(BC) = (AB)C$.

6. Show that the matrix $A = \begin{bmatrix} 1 & 0 & 2 \\ 0 & 2 & 1 \\ 2 & 0 & 3 \end{bmatrix}$ satisfies the equation $A^3 - 6A^2 + 7A + 2I = 0$, where I is the identity matrix of order 3.

7. If $A = \begin{bmatrix} 3 & -4 \\ 1 & 1 \end{bmatrix}$, show that $A^n = \begin{bmatrix} 1+2n & -4n \\ n & 1-2n \end{bmatrix}$, where n is any positive integer.

8. If $I = \begin{bmatrix} 1 & 0 \\ 0 & 1 \end{bmatrix}$, $C = \begin{bmatrix} 0 & 1 \\ 0 & 0 \end{bmatrix}$, show that $(aI + bC)^3 = a^3I + 3a^2bC$.

9. Find the value of x such that $[1 \quad x \quad 1] \begin{bmatrix} 1 & 3 & 2 \\ 2 & 5 & 1 \\ 15 & 3 & 2 \end{bmatrix} \begin{bmatrix} 1 \\ 2 \\ x \end{bmatrix} = 0$.

[**Ans.** $x = -2, -14$]

10. If $A = \begin{bmatrix} 3 & -2 \\ 4 & -2 \end{bmatrix}$ and $I = \begin{bmatrix} 1 & 0 \\ 0 & 1 \end{bmatrix}$, then find k so that $A^2 = kA - 2I$.

[**Ans.** $k = 1$]

11. Find the matrix X so that $X \begin{bmatrix} 1 & 2 & 3 \\ 4 & 5 & 6 \end{bmatrix} = \begin{bmatrix} -7 & -8 & -9 \\ 2 & 4 & 6 \end{bmatrix}$.

$$\left[\textbf{Ans.} \begin{bmatrix} 1 & -2 \\ 2 & 0 \end{bmatrix}\right]$$

12. Given $A = \begin{bmatrix} 3 & -1 & 0 \\ 2 & 4 & 5 \\ 1 & 3 & 6 \end{bmatrix}$ and $B = \begin{bmatrix} -2 & 4 & 3 \\ 5 & 1 & 7 \\ 2 & 3 & 8 \end{bmatrix}$. Verify that

(*i*) $(A + B)' = A' + B'$; (*ii*) $(AB)' = B'A'$; (*iii*) $(2A)' = 2A'$.

13. Show that $\begin{bmatrix} 0 & a & b \\ -a & 0 & -c \\ -b & c & 0 \end{bmatrix}$ is skew-symmetric matrix.

14. Find x any y so that the matrix $A = \frac{1}{3}\begin{bmatrix} 1 & 2 & 2 \\ 2 & 1 & -2 \\ x & 2 & y \end{bmatrix}$ may satisfy the condition $A'A = I_3 = AA'$.

[**Ans.** $x = -2, y = -1$]

15. If A and B are square matrices of the same order and A is skew-symmetric, prove that $B'AB$ is also a skew-symmetric.

16. Express $\begin{bmatrix} 2 & 4 & -1 \\ 3 & 5 & 8 \\ 1 & -2 & 1 \end{bmatrix}$ as the sum of a symmetric and a skew-symmetric matrix.

$$\left[\textbf{Ans. } \begin{bmatrix} 2 & \frac{7}{2} & 0 \\ \frac{7}{2} & 5 & 3 \\ 0 & 3 & 1 \end{bmatrix} + \begin{bmatrix} 0 & \frac{1}{2} & -1 \\ -\frac{1}{2} & 0 & 5 \\ 1 & -5 & 0 \end{bmatrix}\right]$$

2.4. DETERMINANTS

Let $A = [a_{ij}]$ be a square matrix. We can associate with the square matrix A a determinent which is formed by exactly the same array of elements of the matrix A. A determinent formed by the same array of elements of the square matrix X and is called the determinent of the square matrix A and is denoted by the symbol det. A or $|A|$. It should be remembered that the determinent of a square matrix will be a scalar quantity *i.e.* with a determinent we associate some value, whereas a matrix is essentially an arrangement of numbers and has no value.

Determinant of order 1. Let $A = [a_{11}]_{1 \times 1}$ matrix. Then the determinant of A is the number a_{11} itself *i.e.*, $|a_{11}| = a_{11}$. For example, $|3| = 3$, $|-5| = -5$ and $|0| = 0$.

Determinant of order 2. Let $A = \begin{bmatrix} a_{11} & a_{12} \\ a_{21} & a_{22} \end{bmatrix}$ be a 2×2 matrix, then $|A| = \begin{vmatrix} a_{11} & a_{12} \\ a_{21} & a_{22} \end{vmatrix} = a_{11}a_{22} - a_{12}a_{21}$

i.e., the determinant of a 2×2 matrix is obtained by taking the product of the entries on the main diagonal and subtracting from it the product of the entries in the other diagonal.

For example, let a matrix $A = \begin{bmatrix} 5 & 2 \\ 7 & 3 \end{bmatrix}$ so that

$$|A| = \begin{vmatrix} 5 & 2 \\ 7 & 3 \end{vmatrix}$$

The value of $\begin{vmatrix} 5 & 2 \\ 7 & 3 \end{vmatrix} = 5 \times 3 - 7 \times 2 = 15 - 14 = 1$.

Determinant of order 3. If $A = \begin{bmatrix} a_{11} & a_{12} & a_{13} \\ a_{21} & a_{22} & a_{23} \\ a_{31} & a_{32} & a_{33} \end{bmatrix}$ is a square matrix of order 3, then the expression

$$a_{11}\, a_{22}\, a_{33} + a_{12}\, a_{23}\, a_{31} + a_{13}\, a_{23}\, a_{21} - a_{11}\, a_{23}\, a_{32} - a_{22}\, a_{13}\, a_{31} - a_{12}\, a_{21}\, a_{33}$$

is defined as the determinant of A *i.e.*;

$$|A| = \begin{vmatrix} a_{11} & a_{12} & a_{13} \\ a_{21} & a_{22} & a_{23} \\ a_{31} & a_{32} & a_{33} \end{vmatrix}$$

$$= a_{11}(a_{22}\, a_{33} - a_{23}\, a_{32}) - a_{12}(a_{21}\, a_{33} - a_{31}\, a_{23}) + a_{13}(a_{32}\, a_{21} - a_{22}\, a_{31})$$

$$= a_{11}\begin{vmatrix} a_{22} & a_{23} \\ a_{32} & a_{33} \end{vmatrix} - a_{12}\begin{vmatrix} a_{21} & a_{23} \\ a_{31} & a_{32} \end{vmatrix} + a_{13}\begin{vmatrix} a_{21} & a_{22} \\ a_{31} & a_{32} \end{vmatrix}$$

$$= (-1)^{1+1}\, a_{11}\begin{vmatrix} a_{22} & a_{23} \\ a_{32} & a_{33} \end{vmatrix} + (-1)^{1+2}\, a_{12}\begin{vmatrix} a_{21} & a_{23} \\ a_{31} & a_{33} \end{vmatrix} + (-1)^{1+3}\, a_{13}\begin{vmatrix} a_{21} & a_{22} \\ a_{31} & a_{32} \end{vmatrix}$$

Thus the determinant of a square matrix of order 3 is the sum of the product of the element a_{ij} in first row with $(-1)^{i+j}$ times the determinant of a 2×2 submatrix obtained by leaving the first row and column passing through the element.

The above expression of $|A|$ is known as the expansion along first row.

For example, if $$|A| = \begin{vmatrix} 1 & 6 & 7 \\ 2 & 3 & 0 \\ 0 & 1 & 4 \end{vmatrix}$$

then $$|A| = (-1)^{1+1}\, 1\begin{vmatrix} 3 & 0 \\ 1 & 4 \end{vmatrix} + (-1)^{1+2}\, 6\begin{vmatrix} 2 & 0 \\ 0 & 4 \end{vmatrix} + (-1)^{1+3}\, 7\begin{vmatrix} 2 & 3 \\ 0 & 1 \end{vmatrix}$$

$$= 1(12 - 0) - 6(8 - 0) + 7(2 - 0) = 12 - 48 + 14 = 26 - 48 = -22.$$

There are three rows and three column in a square matrix of order 3. The expression for the determinant of order 3 can be arranged in various forms to otain the expansion of $|A|$ along any of its rows or columns. In fact, to expand $|A|$ about a row or column multiply each element a_{ij} in ith row with $(-1)^{i+j}$ times the determinant of the sub-matrix obtained by leaving the row and column passing through that element. For example,

$$|A| = \begin{vmatrix} a_{11} & a_{12} & a_{13} \\ a_{21} & a_{22} & a_{23} \\ a_{31} & a_{32} & a_{33} \end{vmatrix}$$

$$= (-1)^{2+1} a_{21}\begin{vmatrix} a_{12} & a_{13} \\ a_{32} & a_{33} \end{vmatrix} + (-1)^{2+2}\, a_{22}\begin{vmatrix} a_{11} & a_{13} \\ a_{31} & a_{33} \end{vmatrix} + (-1)^{2+3}\, a_{23}\begin{vmatrix} a_{11} & a_{12} \\ a_{31} & a_{32} \end{vmatrix}$$

is the expansion of $|A|$ about second row. The expansion of $|A|$ about 3rd column is

$$|A| = (-1)^{1+3} a_{13}\begin{vmatrix} a_{21} & a_{22} \\ a_{31} & a_{32} \end{vmatrix} + (-1)^{2+3}\, a_{23}\begin{vmatrix} a_{21} & a_{12} \\ a_{31} & a_{32} \end{vmatrix} + (-1)^{3+3} a_{33}\begin{vmatrix} a_{11} & a_{12} \\ a_{21} & a_{22} \end{vmatrix}.$$

Determinant of order 4. (*Chio's Condensation Method*) chio's condensation is a method of evaluating an $n \times n$ determinant in terms of $(n-1) \times (n-1)$ determinants.

$$A = \begin{vmatrix} a_{11} & a_{12} & \cdots & a_{1n} \\ a_{21} & a_{22} & \cdots & a_{2n} \\ \vdots & \vdots & \cdots & \vdots \\ a_{n1} & a_{n2} & \cdots & a_{nn} \end{vmatrix} = \frac{1}{a_{11}^{n-2}} \begin{Vmatrix} \begin{vmatrix} a_{11} & a_{12} \\ a_{21} & a_{22} \end{vmatrix} & \begin{vmatrix} a_{11} & a_{13} \\ a_{21} & a_{23} \end{vmatrix} & \cdots & \begin{vmatrix} a_{11} & a_{1n} \\ a_{21} & a_{2n} \end{vmatrix} \\ \begin{vmatrix} a_{11} & a_{12} \\ a_{31} & a_{32} \end{vmatrix} & \begin{vmatrix} a_{11} & a_{13} \\ a_{31} & a_{33} \end{vmatrix} & \cdots & \begin{vmatrix} a_{11} & a_{1n} \\ a_{31} & a_{3n} \end{vmatrix} \\ \vdots & \vdots & \cdots & \vdots \\ \begin{vmatrix} a_{11} & a_{12} \\ a_{n1} & a_{n2} \end{vmatrix} & \begin{vmatrix} a_{11} & a_{13} \\ a_{n1} & a_{n3} \end{vmatrix} & \cdots & \begin{vmatrix} a_{11} & a_{1n} \\ a_{n1} & a_{nn} \end{vmatrix} \end{Vmatrix}$$

Note: First element (*i.e.* a_{11}) must be a non-zero number.

2.5. MINOR AND COFACTORS

2.5.1. Minor of an Element of a Determinant

The determinant obtained by deleting the ith row and jth column passing through the element a_{ij} is called the minor of element a_{ij} in the determinant $|a_{ij}|$ of order n and is denoted by M_{ij}.

For example, given the 3×3 determinant $\begin{vmatrix} a_{11} & a_{12} & a_{13} \\ a_{21} & a_{22} & a_{23} \\ a_{31} & a_{32} & a_{33} \end{vmatrix}$.

The minor of a_{11} is $\quad M_{11} = \begin{vmatrix} a_{22} & a_{23} \\ a_{32} & a_{33} \end{vmatrix}$

The minor of a_{12} is $\quad M_{12} = \begin{vmatrix} a_{21} & a_{23} \\ a_{31} & a_{33} \end{vmatrix}$, and so on.

2.5.2. Cofactor of an Element of a Determinant

The cofactor of element a_{ij} is $(-1)^{i+j}$ times the determinant obtained by deleting the ith row and jth column passing through a_{ij}. We shall denote the cofactor of an element by the corresponding capital letter with this notation.

$$\text{Cofactor of } a_{ij} = A_{ij} = (-1)^{i+j} M_{ij}.$$

For example, the cofactor of a_{12} in the 3×3 determinant $\begin{vmatrix} a_{11} & a_{12} & a_{13} \\ a_{21} & a_{22} & a_{23} \\ a_{31} & a_{32} & a_{33} \end{vmatrix}$

is
$$A_{12} = (-1)^{1+2} \begin{vmatrix} a_{21} & a_{23} \\ a_{31} & a_{33} \end{vmatrix} = -\begin{vmatrix} a_{21} & a_{23} \\ a_{31} & a_{33} \end{vmatrix}$$

Remark. It is a fact that determinant of a matrix is unique and does not depend on the row or column chosen for its evaluation.

2.6. PROPERTIES OF DETERMINANTS

We now state some useful properties of determinants of order three only. These properties help a good deal in the evaluation of determinants. We use the notations R_i and C_j to denote respectively the ith row and jth column of a determinant.

Property 1. The value of a determinant remains unchanged if its rows and columns are interchanged. That is,

$$\begin{vmatrix} a_{11} & a_{12} & a_{13} \\ a_{21} & a_{22} & a_{23} \\ a_{31} & a_{32} & a_{33} \end{vmatrix} = \begin{vmatrix} a_{11} & a_{21} & a_{31} \\ a_{12} & a_{22} & a_{32} \\ a_{13} & a_{23} & a_{33} \end{vmatrix}$$

Remark. In terms of matrices, if A is a square matrix, then $|A| = |A'|$.

Property 2. If two rows or columns of a determinant are interchanged, then the sign of the determinant is changed. That is

$$\begin{vmatrix} a_{11} & a_{12} & a_{13} \\ a_{21} & a_{22} & a_{23} \\ a_{31} & a_{32} & a_{33} \end{vmatrix} = \begin{vmatrix} a_{11} & a_{12} & a_{13} \\ a_{31} & a_{32} & a_{33} \\ a_{21} & a_{22} & a_{23} \end{vmatrix}$$

Remark. The notation $R_i \leftrightarrow R_j$ ($C_i \leftrightarrow C_j$) is used to represent change of ith and jth row (column).

Property 3. If any two rows or columns of a determinant are identical, then its value is zero. That is,

$$\begin{vmatrix} a_1 & b_1 & c_1 \\ a_1 & b_1 & c_1 \\ a_2 & b_2 & c_2 \end{vmatrix} = 0$$

Property 4. If each element of a row or column of a determinant is multiplied by a constant k, then its value gets multiplied by k. That is,

$$\begin{vmatrix} a_{11} & a_{12} & a_{13} \\ ka_{21} & ka_{22} & ka_{23} \\ a_{31} & a_{32} & a_{33} \end{vmatrix} = k\begin{vmatrix} a_{11} & a_{12} & a_{13} \\ a_{21} & a_{22} & a_{23} \\ a_{31} & a_{32} & a_{33} \end{vmatrix}$$

Remark. The notation $R_1 \to kR_i$, ($C_i \to kC_j$) is used to represent multiplication of each element of ith row (column) by the constant k.

Property 5. If some or all the elements of a row (or column) of a determinant are expressed as sum of two (or more) terms, then the determinant can be expressed as sum of two or more determinents. That is,

$$\begin{vmatrix} a_{11} & a_{12} & a_{13} \\ a_{21} & a_{22} & a_{23} \\ a_{31}+b_1 & a_{32}+b_2 & a_{33}+b_3 \end{vmatrix} = \begin{vmatrix} a_{11} & a_{12} & a_{13} \\ a_{21} & a_{22} & a_{23} \\ a_{31} & a_{32} & a_{33} \end{vmatrix} + \begin{vmatrix} a_{11} & a_{12} & a_{13} \\ a_{21} & a_{22} & a_{23} \\ b_1 & b_2 & b_3 \end{vmatrix}$$

Property 6. If to any row or column of a determinant, a multiple of another row or column is added, the value of the determinant remains the same. That is,

$$\begin{vmatrix} a_{11} & a_{12} & a_{13} \\ a_{21} & a_{22} & a_{23} \\ a_{31} & a_{32} & a_{33} \end{vmatrix} = \begin{vmatrix} a_{11}+ka_{31} & a_{12}+ka_{23} & a_{13}+ka_{33} \\ a_{21} & a_{22} & a_{23} \\ a_{31} & a_{32} & a_{33} \end{vmatrix}.$$

SOLVED EXAMPLES

Example 2.8. Expand the following determinants.

(*i*) $\begin{vmatrix} 2+3i & 4 \\ 1 & 2-3i \end{vmatrix}$; (*ii*) $\begin{vmatrix} \sec\theta & \tan\theta \\ \tan\theta & \sec\theta \end{vmatrix}$; (*iii*) $\begin{vmatrix} 6 & -3 & 2 \\ 2 & -1 & 2 \\ -10 & 5 & 2 \end{vmatrix}$

Solution. (*i*) $\begin{vmatrix} 2+3i & 4 \\ 1 & 2-3i \end{vmatrix} = (2+3i)(2-3i) - (1)(4) = 2^2 + 3^2 - 4 = 9.$

(*ii*) $\begin{vmatrix} \sec\theta & \tan\theta \\ \tan\theta & \sec\theta \end{vmatrix} = \sec^2\theta - \tan^2\theta = 1.$

(*iii*) We have $\begin{vmatrix} 6 & -3 & 2 \\ 2 & -1 & 2 \\ -10 & 5 & 2 \end{vmatrix}$

Expanding in terms of elements along the first row, we get

$$\begin{vmatrix} 6 & -3 & 2 \\ 2 & -1 & 2 \\ -10 & 5 & 2 \end{vmatrix} = 6\begin{vmatrix} -1 & 2 \\ 5 & 2 \end{vmatrix} - (-3)\begin{vmatrix} 2 & 2 \\ -10 & 2 \end{vmatrix} + 2\begin{vmatrix} 2 & -1 \\ 10 & 5 \end{vmatrix}$$

$$= 6(-2-10) + 3(4+20) + 2(10-10) = -72 + 72 + 2(0) = 0.$$

Example 2.9. Find the mirors and cofactor of elements of the determinant $\begin{vmatrix} 2 & -3 & 5 \\ 6 & 0 & 4 \\ 1 & 5 & -7 \end{vmatrix}$.

Solution. We have $M_{11} = \begin{vmatrix} 0 & 4 \\ 5 & -7 \end{vmatrix} = 0 - 20 = -20, A_{11} = -20$

$$M_{12} = \begin{vmatrix} 6 & 4 \\ 1 & -7 \end{vmatrix} = -42 - 4 = -46, A_{12} = 46$$

$$M_{13} = \begin{vmatrix} 6 & 0 \\ 1 & 5 \end{vmatrix} = 30 - 0 = 30, A_{13} = 30$$

$$M_{21} = \begin{vmatrix} -3 & 5 \\ 5 & -7 \end{vmatrix} = 21 - 25 = -4, A_{21} = 4$$

$$M_{22} = \begin{vmatrix} 2 & 5 \\ 1 & -7 \end{vmatrix} = -14 - 15 = -19, A_{22} = -19$$

$$M_{23} = \begin{vmatrix} 2 & -3 \\ 1 & 5 \end{vmatrix} = 10 + 3 = 13, A_{23} = -13$$

$$M_{31} = \begin{vmatrix} -3 & 5 \\ 0 & 4 \end{vmatrix} = -12 - 0 = -12, A_{31} = -12$$

$$M_{32} = \begin{vmatrix} 2 & 5 \\ 6 & 4 \end{vmatrix} = 8 - 30 = -22, A_{32} = 22$$

$$M_{33} = \begin{vmatrix} 2 & -3 \\ 6 & 0 \end{vmatrix} = 0 + 18 = 18, A_{33} = 18.$$

Example 2.10. Evaluate the following determinants:

$$(i) \begin{vmatrix} 265 & 240 & 219 \\ 240 & 225 & 198 \\ 219 & 198 & 181 \end{vmatrix}; \qquad (ii) \begin{vmatrix} 1^2 & 2^2 & 3^2 & 4^2 \\ 2^2 & 3^2 & 4^2 & 5^2 \\ 3^2 & 4^2 & 5^2 & 6^2 \\ 4^2 & 5^2 & 6^2 & 7^2 \end{vmatrix}.$$

Solution. (*i*) Applying $R_1 \to R_1 + R_3 - 2R_2$, we get

$$\Delta = \begin{vmatrix} 4 & -12 & 4 \\ 240 & 225 & 198 \\ 219 & 198 & 181 \end{vmatrix}$$

Taking 4 common from first row

$$= 4\begin{vmatrix} 1 & -3 & 1 \\ 240 & 225 & 198 \\ 219 & 198 & 181 \end{vmatrix}$$

Taking 3 commcn from second column

$$= 4 \times 3\begin{vmatrix} 1 & -1 & 1 \\ 240 & 75 & 198 \\ 219 & 66 & 181 \end{vmatrix}$$

Applying $C_2 \to C_2 + C_1$ and $C_3 \to C_3 - C_1$

$$= 12\begin{vmatrix} 1 & 0 & 0 \\ 240 & 315 & -42 \\ 219 & 285 & -38 \end{vmatrix}$$

Expanding the determinants along R_1

$$\Delta = 12\begin{vmatrix} 315 & -42 \\ 285 & -38 \end{vmatrix}$$

Taking out 21 from R_1 and 19 from R_2, we get

$$= 12 \times 21 \times 19\begin{vmatrix} 15 & -2 \\ 15 & -2 \end{vmatrix}$$

$$= 12 \times 21 \times 19(0) \qquad [\because R_1 \text{ and } R_2 \text{ are identical}]$$

$$= 0$$

(*ii*) The given determinant $= \begin{vmatrix} 1 & 4 & 9 & 16 \\ 4 & 9 & 16 & 25 \\ 9 & 16 & 25 & 36 \\ 16 & 25 & 36 & 49 \end{vmatrix}$

Applying $C_2 \to C_2 - 4C_1$, $C_3 \to C_3 - 9C_1$ and $C_4 \to C_4 - 16C_1$ respectively.

$$= \begin{vmatrix} 1 & 0 & 0 & 0 \\ 4 & -7 & -20 & -39 \\ 9 & -20 & -56 & -108 \\ 16 & -39 & -108 & -207 \end{vmatrix}$$

Expanding along R1, we get $\Delta = (-1)^3\begin{vmatrix} 7 & 20 & 39 \\ 20 & 56 & 108 \\ 39 & 108 & 207 \end{vmatrix}$

Applying $C_2 \to C_2 - 3C_1$ and $C_3 \to C_3 - 2C_2$

$$= -\begin{vmatrix} 7 & -1 & -1 \\ 20 & -4 & -4 \\ 39 & -9 & -9 \end{vmatrix} \qquad [\because\ C_2 \text{ and } C_3 \text{ identical}]$$

Example 2.11. Expand the following determinant using Chio's method:

$$\begin{vmatrix} \mathbf{0} & \mathbf{1} & \mathbf{0} & \mathbf{1} \\ \mathbf{-1} & \mathbf{0} & \mathbf{-1} & \mathbf{3} \\ \mathbf{0} & \mathbf{1} & \mathbf{0} & \mathbf{2} \\ \mathbf{-1} & \mathbf{-3} & \mathbf{-2} & \mathbf{0} \end{vmatrix}.$$

Solution. Since Ist element (*i.e.* a_{11}) is zero, we have to interchange 1st and 2nd column, with a preceeding minue sign

$$= -\begin{vmatrix} 1 & 0 & 0 & 1 \\ 0 & -1 & -1 & 3 \\ 1 & 0 & 0 & 2 \\ -3 & -1 & -2 & 0 \end{vmatrix}$$

Using chio's method we have

$$= -\frac{1}{(1)^{4-2}}\begin{vmatrix} \begin{vmatrix} 1 & 0 \\ 0 & -1 \end{vmatrix} & \begin{vmatrix} 1 & 0 \\ 0 & -1 \end{vmatrix} & \begin{vmatrix} 1 & 1 \\ 0 & 3 \end{vmatrix} \\ \begin{vmatrix} 1 & 0 \\ 1 & 0 \end{vmatrix} & \begin{vmatrix} 1 & 0 \\ 1 & 0 \end{vmatrix} & \begin{vmatrix} 1 & 1 \\ 1 & 2 \end{vmatrix} \\ \begin{vmatrix} 1 & 0 \\ -3 & -1 \end{vmatrix} & \begin{vmatrix} 1 & 0 \\ -3 & -2 \end{vmatrix} & \begin{vmatrix} 1 & 1 \\ -3 & 0 \end{vmatrix} \end{vmatrix} = -\frac{1}{1}\begin{vmatrix} -1 & -1 & 3 \\ 0 & 0 & 1 \\ -1 & -2 & 3 \end{vmatrix}$$

$$= -\frac{1}{(-1)^{3-2}}\begin{vmatrix} \begin{vmatrix} -1 & -1 \\ 0 & 0 \end{vmatrix} & \begin{vmatrix} -1 & 3 \\ 0 & 1 \end{vmatrix} \\ \begin{vmatrix} -1 & -1 \\ -1 & -2 \end{vmatrix} & \begin{vmatrix} -1 & 3 \\ -1 & 3 \end{vmatrix} \end{vmatrix} = \frac{1}{1}\begin{vmatrix} 0 & -1 \\ 1 & 0 \end{vmatrix}$$

$$= 1[(0 \times 0) - (1 \times (-1)] = 1(1) = 1$$

Example 2.12. Prove that

$$\begin{vmatrix} 1+a^2-b^2 & 2ab & -2b \\ 2ab & 1-a^2+b^2 & 2a \\ 2b & -2a & 1-a^2-b^2 \end{vmatrix} = (1+a^2+b^2)^3$$

Solution. Let $$\Delta = \begin{vmatrix} 1+a^2-b^2 & 2ab & -2b \\ 2ab & 1-a^2+b^2 & 2a \\ 2b & -2a & 1-a^2-b^2 \end{vmatrix}$$

Applying $R_1 \to R_1 + bR_3$ and $R_2 \to R_2 - aR_3$, we get

$$= \begin{vmatrix} 1+a^2-b^2 & 0 & -b(1+a^2+b^2) \\ 0 & 1+a^2+b^2 & a(1+a^2+b^2) \\ 2b & -2a & 1-a^2-b^2 \end{vmatrix}$$

$$= (1+a^2+b^2)^2 \begin{vmatrix} 1 & 0 & -b \\ 0 & 1 & a \\ 2b & -2a & 1-a^2-b^2 \end{vmatrix}$$

Expanding along first row, we get

$$= (1+a^2+b^2)^2 \left[\begin{vmatrix} 1 & a \\ -2a & 1-a^2-b^2 \end{vmatrix} - b \begin{vmatrix} 0 & 1 \\ 2b & -2a \end{vmatrix} \right]$$

$$= (1+a^2+b^2)^2\,[(1-a^2-b^2+2a^2) - b(-2b)]$$

$$= (1+a^2+b^2)\,(1+a^2+b^2) = (1+a^2+b^2)^3$$

EXERCISE 2.2

1. Without expanding evaluate

(*i*) $\begin{vmatrix} 91 & 92 & 93 \\ 94 & 95 & 96 \\ 97 & 98 & 99 \end{vmatrix}$ (*ii*) $\begin{vmatrix} 3 & 1 & 43 \\ 2 & 7 & 35 \\ 1 & 3 & 17 \end{vmatrix}$ [**Ans.** (*i*) 0, (*ii*) 0]

2. Without expanding the determinant, prove that

$$\begin{vmatrix} a-b & b-c & c-a \\ b-c & c-a & a-b \\ c-a & a-b & b-c \end{vmatrix} = 0$$

3. Evaluate using chio's method:

(*a*) $\begin{vmatrix} 1 & 2 & 3 & 4 \\ 2 & 0 & 1 & -1 \\ 1 & -2 & 3 & 0 \\ 3 & 1 & -1 & 2 \end{vmatrix}$ (*b*) $\begin{vmatrix} 1 & 2 & 2 & 1 \\ 1 & 2 & 4 & 2 \\ 2 & 7 & 5 & 2 \\ -1 & 4 & -6 & 3 \end{vmatrix}$ (*c*) $\begin{vmatrix} 0 & -1 & 1 & -2 \\ 2 & -1 & 0 & 1 \\ 3 & 2 & -1 & 4 \\ 1 & 0 & 0 & 1 \end{vmatrix}$

[**Ans.** –100, –42, 2]

4. Without expanding the determinant show that $\left(1+\frac{1}{a}+\frac{1}{b}+\frac{1}{c}\right)$ is a factor of the following determinant:

$$\begin{vmatrix} 1+a & 1 & 1 \\ 1 & 1+b & 1 \\ 1 & 1 & 1+c \end{vmatrix}.$$

5. Show that $\begin{vmatrix} a & a+b & a+b+c \\ 2a & 3a+2b & 4a+3b+2c \\ 3a & 6a+3b & 10a+6b+3c \end{vmatrix} = a^3.$

6. Using properties of determinants, prove that $\begin{vmatrix} x+4 & 2x & 2x \\ 2x & x+4 & 2x \\ 2x & 2x & x+4 \end{vmatrix} = (5x+4)(4-x)^2$

7. Show that $\begin{vmatrix} a & b & c \\ a^2 & b^2 & c^2 \\ a^3 & b^3 & c^3 \end{vmatrix} = abc\,(a-b)(b-c)(c-a)$

8. If x, y, z are all different, and if $\begin{vmatrix} x & x^2 & 1+x^3 \\ y & y^2 & 1+y^3 \\ z & z^2 & 1+z^3 \end{vmatrix} = 0$, then prove that $1 + xyz = 0$.

9. Prove that $\begin{vmatrix} a-b-c & 2a & 2a \\ 2b & b-c-a & 2b \\ 2c & 2c & c-a-b \end{vmatrix} = (a+b+c)^3.$

10. Prove that $\begin{vmatrix} \alpha & \beta & \gamma \\ \alpha^2 & \beta^2 & \gamma^2 \\ \beta+\gamma & \gamma+\alpha & \alpha+\beta \end{vmatrix} = (\beta-\gamma)(\gamma-\alpha)(\alpha-\beta)(\alpha+\beta+\gamma)$

11. Show that $\begin{vmatrix} (b+c)^2 & a^2 & a^2 \\ b^2 & (c+a)^2 & b^2 \\ c^2 & c^2 & (a+b)^2 \end{vmatrix} = 2abc\,(a+b+c)^3.$

12. Prove that $\begin{vmatrix} 1 & a & a^2-bc \\ 1 & b & b^2-ca \\ 1 & c & c^2-ab \end{vmatrix} = 0.$

13. Prove that

$$\Delta = \begin{vmatrix} \sin^2 A & \sin A\cos A & \cos^2 A \\ \sin^2 B & \sin B\cos B & \cos^2 B \\ \sin^2 C & \sin C\cos C & \cos^2 C \end{vmatrix}$$

$$= -\sin(A-B)\sin(B-C)\sin(C-A) \text{ where } A+B+C=\pi.$$

2.7. AREA OF A TRIANGLE USING DETERMINANTS

Suppose we are given three points in the cartesian plane as (x_1, y_1), (x_2, y_2) and (x_3, y_3). The area of the triangle obtained by joining these points is given by

$$\Delta = \frac{1}{2}[x_1(y_2 - y_3) + x_2(y_3 - y_1) + x_3(y_1 - y_2)]$$

where Δ represents the area of the triangle and (x_1, y_1), (x_2, y_2) and (x_3, y_3) represents the vertices of the triangle.

The formula for finding area could be represented in the form of determinants as given below

$$\Delta = \frac{1}{2}\begin{vmatrix} x_1 & y_1 & 1 \\ x_2 & y_2 & 1 \\ x_3 & y_3 & 1 \end{vmatrix}$$

Remarks:

1. As we know the value of a determinant can either be negative or a positive value but since we are talking about area and it can never be taken as a negative value, therefore we take the absolute value of the determinant so obtained.
2. If the area of the triangle is already given, then we make use of both the positive and negative values of the determinant.
3. If three points are collinear, we would be left with a straight line instead of a triangle and as the area enclosed by a straight line is zero hence the value of the determinant will also be zero.

SOLVED EXAMPLES

Example 2.13. Find out the area of the triangle whose vertices are given by (3, 8), (– 4, 2) and (5, 1).

Solution. The area of triangle is given by

$$\Delta = \frac{1}{2}\begin{vmatrix} 3 & 8 & 1 \\ -4 & 2 & 1 \\ 5 & 1 & 1 \end{vmatrix}$$

$$= \frac{1}{2}[3(2-1) - 8(-4-5) + 1(-4-10)]$$

$$= \frac{1}{2}(3 + 72 - 14)] = \frac{61}{2}$$

Example 2.14. Find the values of *k* if area of triangle is 4 sq. units and vertices are (*k*, 0), (4, 0) and (0, 2).

Solution. We know that area of triangle is given by

$$\Delta = \frac{1}{2}\begin{vmatrix} x_1 & y_1 & 1 \\ x_2 & y_2 & 1 \\ x_3 & y_3 & 1 \end{vmatrix}$$

Given that, the area of triangle is 4 square units

We know that, the area is always positive.

A triangle can have both positive and negative signs.

$$\therefore \qquad \Delta = \pm 4$$

Now, substitute the values,

$$x_1 = k, y_1 = 0, x_2 = 4, y_2 = 0, x_3 = 0, y_3 = 2$$

$$\pm 4 = \frac{1}{2}\begin{vmatrix} k & 0 & 1 \\ 4 & 0 & 1 \\ 0 & 2 & 1 \end{vmatrix}$$

$$\pm 4 = \frac{1}{2}\left(k\begin{vmatrix} 0 & 1 \\ 2 & 1 \end{vmatrix} - 0\begin{vmatrix} 4 & 1 \\ 0 & 1 \end{vmatrix} + 1\begin{vmatrix} 4 & 0 \\ 0 & 2 \end{vmatrix}\right)$$

$$\pm 4 = \frac{1}{2}[k(0-2) - 0(4-0) + 1(8-0)]$$

$$\pm 4 \times 2 = \{k(-2) - 0 + 1(8)\}$$

$$\pm 8 = -2k + 8$$

Therefore, we get

$$8 = -2k + 8 \text{ (or)}$$

$$-8 = -2k + 8$$

Now, solve both the equations, we get:

Solving

$$8 = -2k + 8$$

$$8 - 8 = -2k$$

$$0 = -2k$$

$$k = \frac{0}{-2} = 0$$

Solving

$$-8 = -2k + 8$$

$$-8 - 8 = -2k$$

$$-16 = -2k$$

$$k = \frac{-16}{-2} = 8$$

Hence the required value of k is either $k = 0$ or $k = 8$.

EXERCISE 2.3

1. Find the area of the triangle with vertices at the point given in each of the following
   (*i*) (2, 7), (1, 1), (10, 8) (*ii*) (– 2, – 3), (3, 2), (– 1, – 8) **[Ans.** (*a*) 47/2 (*b*) 15]
2. Find the equation of the line joining A(1, 3) and B(0, 0) using det. Find K if D, (K, 0) is a point such then area of ΔABC is 3 square unit. **[Ans.** $y = 3x$ and $k = \pm 2$]
3. Find values of K if area of triangle is 35 square units and vertices are (2, – 6), (5, 4), (K, 4). **[Ans.** K = 12]
4. Find equation of line joining (3, 1) and (9, 3) using determinants. **[Ans.** $x - 3y = 0$]
5. Show that points A(a, $b + c$), B(b, $c + a$), C(c, $a + b$) are collinear. **[Ans.** $x - 3y = 0$]

2.8. ADJOINT OF A SQUARE MATRIX

Let $A = [a_{ij}]$ be a square matrix of order n and let A_{ij} be cofactor of a_{ij} in A. Then the transpose of the matrix of cofactors of elements of A is called the adjoint of A and is denoted by adj A.

Thus, $\text{adj } A = [A_{ij}]' = A_{ji}$ = cofactor of a_{ji} in A

$$\text{If } A = \begin{bmatrix} a_{11} & a_{12} & a_{13} \\ a_{21} & a_{22} & a_{23} \\ a_{31} & a_{32} & a_{33} \end{bmatrix}, \text{ then, adj } A = \begin{bmatrix} A_{11} & A_{12} & A_{13} \\ A_{21} & A_{22} & A_{23} \\ A_{31} & A_{32} & A_{33} \end{bmatrix}' = \begin{bmatrix} A_{11} & A_{21} & A_{31} \\ A_{12} & A_{22} & A_{32} \\ A_{13} & A_{23} & A_{33} \end{bmatrix}$$

where A_{ij} denotes the cofactor of a_{ij} in A.

2.8.1. Properties of the Adjoint of a Matrix

1. If A is a square matrix of order n, then
$$A(\text{adj } A) = |A| I_n = (\text{adj } A) A,$$
where I_n is a square matrix of order n.
2. If A is a square matrix of order n, then
$$\text{adj } (A') = (\text{adj } A)'$$
3. If A and B are two square matrices of the same order, then
$$\text{adj } (AB) = \text{adj } (B) \text{ adj } (A)$$

2.9. INVERSE OF A SQUARE MATRIX

Definition. Let A be a n-rowed square matrix. If there exists an n-rowed square matrix B such that
$$AB = BA = \mathrm{I}$$
where I is the identity matrix of order n, then the matrix A is said to be invertible and B is called the inverse (or reciprocal) of A.

Note 1. Only square matrices can have inverse.

Note 2. From the definition, it is clear that if B is the inverse of A, then A is the inverse of B.

Note 3. Inverse of A is denoted by A^{-1}, thus $B = A^{-1}$ and
$$AA^{-1} = A^{-1} A = \mathrm{I}$$

Singular and Non-Singular Matrices: A square matrix A is said to be singular if $|A| = 0$, and it is called non-singular if $|A| \neq 0$. For example, if

$$A = \begin{bmatrix} 3 & 1 & 2 \\ 2 & -3 & -1 \\ 1 & 2 & 1 \end{bmatrix}$$

then $$|A| = 3(-3+2) - 1(2+1) + 2(4+3) = 8$$

Since, $|A| \neq 0$. A is non-singular. Further, if

$$B = \begin{bmatrix} 2 & 0 & 0 \\ 3 & -1 & 4 \\ 5 & -2 & 8 \end{bmatrix}$$

then $$|B| = 2(-8+8) + 0 + 0 = 0$$

Since, $|B| = 0$, B is singular.

Theorem: The ncessary and sufficient condition for a square matrix A to possess inverse is that $|A| \neq 0$ *i.e.* A is non-singular.

Note: From the above theorem, we conclude that A has inverse if and only if $|A| \neq 0$ and then

$$A^{-1} = \frac{\text{Adj. } A}{|A|}$$

2.9.1. Properties of the Inverse of a Matrix

1. A square matrix is invertible if and only if it is non-singular.
2. The inverse of the inverse is the original matrix itself, *i.e.*
$$(A^{-1})^{-1} = A$$
3. The inverse of the transpose of a matrix is the transpose of its inverse *i.e.*, $(A')^{-1} = (A^{-1})'$
4. (Reverse law). If A and B are two invertible matrices of the same order, then AB is also invertible and moreover.
$$(AB)^{-1} = B^{-1}A^{-1}.$$

SOLVED EXAMPLES

Example 2.15. Compute the adjoint of the matrix

$$\mathbf{A} = \begin{bmatrix} 1 & 2 \\ 3 & 4 \end{bmatrix}$$

and verify that A. (ajd. A) = $|A|\, I$.

Solution. Given
$$A = \begin{bmatrix} 1 & 2 \\ 3 & 4 \end{bmatrix}$$

$\therefore$
$$|A| = \begin{vmatrix} 1 & 2 \\ 3 & 4 \end{vmatrix} = 1(4) - 2(3) = -2$$

We have
$$A_{11} = 4 \qquad A_{12} = -3$$
$$A_{21} = -2 \qquad A_{22} = 1$$

$\therefore$
$$\text{Adj. } A = \text{Transpose of } \begin{bmatrix} A_{11} & A_{12} \\ A_{21} & A_{22} \end{bmatrix} = \begin{bmatrix} A_{11} & A_{21} \\ A_{12} & A_{22} \end{bmatrix} = \begin{bmatrix} 4 & -2 \\ -3 & 1 \end{bmatrix}$$

Now
$$A\,.\,(\text{Adj. } A) = \begin{bmatrix} 1 & 2 \\ 3 & 4 \end{bmatrix}\begin{bmatrix} 4 & -2 \\ -3 & 1 \end{bmatrix} = \begin{bmatrix} 4-6 & -2+2 \\ 12-12 & -6+4 \end{bmatrix} = \begin{bmatrix} -2 & 0 \\ 0 & -2 \end{bmatrix}$$

$$= -2\begin{bmatrix} 1 & 0 \\ 0 & 1 \end{bmatrix} = |A|.I = \text{R.H.S.}$$

Hence verified.

Example 2.16. Find the inverse of $A = \begin{bmatrix} 1 & 3 & 3 \\ 1 & 4 & 3 \\ 1 & 3 & 4 \end{bmatrix}$ and verify that $A^{-1}A = I_3$.

Solution. Given
$$A = \begin{bmatrix} 1 & 3 & 3 \\ 1 & 4 & 3 \\ 1 & 3 & 4 \end{bmatrix}$$

$$\therefore \quad |A| = \begin{bmatrix} 1 & 3 & 3 \\ 1 & 4 & 3 \\ 1 & 3 & 4 \end{bmatrix} = 1\begin{vmatrix} 4 & 3 \\ 3 & 4 \end{vmatrix} - 3\begin{vmatrix} 1 & 3 \\ 1 & 4 \end{vmatrix} + 3\begin{vmatrix} 1 & 4 \\ 1 & 3 \end{vmatrix}$$

$$= (16 - 9) - 3(4 - 3) + 3(3 - 4) = 7 - 3 - 3 = 1 \neq 0.$$

$\therefore$ A is non-singular matrix and hence invertible.

$$A_{11} = (-1)^{1+1}\begin{vmatrix} 4 & 3 \\ 3 & 4 \end{vmatrix} = 7;$$

$$A_{12} = (-1)^{1+2}\begin{vmatrix} 1 & 3 \\ 1 & 4 \end{vmatrix} = -1;$$

$$A_{13} = (-1)^{1+3}\begin{vmatrix} 1 & 4 \\ 1 & 3 \end{vmatrix} = -1;$$

$$A_{21} = (-1)^{2+1}\begin{vmatrix} 3 & 3 \\ 3 & 4 \end{vmatrix} = -3;$$

$$A_{22} = (-1)^{2+2}\begin{vmatrix} 1 & 3 \\ 1 & 4 \end{vmatrix} = 1;$$

$$A_{23} = (-1)^{2+3}\begin{vmatrix} 1 & 3 \\ 1 & 3 \end{vmatrix} = 0;$$

$$A_{31} = (-1)^{3+1}\begin{vmatrix} 3 & 3 \\ 4 & 3 \end{vmatrix} = -3;$$

$$A_{32} = (-1)^{3+2}\begin{vmatrix} 1 & 3 \\ 1 & 3 \end{vmatrix} = 0;$$

$$A_{33} = (-1)^{3+3}\begin{vmatrix} 1 & 3 \\ 1 & 4 \end{vmatrix} = 1$$

$$\therefore \quad \text{Adj. } A = \text{Transpose of } \begin{bmatrix} A_{11} & A_{12} & A_{13} \\ A_{21} & A_{22} & A_{23} \\ A_{31} & A_{32} & A_{33} \end{bmatrix} = \begin{bmatrix} A_{11} & A_{21} & A_{31} \\ A_{12} & A_{22} & A_{32} \\ A_{13} & A_{23} & A_{33} \end{bmatrix} = \begin{bmatrix} 7 & -3 & -3 \\ -1 & 1 & 0 \\ -1 & 0 & 1 \end{bmatrix}$$

$$\therefore \quad A^{-1} = \frac{\text{Adj. } A}{|A|} = 1\begin{bmatrix} 7 & -3 & -3 \\ -1 & 1 & 0 \\ -1 & 0 & 1 \end{bmatrix} = 1\begin{bmatrix} 7 & -3 & -3 \\ -1 & 1 & 0 \\ -1 & 0 & 1 \end{bmatrix}$$

Now
$$A^{-1}A = \begin{bmatrix} 7 & -3 & -3 \\ -1 & 1 & 0 \\ -1 & 0 & 1 \end{bmatrix}\begin{bmatrix} 1 & 3 & 3 \\ 1 & 4 & 3 \\ 1 & 3 & 4 \end{bmatrix} = \begin{bmatrix} 7-3-3 & 21-12-9 & 21-9-12 \\ -1+1+0 & -3+4+0 & -3+3+0 \\ -1+0+1 & -3+0+3 & -3+0+4 \end{bmatrix}$$

$$= \begin{bmatrix} 1 & 0 & 0 \\ 0 & 1 & 0 \\ 0 & 0 & 1 \end{bmatrix} = I_3.$$

Hence $A^{-1}A = I_3$ is verified.

Example 2.17. If $A = \begin{bmatrix} 2 & 2 & 1 \\ -2 & 1 & 2 \\ 1 & -2 & 2 \end{bmatrix}$ and $B = \begin{bmatrix} 1 & 3 & 2 \\ 1 & 1 & 1 \\ 2 & -3 & -1 \end{bmatrix}$ be two square matrices of order three, verify that $(AB)^{-1} = B^{-1}A^{-1}$.

Solution.
$$AB = \begin{bmatrix} 2 & 2 & 1 \\ -2 & 1 & 2 \\ 1 & -2 & 2 \end{bmatrix}\begin{bmatrix} 1 & 3 & 2 \\ 1 & 1 & 1 \\ 2 & -3 & -1 \end{bmatrix} = \begin{bmatrix} 6 & 5 & 5 \\ 3 & -11 & -5 \\ 3 & -5 & -2 \end{bmatrix}$$

Now $|AB| = 6(22 - 25) - 5(-6 + 15) + 5(-15 + 33) = 27 \neq 0$

Hence $(AB)^{-1}$ exists and is given by

$$(AB)^{-1} = \frac{1}{|AB|} \text{Adj.}(AB)$$

It can be easily seen that adj. $(AB) = \begin{bmatrix} -3 & -15 & 30 \\ -9 & -27 & 45 \\ 18 & 45 & -81 \end{bmatrix}$

$$\therefore \quad (AB)^{-1} = \frac{1}{27}\begin{bmatrix} -3 & -15 & 30 \\ -9 & -27 & 45 \\ 18 & 45 & -81 \end{bmatrix} \qquad ...(i)$$

Further $|A| = 2(2 + 4) - 2(-4 - 2) + 1(4 - 2) = 27.$

and $|B| = 1(-1 + 3) - 3(-1 - 2) + 2(-3 - 2) = 1$

Hence both A^{-1} and B^{-1} exist and are given by

$$A^{-1} = \frac{1}{|A|}\text{adj. } A \text{ and } B^{-1} = \frac{1}{|B|}\text{adj. } B \text{ respectively.}$$

It can be easily verified that

$$\text{adj. } A = \begin{bmatrix} 6 & -6 & 3 \\ 6 & 3 & -6 \\ 3 & 6 & 6 \end{bmatrix} \quad \text{and} \quad \text{adj. } B = \begin{bmatrix} 2 & -3 & 1 \\ 3 & -5 & 1 \\ -5 & 9 & -2 \end{bmatrix}$$

Thus
$$A^{-1} = \frac{1}{27}\begin{bmatrix} 6 & -6 & 3 \\ 6 & 3 & -6 \\ 3 & 6 & 6 \end{bmatrix} \quad \text{and} \quad B^{-1} = \begin{bmatrix} 2 & -3 & 1 \\ 3 & -5 & 1 \\ -5 & 9 & -2 \end{bmatrix}$$

and hence
$$B^{-1}A^{-1} = \frac{1}{27}\begin{bmatrix} 2 & -3 & 1 \\ 3 & -5 & 1 \\ -5 & 9 & -2 \end{bmatrix}\begin{bmatrix} 6 & -6 & 3 \\ 6 & 3 & -6 \\ 3 & 6 & 6 \end{bmatrix}$$

$$= \frac{1}{27}\begin{bmatrix} -3 & -15 & 30 \\ -9 & -27 & 45 \\ 18 & 45 & -81 \end{bmatrix} = (AB)^{-1} \qquad [\text{Using } (i)]$$

EXERCISE 2.4

1. Find the adjoint of the following matrices:

(i) $\begin{bmatrix} a & b \\ c & d \end{bmatrix}$ (ii) $\begin{bmatrix} \cos\theta & \sin\theta \\ -\sin\theta & \cos\theta \end{bmatrix}$ (iii) $\begin{bmatrix} 1 & 2 & 3 \\ 2 & 3 & 2 \\ 3 & 3 & 4 \end{bmatrix}$

$\left[\textbf{Ans. } (i) \begin{bmatrix} d & -b \\ -c & a \end{bmatrix}, (ii) \begin{bmatrix} \cos\theta & -\sin\theta \\ \sin\theta & \cos\theta \end{bmatrix}, (iii) \begin{bmatrix} 6 & 1 & -5 \\ -2 & -5 & 4 \\ -3 & 3 & -1 \end{bmatrix}\right]$

2. Find the adjoint of the matrix $A = \begin{bmatrix} 6 & -7 & 8 \\ 1 & -3 & 1 \\ 2 & 1 & -4 \end{bmatrix}$ and verify that $(\text{adj. } A)A = |A|I = A(\text{adj. } A)$.

$\left[\textbf{Ans. } \begin{bmatrix} 11 & -20 & 17 \\ 6 & -40 & 2 \\ 7 & -20 & -11 \end{bmatrix}\right]$

3. For the matrix $A = \begin{bmatrix} 1 & -1 & 1 \\ 2 & 3 & 0 \\ 18 & 2 & 10 \end{bmatrix}$, show that $A(\text{ajd. } A) = 0$.

4. For the matrix $A = \begin{bmatrix} \cos\alpha & -\sin\alpha & 0 \\ \sin\alpha & \cos\alpha & 0 \\ 0 & 0 & 1 \end{bmatrix}$, verify that $A(\text{adj. } A) = |A|I$.

5. Prove that the matrix $\begin{bmatrix} -4 & -3 & -3 \\ 1 & 0 & 1 \\ 4 & 4 & 3 \end{bmatrix}$ is its own adjoint.

6. Given the matrix $A = \begin{bmatrix} 1 & 2 & 3 \\ 1 & 3 & 5 \\ 1 & 5 & 12 \end{bmatrix}$. Compute Adj. A and verify that $A(\text{adj. } A) = (\text{adj. } A).A = |A|I$.

$\left[\textbf{Ans. } \text{Adj. } A = \begin{pmatrix} 11 & -9 & 1 \\ -7 & 9 & -2 \\ 2 & -3 & 1 \end{pmatrix}\right]$

7. Find the inverse of the following matrices:

(i) $\begin{bmatrix} 3 & 1 \\ -1 & 2 \end{bmatrix}$ (ii) $\begin{bmatrix} \cos\theta & \sin\theta \\ -\sin\theta & \cos\theta \end{bmatrix}$

(iii) $\begin{bmatrix} 2 & -1 & 3 \\ -5 & 3 & 1 \\ -3 & 2 & 3 \end{bmatrix}$

$\left[\textbf{Ans. } (i) \begin{bmatrix} \frac{2}{7} & \frac{-1}{7} \\ \frac{1}{7} & \frac{3}{7} \end{bmatrix}, (ii) \begin{bmatrix} \cos\theta & -\sin\theta \\ \sin\theta & \cos\theta \end{bmatrix}, (iii) \begin{bmatrix} -7 & -9 & 10 \\ -12 & -15 & 17 \\ 1 & 1 & -1 \end{bmatrix}\right]$

8. Find the inverse of $A = \begin{bmatrix} 1 & 3 & 3 \\ 1 & 4 & 3 \\ 1 & 3 & 4 \end{bmatrix}$ and verify that $A^{-1}A = I_3$.

$\left[\textbf{Ans. } \begin{bmatrix} 7 & -3 & -3 \\ -1 & 1 & 0 \\ -1 & 0 & 1 \end{bmatrix}\right]$

9. If $A = \begin{bmatrix} 2 & 2 & 1 \\ -2 & 1 & 2 \\ 1 & -2 & 2 \end{bmatrix}$ and $B = \begin{bmatrix} 1 & 3 & 2 \\ 1 & 1 & 1 \\ 2 & -3 & -1 \end{bmatrix}$. verify that $(AB)^{-1} = B^{-1}A^{-1}$.

10. If $A = \frac{1}{9}\begin{bmatrix} -8 & 1 & 4 \\ 4 & 4 & 7 \\ 1 & -8 & 4 \end{bmatrix}$, prove that $A^{-1} = A'$, A' being he transpose of A.

11. If $A = \begin{bmatrix} 3 & -3 & 4 \\ 2 & -3 & 4 \\ 0 & -1 & 1 \end{bmatrix}$ (*i*) find A^{-1}, (*ii*) show that $A^3 = A^{-1}$. $\left[\textbf{Ans. } A^{-1} = \begin{pmatrix} 1 & -1 & 0 \\ -2 & 3 & -4 \\ -2 & 3 & -3 \end{pmatrix}\right]$

12. If $A = \begin{bmatrix} 11 & -20 & 17 \\ 6 & -40 & 2 \\ 7 & -20 & -11 \end{bmatrix}$ and $B = \begin{bmatrix} 6 & -7 & 8 \\ 1 & -3 & 1 \\ 2 & 1 & -4 \end{bmatrix}$, compute AB and hence find A^{-1}. $\left[\textbf{Ans. } \frac{1}{80}\begin{bmatrix} 6 & -7 & 8 \\ 1 & -3 & 1 \\ 2 & 1 & -4 \end{bmatrix}\right]$

13. If $A = \begin{bmatrix} 1 & -1 & 1 \\ 2 & -1 & 0 \\ 1 & 0 & 0 \end{bmatrix}$, find A^3 and hence find A^{-1}. $\left[\textbf{Ans. } \begin{bmatrix} 0 & 0 & 1 \\ 0 & -1 & 2 \\ 1 & -1 & 1 \end{bmatrix}\right]$

14. If $A = \begin{bmatrix} 1 & 1 & 2 \\ 1 & 9 & 3 \\ 1 & 4 & 2 \end{bmatrix}$, $B = \begin{bmatrix} 1 & 2 & 0 \\ 2 & 3 & -1 \\ 1 & -1 & 3 \end{bmatrix}$, show that $(AB)^{-1} = B^{-1}A^{-1}$.

15. Compute $(AB)^{-1}$ where $A = \begin{bmatrix} 1 & 3 & 3 \\ 1 & 4 & 3 \\ 1 & 3 & 4 \end{bmatrix}$ and $B^{-1} = \begin{bmatrix} 2 & 1 & 2 \\ 2 & 2 & -1 \\ 1 & 0 & 3 \end{bmatrix}$ $\left[\textbf{Ans. } \begin{bmatrix} 11 & -5 & -4 \\ 13 & -4 & -7 \\ 4 & -3 & 0 \end{bmatrix}\right]$

16. Show that the matrix $A = \begin{bmatrix} 2 & 3 \\ 1 & 2 \end{bmatrix}$, satisfies the equation $A^2 - 4A + I = 0$. Hence find A^{-1}. $\left[\textbf{Ans. } \begin{bmatrix} 2 & -3 \\ -1 & 2 \end{bmatrix}\right]$

17. Given $A = \begin{bmatrix} 2 & -3 \\ -4 & 7 \end{bmatrix}$, compute A^{-1} and show that $2A^{-1} = 9I - A$. $\left[\textbf{Ans. } \frac{1}{2}\begin{bmatrix} 7 & 3 \\ 4 & 2 \end{bmatrix}\right]$

18. Find the matrix X satisfying $X\begin{bmatrix} 5 & 3 \\ -1 & -2 \end{bmatrix} = \begin{bmatrix} 14 & 7 \\ 7 & 7 \end{bmatrix}$. $\left[\textbf{Ans. } \begin{bmatrix} 3 & 1 \\ 1 & -2 \end{bmatrix}\right]$

19. If $A = \begin{bmatrix} 6 & -7 & 8 \\ 1 & -3 & 1 \\ 2 & 1 & -4 \end{bmatrix}$, compute det (adj. A). [**Ans.** 6400]

20. If $A = \begin{bmatrix} 1 & 0 & 2 \\ 0 & 2 & 1 \\ 2 & 0 & 3 \end{bmatrix}$, find A (adj. A). $\left[\textbf{Ans. } \begin{bmatrix} -2 & 0 & 0 \\ 0 & -2 & 0 \\ 0 & 0 & -2 \end{bmatrix}\right]$

21. If $D = \text{diag.}\ [d_1, d_2, d_3]$ where $d_1, d_2, d_3 \neq 0$ then show that $D^{-1} = \text{diag.}\ [d_1^{-1}, d_2^{-1}, d_3^{-1}]$

2.10. SOLUTION OF SYSTEM OF LINEAR EQUATIONS BY DETERMINANTS (CREMER'S RULE)

Consider the system of two linear equations in two unknown x and y:

$$a_1x + b_1y = c_1$$
$$a_2x + b_2y = c_2$$

Solving these equation, we get

$$x = \frac{b_2c_1 - b_1c_2}{a_1b_2 - a_2b_1}$$

$$y = \frac{a_1c_2 - a_2c_1}{a_1b_2 - a_2b_1}$$

Writing determinants, we can express the solution as

$$x = \frac{b_2c_1 - b_1c_2}{a_1b_2 - a_2b_1} = \frac{\begin{vmatrix} c_1 & b_1 \\ c_2 & b_2 \end{vmatrix}}{\begin{vmatrix} a_1 & b_1 \\ a_2 & b_2 \end{vmatrix}} = \frac{D_1}{D}$$

$$y = \frac{a_1c_2 - a_2c_1}{a_1b_2 - a_2b_1} = \frac{\begin{vmatrix} a_1 & c_1 \\ a_2 & c_2 \end{vmatrix}}{\begin{vmatrix} a_1 & b_1 \\ a_2 & b_2 \end{vmatrix}} = \frac{D_2}{D}$$

This is known as Cramer's Rule.

The rule states that if

(*i*) D = the determinant of the coefficients.

(*ii*) D_1 = what becomes of D when the coefficients of x are replaced by c's.

(*iii*) D_2 = what becomes of D when the coefficients of y are replaced by c's then

$$x = \frac{D_1}{D} \quad \text{and} \quad y = \frac{D_2}{D}.$$

The same rule can be applied to solve the equations in three unknowns.

If
$$a_1x + b_1y + c_1z = d_1$$
$$a_2x + b_2y + c_2z = d_2$$
$$a_3x + b_3y + c_3z = d_3$$

then $x = \frac{D_1}{D}, y = \frac{D_2}{D}, z = \frac{D_3}{D}$, where

$$D = \begin{vmatrix} a_1 & b_1 & c_1 \\ a_2 & b_2 & c_3 \\ a_3 & b_3 & c_3 \end{vmatrix}$$

i.e. the determinant of the coefficients of x, y, z.

$$D_1 = \begin{vmatrix} d_1 & b_1 & c_1 \\ d_2 & b_2 & c_2 \\ d_3 & b_3 & c_3 \end{vmatrix}$$

i.e. the determinant obtained from D by replacing a's, the coefficient of x by d's.

$$D_2 = \begin{vmatrix} a_1 & d_1 & c_1 \\ a_2 & d_2 & c_2 \\ a_3 & d_3 & c_3 \end{vmatrix}$$

i.e. the determinant obtained from D by replacing b's, the coefficients of y by d's.

$$D_3 = \begin{vmatrix} a_1 & b_1 & d_1 \\ a_2 & b_2 & d_2 \\ a_3 & b_3 & d_3 \end{vmatrix}$$

i.e. the determinant obtained from D by replacing c's, the coefficients of z by d's.

Cramer's rule can be used in exactly the same way to solve the system of n equations in n unknowns.

Note 1. If $D \neq 0$, then the system has a unique solution and is, therefore consistent.

Note 2. If $D = 0$, then $x \,.\, 0 = D_1, y \,.\, 0 = D_2, z \,.\, 0 = D_3$

(*i*) If $D_1 = D_2 = D_3 = 0$, then the system is **consistent** and has an infinite number of solutions.

(*ii*) If $D_i \neq 0$ for atleast one $i = 1, 2, 3$, then the system has **no solution,** and is therefore, inconsistent.

Note 3. We have assumed in the above discussion that $d_i \neq 0$. for atleast one $i = 1, 2, 3$ and the system is called non-homogeneous.

If, however $d_1 = d_2 = d_3 = 0$, then we have a system of homogeneous linear equations:

$$a_1x + b_2y + c_2z = 0$$
$$a_2x + b_2y + c_2z = 0$$
$$a_3x + b_3y + c_3z = 0$$

This system is always consistent, because $x = y = z = 0$ is a solution which is called **trivial solution.**

SOLVED EXAMPLES

Example 2.18. Solve the following system of equations by Cramer's rule.

$$5x - 7y + z = 11$$
$$6x - 8y - z = 15$$
$$3x + 2y - 6z = 7$$

Solution. We have
$$D = \begin{vmatrix} 5 & -7 & 1 \\ 6 & -8 & -1 \\ 3 & 2 & -6 \end{vmatrix} = 5(48 + 2) + 7(-36 + 3) + 1(12 + 24) = 55$$

Since $D \neq 0$; therefore by Cramer's rule, the solution is given by

$$x = \frac{D_1}{D}, y = \frac{D_2}{D}; z = \frac{D_3}{D}$$

where
$$D_1 = \begin{vmatrix} 11 & -7 & 1 \\ 15 & -8 & -1 \\ 7 & 2 & -6 \end{vmatrix} = 11(48 + 2) + 7(-90 + 7) + 1(30 + 56) = 55$$

$$D_2 = \begin{vmatrix} 5 & 11 & 1 \\ 6 & 15 & -1 \\ 3 & 7 & -6 \end{vmatrix} = 5(-90 + 7) - 11(-36 + 3) + 1(42 - 45) = -55$$

$$D_3 = \begin{vmatrix} 5 & -7 & 11 \\ 6 & -8 & 15 \\ 3 & 2 & 7 \end{vmatrix} = 5(-56 - 30) + 7(42 - 45) + 11(12 + 24) = -55$$

$\therefore$
$$x = \frac{D_1}{D} = \frac{55}{55} = 1; y = \frac{D_2}{D} = \frac{-55}{55} = -1; z = \frac{D_3}{D} = \frac{-55}{55} = -1$$

Hence $x = 1, y = -1$ and $z = -1$.

Example 2.19. **Solve, if consistent, the following system of equations using cramer's rule:**

$$2x - y + z = 4 \quad ...(i)$$
$$x + 3y + 2z = 12 \quad ...(ii)$$
$$3x + 2y + 3z = 16 \quad ...(iii)$$

Solution. We have
$$D = \begin{vmatrix} 2 & -1 & 1 \\ 1 & 3 & 2 \\ 3 & 2 & 3 \end{vmatrix} = 2(9-4) + 1(3-6) + 1(2-9) = 0$$

$$D_1 = \begin{vmatrix} 4 & -1 & 1 \\ 12 & 3 & 2 \\ 16 & 2 & 3 \end{vmatrix} = 4(9-4) + 1(36-32) + 1(24-48) = 0$$

$$D_2 = \begin{vmatrix} 2 & 4 & 1 \\ 1 & 12 & 2 \\ 3 & 16 & 3 \end{vmatrix} = 2(36-32) - 4(3-6) + 1(16-36) = 0$$

$$D_3 = \begin{vmatrix} 2 & -1 & 4 \\ 1 & 3 & 12 \\ 3 & 2 & 16 \end{vmatrix} = 2(48-24) + 1(16-36) + 4(2-9) = 0$$

Since $D = D_1 = D_2 = D_3 = 0$, the system is consistent and has infinite many solutions. To find these solution, consider the first two equations and take $z = k$, we get

$$2x - y = 4 - k$$
$$x + 3y = 12 - 2k$$

Solving these equations, by Cramer's rule, we have

$$D = \begin{vmatrix} 2 & -1 \\ 1 & 3 \end{vmatrix} = 7,$$

$$D_1 = \begin{vmatrix} 4-k & -1 \\ 12-2k & 3 \end{vmatrix} = 24 - 5k$$

$$D_2 = \begin{vmatrix} 2 & 4-k \\ 1 & 12-2k \end{vmatrix} = 20 - 3k$$

Hence, by Cramer's rule, we have

$$x = \frac{D_1}{D} = \frac{24-5k}{7} \text{ and } y = \frac{D_2}{D} = \frac{20-3k}{7}$$

By substituting the values of x, y and z in (*iii*), we get

$$\text{L.H.S.} = 3x + 2y + 3z = \frac{3(24-5k)}{7} + \frac{2(20-3k)}{7} + 3k = 16 = \text{R.H.S}$$

Since Eq. (*iii*) is satisfied $x = \dfrac{24-5k}{7}$, $y = \dfrac{20-3k}{7}$, $z = k$

given the solution for each value of k. Hence, the system has infinite solutions.

Example 2.20. Solve the following system, if possible using Cramer's rule:

$$x - 3y + 4z = 3$$
$$2x - 5y + 7z = 6$$
$$3x - 8y + 11z = 11$$

Solution. We have
$$D = \begin{vmatrix} 1 & -3 & 4 \\ 2 & -5 & 7 \\ 3 & -8 & 11 \end{vmatrix} = 1(-55+56) + 3(22-21) + 4(-16+15) = 0$$

Further,
$$D_1 = \begin{vmatrix} 3 & -3 & 4 \\ 6 & -5 & 7 \\ 11 & -8 & 11 \end{vmatrix} = 3(-55+56) + 3(66-77) + 4(48+55) = 1$$

Since $D = 0$ and $D_1 \neq 0$, the given system is inconsistent.

EXERCISE 2.5

1. Using Cramer's rule, solve the following system of equations:
   (*i*) $2y - 3z = 0, x + 3y = -4, 3x + 4y = 3.$ [**Ans.** $x = 5, y = -3, z = 2$]
   (*ii*) $x - 4y - z = 11, 2x - 5y + 2z = 39, -3x + 2y + z = 1.$ [**Ans.** $x = -1, y = -5, z = 8$]
   (*iii*) $\frac{2}{x} + \frac{3}{y} + \frac{10}{z} = 4$
   $\frac{4}{x} - \frac{6}{y} + \frac{5}{z} = 1$
   $\frac{6}{x} + \frac{9}{y} - \frac{20}{z} = 2$ [**Ans.** $x = 2, y = 3, z = 5$]
2. Using Determinants, solve:
   (*i*) $x - 3y - 8z = -10, 3x + y - 4z = 0, 2x + 5y + 6z = 13.$ [**Ans.** $x = 2k - 1, y = -2k + 3$ and $z = k$].
   (*ii*) $x - y + z = 3, 2x + y - z = 2, -x - 2y + 2z = 1.$ $\left[\textbf{Ans. } x = \frac{5}{3}, y = k - \frac{4}{3}, z = k\right]$
   (*iii*) $2x + y - 2z = 4, x - 2y + z = -2, 5x - 5y + z = -2.$ $\left[\textbf{Ans. } x = \frac{6+3k}{5}, y = \frac{8+4k}{5}, z = k\right]$
3. Using Cramer's rule, solve:
   (*i*) $3x - y + 2z = 3, 2x + y + 3z = 5, x - 2y - z = 1.$ [**Ans.** Inconsistent]
   (*ii*) $x + y + z = 3, 2x - y + z = 2, 3x + 6y + 5z = 20.$ [**Ans.** Inconsistent]
   (*iii*) $x + y + z = 1, 2x + 2y + 2z = 2, 3z + 3y + 3z = 4.$ [**Ans.** Inconsistent]

2.11. SOLUTION OF SYSTEM OF LINEAR EQUATIONS BY MATRICES

Consider the following System of n linear equations in n unknowns $x_1, x_2, ..., x_n$:

$$\left.\begin{array}{l} a_{11}x_1 + a_{12}x_2 + \ldots + a_{1n}x_n = b_1 \\ a_{21}x_1 + a_{22}x_2 + \ldots + a_{2n}x_n = b_2 \\ \ldots\ldots\ldots\ldots\ldots\ldots\ldots\ldots \\ \ldots\ldots\ldots\ldots\ldots\ldots\ldots\ldots \\ a_{n1}x_1 + a_{n2}x_2 + \ldots + a_{nn}x_n = b_n \end{array}\right]$$

Let A be the coefficient matrix *i.e.*,

$$A = \begin{bmatrix} a_{11} & a_{12} & \dots & a_{1n} \\ a_{21} & a_{22} & \dots & a_{2n} \\ \dots & \dots & \dots & \dots \\ a_{n1} & a_{n2} & \dots & a_{nn} \end{bmatrix}_{n\times n}$$

Suppose $$X = \begin{bmatrix} x_1 \\ x_2 \\ x_3 \\ \vdots \\ x_n \end{bmatrix}_{n\times 1} \quad \text{and} \quad b = \begin{bmatrix} b_1 \\ b_2 \\ b_3 \\ \vdots \\ b_n \end{bmatrix}_{n\times 1}$$

This sytem can be represented by the single matrix equation

$$AX = B \qquad \dots(ii)$$

If A is non-singular, *i.e.*, $|A| \neq 0$, so that A^{-1} exists, then from (*ii*), we obtain

$$A^{-1}(AX) = A^{-1}B$$
$$(A^{-1}A)X = A^{-1}B$$
$$IX = A^{-1}B$$
$$X = A^{-1}B$$

which gives the required solution of the given system.

Criterion of consistency. We give the following criterion for the consistency or inconsistency of a system of linear equations given by $AX = B$, where A is a square matrix.

(*i*) If $|A| \neq 0$, then the system is consistent and has a **unique solution.**

(*ii*) If $|A| = 0$, the system of equations has either no solution or an infinite number of solutions. Now find (Adj. A)B

(*a*) If (Adj. $A)B \neq 0$, the system has **no solution** and is therefore **inconsistent.**

(*b*) If (Adj. $A)B = 0$, the system is **consistent** and has **infinite many solutions.**

SOLVED EXAMPLES

Example 2.21. Solve the following system of equations by matrix method:

$$2x - y = 2$$
$$3y + 2z = 16$$
$$5x + 3z = 21$$

Solution. The system can be represented by the matrix equation $AX = B$

where $$A = \begin{bmatrix} 2 & -1 & 0 \\ 0 & 3 & 2 \\ 5 & 0 & 3 \end{bmatrix}, X = \begin{bmatrix} x \\ y \\ z \end{bmatrix}, B = \begin{bmatrix} 2 \\ 16 \\ 21 \end{bmatrix}$$

$$|A| = \begin{vmatrix} 2 & -1 & 0 \\ 0 & 3 & 2 \\ 5 & 0 & 3 \end{vmatrix} = 2(9) + 1(-10) = 8 \neq 0$$

$\therefore$ A is non-singular.

The system has a unique solution given by

$$X = A^{-1}B$$

Now $A_{11} = 9$ $A_{21} = 3$ $A_{31} = -2$

$A_{12} = 10$ $A_{22} = 6$ $A_{32} = -4$

$A_{13} = -15$ $A_{23} = -5$ $A_{33} = 6$

$$\text{Adj. } A = \begin{bmatrix} 9 & 3 & -2 \\ 10 & 6 & -4 \\ -15 & -5 & 6 \end{bmatrix}$$

$$A^{-1} = \frac{\text{Adj. } A}{|A|} = \frac{1}{8}\begin{bmatrix} 9 & 3 & -2 \\ 10 & 6 & -4 \\ -15 & -5 & 6 \end{bmatrix}$$

The solution is given by $X = A^{-1}B$

$$\therefore \quad \begin{bmatrix} x \\ y \\ z \end{bmatrix} = \frac{1}{8}\begin{bmatrix} 9 & 3 & -2 \\ 10 & 6 & -4 \\ -15 & -5 & 6 \end{bmatrix}\begin{bmatrix} 2 \\ 16 \\ 21 \end{bmatrix} = \frac{1}{8}\begin{bmatrix} 24 \\ 32 \\ 16 \end{bmatrix} = \begin{bmatrix} 3 \\ 4 \\ 2 \end{bmatrix}$$

Hence $x = 3$, $y = 4$ and $z = 2$.

Example 2.22. Use matrix method to solve the following system of equations:

$$\mathbf{4x + 2y + 6z = 28}$$
$$\mathbf{3x + y + 2z = 20}$$
$$\mathbf{10x + 5y + 15z = 70}$$

Solution. The given equations can be written in matrix as $AX = B$

where $$A = \begin{bmatrix} 4 & 2 & 6 \\ 3 & 1 & 2 \\ 10 & 5 & 15 \end{bmatrix}, X = \begin{bmatrix} x \\ y \\ z \end{bmatrix}, B = \begin{bmatrix} 28 \\ 20 \\ 70 \end{bmatrix}$$

$$|A| = \begin{vmatrix} 4 & 2 & 6 \\ 3 & 1 & 2 \\ 10 & 5 & 15 \end{vmatrix}$$

$$= 4(15 - 10) - 2(45 - 20) + 6(15 - 10) = 20 - 50 + 30 = 0$$

Since $|A| = 0$, A^{-1} does not exist.

Let us find (Adj. A) B

$A_{11} = 5$ $A_{21} = 0$ $A_{31} = -2$

$A_{12} = -25$ $A_{22} = 0$ $A_{32} = 10$

$A_{13} = 5$ $A_{23} = 0$ $A_{33} = -2$

$$\text{Adj. } A = \begin{bmatrix} 5 & 0 & -2 \\ -25 & 0 & 10 \\ 5 & 0 & -2 \end{bmatrix}$$

$$(\text{Adj. } A)\,B = \begin{bmatrix} 5 & 0 & -2 \\ -25 & 0 & 10 \\ 5 & 0 & -2 \end{bmatrix}\begin{bmatrix} 28 \\ 20 \\ 70 \end{bmatrix} = \begin{bmatrix} 0 \\ 0 \\ 0 \end{bmatrix}$$

Since (Adj. A)$B = 0$, the given system of equations in consistent and has infinite many solutions.

To obtain the finite solution, take $z = k$ and solve first two of the given equations as follows:

$$4x + 2y + 6z = 28$$
$$6x + 2y + 4z = 40$$
$$2x - 2z = 12$$
$$x = 6 + z$$

But $z = k$, $\quad x = 6 + k$

$\therefore \quad y = 20 - 3(6 + k) - 2k = 2 - 5k$

Hence $x = 6 + k$, $y = 2 - 5k$ and $z = k$.

Example 2.23. Test for consistency of the equations

$$\mathbf{4x - 5y - 2z = 2}$$
$$\mathbf{5x - 4y + 2z = -2}$$
$$\mathbf{2x + 2y + 8z = -1}$$

Solution. The system can be represented by the matrix equation $AX = B$ where

$$A = \begin{bmatrix} 4 & -5 & -2 \\ 5 & -4 & 2 \\ 2 & 2 & 8 \end{bmatrix}, X = \begin{bmatrix} x \\ y \\ z \end{bmatrix}, B = \begin{bmatrix} 2 \\ -2 \\ -1 \end{bmatrix}$$

$$|A| = \begin{vmatrix} 4 & -5 & -2 \\ 5 & -4 & 2 \\ 2 & 2 & 8 \end{vmatrix} = 4(-32 - 4) - (-5)(40 - 4) - 2(10 + 8)$$

$$= 144 + 180 - 36 = -180 + 180 = 0$$

Since $|A| = 0$, $\therefore \quad A^{-1}$ does not exist.

Let us find (Adj. A) B

$$A_{11} = -36 \qquad A_{21} = 36 \qquad A_{31} = -18$$
$$A_{12} = -36 \qquad A_{22} = 36 \qquad A_{32} = -18$$
$$A_{13} = 18 \qquad A_{23} = -18 \qquad A_{33} = 9$$

$\therefore$
$$\text{Adj. } A = \begin{bmatrix} -36 & 36 & -18 \\ -36 & 36 & -18 \\ 18 & -18 & 9 \end{bmatrix}$$

$$(\text{Adj. } A)\,B = \begin{bmatrix} -36 & 36 & -18 \\ -36 & 36 & -18 \\ 18 & -18 & 9 \end{bmatrix} \begin{bmatrix} 2 \\ -2 \\ -1 \end{bmatrix} = \begin{bmatrix} -126 \\ -126 \\ 63 \end{bmatrix} \neq 0$$

Since $|A| = 0$ and (Adj. A) $B \neq 0$, the given equation has no solution and, therefore, inconsistent.

Example 2.24. Given that $A = \begin{bmatrix} -4 & 4 & 4 \\ -7 & 1 & 3 \\ 5 & -3 & -1 \end{bmatrix}$ **and** $B = \begin{bmatrix} 1 & -1 & 1 \\ 1 & -2 & -2 \\ 2 & 1 & 3 \end{bmatrix}$ **Find AB. use this result. to solve the following system of linear equations:**

$$\mathbf{x - y + z = 4}$$
$$\mathbf{x - 2y - 2z = 9}$$
$$\mathbf{2x + y + 3z = 1.}$$

Solution. Here,
$$AB = \begin{bmatrix} -4 & 4 & 4 \\ -7 & 1 & 3 \\ 5 & -3 & -1 \end{bmatrix}\begin{bmatrix} 1 & -1 & 1 \\ 1 & -2 & -2 \\ 2 & 1 & 3 \end{bmatrix}$$
$$= \begin{bmatrix} -4+4+8 & 4-8+4 & 7-2+3 \\ -5+6-1 & 7-2+3 & -7-2+9 \\ 5-3-2 & -5+6-1 & 5+6-3 \end{bmatrix}$$
$$= \begin{bmatrix} 8 & 0 & 0 \\ 0 & 8 & 0 \\ 0 & 0 & 8 \end{bmatrix} = 8\begin{bmatrix} 1 & 0 & 0 \\ 0 & 1 & 0 \\ 0 & 0 & 1 \end{bmatrix} = 8I$$

$\Rightarrow \quad |AB| = |8I| = 8^3 |I| = 83 . 1 \Rightarrow |A| . |B| = 512 \neq 0$

$\Rightarrow \quad |B| \neq 0$

$\Rightarrow$ B is invertible and $B^{-1} = \dfrac{1}{8}A$ $\quad \left[\because \; AB = 8I \Rightarrow \dfrac{1}{8}AB = I \Rightarrow \left(\dfrac{1}{8}A\right)B = I\right]$

Now, the given system can be written as
$$\begin{bmatrix} 1 & -1 & 1 \\ 1 & -2 & -2 \\ 2 & 1 & 3 \end{bmatrix}\begin{bmatrix} x \\ y \\ z \end{bmatrix} = \begin{bmatrix} 4 \\ 9 \\ 1 \end{bmatrix} \text{ i.e., } BX = C,$$

where $X = \begin{bmatrix} x \\ y \\ z \end{bmatrix}$ and $C = \begin{bmatrix} 4 \\ 9 \\ 1 \end{bmatrix}$. Since B is invertible, therefore, on premultiplying by B^{-1}, we get
$$B^{-1}(BX) = B^{-1}C \quad \Rightarrow \quad (B^{-1}B)X = B^{-1}C$$

$$\Rightarrow \quad IX = \frac{1}{8}AC \quad \Rightarrow \quad X = \frac{1}{8}\begin{bmatrix} -4 & 4 & 4 \\ -7 & 1 & 3 \\ 5 & -3 & 1 \end{bmatrix}\begin{bmatrix} 4 \\ 9 \\ 1 \end{bmatrix}$$

$$\Rightarrow \quad \begin{bmatrix} x \\ y \\ z \end{bmatrix} = \frac{1}{8}\begin{bmatrix} -16+36+4 \\ -28+9+3 \\ 20-27-1 \end{bmatrix} = \frac{1}{8}\begin{bmatrix} 24 \\ -16 \\ -8 \end{bmatrix} = \begin{bmatrix} 3 \\ -2 \\ 1 \end{bmatrix}$$

Hence $x = 3$, $y = -2$ and $z = -1$.

Example 2.25. Find all integers λ for which the system of equations
$$\mathbf{x + 2y - 3z = 1}$$
$$\mathbf{2x - \lambda y - 3z = 2}$$
$$\mathbf{x + 2y + \lambda z = 3}$$
has a unique solution. Find the solution for $\lambda = 0$.

Solution. Let Δ be the determinant of the matrix of coefficients, then
$$\Delta = \begin{vmatrix} 1 & 2 & -3 \\ 2 & -\lambda & -3 \\ 1 & 2 & \lambda \end{vmatrix} = 1(-\lambda^2 + 6) + 2(2\lambda + 3) - 3(4 + \lambda)$$
$$= -\lambda^2 - 7\lambda - 12 = -(\lambda + 3)(\lambda + 4)$$

Now, the given equations will have a unique solution iff $\Delta \neq 0$.

i.e., iff $-(\lambda + 3)(\lambda + 4) \neq 0$, *i.e.,* $\lambda \neq -3, -4$.

Hence the given system of equations has a unique solution for all integral values of λ except -3 and -4.

In particular, when $\lambda = 0$, the given system of equations becomes

$$x + 2y - 3z = 1$$
$$2x - 3z = 2$$
$$z + 2y = 3$$

which can be written as

$$\begin{bmatrix} 1 & 2 & -3 \\ 2 & 0 & -3 \\ 1 & 2 & 0 \end{bmatrix}\begin{bmatrix} x \\ y \\ z \end{bmatrix} = \begin{bmatrix} 1 \\ 2 \\ 3 \end{bmatrix}$$

i.e., $AX = B$, where
$$A = \begin{bmatrix} 1 & 2 & -3 \\ 2 & 0 & -3 \\ 1 & 2 & 0 \end{bmatrix}$$

Now
$$|A| = \begin{vmatrix} 1 & 2 & -3 \\ 2 & 0 & -3 \\ 1 & 2 & 0 \end{vmatrix} = 1(0 + 6) - 2(0 + 3) - 3(4 - 0)$$
$$= 6 - 6 - 12 = -12 \neq 0.$$

Hence
$$A^{-1} = \frac{1}{|A|}.\text{Adj}.A = -\frac{1}{12}\begin{bmatrix} 6 & -3 & 4 \\ -6 & 3 & 0 \\ -6 & -3 & -4 \end{bmatrix} = -\frac{1}{12}\begin{bmatrix} 6 & -6 & -6 \\ -3 & 3 & -3 \\ 4 & 0 & -4 \end{bmatrix}$$

Thus from $AX = B$, we get on premultiplying by A^{-1},
$$A^{-1}(AX) = A^{-1}B$$

i.e.,
$$X = A^{-1}B = -\frac{1}{12}\begin{bmatrix} 6 & -6 & -6 \\ -3 & 3 & -3 \\ 4 & 0 & -4 \end{bmatrix}\begin{bmatrix} 1 \\ 2 \\ 3 \end{bmatrix} = -\frac{1}{12}\begin{bmatrix} 6-12-18 \\ -3+6-9 \\ 4+0-12 \end{bmatrix}$$

$$\Rightarrow \quad \begin{bmatrix} x \\ y \\ z \end{bmatrix} = -\frac{1}{12}\begin{bmatrix} -24 \\ -6 \\ -8 \end{bmatrix} = \begin{bmatrix} 2 \\ \frac{1}{2} \\ \frac{2}{3} \end{bmatrix}$$

$$\Rightarrow \quad x = 2, y = \frac{1}{2} \text{ and } z = \frac{2}{3}.$$

EXERCISE 2.6

1. Solve the following equations by matrix method

$$5x + 3y + z = 16$$
$$2x + y + 3z = 19$$
$$x + 2y + 4z = 25$$

[**Ans.** $x = 1, y = 2, z = 5$]

2. Find A^{-1} if $A = \begin{bmatrix} -1 & 2 & 5 \\ 2 & -3 & 1 \\ -1 & 1 & 1 \end{bmatrix}$ and hence, solve the system of linear equations:

$$-x + 2y + 5z = 2$$
$$2x - 3y + z = 15$$
$$-x + y + z = -3$$

$\left[\textbf{Ans. } -\frac{1}{7}\begin{bmatrix} -4 & 3 & 17 \\ -3 & 4 & 11 \\ -1 & -1 & -1 \end{bmatrix}\right]$

3. Solve the following system of equations by matrix method:

$$3x - 2y + 3z = 8$$
$$2x + y - z = 1$$
$$4x - 3y + 2z = 4$$

[**Ans.** $x = 1, y = 2, z = 3$]

4. Use matrix method to solve the following system of equations:

(*i*) $x - y + z = 4$
$x + y - 3z = 0$
$x + y + z = 2$

(*ii*) $x + y + z = 2$
$2x - y = 3$
$2y + z = 0$

[**Ans.** (*i*) $x = 2,\ y = -1,\ z = 1$
(*ii*) $x = 1,\ y = -1,\ z = 2$]

5. Which of the following equations are consistent? If consistent, find the solutions:

(*i*) $4x - 3y = 5$
$8x - 6y = 9$

(*ii*) $x + y - 5z = 26$
$x + 2y + z = -4$
$x + 3y + 6z = -29$

(*iii*) $3x - y\ 2z = 2$
$2y + z = -1$
$3x - 5y = 3$

(*iv*) $x - 3y - 8z = -10$
$3x + y - 4z = 0$
$2x + 5y + 6z = 13$

[**Ans.** (*i*) Inconsistent; (*ii*) Consistent : $x = 1, y = 0, z = -5$; (*iii*) Inconsistent;
(*iv*) Consistent, $x = 2K - 1, y = 3 - 2K, z = K, K \in R$]

6. If $A = \begin{bmatrix} 2 & -1 & 3 \\ 1 & 3 & 2 \\ 3 & -4 & -1 \end{bmatrix}$, find A^{-1}. Using A^{-1}, solve the following system of linear equations

$$2x - y + 3z = 13$$
$$x + 3y + 2z = 1$$
$$3z - 4y - z = 8$$

[**Ans.** $x = 1, y = -2, z = 3$]

7. Solve the system of equations:

$$5x + 3y + 7z = 4$$
$$3x + 26y + 2z = 9$$
$$7x + 2y + 11z = 5$$

with the help of matrix inversion.

$\left[\textbf{Ans. } x = \frac{7}{11}, y = \frac{3}{11}, z = 0\right]$

8. If $A = \begin{bmatrix} 2 & -3 & 5 \\ 3 & 2 & -4 \\ 1 & 1 & -2 \end{bmatrix}$, find A^{-1} and hence solve the system

$$2x - 3y + 5z = 11$$
$$2x + 2y - 4z = -5$$
$$x + y - 2z = -3$$

[**Ans.** $x = 1, y = 2, z = 3$]

9. Solve with help of matrices, the simultaneous equations:

$$x + y + z = 3$$
$$x + 2y + 3z = 4$$
$$x + 4y + 9z = 6$$

[**Ans.** $x = 2, y = 1, z = 0$]

10. For what value of t, will the system

$$tx + 3y - z = 1$$
$$x + 2y + z = 2$$
$$-tx + y + 2z = -1.$$

fails to have a unique solution? Will it have any solution for this value of t?

$\left[\textbf{Ans. } -\frac{7}{2}; \text{No}\right]$

11. Show that the system of equations

$$3x - y + 4z = 3$$
$$x + 2y - 3z = -2$$
$$6x + 5y + \lambda z = -3$$

has at least one solution for every real λ. Find the set of solutions when $\lambda = -5$.

$\left[\textbf{Ans. } x = \frac{1}{7}(4 - 5k),\ y = \frac{1}{7}(13k - 9),\ z = k, \text{ where } k \text{ is arbitrary}\right]$

❑❑❑

CHAPTER 3

Vector Algebra

INSIDE THIS CHAPTER

3.1. INTRODUCTION TO VECTORS

The physical quantities are divided into two categories:

Scalar quantities and Vector quantities

Scalars: A quantity that has magnitude only is known as a scalar.

Examples: Each of the quantities mass, length, time, temperature, density, speed etc., is a scalar.

Vectors: A quantities that has magnitude as well as direction is called a vector.

Examples: Each of the quantities force, velocity, acceleration and momentum is a vector.

However, we define a vector as *a directed line segment.*

A directed line segment with **initial point** A and the terminal point B is the vector denoted by $\overrightarrow{AB}$.

The magnitude of $\overrightarrow{AB}$ is denoted by $|\overrightarrow{AB}|$.

Note: We usually denote a vector by a single letter with an arrow on it and its magnitude is denoted by this letter only.

Thus, $$\overrightarrow{AB} = \vec{a} \quad \text{and} \quad |\overrightarrow{AB}| = |\vec{a}| = a$$

3.2. TYPES OF VECTORS

- **Unit Vector:** A vector $\vec{a}$ is called a unit vector if $|\vec{a}| = 1$ and it is denoted by $\hat{a}$.

- **Equal Vectors:** Two vectors $\vec{a}$ and $\vec{b}$ are said to be equal if they have the same magnitude and the same direction regardless of the positions of their initial points.

• **Like Vectors:** Vectors which have the same direction (parallel), same sense (arrow), the magnitude may be different.

• **Unlike Vectors:** Vectors which have the same direction (parallel), opposite sense (arrow), the magnitude may be different.

• **Negative of a Vectors:** A vector having the same magnitude as that of a given vector $\vec{a}$ but opposite direction is called the negative of $\vec{a}$ and is denoted by $-\vec{a}$. Thus, if $\overrightarrow{AB} = \vec{a}$, then $\overrightarrow{BA} = -\vec{a}$.

• **Zero or Null Vector:** A vector whose initial and terminal points coincide is called a zero vector denoted by $\vec{O}$. Clearly, the magnitude of a zero vector is O but it cannot be assigned a definite direction. Thus $\overrightarrow{AA} = \vec{O}$.

Vector other than the null are called *proper vectors*.

• **Co-initial Vectors:** Vectors having the same initial point are called co-initial vectors.

• **Collinear Vectors:** Vectors $\vec{a}$ and $\vec{b}$ are collinear if they have the same direction or are parallel or anti-parallel. Since their magnitudes are different, we can find some scalar λ for which $\vec{a} = \lambda\vec{b}$. If $\lambda > 0$, $\vec{a}$ and $\vec{b}$ are in the same direction, if $\lambda < 0$, $\vec{a}$ and $\vec{b}$ are in opposite directions. Collinear vectors are often called *dependent vectors.*

• **Localized and Free Vectors:** A vector drawn parallel to a given vector through a specified point the initial point is called a localized vector.

If the initial point of vector is not specified then it is said to be free vector.

• **Coplanar Vectors:** Three or more non-zero vectors lying in the same plane (or parallel to the same plane) are said to be coplanar vectors.

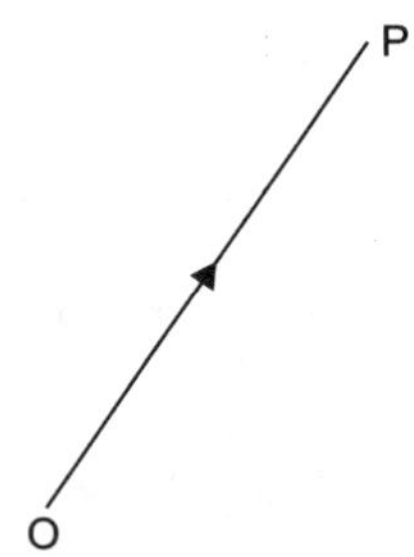

• **Position Vector of a Point:** When a vector $\overrightarrow{OP}$ is used to specify the position of a point P relative to another point O, it is called the position vector of P relative to the point O. The point O is arbitrarily chosen and is called the *origin of reference*.

Thus position vector of P briefly written as **P.V. of *P* is the vector $\overrightarrow{OP}$ with tip at *P* and tail at *O*.**

If the position vector of the point P with an assigned origin O is the vector $\vec{a}$, then it is often referred to as the point $P(\vec{a})$.

3.3. VECTOR ADDITION

Let $\vec{a}$ and $\vec{b}$ be any two vectors. Take any point O and draw segments $\overrightarrow{OA}$ and $\overrightarrow{AB}$ such that $\overrightarrow{OA} = \vec{a}$ and $\overrightarrow{AB} = \vec{b}$. Join OB. Then $\overrightarrow{OB}$ is called the **sum** or **resultant** of $\vec{a}$ and $\vec{b}$.

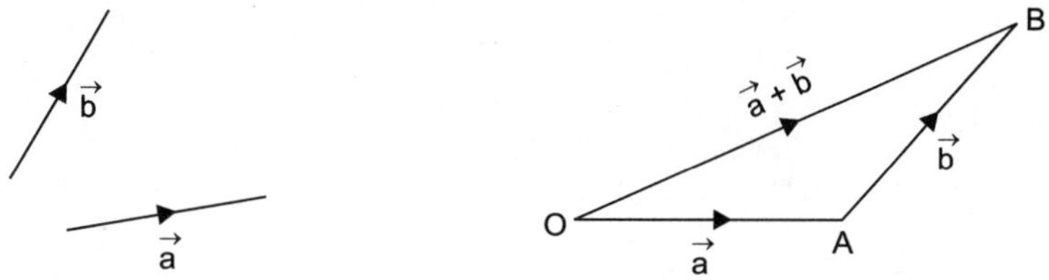

$$\therefore \quad \vec{a} + \vec{b} = \overrightarrow{OA} + \overrightarrow{AB} = \overrightarrow{OB}$$

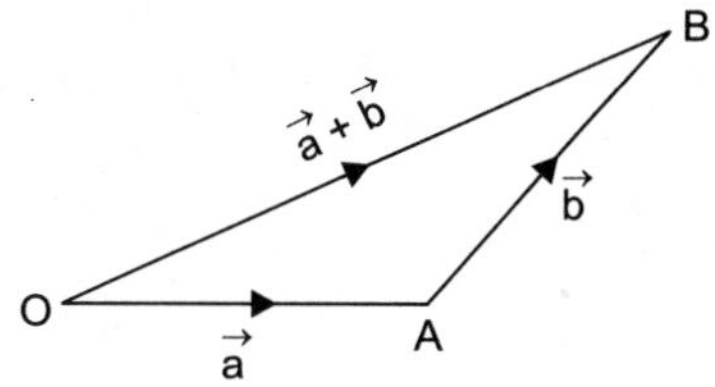

Triangle Law of Addition of Vectors: In a ΔOAB, if $\overrightarrow{OA}$ and $\overrightarrow{AB}$ represent $\vec{a}$ and $\vec{b}$ respectively, then $\overrightarrow{OB}$ represents $\vec{a} + \vec{b}$.

This is known as triangle law of addition of vectors.

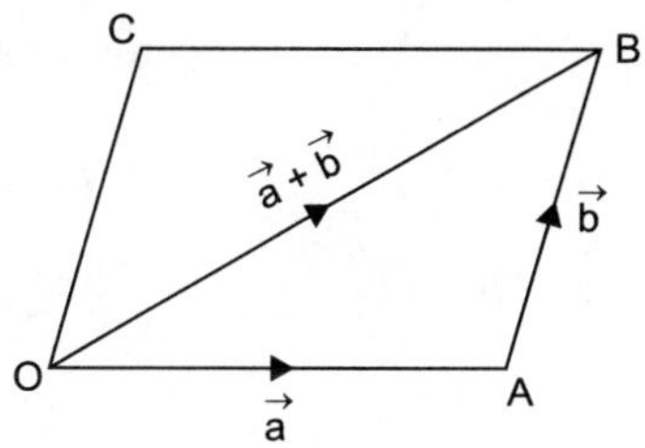

Parallelogram Law of Addition of Vectors: In a parallelogram $OABC$, if $\overrightarrow{OA}$ and $\overrightarrow{AB}$ represent $\vec{a}$ and $\vec{b}$ respectively, then $\overrightarrow{OB}$ represents $\vec{a} + \vec{b}$.

This is known as parallelogram law of addition of vectors.

3.4. LAWS OF ADDITION OF VECTORS

Theorem 1: (Cumulative law): *Vector addition is commutative.*

i.e., $$\vec{a} + \vec{b} = \vec{b} + \vec{a}$$

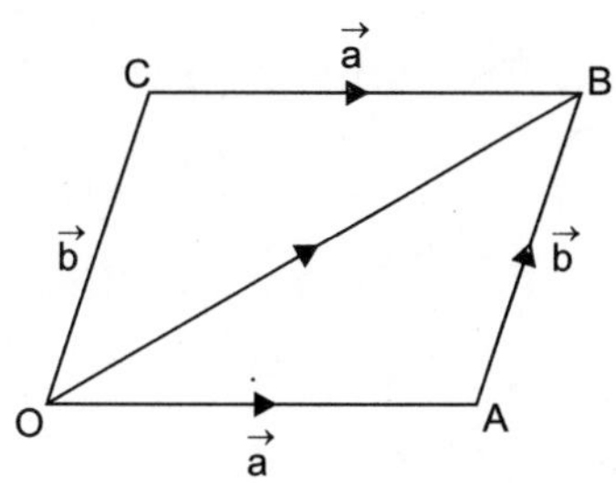

Proof: Let $\vec{a}$ and $\vec{b}$ be the given vectors represented by $\overrightarrow{OA}$ and $\overrightarrow{OB}$ respectively. Complete the parallelogram OABC.

Then, $$\overrightarrow{OC} = \overrightarrow{AB} = \vec{b}$$

and $$\overrightarrow{CB} = \overrightarrow{OA} = \vec{a}$$

$$\therefore \quad \vec{a} + \vec{b} = \overrightarrow{OA} + \overrightarrow{AB} = \overrightarrow{OB}$$

and $$\vec{b} + \vec{a} = \overrightarrow{OC} + \overrightarrow{CB} = \overrightarrow{OB}$$

Hence, $$\vec{a} + \vec{b} = \vec{b} + \vec{a}$$

Theorem 2: (Associative law): *Vector addition is associative*

i.e., $$(\vec{a} + \vec{b}) + \vec{c} = \vec{a} + (\vec{b} + \vec{c})$$

Proof: Let $\overrightarrow{OA} = \vec{a}$, $\overrightarrow{AB} = \vec{b}$ and $\overrightarrow{BC} = \vec{c}$

Join OB, OC and AC

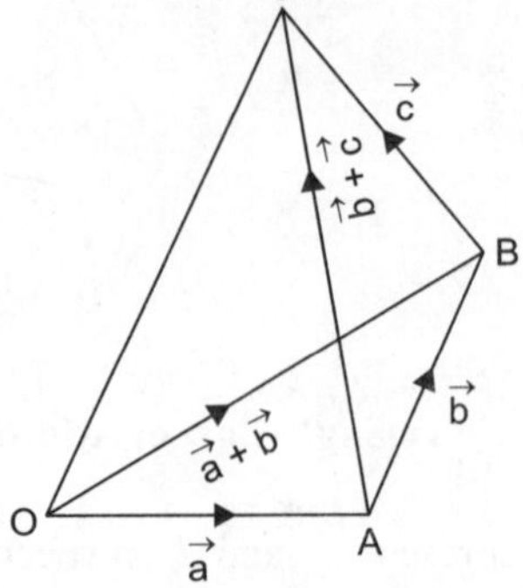

$$(\vec{a}+\vec{b})+\vec{c} = (\overrightarrow{OA}+\overrightarrow{AB})+\overrightarrow{BC}$$

$$= (\overrightarrow{OB}+\overrightarrow{BC}) \qquad [\because \overrightarrow{OA}+\overrightarrow{AB}=\overrightarrow{OB}]$$

$$= \overrightarrow{OC}$$

$$\vec{a}+(\vec{b}+\vec{c}) = \overrightarrow{OA}+(\overrightarrow{AB}+\overrightarrow{BC})$$

$$= \overrightarrow{OA}+\overrightarrow{AC} \qquad [\because \overrightarrow{AB}+\overrightarrow{BC}=\overrightarrow{AC}]$$

$$= \overrightarrow{OC}$$

$$\therefore \quad (\vec{a}+\vec{b})+\vec{c} = \vec{a}+(\vec{b}+\vec{c})$$

Theorem 3: (Existence of Additive Identity): For any vector $\vec{a}$, prove that:

$$\vec{a}+\vec{0} = \vec{0}+\vec{a}=\vec{a}$$

Proof: Let $\overrightarrow{OA}=\vec{a}$. Then, $\vec{a}+\vec{0}=\overrightarrow{OA}+\overrightarrow{AA}=\overrightarrow{OA}=\vec{a}$

and,
$$\vec{0}+\vec{a} = \overrightarrow{OO}+\overrightarrow{OA}=\overrightarrow{OA}=\vec{a}$$

$$\therefore \quad \vec{a}+\vec{0} = \vec{0}+\vec{a}=\vec{a}$$

Remark: The vector $\vec{0}$ is called the *additive identity* for vectors.

Theorem 4: (Existence of Additive Inverse): For any vector $\vec{a}$, prove that:

$$\vec{a}+(-\vec{a}) = (-\vec{a})+\vec{a}=\vec{0}$$

Proof: Let $\overrightarrow{OA}=\vec{a}$. Then, $\overrightarrow{AO}=-\vec{a}$

$$\therefore \quad (\vec{a})+(-\vec{a}) = \overrightarrow{OA}+\overrightarrow{AO}=\overrightarrow{OO}=\vec{0}$$

and
$$(-\vec{a})+(\vec{a}) = \overrightarrow{AO}+\overrightarrow{OA}=\overrightarrow{AA}=\vec{0}$$

Hence,
$$\vec{a}+(-\vec{a}) = (-\vec{a})+\vec{a}=\vec{0}$$

Remark: The vector $-\vec{a}$ is called the *additive inverse* of $\vec{a}$.

3.5. SUBTRACTION OF TWO VECTORS

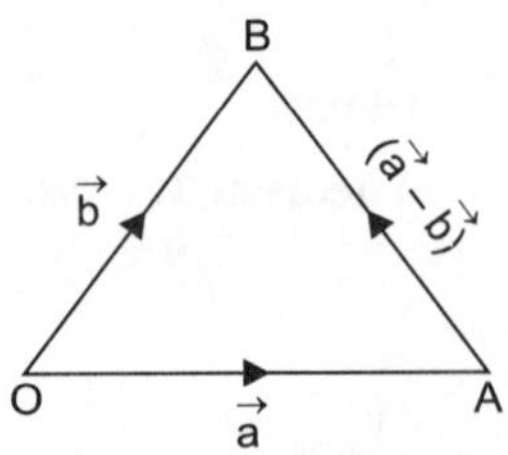

For any two vectors $\vec{a}$ and $\vec{b}$, we define

$$\vec{a}-\vec{b} = \vec{a}+(-\vec{b})$$

Let $\overrightarrow{OA}=\vec{a}$ and $\overrightarrow{OB}=\vec{b}$. Then $\overrightarrow{BO}=-\vec{b}$.

$$\therefore \qquad (\vec{a} - \vec{b}) = \vec{a} + (-\vec{b}) = \overrightarrow{OA} + \overrightarrow{BO} = \overrightarrow{BO} + \overrightarrow{OA} = \overrightarrow{BA}$$

Thus $\qquad (\overrightarrow{OA} - \overrightarrow{OB}) = \overrightarrow{BA}$

Similarly, $\qquad (\overrightarrow{OB} - \overrightarrow{OA}) = \overrightarrow{AB}$

SOLVED EXAMPLES

Example 3.1. Find the value of x, y and z. So that the vectors

$$\vec{a} = x\hat{i} - 2\hat{j} + z\hat{k} \quad \text{and} \quad \vec{b} = 3\hat{i} + y\hat{j} - \hat{k}$$

are equal.

Solution. $\vec{a} = \vec{b} \Rightarrow x\hat{i} - 2\hat{j} + z\hat{k} = 3\hat{i} + y\hat{j} - \hat{k}$

$$\Rightarrow \qquad x = 3,\ y = -2 \text{ and } z = -1$$

Example 3.2. Let $\vec{a} = \hat{i} + 2\hat{j}$ and $\vec{b} = 2\hat{i} + \hat{j}$. Is $|\vec{a}| = |\vec{b}|$? Are the vectors $\vec{a}$ and $\vec{b}$ equal?

Solution. We have

$$|\vec{a}| = \sqrt{1^2 + 2^2} = \sqrt{5}$$

and

$$|\vec{b}| = \sqrt{2^2 + 1^2} = \sqrt{5}$$

So, $|\vec{a}| = |\vec{b}|$. But the two vectors are not equal since their corresponding components are distinct.

Example 3.3. Find a unit vector in the direction of the vector $\vec{a} = \hat{i} + 2\hat{j} + 3\hat{k}$.

Solution. $\qquad \vec{a} = \hat{i} + 2\hat{j} + 3\hat{k}$

$$\Rightarrow \qquad |\vec{a}| = \sqrt{1^2 + 2^2 + 3^2} = \sqrt{14}$$

The unit vector in the direction of $\vec{a}$ is given by

$$\hat{a} = \frac{\vec{a}}{|\vec{a}|} = \frac{\hat{i} + 2\hat{j} + 3\hat{k}}{\sqrt{14}} = \frac{1}{\sqrt{14}}\hat{i} + \frac{2}{\sqrt{14}}\hat{j} + \frac{3}{14}\hat{k}$$

Example 3.4. Find a unit vector in the direction of the sum of the vectors,

$$\vec{a} = 2\hat{i} + 2\hat{j} - 5\hat{k} \quad \text{and} \quad \vec{b} = 2\hat{i} + \hat{j} + 3\hat{k}.$$

Solution. Let $\vec{c} = \vec{a} + \vec{b}$. Then

$$\vec{c} = \vec{a} + \vec{b} = (2\hat{i} + 2\hat{j} - 5\hat{k}) + (2\hat{i} + \hat{j} + 3\hat{k}) = 4\hat{i} + 3\hat{j} - 2\hat{k}$$

$$|\vec{c}| = \sqrt{4^2 + 3^2 + (-2)^2} = \sqrt{29}$$

Hence, the required vector is given by

$$\hat{c} = \frac{\vec{c}}{|\vec{c}|} = \frac{4\hat{i} + 3\hat{j} - 2\hat{k}}{\sqrt{29}} = \frac{4}{\sqrt{29}}\hat{i} + \frac{3}{\sqrt{29}}\hat{j} - \frac{2}{\sqrt{29}}\hat{k}$$

Example 3.5. **If $\vec{a} + 5\vec{b} = \vec{c}$ and $\vec{a} - 7\vec{b} = 2\vec{c}$, show that $\vec{a}$ has the same direction as $\vec{c}$ and opposite to that of $\vec{b}$.**

Solution. Given that

$$\vec{a} + 5\vec{b} = \vec{c} \quad \ldots(i)$$

$$\vec{a} - 7\vec{b} = 2\vec{c} \quad \ldots(ii)$$

Eliminating $\vec{b}$ from (*i*) and (*ii*), we get

$$7\vec{a} + 35\vec{b} = 7\vec{c}$$

$$5\vec{a} - 35\vec{b} = 10\vec{c}$$

$$\Rightarrow \quad 12\vec{a} = 17\vec{c}$$

$$\therefore \quad \vec{a} = \frac{17}{12}\vec{c} = m\vec{c} \quad \ldots(iii)$$

where $m = \frac{17}{12} > 0$

$\therefore$ Vector $\vec{a}$ has the same direction as $\vec{c}$

From (*iii*), $\vec{c} = \frac{12}{17}\vec{a}$

Putting $\vec{c}$ in (*i*)

$$\vec{a} + 5\vec{b} = \frac{12}{17}\vec{a} \quad \text{or} \quad \vec{a} - \frac{12}{17}\vec{a} = -5\vec{b}$$

or $$\frac{5}{17}\vec{a} = -5\vec{b} \quad \text{or} \quad \vec{a} = -17\vec{b}$$

$$\therefore \quad \vec{a} = n\vec{b}, \quad \text{where } n = -17 < 0$$

$\therefore$ Vector $\vec{a}$ has a direction opposite to that of $\vec{b}$.

Example 3.6. **If D is the mid-point of BC of a ΔABC, then prove that $\overrightarrow{AB} + \overrightarrow{AC} = 2\overrightarrow{AD}$.**

Solution. By triangle law of addition of vectors

$$\overrightarrow{AB} + \overrightarrow{BD} = \overrightarrow{AD} \quad \text{and} \quad \overrightarrow{AC} + \overrightarrow{CD} = \overrightarrow{AD}$$

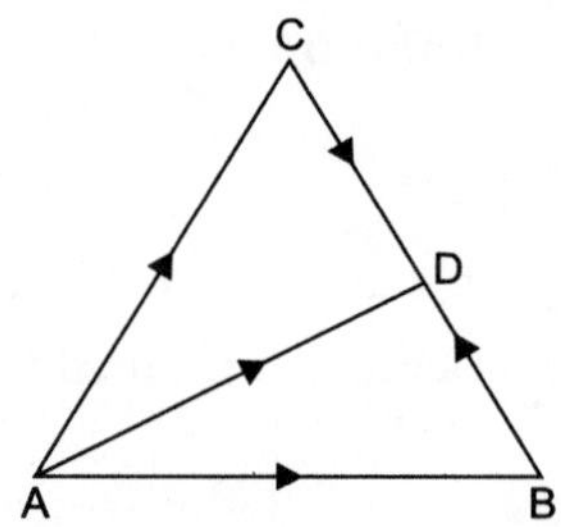

Adding $$\overrightarrow{AB} + \overrightarrow{AC} + \overrightarrow{BD} + \overrightarrow{CD} = 2\overrightarrow{AD} \quad \ldots(i)$$

But $$\overrightarrow{BD} + \overrightarrow{CD} = \vec{O}$$

(Since $\overrightarrow{BD}$ and $\overrightarrow{CD}$ are equal in magnitude and opposite in direction)

$\therefore$ From (*i*), $$\overrightarrow{AB} + \overrightarrow{AC} = 2\overrightarrow{AD}$$

Example 3.7. If $\vec{a}$ and $\vec{b}$ are non-collinear (or non-parallel) vectors and $x\vec{a} + y\vec{b} = \vec{0}$; then show that $x = 0$ and $y = 0$.

Solution. $$x\vec{a} + y\vec{b} = \vec{0} \qquad \text{(given)} \quad \ldots(i)$$

If possible let $x \neq 0$

$\therefore$ From (*i*), $$x\vec{a} = -y\vec{b}$$

Dividing by $x \neq 0$

$$\vec{a} = \frac{-y}{x}\vec{b} \quad \text{or} \quad \vec{a} = m\vec{b}, \quad \text{where } m = \frac{-y}{x}$$

$\therefore$ Vectors $\vec{a}$ and $\vec{b}$ are collinear (or parallel) vectors.

But this is contrary to given

$\therefore$ Our supposition is wrong.

$\therefore$ $x = 0$. Similalrly $y = 0$

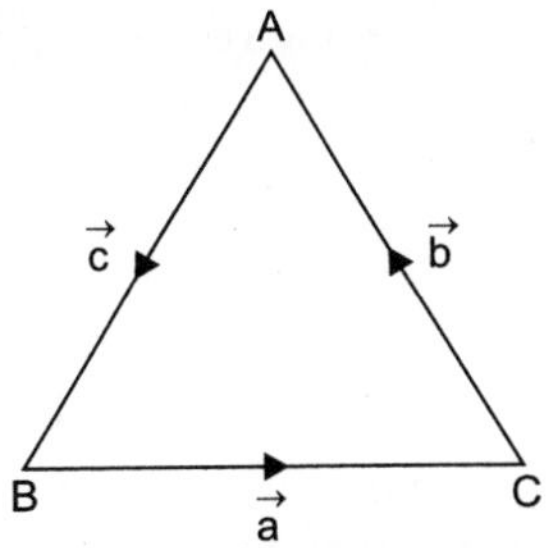

Example 3.8. If $\vec{a}, \vec{b}, \vec{c}$ be the vectors represented by the sides of a triangle taken in order, then prove that $\vec{a} + \vec{b} + \vec{c} = \vec{0}$.

Solution. Let ABC be a triangle such that $\overrightarrow{BC} = \vec{a}$, $\overrightarrow{CA} = \vec{b}$ and $\overrightarrow{AB} = \vec{c}$.

Then

$$\vec{a} + \vec{b} + \vec{c} = \overrightarrow{BC} + \overrightarrow{CA} + \overrightarrow{AB}$$

$$= \overrightarrow{BA} + \overrightarrow{AB} \qquad (\because \text{ triangle law of addition of vectors})$$

$$= \overrightarrow{BB} \qquad \text{(By triangle law)}$$

$$= \vec{0} \qquad \text{(By def. of null vectors)}$$

Hence, $$\vec{a} + \vec{b} + \vec{c} = \vec{0}.$$

Example 3.9. If $\vec{a}$ and $\vec{b}$ are the vectors determined by the two adjacent sides of a regular hexagon, what are the vectors determined by the other sides taken in order?

Solution. Let $ABCDEF$ be a regular hexagon such that $\overrightarrow{AB} = \vec{a}$ and $\overrightarrow{BC} = \vec{b}$. Since AD is parallel to BC such that $AD = 2BC$. Therefore

$$\overrightarrow{AD} = 2\overrightarrow{BC} = 2\vec{b}$$

In ΔABC, we have $$\overrightarrow{AB} + \overrightarrow{BC} = \overrightarrow{AC}$$

$\Rightarrow$ $$\vec{a} + \vec{b} = \overrightarrow{AC}$$

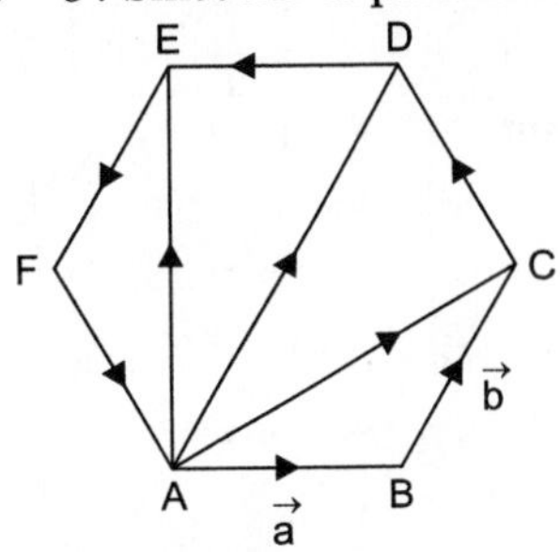

Fig. 9.11

In ΔACD, we have $\overrightarrow{AC} + \overrightarrow{CD} = \overrightarrow{AD}$

$\Rightarrow$ $\overrightarrow{CD} = \overrightarrow{AD} - \overrightarrow{AC} = 2\vec{b} - (\vec{a} + \vec{b}) = \vec{b} - \vec{a}$

Clearly, $\overrightarrow{DE} = -\overrightarrow{AB} = -\vec{a},\ \overrightarrow{EF} = -\overrightarrow{BC} = -\vec{b}$

and $\overrightarrow{FA} = -\overrightarrow{CD} = -(\vec{b} - \vec{a}) = \vec{a} - \vec{b}$

Example 3.10. For any two vectors $\vec{a}$ and $\vec{b}$, prove that:

(i) $|\vec{a} + \vec{b}| \le |\vec{a}| + |\vec{b}|$ ***(ii)*** $|\vec{a} - \vec{b}| \le |\vec{a}| + |\vec{b}|$

(iii) $|\vec{a} - \vec{b}| \ge |\vec{a}| - |\vec{b}|$

Solution. (*i*) **Case I:** *When $\vec{a}, \vec{b}$ are non-collinear vectors*

Let the vectors $\vec{a}$ and $\vec{b}$ be represented by sides $\overrightarrow{OA}$ and $\overrightarrow{AB}$ of a triangle OAB. Then

$$\overrightarrow{OA} + \overrightarrow{AB} = \overrightarrow{OB} \quad \text{(by triangle law of addition)}$$

$$\Rightarrow \quad \vec{a} + \vec{b} = \overrightarrow{OB}$$

Clearly, $|\vec{a}| = OA$, $|\vec{b}| = AB$ and $|\vec{a} + \vec{b}| = OB$

Since the sum of two sides of a triangle is always greater than the third sides, therefore

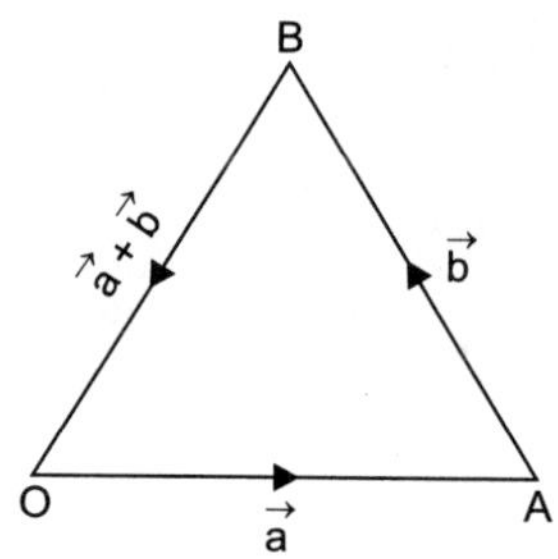

$$OA + AB > OB$$

$$\Rightarrow \quad |\overrightarrow{OB}| < |\overrightarrow{OA}| + |\overrightarrow{AB}|$$

$$\Rightarrow \quad |\vec{a} + \vec{b}| < |\vec{a}| + |\vec{b}|$$

Case II: *When $\vec{a}, \vec{b}$ are collinear vectors*

Let $\overrightarrow{OA} = \vec{a}$, $\overrightarrow{AB} = \vec{b}$. Then

$$\vec{a} + \vec{b} = \overrightarrow{OA} + \overrightarrow{AB} = \overrightarrow{OB} \quad \text{(by triangle law of addition)}$$

Clearly, $OA = |\vec{a}|,\ AB = |\vec{b}|$

and $OB = |\vec{a} + \vec{b}|$

Now, $OB = OA + AB$

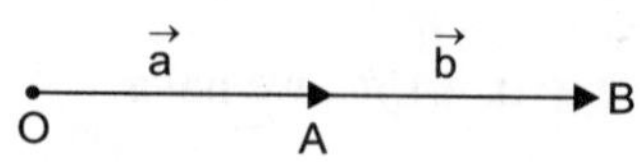

$$\Rightarrow \quad |\overrightarrow{OB}| = |\overrightarrow{OA}| + |\overrightarrow{AB}|$$

$$\Rightarrow \quad |\vec{a} + \vec{b}| = |\vec{a}| + |\vec{b}|$$

Hence, in general $|\vec{a} + \vec{b}| \le |\vec{a}| + |\vec{b}|$.

(*ii*) We have

$$|\vec{a} - \vec{b}| = |\vec{a} + (-\vec{b})| \le |\vec{a}| + |-\vec{b}| \qquad \text{[Using } (ii)]$$

$$\Rightarrow \quad |\vec{a} - \vec{b}| \le |\vec{a}| + |\vec{b}| \qquad (\because |-\vec{b}| = |\vec{b}|)$$

(*iii*) We have

$$|\vec{a}| = |\vec{a} - \vec{b} + \vec{b}| \le |\vec{a} - \vec{b}| + |\vec{b}|$$

$$\Rightarrow \quad |\vec{a}| - |\vec{b}| \le |\vec{a} - \vec{b}| \quad \text{or} \quad |\vec{a} - \vec{b}| \ge |\vec{a}| - |\vec{b}|$$

EXERCISE 3.1

1. If $|\vec{a}| = 3$, what is $|5\vec{a}|$ and $|-2\vec{a}|$? **[Ans.** 15; 6]
2. Let $\vec{a} = 3\hat{i} + 2\hat{j}$ and $\vec{b} = 2\hat{i} + 3\hat{j}$. Is $|\vec{a}| = |\vec{b}|$? Is $\vec{a} = \vec{b}$. **[Ans.** Yes; No]
3. Find a vector in the direction of the vector $\vec{a} = 3\hat{i} + \hat{j}$ that has magnitude 5 units. $\left[\textbf{Ans.} \frac{15}{\sqrt{10}}\hat{i} + \frac{5}{\sqrt{10}}\hat{j}\right]$
4. If $\overrightarrow{PO} + \overrightarrow{OQ} = \overrightarrow{QO} + \overrightarrow{OR}$ show that the points P, Q, R are collinear.
5. If the sum of two unit vectors in a unit vector, prove that the magnitude of their difference is $\sqrt{3}$.
6. Prove that the medians of a triangle meet in a point which is a point of trisection of the medians.
7. If the mid-points of the consecutive sides of a quadrilateral are joined, prove that the resulting quadrilateral is a parallelogram.
8. If $\vec{a}, \vec{b}$ are any two vectors then give the geometrical interpretation of the relation $|\vec{a} + \vec{b}| = |\vec{a} - \vec{b}|$.
9. Show that the three points having position vectors $\vec{a} - 2\vec{b} + 3\vec{c}, -2\vec{a} + 3\vec{b} + 2\vec{c}, -8\vec{a} + 13\vec{b}$ are collinear whatever be $\vec{a}, \vec{b}, \vec{c}$.
10. If P_1, P_2, P_3, P_4 are points in a plane or space and O the origin of vectors, show that P_4 coinsides with 0 iff

$$\overrightarrow{OP} + \overrightarrow{P_1P_2} + \overrightarrow{P_2P_3} + \overrightarrow{P_3P_4} = \vec{0}$$

3.6. SECTION FORMULAE

(*a*) Internal Division: Let A and B be two points with position vectors $\vec{a}$ and $\vec{b}$ respectively and let c be a point dividing AB internally in the ratio $m : n$. Then, the position vector of c is given by $\overrightarrow{OC} = \dfrac{m\vec{b} + n\vec{a}}{m+n}$.

Remark 1. If C is the mid-point of AB, then it divides AB in the ratio $1 : 1$. Therefore, position vector of C is given by

$$\frac{1 \cdot \vec{a} + 1 \cdot \vec{b}}{1+1} = \frac{\vec{a} + \vec{b}}{2}$$

Thus, the position vector of the mid-point of AB is $\frac{1}{2}(\vec{a} + \vec{b})$.

(b) External Division: Let A and B be two points with position vectors $\vec{a}$ and $\vec{b}$ respectively and let C be a point dividing AB externally in the ratio $m : n$. Then, the position vector of C is given by

$$\vec{OC} = \frac{m\vec{b} - n\vec{a}}{m - n}$$

SOLVED EXAMPLES

Example 3.11. Consider two points P and Q with position vectors $\vec{OP} = 3\vec{a} - 2\vec{b}$ and $\vec{OQ} = \vec{a} + \vec{b}$. Find the position vector of a point R which divides the line joining P and Q in the ratio 2 : 1: (*i*) internally and (*ii*) externally

Solution. (*i*) The position vector of a point R dividing the join of P and Q internally in the ratio 2 : 1 is

$$\vec{OR} = \frac{2(\vec{a} + \vec{b}) + (3\vec{a} - 2\vec{b})}{2 + 1} = \frac{5\vec{a}}{3}$$

(*ii*) The position vector of the point R dividing the join of P and Q externally in the ratio 2 : 1 is

$$\vec{OR} = \frac{2(\vec{a} + \vec{b}) - (3\vec{a} - 2\vec{b})}{2 - 1} = 4\vec{b} - \vec{a}$$

Example 3.12. Show that the points $A(2i - j + k)$, $B(i - 3j - 5\hat{k})$, $(3i - 4j - 4\hat{k})$ are the vertices of a right angled triangle.

Solution. We have

$$\vec{AB} = (1-2)\hat{i} + (-3+1)\hat{j} + (-5-1)\hat{k} = -\hat{i} - 2\hat{j} - 6\hat{k}$$

$$\vec{BC} = (3-1)\hat{i} + (-4+3)\hat{j} + (-4+5)\hat{k} = 2\hat{i} - \hat{j} + \hat{k}$$

and

$$\vec{CA} = (2-3)\hat{i} + (-1+4)\hat{j} + (1+4)\hat{k} = -\hat{i} + 3\hat{j} + 5\hat{k}$$

Further, note that

$$|\vec{AB}|^2 = 41 = 6 + 35 = |\vec{BC}| + |\vec{CA}|$$

Hence, the triangle is a right angled triangle.

EXERCISE 3.2

1. Find the position vectors of the points which divide the join of the points $2\vec{a} - 3\vec{b}$ and $3\vec{a} - 2\vec{b}$ internally and externally in the ratio 2 : 3.

[**Ans.** $\frac{12}{5}\vec{a} - \frac{13}{5}\vec{b}; -5\vec{b}$]

2. If $\vec{a}$ and $\vec{b}$ are the position vectors of A and B respectively, find the position vector of a point C on BA produced such that $BC = 1.5\,BA$. $\left[\textbf{Ans.}\ \frac{1}{2}(3\vec{a} - \vec{b})\right]$

3. Four points A, B, C, D with position vectors $\vec{a}, \vec{b}, \vec{c}, \vec{d}$ respectively are such that $3\vec{a} - \vec{b} + 2\vec{c} - 4\vec{d} = \vec{0}$. Show that the four points are coplanar. Also find the position vector of the point of intersection of line segments AC and BD.

4. Let $\vec{a}, \vec{b}, \vec{c}$ be the position vectors of three distinct points A, B, C. If there exist scalars x, y, z (not all zero) such that $x\vec{a} + y\vec{b} + z\vec{c} = \vec{0}$ and $x + y + z = 0$, then show that A, B and C lie on a line.

3.7. SCALAR (OR DOT) PRODUCT OF TWO VECTORS

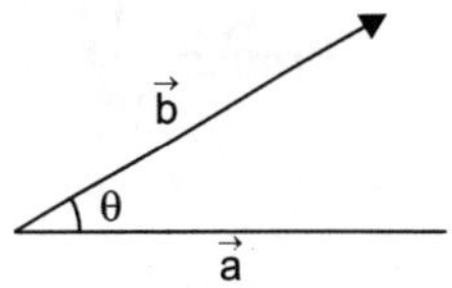

The scalar or dot product of two non-zero vectors $\vec{a}$ and $\vec{b}$ denoted by $\vec{a} \cdot \vec{b}$, is defined as $\vec{a} \cdot \vec{b} = |\vec{a}||\vec{b}|\cos\theta$, where $0 \le \theta \le \pi$.

The scalar product of two vectors is also called the *dot product* of two vectors, under consideration. The scalar product $\vec{a} \cdot \vec{b}$ is read as $\vec{a}$ dot $\vec{b}$. The scalar product of two vectors is a scalar quantity and not a vector quantity.

The scalar product $\vec{a} \cdot \vec{b}$ is positive or negative or zero according as the angle θ between $\vec{a}$ and $\vec{b}$ is acute, obtuse or right respectively.

Remarks:

1. $\vec{a} \cdot \vec{b}$ is a real number.

2. Let $\vec{a}$ and $\vec{b}$ are perpendicular to each other *i.e.*, $\vec{a} \cdot \vec{b} = 0 \Leftrightarrow \vec{a} \perp \vec{b}$

3. If $\theta = 0$, then $\vec{a} \cdot \vec{b} = |\vec{a}||\vec{b}|\cos 0 = |\vec{a}||\vec{b}|$

   In particular, $\vec{a} \cdot \vec{a} = |\vec{a}||\vec{a}|\cos 0 = |\vec{a}|^2 \cdot 1 = |\vec{a}|^2$

4. If $\theta = \pi$, then $\vec{a} \cdot \vec{b} = |\vec{a}||\vec{b}|\cos\pi = -|\vec{a}||\vec{b}|$

5. The angle between two non-zero vectors $\vec{a}$ and $\vec{b}$ is given by

$$\cos\theta = \frac{\vec{a} \cdot \vec{b}}{|\vec{a}||\vec{b}|} \quad \text{or} \quad \theta = \cos^{-1}\left(\frac{\vec{a} \cdot \vec{b}}{|\vec{a}||\vec{b}|}\right)$$

6. For mutually perpendicular unit vectors $\hat{i}$, $\hat{j}$ and $\hat{k}$, we have

$$\hat{i}\cdot\hat{i} = |\hat{i}|\cdot|\hat{i}|\cos 0° = 1\times 1\times 1 = 1$$

$\therefore$ $\hat{i}\cdot\hat{i} = 1$. Similarly $\hat{j}\cdot\hat{j} = 1$ and $\hat{k}\cdot\hat{k} = 1$

But $\hat{i}\cdot\hat{j} = |\hat{i}||\hat{j}|\cos 90° = 1\times 1\times 0 = 0$

Similarly $\hat{j}\cdot\hat{i} = 0,\ \hat{j}\cdot\hat{k} = 0,\ \hat{k}\cdot\hat{j} = 0,$

$\hat{i}\cdot\hat{k} = 0$ and $\hat{k}\cdot\hat{i} = 0$

3.8. PROPERTIES OF SCALAR (DOT) PRODUCT

Property I. Scalar product is commutative *i.e.*, if $\vec{a}$ and $\vec{b}$ be any two vectors then $\vec{a}\cdot\vec{b} = \vec{b}\cdot\vec{a}$.

Property II. Scalar product is distributive over addition *i.e.*, if $\vec{a}, \vec{b}$ and $\vec{c}$ be any three vectors then

(*a*) $\vec{a}\cdot(\vec{b}\cdot\vec{c}) = \vec{a}\cdot\vec{b}\cdot\vec{a}\cdot\vec{c}$ (Left distributive law)

(*b*) $(\vec{b}+\vec{c})\cdot\vec{a} = \vec{b}\cdot\vec{a} = \vec{c}\cdot\vec{a}$ (Right distributive law)

Property III. For any scalar m, $(m\vec{a})\cdot\vec{b} = m(\vec{a}\cdot\vec{b}) = \vec{a}\cdot(m\vec{b})$

3.9. SCALAR PRODUCT IN TERMS OF COMPONENTS

Let $\vec{a} = a_1\hat{i} + a_2\hat{j} + a_3\hat{k}$ and $\vec{b} = b_1\hat{i} + b_2\hat{j} + b_3\hat{k}$

then
$$\vec{a}\cdot\vec{b} = (a_1\hat{i} + a_2\hat{j} + a_3\hat{k})\cdot(b_1\hat{i} + b_2\hat{j} + b_3\hat{k})$$
$$= a_1b_1\hat{i}\cdot\hat{i} + a_1b_2\hat{i}\cdot\hat{j} + a_1b_3\hat{i}\cdot\hat{k} + a_2b_2\hat{j}\cdot\hat{i} + a_2b_2\hat{j}\cdot\hat{j} + a_2b_3\hat{j}\cdot\hat{k} + a_3b_1\hat{k}\cdot\hat{i} + a_2b_2\hat{k}\cdot\hat{j}$$
$$= a_1b_1 + a_2b_2 + a_3b_3 \quad [\because\ \hat{i}\cdot\hat{i} = \hat{j}\cdot\hat{j} = \hat{k}\cdot\hat{k} = 1 \text{ and } \hat{i} + \hat{j} = \hat{j}\cdot\hat{k} = \hat{k}\cdot\hat{i} = 0]$$

Therefore, the scalar product of two vectors is equal to the sum of the product of their corresponding rectangular components.

Corollary 1. To find the length or magnitude of $\vec{a} = a_1\hat{i} + a_2\hat{j} + a_3\hat{k}$, we note that

$$|\vec{a}|^2 = \vec{a}\cdot\vec{a} = (a_1\hat{i} + a_2\hat{j} + a_3\hat{k})\cdot(a_1\hat{i} + a_2\hat{j} + a_3\hat{k}) = a_1^2 + a_2^2 + a_3^2$$

$$\therefore \quad |\vec{a}| = \sqrt{a_1^2 + a_2^2 + a_3^2}$$

Corollary 2. Let $\vec{a} = a_1\hat{i} + a_2\hat{j} + a_3\hat{k}, b = b_1\hat{i} + b_2\hat{j} + b_3\hat{k}$. Then $\vec{a}$ and $\vec{b}$ are perpendicular iff $\vec{a}\cdot\vec{b} = \vec{0}$

i.e., iff $a_1b_1 + a_2b_2 + a_3b_3 = 0$. Thus $\vec{a}$ and $\vec{b}$ are perpendicular iff $a_1b_2 + a_2b_2 + a_3b_3 = 0$. Also note that $\vec{a}$ and $\vec{b}$ are parallel (collinear) iff $\frac{a_1}{b_1} = \frac{a_2}{b_2} = \frac{a_3}{b_3}$.

3.10. ANGLE BETWEEN TWO VECTORS

Let θ be the angle between two (non-zero) vectors $\vec{a}$ and $\vec{b}$, then

$$\vec{a}\cdot\vec{b} = ab\cos\theta \quad \text{where} \quad a = |\vec{a}| \text{ and } b = |\vec{b}|$$

$$\Rightarrow \qquad \cos\theta = \frac{\vec{a}\cdot\vec{b}}{ab} \qquad ...(i)$$

Let $\vec{a} = a_1\hat{i} + a_2\hat{j} + a_3\hat{k}$ and $\vec{b} = b_1\hat{i} + b_2\hat{j} + b_3\hat{k}$

then $\vec{a}\cdot\vec{b} = a_1b_1 + a_2b_2 + a_3b_3$

and $a = |\vec{a}| = \sqrt{a_1^2 + a_2^2 + a_3^2}$, $|\vec{b}| = \sqrt{b_1^2 + b_2^2 + b_3^2}$

From Eq. (*i*), we get $\cos\theta = \dfrac{a_1b_1 + a_2b_2 + a_3b_3}{\sqrt{a_1^2 + a_2^2 + a_3^2}\,\sqrt{b_1^2 + b_2^2 + b_3^2}}$

$$\Rightarrow \qquad \theta = \cos^{-1}\left(\frac{a_1b_1 + a_2b_2 + a_3b_3}{\sqrt{a_1^2 + b_2^2 + a_3^2}\,\sqrt{b_1^2 + b_2^2 + b_3^2}}\right)$$

3.11. COMPONENTS OF A VECTOR ALONG AND PERPENDICULAR TO VECTOR

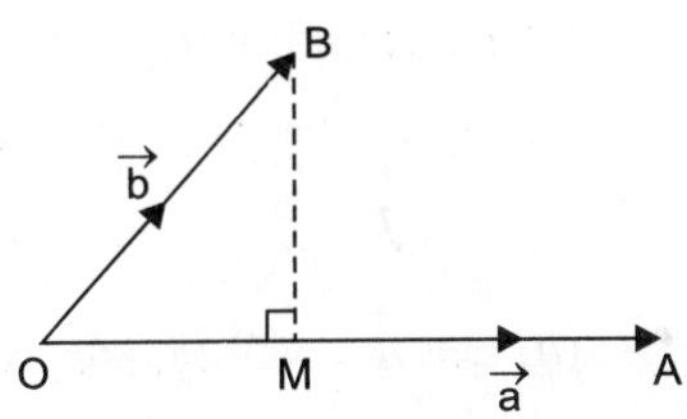

Let $\vec{a}$ and $\vec{b}$ be two vectors represented by $\overrightarrow{OA}$ and $\overrightarrow{OB}$, respectively and let θ be the angle between $\vec{a}$ and $\vec{b}$. Then

$$\vec{b} = \overrightarrow{OM} + \overrightarrow{MB}$$

Also $\overrightarrow{OM} = (OM)\hat{a} = (OB\cos\theta)\hat{a} = (|\vec{b}|\cos\theta)\hat{a}$

$$= \left[|\vec{b}|\frac{\vec{a}\cdot\vec{b}}{|\vec{a}||\vec{b}|}\right]\hat{a} = \left(\frac{\vec{a}\cdot\vec{b}}{|\vec{a}|}\right)\hat{a} = \frac{\vec{a}\cdot\vec{b}}{|\vec{a}||\vec{a}|}\vec{a} = \left(\frac{\vec{a}\cdot\vec{b}}{|\vec{a}|^2}\right)\vec{a}$$

Also $\vec{b} = \overrightarrow{OM} + \overrightarrow{MB} \quad \Rightarrow \quad \overrightarrow{MB} = \vec{b} - \overrightarrow{OM} = \vec{b} - \left(\dfrac{\vec{a}\cdot\vec{b}}{|\vec{a}|^2}\right)\vec{a}$

Thus, the component of $\vec{b}$ along and perpendicular to $\vec{a}$ are $\left(\dfrac{\vec{a}\cdot\vec{b}}{|\vec{a}|^2}\right)\vec{a}$ and $\vec{b} - \left(\dfrac{\vec{a}\cdot\vec{b}}{|\vec{a}|^2}\right)\vec{a}$ respectively.

SOLVED EXAMPLES

Example 3.13. **Find $\vec{a}\cdot\vec{b}$ when:**

(*i*) $\vec{a}=3\hat{i}-2\hat{j}+\hat{k},\ \vec{b}=2\hat{i}+3\hat{j}$ (*ii*) $\vec{a}=(2,3,-1)$ and $\vec{b}=(-1,2,3)$

Solution. (*i*) We have

$$\vec{a}\cdot\vec{b} = (3\hat{i}-2\hat{j}+\hat{k})\cdot(2\hat{i}+3\hat{j})$$
$$= 3(2)+(-2)(3)+1(0) = 6-6+0 = 0$$

(*ii*) We have

$$\vec{a}\cdot\vec{b} = (2\hat{i}+3\hat{j}-\hat{k})\cdot(-\hat{i}+2\hat{j}+3\hat{k})$$
$$= (2)(-1)+(3)(2)+(-1)(3) = -2+6-3 = 1$$

Example 3.14. **Find the angle between the vectors:**

(*i*) (1, – 2, 3) and (3, – 2, 1) (*ii*) $3\hat{i}-2\hat{j}-6\hat{k}$ and $4\hat{i}-\hat{j}+8\hat{k}$

Solution. (*i*) Let $\vec{a}=\hat{i}-2\hat{j}+3\hat{k}$ and $\vec{b}-3\hat{i}-2\hat{j}+\hat{k}$ and θ be the angle between them, then

$$\cos\theta = \frac{\vec{a}\cdot\vec{b}}{|\vec{a}||\vec{b}|} \qquad \ldots(i)$$

Now

$$\vec{a}\cdot\vec{b} = (\hat{i}-2\hat{j}+3\hat{k})\cdot(3\hat{i}-2\hat{j}+\hat{k})$$
$$= (1)(3)+(-2)(-2)+(3)(1) = 3+4+3 = 10$$
$$|\vec{a}| = \sqrt{1^2+(-2)^2+3^2} = \sqrt{1+4+9} = \sqrt{14}$$
$$|\vec{b}| = \sqrt{3^2+(-2)^2+1^2} = \sqrt{9+4+1} = \sqrt{14}$$

Hence, from Eq. (*i*).

$$\cos\theta = \frac{10}{\sqrt{14}\sqrt{14}} = \frac{10}{14} = \frac{5}{7}$$

or

$$\theta = \cos^{-1}\left(\frac{5}{7}\right)$$

(*ii*) Let $\vec{a}=3\hat{i}-2\hat{j}-6\hat{k}$ and $\vec{b}=4\hat{i}-\hat{j}+8\hat{k}$ and let θ be the angle between them, then

$$\cos\theta = \frac{\vec{a}\cdot\vec{b}}{|\vec{a}||\vec{b}|} \qquad \ldots(i)$$

Now

$$\vec{a}\cdot\vec{b} = (3\hat{i}-2\hat{j}-6\hat{k})\cdot(4\hat{i}-\hat{j}+8\hat{k})$$
$$= (3)(4)+(-2)+(-1)+(-6)(8) = 12+2-48 = -34$$
$$|\vec{a}| = \sqrt{3^2+(-2)^2+(-6)^2} = \sqrt{49} = 7$$
$$|\vec{b}| = \sqrt{4^2+(-1)^2+8^2} = \sqrt{81} = 9$$

Hence, from Eq. (*i*)

$$\cos\theta = \frac{-34}{7\times 9} = \frac{-34}{63}$$

⇒

$$\theta = \cos^{-1}\left(\frac{-34}{63}\right) = \pi-\cos^{-1}\left(\frac{34}{63}\right)$$

Example 3.15. Find the value λ so that the vector $\vec{a} = 2\hat{i} + 3\hat{j} - \hat{k}$ and $\vec{b} = -4\hat{i} - 6\hat{j} + \lambda\hat{k}$ are (*i*) parallel, (*ii*) perpendicular to each other.

Solution. (*i*) We know that the vectors $\vec{a} = a_1\hat{i} + a_2\hat{j} + a_3\hat{k}$ and $\vec{b} = b_1\hat{i} + b_2\hat{j} + b_3\hat{k}$ are parallel iff $\vec{a} = \lambda\vec{b}$

$$\Leftrightarrow \quad (a_1\hat{i} + a_2\hat{j} + a_3\hat{k}) = \lambda(b_1\hat{i} + b_2\hat{j} + b_3\hat{k})$$

$$\Leftrightarrow \quad a_1 = \lambda b_1, \quad a_2 = \lambda b_2, \quad a_3 = \lambda b_3$$

$$\Leftrightarrow \quad \frac{a_1}{b_1} = \frac{a_2}{b_2} = \frac{a_3}{b_3} \ (= 1)$$

So, given vectors $\vec{a} = 2\hat{i} + 3\hat{j} - \hat{k}$ and $\vec{b} = -4\hat{i} - 6\hat{j} + \lambda\hat{k}$ will be parallel iff

$$\frac{2}{-4} = \frac{3}{-6} = \frac{1}{-\lambda} \quad \Rightarrow \quad 1 = 2$$

(*ii*) $\vec{a}$ and $\vec{b}$ are perpendicular to each other iff $\vec{a}.\vec{b} = 0$

i.e., iff (2) (– 4) + 3(– 6) + (– 1) (λ) = 0

$$\Rightarrow \quad \lambda = -8 - 18 = -26.$$

Example 3.16. Let $\vec{a} = 4\hat{i} + 5\hat{j} - \hat{k}$, $\vec{b} = \hat{i} - 4\hat{j} + 5\hat{k}$ and $\vec{c} = 3\hat{i} + \hat{j} - \hat{k}$. Find a vector $\vec{d}$ which is perpendicular to both $\vec{a}$ and $\vec{b}$ satisfying $\vec{d}.\vec{c} = 21$.

Solution. Let $\vec{d} = x\hat{i} + y\hat{j} + z\hat{k}$. Since $\vec{d}$ is perpendicular to both $\vec{a}$ and $\vec{b}$. Therefore,

$$\vec{d}.\vec{a} = 0 \quad \text{and} \quad \vec{d}.\vec{b} = 0$$

Now, $\vec{d}.\vec{a} = 0 \Rightarrow (x\hat{i} + y\hat{j} + z\hat{k})\cdot(4\hat{i} + 5\hat{j} - \hat{k}) = 0 \Rightarrow 4x + 5y - z = 0$ ...(*i*)

$\vec{d}.\vec{b} = 0 \Rightarrow (x\hat{i} + y\hat{j} + z\hat{k})\cdot(\hat{i} - 4\hat{j} + 5\hat{k}) = 0 \Rightarrow x - 4y + 5z = 0$ ...(*ii*)

$\vec{d}.\vec{c} = 21 \Rightarrow (x\hat{i} + y\hat{j} + z\hat{k})\cdot(3\hat{i} + \hat{j} - \hat{k}) = 21 \Rightarrow 3x + y - z = 21$ ...(*iii*)

Solving Eqs. (*i*), (*ii*) and (*iii*) for *x*, *y* and *z*, we get

$$x = 7, y = -7, z = -7$$

Hence $\vec{d} = 7\hat{i} - 7\hat{j} - 7\hat{k}$.

Example 3.17. If with reference to a right handed system of mutually perpendicular with vector $\hat{i}, \hat{j}, \hat{k}$, we have $\vec{\alpha} = 3\hat{i} - \hat{j}$ and $\vec{\beta} = 2\hat{i} + \hat{j} - 3\hat{k}$. Express β in the form of $\vec{\beta} = \vec{\beta_1} + \vec{\beta_2}$ where $\vec{\beta_1}$ is parallel to $\vec{\alpha}$ and $\vec{\beta_2}$ is perpendicular to $\vec{\alpha}$.

Solution. Since $\vec{\beta_1}$ is parallel to $\vec{\alpha}$. Therefore

$$\vec{\beta_1} = \lambda\vec{\alpha} \text{ for some scalar } \lambda.$$

$$\Rightarrow \qquad \vec{\beta_1} = \lambda(3\hat{i} - \hat{j}) \qquad \text{...(i)}$$

It is given that $\vec{\beta} = \vec{\beta_1} + \vec{\beta_2}$

$$\Rightarrow \qquad \vec{\beta_2} = \vec{\beta} - \vec{\beta_1}$$

$$\Rightarrow \qquad \vec{\beta_2} = (2\hat{i} + \hat{j} - 3\hat{k}) - \lambda(3\hat{i} - \hat{j})$$

$$\Rightarrow \qquad \vec{\beta_2} = 3(-3\lambda + 2)\hat{i} + (\lambda + 1)\hat{j} - 3\hat{k} \qquad \text{...(ii)}$$

It is also given that $\vec{\beta_2}$ is $\perp$ to $\vec{\alpha}$. Therefore

$$\vec{\beta_2} . \vec{\alpha} = 0$$

$$\Rightarrow \quad [(-3\lambda + 2)\hat{i} + (\lambda + 1)\hat{j} - 3\hat{k})] . (-3\hat{i} - \hat{j}) = 0$$

$$\Rightarrow \qquad 3(-3\lambda + 2) - (1 + 1) = 0 \;\Rightarrow\; -10\,\lambda + 5 = 0 \;\Rightarrow\; \lambda = \frac{1}{2}$$

Putting the value of l in Eqs. (*i*) and (*ii*), we get

$$\vec{\beta_1} = \frac{1}{2}(3\hat{i} - \hat{j}) \quad \text{and} \quad \vec{\beta_2} = \frac{1}{2}(\hat{i} + 3\hat{j} - 6\hat{k})$$

Hence $\beta = \beta_1 + \beta_2 = \frac{1}{2}(3\hat{i} - \hat{j}) + \frac{1}{2}(\hat{i} + 3\hat{j} - 6\hat{k})$

Geometrical Problems Based on Dot Product

3.12. APPLICATION OF SCALAR (OR DOT) PRODUCT

Work done by a Force. A force acting on a particle does not work when the particle is displaced in a direction which is not perpendicular to the force.

Let $\vec{F}$ be a force acting on a particle at A. The work is said to b done by $\vec{F}$ if the particle undergoes a displacement by the action of the force and measured by the product of the magnitude of $\vec{F}$ and the distance covered by the particle in the direction of $\vec{F}$.

Suppose the force $\vec{F}$ acting on a particle displaces it from A to B. Let $\vec{AB} = \vec{d}$.

Now, the distance covered by the particle in the direction of

$$\vec{F} = \text{the projection of } \vec{d} \text{ along } \vec{F} = \frac{\vec{d} \cdot \vec{F}}{|\vec{F}|}$$

$$\therefore \quad \text{Work done by the force} \qquad \vec{F} = \left(\frac{\vec{d} \cdot \vec{F}}{|\vec{F}|}\right)|\vec{F}| = \vec{d} \cdot \vec{F} = \vec{F} \cdot \vec{d} \text{ units.}$$

Remark. If the dot product of the force vector and the displacement vector is 0, then it means that the displacement is at right angles to the force.

SOLVED EXAMPLES

Example 3.18. Constant forces $2\hat{i} - 5\hat{j} + 6\hat{k}$ and $-\hat{i} + 2\hat{j} - \hat{k}$ act on a particle. Determine the work done with position vector $4\hat{i} - 3\hat{j} - 2\hat{k}$ to a point *B* with position vector $6\hat{i} + \hat{j} - 3\hat{k}$.

Solution. Here

$$\vec{F} = \vec{F_1} + \vec{F_2}$$
$$= (2\hat{i} - 5\hat{j} + 6\hat{k}) + (-\hat{i} + 2\hat{j} - \hat{k}) = (\hat{i} - 3\hat{j} + 5\hat{k})$$
$$\vec{OA} = 4\hat{i} - 3\hat{j} - 2\hat{k}, \vec{OB} = 6\hat{i} + \hat{j} - 3\hat{k}$$

$$\text{Displacement } \vec{S} = \vec{AB} = \vec{OB} - \vec{OA}$$
$$= (6\hat{i} + \hat{j} - 3\hat{k}) - (4\hat{i} - 3\hat{j} - 2\hat{k})$$
$$= (6-4)\hat{i} + (1+3)\hat{j} + (-3+2)\hat{k} = 2\hat{i} + 4\hat{j} - \hat{k}$$

$$\text{Work done} = \vec{F} \cdot \vec{S} = (\hat{i} - 3\hat{j} + 5\hat{k}) \cdot (2\hat{i} + 4\hat{j} - \hat{k})$$
$$= (1)(2) + (-3) + (4) + 5(-1) = 2 - 12 - 5 = -15$$

Example 3.19. A particle on by constant forces $4\hat{i} + \hat{j} - 3\hat{k}$ and $3\hat{i} + \hat{j} - \hat{k}$ is displaced from the point $\hat{i} + 2\hat{j} + 3\hat{k}$ to the point $5\hat{i} + 4\hat{j} + \hat{k}$. Find the total work done by the forces.

Solution. Total force $\vec{F} = \vec{F_1} + \vec{F_2} = 4\hat{i} + \hat{j} - 3\hat{k} + 3\hat{i} + \hat{j} - \hat{k} = 7\hat{i} + 2\hat{j} - 4\hat{k}$

(Displacement) $\vec{S} = \vec{AB} = \vec{OB} - \vec{OA} = (5\hat{i} + 4\hat{j} + \hat{k}) - (\hat{i} + 2\hat{j} + 3\hat{k}) = 4\hat{i} + 2\hat{j} - 2\hat{k}$

$$\text{Total work done} = \vec{F} \cdot \vec{S} = (7\hat{i} + 2\hat{j} - 4\hat{k}) \cdot (4\hat{i} + 2\hat{j} - 2\hat{k})$$
$$= (7)(4) + (2)(2) + (-4)(-2) = 28 + 4 + 8 = 40 \text{ units.}$$

Example 3.20. (*a*) Find the work done in moving an objct along the vector $3\hat{i} + 2\hat{j} - 5\hat{k}$ if the force appplied is $\vec{F} = 2\hat{i} - \hat{j} - \hat{k}$.

(*b*) A particle acted on by constant forces $3\hat{i} + 2\hat{j} + 5\hat{k}$ and $2\hat{i} + \hat{j} - 3\hat{k}$ is displaced from the point $2\hat{i} - \hat{j} - 3\hat{k}$ to the point $4\hat{i} - 3\hat{j} + 7\hat{k}$. Find the total work done by the forces.

Solution. (*a*) $\vec{F} = 2\hat{i} - \hat{j} - \hat{k}$

Displacement veector = $\vec{S} = 3\hat{i} + 2\hat{j} - 5\hat{k}$

$$\text{Work done} = \vec{F} \cdot \vec{S} = (2\hat{i} - \hat{j} - \hat{k}) \cdot (3\hat{i} + 2\hat{j} - 5\hat{k})$$
$$= (2)(3) + (-1)(2) + (-1)(-5) = 6 - 2 + 5 = 9 \text{ units}$$

(*b*) Sum of two forces *i.e.*, $\vec{F} = (3\hat{i} + 2\hat{j} + 5\hat{k}) + (2\hat{i} + \hat{j} - 3\hat{k}) = 5\hat{i} + 3\hat{j} + 2\hat{k}$

Displacement vector, $\vec{d} = (4\hat{i} - 3\hat{j} + 7\hat{k}) - (2\hat{i} - \hat{j} - 3\hat{k}) = 2\hat{i} - 2\hat{j} + 10\hat{k}$

$$\text{Work done} = \vec{F} \cdot \vec{d} = (5\hat{i} + 3\hat{j} + 2\hat{k}) \cdot (2\hat{i} - 2\hat{j} + 10\hat{k})$$
$$= (5)(2) + (3)(-2) + (2)(10) = 10 - 6 + 20 = 24 \text{ units}$$

EXERCISE 3.3

1. If $\vec{a}$ and $\vec{b}$ are two vectors such that $|\vec{a}| = 0$, $|\vec{b}| = 15$ and $\vec{a}\cdot\vec{b} = 75\sqrt{2}$, find the angle between $\vec{a}$ and $\vec{b}$. [**Ans.** 45°]
2. Prove that the three vectors $3\hat{i}+\hat{j}+2\hat{k}, \hat{i}-\hat{j}-\hat{k}$ and $\hat{i}+5\hat{j}-4\hat{k}$ are at right angles to one another.
3. Find the values of λ for which the vectors $\vec{a} = 3\hat{i}+2\hat{j}+9\hat{k}$ and $\vec{b} = \hat{i}+\lambda\hat{j}+3\hat{k}$ are:

   (*i*) perpendicular (*ii*) parallel $\left[\textbf{Ans.} -15, \frac{2}{3}\right]$
4. Find the angles of the triangle whose vertices are *A* (– 1, – 2, – 3), *B* (2, 0, 3) and *C* (4, 6, 0). [**Ans.** $\angle A = 45°$, $\angle B = 90°$, $\angle C = 45°$]
5. Dot product of a vector with vectors $3\hat{i}-5\hat{j}, 2\hat{i}+7\hat{j}$ and $\hat{i}+\hat{j}+\hat{k}$ are respectively – 1, 6 and 5. Find the vector. [**Ans.** $3\hat{i}+0\hat{j}+2\hat{k}$]
6. If $\vec{a}$ and $\vec{b}$ be unit vectors such that $|\vec{a}+\vec{b}| = \sqrt{3}$, then find the value of $(2\vec{a}-5\vec{b})\cdot(3\vec{a}+\vec{b})$. $\left[\textbf{Ans.} \frac{-11}{2}\right]$
7. Find $|\vec{a}|$ and $|\vec{b}|$, if $(\vec{a}-\vec{b})\cdot(\vec{a}+\vec{b}) = 27$ and $|\vec{a}| = 2|\vec{b}|$. [**Ans.** 6, 3]
8. Find the projection of the vector $7\hat{i}+\hat{j}-4\hat{k}$ on $2\hat{i}+6\hat{j}+3\hat{k}$. $\left[\textbf{Ans.} \frac{8}{7}\right]$
9. If two vectors $\vec{a}$ and $\vec{b}$ are such that $|\vec{a}| = 3$, $|\vec{b}| = 2$ and $\vec{a}\cdot\vec{b} = 6$, find $|\vec{a}\cdot\vec{b}|$ and $|\vec{a}-\vec{b}|$. [**Ans.** 5; 1]
10. Find a unit vector perpendicular to both the vectors $4\hat{i}-\hat{j}+3\hat{k}$ and $-2\hat{i}+\hat{j}-2\hat{k}$. $\left[\textbf{Ans.} \pm\frac{1}{3}(-\hat{i}+2\hat{j}+2\hat{k})\right]$
11. In a right angled Δ*AOB* right angled at *O*, if *P* and *Q* are points of trisection of *AB*, prove that
$$OP^2 + OQ^2 = \frac{5}{9}AB^2$$
12. Prove that in any triangle *ABC*:

    (*i*) $\cos B = \dfrac{c^2+a^2-b^2}{2ca}$ (*ii*) $\cos C = \dfrac{a^2+b^2-c^2}{2ab}$
13. Prove that the middle point of the hypotenuse of straight angled triangle is equidistance from its vertices.
14. Prove that the right bisectors of the sides of a triangle are concurrent.
15. If *D* is the mid-point of the side *BC* of a triangle *ABC*, show that $AB^2 + AC^2 = 2(AD^2 + BD^2)$.
16. If the median of the base of a triangle is perpendicular to the base, then prove that the triangle is isosceles.
17. If $\vec{a}+\vec{b}+\vec{c} = 0$ and $|\vec{a}| = 6$, $|\vec{b}| = 5$, $|\vec{c}| = 7$, find the angle between $\vec{b}$ and $\vec{c}$. $\left[\textbf{Ans.} \cos^{-1}\left(\frac{19}{35}\right)\right]$
18. Find the work done by the force $\vec{F} = \hat{i}+2\hat{j}+\hat{k}$ acting on a particle, if the particle is displaced from the point with position vector $2\hat{i}+\hat{j}+\hat{k}$ to the point with the position vector $3\hat{i}+2\hat{j}+4\hat{k}$. [**Ans.** 6 units]

19. Forces of magnitude 5 and 3 units acting in the direction $6\hat{i}+2\hat{j}+3\hat{k}$ and $3\hat{i}-2\hat{j}+6\hat{k}$ respectively act on a particle which is displaced from the point (2, 2, – 1) to (4, 3, 1). Find the work done by the forces.

$$\left[\textbf{Ans. } \frac{148}{7} \text{ units}\right]$$

20. Forces F_1, F_2, F_3 of magnitude 5, 3, 1 units acting in the directions $6\hat{i}+2\hat{j}+3\hat{k}, 3\hat{i}-2\hat{j}+6\hat{k}, 2\hat{i}-3\hat{j}-6\hat{k}$ respectively act on a particl which is displaced from the point (2, – 1, – 3) to (5, – 1, 1). Find the work done by the forces $\vec{F_1}, \vec{F_2}, \vec{F_3}$ and $\vec{F}=\vec{F_1}+\vec{F_2}+\vec{F_3}$.

$$\left[\textbf{Ans. } \frac{150}{7} \text{ units}; \frac{99}{7} \text{ units}; \frac{18}{7} \text{ units}; 33 \text{ units}\right]$$

3.13. VECTOR (OR CROSS) PRODUCT OF TWO VECTORS

The vector product of two proper vectors $\vec{a}$ and $\vec{b}$ (in an assigned order), denoted by $\vec{a}\times\vec{b}$, is a vector perpendicular to both of them in a direction which corresponds to the advance of a right-handed screw due to rotation which carries the first vector to the second and having modulus equal to the product of their modulii and the sin of the angle between them.

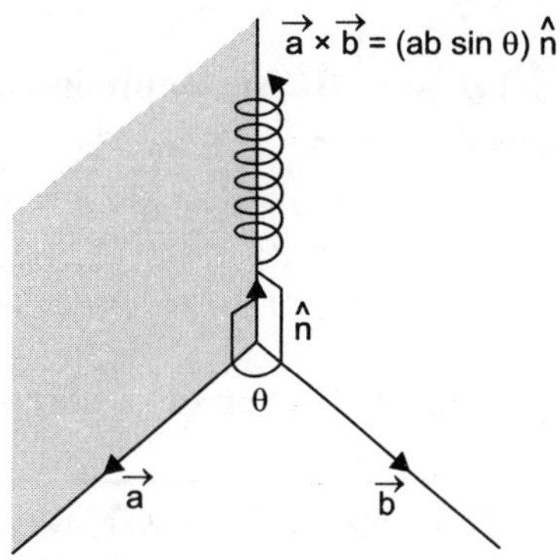

In other words, $\vec{a}\times\vec{b} = |\vec{a}||\vec{b}|\sin\theta\,\hat{\eta}$

where θ is the angle between $\vec{a}$ and $\vec{b}$ and $\hat{\eta}$ is a unit vector perpendicular to the plane of $\vec{a}$ and $\vec{b}$ such that $\vec{a}, \vec{b}, \hat{\eta}$ form a right handed system.

The vector product of two vectors is also called the *cross product of the vectors*, under consideration. The vector product $\vec{a}\times\vec{b}$, is read as "$\vec{a}$ cross $\vec{b}$".

Remarks:

1. $\vec{a}\times\vec{b}$ is a vector.

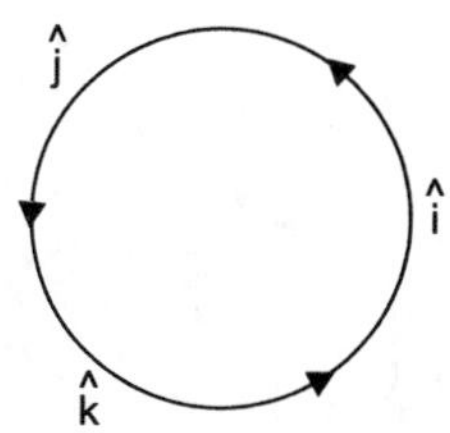

2. If $\vec{a}$ and $\vec{b}$ be unit vectors then $|\vec{a}\times\vec{b}| = \sin\theta$. Thus, we find that the modulus of the vector product of two unit vectors is the sine of the angle between their directions.
3. If $\vec{a}$ and $\vec{b}$ are perpendicular, than the three vectors $\vec{a}, \vec{b}$ and $\vec{a}\times\vec{b}$ form a right handed triad of mutually perpendicular vectors.
4. If $\vec{a}$ is parallel to $\vec{b}$, then $\theta = 0$ or π *i.e.*, $\sin\theta = 0$ and we obtain $\vec{a}\times\vec{b} = 0$. Conversely, if $\vec{a}\times\vec{b} = 0$ *i.e.*, $|\vec{a}||\vec{b}|\sin\theta\,\hat{\eta} = \vec{0}$ then either $\vec{a}$ or $\vec{b}$ (or both) is a zero vector or else they are parallel.
5. If $\vec{a}=\vec{b}$ then $\theta = 0$ *i.e.*, $\sin\theta = 0$ and we get $\vec{a}\times\vec{b} = \vec{0}$.

6. Since $\hat{i}$, $\hat{j}$ and $\hat{k}$ are unit vectors along three co-ordinate axes, these form a right handed triad of mutually perpendicular unit vectors, we have

$$\hat{i}\times\hat{j} = \hat{k} = -\hat{j}\times\hat{i},\ \hat{j}\times\hat{k} = \hat{i} = -\hat{k}\times\hat{j} \text{ and } \hat{k}\times\hat{i} = \hat{j} = -\hat{i}\times\hat{k}$$

Also as $\vec{a}\times\vec{a} = \vec{0}$ for any vector $\vec{a}$, we have $\hat{i}\times\hat{i} = \hat{j}\times\hat{j} = \hat{k}\times\hat{k} = \vec{0}$

3.14. PROPERTIES OF VECTOR (CROSS) PRODUCT

Property I. Vector product is not commutative and $\vec{a}\times\vec{b} = -(\vec{b}\times\vec{a})$.

Property II. Vector Product is not associative.

$$\therefore \qquad (\vec{a}\times\vec{b})\times\vec{c} \neq \vec{a}\times(\vec{b}\times\vec{c})$$

Property III. Vector product is distributive over addition *i.e.*, if $\vec{a}, \vec{b}$ and $\vec{c}$ be any three vectors, then

(*a*) $\vec{a}\times(\vec{b}+\vec{c}) = \vec{a}\times\vec{b}+\vec{a}\times\vec{c}$ [Left distributive law]

(*b*) $(\vec{b}+\vec{c})\times\vec{a} = \vec{b}\times\vec{a}+\vec{c}\times\vec{a}$ [Right distributive law]

Property IV. For any scalar *m*, $(m\vec{a})\times\vec{b} = m(\vec{a}\times\vec{b}) = \vec{a}\times(m\vec{b})$.

3.15. VECTOR PRODUCT IN TERMS OF COMPONENTS

Let $\vec{a} = a_1\hat{i}+a_2\hat{j}+a_3\hat{k}$ and $\vec{b} = b_1\hat{i}+b_2\hat{j}+b_3\hat{k}$ be two vectors. Then,

$$\vec{a}\times\vec{b} = (a_1\hat{i}+a_2\hat{j}+a_3\hat{k})\times(b_1\hat{i}+b_2\hat{j}+b_3\hat{k})$$

$$= a_1\hat{i}\times(b_1\hat{i}+b_2\hat{j}+b_3\hat{k}) + a_2\hat{j}\times(b_1\hat{i}+b_2\hat{j}+b_3\hat{k}) + a_3\hat{k}\times(b_1\hat{i}+b_2\hat{j}+b_3\hat{k})$$

[Using right distributive law]

$$= (a_1\hat{i})\times(b_1\hat{i}) + (a_1\hat{i})\times(b_2\hat{j}) + (a_1\hat{i})\times(b_3\hat{k}) + ((a_2\hat{j})\times(b_1\hat{i}) + (a_2\hat{j})\times(b_2\hat{j}) + (a_2\hat{j})\times(b_3\hat{k})) + ((a_3\hat{i})\times(b_1\hat{i}) + (a_3\hat{k})\times(b_2\hat{j}) + (a_3\hat{k})\times(b_3\hat{k}))$$

[Using left distributive law]

$$= (a_1b_1\hat{i}\times\hat{i} + a_1b_2\hat{j}\times\hat{j} + a_1b_3\hat{k}\times\hat{k}) + (a_2b_1\hat{j}\times\hat{i} + a_2b_2\hat{j}\times\hat{j} + a_2b_3\hat{j}\times\hat{k}) + (a_3b_1\hat{k}\times\hat{i} + a_3b_2\hat{k}\times\hat{j} + a_3b_3\hat{k}\times\hat{k})$$

$$= a_1b_1\vec{0} + a_1b_2\hat{k} + a_1b_3(-\hat{j}) + a_2b_1(-\hat{k}) + a_2b_2\vec{0} + a_2b_3\hat{i} + a_3b_1\hat{j} + a_3b_2(-\hat{i}) - a_3b_3\vec{0}$$

$$= (a_2b_3 - a_3b_2)\hat{i} - (a_1b_3 - a_3b_1)\hat{j} + (a_1b_2 - a_2b_1)\hat{k} = \begin{vmatrix} \hat{i} & \hat{j} & \hat{k} \\ a_1 & a_2 & a_3 \\ b_1 & b_2 & b_3 \end{vmatrix}.$$

Thus, $$(a_1\hat{i}+a_2\hat{j}+a_3\hat{k})\times(b_1\hat{i}+b_2\hat{j}+b_3\hat{k}) = \begin{vmatrix} \hat{i} & \hat{j} & \hat{k} \\ a_1 & a_2 & a_3 \\ b_1 & b_2 & b_3 \end{vmatrix}.$$

3.16. VECTORS NORMAL TO THE PLANE OF TWO GIVEN VECTORS

If $\vec{a}$ and $\vec{b}$ are non-zero, non-parallel vectors, then a unit vector perpendicular to the plane of the vector $\vec{a}$ and $\vec{b}$ is given by $\pm \dfrac{\vec{a} \times \vec{b}}{|\vec{a} \times \vec{b}|}$.

Proof. Let θ be the angle between the vectors $\vec{a}$ and $\vec{b}$. Since $\vec{a}$ and $\vec{b}$ are non-parallel vectors, the cross product, $\vec{a} \times \vec{b}$ is non-zero.

$$\therefore \qquad |\vec{a} \times \vec{b}| \neq 0$$

The vector $\vec{a} \times \vec{b}$ is perpendicular to the plane of $\vec{a}$ and $\vec{b}$.

The vector $-(\vec{a} \times \vec{b})$ is also perpendicular to the plane of $\vec{a}$ and $\vec{b}$.

$\therefore \quad \dfrac{\vec{a} \times \vec{b}}{|\vec{a} \times \vec{b}|}$, being a scalar multiple of $\pm(\vec{a} \times \vec{b})$ is also perpendicular to $\vec{a}$ and $\vec{b}$ and its magnitude is one.

$\therefore \quad$ Using vector perpendicular to the plane of vector $\vec{a}$ and $\vec{b}$ is $\dfrac{\vec{a} \times \vec{b}}{|\vec{a} \times \vec{b}|}$.

Remark 1. It is important to observe that the unit vector along $\vec{a} \times \vec{b}$ is equal to $\pm \dfrac{\vec{a} \times \vec{b}}{|\vec{a} \times \vec{b}|}$, which is the perpendicular to the plane vector $\vec{a}$ and $\vec{b}$.

Remark 2. Unit vector of length k(¹ 0) and perpendicular to the plane of vectors $\vec{a}$ and $\vec{b}$ is given by $k\left(\pm \dfrac{\vec{a} \times \vec{b}}{|\vec{a} \times \vec{b}|}\right)$ *i.e.*, by $\left(\pm \dfrac{\vec{a} \times \vec{b}}{|\vec{a} \times \vec{b}|}\right) \cdot (\vec{a} \times \vec{b})$.

3.17. ANGLE BETWEEN TWO VECTORS

Let θ be the angle between two (proper) vectors $\vec{a}, \vec{b}$ and $\hat{n}$ be a unit vector perpendicular to the plane of $\vec{a}, \vec{b}$ and positive for a right handed rotation from $\vec{a}$ to $\vec{b}$, then

$$\vec{a} \times \vec{b} = |\vec{a}||\vec{b}| \sin\theta \, \hat{n}$$

$$\Rightarrow \qquad |\vec{a} \times \vec{b}| = |\vec{a}||\vec{b}| \sin\theta \qquad [\because |\hat{n}| = 1 \text{ and } 0 \leq \theta \leq \pi \Rightarrow \sin\theta \geq 0]$$

$$\Rightarrow \qquad \sin\theta = \frac{|\vec{a} \times \vec{b}|}{|\vec{a}||\vec{b}|} \qquad \ldots(i)$$

Let $\vec{a} = a_1\hat{i} + a_2\hat{j} + a_3\hat{k}$ and $\vec{b} = b_1\hat{i} + b_2\hat{j} + b_3\hat{k}$, then

$$\vec{a} \times \vec{b} = (a_2b_3 - a_3b_2)\hat{i} + (a_3b_1 - a_1b_3)\hat{j} + (a_1b_2 - a_2b_1)\hat{k}$$

$$|\vec{a} \times \vec{b}| = \sqrt{(a_2b_3 - a_3b_2)^2 + (a_3b_1 - a_1b_3)^2 + (a_1b_2 - a_2b_1)^2}$$

Also, $$|\vec{a}| = \sqrt{a_1^2 + a_2^2 + a_3^2} \text{ and } |\vec{b}| = \sqrt{b_1^2 + b_2^2 + b_3^2}$$

From Eq. (*i*) we get $$\sin\theta = \frac{\sqrt{(a_2b_3 - a_3b_2)^2 + (a_3b_1 - a_1b_3)^2 + (a_1b_2 - a_2b_1)^2}}{\sqrt{a_1^2 + a_2^2 + a_3^2}\sqrt{b_1^2 + b_2^2 + b_3^2}}.$$

SOLVED EXAMPLES

Example 3.21. Find $|\vec{a} \times \vec{b}|$, if $\vec{a} = 2\hat{i} + 3\hat{j} + 6\hat{k}$ and $\vec{b} = 3\hat{i} - 6\hat{j} + 2\hat{k}$.

Solution. We have $$|\vec{a} \times \vec{b}| = \begin{vmatrix} \hat{i} & \hat{j} & \hat{k} \\ 2 & 3 & 6 \\ 3 & -6 & 2 \end{vmatrix} = (6+36)\hat{i} - (4-18)\hat{j} + (-12-9)\hat{k} = 42\hat{i} + 14\hat{j} - 21\hat{k}$$

Hence $$|\vec{a} \times \vec{b}| = \sqrt{(42)^3 + (14)^2 + (-21)^2} = \sqrt{2401} = 49$$

Example 3.22. (*a*) If $\vec{a} = 3\hat{i} - \hat{j} - 2\hat{k}\ \vec{b} = 2\hat{i} + 3\hat{j} - 5\hat{k}$, find $(\vec{a} + 2\vec{b}) \times (2\vec{a} - \vec{b})$.

(*b*) Calculate the product $[(\hat{i} - 2\hat{j} + 3\hat{k}) \times (2\hat{i} + \hat{j} - 3\hat{k})] \times (-3\hat{i} + \hat{j} + 2\hat{k})$.

Solution. (*a*) $$(\vec{a} + 2\vec{b}) = (3\hat{i} - \hat{j} - 2\hat{k}) + 2(2\hat{i} + 3\hat{j} - 5\hat{k}) = 7\hat{i} + 5\hat{j} + 0\hat{k}$$

$$(2\vec{a} - \vec{b}) = 2(3\hat{i} - \hat{j} - 2\hat{k}) - (2\hat{i} + 3\hat{j} - 5\hat{k}) = 4\hat{i} - 5\hat{j} - 5\hat{k}$$

$$\therefore \quad (\vec{a} + 2\vec{b}) \times (2\vec{a} - \vec{b}) = \begin{vmatrix} \hat{i} & \hat{j} & \hat{k} \\ 7 & 5 & 0 \\ 4 & -5 & -5 \end{vmatrix}$$

$$= (-25+0)\hat{i} + (0+35)\hat{j} + (-35-20)\hat{k} = -25\hat{i} + 35\hat{j} - 55\hat{k}$$

(*b*) $$(\hat{i} + 2\hat{j} + 3\hat{k}) \times (2\hat{i} + \hat{j} - 3\hat{k}) = \begin{vmatrix} \hat{i} & \hat{j} & \hat{k} \\ 1 & -2 & 3 \\ 2 & 1 & -3 \end{vmatrix}$$

$$= (6-3)\hat{i} - (-3-6)\hat{j} + (1+4)\hat{k} = 3\hat{i} + 9\hat{j} + 5\hat{k}$$

$$\therefore \quad [(\hat{i} - 2\hat{j} + 3\hat{k}) \times (2\hat{i} + \hat{j} - 3\hat{k})] \times (-3\hat{i} + \hat{j} + 2\hat{k}) = (3\hat{i} + 9\hat{j} + 9\hat{k}) \times (-3\hat{i} + \hat{j} + 2\hat{k})$$

$$= \begin{vmatrix} \hat{i} & \hat{j} & \hat{k} \\ 3 & 9 & 5 \\ -3 & 1 & 2 \end{vmatrix}$$

$$= (18-5)\hat{i} - (6+15)\hat{j} + (3+27)\hat{k} = 13\hat{i} - 21\hat{j} + 30\hat{k}.$$

Example 3.23. Find a unit perpendicular to each of the vector $(\vec{a}+\vec{b})$ and $(\vec{a}-\vec{b})$ where $\vec{a}=\hat{i}+\hat{j}+\hat{k}$, $\vec{b}=\hat{i}+2\hat{j}+3\hat{k}$.

Solution. We have $\vec{a}+\vec{b} = 2\hat{i}+3\hat{j}+4\hat{k}$ and $\vec{a}-\vec{b} = -\hat{j}-2\hat{k}$

A vector which is perpendicular to both $\vec{a}+\vec{b}$ and $\vec{a}-\vec{b}$ is given by

$$(\vec{a}+\vec{b})\times(\vec{a}-\vec{b}) = \begin{vmatrix} \hat{i} & \hat{j} & \hat{k} \\ 2 & 3 & 4 \\ 0 & -1 & -2 \end{vmatrix} = -2\hat{i}+4\hat{j}-2\hat{k} \ (= \vec{c}\text{, say})$$

Now, $$|\vec{c}| = 4\sqrt{4+16+4} = \sqrt{24} = 2\sqrt{6}$$

Therefore, the required unit vectors is

$$\frac{\vec{c}}{|\vec{c}|} = -\frac{1}{\sqrt{6}}\hat{i}+\frac{1}{\sqrt{6}}\hat{j}-\frac{1}{\sqrt{6}}\hat{k}.$$

Example 3.24. Find the area of the triangle whose vertices are $A(3, -1, 2)$, $B(1, -1, -3)$ and $C(4, -3, 1)$.

Solution. Let $\vec{a}, \vec{b}$ and $\vec{c}$ be the position vectors of points A, B and C respectively. Then,

$$\vec{a} = 3\hat{i}-\hat{j}+2\hat{k},\ \vec{b}=\hat{i}-\hat{j}-3\hat{k} \text{ and } \vec{c}=4\hat{i}-3\hat{j}+\hat{k}$$

We have $$\text{Area of } \Delta ABC = \frac{1}{2}|\overrightarrow{AB}\times\overrightarrow{AC}|$$

$$\overrightarrow{AB} = \text{Position vector of } B - \text{Position vector of } A$$

$$= \vec{b}-\vec{a} = (\hat{i}-\hat{j}-3\hat{k})-(3\hat{i}-\hat{j}+2\hat{k}) = -2\hat{i}+0\hat{j}-5\hat{k}$$

$$\overrightarrow{AC} = \text{Position vector of } C - \text{Position vector of } A$$

$$= \vec{c}-\vec{a} = (4\hat{i}-3\hat{j}+\hat{k})-(3\hat{i}-\hat{j}+2\hat{k}) = \hat{i}-2\hat{j}-\hat{k};$$

$\therefore$ $$\overrightarrow{AB}\times\overrightarrow{AC} = \begin{vmatrix} \hat{i} & \hat{j} & \hat{k} \\ -2 & 0 & -5 \\ 1 & -2 & -1 \end{vmatrix} = (0-10)\hat{i}-(2+5)\hat{j}+(4-0)\hat{k}$$

$$= 10\hat{i}-7\hat{j}+4\hat{k}$$

$\Rightarrow$ $$|\overrightarrow{AB}\times\overrightarrow{AC}| = \sqrt{(-10)^2+(-7)^2+(4)^2} = \sqrt{165}$$

So, area of $$\Delta ABC = \frac{1}{2}|\overrightarrow{AB}\times\overrightarrow{AC}| = \frac{1}{2}\sqrt{165}.$$

Example 3.25. Find the area of a parallelogram whose adjacent sides are given by the vectors $\vec{a} = \hat{i} + 2\hat{j} + 3\hat{k}, \vec{b} = -3\hat{i} - 2\hat{j} + \hat{k}$.

Solution. We have $\vec{a} = \hat{i} + 2\hat{j} + 3\hat{k}, \vec{b} = -3\hat{i} - 2\hat{j} + \hat{k}$

$$\therefore \quad \vec{a} \times \vec{b} = \begin{vmatrix} \hat{i} & \hat{j} & \hat{k} \\ 1 & 2 & 3 \\ -3 & -2 & 1 \end{vmatrix} = (2+6)\hat{i} + (1+9)\hat{j} + (-2+6)\hat{k} = 8\hat{i} - 10\hat{j} + 4\hat{k}$$

Area of the parallelogram with adjacent sides $\vec{a}$ and $\vec{b}$

$$= |\vec{a} \times \vec{b}| = \sqrt{8^2 + (-10)^2 + (4)^2} = 6\sqrt{5} \text{ sq. units.}$$

3.18. MOMENT OF A FORCE ABOUT A POINT

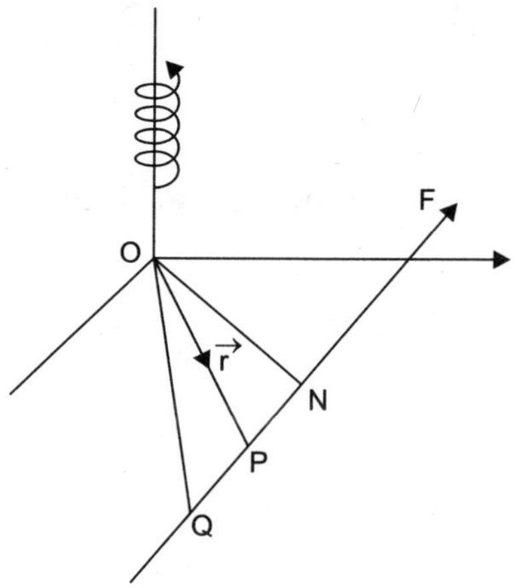

The vector moment or torque $\vec{M}$ of a force $\vec{F}$ about any given point O is in magnitude equal to F times the perpendicular distance of O from the line of action of F.

Thus, $$|\vec{M}| = F\,.\,ON = F\,.\,OP \sin\theta$$

If $\vec{r}$ be the position vector of any point P on the line of action of force $\vec{F}$, then

$$|\vec{r} \times \vec{F}| = F.\ OP \sin\theta$$

$$|\vec{M}| = |\vec{r} \times \vec{F}|$$

The direction of vector moment is to be regarded positive relative to a rotation from $\vec{r}$ to $\vec{F}$. Thus in the figure the direction of the vector moment is along the normal drawn through O on the plane of the paper and pointing towards the reader.

$\therefore \quad \vec{M} = \vec{r} \times \vec{F}$ and its magnitude is $F\,.\,ON$ or $Fr \sin\theta$.

As a convention of moment of a force $\vec{F}$ about a point is regarded as + ve when the rotation of the force about the point is anti-clockwise, otherwise negative.

Moment is independent of the position P on the line of action of F.

We have shown that moment of force $\vec{F}$ about a point O is $\vec{r} \times \vec{F}$ where $\vec{r}$ is the position vector of any point P on the line of action of $\vec{F}$ relative to O.

If in place of P, we choose any other point Q on the line of action of $\vec{F}$ whose position vector relative to O be $\vec{q}$, then moment is $\vec{M} = \vec{q} \times \vec{F}$.

But $\quad \vec{q} = \overrightarrow{OQ} = \overrightarrow{OP} + \overrightarrow{PQ} = \vec{r} + \overrightarrow{PQ}$

$\therefore \quad \vec{M} = \vec{q} \times \vec{F} = (\vec{r} + \overrightarrow{PQ}) \times \vec{F} = \vec{r} \times \vec{F} + \overrightarrow{PQ} \times \vec{F} = \vec{r} \times \vec{F} + 0 = \vec{r} \times \vec{F}$

$\because \quad \overrightarrow{PQ} \times \vec{F} = 0$ as both $\overrightarrow{PQ}$ and $\vec{F}$ are in the same line.

Working rule. In order to get the vector moment of a force $\vec{F}$ about a point O, find the vector product $\vec{r} \times \vec{F}$ where $\vec{r}$ is the position vector of any point P on the line of action of the force relative to origin O.

The algebraic sum of the moments of a system of force about any point is equal to the moment of their resultant about the same point.

Let the system of concurrent force $\vec{F_1}, \vec{F_2}, \vec{F_3}$ ... which all pass through the point P, have the resultant $\vec{F}$.

Then $\quad \vec{F} = \vec{F_1} + \vec{F_2} + \vec{F_3}$

Also $\vec{F}$ passes through the same P.

Now the vector moment $\vec{M}$ of $\vec{F}$ about any given point O (say)

$= \vec{r} \times \vec{F}$, where $\vec{r}$ is the position vector of P referred to O.

$= \vec{r} \times (\vec{F_1} + \vec{F_2} + \vec{F_3} + ...) = \vec{r} \times \vec{F_1} + \vec{r} \times \vec{F_2} + \vec{r} \times \vec{F_3} + ...$

= The algebraic sum of the moments of the system of concurrent force $\vec{F_1}, \vec{F_2}, \vec{F_3}$ ... about O.

3.19. MOMENT OF A COUPLE

A system consisting of a pair of equal unlike parallel force is called a couple.

Let a couple consists of a pair of equal unlike parallel to forces $\vec{F}$ and $-\vec{F}$. Let the lines of action of the force $\vec{F}$ and $-\vec{F}$ of the couple be the two parallel lines as shown in Fig. 3.26, let A, C be any two points on these parallel lines.

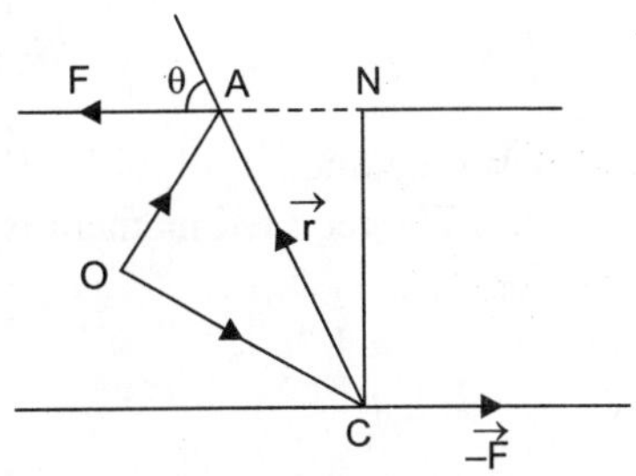

The moment of couple about a given point P is defined as the sum of moment of $\vec{F}$ and $-\vec{F}$ about O.

$\therefore$ Moment of couple about O = Moment of $\vec{F}$ about O + moment of $-\vec{F}$ about O

$$= \overrightarrow{OA} \times \vec{F} + \overrightarrow{OC} \times (-\vec{F})$$

$$= (\overrightarrow{OA} - \overrightarrow{OC}) \times \vec{F} = \overrightarrow{CA} \times \vec{F}$$

We see that the sum of the moments of the forces of the given couple is independent of O. Therefore, moment of the couple is same about any point and equal $\overrightarrow{CA} \times \vec{F}$.

Again $$|\vec{M}| = |\overrightarrow{CA} \times \vec{F}|$$

$$= |\vec{F}||\overrightarrow{CA}||\sin\theta|, \text{ where } \theta \text{ is the angle between } \overrightarrow{CA} \text{ and } \vec{F}$$

$$= |\vec{F}|CN$$

Any line perpendicular to the plane of a couple is called the axis of the couple. So the axis of couple is parallel to its moment vector.

Hence, the moment of a couple is a vector perpendicular to the plane of the couple and its magnitude is the product of the magnitude of either force with the perpendicular distance between the line of the forces.

SOLVED EXAMPLES

Example 3.26. Find the moment about (1, – 1, – 2) of the force $3\hat{i} + 4\hat{j} - 5\hat{k}$ acting at (1, 0, – 2).

Solution. $$\vec{F} = 3\hat{i} + 4\hat{j} - 5\hat{k}$$

$$\vec{r} = (\hat{i} - 2\hat{k}) - (\hat{i} - \hat{j} - \hat{k}) = \hat{j} - \hat{k}$$

The required moment is given by

$$= \vec{r} \times \vec{F} = (\hat{i} - 2\hat{k}) - (\hat{i} - \hat{j} - \hat{k}) = \hat{j} - \hat{k}$$

$$= \begin{vmatrix} \hat{i} & \hat{j} & \hat{k} \\ 0 & 1 & -1 \\ 3 & 4 & -5 \end{vmatrix} = (-5+4)\hat{i} - (0+3)\hat{j} + (0-3)\hat{k}.$$

Example 3.27. The force represent by $3\hat{i} + 2\hat{k}$ is acting through the point $5\hat{i} + 4\hat{j} - 3\hat{k}$. Find its moment about the point $\hat{i} + 3\hat{j} + \hat{k}$.

Solution. Let A be the point $5\hat{i} + 4\hat{j} - 3\hat{k}$ and B be the point $\hat{i} + 3\hat{j} + \hat{k}$

Then $$\overrightarrow{BA} = \vec{r} = (5\hat{i} + 4\hat{j} - 3\hat{k}) - (\hat{i} + 3\hat{j} + \hat{k}) = 4\hat{i} + \hat{j} - 4\hat{k}$$

Also the force $$\vec{F} = 3\hat{i} + 2\hat{k}$$

∴ The required moment is given by

$$= \vec{r} \times \vec{F} = (4\hat{i} + \hat{j} - 4\hat{k}) \times (3\hat{i} + 2\hat{k})$$

$$= \begin{vmatrix} \hat{i} & \hat{j} & \hat{k} \\ 4 & 1 & -4 \\ 0 & 3 & 2 \end{vmatrix} = \hat{i}(2+12) - \hat{j}(8-0) + \hat{k}(12-0) = 14\hat{i} - 8\hat{j} + 12\hat{k}.$$

Example 3.28. Two unlike force of equal magnitudes $\hat{i} + 2\hat{k}$ and $-\hat{j} - 2\hat{k}$ are acting at the point whose position vectors are given by $\hat{i} + \hat{j} + \hat{k}$ respectively. Find the moment of the couple formed by these forces.

Solution.

$$\vec{r_1} = \hat{i} + \hat{j} + \hat{k}$$

$$\vec{r_2} = \hat{i} + 2\hat{j} + 3\hat{k}$$

$$\vec{r_1} - \vec{r_2} = (\hat{i} + \hat{j} + \hat{k}) - (\hat{i} + 2\hat{j} + 3\hat{k}) = (1-1)\hat{i} + (1-2)\hat{j} + (1-3)\hat{k} = -\hat{j} - 2\hat{k}$$

$$\vec{F} = \hat{j} + 2\hat{k}$$

Now Moment of the couple $= (\vec{r_1} - \vec{r_2}) \times \vec{F} = (-\hat{j} + 2\hat{k}) \times (\hat{j} + 2\hat{k})$

$$= \begin{vmatrix} \hat{i} & \hat{j} & \hat{k} \\ 0 & -1 & -2 \\ 0 & 1 & 2 \end{vmatrix} = 0.$$

EXERCISE 3.4

1. If $\vec{a} = 3\hat{k} + 4\hat{j}$ and $\vec{b} = \hat{i} + \hat{j} - \hat{k}$, then find $|\vec{a} \times \vec{b}|$. **[Ans.** $\sqrt{74}$ **]**
2. Show that $\vec{a} \times (\vec{a} + \vec{b}) + \vec{b} \times (\vec{c} + \vec{a}) + \vec{c} \times (\vec{a} + \vec{b}) = \vec{0}$.
3. If $\vec{a} = \hat{i} + 2\hat{j} + 3\hat{k}$, $\vec{b} = 2\hat{i} + \hat{j}$, $\vec{c} = 3\hat{i} - 4\hat{j} - 5\hat{k}$, then verify that $\vec{a} \times (\vec{b} + \vec{c}) = (\vec{a} \times \vec{b}) + (\vec{a} \times \vec{c})$.
4. Find a unit vector perpendicular toboth the vector $\hat{i} - 2\hat{j} + 3\hat{k}$ and $\hat{i} + 2\hat{j} - \hat{k}$. $\left[\textbf{Ans.} \frac{1}{\sqrt{3}}(-\hat{i} + \hat{j} + \hat{k})\right]$
5. Given $|\vec{a}| = 10$, $|\vec{b}| = 2$ and $\vec{a} \cdot \vec{b} = 12$, find $|\vec{a} \times \vec{b}|$.
6. If $\vec{a} = \hat{i} - 2\hat{j} + 3\hat{k}$, $\vec{b} = 2\hat{i} + 3\hat{j} - 5k$ then find $\vec{a} \times \vec{b}$. Verify that $\vec{a}$ and $\vec{a} \times \vec{b}$ are perpendicular to each other. **[Ans.** $\hat{i} + 11\hat{j} + 7\hat{k}$ **]**
7. Find a vactor of magnitude 9 which is perpendicular to both the vectors $4\hat{i} - \hat{j} + 3\hat{k}$ and $-2\hat{i} + \hat{j} - 2k$.
8. Find a unit vector perpendicular to the plane *ABC* where *A*, *B*, *C* are the point (3, – 1, 2) (1, – 1, – 3) and (4, – 3, 1) respectively. $\left[\textbf{Ans.} \frac{1}{\sqrt{165}}(-10\hat{i} - 7\hat{j} + 4\hat{k})\right]$
9. If $\vec{a}, \vec{b}, \vec{c}$ are three vectors and $\vec{a}$ is not zero vector, $\vec{a} \cdot \vec{b} = \vec{a} \cdot \vec{c}$ and $\vec{a} \times \vec{c} = \vec{b} \times \vec{c}$, then prove that $\vec{b} = \vec{c}$.
10. Find the value of a so that angle between the vectors $\hat{j} + \hat{k}$ and $\hat{i} - \hat{j} + a\hat{k}$ may be $\left(\frac{\pi}{3}\right)^c$. **[Ans.** -4**]**
11. Find the value of a for which the vectors $3\hat{i} + 2\hat{j} + 9\hat{k}$ and $\hat{i} + a\hat{j} + 3\hat{k}$ are (*i*) perpendicular, (*ii*) parallel. $\left[\textbf{Ans.} -15\ \frac{2}{3}\right]$
12. Find the area of a parallelogram having diagonals $3\hat{i} + \hat{j} - 2\hat{k}$ and $\hat{i} - 3\hat{j} + 4\hat{k}$. **[Ans.** $5\sqrt{3}$ sq. unit**]**

13. If $\vec{a} = 2\hat{i} - 3\hat{j} + \hat{k}, \vec{b} = -\hat{i} + 4\hat{k}, \vec{c} = 2\hat{j} - \hat{k}$, find the area of the parallelogram having diagonals $\vec{a} + \vec{b}$ and $\vec{b} + \vec{c}$. $\left[\textbf{Ans. } \frac{1}{2}\sqrt{21} \text{ sq. units}\right]$

14. Let $\vec{a} = \hat{i} - \hat{j}, \vec{b} = 3\hat{j} - \hat{k}$ and $\vec{c} = 7\hat{j} - \hat{k}$. Find the vector $\vec{d}$ which is perpendicular to both $\vec{a}$ and $\vec{b}$ and $\vec{c}.\vec{d} = 1$. $\left[\textbf{Ans. } \frac{1}{4}(\hat{i} + \hat{j} + 3\hat{k})\right]$

15. For any two vectors $\vec{a}$ and $\vec{b}$, show that $(1 + |\ (1+|\vec{a}|^2)(1+|\vec{b}|^2) = (\vec{a}.\vec{b})^2 + |\vec{a} + \vec{b} + (\vec{a} \times \vec{b})|^2$.

16. If $|\vec{a}| = 2, |\vec{b}| = 7$, and $\vec{a} \times \vec{b} = 3\hat{i} + 2\hat{j} + 6\hat{k}$, find the angle between $\vec{a}$ and $\vec{b}$. $\left[\textbf{Ans. } \frac{\pi}{6}\right]$

17. If $\vec{p}$ and $\vec{q}$ are units vectors forming an angle of 30°, find the area of the parallelogram having a $\vec{a} = \vec{p} + 2\vec{q}$ and $\vec{b} = 2\vec{p} + \vec{q}$ as its diagonals. $\left[\textbf{Ans. } \frac{3}{4} \text{ sq. units}\right]$

18. Using vectors, find the area of ΔABC with vertices $A(1, 2, 3), B(2, -1, 4)$ ans $C(4, 5, -1)$.

$\left[\textbf{Ans. } \frac{1}{2}\sqrt{274} \text{ sq. units}\right]$

19. If $\vec{a} \times \vec{b} = \vec{b} \times \vec{c} \neq \vec{0}$, then show that $\vec{a} + \vec{c} = k\vec{b}$ where k is a scalar.

20. If $\vec{a}, \vec{b}, \vec{c}$ are three non-zero vectors such that $\vec{a}\ (\vec{b} \times \vec{c}) = 0$ and $\vec{b}$ and $\vec{c}$ are not parallel vectors, prove that $\vec{a} = \lambda\vec{b} + \mu\vec{c}$, where λ and μ are scalars.

21. If $\vec{a}, \vec{b}, \vec{c}$ are the position vectors of the vertices A, B, C of a triangle ABC, show that area of triangle ABC is $\frac{1}{2}|\vec{a} \times \vec{b} + \vec{b} \times \vec{c} + \vec{c} \times \vec{a}|$.

22. Prove by vector method, that the parallelograms upon the same base and between the same parallels are equal in area.

23. Prove vector method that, $\sin(A - B) = \sin A \cos B - \cos A \sin B$.

24. Given $\vec{A} = \hat{i} + 2\hat{j} + 3\hat{k}, \vec{B} = -\hat{i} + 2\hat{j} + \hat{k}$ and $\vec{C} = 3\hat{i} + \hat{j}$, find a unit vector in the direction of the resultant of these vectors. Also find a vector $\vec{D}$ which is normal to both $\vec{A}$ and $\vec{B}$. What is the inclination of $\vec{D}$ on $\vec{C}$?

$\left[\textbf{Ans. } \frac{3}{5\sqrt{2}}\hat{i} + \frac{1}{\sqrt{2}}\hat{j} + \frac{4}{5\sqrt{2}}\hat{k};\ -4\hat{i} - 4\hat{j} + 4\hat{k};\ \cos^{-1}\left(\frac{-4}{\sqrt{30}}\right)\right]$

❑❑❑

UNIT - II

COORDINATE GEOMETRY

CHAPTER 4

Two Dimensional Geometry

INSIDE THIS CHAPTER

4.1. INTRODUCTION

Analytic or Co-ordinate Geometry is the branch of Mathematics that came into existance by combining two branches—*Algebra* and *Geometry*. So, it is also known as *Algebraic geometry*. This concept was discovered by *Rene Descartes* with another French Mathematician '*Pierre de Femat*'.

In co-ordinate geometry, a point in a place is represented by an ordered pair of real numbers called co-ordinates of point.

4.1.1. Rectangular Axis

Let $X'OX$ and $Y'OY$ be two perpendi-cular fixed straight lines intersect at origin O. Then

(*i*) $X'OX$ is called x-axis (horizontal)

(*ii*) $Y'OY$ is called y-axis (vertical)

(*iii*) The point of intersection O is called origin.

(*iv*) $X'OX$ and $Y'OY$ taken together are called co-ordinates axes. This is shown in Fig. 4.1.

4.1.2. Co-ordinates of a Point

Let $X'OX$ and $Y'OY$ be co-ordinates axes. Let Q be any point in the plane. Draw QM and QN perpendiculars on X-aixs and Y-axis respectively from Q. Then

(*i*) The length NQ (or OM) is called x-coordinates or *abscissa* of point Q.

(*ii*) The length MQ (or ON) is called y-coordinate or ordinate of point Q.

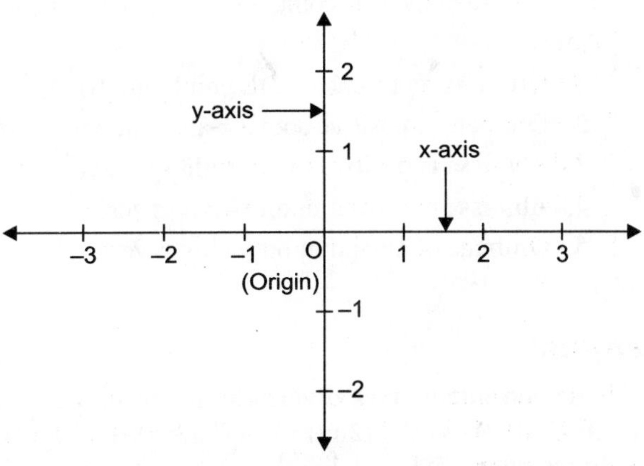

Fig. 4.1

4.1.3. Point Representation

The point $Q(x, y)$ is represented by ordered pair of x and y in which x-coordinate is written first and then y-coordinate, separated by comma (,) and written with parenthesis ( ). This is shown in Fig. 4.2.

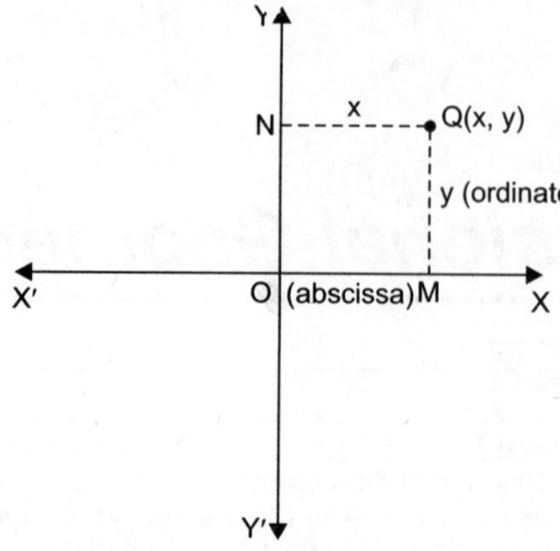

Fig. 4.2

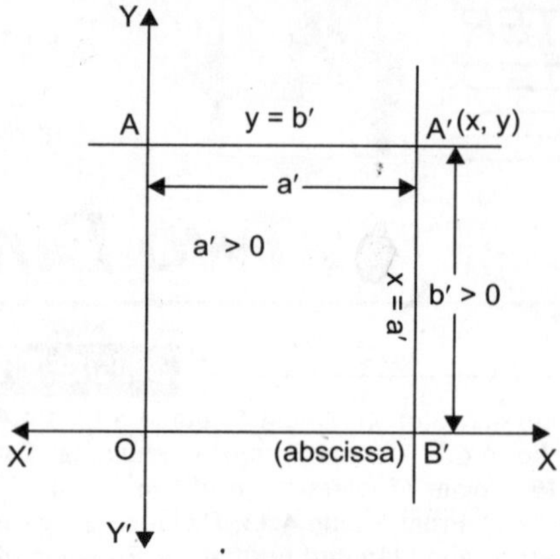

Fig. 4.3

4.2. STRAIGHT LINE

"A straight line can be defined as the shortest distance between two points."

4.2.1. Equation of a Straight Line Parallel to x-axis

In the following Fig. 4.3, AA' is the line parallel to x-aixs, then it is always at a constant distance *i.e.*, b' from x-axis. Let (x, y) be any point on this straight line and its ordinate is constant.

Thus, Equation of straight line parallel to x-axis is
$$y = b'$$

When $b^2 = 0$, the line coincides with x-axis and get $y = 0$, equation of x-axis.

4.2.2. Equation of a Straight Line Parallel to y-axis

In the above Fig. 4.3, $A'B'$ is the line parallel to y-axis, then it is always at a constant distance *i.e.*, a' from y-axis. Then abcissa is always constant so

Equation of straight line parallel to y-axis be
$$x = a'$$

If $a' = 0$, then the line coincides with y-axis and get $x = 0$, equation of the y-axis.

Notes:

1. Origin is represented with point $O(0, 0)$.
2. Ordinate is positive above x-axis and vice-versa.
3. Abscissa is positive to the right of y-axis and vice versa.
4. Abscissa of any point on y-axis is zero.
5. Ordinate of any point on y-axis is zero.

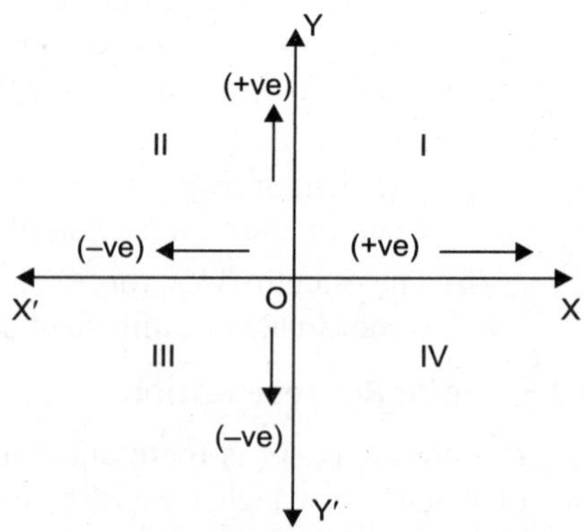

Fig. 4.4

Quadrants

The coordinate axes divide the plane into four quadrants *i.e.*, XOY (1st), YOX' (2nd), $X'OY'$ (3rd) and $Y'OX$ (Fourth) quadrants shown in Fig. 4.4.

Thus, if point $Q(x, y)$ lies in

(*i*) First quadrant, then $x > 0, y > 0$ or (+, +)

(*ii*) Second quadrant, then $x < 0, y > 0$ or (–, +)

(*iii*) Third quadrant, then $x < 0, y < 0$ or (–, –)

(*iv*) Fourth quadrant, then $x > 0, y < 0$ or (+, –)

4.3. SLOPE A LINE

In the coordinate plane, a line AB which is not parallel to x-axis, then it intersects x-aixs. Such a line makes two angles with x-axis, which are supplementary. In order to define, we select an angle 'θ' which is made going anti-clockwise direction from x-axis. This angle 'θ' will have two values between 0° and 180° and is called angle of inclination.

Definition. The tangent of the angle which a line makes with the positive direction of x-axis is called slope of gradient of line. It is generally denoted by 'm' as shown in Fig. 4.5.

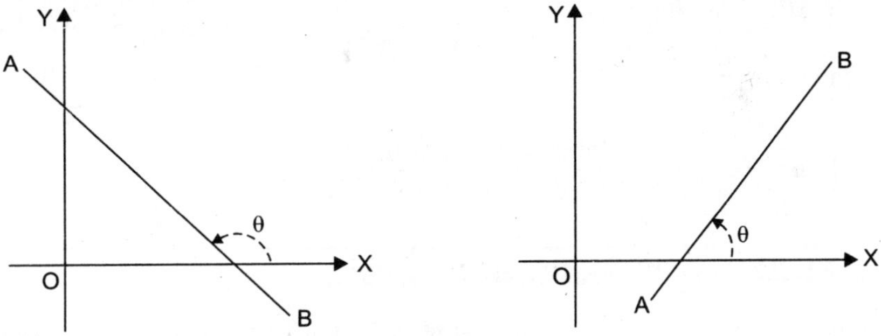

Fig. 4.5

Thus, if a line makes an angle θ with x-axis, then its slope = tan θ

i.e., $$m = \tan\theta$$

If a line is parallel to x-axis, then θ = 0

∴ $$m = \tan 0° = 0$$

If a line is parallel to y-axis, *i.e.*, it is perpendicular to x-axis so that θ = 90°

∴ $m = \tan 90° = \infty$.

Equation of a Straight Line Passing Through Origin and Having Slope 'm'

Let OA be the straight line passing through origin O making an angle θ with the +ve direction of x-axis as shown in Fig. 4.6.

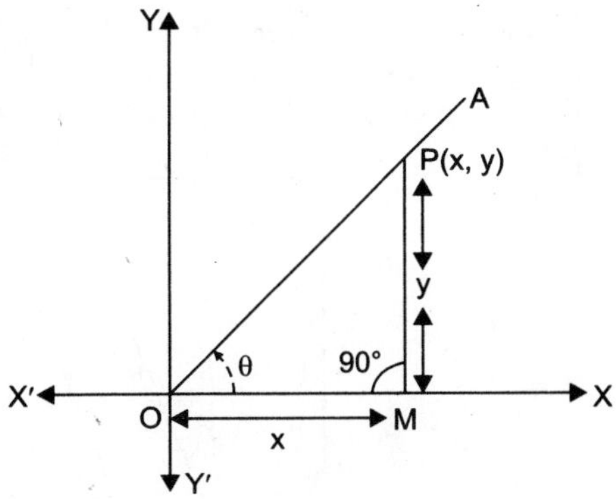

Fig. 4.6

Then its slope $m = \tan\theta$.

Let $P(x, y)$ be any point on the line. Draw $PM \perp x$-axis so that $OM = x$, $MP = y$.

In right angled ΔOMP, we have

$$\tan\theta = \frac{PM}{OM} \quad \text{or} \quad \tan\theta = \frac{y}{x} = m$$

i.e., $$y = mx$$

which is required equation of line.

Slope of a Line Joining Two given Points $P(x_1, y_1)$ and $Q(x_2, y_2)$

Let $P(x_1, y_1)$ and $Q(x_2, y_2)$ be two given points and 'θ' be the inclination of line PQ with x-axis as shown in Fig. 4.7.

So that $m = \tan\theta$

From P, draw $PM \perp x$-axis and from Q, draw $QN \perp x$-axis. Also draw $PK \perp QN$. Let PQ meets OX at A.

$$\angle KPQ = \angle XAQ = \theta$$
$$PK = MN$$
$$= ON - OM = x_2 - x_1$$
$$KQ = NQ - NK$$
$$= NQ - PM = y_2 - y_1$$

From right angle ΔPQK, we have

$$\tan\theta = \frac{QK}{PK} = \frac{y_2 - y_1}{x_2 - x_1}$$

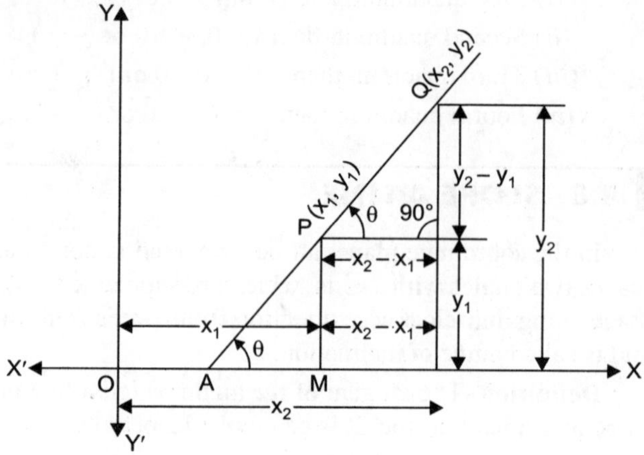

Fig. 4.7

$$\therefore \quad \boxed{m = \frac{y_2 - y_1}{x_2 - x_1} = \frac{\text{Difference of ordinates}}{\text{Difference of abcissa}}}$$

4.4. PARALLEL AND PERPENDICULAR LINES

Two lines (neither parallel to y-axis) are parallel if there slopes are equal as shown in Fig. 4.8. Let m_1 and m_2 be the slopes of two lines AB and CD respectively and 'θ_1' and 'θ_2' be the inclinations of lines AB and CD respectively.

Two lines are parallel off	$m_1 = m_2$
where	$m_1 = \tan\theta_1$
	$m_2 = \tan\theta_2$

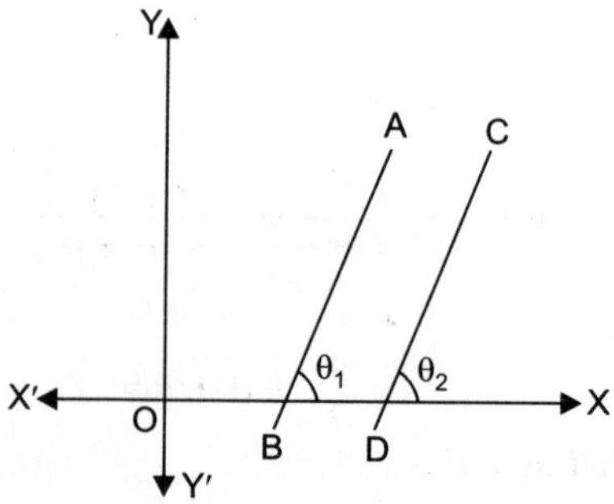

Fig. 4.8

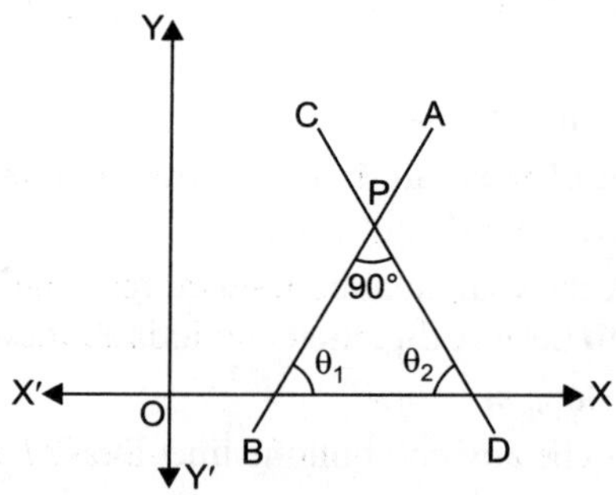

Fig. 4.9

Two lines (neither parallel to y-axis) are perpendicular. If their slopes $m_1 m_2 = -1$ as shown in Fig. 4.9.

Two lines are perpendicular off $m_1 m_2 = -1$

where
$$m_1 = \tan\theta_1$$
$$m_2 = \tan\theta_2$$

SOLVED EXAMPLES

Example 4.1. Find the slope of line passing through the point:
(*i*) (4, –6), (–2, –5) (*ii*) (0, –4), (–6, 2).

Solution. (*i*) $$\text{Slope of line} = \frac{y_2 - y_1}{x_2 - x_1}$$

[∵ Here given $P(x, y) = (4, -6)$; $Q(x_2, y_2) = (-2, -5)$]

$$= \frac{-5+6}{-2-4} = \frac{1}{-6} = \frac{-1}{6}$$

(*ii*) Slope of line with given point $P(x_1, y_1) = (0, -4)$; $Q(x_2, y_2) = (-6, 2)$

$$\text{Slope of line} = \frac{y_2 - y_1}{x_2 - x_1} = \frac{2+4}{-6-0} = \frac{6}{-6} = -1$$

Example 4.2. Show that line joining (2, –5) and (–2, 5) is perpendicular to the line joining (6, 3) and (1, 1).

Solution. Let the given points be $A(2, -5)$, $B(-2, 5)$, $C(6, 3)$ and $D(1, 1)$

∴ Slope of line through A and B

$$m_1 = \frac{y_2 - y_1}{x_2 - x_1} = \frac{5+5}{-2-2} = \frac{-10}{4} = \frac{-5}{2}$$

∴ Slope of line through C and D

$$m_2 = \frac{x_2 - y_1}{x_2 - x_1} = \frac{1-3}{1-6} = \frac{-2}{-5} = \frac{2}{5}$$

Since $$m_1 m_2 = \left(\frac{-5}{2}\right)\left(\frac{2}{5}\right) = -1$$

∴ The lines are perpendicular.

Example 4.3. Show that the points (1, –1), (2, 1) and (4, 5) are collinear.

Solution. Since points are collinear if they have same slope.
Let $A(1, -1)$, $B(2, 1)$, $C(4, 5)$ be given points.

∴ $$\text{Slope of } AB = \frac{1+1}{2-1} = \frac{2}{1} = 2$$

$$\text{Slope of } BC = \frac{5-1}{4-2} = \frac{4}{2} = 2$$

Since Slope of AB = Slope of BC

∴ Point A, B and C are collinear.

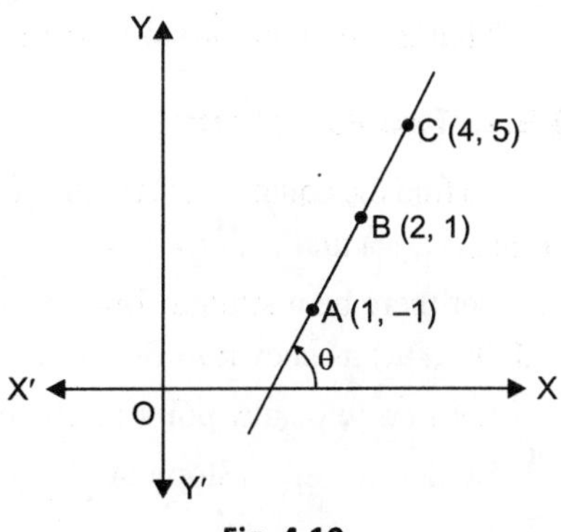

Fig. 4.10

EXERCISE 4.1

1. Find the equation of line, which is parallel to x-axis or ⊥ to y-axis at a distance of (*i*) 3 units above it, (*ii*) 5 units below it. [**Ans.** (*i*) $y = 3$, (*ii*) $y = -5$]

2. Find the equation of line, which is parallel to y-axis or ⊥ to x-axis at a distance of (i) 2 units to right of it, (ii) 8 units to left of it. **[Ans.** (i) $x = 2$, (ii) $x = -8$]
3. Determine x so that slope of the line joining (2, 5) and (x, 3) is 2. **[Ans.** 1]
4. Find the slope of line perpendicular to line joining the points (6, 4) and (2, 12). $\left[\textbf{Ans. } \frac{1}{2}\right]$
5. Show that points (1, 1), (2, 3) and (3, 5) lie on same line.
6. Show that the line passing through (–2, 6) and (4, 8) is perpendicular to line through (8, 12) and (4, 24).
7. State whether two lines are parallel, perpendicular or neither parallel nor perpendicular (i) through (8, 2) and (–5, 3); through (16, 6) and (3, 15).
8. The points (h, 0), (0, k) and (a, b) are collinear then prove that $\frac{a}{h} + \frac{b}{h} = 1$.
9. Find k such that line through (3, k), (2, 7) is perpendicular to the line through (–1, 4) and (1, 8). $\left[\textbf{Ans. } \frac{13}{2}\right]$

4.5. POINTS-SLOPE FORM

To find equation of straight line passing through (x_1, y_1) and having slope 'm'. Let $A(x_1, y_1)$ be the given point and 'm' be the given slope. Let $P(x, y)$ be any point on straight line as shown in Fig. 4.11.

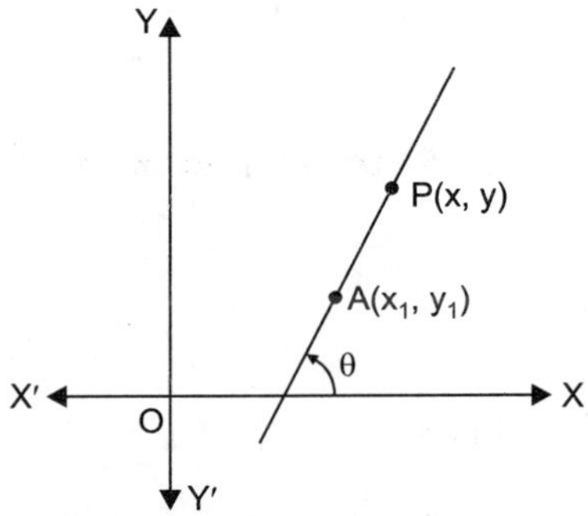

Fig. 4.11

Slope of straight line A and $P = \dfrac{y - y_1}{x - x_1}$

Now, by the definition $\quad m = \dfrac{y - y_1}{x - x_1} \qquad [\because \; x - x_1 \neq 0]$

$$\Rightarrow \qquad y - y_1 = m(x - x_1)$$

which is required equation.

That is called point-slope form.

4.5.1. Two Point Form

To find the equation of straight line passing through two given points (x_1, y_1) and (x_2, y_2).

Let there be a straight line passing through points $C(x_1, y_1)$ and $D(x_2, y_2)$ as shown in Fig. 4.12.

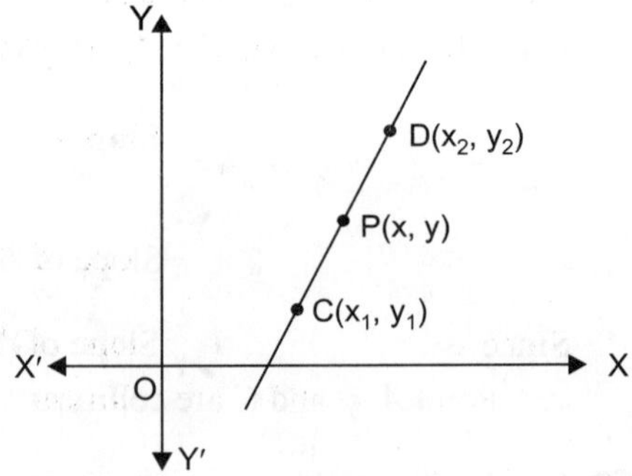

Fig. 4.12

Let $P(x, y)$ be any point on the line then

By definition $\quad$ Slope of CP = Slope of CD

$$i.e., \qquad \frac{y - y_1}{x - x_1} = \frac{y_2 - y_1}{x_2 - x_1}$$

$$\Rightarrow \qquad (y - y_1) = \frac{y_2 - y_1}{x_2 - x_1}(x - x_1)$$

be required equation of line.

SOLVED EXAMPLES

Example 4.4. Find the equation of straight line passing through (3, –5) and parallel to the line joining the points (1, 2) and (–3, 4).

Solution. Let m be the slope of line passing through (3, –5)

Then its equation is $y-(-5) = m(3-x)$ ... (*i*) $[\because \; y - y_1 = m(x - x_1)]$

$$= \frac{y_2 - y_1}{x_2 - x_1} = \frac{4-2}{-3-1} = \frac{2}{-4} = \frac{-1}{2}$$

Since the line with slope $\frac{-1}{2}$ is parallel to the required line. Thus, their slopes are equal.

$$\therefore \qquad m = -\frac{1}{2}$$

From Eq. (*i*) becomes

$$y + 5 = -\frac{1}{2}(x-3)$$

$$\Rightarrow \qquad 2y + 10 = -x + 3 \qquad \Rightarrow \qquad x + 2y + 7 = 0$$

Example 4.5. Find the equation of line passing through (0, –1) and perpendicular to line joining the points (0, 1) and (3, 5).

Solution. Let m be the slope of line passing through point (0, –1).

Then its equation is $y-(-1) = m(x-0)$

Slope of line joining the points (0, 1) and (3, 5) is

$$= \frac{y_2 - y_1}{x_2 - x_1} = \frac{5-1}{3-0} = \frac{4}{3}$$

Since the with slope $\frac{4}{3}$ is perpendicular to the required line. Then their slopes exist like

So $\qquad m_1 m_2 = -1$

$$\therefore \qquad m\left(\frac{4}{3}\right) = -1$$

$$m = \frac{-3}{4}$$

$\therefore$ From equation (*i*) $\qquad y + 1 = \frac{-3}{4}(x-0)$

$$\Rightarrow \qquad 4y + 4 = -3x$$

$\Rightarrow \qquad 3x + 4y + 4 = 0$ be required equation of line.

Example 4.6. Find the equation of sides of the triangle whose vertices are (–1, 8), (4, –2) and (–5, –3).

Solution. Let the vertices of triangle be $A(-1, 8)$, $B(4, -2)$ and $C(-5, -3)$

(*i*) The equation of side AB is:

$$y - 8 = \frac{-2-8}{4-(-1)}[x-(-1)] \qquad \left[\because \text{ Using } y - y_1 = \frac{y_2 - y_1}{x_2 - x_1}(x - x_1)\right]$$

$$\Rightarrow \quad y-8=\frac{-10}{5}(x+1)$$

$$\Rightarrow \quad y-8=-2(x+1)$$

$$\Rightarrow \quad y-8=-2x-2 \quad \Rightarrow \quad 2x+y-6=0$$

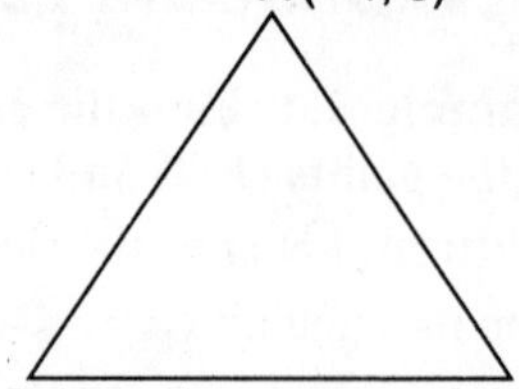

Fig. 4.13

(*ii*) the equation of side *BC* is

$$y-(-2)=\frac{-3-(-3)}{-5-4}(x-4)$$

$$\left[\because \text{ Using } y-y_1=\frac{y_2-y_1}{x_2-x_1}(x-x_1)\right]$$

$$\Rightarrow \quad y+2=\frac{-1}{-9}(x-4)$$

$$\Rightarrow \quad y+2=\frac{1}{9}(x-4)$$

$$\Rightarrow \quad 9y+18=x-4 \quad \Rightarrow \quad x-9y-22=0$$

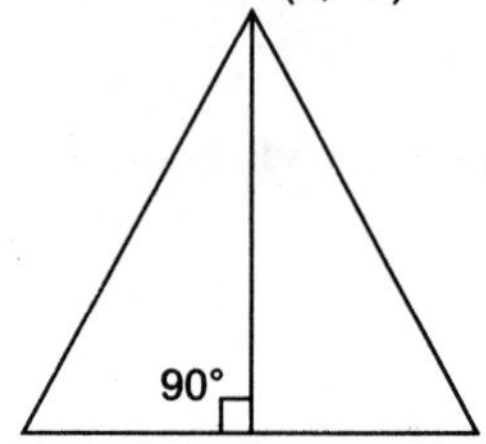

Fig. 4.14

(*iii*) The equation of the side *CA* is

$$y-(-3)=\frac{8-(-3)}{-1-(-5)}[x-(-5)]$$

$$\left[\because \text{ Using } y-y_1=\frac{y_2-y_1}{x_2-x_1}(x-x_1)\right]$$

$$\Rightarrow \quad y+3=\frac{11}{4}(x+5)$$

$$\Rightarrow \quad 4y+12=11x+55$$

$$\Rightarrow \quad 11x-4y+43=0$$

EXERCISE 4.2

1. Find the straight line that passes through
   (*i*) Point (2, 3) and gradient –1
   (*ii*) Point (–1, –2) and slope $\frac{2}{3}$. **[Ans.** (*i*) $y+x-5=0$, (*ii*) $3y-2x+4=0$]
2. Find the equation of line passes through $(2, 2\sqrt{3})$ and is inclined with *x*-axis at an angle of 75°.
   [**Ans.** $(2+\sqrt{3})x-y-4=0$]
3. Find the equation of line which intersects the *y*-axis at a distance of 2 units above the origin and making an angle of 30° with positive direction of *x*-axis. [**Ans.** $x-\sqrt{3}y+2\sqrt{3}=0$]
4. Find the equation of the right bisector of segment joining (1, 1) and (2, 3). [**Ans.** $2x+4y=11$]
5. Find the equation of a line passing through the point (2, 1) and parallel to line joining the points (1, 3) and (–3, 1). [**Ans.** $x-2y=0$]
6. Find the equation of line passing through (–3, 5) and perpendicular to the line through points (2, 5) and (–3, 6). [**Ans.** $5x-y=-20$]

7. Find the equation of straight line passing through points
   (*i*) (0, –4) and (–6, 2) [**Ans.** $x + y + 4 = 0$]
   (*ii*) (–1, 1) and (2, –4) [**Ans.** $5x + 3y + 2 = 0$]
   (*iii*) (–1, –3) and (–4, –5) [**Ans.** $y - x + 1 = 0$]
   (*iv*) (1, 2) and (0, 5) [**Ans.** $y + 3x - 5 = 0$]
8. The vertices of triangle are points $A(2, 1)$, $B(-2, 3)$ and $C(4, 5)$. Find the equations of its sides.
   [**Ans.** $x + 2y - 4 = 0, x - 3y + 11 = 0, 2x - y - 3 = 0$]
9. The vertices of a triangle are $P(4, 10)$, $Q(9, -4)$ and $R(-1, -2)$. Find the equation of
   (*i*) side BC. [**Ans.** $x + 5y + 11 = 0$]
   (*ii*) the median bisecting BC. [**Ans.** $x = 4$]
10. If $P(2, -2)$, $Q(1, 1)$ and $R(-1, 0)$ be vertices of a ΔPQR and $PD \perp QR$, find equation of altitude PD.
   [**Ans.** $2x + y - 2 = 0$]

4.6. INTERCEPT FORM

Let a straight line M meets x-axis at A and y-axis at B then:
(*i*) OA is called intercept of the line on x-axis.
(*ii*) OB is called intercept of the line on y-axis.

The intercept OA on x-axis is +ve of A lies to right of O and negative of A lies to left of O.
The intercept OB on y-axis is positive of B lies above O and negative of B lies below O.

∴ Equation of line M passing through points A $(a, 0)$ and B $(0, b)$ is

$$y - 0 = \frac{b-0}{0-a}(x-a) \qquad \left[\because \text{ Using } y - y_1 = \frac{y_2 - y_1}{x_2 - x_1}(x - x_1)\right]$$

$$y = \frac{-b}{a}(x-a)$$

$$\frac{y}{b} = \frac{-1}{a}(x-a) \Rightarrow \frac{y}{b} = \frac{-x}{a} + \frac{a}{a}$$

$$\Rightarrow \boxed{\frac{x}{a} + \frac{y}{b} = 1} \text{ which is required equation.}$$

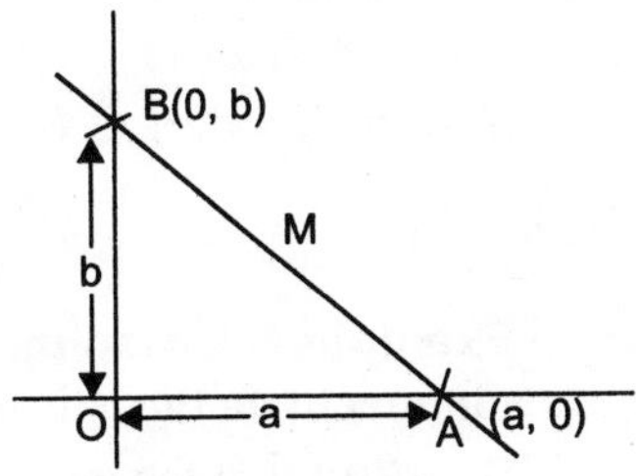

Fig. 4.15

Aid to Memory

$$\boxed{\frac{x}{\text{Intercept on } x\text{-axis}} + \frac{y}{\text{Intercept on } y\text{-axis}} = 1}$$

4.6.1. Slope Intercept Form

To find the equation of straight line, which cuts of a given intercept 'C' on y-axis and having slope 'm' as shown in Fig. 4.16.

Thus equation of straight line passing through point $(0, C)$ and having slope 'm' is

$$y - C = M(x - 0)$$

[Using $y - y_1 = m(x - x_1)$]

$$y = mx + C$$

be required equation of line.

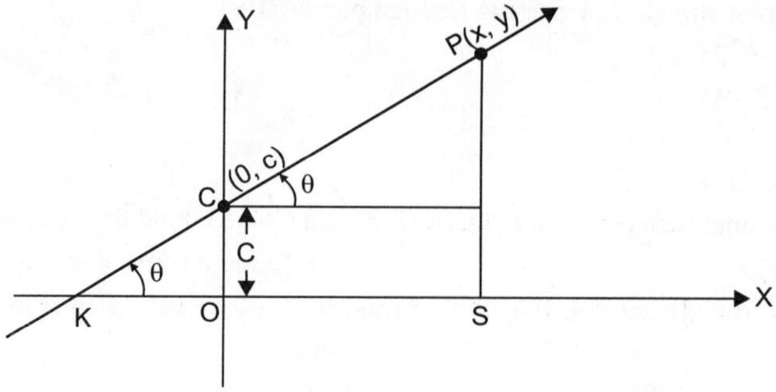

Fig. 4.16

Notes:

1. To find equation of the straight line, which cuts off a given intercept 'C' on x-axis and having slope 'm' is

$$\boxed{y = m\,(x - c)}$$

2. When the line passes through origin $C\,(0, 0)$. So intercept 'C' = 0, then euation becomes

$$\boxed{y = mx}$$

SOLVED EXAMPLES

Example 4.7. Find the equation of straight line with slope – 3 and y-intercept 2.

Solution. Here $\quad$ 'm' = – 3, 'C' = 2

∴ The equation of straight line is

$$y = mx + C$$

Now using values of Eq. (i), we get

$$y = -3x + 2$$

⇒ $\quad y + 3x - 2 = 0$ be required equation.

Example 4.8. Find the equation of a straight line passing through the point (2, 2) such that sum of its intercept on the axis is 9.

Solution. Let the intercepts be 'a' and 'b' so that

$$a + b = 9$$

$$b = 9 - a \qquad \ldots (i)$$

∴ Equation of straight line is $\quad \dfrac{x}{a} + \dfrac{y}{b} = 1$

Now using Eq. (i), we get

$$\frac{x}{a} + \frac{y}{9-a} = 1$$

Since is passes through point (2, 2)

∴ $$\frac{2}{a} + \frac{2}{9-a} = 1$$

⇒ $$2(9 - a) + 2a = a\,(9 - a)$$

⇒ $$18 - 2a + 2a = 9a - a^2$$

⇒ $$a^2 - 9a + 18 = 0$$

$\Rightarrow$ $a^2 - 6a - 3a + 18 = 0$

$\Rightarrow$ $a(a-6) - 3(a-6) = 0$

$\Rightarrow$ $(a-6)(a-3) = 0$

$\Rightarrow$ $a = 6, 3$

When $a = 6$, then from Eq. (*i*) $b = 9 - 6 = 3$

When $a = 3$, then from Eq. (*i*), $b = 9 - 3 = 6$

So when $a = 6$, $b = 3$, then equation of straight line is

$$\frac{x}{6} + \frac{y}{3} = 1$$

$\Rightarrow$ $x + 2y = 6$ $\left[\because \text{ Using } \frac{x}{a} + \frac{y}{b} = 1\right]$

So, when $a = 3$, $b = 6$, then equation of straight line is

$$\frac{x}{3} + \frac{y}{6} = 1$$

$\Rightarrow$ $2x + y = 6$ $\Rightarrow$ $2x + y - 6 = 0$

EXERCISE 4.3

1. Find the equation of lines having following intercepts:
   (*i*) $a = -3, b = 1$ (*ii*) $a = 3, b = -2$
   [Ans. (*i*) $x - 3y + 3 = 0$, (*ii*) $2x - 3y - 6 = 0$]
2. Find the equation of the line:
   (*i*) Passing through $(-2, 3)$ and having slope -4
   (*ii*) Passing through $(\sqrt{2}, 2\sqrt{2})$ and having slope $\frac{2}{3}$. **[Ans.** (*i*) $4x + y + 5 = 0$, (*ii*) $2x - 3y = 4\sqrt{2} = 0$]
3. Find the equation of straigh line with slope 3 and y-inercept is -2. **[Ans.** $y - 3x + 2 = 0$]
4. Find the equation of the line which cuts off intercept 4 on x-axis and makes an angle $\frac{\pi}{3}$ with positive direction of x-axis. $\left[\textbf{Ans. } y = \sqrt{3}(x-4)\right]$
5. Find is equation of line that cuts off equal intercepts on co-ordinate axes and passes through the point:
   (*i*) (5, 6) (*ii*) (2, 3) **[Ans.** (*i*) $x + y - 11 = 0$, (*ii*) $x + y - 5 = 0$]

4.7. NORMAL FORM/PERPENDICULAR FORM

To find the equation of a straight line terms of 'p', the length of the perpendicular from the origin upon it and 'α' the angle which perpendicular makes with x-axis as shown in Fig. 4.17.

The equation of a straight line in perpendicular form

$$x \cos\alpha + y \sin\alpha = p$$

Note: p is always positive and α lies between 0° and 360°.

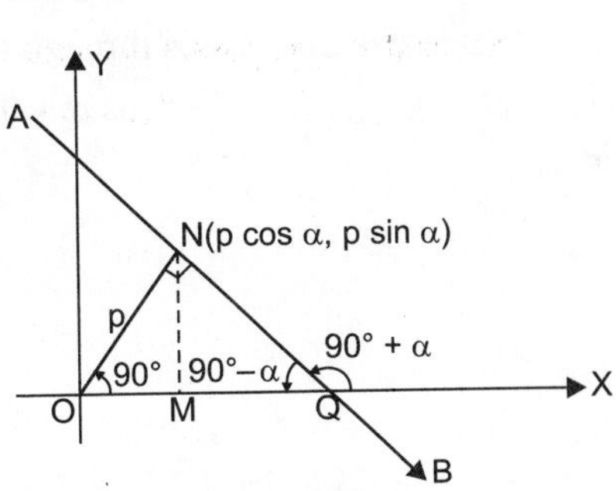

Fig. 4.17

4.7.1. Symmetric Form/Distance Form

Let $A(x_1, y_1)$ is the given point. Let $P(x, y)$ be any point on the straight line, which makes an angle θ with positive direction of x-axis. $AP = r$ as shown in Fig. 4.18.

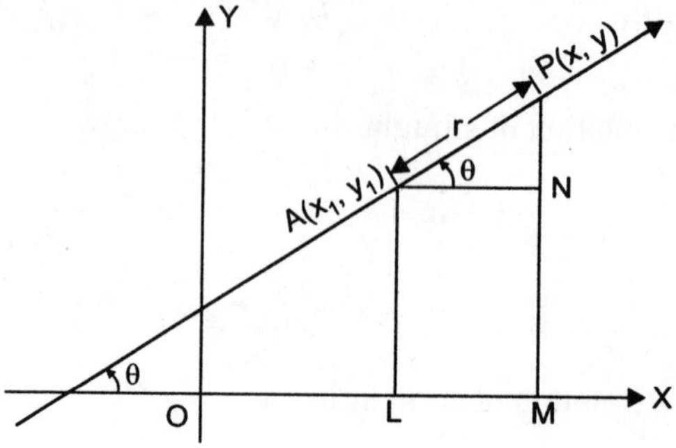

Fig. 4.18

So, the equation of straight line through the point $A(x_1, y_1)$ and making an angle θ with the positive direction of x-axis in the form

$$\frac{x - x_1}{\cos\theta} = \frac{y - y_1}{\sin\theta} = r$$

where 'r' is the distance of any point $P(x, y)$ from $P_1(x_1, y_1)$.

Notes: 1. From above, we have

$$x - x_1 = r\cos\theta \quad \text{and} \quad y - y_1 = r\sin\theta$$
$$x = x_1 + r\cos\theta \quad \text{and} \quad y = y_1 + r\sin\theta$$

which are parametric equations of straight line, θ being parameter.

2. Any point P on $\frac{x - x_1}{\cos\theta} = \frac{y - y_1}{\sin\theta}$ is $(x_1 + r\cos\theta, y_1 + r\sin\theta)$, where '$r$' is distance of P from $A(x_1, y_1)$.

SOLVED EXAMPLES

Example 4.9. Find the equation of the line whose x-intercept is 5 and distance from origin is 3.

Solution. Since the x-intercept is 5 *i.e.*, line passes through point (5, 0) and $p = 3$ (given)

∴ Equation of line in normal form is

$$x\cos\alpha + y\sin\alpha = 3 \quad \text{...(i)}$$

Since this line passes through the point (5, 0)

$$5\cos\alpha + 0\cdot\sin\alpha = 3$$

$$\Rightarrow \quad \cos\alpha = \frac{3}{5}$$

$$\therefore \quad \sin^2\alpha = 1 - \cos^2\alpha = 1 - \left(\frac{3}{5}\right)^2 = 1 - \frac{9}{25} = \frac{16}{25}]$$

$$\sin\alpha = \pm\frac{4}{5}$$

$\therefore$ From equation (*i*), we get

$$x\cdot\left(\frac{3}{5}\right)+y\left(\pm\frac{4}{5}\right)=3 \quad\Rightarrow\quad 3x\pm 4y=15$$

$\Rightarrow$ $3x \pm 4y - 15 = 0$ which are required equation of lines.

Example 4.10. If '*p*' be measure of the perpendicular segment form the origin on the line whose intercepts on the axes '*a*' and '*b*', show that:

$$\frac{1}{p^2}=\frac{1}{a^2}+\frac{1}{b^2}.$$

Solution. Let the straight line *CD* cuts the axes in *C* and *D* such that $OC = a$ and $OD = b$ as shown in Fig. 4.19.

Let $OM = p$ be the measure of the perpendicular segment from *O* on *CD*.

Let $\angle COM = \alpha$, then $\angle MOD = 90° - \alpha$

In right angle, ΔOCM, $\frac{p}{a} = \cos\alpha$ ...(*i*)

In right angle, ΔOMD, $\frac{p}{b} = \cos(90° - \alpha)$

i.e., $\frac{p}{b} = \sin\alpha$ ...(*ii*)

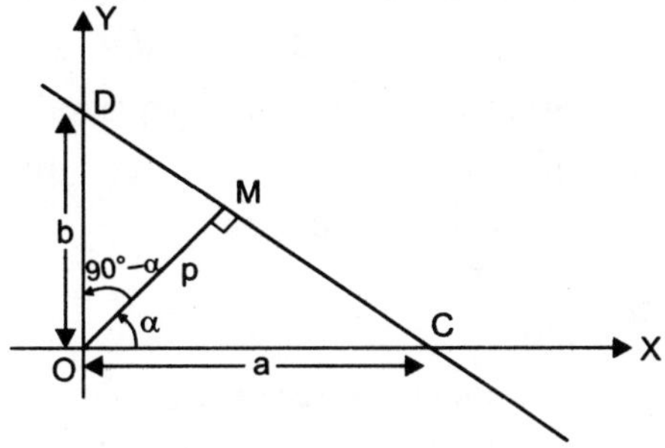

Fig. 4.19

Squaring and adding (*i*) and (*ii*), we get

$$\frac{p^2}{a^2}+\frac{p^2}{b^2}=\cos^2\alpha+\sin^2\alpha$$

[$\because$ $\cos^2\alpha + \sin^2\alpha = 11$]

$$\Rightarrow \quad \frac{p^2}{a^2}+\frac{p^2}{b^2}=1$$

$$\Rightarrow \quad \frac{1}{p^2}=\frac{1}{a^2}+\frac{1}{b^2}$$ which is true.

EXERCISE 4.4

1. Find the perpendicular form of equation of straight line for given follwoing values:

   (*i*) $p = 1, \alpha = 90°$ (*ii*) $p = 2, \alpha = 150°$ [**Ans.** (*i*) $y = 1$, (*ii*) $\sqrt{3}x - y + 4 = 0$]

2. Find the equation of line such that perpendicular distance of a line form the origin is 5 units and its slope is – 1.

   [**Ans.** $x + y - 5\sqrt{2} = 0, x + y + 5\sqrt{2} = 0$]

3. Find the coordinates of a point at a distance 6 units from the point (2, 4) and lying on the straight line passing through (2, 4) and making an angle $\frac{\pi}{3}$ with *x*-axis. [**Ans.** $5, 4+3\sqrt{3}$]

4. Find the locus of middle points of variable line $x\cos\alpha + y\sin\alpha - p = 0$ intercepted by axes given that premains constant. [**Ans.** $\frac{1}{x^2}+\frac{1}{y^2}=\frac{4}{p^2}$]

5. Find the coordinates of two points which are $2\sqrt{2}$ units from the point *P* (3, 4) and lie on straight line passing through this point and inclined at angle of 135° to the *x*-axis. [**Ans.** (1, 6), (5, 2)]

4.8. GENERAL FORM OF STRAIGHT LINE

From all the forms, we observe that equations of the straight lines are of first degree in x and y. The most common form of the equation of first degree in x and y be:

$$Ax + By + C = 0 \qquad \ldots(i)$$

In Eq. (i), it involves three constants *i.e.*, A, B and C but actually, it involves only two constants *i.e.*, $\frac{A}{C}$ and $\frac{B}{C}$.

SOLVED EXAMPLES

Example 4.11. Reduce each of the following into slope intercept form:

(i) $7x + 3y - 6 = 0$ (ii) $6x + 3y - 5 = 0$.

Solution. (i) The given equation is

$$7x + 3y - 6 = 0$$

This can be written as

$$3y = -7x + 6$$

$$\Rightarrow \quad y = \frac{-7}{3}x + 2$$

Now this is the required form *i.e.*, $y = mx + c$.

(ii) The slope intercept form of line is $y = mx + c$

Now the given equation is

$$6x + 3y - 5 = 0$$

$$\Rightarrow \quad 3y = -6x + 5$$

$$\Rightarrow \quad y = -2x + \frac{5}{3}$$

which is required slope intercept form.

Example 4.12. Reduce the equation $x - \sqrt{3}y + 2 = 0$ to the normal form and hence find the values of α and p.

Solution. The given equation is

$$x - \sqrt{3y} + 2 = 0$$

$$\Rightarrow \quad x - \sqrt{3}y = -2 \quad \Rightarrow \quad -x + \sqrt{3}y = 2$$

Dividing both sides by $\sqrt{(\text{Coefficient of } x)^2 + (\text{Coefficient of } y)^2}$ *i.e.*, $\sqrt{(-1)^2 + (\sqrt{3})^2}$

$$\frac{-x}{2} + \frac{\sqrt{3}}{2}y = 1$$

Which is required form.

Now, comparing it with $x \cos\alpha + y \sin\alpha = p$, we get

$$\therefore \quad \cos\alpha = \frac{-1}{2}, \sin\alpha = \frac{\sqrt{3}}{2}, p = 1$$

$\sin\alpha$ is positive and $\cos\alpha$ is negative in II quadrant.

$$\therefore \qquad \tan\alpha = \frac{\sin\alpha}{\cos\alpha} + \frac{\frac{\sqrt{3}}{2}}{\frac{-1}{2}} = -\sqrt{3}$$

$$= -\tan + 60° = \tan(180° - 60°) = \tan 120°$$

$$\therefore \qquad \alpha = 120°$$

Hence, required form is $x\cos 120° + y\sin 120° = 1$ where $\alpha = 120°$, $p = 1$.

EXERCISE 4.5

1. Reduce the equation $3x + 5y + 5 = 0$ to intercept form and find its intercept on the axes.

$$\left[\textbf{Ans. } \frac{x}{\frac{-5}{3}} + \frac{y}{\frac{-5}{4}} = 1, a = \frac{-5}{3}, b = \frac{-5}{4}\right]$$

2. Reduce each of following to the perpendicular form find p:
   (*i*) $x + y - 2 = 0$ (*ii*) $x - 4 = 0$

$$\left[\textbf{Ans. } \frac{x}{\sqrt{2}} + \frac{y}{\sqrt{2}} = \sqrt{2}, p = \sqrt{2}, (ii)\ x = 4, p = 4\right]$$

3. Find the angle made by line $x\cos 30° + y\sin 30° + \sin 120°$ with positive direction of x-axis. [**Ans.** 120°]
4. Show that the origin is equidistant from the three straight lines $4x + 3y + 10 = 0$, $5x - 12y + 26 = 0$ and $7x + 24y - 50 = 0$.

4.9. ANGLE BETWEEN TWO LINES

To find the angle between straight line l_1 and l_2 given respectively by:

$$y = m_1 x + c_1 \quad \text{and} \quad y = m_2 x + c_2$$

Let θ_1, θ_2 be inclinations of the given lines with x-axis, so that

$$\tan\theta_1 = m_1 \quad \text{and} \quad m_2 = \tan\theta_2$$

Let θ be the required angle.

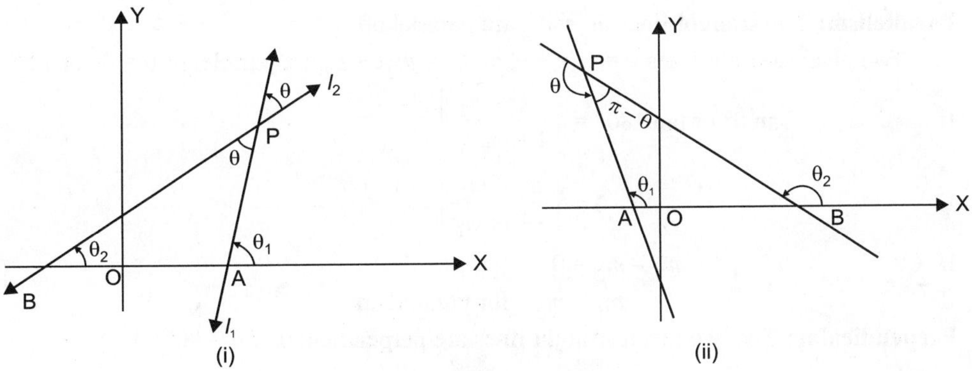

Fig. 4.20

In Fig. 4.20(*i*), $\theta_1 = \theta_2 + \theta$

[∵ Exterior angle = sum of two opposite interior angles]

$$\theta = \theta_1 - \theta_2$$

$\therefore$ $$\tan\theta = \tan(\theta_1 - \theta_2) = \frac{\tan\theta_1 - \tan\theta_2}{1+\tan\theta_1\tan\theta_2}$$

$\Rightarrow$ $$\tan\theta = \frac{m_1 - m_2}{1+m_1m_2}$$

In Fig. 4.20(*ii*),
$$\theta_2 = \theta_1 + (\pi - \theta) \qquad [\because \text{ Here } \theta_1 \text{ and } \pi - \theta \text{ are interior angles}]$$
$$\theta = \theta_1 - \theta_2 + \pi$$
$$\tan\theta = \tan[\pi + (\theta_1 - \theta_2)] = \tan(\theta_1 - \theta_2) = \frac{\tan\theta_1 - \tan\theta_2}{1+\tan\theta_1\tan\theta_2}$$
$$\tan\theta = \frac{m_1 - m_2}{1+m_1m_2} \qquad \text{...}(i)$$

Also $$\tan(\pi - \theta) = -\tan\theta = -\frac{m_1 - m_2}{1+m_1m_2}$$

Thus, if θ is the angle between the lines l_1 and l_2 then
$$\tan\theta = \pm\frac{m_1 - m_2}{1+m_1m_2}$$

Hence, $$\tan\theta = \frac{m_1 - m_2}{1+m_1m_2}$$

Aid to Memory:

- If θ is acute, then $\tan\theta$ is +ve so that the acute angle between the lines is given by $\tan\theta = \left|\frac{m_1 - m_2}{1+m_1m_2}\right|$.
- If is obtuse, then that θ is –ve so that obtuse angle between the line is given by $\tan\theta = -\left|\frac{m_1 - m_2}{1+m_1m_2}\right|$.
- Whenever we are to find the angle between two straight lines, we have to find acute angle between them by discarding the negative sign.

Conditions:

(*i*) **Parallelism:** Two straight lines $m_1 = m_2$ are parallel off

$\therefore$ Two given straight lines $y = m_1x + c$, and $y = m_2x + c_2$ are parallel of $\theta = 0°$ or $180°$.

If $$\text{Tan } 0° \text{ or } \tan 180° = \left|\frac{m_1 - m_2}{1+m_1m_2}\right|$$

If $$0 = \left|\frac{m_1 - m_2}{1+m_1m_2}\right|$$

If $$m_1 - m_2 = 0$$

If $$m_1 = m_2 \quad \text{for parallelism.}$$

(*ii*) **Perpendicular:** The two given straight lines are perpendicular if $\theta = 90°$.

If $$\tan 90° = \left|\frac{m_1 - m_2}{1+m_1m_2}\right|$$

If $$\frac{\sin 90°}{\cos 90°} = \left|\frac{m_1 - m_2}{1+m_1m_2}\right|$$

If $$\frac{1}{0} = \left|\frac{m_1 - m_2}{1+m_1m_2}\right|$$

Cross multiplying, we get

If $\qquad 1 + m_1 m_2 = 0$

If $\qquad m_1 m_2 = -1,$ for perpendicular

(*iii*) Two straight lines are coincident if $\theta = 0$

i.e., if $\qquad m_1 = m_2$ for coincident

SOLVED EXAMPLES

Example 4.13. Find the angle between the straight line $y - \sqrt{3}x - 5 = 0$ and $\sqrt{3}y - x + 6 = 0$.

Solution. Here
$$m_1 = \frac{-\text{Coefficient of } x}{\text{Coefficient of } y} = -\frac{\sqrt{3}}{1} = \sqrt{3}$$
$$m_2 = -\frac{-1}{\sqrt{3}} = \frac{1}{\sqrt{3}}$$

Let θ be angle between the given lines

Then
$$\tan\theta = \left|\frac{m_1 - m_2}{1 + m_1 m_2}\right| = \left|\frac{\sqrt{3} - \frac{1}{\sqrt{3}}}{1 + (\sqrt{3})\left(\frac{1}{\sqrt{3}}\right)}\right| = \left|\frac{3-1}{2\sqrt{3}}\right| = \left|\frac{2}{2\sqrt{3}}\right| = \frac{1}{\sqrt{3}}$$

$\therefore$
$$\tan\theta = \tan 30°$$
$$\theta = 30° \text{ or } \frac{\pi}{6}.$$

Example 4.14. The acute angle between two lines is 45° and slope of one of them is $\frac{1}{2}$. Find slope of other line.

Solution. Here
$$\theta = 45°, \quad m_1 = \frac{1}{2} \text{ (given)}$$

Using
$$\tan\theta = \left|\frac{m_1 - m_2}{1 + m_1 m_2}\right|, \text{ we get}$$
$$\tan 45° = \left|\frac{\frac{1}{2} - m_2}{1 + \left(\frac{1}{2}\right)(m_2)}\right|$$
$$\pm 1 = \left|\frac{1 - 2m_2}{2 + m_2}\right|$$

$\Rightarrow \quad \frac{1-2m_2}{2+m_2} = 1 \quad$ or $\quad \frac{1-2m_2}{2+m_2} = 1$

$\Rightarrow \quad 1 - 2m_2 = 2 + m_2 \quad$ or $\quad 1 - 2m_2 = -2 - m_2$

$\Rightarrow \quad 1 - 2 = 3m_2 \quad$ or $\quad -m_2 = -2 - 1$

$\Rightarrow \quad -1 = 3m_2 \quad$ or $\quad -m_2 = -3$

$\Rightarrow \quad m_2 = \frac{-1}{3} \quad$ or $\quad m_2 = 3$

Hence, slope of other line is -3 or $\frac{1}{3}$.

Example 4.15. Find the equations of the straight lines which pass through (4, 5) and make angle 45° with straight line $2x + y + 1 = 0$.

Solution. Let equation of straight line passing through point (4, 5) and having slope 'm' is

$$y - 5 = m(x - 4) \quad [\text{Using } y - y_1 = m(x - x_1)]$$

$$y - 5 = m(x - 4) \quad \ldots(i)$$

Here $m_1 = m$

The other given line is $2x + y + 1 = 0$

$$\therefore \quad m_2 = \frac{-\text{Coefficient of } x}{\text{Coefficient of } y} = \frac{-2}{1} = -2$$

Now $m_1 = m,\ m_2 = -2,\ \theta = 45°$

$$\therefore \quad \tan 45° = \left|\frac{m+2}{1+(m)(-2)}\right|$$

$$\pm 1 = \frac{m+2}{1-2m}$$

Taking +ve value | or | Taking – ve value

$\Rightarrow \quad 1 - 2m = m + 2$ or $-1(1 - 2m) = m + 2$

$\Rightarrow \quad 1 - 2 = 3m$ or $-1 + 2m = m + 2$

$\Rightarrow \quad m = \frac{-1}{3}$ or $m = 3$

When $m = \frac{-1}{3}$, from Eq. (*i*)

$$y - 5 = \frac{-1}{3}(x - 4)$$

$$\Rightarrow \quad 3y - 15 = -x + 4$$

$$\Rightarrow \quad x + 3y - 19 = 0$$

When $m = 3$, from Eq. (*i*)

$$y - 5 = 3(x - 4)$$

$$\Rightarrow \quad 3x - y - 7 = 0$$

EXERCISE 4.6

1. Find the acute angles between straight lines:

   (*i*) $2x - y + 3\ \ 0$ and $x + y - 2 = 0$

   (*ii*) $\sqrt{3}x + y - 1 = 0$ and $x + \sqrt{3}y - 1 = 0$ **[Ans.** (*i*) $\tan^{-1} 3$, (*ii*) 30°]

2. The slope of a line is double of the slope of another line. If tangent of anle between them is $\frac{1}{3}$, find slopes of lines. $\left[\textbf{Ans.}\ \frac{1}{2} \text{ or } 1, \text{ or } 2 \text{ or } \frac{-1}{2}, -1 \text{ or } -1, -2\right]$

3. Find the equations of line passing through (4, 5) and making an angle of $\frac{\pi}{4}$ with the line $2x - y + 7 = 0$.

   [Ans. $3x + y - 17 = 0$; $x - 3y + 11 = 0$]

4. A line passing through the point $(K, 2K)$ and $(-2, 3)$ is perpendicular to the line $4x + 3y + 5 = 0$; find the value of K. $\left[\textbf{Ans. } \frac{18}{5}\right]$

5. Find the equation of line that is perpendicular to $3x + 2y = 8$ and passes through the mid point of line joining $(5, -2)$ and $(2, 2)$. [**Ans.** $2x - 3y - 7 = 0$]

4.10. POINTS OF INTERSECTION OF TWO STRAIGHT LINES

To find the co-ordinates of the point of intersection of two straight line:

$$a_1x + b_1y + c_1 = 0 \quad \ldots(i)$$

$$a_2x + b_2y + c_2 = 0 \quad \ldots(ii)$$

Solving Eqs. (*i*) and (*ii*) by cross multiplication, we get

$$\frac{x}{b_1c_2 - b_2c_1} = \frac{y}{c_1a_2 - c_2a_1} = \frac{1}{a_1b_2 - a_2b_1}$$

$$\Rightarrow \quad x = \frac{b_1c_2 - b_2c_1}{a_1b_2 - a_2b_1} \quad \text{and} \quad y = \frac{c_1a_2 - c_2a_1}{a_1b_2 - a_2b_1}$$

Remarks:

1. When $\frac{a_1}{a_2} \neq \frac{b_1}{b_2}$ *i.e.*, $a_1b_2 - a_2b_1 \neq 0$, that is the condition for the lines to be intersecting.

2. When $\frac{a_1}{a_2} = \frac{b_1}{b_2} \neq \frac{c_1}{c_2}$ in this case lines are parallel and there is no common point on both the lines.

3. When $\frac{a_1}{a_2} = \frac{b_1}{b_2} = \frac{c_1}{c_2}$ = say (k), in this case, lines are coincident and there are infinitely many points lie on both the lines.

4.11. CONDITION OF CONCURRENCY OF THREE LINES

Three lines are said to be concurrent of they meet a point (*i.e.*, they pass through a common point). To find the condition that three straight lines may meet in a single point.

The given straight lines are

$$a_1x + b_1y + c_1 = 0 \quad \ldots(i)$$

$$a_2x + b_2y + c_2 = 0 \quad \ldots(ii)$$

$$a_3x + b_3y + c_3 = 0 \quad \ldots(iii)$$

Solving Eq. (*i*) and (*ii*) by cross-multiplication, we get

$$\frac{x}{b_1c_2 - b_2c_1} = \frac{y}{c_1a_2 - c_2a_1} = \frac{1}{a_1b_2 - a_2b_1}$$

$$\therefore \quad x = \frac{b_1c_2 - b_2c_1}{a_1b_2 - a_2b_1} \quad \text{and} \quad y = \frac{c_1a_2 - c_2a_1}{a_1b_2 - a_2b_1}$$

where $a_1b_2 - a_2b_1 \neq 0$.

$\therefore$ The coordinates of the point of intersection of (*i*) and (*ii*) are:

$$\left(\frac{b_1c_2 - b_2c_1}{a_1b_2 - a_2b_1}, \frac{c_1a_2 - c_2a_1}{a_1b_2 - a_2b_1}\right), \text{ where } a_1b_2 - a_2b_1 \neq 0$$

Substituting these in Eq. (*iii*), we get

$$a_3\left(\frac{b_1c_2 - b_2c_1}{a_1b_2 - a_2b_1}\right) + b_3\left(\frac{c_1a_2 - c_2a_1}{a_1b_2 - a_2b_1}\right) + c_3 = 0$$

$$\Rightarrow \quad a_3(b_1c_2 - b_2c_1) + b_3(c_1a_2 - c_2a_1) + c_3(a_1b_2 - a_2b_1) = 0$$

which is required solution.

SOLVED EXAMPLES

Example 4.16. Find the points of intersection of the following lines:

(*i*) $x - 4y - 3 = 0$; $6x - y - 11 = 0$ **(*ii*) $x - 3y + 4 = 0$; $7x + 4y + 8 = 0$**

Solution. (*i*) The given straight lines are

$$x - 4y - 3 = 0 \quad \ldots(i)$$

$$6x - y - 11 = 0 \quad \ldots(ii)$$

Solving by cross-multiplication, we have

$$\frac{x}{44-3} = \frac{y}{-18+11} = \frac{1}{-1+24}$$

$$\frac{x}{41} = \frac{y}{-7} = \frac{1}{23} \quad \Rightarrow \quad x = \frac{41}{23}, y = \frac{-7}{23}$$

Hence, point of intersection are $\left(\frac{41}{23}, \frac{-7}{23}\right)$.

(*ii*) The given straight lines are

$$x - 3y + 4 = 0 \quad \ldots(i)$$

$$7x + 4y + 8 = 0 \quad \ldots(ii)$$

Solving by cross-multiplication; we get

$$\frac{x}{-24-16} = \frac{y}{28-8} = \frac{1}{4+21}$$

$$\frac{x}{-40} = \frac{y}{20} = \frac{1}{25}$$

$$\Rightarrow \quad x = \frac{-40}{25} = \frac{-8}{5} \quad \text{and} \quad y = \frac{20}{25} = \frac{4}{5}$$

Hence, point of intersection are $\left(\frac{-8}{5}, \frac{4}{5}\right)$.

Example 4.17. Find the coordinates of the foot of perpendicular from a point (3, 13) on the line $2x + 3y - 32 = 0$.

Solution. The given line is $2x + 3y - 32 = 0 \quad \ldots(i)$

Equation of any line ⊥ to the given line is

$$3x - 2y + K = 0 \quad \ldots(ii)$$

Since it passes through (3, 13)

$$\therefore \quad 3(3) - 2(13) + K = 0$$

$$\Rightarrow \quad 9 - 26 + K = 0$$

$$\Rightarrow \quad K = 17$$

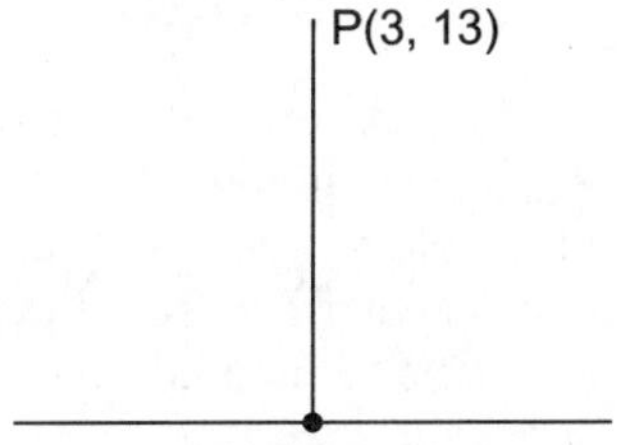

Fig. 4.21

Put K in Eq. (ii), we get

$$3x - 2y + 17 = 0 \quad \ldots(iii)$$

Solving (i) and (iii), we get

$$\frac{x}{64-51} = \frac{y}{34+96} = \frac{1}{9+4}$$

$$\frac{x}{13} = \frac{y}{130} = \frac{1}{13}$$

$$\Rightarrow \quad \frac{x}{13} = \frac{1}{13} \text{ or } \frac{y}{130} = \frac{1}{13}$$

$$\Rightarrow \quad x = 1 \text{ or } y = 10$$

∴ Hence required foot of the perpendicular is (1, 10).

Example 4.18. Find the image of the point (3, 8) in the line $x + 3y - 7 = 0$.

Solution. Let D be point (3, 8) and AB be line.

$$x + 3y - 7 = 0 \quad \ldots(i)$$

Let D' be image of the point D (3, 8) in the line (i)

Equation of any line ⊥ to (i) is

$$3x - y + K = 0 \quad \ldots(ii)$$

Since it passes through (3, 8)

$$\therefore \quad 9 - 8 + K = 0$$

$$K = -1$$

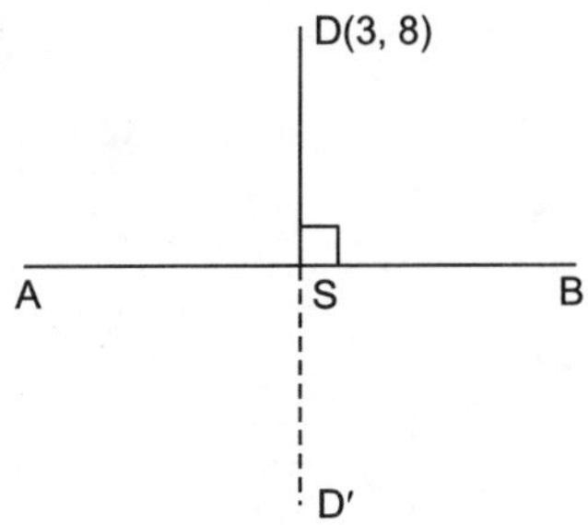

Fig. 4.22

From (ii), equation of DD' is

$$3x - y - 1 = 0 \quad \ldots(iii)$$

Solving Eqs. (i) and (iii), we get

$$\frac{x}{7+3} = \frac{y}{21-1} = \frac{1}{9+1}$$

$$\therefore \quad \frac{x}{10} = \frac{y}{20} = \frac{1}{10}$$

$$\frac{x}{10} = \frac{1}{10} \text{ or } \frac{y}{20} = \frac{1}{10}$$

∴ Two lines AB and DD' intersect at point S *i.e.*, (1, 2)

S is the mid point of DD'

∴ Let (α, β) be coordinates of D'

$$\therefore \quad \frac{3+\alpha}{2} = 1 \text{ or } \frac{\beta+8}{2} = 2$$

$$\alpha = -1 \text{ or } \beta = -4$$

∴ Coordinates of D' (*i.e.*, image of point D(3, 8) are (– 1, – 4)).

4.12. DISTANCE OF A POINT FROM A LINE

The distance of a point from a line is the length of the perpendicular drawn from the point to the line. Let $L: Ax + By + C = 0$ be a line, whose distance from the point $P(x_1, y_1)$ is d. Draw a perpendicular PM from the point P to the line (Fig. 4.23). If the line meets the x and y-axis at the point Q and R, respectively; then coordinates of the points are $Q\left(\frac{-C}{A}, 0\right)$ and $R\left(0, \frac{-C}{B}\right)$. Thus, the area of the triangle PQR is given by

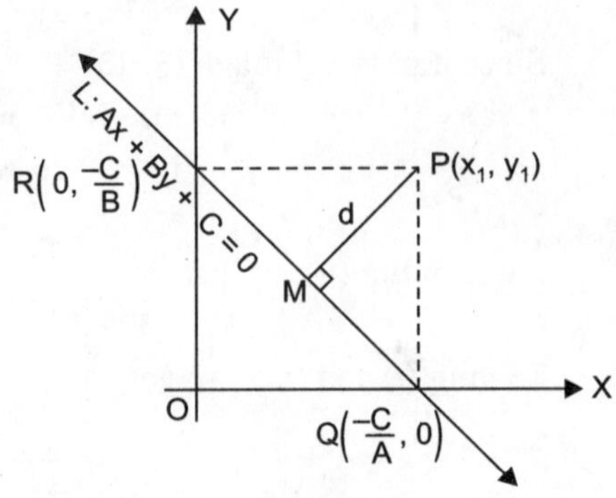

Fig. 4.23

$$\text{area } (\Delta PQR) = \frac{1}{2} PM \cdot QR$$

or
$$PM = \frac{2 \text{ area } (\Delta PQR)}{QR} \quad \ldots(i)$$

Also,
$$\text{area } (\Delta PQR) = \frac{1}{2}\left| x_1\left(0 + \frac{C}{B}\right) + \left(-\frac{C}{A}\right)\left(-\frac{C}{B} - y_1\right) + 0(y_1 - 0)\right|$$

$$= \frac{1}{2}\left| x_1 \frac{C}{B} + y_1 \frac{C}{A} + \frac{C^2}{AB}\right|$$

or
$$2 \text{ area } (\Delta PQR) = \left|\frac{C}{AB}\right| \cdot | Ax_1 + By_1 + C |$$

and
$$QR = \sqrt{\left(0 + \frac{C}{A}\right)^2 + \left(\frac{C}{B} - 0\right)^2} = \left|\frac{C}{AB}\right| \sqrt{A^2 + B^2}$$

Substituting the values of area (ΔPQR) and QR in (i), we get

$$PM = \frac{| Ax_1 + By_1 + C |}{\sqrt{A^2 + B^2}}$$

Thus, the perpendicular distance (d) of a line $Ax + By + C$ from a point (x_1, y_1) is given by

$$\boxed{d = \frac{| Ax_1 + By_1 + C |}{\sqrt{A^2 + B^2}}}$$

4.13. DISTANCE BETWEEN TWO PARALLEL LINES

Two parallel lines can be taken in the form

$$y = mx + c_1 \quad \ldots(i)$$

(∵ Slopes of two parallel lines are equal)

and
$$y = mx + c_2 \quad \ldots(ii)$$

Thus, the distance d between two parallel $y = mx + c_1$ and $y = mx + c_2$ is given by

$$d = \frac{|c_1 - c_2|}{\sqrt{1+m^2}} \quad \text{or} \quad d = \frac{|c_1 - c_2|}{\sqrt{A^2 + B^2}}$$

($\because$ A and B will be same for parallel lines)

SOLVED EXAMPLES

Example 4.19. Find the distance of the point (– 1, 1) from the line 12(x + 6) = 5(y – 2).

Solution. The given equation of the line is $12(x + 6) = 5(y - 2)$

$$\Rightarrow \quad 12x - 5y + 82 = 0 \quad \ldots(i)$$

Comparing (*i*) with general equation of line $Ax + By + C = 0$, we obtain $A = 12$, $B = -5$ and $C = 82$

The given point is $(x_1, y_1) = (-1, 1)$.

$\therefore$ The distance of point (– 1, 1) from the given line

$$= \frac{|12(-1) + (-5)(1) + 82|}{\sqrt{(12)^2 + (-5)^2}} = \frac{|-12 - 5 + 82|}{\sqrt{169}} = \frac{|65|}{13} = 5 \text{ units}$$

Example 4.20. Find the distance between the parallel lines $15x + 8y - 34 = 0$ and $15x + 8y + 31 = 0$.

Solution. The given parallel lines are $15x + 8y - 34 = 0$ and $15x + 8y + 31 = 0$.

Here, $A = 15$, $B = 8$, $C = -34$ and $C_2 = 31$

$\therefore$ The distance between the parallel lines is

$$d = \frac{|C_1 - C_2|}{\sqrt{A^2 + B^2}} = \frac{|-34 - 31|}{\sqrt{(15)^2 + (8)^2}}$$

$$= \frac{|-65|}{17} = \frac{65}{17} \text{ units}$$

EXERCISE 4.7

1. Find the distance of the point (3, – 5) from the line $3x - 4y - 26 = 0$. **[Ans.** 3/5]
2. Find the distance between the parallel lines $3x - 4y + 7 = 0$ and $3x - 4y + 5 = 0$. **[Ans.** 2/5]
3. Show that the path of a moving point such that its distances from two lines $3x - 2y = 5$ and $3x + 2y = 5$ are equal is a straight line.
4. If p and q are the lengths of perpendiculars from the origin to the lines $x \cos \theta - y \sin \theta = k \cos 2\theta$ and $x \sec \theta + y \operatorname{cosec} \theta = k$, respectively, prove that

$$p^2 + 4q^2 = k^2$$

4.14. THE CIRCLE

Definition: A circle is the locus of a point, which moves in a plane such that it remains at a constant distance from a fixed point.

- The fixed point is called the **centre** of circle.
- The distance from the centre to a point on circle is called **radius** of that circle. This is shown in Fig. 4.24.

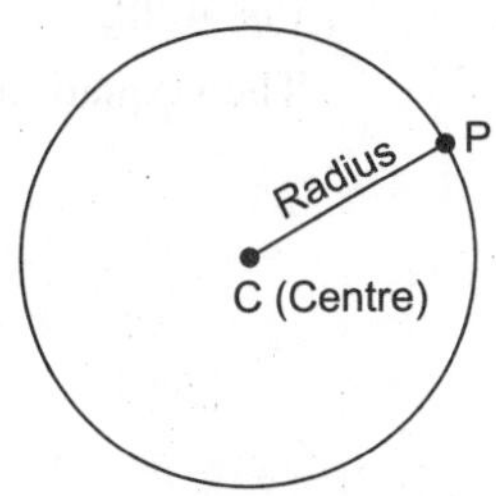

Fig. 4.24

4.14.1. Standard Form of the Equation of a Circle

To find the equation of a circle whose centre is the point C(*h*, *k*) and whose radius is *r*. Let P(*x*, *y*) be any point on the circle as shown in Fig. 4.25.

$\therefore \quad |CP| = r = \text{radius}$

$$\sqrt{(x-h)^2+(y-k)^2} = r$$

Squaring on both sides

$$(x-h)^2 + (y-k)^2 = r^2 \qquad ...(i)$$

So, this is required equation of a circle whose centre is *C*(*h*, *k*) and radius be '*r*'.

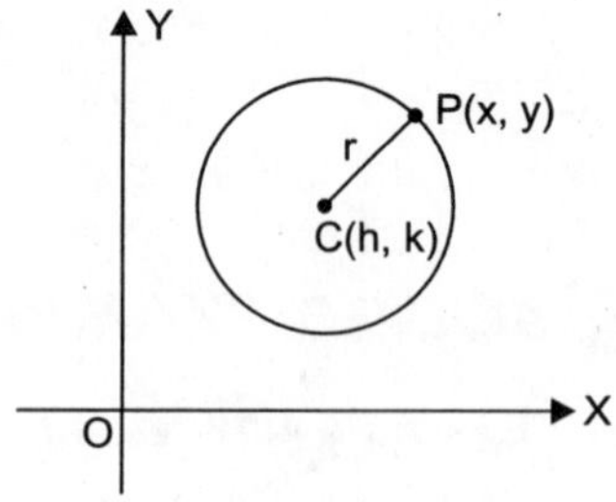

Fig. 4.25

Special Cases:

1. If centre of the circle is at origin, then $h = 0$, $k = 0$.
   Put these value in Eq. (*i*)
   $$(x-0)^2 + (y-0)^2 = r^2$$
   $$x^2 + y^2 = r^2$$
2. (*a*) If a circle passes through the origin and its centre lies on *x*-axis, then its equation is of the form according to Fig. 4.26
   $$(x-l)^2 + (y-0)^2 = 12$$
   [$\because$ On *x*-axis, *y*-coordinate is always zero]

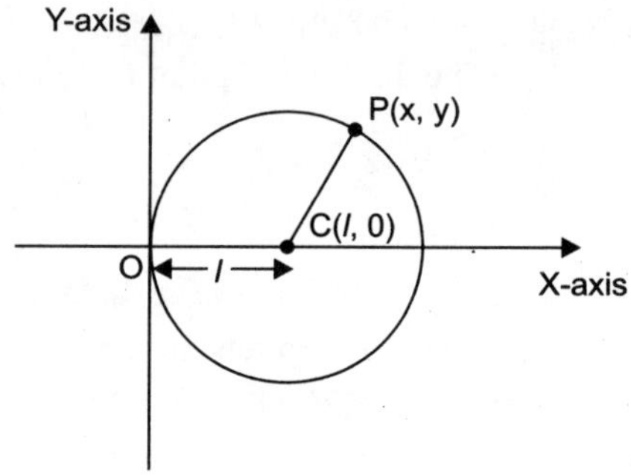

Fig. 4.26

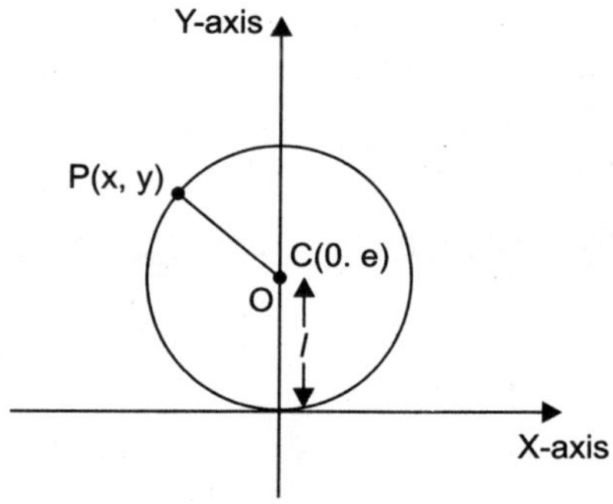

Fig. 4.27

   (*b*) If circle passes through origin and its centre lies on *y*-axis, then equation is of the form according to Fig. 4.27
   $$(x-0)^2 + y - l)^2 = l^2$$
   [$\because$ on *y*-axis, *x*-coordinate is always zero]

3. If a circle touches *x*-axis in first quadrant as shown in Fig. 4.28.
   Then equation is of the form
   $$(x-h)^2 + (y-k)^2 = k^2 \quad \text{where } h, k > 0$$

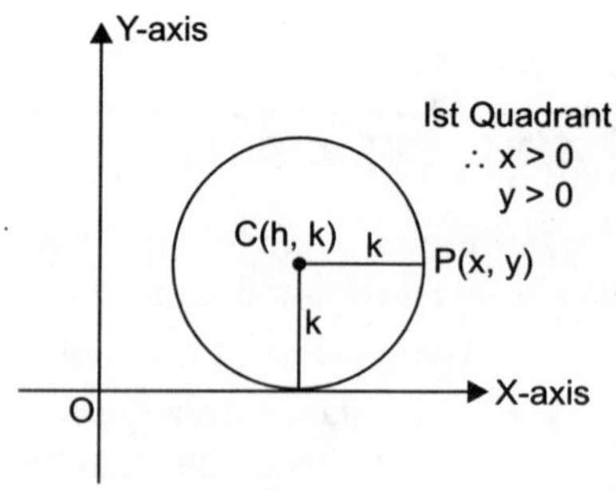

Fig. 4.28

4. If a circle touches y-axis in first quadrant as shown in Fig. 4.29 then equation is of the form

$$(x-h)^2+(y-k)^2=h^2 \quad \text{where } h, k>0$$

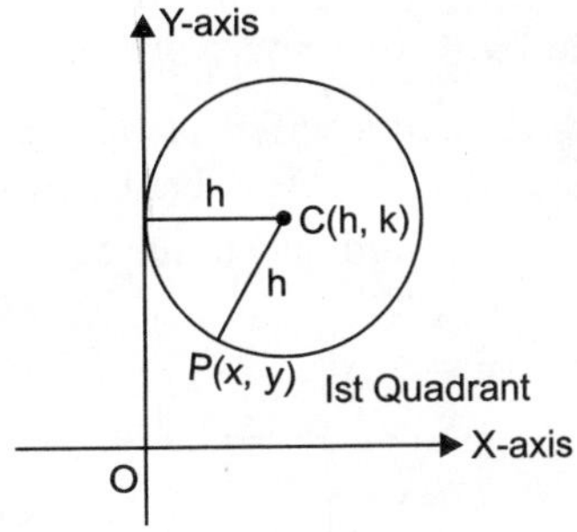

Fig. 4.29

5. If a circle touches both the axis in 1st quadrant as shown in Fig. 4.30.
 The equation is of the form:

$$(x-l)^2+(y-l)^2=l^2 \quad \text{where } l>0$$

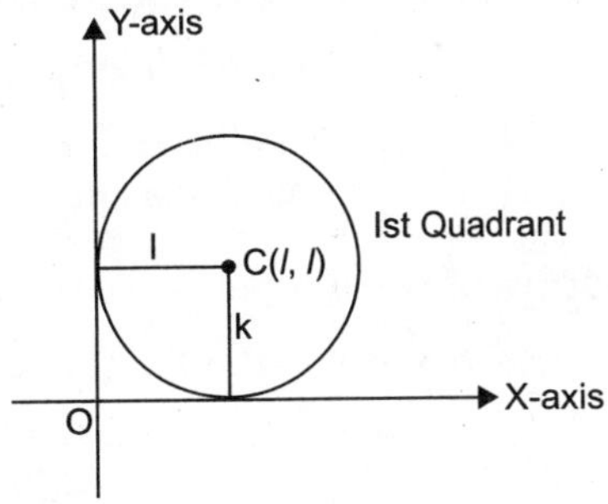

Fig. 4.30

To Obtain the Equation of a Circle in the Form

$$x^2+y^2+2gx+2fy+c=0$$

As we know, the equation of circle with centre (h, k) and radius 'r' is

$$(x-h)^2+(y-k)^2=r^2$$

$$\Rightarrow \quad x^2+h^2-2hx+y^2+k^2-2yk=r^2 \qquad [\because \ (a-b)^2=a^2+b^2-2ab]$$

$$\Rightarrow x^2+y^2-2hx-2ky+(h^2+k^2-r^2)=0$$

This can be written in the form

$$x^2+y^2+2gx+2fy+c=0 \qquad \text{where } h=-g, k=-f \text{ and } h^2+k^2-r^2=c$$

General Form

To prove the equation $x^2+y^2+2gx+2fy+c=0$ represents a circle and find its centre and radius. The given equation is

$$\Rightarrow \quad x^2+y^2+2gx+2fy+c=0 \qquad \ldots(ii)$$

$$\Rightarrow \quad (x^2+2gx)+(y^2+2fy)=-c$$

Adding g^2+f^2 on both sides

$$\Rightarrow (x^2+2gx)+(y^2+2fy)+(g^2+f^2)=g^2+f^2-c$$

$$\Rightarrow (x^2+g^2+2gx)+(y^2+f^2+2fy)=g^2+f^2-c$$

$$\Rightarrow (x+g)^2+(y+f)^2=g^2+f^2-c$$

$$\Rightarrow [x-(-g)]^2+[y-(-f)]^2=(\sqrt{g^2+f^2-c})^2$$

Comparing it with central form

$$(x-h)^2+(y-k)^2=r^2$$

$$\therefore \qquad h=-g, \quad k=-f, \quad r=\sqrt{g^2+f^2-c}$$

So, equation (ii) represents a circle whose centre is $(-g, -f)$ and radius is $\sqrt{g^2+f^2-c}$, $(g^2+f^2)\geq c$.

Rules to Find the Centre and Radius of a Circle

(*i*) Make the coefficient of x^3 and y^2 each equal to 1 by dividing throughout (if necessary) by coefficient of $x^2 + y^2$ [R.H.S. being zero]

(*ii*) The co-ordinate of the centre are

$$\left(-\frac{1}{2}\text{ coefficient of } x, -\frac{1}{2}\text{ coefficient of } y\right)$$

(*iii*) Radius $= \sqrt{\left(\frac{-1}{2}\text{ coefficient of } x\right)^2 + \left(-\frac{1}{2}\text{ coefficient of } y\right)^2 - (\text{constant term})}$

Characteristics of the Equation of a Circle

(*i*) It must be a second degree equation in x and y.

(*ii*) The coefficients of x^2 and y^2 must be equal (*i.e.,* 1).

(*iii*) There must not contain term involving the xy product.

Nature of the Circle

1. If $g^2 + f^2 - c > 0$ *i.e.,* $c < g^2 + f^2$, the radius of circle is real and hence **circle is real.**
2. If $g^2 + f^2 - c = 0$, the radius of the circle is zero and it is called **point circle.**
3. If $g^2 + f^2 - c < 0$, the radius of the circle is imaginary but its centre is real. It is called **imaginary circle** or **virtual circle.**

Circle under Given Conditions

Equation of circle through given three point.

Let the required equation of a circle is

$$x^2 + y^2 + 2gx + 2fy + c = 0$$

Suppose it passes through three points (x_1, y_1), (x_2, y_2), (x_3, y_3)

∴ These three points will satisfy the required equation.

∴ $$x_1^2 + x_2^2 + 2gx_1 + 2fy + c = 0 \quad ...(i)$$

$$x_2^2 + y_2^2 + 2gx_2 + 2fy_2 + c = 0 \quad ...(ii)$$

$$x_3^2 + y_3^2 + 2gx_3 + 2fy_3 + c = 0 \quad ...(iii)$$

Now, we will solve these three equations and get the values of g, f and c. After putting values of g, f and c in (*i*), we will get required equation of circle.

Diameter Form

To find the equation of circle on the join of (x_1, y_1) and (x_2, y_2) as diameter.

Let $A(x_1, y_1)$ and $B(x_2, y_2)$ be the point of diameter AB and $P(x, y)$ be any point on the circle. Join AP and BP.

∴ $$\angle APB = 90°$$

Now, $$m_1 = \text{Slope of } PA = \frac{\text{Different of ordinates}}{\text{Difference of absissae}} = \frac{y - y_1}{x - x_1}$$

and $$m_2 = \text{The slope of } PB = \frac{\text{Different of ordinates}}{\text{Difference of absissae}} = \frac{y - y_2}{x - x_2}$$

$\therefore \quad m_1 m_2 = -1$

gives $\left(\frac{y-y_1}{x-x_1}\right)\left(\frac{y-y_2}{x-x_2}\right) = -1$

$\Rightarrow \quad (y-y_1)(y-y_2) = -(x-x_1)(x-x_2)$

$\Rightarrow \quad (x-x_1)(x-x_2) + (y-y_1)(y-y_2) = 0$ which is the required equation of circle.

SOLVED EXAMPLES

Example 4.21. Find the equation of a circle whose circle is at (0, 2) and radius = 2.

Solution. Let the required equation of circle be

$$(x-h)^2 + (y-k)^2 = r^2 \qquad ...(i)$$

Given, centre = $(h, k) = (0, 2)$

r = radius = 2

$\therefore$ From Eq. (*i*) $(x-0)^2 + (y-2)^2 = (2)^2$

$\Rightarrow \quad x^2 + y^2 + 4 - 4y = 4$

$\Rightarrow \quad x^2 + y^2 - 4y = 0$ be the required equation.

Example 4.22. Find centre and radius of following circles

(*i*) $x^2 + y^2 - 12x + 6y + 45 = 0$ **(*ii*) $(x+5)^2 + (y-3)^2 = 36$**

(*iii*) $2x^2 + 2y^2 - 6x + 8y - 4 = 0$ **(*iv*) $4(x^2 + y^2) - 10x + 5y + \frac{5}{4} = 0$.**

Solution. (*i*) The equation of a given circle is

$$x^2 + y^2 - 12x + 6y + 45 = 0$$

Comparing with $x^2 + y^2 + 2gx + 2fy + c = 0$

$\therefore \quad 2g = -12 \Rightarrow g = -6$

$2f = 6 \Rightarrow f = 3$

$c = 45$

$\therefore$ Co-ordinate of centre are $= (-g, -f) = (6, -3)$

and Radius of circle $= \sqrt{g^2 + f^2 - c} = \sqrt{(6)^2 + (-3)^2 - 45} = \sqrt{36 + 9 - 45} = 0$ units

(*ii*) The equation of given circle is $(x+5)^2 + (y-3)^2 = 36 = (6)^2$

Comparing it with $(x-h)^2 + (y-k)^2 = r^2$

$\therefore$ Centre of circle is $= (h, k) = (-5, 3)$

$\therefore$ Radius = 6 units

(*iii*) The equation of given circle is

$$2x^2 + 2y^2 - 6x + 8y - 4 = 0 \qquad ...(i)$$

[$\because$ As we know, to make the equation of circle, coefficient of x^2 and y^2 must be unity]

$\therefore$ Dividing Eq. (*i*) by 2

$\therefore \quad x^2 + y^2 - 3x + 4y - 2 = 0$

Now comparing it with equation

$$x^2 + y^2 + 2gx + 2fy + c = 0$$

$\therefore \quad 2g = -3 \Rightarrow g = \frac{-3}{2}$ and $2f = 4$

$\Rightarrow \quad f = 2, \quad c = -2$

$\therefore$ Coordinate of centre $= (-g, -f) = \left[-\left(\frac{-3}{2}\right), -2\right] = \left(\frac{3}{2}, -2\right)$

$$\text{Radius} = \sqrt{g^2 + f^2 - c} = \sqrt{\left(\frac{3}{2}\right)^2 + (-2)^2 - (-2)}$$

$$= \sqrt{\frac{9}{4} + \frac{4}{1} + 2} = \sqrt{\frac{9 + 16 + 8}{4}} = \frac{\sqrt{33}}{2} \text{ units}$$

(*iv*) The equation of given circle is $4(x^2 + y^2) - 10x + 5y + \frac{5}{4} = 0$ ...(i)

Dividing Eq. (*i*) by 4 $x^2 + y^2 - \frac{10}{4}x + \frac{5}{4}y + \frac{5}{4} = 0$

Comparing it with

$$x^2 + y^2 + 2gx + 2fy + c = 0$$

$\therefore$ $2g = \frac{-10}{4} \Rightarrow g = \frac{-10}{4 \times 2} = \frac{-5}{4}$

$\therefore$ $2f = \frac{5}{4} \Rightarrow f = \frac{5}{8}$

$$c = \frac{5}{4}$$

$\therefore$ Cenrte $= (-g, -f) = \left(\frac{5}{4}, \frac{-5}{8}\right)$

$$\text{Radius} = \sqrt{g^2 + f^2 - c} = \sqrt{\left(\frac{5}{4}\right)^2 + \left(\frac{-5}{8}\right)^2 - \frac{5}{4}}$$

$$= \sqrt{\frac{25}{16} + \frac{25}{64} - \frac{5}{4}} = \sqrt{\frac{45}{64}} = \frac{3\sqrt{5}}{8} \text{ unit.}$$

Example 4.23. Find the equation of the circle whose centre is (2, –1) and passes through (3, 6).

Solution. Given $C(h, k) = (2, -1)$ and $P(x, y) = (3, 6)$

$\therefore$ Radius $= |CP|$

$$= \sqrt{(3-2)^2 + (6+1)^2}$$

$$= \sqrt{1 + 49} = \sqrt{50} = 5\sqrt{2}$$

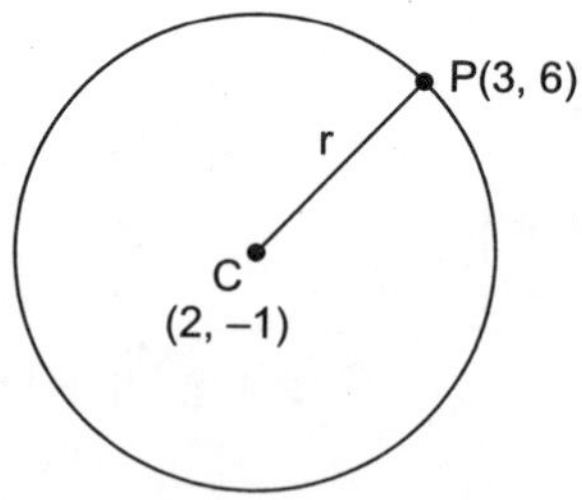

Fig. 4.31

Let the equation of circle be

$$(x - h)^2 + (y - k)^2 = r^2$$

Now put $C(h, k) = C(2, -1)$ and $r = 5\sqrt{2}$ in Eq. (*i*)

$$(x - 2)^2 + (y + 1)^2 = (5\sqrt{2})^2$$

$\Rightarrow x^2 + 4 - 4x + y^2 + 1 + 2y = 50$

$\Rightarrow x^2 + y^2 - 4x + 2y - 45 = 0$ be required equation of circle.

Example 4.24. Find the equation of a circle whose centre is (4, 5) and which passes through centre of the circle, $x^2 + y^2 + 4x - 6y = 12$.

Solution. Equation of given circle is

$$x^2 + y^2 + 4x - 6y - 12 = 0$$

Comparing it with

$$x^2 + y^2 + 2gx + 2fy + c = 0$$

$\therefore$ Centre = $(-g, -f) = (-2, 3)$

$\therefore$ Equation of required circle whose centre is (4, 5) and passes through (–2, 3) is

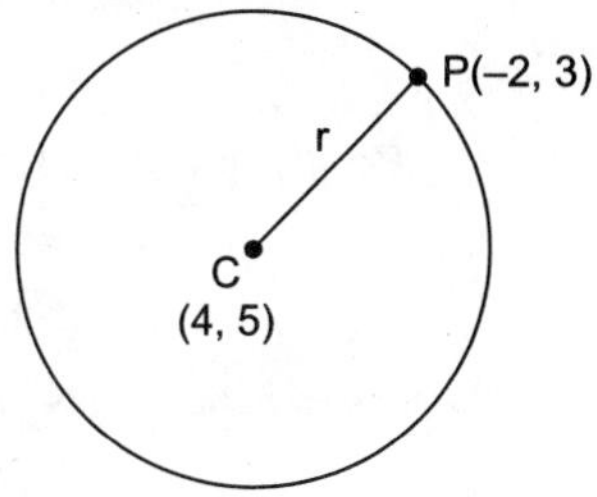

Fig. 4.32

$\because$ $r = |CP|$

$$= \sqrt{(4+2)^2 + (5-3)^2} = \sqrt{36+4} = \sqrt{40} = 2\sqrt{10} \text{ units}$$

$$(x-4)^2 + (y-5)^2 = (2\sqrt{10})^2$$

$\Rightarrow$ $x^2 + 16 - 8x + y^2 + 25 - 10y = 40$

$\Rightarrow$ $x^2 + y^2 - 8x - 10y + 1 = 0$ be required equation of circle.

Example 4.25. Find the equation of a circle passing through the points (5, 7), (6, 6) and (2, –2). Also find its centre and radius.

Solution. Let equation of required circle be

$$x^2 + y^2 + 2gx + 2fy + c = 0 \quad ...(i)$$

Since 1 passes through (5, 7). So, it satisfy Eq. (*i*)

$$25 + 49\ 10g + 14f + c = 0 \Rightarrow 74 + 10g + 14f + c = 0 \quad ...(ii)$$

Equation (*i*) passes through (6, 6)

$\therefore$ $36 + 36 + 12g + 12f + c = 0 \Rightarrow 72 + 15g + 12f + c = 0 \quad ...(iii)$

Also Eq. (*i*) passes through (2, –2)

$\therefore$ $4 + 4 + 4g - 4f + c = 0 \Rightarrow 8 + 4g - 4f + c = 0 \quad ...(iv)$

$(iii) - (ii) \Rightarrow -2 + 2g - 2f = 0 \Rightarrow g - f = 1$

$(iv) - (iii) \Rightarrow -64 - 8g - 16f = 0 \Rightarrow g + 2f + 8 = 0 \quad ...(v)$

$(v) - (vi) \Rightarrow -3f - 9 = 0 \Rightarrow f = 3 \quad ...(vi)$

Put the value of f = –3 in Eq. (*v*)

$$g + 3 = 1 \Rightarrow g = -2$$

Putting g = –2, f = –3 in Eq. (*v*)

$\Rightarrow$ $8 - 8 + 12 + c = 0 \Rightarrow c = -12$

Putting value of g, f, c in Eq. (*i*)

$$x^2 + y^2 - 4x - 6y - 12 = 0$$

Comparing it with

$$x^2 + y^2 + 2gx + 2fy + c = 0$$

$\therefore$ $2g = -4 \Rightarrow g = -2$

$2f = -6 \Rightarrow f = -3 \quad \therefore \quad c = -12$

Centre = $(-g, -f) = (2, 3)$

Radius = $\sqrt{g^2 + f^2 - c} = \sqrt{4+9+12} = \sqrt{25} = 5$ units

Example 4.26. Find the equation of circle which passes through the points (3, 7), (5, 5) and whose centre lies on the line $x - 4y - 1 = 0$.

Solution. Let the equation of required circle be

$$x^2 + y^2 + 2gx + 2fy + c = 0 \quad ...(i)$$

$\because$ Eq. (*i*) passes through (5, 5), so it satisfy the Eq. (*i*)

$\therefore$ $$25 + 25 + 10g + 10f + c = 0$$

$\Rightarrow$ $$50 + 10g + 10f + c = 0 \quad ...(ii)$$

$\therefore$ Eq. (*ii*) passes through (3, 7)

$\therefore$ $$9 + 49 + 6g + 14f + c = 0$$

$\Rightarrow$ $$58 + 6g + 14f + c = 0 \quad ...(iii)$$

Note: Now centre $(-g, -f)$ lies on line $x - 4y - 1 = 0$

$\therefore$ $$-g + 4f - 1 = 0$$

(*ii*) – (*iii*) $\Rightarrow$ $$4g - 4f - 8 = 0 \quad ...(iv)$$

$$g - f - 2 = 0$$

(*iv*) + (*v*) $\Rightarrow$ $$3f - 3 = 0 \quad \Rightarrow \quad f = 1 \quad ...(v)$$

Put $f = 1$ in Eq. (*v*) $$g - 1 - 2 = 0$$

$$g = 3$$

Put $f = 1$, $g = 3$ in Eq. (*ii*)

$$50 + 10(3) + 10(1) + c = 0 \quad \Rightarrow \quad c = -90$$

Putting value of g, f and c in Eq. (*i*)

$x^2 + y^2 + 6x + 2y - 90 = 0$ is required equation of circle.

Example 4.27. Prove that the points (1, – 6), (5, 2), (7, 0) and (– 1, – 4) are concylic. Find radius of circle.

Solution. Let equation of circle be

$$x^2 + y^2 + 2gx + 2fy + c = 0 \quad ...(i)$$

Since Eq. (*i*) passes through the point (1, – 6)

$\therefore$ $$1 + 36 + 2g - 12f + c = 0$$

$\Rightarrow$ $$37 + 2g - 12f + c = 0 \quad ...(ii)$$

Since Eq. (*i*) passes through the point (5, 2)

$\therefore$ $$25 + 4 + 10g + 4f + c = 0$$

$\Rightarrow$ $$29 + 10g + 4f + c = 0 \quad ...(iii)$$

Since Eq. (*i*) passes through (7, 0)

$\therefore$ $$49 + 0 + 14g + 0 + c = 0$$

$\Rightarrow$ $$49 + 14g + c = 0 \quad ...(iv)$$

Eqs. (*iii*) – (*ii*) $\Rightarrow$ $$-8 + 8g + 16f = 0$$

$$-1 + g + 2f = 0 \quad ...(v)$$

(*iv*) – (*iii*) $\Rightarrow$ $$20 + 4g - 4f = 0$$

$$5 + g - f = 0 \quad ...(vi)$$

(*vi*) – (*v*) $\Rightarrow$ $$6 - 3f = 0$$

$$3f = 6 \quad \Rightarrow \quad f = 2$$

Put $f = 2$ in Eq. (*vi*) $$5 + g - 2 = 0$$

$$g = -3$$

Put value of g, f in Eq. (*iii*)

$$29 + 10(-3) + 4(2) + c = 0$$

$\Rightarrow$ $$29 - 30 + 8 + c = 0$$

$\Rightarrow$ $$c = -7$$

Put value of g, f, c in Eq. (*i*)

$$x^2 + y^2 + 2(-3)x + 2(2)y - 7 = 0$$

$$x^2 + y^2 - 6x + 4y - 7 = 0$$

Putting the coordinates of fourth point (– 1, – 4) in Eq. (*vii*)

$\Rightarrow$ $(-1)^2 + (-4)^2 - 6(-1) + 4(-4) - 7 = 0$

$\Rightarrow$ $1 + 16 + 6 - 16 - 7 = 0$

$\Rightarrow$ $0 = 0$ which is true.

Hence, given four points are concyclic

$\therefore$ Radius $= \sqrt{g^2 + f^2 - c} = \sqrt{9 + 4 + 7} = \sqrt{20} = 2\sqrt{5}$ unit

EXERCISE 4.8

1. Find equation of following circle is:

   (*i*) Centre (1, 1) and radius $\sqrt{2}$ **[Ans.** $x^2 + y^2 - 2x - 2y = 0$]

   (*ii*) Centre (– 3, 2) and radius 4 **[Ans.** $x^2 + y^2 + 6x - 4y - 3 = 0$]

   (*iii*) Centre $\left(\frac{1}{3}, \frac{1}{4}\right)$ and radius $\frac{1}{5}$ $\left[\textbf{Ans. } x^2 + y^2 - \frac{2}{3}x - \frac{y}{2} + \frac{481}{3,600} = 0\right]$

   (*iv*) Centre (a, a) and radius $\sqrt{2}a$ **[Ans.** $x^2 + y^2 - 2ax - 2ay = 0$]

2. Find centre and radius of following circles:

   (*i*) $(x - 3)^2 + y^2 = 4$ **[Ans.** Centre = (3, 0), Radius = 2]

   (*ii*) $x^2 + y^2 + 8x + 10y - 8 = 0$ **[Ans.** (– 4, – 5), 7]

   (*iii*) $2x^2 + 2y^2 - x = 0$ $\left[\textbf{Ans.} \left(\frac{1}{4}, 0\right), \frac{1}{4}\right]$

   (*iv*) $3x^2 + 3y^2 + 6x - 4y - 1 = 0$ $\left[\textbf{Ans.} \left(-1, \frac{2}{3}\right), \frac{4}{3}\right]$

3. Find equation of a circle having centre (2, 2) and passes through point (4, 5).
   [Ans. $x^2 + y^2 - 4x - 4y - 5 = 0$]

4. Find equation of a circle with radius 5 whose centre lies in x-axis and passes through point (2, 3).
   [Ans. $x^2 + y^2 + 4x - 21$]

5. Find equation of circle having centre (3, – 2) and which passes through centre of circle $4x^2 + 4y^2 + 8x + 2y + 1 = 0$. **[Ans.** $16(x^2 + y^2) - 96x + 64y - 97 = 0$]

6. Find equation of circle cocentric (same centre) with circle $x^2 + y^2 + 4x + 6y + 11 = 0$ and passing through the point (5, 4). **[Ans.** $x^2 + y^2 + 4x + 6y - 85 = 0$]

7. (*i*) Find the equation of circle, passes through points (2, – 3) and (3, – 2) and has its centre on line $2x - 3y = 8$.
   [Ans. $5x^2 + 5y^2 - 16x + 16y + 15 = 0$]

   (*ii*) Find equation of circle, which passes through the points (2, 3) and (– 1, 1) and has its centre on line $x - 3y - 11 = 0$. **[Ans.** $x^2 + y^2 + 4x - 21 = 0; x^2 + y^2 - 12x + 11 = 0$]

8. (*i*) Find equation of circle whose centre is (2, – 3) and which passes through the intersection of straight lines $3x + 2y = 11$ and $2x + 3y = y$. **[Ans.** $x^2 + y^2 - 4x + 6y + 3 = 0$]

   (*ii*) Find equation of the circle passing through the point (2, 4) and has its centre at the intersection of $x - y + 4 = 0$ and $2x + 3y + 7 = 0$. **[Ans.** $x^2 + y^2 - 2x + 6y + 3 = 0$]

9. Show that points (5, 5), (6, 4), (– 2, 4) and (7, 1) are concyclic, and find its equation centre and radius.

[**Ans.** $x^2 + y^2 - 4x - 2y - 20 = 0$, (2, 1), 5]

10. Find the equation of circle when end point of a diameter are:

(*i*) (3, 2) and (2, 5) [**Ans.** $x^2 + y^2 - 5x - 7y + 16 = 0$]

(*ii*) (– 1, 2) and (3, – 4) [**Ans.** $x^2 + y^2 - 2x + 2y - 11 = 0$]

(*iii*) (2, 3) and (– 1, – 3) [**Ans.** $x^2 + y^2 - x - 11 = 0$]

11. Find equation of circle which passes through origin and makes intercepts 4, and 2 on *x*- and *y*-axis respectively.

12. One end of diameter of circle $x^2 + y^2 - 3x + 5y - 4 = 0$ is (2, 1). Find the coordinates of other end.

$$\left[\textbf{Ans.} \left(\frac{-1}{2}, \frac{3}{2}\right)\right]$$

4.15. PARABOLA

Definition: A parabola is set of all points in a plane, equidistant from a fixed point and a fixed line in that plane in which:

(*i*) The fixed point is called **Focus**.

(*ii*) The fixed line is called **Directix**.

(*iii*) The line through the focus and perpendicular to the directix is called **Axis** of the parabola.

(*iv*) The point, where the parabola intersects the axis (*i.e.*, *x*-axis and *y*-axis) is called **vertex**.

4.16. EQUATION OF PARABOLA IN STANDARD FORM

∴ Equation of parabola is

$$y^2 = 4ax, (a > 0)$$

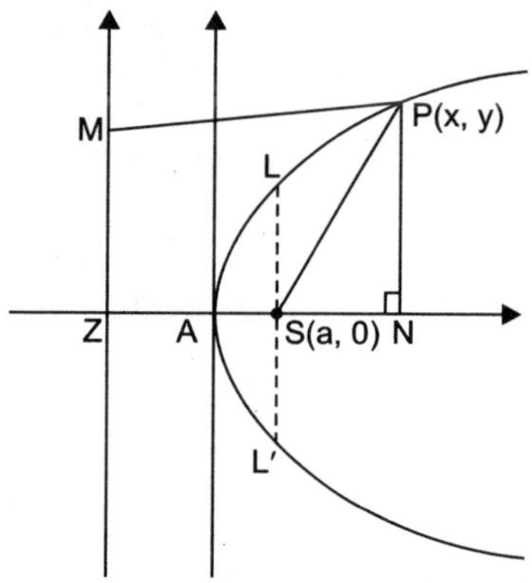

Fig. 4.33

Observations

(*i*) Co-ordinates of focus *S* are (*a*, 0)

(*ii*) Equation of directrix *ZM* $\Rightarrow x + a = 0$

(*iii*) Equation of axis is $y = 0$

(*iv*) Co-ordinates of vertex *A* are (0, 0)

Focal Distance

The distance of a point '*P*' on parabola from the focus (*S*) is called focal distance of the point.

From the Fig 4.33.

To find the focal distance of any point *P*(*x*, *y*) on parabola $y^2 = 4ax$ is

$$= SP = MP = ZN = ZA + AN = a + x$$

Focal distance of $P = x + a$

Latus Rectum

A final chord (chord passing through focus) perpendicular to axis of parabola is called *latus rectum of parabola*. In Fig. 12.9 *L′L* is latus-rectum.

∴ Equation of latus rectum is $x - a = 0$

SOLVED EXAMPLES

Example 4.28. Find vertex, focus directrix, latus rectum and axis of following parabola:

(i) $y^2 = 4x$ (ii) $y^2 = -8x$ (iii) $3x^2 = 8y$ (iv) $x^2 = -24y$

Solution. (i) The given parabola is

$$y^2 = 4x$$

Compare it $y^2 = 4ax$

$\therefore$ $4a = 4 \Rightarrow a = 1$

Hence, Vertex = (0, 0)

Focus = $(a, 0) = (1, 0)$

Equation of directix $\Rightarrow x = -a \Rightarrow x = -1 \Rightarrow x + 1 = 0$

Latus rectum = $4a = 4(1) = 4$ units

Equation of axis $y = 0$

(ii) The given parabola is $y^2 = -8x$

Compare it with $y^2 = -4ax$

$\therefore$ $-4a = -8 \Rightarrow a = 2$

Hence, Vertex = (0, 0)

Focus = $(-a, 0) = (-2, 0)$

Equation of directix $x = a \Rightarrow x = 2 \Rightarrow x - 2 = 0$

Latus rectum = $4a = 4(2) = 8$ units

Equation of axis $y = 0$

(iii) The given parabola is $3x^2 = 8y$

$\Rightarrow$ $x^2 = \frac{8}{3}y$

Compare it with $x^2 = 4ay$

$\therefore$ $4a = \frac{8}{3} \Rightarrow a = \frac{8}{3 \times 4} = \frac{2}{3}$

Hence, Vertex = (0, 0)

Focus = $(0, a) = \left(0, \frac{2}{3}\right)$

Equation of directrix $\Rightarrow y = -a \Rightarrow y = \frac{-2}{3} \Rightarrow 3y + 2 = 0$

Latus rectum = $4a \Rightarrow 4\left(\frac{2}{3}\right) = \frac{8}{3}$

Equation of axis $\Rightarrow x = 0$

(iv) The given parabola is $x^2 = -24y$

Compare it with $x^2 = -4ay$

$\therefore$ $-4a = -24 \Rightarrow a = 6$

Hence, Vertex = (0, 0)

Focus = $(0, -a) = (0, -6)$

Equation of directrix $\Rightarrow y = a \Rightarrow y - 6 = 0$

Latus rectum = $4a \Rightarrow 4 \times 6 = 24$ units

Equation of axis $\Rightarrow x = 0$

Example 4.29. Find the equation of parabola with vertex at (0, 0) and focus is at (0, 4).

Solution. Since vertex is (0, 0) and focus is at (0, 4) which is on y-axis *i.e.*, symmetric about y-axis on upward direction.

$\therefore$ Equation of the parabola is of the form

$$x^2 = 4ay$$

$$\Rightarrow \quad x^2 = 4(4)y \quad \Rightarrow \quad x^2 = 16y$$

is required equation of parabola.

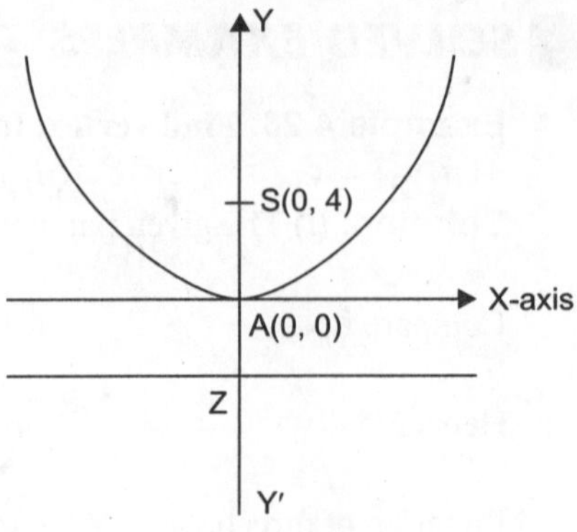

Fig. 4.34

Example 4.30. Find the equation of parabola which has vertex (0, 0) and is symmetric about y-axis and passes throuh the point (– 2, – 3).

Solution. Since the required parabola is symmetric about y-axis

$\therefore$ Equation of parabola may be $x^2 = 4ay$ or $x^2 = -4ay$

But, it passes through point (– 2, – 2) that lies in 3rd quadrant. So, required parabola is symmetric about y-axis in downward direction.

$\therefore$ Equation of required parabola is $\quad x^2 = -4ay \quad \ldots(i)$

Since it passes through (– 2, – 3)

$$\therefore \quad (-2)^2 = -4(-3)$$

$$4 = 12a$$

$$a = \frac{4}{12} = \frac{1}{3}$$

Put $a = \frac{1}{3}$ in Eq. (i) $\quad x^2 = -4\left(\frac{1}{3}\right)y$

$\Rightarrow \quad 3x^2 = -4y$, is required equation.

Example 4.31. Find the equation of the parabola whose focus is (0, – 1) and the directix is $x + y = 1$.

Solution. Let $P(x, y)$ be any point on the parabola whose focus is S (0, – 1) and directix is

$$x + y - 1 = 0 \quad \ldots(i)$$

Join SP and draw $PM \perp 1$ then by definition of parabola

$$|SP| = |PM|$$

$$\sqrt{(x-0)^2 + (y+1)^2} = |PM|$$

$$|PM| = \sqrt{x^2 + (y+1)^2} \quad \ldots(ii)$$

$|PM|$ = Length of $\perp$ from $P(x, y)$ to directix (1)

$$= \left|\frac{x+y-1}{\sqrt{(1)^2 + (1)^2}}\right| = \left|\frac{x+y-1}{\sqrt{2}}\right|$$

From Eq. (ii) $\quad \sqrt{x^2 + (y+1)^2} = \left|\frac{x+y-1}{\sqrt{2}}\right| \quad \ldots(iii)$

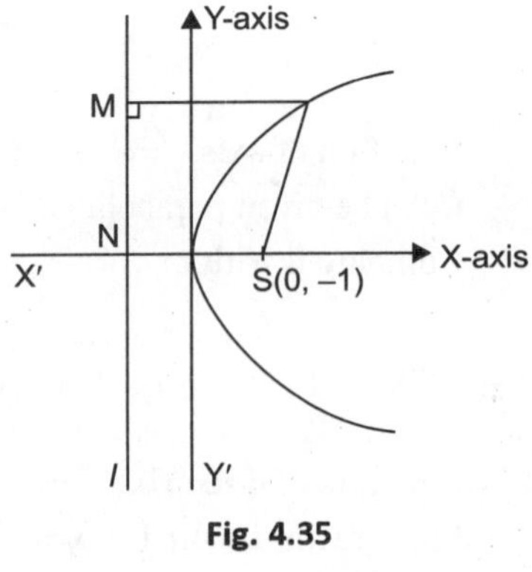

Fig. 4.35

Squaring both sides $$x^2 + (y+1)^2 = \frac{1}{2}(x+y-1)^2$$

$$\Rightarrow \quad x^2 + y^2 + 2y + 1 = \frac{1}{2}(x^2 + y^2 + 2 + 2xy - 2x - 2y)$$

$$\Rightarrow \quad 2x^2 + 2y^2 + 4y + 2 = x^2 + y^2 + 1 + 2xy - 2x - 2y$$

$$\Rightarrow \quad x^2 + y^2 - 2xy + 2x + 6y + 1 = 0; \text{ which is required equation of parabola}$$

EXERCISE 4.9

1. Find the following parabola, find focus latus rectum equation of axis and equation of directix:

   (*i*) $y^2 = 18x$ $\left[\textbf{Ans.}\left(\frac{9}{2},0\right), 18;\ y=0;\ 2x+9=0\right]$

   (*ii*) $3y^2 = -16x$ $\left[\textbf{Ans.}\left(\frac{-4}{3},0\right), \frac{16}{3};\ y=0;\ 3x-4=0\right]$

   (*iii*) $2x^2 = 5y$ $\left[\textbf{Ans.}\left(0,\frac{5}{8}\right), \frac{5}{2};\ x=0;\ 8y+5=0\right]$

   (*iv*) $x^2 = -4y$ [**Ans.** $(0, -1), 4, x = 0; y - 1 = 0$]
2. Find the equation of parabola having focus at (2, 0) and directix is $x + 2 = 0$. [**Ans.** $y^2 = 8x$]
3. Find the equation of parabola having focus at (3, – 4) and directix is $x + y - 2 = 0$. [**Ans.** $x^2 - 2xy + 2y - 8x + 20y + 46 = 0$]
4. Find the equation of parabola whose vertex is (2, 1) and directix is $x - y + 1 = 0$. [**Ans.** $x^2 + y^2 + 2xy - 14x + 2y + 17 = 0$]
5. Find the focus, vertex, equation of directix and axis of the parabola $x = y^2 - 2y + 3$. $\left[\textbf{Ans.}\left(\frac{9}{4},1\right) : (2,1);\ 4x-7=0;\ y=1\right]$
6. Find the equation of parabola with vertex at origin and passing through (– 3, 7) and axis along the x-axis. [**Ans.** $3y^2 = -4y$]
7. Find the coordinates of a point on parabola $y^2 = 18x$, where ordinate is equal to three times the abscissa. [**Ans.** 2, 6]
8. Find the equation of parabola with vertex (2, 1) and focus (1, – 1). [**Ans.** $4x^2 + y^2 - 4xy + 8x + 46y - 71 = 0$]

4.17. EQUATION OF AN ELLIPSE IN THE STANDARD FORM

$$\frac{x^2}{a^2} + \frac{y^2}{b^2} = 1$$

Proof. Let S be the focus, ZM the directrix and e (< 1) the eccentricity of the ellipse.

Draw $SZ \perp ZM$. Divide ZS internally at A and externally at A' in the ratio $1 : e$ so that

$$\frac{ZA}{SA} = \frac{1}{e} \Rightarrow AS = eZA \qquad \ldots(i)$$

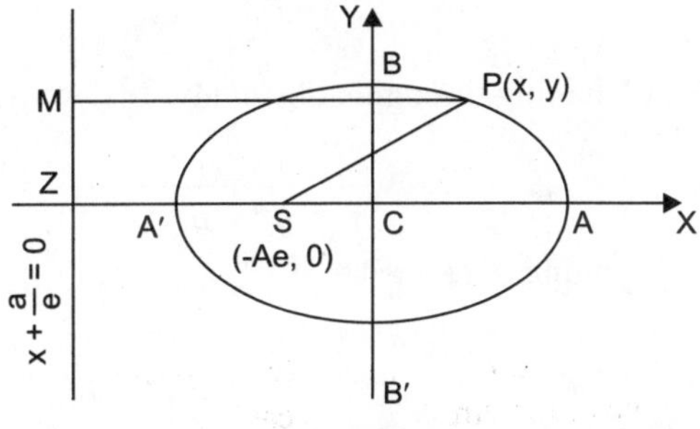

Fig. 4.36

and
$$\frac{ZA'}{A'S} = \frac{1}{e}$$

$\Rightarrow$ $$A'S = eZA \qquad ...(ii)$$

Thus, by definition A and A' lie in the ellipse. Let C be the middle point of AA' and $AA' = 2a$, then
$$CA = CA' = a$$

Adding (*i*) and (*ii*), we get
$$AS + A'S = e\,(ZA + ZA')$$
$$\Rightarrow \quad AA' = (CZ + CA + CZ - CA')$$
$$\Rightarrow \quad 2a = e \cdot 2CZ \qquad [\because \; CA = CA', AA' = 2a]$$
$$\Rightarrow \quad CZ = a/e \qquad ...(iii)$$

Again, subtracting (*ii*) from (*i*), we get
$$AS - A'S = e\,(ZA - ZA')$$
$$\Rightarrow \quad (AC + CS) - (A'C - CS) = e \cdot AA \quad \Rightarrow \quad 2CS = e \cdot AA'$$
$$\therefore \quad CS = ae \qquad ...(iv)$$

Take C as origin, AA' as the axis of x and CY a line through C and perpendicular to AA' as the axis of y. Then Co-ordinates of S are $(-\,ae, 0)$ and the equation of ZM is
$$x = -\frac{a}{e} \quad \Rightarrow \quad x + -\frac{a}{e} = 0 \qquad ...(v)$$

[$\because$ ZM is the line parallel to y-axis and at distance $\frac{a}{e}$ to the left of it]

Let $P(x, y)$ by any point on the ellipse. Join SP and from P, draw $PM \perp$ on ZM. Then by definition
$$\frac{SP}{PM} = e \quad \Rightarrow \quad SP = ePM \qquad ...(vi)$$

But SP = Distance between the point $P\,(x, y)$ and $S\,(-\,ae, 0) = \sqrt{(x + ae)^2 + y^2}$

and PM = perpendicular from $P\,(x, y)$ on the direction
$$\left(x + \frac{a}{e} = 0\right) \text{ is} = x + \frac{a}{e}$$

Substituting for SP and PM in (*iv*), we get
$$\sqrt{(x + ae)^2 + y^2} = e\left(x + \frac{a}{e}\right) = ex + a$$

Squaring both sidse,
$$x^2 + 2aex + a^2e^2 + y^2 = e^2x^2 + 2aex + a^2$$
$$\Rightarrow \quad x^2\,(1 - e^2) + y^2 = a^2\,(1 - e^2)$$

Dividing throughout by $a^2\,(1 - e^2)$, we get
$$\frac{x^2}{a^2} + \frac{y^2}{a^2(1 - e^2)} = 1$$

Putting $a^2\,(1 - e^2) = b^2$
$$\frac{x^2}{a^2} + \frac{y^2}{b^2} = 1$$

which is the required equation of the ellipse.

This is the standard form of the ellipse.

4.18. SHAPE OF THE ELLIPSE WHOSE EQUATION IS GIVEN BY

$$\frac{x^2}{a^2} + \frac{y^2}{b^2} = 1, a > b > 0 \qquad ...(i)$$

1. The ellipse is symmetrical about x-axis as well about y-axis. Also the ellipse is symmetrical about the origin. Thus, origin is the **centre** of ellipse.
2. An ellipse has two axes. Putting $y = 0$ in (i), we have

$$\frac{x^2}{a^2} = 1 \Rightarrow x^2 = a^2 \Rightarrow x = \pm a$$

$\therefore$ x-axis meets ellipse in two points A (a, 0) and A'($-$ a, 0)

Putting $x = 0$ in (1), we have

$$\frac{y^2}{b^2} = 1 \Rightarrow y^2 = b^2 \Rightarrow y = -ab$$

$\therefore$ y-axis meets ellipse in two points B(0, b) and B'(0, $-$ b) thus ellipse has four vertices namely, A, A', B, B'.

Obviously $AA' = 2a$ and $BB' = 2b$.

Since $a > b$, therefore $AA' > BB'$.

AA' is called the major axis and BB' is called the **minor axis** of the ellipse

3. $\frac{x^2}{a^2} + \frac{y^2}{b^2} = 1 \Rightarrow y^2 = b^2\left(1 - \frac{x^2}{a^2}\right) = \frac{b^2}{a^2}(a^2 - x^2)$

Since $y^2 \geq 0$

$\therefore$ $$a^2 - x^2 \geq 0 \Rightarrow x^2 \leq a^2$$

$\Rightarrow$ $$|x| \leq a \Rightarrow -a \leq x \leq a$$

Similarly, $-b \leq y \leq b$

Thus the ellipse lies completely within the rectangle formed by the line $x = \pm a, y = \pm b$. Hence, an ellipse is a bounded curve.

4. The symmetry of the ellipse that the ellipse $\frac{x^2}{a^2} + \frac{y^2}{b^2} = 1$, has two foci and two directrices.

A rough sketch of the ellipse is given 4.37.

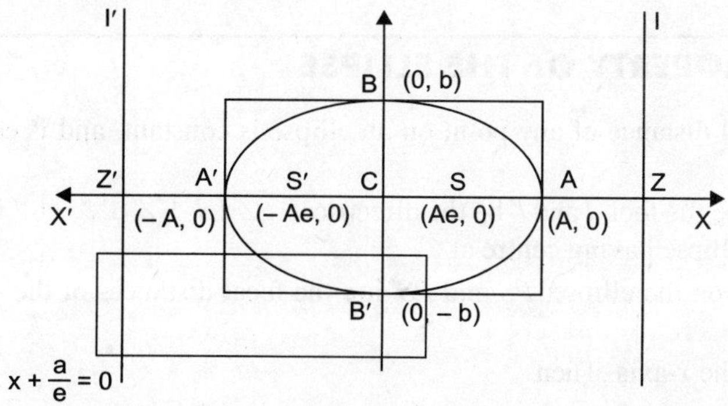

Fig. 4.37

4.19. LENGTH OF LATUS RECTUM OF ELLIPSE

$$\frac{x^2}{a^2}+\frac{y^2}{b^2}=1\ (a>b>0)$$

The equation of ellipse is $\dfrac{x^2}{a^2}+\dfrac{y^2}{b^2}=1$ ...(*i*)

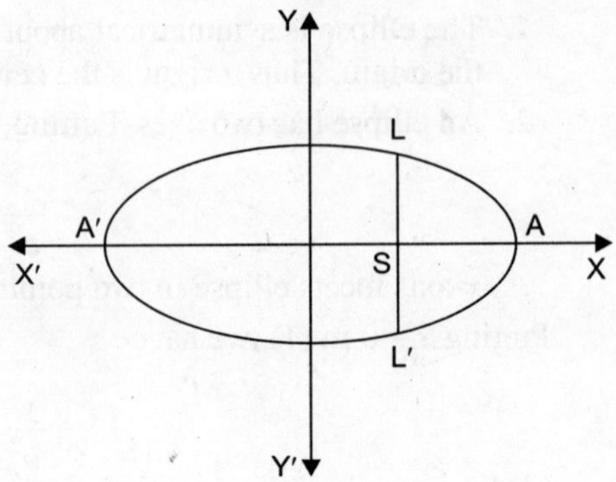

Fig. 4.38

Let LSL' be the latus rectum through S.

Let $SL = l$

∴ The point (ae, l) lies on (*i*)

$$\therefore \quad \frac{a^2e^2}{a^2}+\frac{l^2}{b^2}=1$$

$$\Rightarrow \quad \frac{l^2}{b^2}=1-e^2$$

$$\therefore \quad \frac{l^2}{b^2}=\frac{b^2}{a^2} \qquad [\because\ b^2=a^2(1-e^2)]$$

$$\Rightarrow \quad l^2=\frac{b^4}{a^2} \Rightarrow l=\frac{b^2}{a}$$

∴ Length of latus rectum $= 2l = \dfrac{2b^2}{a}$

Corollary. The ends of latus rectum through S are $\left(as, \dfrac{b^2}{a}\right), \left(ae, -\dfrac{b^2}{a}\right)$

Also the ends of latus rectum through S′ are $\left(-ae, \dfrac{b^2}{a}\right), \left(-ae, -\dfrac{b^2}{a}\right)$

The equation of latus rectum through S is

$$x = ae \quad \Rightarrow \quad x - ae = 0$$

and the equation of latus rectum through S' is

$$x = -ae \Rightarrow x + ae = 0$$

4.20. FOCAL PROPERTY OF THE ELLIPSE

The sum of the focal distance of any point on an ellipse is constant, and is equal to the length of the major axis of the ellipse.

Proof. Let S and S' be the foci, l and l' be the directrices. $SZ \perp l$, $S'Z' \perp l' \cdot AA'$ is the major axis and BB' is the minor axis of the ellipse having centre at C.

$P(x, y)$ be any point on the ellipse, PS and PS' are the focal distances of the point. Draw $PM \perp l$ and $PM' \perp l'$.

Also draw $PN \perp$ to the x-axis. Then,

$$PS = ePM \quad (e \text{ is the eccentricity of the ellipse})$$

$$\Rightarrow \qquad PS = e \cdot ZN = e \cdot \left(\frac{a}{e} - x\right)$$

$$\Rightarrow \qquad PS = a - ex \qquad \ldots(i)$$

$$\Rightarrow \qquad PS' = ePM' = eNZ' = e(CZ' + CN)$$

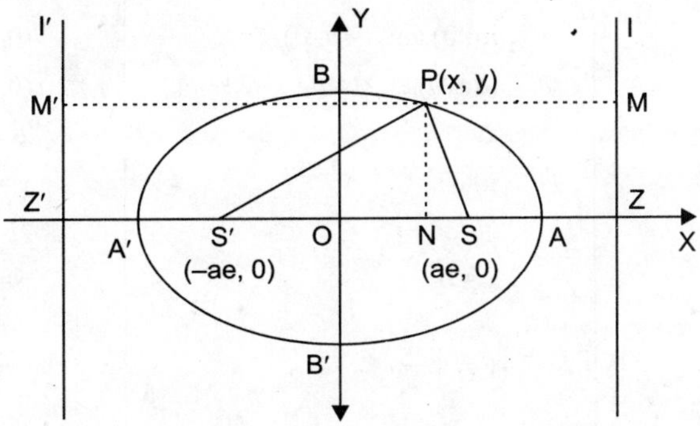

Fig. 4.39

Adding (i) and (ii), we get

$$PS + PS' = (a - ex) + (a + ex) = 2a$$

which is the length of major axis of the ellipse.

4.21. OTHER FORM OF ELLIPSE

If the major axis of the ellipse is along the y-axis then the equation of the ellipse takes the form

$$\frac{x^2}{b^2} + \frac{y^2}{a^2} = 1, \text{ for } a^2 > b^2$$

The ellipse is shown as Fig. 4.40.

(*i*) This is symmetrical about the co-ordinate axis centre = $O(0, 0)$

(*ii*) The vertices are at $(0, \pm a)$

(*iii*) The foci at $(0, \pm ae)$

(*iv*) The end point of the minor ais are $(\pm b, 0)$

(*v*) Equations of the directrices are $y = \pm \dfrac{a}{e}$

(*vi*) Length of the latus rectum = $\dfrac{2b^2}{a}$

(*vii*) Equations of the latus rectum are $y = \pm ae$.

(*viii*) Eccentricity is given by $b^2 = a^2 (1 - e^2)$

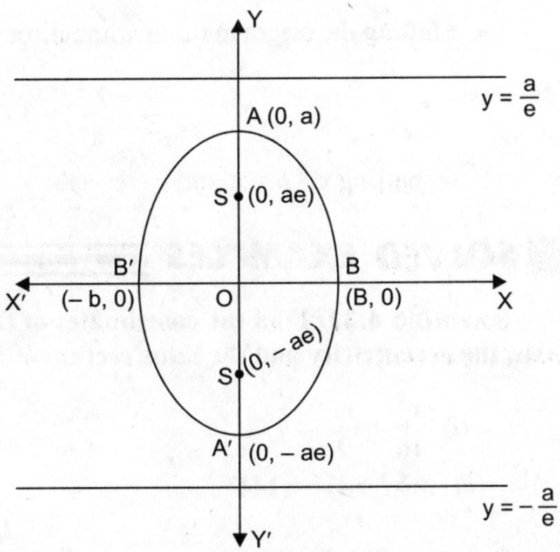

Fig. 4.40

4.21.1 Important Results of Different Ellipses

Equation of standard form	$\frac{x^2}{a^2}+\frac{y^2}{b^2}=1\ a>b>0$	$\frac{x^2}{b^2}+\frac{y^2}{a^2}=1\ a>b>0$
Co-ordinates of centre	(0, 0)	(0, 0)
Co-ordinates of vertices	$(a, 0)$ and $(-a, 0)$	$(0, a)$ and $(0, -a)$
Co-ordinates of foci	$(ae, 0)$ and $(-ae, 0)$	$(0, ae)$ and $(0, -ae)$
Length of major axis	$2a$	$2a$
Equation of major axis	$y = 0$	$x = 0$
Length of minor axis	$2b$	$2b$
Equation of minor axis	$x = 0$	$y = 0$
Equation of directrices	$x = \pm\frac{a}{e}$	$y = \pm\frac{a}{e}$
Length of latus rectum	$x = \pm ae$	$y = \pm ae$
Equations of latus-rectum	$x = \pm ae$	$y = \pm ae$
Eccentricity is given by	$b^2 = a^2(1 \cdot e^2)$	$b^2 = a^2(1 \cdot e^2)$

Note:

1. **(An important property of an ellipse):** The sum of the focal distances of any point on an ellipse is constant and is equal to length of the major axis of the ellipse.
2. If the centre of the ellipse is at point (h, k) and the directions of the axis are parallel to the co-ordinate axes, then its equation is

$$\frac{(x-h)^2}{a^2}+\frac{(y-k)^2}{b^2}=1$$

Shifting the origin at (h, k) without rotatin the coordinate axes, the equation can be written as

$$\frac{x^2}{a^2}+\frac{Y^2}{b^2}=1 \text{ (in standard form)}$$

by putting $x - h = X$ and $y - k = Y$.

SOLVED EXAMPLES

Example 4.32. Find the coordinates of the foci, the vertices, the length of major axis, the minor axis, the eccentricity and the latus rectum of the ellipse.

(i) $\frac{x^2}{16}+\frac{y^2}{9}=1$ ***(ii)*** $\frac{x^2}{4}+\frac{y^2}{25}=1$

(iii) $36x^2 + 4y^2 = 144$ ***(iv)*** $4x^2 + 9y^2 = 36$

Solution. (*a*) The given equation is $\frac{x^2}{16}+\frac{y^2}{9}=1$.

Here, the denominator of $\frac{x^2}{16}$ is greater than the denominator of $\frac{y^2}{9}$.

Therefore, the major axis is along the x-axis, while the minor axis is along the y-axis.

On comparing the given equatio with $\frac{x^2}{a^2}+\frac{y^2}{b^2}=1.$

We obtain $a = 4$ and $b = 3$

$$\therefore \qquad C=\sqrt{a^2-b^2}=\sqrt{16-9}=\sqrt{7}$$

Therefore,

The coordianates of the foci are $(\pm\sqrt{7}, 0)$

The coordianates of the vertices are $(\pm 4, 0)$

Length of the major axis = $2a = 8$

Length of the minor axis = $2b = 6$

$$\text{Eccentricity,} \qquad e=\frac{c}{a}=\frac{\sqrt{7}}{4}$$

$$\text{Length of latus rectum} = \frac{2b^2}{a}=\frac{2\times 9}{4}=\frac{9}{2}$$

(b) The given equation is $\frac{x^2}{4}+\frac{y^2}{25}=1$

Here, the denominator of $\frac{y^2}{25}$ is greater than the denominator of the $\frac{x^2}{4}$.

Therefore, the major axis is along the y-axis, while the minor axis is along the x-axis.

On comparing the given equation with $\frac{x^2}{b^2}+\frac{y^2}{a^2}=1$, we obtain $b = 2$ and $a = 5$.

$$\therefore \qquad C=\sqrt{a^2-b^2}=\sqrt{25-4}=\sqrt{21}$$

Therefore

The coordinates of the foci are $(0, \pm\sqrt{21})$

The coordinates of the vertices are $(0, \pm 5)$

Length of major axis = $2a = 10$

Length of minor axis = $2b = 4$

$$\text{Eccentricity,} \qquad e=\frac{c}{a}=\frac{\sqrt{21}}{5}$$

$$\text{Length of latus rectum} = \frac{2b^2}{a}=\frac{2\times 4}{5}=\frac{8}{5}$$

(c) The given equation is $36x^2 + 4y^2 = 144$

It can be written as

$$\frac{x^2}{4}+\frac{y^2}{36}=1 \quad \text{or} \quad \frac{x^2}{2^2}+\frac{y^2}{6^2}=1$$

Here, the denominator of $\frac{y^2}{36}$ is greater than the denominator of $\frac{x^2}{4}$

Therefore, the major axis is along the y-axis, while the minor axis is along the x-axis.

On comparing the given equation with $\frac{x^2}{b^2}+\frac{y^2}{a^2} = 1$ we obtain $b = 2$ and $a = 6$

$\therefore$ $$C= \sqrt{a^2-b^2} = \sqrt{36-4} = \sqrt{32} = 4\sqrt{2}$$

Therefore

The coordinates of the foci are $(0, \sqrt{2})$

The coordinates of the vertices are $(0, \pm 6)$

Length of major axis $= 2a = 12$

Length of minor axis $= 2b = 4$

Eccentricity, $$e = \frac{c}{a} = \frac{4\sqrt{2}}{6} = \frac{2\sqrt{2}}{3}$$

Length of latus rectum $= \frac{2b^2}{a} = \frac{2\times 4}{6} = \frac{4}{3}$

(*d*) The given equation is $4x^2 + 9y^2 = 36$

It can be written as $\frac{x^2}{9}+\frac{y^2}{4} = 1$

Here, the denominator of $\frac{x^2}{9}$ is greater than $\frac{y^2}{4}$.

Therefore, the major axis is along the x-axis while the minor axis is along the y-axis.

On comparing the given equation with $\frac{x^2}{a^2}+\frac{y^2}{b^2} = 1$

We obtain $a = 3$ and $b = 2$

$\therefore$ $$c= \sqrt{a^2-b^2} = \sqrt{9-4} = \sqrt{5}$$

Therefore,

The coordinates of the foci are $(\pm\sqrt{5}, 0)$

The coordinates of the vertices are $(\pm 3, 0)$

Length of major axis $= 2a = 6$

Length of minor axis $= 2b = 4$

Eccentricity, $$e = \frac{c}{a} = \frac{\sqrt{5}}{3}$$

Example 4.33. Find the equation of the ellipse satisfying the given conditions:

(*a*) Vertices (± 13, 0), Foci (± 5, 0)

(*b*) Vertices (0, ± 5), Foci (0, ± 4).

Solution. Vertices $(\pm 13, 0)$, Foci $(\pm 5, 0)$

Here, vertices are on the x-axis.

Therefore, the equation of the ellipse will of the form $\frac{x^2}{a^2}+\frac{y^2}{b} = 1$, where a is the semi-major axis.

Accordingly, $a = 13$ and $c = 5$

It is known that $a^2 = b^2 + c^2$

$\therefore$ $169 = b^2 + 25 \Rightarrow b = 12$

Hence, the equation of the ellipse is $\dfrac{x^2}{169} + \dfrac{y^2}{144} = 1$

(*b*) Vertices (0, ± 5), Foci (0, ± 4)

Here, the vertices are on the *y*-axis

Therefore, the equation of the ellipse will be of the form $\dfrac{x^2}{b^2} + \dfrac{y^2}{a^2} = 1$ where *a* is the semi-major axis.

Accordingly, $a = 5$ and $c = 4$

It is known that $a^2 = b^2 + c^2$

$\therefore$ $5^2 = b^2 + 4^2 \Rightarrow b = 3$

Hence, the equation of the ellipse is $\dfrac{x^2}{9} + \dfrac{y^2}{25} = 1.$

Example 4.34. Find the equation of the ellipse satisfying the given conditions:

(*a*) Foci (± 5, 0) and eccentrcity 5/8

(*b*) Major axis = 8 and $e = \dfrac{1}{2}$

Solution. Here the foci are on the *x*-axis.

Therefore the equation of the ellipse will be of the form $\dfrac{x^2}{a^2} + \dfrac{y^2}{b^2} = 1$ where *a* is the semi-major axis.

Accordingly, $ae = 5 \therefore a\left(\dfrac{5}{8}\right) = 5 \Rightarrow a = 8$

Also $b^2 = a^2(1 - e^2) = 64\left(1 - \dfrac{25}{64}\right) = 64 \times \dfrac{39}{64} = 39$

Hence, the equation of the ellipse is $\dfrac{x^2}{64} + \dfrac{y^2}{39} = 1$

(*b*) Let the equation of the ellipse be $\dfrac{x^2}{a^2} + \dfrac{y^2}{b^2}$

Major axis $2a = 8 \Rightarrow a = 4$

Also $e = \dfrac{1}{2}$

$\therefore$ $b^2 = a^2(1 - e^2) = 16\left(1 - \dfrac{1}{4}\right) = 16 \times \dfrac{3}{4} = 12$

Hence, the equation of the ellipse is $\dfrac{x^2}{16} + \dfrac{y^2}{12} = 1.$

EXERCISE 4.10

Find the length of major and minor axis, the coordinates of foci and vertices the eccentricity of the ellipse.

1. $\frac{x^2}{169}+\frac{y^2}{25}=1$ [**Ans.** 36, 10 (± 12, 0), (± 13, 0)]
2. $\frac{x^2}{100}+\frac{y^2}{25}=1$ [**Ans.** 20, 10 $(\pm 5\sqrt{3}, 0)$, (± 10, 0)]
3. $\frac{x^2}{225}+\frac{y^2}{289}=1$ [**Ans.** 34, 30 (0, ± 8), (0, ± 17)]
4. $16x^2 + y^2 = 16$ [**Ans.** 8, 2,$(0, \pm\sqrt{15})$, (0, ± 4)]
5. $36x^2 + 4y^2 = 144$ [**Ans.** 12, 4 $(0, \pm 4\sqrt{2})$, (0, ± 6)]
6. $4x^2 + 9y^2 = 36$ [**Ans.** 6, 4 $(\pm\sqrt{5}, 0)$, (± 3, 0)]

Find the equation of the ellipse satisfying the given conditions:

7. Vertices (± 5, 0), Foci (± 4, 0) $\left[\textbf{Ans. } \frac{x^2}{25}+\frac{y^2}{9}=1\right]$
8. Foci (0, ± 5), vertices (0, ± 13) $\left[\textbf{Ans. } \frac{x^2}{144}+\frac{y^2}{169}=1\right]$
9. Foci (0, ± 4); $e = \frac{4}{5}$ $\left[\textbf{Ans. } \frac{x^2}{9}+\frac{y^2}{25}=1\right]$
10. Latus rectum = 8 and $e = \frac{1}{\sqrt{2}}$ $\left[\textbf{Ans. } \frac{x^2}{64}+\frac{y^2}{32}=1\right]$
11. Axis along co-ordinate axis, passing through (4, 3)and (– 1, 4). $[7x^2 + 15y^2 = 247]$
12. End of major axis (± 3, 0) ends of minor axis (0, ± 2). $\left[\textbf{Ans. } \frac{x^2}{9}+\frac{y^2}{4}=1\right]$
13. Ends of major axsi $(0, \pm\sqrt{5})$ ends of minor axis (± 1, 0). $\left[\textbf{Ans. } \frac{x^2}{1}+\frac{y^2}{5}=1\right]$
14. Length of major axis 26, foci (± 5, 0). $\left[\textbf{Ans. } \frac{x^2}{169}+\frac{y^2}{144}=1\right]$
15. Length of minor axis 16, foci (0, ± 6). $\left[\textbf{Ans. } \frac{x^2}{64}+\frac{y^2}{100}=1\right]$

4.22. HYPERBOLA

A hyperbola is the set of all points in a plane whose distances for a fixed point in the plane and their distances from a fixed line in the plane bear a constant ratio, greater that one.

The fixed point is called *focus*, the fixed line *directix* and the constant ratio e (> 1), the *eccentricity* of the hyperbola.

4.23. TO FIND THE STANDARD EQUATION OF A HYPERBOLA

Article. *To find the equation of a hyperbola in the form*

$$\frac{x^2}{a^2} - \frac{y^2}{b^2} = 1$$

Let S be the focus, ZM the directrix and e (> 1) the eccentricity of the hyperbola.

From S draw SZ perpendicular to the directix, ZM. Divide SZ internally at A and externally at A' in the ratio $e : 1$, so that

$$\frac{SA}{AZ} = \frac{e}{1} \quad \text{or} \quad SA = eAZ \qquad \ldots(i)$$

and $$\frac{SA'}{A'Z} = \frac{e}{1} \quad \text{or} \quad SA' = eA'Z \qquad \ldots(ii)$$

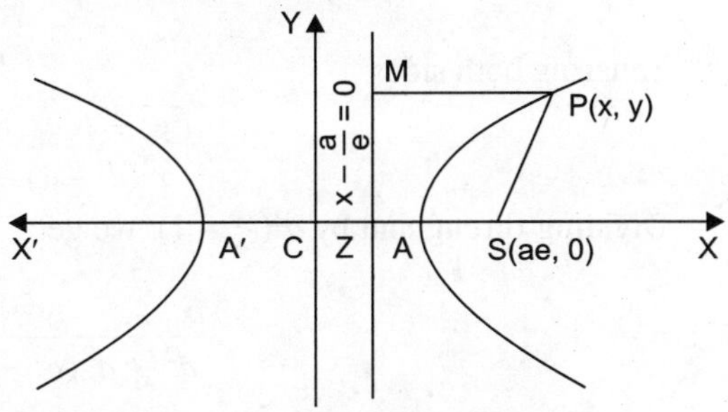

Fig. 4.41

Thus, by definition, A and A' lie on the hyperbola. Let C be the middle point of AA' and $AA' = 2a$, then

$$CA = CA' = a$$

Adding (*i*) and (*ii*), we get

$$SA + SA' = e\,(AZ + A'Z)$$

$$\Rightarrow \quad (CS - CA) + (CS + CA') = e \cdot AA'$$

$$\Rightarrow \quad 2CS = e \cdot 2a \qquad [\because\ CA = CA', AA' = 2a]$$

$$\Rightarrow \quad CS = ae \qquad \ldots(iii)$$

Now subtracting (*i*) from (*ii*), we have

$$SA' - SA = e\,(A'Z - AZ)$$

$$\Rightarrow \quad AA' = e\,[(CA' + CZ) - (CA - CZ)]$$

$$\Rightarrow \quad 2a = e \cdot 2CZ \qquad [\because\ CA = CA', AA' = 2a]$$

$$\Rightarrow \quad CZ = \frac{a}{e} \qquad \ldots(iv)$$

Take C as origin, AA' as the x-axis and CY, a line through C and $\perp AA'$ as the y-axis.

$\therefore$ From (*iii*) and (*iv*), the coordinates of S are $(ae, 0)$ and the equation of ZM is

$$x = \frac{a}{e}$$

$$\Rightarrow \quad x - \frac{a}{e} = 0 \qquad \ldots(v)$$

Let $P\,(x, y)$ be any point on the hyperbola. Join SP. Draw $PM \perp$ on ZM.

Then by def

$$\frac{SP}{PM} = e \quad \text{or} \quad SP = ePM \qquad \ldots(vi)$$

But SP = Distance between $S(ae, 0)$ and $P(x, y)$

$$= \sqrt{(x - ae)^2 + y^2}$$

and $PM =$ perpendicular from $P(x, y)$ on (v)

$$= \frac{x - \frac{a}{e}}{1} = x - \frac{a}{e}$$

Substituting the values of SP and PM in (vi)

$$\sqrt{(x - ae)^2 + y^2} = e\left(x - \frac{a}{e}\right) = ex - a$$

Squaring both sides,

$$x^2 + a^2e^2 - 2aex + y^2 = e^2x^2 - 2aex + a^2$$

$$\Rightarrow \quad x^2(e^2 - 1) - y^2 = a^2(e^2 - 1)$$

Dividing throughout by $a^2(e^2 - 1)$, we get

$$\Rightarrow \quad \frac{x^2}{a^2} - \frac{y^2}{a^2(e^2 - 1)} = 1$$

As $a^2(e^2 - 1) = b^2$, the required equation is

$$\frac{x^2}{a^2} - \frac{y^2}{b^2} = 1$$

This is the standard equation of the hyperbola.

4.24. SHAPE OF HYPERBOLA $\frac{x^2}{a^2} - \frac{y^2}{b^2} = 1, a > b$

where $b^2 = a^2(e^2 - 1)$ and $e > 1$

1. **Symmetry.** The hyperbola is symmetrical about the co-ordinate axes and also symmetrial about the origin. The origin is the **centre** of the hyperbola.
2. The hyperbola has two axis.
   Putting $y = 0$ in (i), we have

$$\frac{x^2}{a^2} = 1 \Rightarrow x^2 = a^2; x = \pm a$$

$\therefore$ x-axis meets hyperbola in two points $A(a, 0)$ and $A'(-a, 0)$. The hyperbola does not intersect y-axis.

$\therefore$ AA' is called the transverse axis. If we take two points $B = (0, b)$ and $B' = (0, -b)$ on the y-axis, then BB' is called the *conjugate axis*.

Length of transverse axis $= |AA'| = 2a$

Length of conjugate axis $= |BB'| = 2b$

The points A and A' are called *vertices of the hyperbola.*

3. As in the case of an ellipse, every chord of the hyperbola which passes through the origin is bisected at it. Hence, the origin is the centre of the hyperbola and the hyperbola is also a central conic.
4. Let $A'A$ and $B'B$ be the axes of the hyperbola

$$\frac{x^2}{a^2} - \frac{y^2}{b^2} = 1$$

Complete the rectangle by drawing lines through $A'A$, B' and B parallel to the axes, as shown in Fig. 4.7.

Obviously the slope of the diagonal is $\frac{b}{a}$ and the slope of the other diagonal is $-\frac{b}{a}$. The equation of the diagonals are $y = \frac{b}{a}$ and $y = -\frac{b}{a}x$. These are equations of asymptotes of the hyperbola.

A rought sketch of the hyperbola is given in the Fig. 4.42.

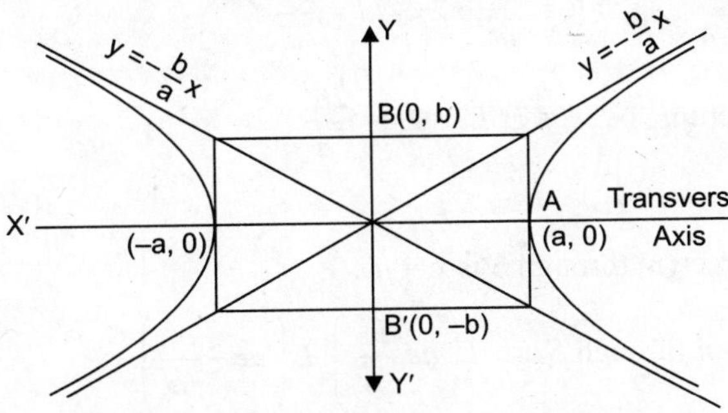

Fig. 4.42

Note. The hyperbola has a second focus and a second directix.

$\therefore$ The hyperbola has two foci and two directices (See Fig. 4.43)

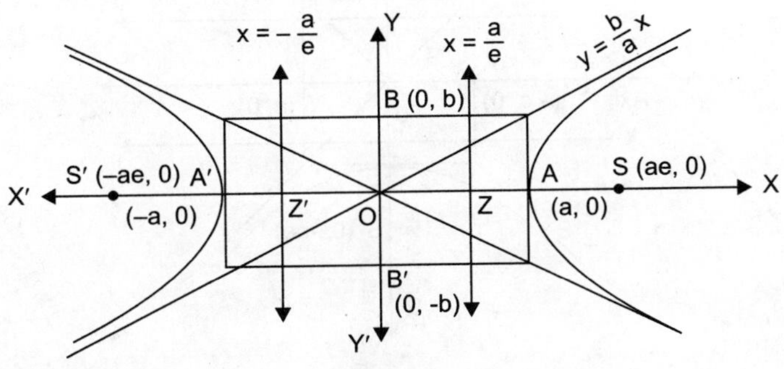

Fig. 4.43

4.25. LENGTH OF LATUS RECTUM

Length of Latus rectum of the hyperbola

$$\frac{x^2}{a^2} - \frac{y^2}{b^2} = 1$$

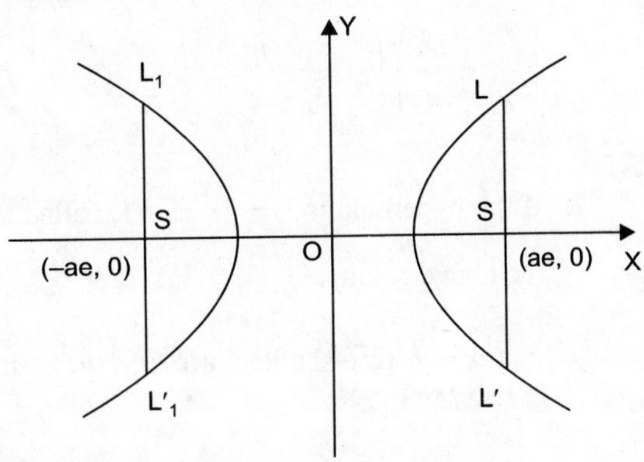

Fig. 4.44

Solution. The equation of the hyperbola is

$$\frac{x^2}{a^2} - \frac{y^2}{b^2} = 1 \quad \ldots(i)$$

Let LSL' be the latus rectum through S.

Let $SL = l$

$\therefore$ The point $L(ae, l)$ lies on the hyperbola (i)

$\therefore$ The hyperbola has two foci and two directices (See Fig. 4.44)

$$\therefore \quad \frac{a^2e}{a^2} - \frac{l^2}{b^2} \Rightarrow e^2 - \frac{l^2}{b^2} = 1$$

$$\Rightarrow \quad \frac{l^2}{b^2} = e^2 - 1 = \frac{b^2}{a^2} \qquad [\because \ b^2 = a^2(e^2 - 1)]$$

$$\Rightarrow \quad l^2 = \frac{b^4}{a^2} \Rightarrow l = \frac{b^2}{a}$$

$\therefore$ The length of lotus rectum $= LSL' = 2l = \dfrac{2b^2}{a}$

Corollary

1. The equation of latus rectym through S is $x = ae$.
2. The ends of latus rectum through S are $L\left(ae, \dfrac{b^2}{a}\right), L'\left(ae - \dfrac{b^2}{a}\right)$.

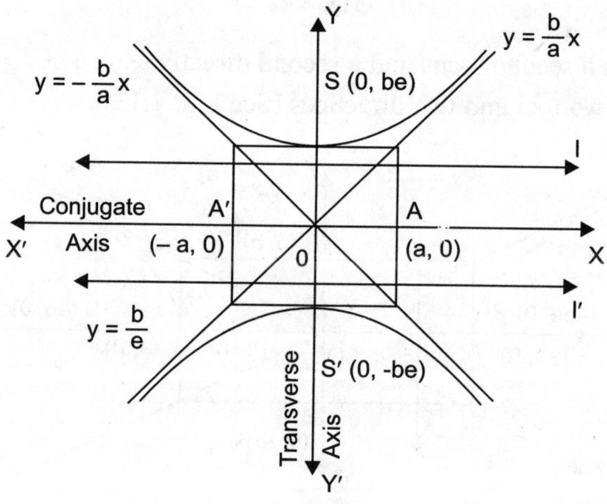

Fig. 4.45

Note:

1. The equation of the latus rectum through S' is $x = -ar$ and ends of latus recta are $\left(-ae, \dfrac{b^2}{a}\right), \left(-ae, -\dfrac{b^2}{a}\right)$.

2. If the hyperbola is $\dfrac{y^2}{b^2} - \dfrac{x^2}{a^2} = 1$, called the *conjugate hyperbola*, a and b will be interchanged all the above results.

$\therefore$ $a^2 = b^2(e^2 - 1)$. Foci are $(0, \pm be)$, directrices are $y = \pm \dfrac{b}{e}$ and length of latus rectum $= \dfrac{2a^2}{b}$

4.26. RECTANGULAR HYPERBOLA

If the lengths of the transverse and conjugate axes a hyperbola be equal, then it is called a *Rectangular hyperbola.*

In case of a rectangular hyperbola

$$2a = 2b \Rightarrow a = b$$

$$\therefore \quad \frac{x^2}{a^2} - \frac{y^2}{a^2} = 1 \Rightarrow x^2 - y^2 = a^2$$

which is the equation of rectangular hyperbola.

Also $\quad b^2 = a^2(e^2 - 1) \Rightarrow 1 = e^2 - 1 \quad [\because a = b]$

$$\Rightarrow \quad e^2 = 2 \Rightarrow e = \sqrt{2}$$

$\therefore$ Eccentricity of a rectangular hyperbola is $\sqrt{2}$.

4.27. CONJUGATE HYPERBOLA

The hyperbola whose transverse and conjugate axes are respectively the conjugate and transverse axes of a given hyperbola is called the *conjugate hyperbola* of the gien hyperbola.

If the hyperbola is $\frac{x^2}{a^2} - \frac{y^2}{b^2} = 1$, then its conjugate hyperbola is $-\frac{x^2}{a^2} + \frac{y^2}{b^2} = 1$ or $\frac{x^2}{a^2} - \frac{y^2}{b^2} = -1$.

Important Results of Different Hyperbola

Equation of standard form	*Hyperbola* $\frac{x^2}{a^2} - \frac{y^2}{b^2} = 1$	*Conjugate hyperbola* $\frac{x^2}{a^2} - \frac{y^2}{b^2} = -1$
Co-ordinate of the centre	(0, 0)	(0, 0)
Co-ordinate of the vertices	$(\pm a, 0)$	$(0, \pm b)$
Equation of transverse axis	$y = 0$	$x = 0$
Length of transverse axis	$2a$	$2b$
Conjugate axis equation	$x = 0$	$y = 0$
Co-ordinate of foci	$(\pm ae, 0)$	$(0, \pm be)$
Equations of directices	$x = \pm \frac{a}{e}$	$y = \pm \frac{a}{e}$
Length of latus rectum	$\frac{2b^2}{a}$	$\frac{2a^2}{b}$
Eccentricity is given by	$b^2 = a^2(e^2 - 1)$	$a^2 = b^2(e^2 - 1)$
Late a recta equation	$x = \pm ae$	$y = \pm be$

Note 1. If the centre of the hyperbola is at (h, k) and the direction of the axes are respectively parallel to the coordinate axes, then equation of the hyperbola is

$$\frac{(x-h)^2}{a^2} - \frac{(y-k)^2}{b^2} = 1 \quad \ldots(i)$$

Shifting the origin at (h, k) without rotating the coordinate axes, the equation (i) reduces to $\frac{x^2}{a^2} - \frac{y^2}{b^2} = 1$ (standard form) by putting $x - h = X$ and $y - k = Y$

Note 2.

(i) If the centre of the hyperbola is at the oriring (0, 0) and the foci are on the x-axis, then the equation of the hyperbola is

$$\frac{x^2}{a^2} - \frac{y^2}{b^2} = 1$$

(ii) If we devote the distance between the two foci by $2c$, the length of the transverse axis by $2a$ and the length of the conjugate axis by $2b$, then

(a) $2c = 2ae \Rightarrow e = \frac{c}{a}$

Thus ratio $e = \frac{c}{a}$ is called the eccentricity of the hyperbola.

(b) a, b, c are related as $c^2 = a^2 - b^2$

(c) In terms of the eccentricity, the foci are at a distance of ae from the centre

(d) In finding the equation of the hyperbola, as

$$\frac{x^2}{a^2} - \frac{y^2}{b^2} = 1$$

we placed the foci on x-axis at $(ae, 0)$ and $(0, -ae)$.

However, if we put the foci on y-axis at $(0, ae)$ and $(0, -ae)$ then we can derive the equation of the hyperbola as

$$\frac{x^2}{a^2} - \frac{y^2}{b^2} = 1$$

Here the transverse axis is along y-axis and the conjugae axis along x-axis.

SOLVED EXAMPLES

Example 4.35. For the following hyperbolas, find the lengths of transverse an conjugate axis, eccentricity and the coordinates of foci and vertices:

(a) $16x^2 - 9y^2 = 144$ **(b) $2x^2 - 3y^2 - 6 = 0$**

(c) $y^2 - 16x^2 = 16$

Solution. (a) Dividing the equation by 144 on both sides, we have

$$\frac{x^2}{9} - \frac{y^2}{16} = 1$$

Comparing the equation $\frac{x^2}{9} - \frac{y^2}{16} = 1$ with the standard equation $\frac{x^2}{a^2} - \frac{y^2}{b^2} = 1$

Here, $a = 3$, $b = 4$ and $c = \sqrt{a^2 + b^2} = \sqrt{9 + 16} = 5$

Therefore

The length of transverse axis = $2a = 2 \times 3 = 6$

The length of conjugate axis = $2b = 2 \times 4 = 8$

Eccentricity, $$e = \frac{c}{a} = \frac{5}{3}$$

$\therefore$ $$ae = 3 \times \frac{5}{3} = 5$$

$\therefore$ Coordinates of foci are $(\pm ae, 0) = (\pm 5, 0)$

Coordinates of vertices are $(\pm a, 0) = (\pm 3, 0)$

(*b*) Given hyperbola is $\quad 2x^2 - 3y^2 - 6 = 0$

$\Rightarrow \quad 2x^2 - 3y^2 = 6$

Dividing the equation by 6 on both sides, we have

$$\frac{x^2}{3} - \frac{y^2}{2} = 1$$

Here $a = \sqrt{3}$, $b = \sqrt{2}$ and $c = \sqrt{a^2 + b^2} = \sqrt{3+2} = \sqrt{5}$

Therefore,

The length of transverse axis = $2a = 2\sqrt{3}$

The length of conjugate axis = $2b = 2\sqrt{2}$

Vertices are $(\pm a, 0) = (\pm\sqrt{3}, 0)$

Eccentricity, $e = \frac{c}{a} = \frac{\sqrt{5}}{\sqrt{3}}$

$\therefore$ $$ae = \sqrt{3} \times \frac{\sqrt{5}}{\sqrt{3}} = \sqrt{5}$$

$\therefore$ Coordinates of foci are $(\pm\sqrt{5}, 0)$.

(*c*) Dividing the equation by 16 on both sides, we get

$$\frac{y^2}{16} - \frac{x^2}{1} = 1 \Rightarrow \frac{x^2}{1} - \frac{y^2}{16} = -1$$

It is of the form $\frac{x^2}{a^2} - \frac{y^2}{b^2} = -1$ with $a^2 = 1$, $b^2 = 16$

$\therefore$ $a = 1$, $b = 4$ and $c = \sqrt{a^2 + b^2} = \sqrt{1+16} = \sqrt{17}$

Therefore,

The length of transverse axis = $2b = 8$

The length of conjugate axis = $2a = 2$

Eccentricity, $e = \frac{c}{a} = \frac{\sqrt{17}}{1}$

$\therefore$ $$ae = 1 \times \sqrt{17} = \sqrt{17}$$

$\therefore$ Coordinates of foci are $(0, \pm\sqrt{17})$

Coordinates of vertices are $(0, \pm 4)$.

Example 4.36. Find the eqiation of the hyperbola satisfying the given conditions:

(*a*) Vertices $(\pm 5, 0)$, foci $(\pm 7, 0)$

(*b*) Vertices $(0, \pm 3)$, foci $(0, \pm 5)$

(*c*) Length of transverse axis is 6 and latus rectum is $\frac{32}{3}$

(*d*) Length of conjugate axis is 24 and foci $(0, \pm 13)$

(*e*) Length of transverse axis is 8 and foci $(\pm 5, 0)$

(*f*) Foci $(\pm 4, 0)$, the latus rectum is of length 12

(*g*) Vertices $(\pm 7, 0)$, $e = \frac{4}{3}$

(*h*) Length of transverse axis is 7 and it passes through $(5, -2)$.

Solution. (*a*) Here, the vertices are on the *x*-axis

Therefore, the equation of the hyperbola is of the form

$$\frac{x^2}{a^2} - \frac{y^2}{b^2} = 1$$

Since the vertices are $(\pm 5, 0)$, $a = 5$

Since the foci are $(\pm 7, 0)$

$\therefore$ $$ae = 7 \Rightarrow 5e = 7 \Rightarrow e = \frac{7}{5}$$

$\therefore$ $$b^2 = a^2(e^2 - 1) = 25\left(\frac{49}{25} - 1\right) = 25 \times \frac{24}{25} = 24$$

Therefore, the equation of the hyperbola is

$$\frac{x^2}{a^2} - \frac{y^2}{b^2} = 1$$

$\Rightarrow$ $$\frac{x^2}{5^2} - \frac{y^2}{24} = 1 \Rightarrow \frac{x^2}{25} - \frac{y^2}{24} = 1$$

(*b*) Here, the vertices are on *y*-axis

Therefore, the equation of the hyperbola is of the form

$$\frac{y^2}{a^2} - \frac{x^2}{b^2} = 1$$

Since the vertices are $(0, \pm 3)$, $a = 3$

Since the foci are $(0, \pm 5)$

$\therefore$ $$ae = 5 \Rightarrow 3e = 5 \Rightarrow e = \frac{5}{3}$$

$\therefore$ $$b^2 = a^2(e^2 - 1) = 9\left(\frac{25}{9} - 1\right) = 9 \times \frac{16}{9} = 16$$

$\therefore$ The equation of the hyperbola is

$$\frac{y^2}{9} - \frac{x^2}{16} = 1$$

(*c*) Transverse axis = $2a$

$\therefore$ $$2a = 6 \Rightarrow a = 3;\ a^2 = 9$$

$$\text{Latus rectum} = \frac{2b^2}{a} = \frac{32}{3} \quad \text{(given)}$$

$\Rightarrow$ $$b^2 = \frac{32a}{3\times 2} = \frac{16}{3}a = \frac{16\times 3}{3} = 16$$

Hence, the required equation of the hyperbola is

$$\frac{x^2}{9} - \frac{y^2}{16} = 1$$

(*d*) Since the foci are along y-axis the equation of the hyperbola is of the form

$$\frac{y^2}{a^2} - \frac{x^2}{b^2} = 1$$

Coordinates of foci = $(0, \pm c) = (0, \pm 13)$

$\therefore$ $c = 13$

Length of conjugate axis = $2b = 24$

$\Rightarrow$ $$b = 12$$

From the relation $c^2 = a^2 + b^2$, we get

$$a^2 = c^2 - b^2 = (13)^2 - (12)^2 = 25$$

$\therefore$ $$a = 5$$

Hence, the equation of the hyperbola is $\dfrac{y^2}{25} - \dfrac{x^2}{144} = 1$.

(*e*) Since the foci are along x-axis the equation of the hyperbola is of the form

$$\frac{x^2}{a^2} - \frac{y^2}{b^2} = 1$$

Co-ordinates of foci = $(\pm c, 0) = (\pm 5, 0)$

$\Rightarrow$ $$c = 5$$

Length of transverse axis = $2a = 8 \Rightarrow a = 4$

From the relation $$b^2 = c^2 - a^2 = 5^2 - 4^2 = 9$$

$\Rightarrow$ $$b = 3$$

Hence, the equation of the hyperbola is

$$\frac{x^2}{16} - \frac{y^2}{9} = 1$$

(*f*) Since the foci are along x-axis the equation of the hyperbola is of the form

$$\frac{x^2}{a^2} - \frac{y^2}{b^2} = 1$$

Foci are $(\pm c, 0) = (\pm 4, 0)$

$\therefore \quad c = 4 \Rightarrow c^2 = 16$

$$\text{Latus rectum} = \frac{2b^2}{a} = 12 \quad \text{(given)}$$

$\Rightarrow \quad b^2 = 6a$

Now $\quad c^2 = a^2 + b^2 \Rightarrow 16 = a^2 + 6a$

$\Rightarrow \quad a^2 + 6a - 16 = 0 \Rightarrow (a + 8)(a - 2) = 0$

$\Rightarrow \quad a = -8, 2$

$\therefore \quad a = 2 \qquad (\because$ a being length can't be – ve)

$\therefore \quad b^2 = 6a = 12$

Hence, equation of the hyperbola is $\dfrac{x^2}{4} - \dfrac{y^2}{12} = 1.$

(*g*) Since the foci are along *x*-axis the equation of the hyperbola is of the form

$$\frac{x^2}{a^2} - \frac{y^2}{b^2} = 1$$

Vertices are $(\pm a, 0) = (\pm 7, 0)$

$\therefore \quad a = 7$

Eccentricity, $\quad e = \dfrac{c}{a} = \dfrac{4}{3}$ (given)

$\Rightarrow \quad c = \dfrac{4}{3}a = \dfrac{4}{3} \times 7 = \dfrac{28}{3}$

Now from the relation, $b^2 = c^2 - a^2$, we get

$$b^2 = \left(\frac{28}{3}\right)^2 - 7^2 = \frac{789}{9} - 49 = \frac{343}{9}$$

Hence, the equation of the hyperbola is $\dfrac{x^2}{49} - \dfrac{y^2}{\left(\dfrac{343}{9}\right)} = 1.$

$\Rightarrow \quad \dfrac{x^2}{49} - \dfrac{9y^2}{343} = 1$

(*h*) Let the equation of the hyperbola be

$$\frac{x^2}{a^2} - \frac{y^2}{b^2} = 1$$

$\therefore \quad$ Transverse axis $= 2a = 7 \;\therefore\; a = \dfrac{7}{2}$

$\therefore \quad$ Equation of the hyperbola is $\dfrac{x^2}{\dfrac{49}{4}} - \dfrac{y^2}{b^2} = 1$

It passes through the point $(5, -2)$

$\therefore \quad \dfrac{25}{\dfrac{49}{4}} - \dfrac{4}{b^2} = 1 \;\therefore\; \dfrac{100}{49} - \dfrac{4}{b^2} = 1$

$$\Rightarrow \qquad \frac{51}{49} = \frac{4}{b^2} \quad \therefore \qquad b^2 = \frac{196}{51}$$

The equation of the hyperbola is

$$\frac{x^2}{\frac{49}{4}} - \frac{y^2}{\frac{196}{51}} = 1 \Rightarrow \frac{4x^2}{49} - \frac{51y^2}{196} = 1$$

EXERCISE 4.11

Find the coordinates of the foci and the vertices, the eccentricity and the length of latus rectum of the hyperbolas:

1. $\frac{x^2}{16} - \frac{y^2}{9} = 1$ $\left[\textbf{Ans.} \text{ Foci } (\pm 5, 0); \text{ Vertices } (\pm 4, 0); e = \frac{5}{4}; \text{ Latus rectum} = \frac{a}{2}\right]$

2. $\frac{y^2}{9} - \frac{x^2}{27} = 1$ [**Ans.** Foci $(0, \pm 6)$; Vertices $(0, \pm 3)$; $e = 2$; Latus rectum = 18]

3. $9y^2 - 4x^2 = 36$ $\left[\textbf{Ans.} \text{ Foci } (0, \pm\sqrt{13}); \text{ Vertices } (0, \pm 2); e = \pm\frac{\sqrt{13}}{2}; \text{ Latus rectum} = 9\right]$

4. $16x^2 - 9y^2 = 576$ $\left[\textbf{Ans.} \text{ Foci } (\pm 10, 0); \text{ Vertices } (\pm 6, 0); e = \frac{5}{3}; \text{ Latus rectum} = \frac{64}{3}\right]$

5. $5y^2 - 9x^2 = 36$ $\left[\textbf{Ans.} \text{ Foci } \left(0, \pm\frac{2\sqrt{14}}{\sqrt{5}}\right); \text{ Vertices } \left(0, \pm\frac{6}{\sqrt{5}}\right); e = \frac{\sqrt{14}}{3}; \text{ Latus rectum} = \frac{4\sqrt{5}}{3}\right]$

6. $49y^2 - 16x^2 = 784$ $\left[\textbf{Ans.} \text{ Foci } (0, \pm\sqrt{65}); \text{ Vertices } (0, \pm 4); e = \frac{\sqrt{65}}{4}; \text{ Latus rectum} = \frac{49}{2}\right]$

Find the equation of the hyperbola satisfying the given conditions:

7. Vertices $(\pm 2, 0)$, foci $(\pm 3, 0)$. $\left[\textbf{Ans.} \frac{x^2}{4} - \frac{y^2}{5} = 1\right]$

8. Vertices $(0, \pm 5)$, foci $(0, \pm 8)$. $\left[\textbf{Ans.} \frac{y^2}{25} - \frac{x^2}{39} = 1\right]$

9. Foci $(\pm 5, 0)$, the transverse axis of length 8. $\left[\textbf{Ans.} \frac{x^2}{16} - \frac{y^2}{9} = 1\right]$

10. Foci $(0, \pm 13)$, the conjugate axis of the length 24. $\left[\textbf{Ans.} \frac{y^2}{25} - \frac{x^2}{144} = 1\right]$

11. Foci $(\pm 4, 0)$, the latus rectum is of length 12. $\left[\textbf{Ans.} \frac{x^2}{4} - \frac{y^2}{12} = 1\right]$

12. Foci $(0, \pm\sqrt{10})$, passing through (2, 3). $\left[\textbf{Ans.} \frac{y^2}{5} - \frac{x^2}{5} = 1\right]$

❑❑❑

CHAPTER 5

Three Dimensional Coordinate System

INSIDE THIS CHAPTER

5.1. INTRODUCTION

Let $X'OX$, $Y'OY$ and $Z'OZ$ are three mutually perpendicular lines intersecting at O. Point O is called origin. The lines $X'OX$, $Y'OY$ and $Z'OZ$ are called x axis, y-axis and z-axis respectively. These are also called the rectangular axis of coordinates. The planes formed by these coordinates axis are called the rectangular coordinate planes.

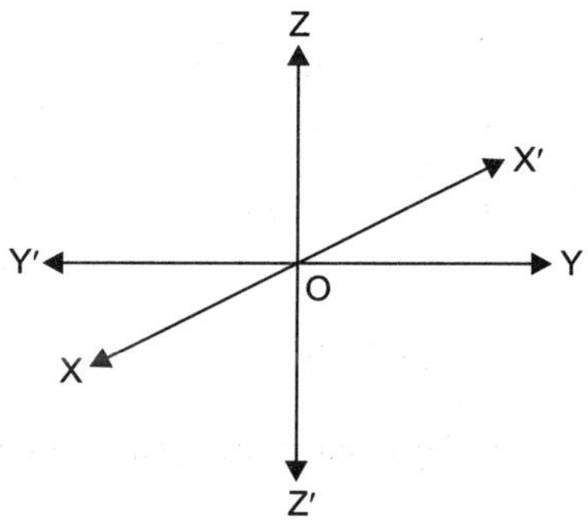

Let P be any point on the three dimensional coordinate system. The coordinates of P are the perpendicular distance from P on three mutually perpendicular coordinate planes, YOZ, ZOX and XOY respectively.

5.2. DISTANCE FORMULA

The distance between the point $P(x_1, y_1, z_1)$ and $Q(x_2, y_2, z_2)$ is given by

$$PQ = \sqrt{(x_2 - x_1)^2 + (y_2 - y_1)^2 + (z_2 - z_1)^2}$$

SOLVED EXAMPLES

Example 5.1. Find the distance between the points $P(2, 3, 4)$ and $Q(3, -2, 1)$.

Solution. The distance PQ is given by

$$PQ = \sqrt{(x_2 - x_1)^2 + (y_2 - y_1)^2 + (z_2 - z_1)^2}$$

$$= \sqrt{(3-2)^2 + (-2-3)^2 + (1-4)^2}$$

$$= \sqrt{1^2 + (-5)^2 + (-3)^2}$$

$$= \sqrt{1 + 25 + 9} = \sqrt{35} \text{ units}$$

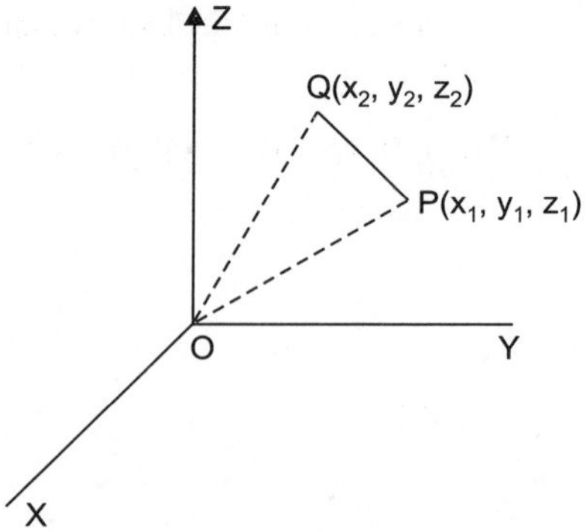

Example 5.2. Using distance formula, prove that following points are collinear:

(a) $P(4, -3, -1)$, $Q(5, -7, 6)$, $R(3, 1, -8)$

(b) $P(0, 7, -7)$, $Q(1, 4, -5)$, $R(-1, 10, -9)$

Solution. (*a*)

$$PQ = \sqrt{(5-4)^2 + (-7+3)^2 + (6+1)^2} = \sqrt{1+16+49} = \sqrt{66}$$

$$QR = \sqrt{(3-5)^2 + (1+7)^2 + (-8-6)^2} = \sqrt{4+64+196} = \sqrt{264} = 2\sqrt{66}$$

$$PR = \sqrt{(3-4)^2 + (1+3)^2 + (-8+1)^2} = \sqrt{1+16+49} = \sqrt{66}$$

Since $PQ + PR = QR$. Therefore given points are collinear

(*b*)

$$PQ = \sqrt{(1-0)^2 + (4-7)^2 + (-5+7)^2} = \sqrt{1+9+4} = \sqrt{14}$$

$$QR = \sqrt{(-1-1)^2 + (10-4)^2 + (-9+5)^2} = \sqrt{4+36+16} = \sqrt{56} = 2\sqrt{14}$$

$$PR = \sqrt{(-1-0)^2 + (10-7)^2 + (-9+7)^2} = \sqrt{1+9+4} = \sqrt{14}$$

Since $PQ + PR = QR$. Therefore given points are collinear.

5.3. SECTION FORMULA

Let $P(x, y, z)$ and $Q(x_2, y_2, z_2)$ be two points and let $R(x, y, z)$ be a point on PQ dividing it in the ratio $m : n$, then

$$x = \frac{mx_2 + nx_1}{m+n}$$

$$y = \frac{my_2 + ny_1}{m+n}$$

$$z = \frac{mz_2 + nz_1}{m+n}$$

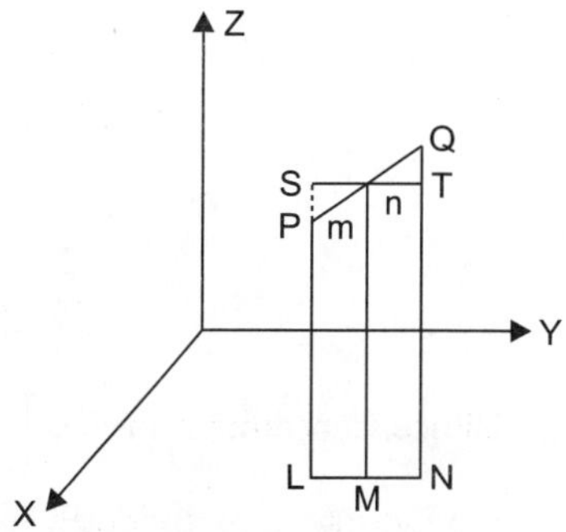

Corollary 1: If the point R divides PQ externally in the ratio $m : n$, then its coordinates are obtained by replacing n by $-n$ and the coordinates of R will be

$$\left(\frac{mx_2 + nx_1}{m-n}, \frac{my_2 - ny_1}{m-n}, \frac{mz_2 + nz_1}{m-n}\right)$$

Corollary 2: If the point R is the mid point of PQ, then the coordinates of R will be

$$\left(\frac{x_1 + x_2}{2}, \frac{y_1 + y_2}{2}, \frac{z_1 + z_2}{2}\right)$$

Centroid of a Triangle

Let ABC be a triangle whose vertices are $A(x_1, y_1, z_1)$, $B(x_2, y_2, z_2)$ and $C(x_2, y_2, z_2)$

AD is the median of the triangle

$\therefore$ Coordinates of point

$$D = \left(\frac{x_2 + x_3}{2}, \frac{y_2 + y_3}{2}, \frac{z_2 + z_3}{2}\right)$$

Let, G be the centroid of ΔABC, we know that centroid divides a median in the ratio 2 : 1.

$\therefore \qquad AG : GD = 2 : 1$

Coordinate of centroid G are

$$\left(\frac{\frac{2 \cdot x_2 + x_3}{2} + 1 \cdot x_1}{2+1}, \frac{\frac{2 \cdot y_2 + y_3}{2} + 1 \cdot y_1}{2+1}, \frac{\frac{2 \cdot z_2 + z_3}{2} + 1 \cdot z_1}{2+1} \right)$$

$$\left(\frac{x_1 + x_2 + x_3}{3}, \frac{y_1 + y_2 + y_3}{3}, \frac{z_1 + z_2 + z_3}{3} \right)$$

SOLVED EXAMPLES

Example 5.3. Find the coordinates of the point which divides the line segment joining the points (– 2, 3, 5) and (1, – 4, 6) in the ratio:

(*i*) 2 : 3 internally **(*ii*) 2 : 3 externally**

Solution. Let $P(-2, 3, 5)$ and $Q(1, -4, 6)$ be the given points

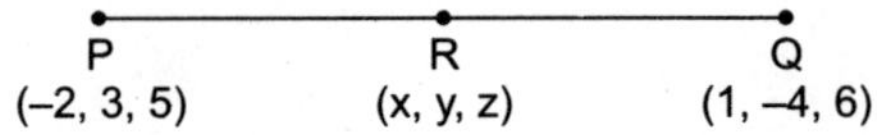

(*i*) Let $R(x, y, z)$ be the point which divide PQ internally in the ratio 2 : 3

Coordinates of R

$$x = \frac{2 \times 1 + 3 \times -2}{2+3} = \frac{-4}{5}$$

$$y = \frac{2 \times -4 + 3 \times 3}{2+3} = \frac{1}{5}$$

$$z = \frac{2 \times 6 \times 3 \times 5}{2+3} = \frac{27}{5}$$

Hence, coordinates of R is $\left(\frac{-4}{5}, \frac{1}{5}, \frac{27}{5} \right)$.

(*ii*) Let $R(x, y, z)$ divide PQ externally in the ratio 2 : 3, coordinates of R

$$x = \frac{2 \times 1 - 3 \times (-2)}{2-3} = \frac{2+6}{-1} = -8$$

$$y = \frac{2 \times (-4) - 3 \times 3}{2-3} = \frac{-17}{-1} = 17$$

$$z = \frac{2 \times 6 - 3 \times 5}{2-3} = \frac{-3}{-1} = 3$$

Hence, coordinates of R is (– 8, 17, 3).

Example 5.4. Given that P(3, 2, – 4), *Q*(5, 4, – 6) and *R*(9, 8, – 10) are collinear. Find the ratio in which *Q* divides *PR*.

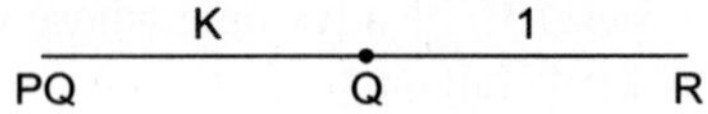

Solution. Let $Q(5, 4, -6)$ divide PR in the ratio $k : 1$

Then coordinates of Q are

$$5 = \frac{k \times 9 + 1 + 3}{k+1} \quad \Rightarrow \quad 5k + 5 = 9k + 3$$

$\Rightarrow \quad 4k = 2 \quad \therefore \quad k = \frac{1}{2}$

Hence, point Q divides PR in the ratio 1 : 2

Example 5.5. Find the third vertex of the triangle, whose centroid is origin and two vertices are (2, 4, 6) and (0, – 2, – 5).

Solution. Let ABC be a triangle whose two vertices be $A(2, 4, 6)$ and $B(0, -2, -5)$.

Let coordinates of C be (x, y, z).

Centroid of the triangle is $O(0, 0, 0)$.

Extend OC upto D, then D will be the mid-point of AB.

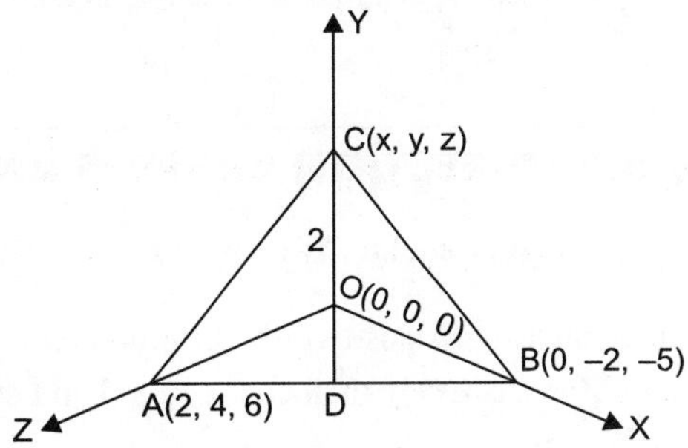

$\therefore \quad$ Centroid of $D = \left(\frac{2+0}{2}, \frac{4-2}{2}, \frac{6-5}{2}\right) = \left(1, 1, \frac{1}{2}\right)$

We know that centroid divides the median in the ratio 2 : 1

$\therefore \quad CO : OD = 2 : 1$

Coordinate of centroid O are,

$$(0, 0, 0) = \left(\frac{2\times1+1\times x}{2+1}, \frac{2\times1+1\times y}{2+1}, \frac{2\times\frac{1}{2}+1\times z}{2+1}\right)$$

$$(0, 0, 0) = \left(\frac{2+x}{3}, \frac{2+y}{3}, \frac{1+z}{3}\right)$$

$$\frac{2+x}{3} = 0 \Rightarrow x = -2$$

$$\frac{2+y}{3} = 0 \Rightarrow y = -2$$

$$\frac{1+z}{3} = 0 \Rightarrow z = -1$$

Hence, coordinates of C is $(-2, -2, -1)$.

EXERCISE 5.1

1. Find the distance between following pair of points:
   (*a*) $A(1, -1, 0)$ and $B(2, 1, 2)$ **[Ans.** 3]
   (*b*) $A(1, -1, 3)$ and $B(2, 3, -5)$ **[Ans.** 9]
   (*c*) $A(3, 2, -1)$ and $B(-1, -1, -1)$ **[Ans.** 5]
   (*d*) $A(-2, 3, 1)$ and $B(2, 1, 2)$ **[Ans.** $\sqrt{21}$]
2. Using distance formula show that following points are collinear
   (*a*) $A(3, -5, 1), B(-1, 0, 8), C(7, -10, -6)$
   (*b*) $A(3, -2, 4), B(1, 1, 1), C(-1, 4, -2)$
3. Determine the point of y axis which is equidistant from (3, 1, 2) and (5, 5, 2). **[Ans.** 0, 5, 0]
4. Show that the points $A(0, 1, 2)$, $B(2, -1, 3)$ and $C(1, -3, 1)$ are vertices of isosceles right angled triangle.
5. Find the coordinates of the point which divides the join of $A(2, -1, 4)$ and $B(4, 3, 2)$ in the ratio 2 : 3; (*i*) internally and (*ii*) externally. **[Ans.** 2, – 9, 8]

6. Find the ratio in which the line segment joining the point (2, – 1, 3) and (– 1, 2, 1) is divided by the plane $x + y + z = 5$. **[Ans.** (1 : 3 externally)]
7. Three vertices of a parallelogram *ABCD* are *A*(3, – 1, 2), *B*(1, 2, – 4) and *C*(– 1, 1, 2). Find the co-ordinates of the fourth vertex. **[Ans.** (1, – 2, 8)]
8. If the origin is the centroid of the triangle *PQR* with vertices *P*(2*a*, 2, 6), *Q*(– 4, 3*b*, – 10) and *R*(8, 14, 2*c*) then find the value of *a*, *b* and *c*. **[Ans.** – 2, – 16/3, 2]
9. Find the centroid of a triangle mid points of whose sides are (1, 2, – 3), (3, 0, 1) and (1, 1, – 4). **[Ans.** *A*(1, – 1, 0), *B*(– 3, 3, – 8) and *C*(5, 1, 2)]

5.4. DIRECTION COSINES AND DIRECTION RATIOS

Direction cosine: If α, β and γ are the angles which vector $\overrightarrow{OP}$ makes with positive directions of coordinates axis *OX, OY* and *OZ* respectively then cos α, cos β and cos γ are known as the direction cosines of $\overrightarrow{OP}$ and are denoted by *l, m, n* respectively.

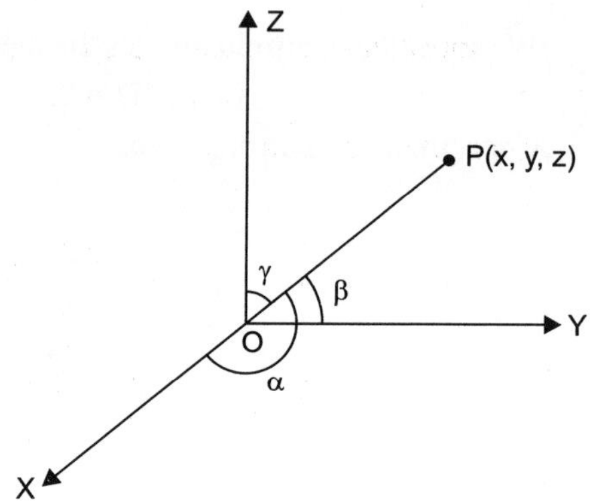

The angles of α, β and γ are called direction angles of $\overrightarrow{OP}$.

Direction ratios: Let *l*, *m*, *n* are direction cosines of a vector $\overrightarrow{OP}$ and *a, b, c* are three numbers such that

$$\frac{l}{a} = \frac{m}{b} = \frac{n}{c}$$

Then *a, b, c* are called direction ratios of $\overrightarrow{OP}$.

Direction Cosines of the Vector $\vec{r} = a\hat{i} + b\hat{j} + c\hat{k}$

Let α, β, γ be the direction angles and *l, m, n* are direction ratio of $\vec{r}$

We know that $l = \cos\alpha$, $m = \cos\beta$, $n = \cos\gamma$

We have
$$\vec{r}\cdot\hat{i} = (a\hat{i} + b\hat{j} + c\hat{k})\cdot\hat{i} = a$$
$$\vec{r}\cdot\hat{j} = (a\hat{i} + b\hat{j} + c\hat{k})\cdot\hat{j} = b$$
$$\vec{r}\cdot\hat{k} = (a\hat{i} + b\hat{j} + c\hat{k})\cdot\hat{j} = c$$

Now since
$$\vec{r}\cdot\hat{i} = a \quad\Rightarrow\quad |\vec{r}|\cdot|\hat{i}|\cos\alpha = a \quad\Rightarrow\quad \cos\alpha = \frac{a}{|\vec{r}|}$$

Now
$$l = \cos\alpha = \frac{a}{|\vec{r}|} = \frac{a}{\sqrt{a^2 + b^2 + c^2}}$$

Similarly
$$m = \cos\beta = \frac{b}{\sqrt{a^2 + b^2 + c^2}} \quad\text{and}\quad n = \cos\gamma = \frac{c}{\sqrt{a^2 + b^2 + c^2}}$$

Note: 1. $l^2 + m^2 + n^2 = 1$

Proof: $l^2 + m^2 + n^2$

$$\Rightarrow \quad \frac{a^2}{a^2+b^2+c^2} + \frac{b^2}{a^2+b^2+c^2} + \frac{c^2}{a^2+b^2+c^2}$$

$$\Rightarrow \quad \frac{a^2+b^2+c^2}{a^2+b^2+c^2} = 1$$

2. If $A(x_1, y_1, z_1)$ and $B(x_2, y_2, z_2)$ are two points on a line, then its direction ratios are proportional to $x_2 - x_1$, $y_2 - y_1, z_2 - z_1$.

3. Two parallel vectors has proportional direction ratios *i.e.*, $\frac{a_1}{a_2} = \frac{b_1}{b_2} = \frac{c_1}{c_2}$

4. If two lines having direction ratios a_1, b_1, c_1 and a_2, b_2, c_2 are perpendicular then $a_1a_2 + b_1b_2 + c_1c_2 = 0$

SOLVED EXAMPLES

Example 5.6. Find the direction cosines of a line whose direction ratios are 1, 3, – 5.

Solution. Let $a = 1, b = 3, c = -5$

then
$$l = \frac{a}{\sqrt{a^2+b^2+c^2}} = \frac{1}{\sqrt{1+9+25}} = \frac{1}{\sqrt{35}}$$

$$m = \frac{b}{\sqrt{a^2+b^2+c^2}} = \frac{3}{\sqrt{35}}$$

$$n = \frac{c}{\sqrt{a^2+b^2+c^2}} = \frac{-5}{\sqrt{35}}$$

Example 5.7. Can a vector have direction angles 45°, 60° and 120°.

Solution.
$$l = \cos 45° = \frac{1}{\sqrt{2}}$$

$$m = \cos 60° = \frac{1}{2}$$

$$n = \cos 120° = \cos (90° + 30°) = -\sin 30° = -\frac{1}{2}$$

Now
$$l^2 + m^2 + n^2 = \frac{1}{2} + \frac{1}{4} + \frac{1}{4} = 1$$

Yes, a vector have direction angles 45°, 60° and 120°.

EXERCISE 5.2

1. Find the direction cosines of the following vectors:

(*a*) $-2i + 6\hat{j} - 3\hat{k}$ (*b*) $3\hat{i} - 4\hat{k}$ (*c*) $-2i + 6\hat{j} - 3\hat{k}$ (*d*) $5\hat{i} + \hat{k}$

2. Find the direction cosines of a line segment whose direction ratios are

(*a*) 1, – 4, 2 (*b*) 2, 1, 1 (*c*) 3, – 1, 1 (*d*) 1, 0, – 1

3. Find direction ratio and direction cosine of line joining points:
   (*a*) $A(2, 1, 4)$ and $B(-2, -1, 2)$ (*b*) $A(1, 3, 1)$ and $B(5, 2, 1)$
   (*c*) $A(1, 1, 0)$ and $B(0, 0, 1)$ (*d*) $A(2, -2, -2)$ and $B(1, 0, -2)$
4. Find the angle at which following vectors are inclined to coordinate axis
   (*a*) $i + j + k$ (*b*) $\hat{j} - \hat{k}$ (*c*) $4\hat{i} + 8\hat{j} + \hat{k}$ (*d*) $2\hat{i} + \hat{j} + 3\hat{k}$

5.5. STRAIGHT LINE

In this topic we will study the cartesian equation of straight line under different conditions

1. The cartesian equation of a straight line passing through a fixed point (x_1, y_1, z_1) and having direction ratios proportional to a, b, c is given by

$$\frac{x - x_1}{a} = \frac{y - y_1}{b} = \frac{z - z_1}{c} \quad \text{(symmetrical form of line)}$$

Note:

(*a*) The coordinate of any point on line $\frac{x - x_1}{a} = \frac{y - y_1}{b} = \frac{z - z_1}{c}$ are $(x_1 + \lambda a, y_1 + \lambda b, z_1 + \lambda c)$ where λ is the parameter.

(*b*) Since the direction cosines of a line are also direction ratios therefore the equation of the line passing through (x_1, y_1, z_1) and having direction cosines l, m, n is $\frac{x - x_1}{l} = \frac{y - y_1}{m} = \frac{z - z_1}{n}$.

2. The cartesian equation of a line passing through two given points (x_1, y_1, z_1) and (x_2, y_2, z_2) is given

$$\frac{x - x_1}{x_2 - x_1} = \frac{y - y_1}{y_2 - y_1} = \frac{z - z_1}{z_2 - z_1} \quad \text{(two point form)}$$

5.6. VECTOR EQUATION OF A LINE

The vector equation of a line passing through a fixed point with position vector $\vec{a}$ and parallel to a given vector $\vec{b}$ is $\vec{r} = \vec{a} + \lambda \vec{b}$, where λ is scalar.

Here $\vec{r}$ is a position vector of any point $P(x, y, z)$ on the line. Therefore $\vec{r} = x\hat{i} + y\hat{j} + z\hat{k}$. The position vector of any point on the line is taken as $\vec{a} + \lambda \vec{b}$.

1. The vector equation of a line passing through two points with position vector $\vec{a}$ and $\vec{b}$ is

$$\vec{r} = \vec{a} + \lambda(\vec{b} - \vec{a})$$

Reduction From Cartesian to Vector Form and Vice-versa

Let the Cartesian form of equation is

$$\frac{x - x_1}{a} = \frac{y - y_1}{b} = \frac{z - z_1}{c}$$

Given equation of line passes through (x_1, y_1, z_1) and its direction ratio are proportional to a, b, c.
The vector equation is given by

$$\vec{r} = (x_1\hat{i} + y_1\hat{j} + z_1\hat{k}) + \lambda(a\hat{i} + b\hat{j} + c\hat{k})$$

Conversely if the vector equation of a line is

$$\vec{r} = \vec{a} + \lambda \vec{b}$$

where $\vec{a} + \lambda \vec{b} = x_1\hat{i} + y_1\hat{j} + z_1\hat{k}$ and $\vec{b} = a\hat{i} + b\hat{j} + c\hat{k}$

then Cartesian equation of line is given by

$$\frac{x - x_1}{a} = \frac{y - y_1}{b} = \frac{z - z_1}{c}$$

SOLVED EXAMPLES

Example 5.8. Find the cartesian equation of a line passing through the points *A*(1, 4, 8) and *B*(4, 9, 3).

Solution. The cartesian equation of line passing through *A*(1, 4, 3) and *B*(4, 9, 3) is given by

$$\frac{x-1}{4-1} = \frac{y-4}{9-4} = \frac{z-8}{3-8}$$

$$\frac{x-1}{3} = \frac{y-4}{5} = \frac{z-8}{-5}$$

Example 5.9. The points *A*(6, 7, 12), *B*(4, 5, 6) and *C*(3, 4, 1) are three points of parallelogram *ABCD*. Find cartesian equation of *AB* and *BC* and also find coordinate of *D*. Also reduce the equation in vector form.

Solution. Cartesian equation of *AB* is given by

$$\frac{x-6}{4-6} = \frac{y-7}{5-7} = \frac{z-12}{6-12}$$

$$\Rightarrow \quad \frac{x-6}{-2} = \frac{y-7}{-2} = \frac{z-12}{-6}$$

$$\Rightarrow \quad \frac{x-6}{1} = \frac{y-7}{1} = \frac{z-12}{3}$$

Since *BC* passes through *B*(4, 5, 6) and *C*(3, 4, 1) therefore equation of *BC* is given by

$$\frac{x-4}{3-4} = \frac{y-5}{4-5} = \frac{z-6}{1-6}$$

$$\Rightarrow \quad \frac{x-4}{-1} = \frac{y-5}{-1} = \frac{z-6}{-5}$$

$$\Rightarrow \quad \frac{x-4}{1} = \frac{y-5}{1} = \frac{z-6}{5}$$

Let coordinate *D* be (*x*, *y*, *z*). Since ABCD is a parallelogram therefore diagonal AC and BD bisect each other. Therefore AC and BD must have the same mid point.

The coordinate of mid point of AC are $\left(\frac{6+3}{2}, \frac{7+4}{2}, \frac{12+1}{2}\right)$ *i.e.*, $\left(\frac{9}{2}, \frac{11}{2}, \frac{13}{2}\right)$ and coordinates of midpoitns of BD are $\left(\frac{x+4}{2}, \frac{y+5}{2}, \frac{z+6}{2}\right)$. Since the mid-points coincide therefore

$$\frac{x+4}{2} = \frac{9}{2} \quad \Rightarrow \quad x = 9 - 4 = 5$$

$$\frac{y+5}{2} = \frac{11}{2} \qquad \Rightarrow \qquad y = 11 - 5 = 6$$

$$\frac{z+6}{2} = \frac{13}{2} \qquad \Rightarrow \qquad z = 13 - 6 = 7$$

The coordinates of D are (5, 6, 7).

The vectors equation of line is $\vec{r} = (6\hat{i} + 7\hat{j} + 12\hat{k}) + \lambda(\hat{i} + \hat{j} + 3\hat{k})$.

Example 5.10. The cartesian equation of a line are $6x - 2 = 3y + 1 = 2z - 2$. Find the direction ratio of line. Also find the vectors equation of the line.

Solution. In symmetrical form of line the coefficient of x, y, z are unity

$$\therefore \qquad 6x - 2 = 3y + 1 = 2z - 2$$

$$\Rightarrow \qquad 6\left(x - \frac{1}{3}\right) = 6\left(x + \frac{1}{3}\right) = 2(z-1)$$

$$\Rightarrow \qquad \frac{x - \frac{1}{3}}{1} = \frac{y + \frac{1}{3}}{2} = \frac{z-1}{3}$$

Therefore the direction ratio of line are 1, 2, 3.

The vector equation is given by $\vec{r} = \left(\frac{-1}{3}\hat{i} + \frac{1}{3}\hat{j} + \hat{k}\right) + \lambda(\hat{i} + 2\hat{j} + 3\hat{k})$.

Example 5.11. Find the vector equation of a line which passes through the point with position vector $3\hat{i} - 2\hat{j} + \hat{k}$ and in direction $2\hat{i} + \hat{j} - 3\hat{k}$.

Solution. Here $\vec{a} = 3\hat{i} - 2\hat{j} + \hat{k}$ and $\vec{b} = 2\hat{i} + \hat{j} - 3\hat{k}$

So the vector equation of the required line is

$$\vec{r} = \vec{a} + \lambda\vec{b}$$

$$\vec{r} = (3\hat{i} - 2\hat{j} + \hat{k}) + \lambda(2\hat{i} + \hat{j} - 3\hat{k})$$

Example 5.12. Find the vector equation of line which passes through the position vectors $2\hat{i} + 3\hat{j} + 4\hat{k}$ and $\hat{i} + 2\hat{j} + \hat{k}$.

Solution. The vector equation of line passing through

$\vec{a} = 2\hat{i} + 3\hat{j} + 4\hat{k}$ and $\vec{b} = \hat{i} + 2\hat{j} + \hat{k}$ is given by

$$\vec{r} = \vec{a} + \lambda(\vec{b} - \vec{a})$$

$$\vec{r} = (2\hat{i} + 3\hat{j} + 4\hat{k}) + \lambda[(\hat{i} + 2\hat{j} + \hat{k}) - (2\hat{i} + 3\hat{j} + 4\hat{k})]$$

$$= (2\hat{i} + 3\hat{j} + 4\hat{k}) + \lambda(-\hat{i} - \hat{j} - 3\hat{k})$$

Example 5.13. Find the vector equation of the line through $A(-3, 9, -7)$ and $B(7, -7, 6)$.

Solution. The vector equation of line passing through the position vector $\vec{a}$ and $\vec{b}$ is given by

$$\vec{r} = \vec{a} + (\vec{b} - \vec{a}) \text{ where } \lambda \text{ is a scalar}$$

$$\vec{a} = -3\hat{i} + 9\hat{j} - 7\hat{k}$$

$$\vec{b} = 7\hat{i} - 7\hat{j} + 6\hat{k}$$

Therefore the equation of the line is given by

$$\vec{r} = (-3\hat{i} + 9\hat{j} - 7\hat{k}) + \lambda[(7\hat{i} - 7\hat{j} + 6\hat{k}) - (-3\hat{i} + 9\hat{j} - 7\hat{k})]$$

$$\vec{r} = (-3\hat{i} + 9\hat{j} - 7\hat{k}) + \lambda(10\hat{i} - 16\hat{j} + 13\hat{k})$$

EXERCISE 5.3

1. Prove that points (1, 2, 3), (4, 0, 4), (– 2, 4, 2) and (7, – 2, 5) are collinear.
2. Find the cartesian equation of a line passing through (2, – 1, 2) and parallel to the line whose equation are $\frac{x-3}{1} = \frac{y-1}{2} = \frac{z+1}{-2}$. Also find the vector equation of line.
   $\left[\textbf{Ans. } \frac{x-2}{1} = \frac{y+1}{2} = \frac{z-2}{-2} \quad \vec{r} = 2\hat{i} - \hat{j} - 2\hat{k} + \lambda(\hat{i} + 2\hat{j} - 2\hat{k})\right]$
3. The cartesian equations of a line are $3x + 1 = 6y - 2z = 1 - z$. Find the fixed point it passes through and its direction ratios.
   $\left[\textbf{Ans. } \left(\frac{-1}{3}, \frac{1}{3}, 1\right) \text{ and } 2, 1, -6\right]$
4. Find the direction cosines of line $\frac{4-x}{2} = \frac{y}{6} = \frac{1-z}{3}$.
   $\left[\textbf{Ans. } \left(\frac{-2}{7}, \frac{6}{7}, \frac{3}{7}\right)\right]$
5. Find the vector equation of line passing through points with position vectors $\hat{i} - 2\hat{j} + 3\hat{k}$ and $3\hat{i} - 4\hat{j} + \hat{k}$. Also find the cartesian equation.
   $\left[\textbf{Ans. } \vec{r} = (i - 2\hat{j} + 3\hat{k}) + \lambda(2\hat{i} - 2\hat{j} - 2\hat{k}) \quad \frac{x-1}{2} = \frac{y+2}{-2} = \frac{z+3}{-2}\right]$

5.7. ANGLE BETWEEN TWO LINES

Cartesian Form: Let

$$\frac{x-x_1}{a_1} = \frac{y-y_1}{b_1} = \frac{z-z_1}{c_1} \quad \text{and} \quad \frac{x-x_2}{a_2} = \frac{y-y_2}{b_2} = \frac{z-z_2}{c_2}$$

be cartesian equation of two lines. Then, θ is the angle between two lines and is given by

$$\cos\theta = \frac{a_1a_2 + b_1b_2 + c_1c_2}{\sqrt{a_1^2 + b_1^2 + c_1^2}\,\sqrt{a_2^2 + b_2^2 + c_2^2}}$$

Condition for perpendicularity: If two lines are perpendicular then $a_1a_2 + b_1b_2 + c_1c_2 = 0$

Condition for parallelism: If two lines are parallel then

$$\frac{a_1}{a_2} = \frac{b_1}{b_2} = \frac{c_1}{c_2}$$

Vector Form: Let the vector equation of two lines be

$$\vec{r} = \vec{a_1} + \lambda_1 \vec{b_1} \quad \text{and} \quad \vec{r} = \vec{a_2} + \lambda_2 \vec{b_2}$$

The angle between these two lines is given by

$$\cos\theta = \frac{\vec{b_1} \cdot \vec{b_2}}{|\vec{b_1}||\vec{b_2}|}$$

Condition for perpendicularity: If lines are perpendicular then $\vec{b_1} \cdot \vec{b_2} = 0$

Condition for parallelism: If lines are parallel then $\vec{b_1} \cdot \lambda \vec{b_2}$ for scalar λ.

Note: A line passing through a point having position vector $\vec{b_1} \cdot \lambda \vec{b_2}$ and perpendicular to lines $\vec{r} = \vec{a_1} + \lambda \vec{b_1}$ and $\vec{r} = \vec{a_2} + \lambda \vec{b_2}$ is parallel to $\vec{b_1} \times \vec{b_2}$ and its vector equation is given by

$$\vec{r} = \vec{a} + \lambda(\vec{b_1} \times \vec{b_2})$$

SOLVED EXAMPLES

Example 5.14. Find the angle between the lines

$$\frac{x+3}{3} = \frac{y+32}{-1}, z = 3 \quad \text{and} \quad \frac{x+1}{2} = \frac{2y+2}{3} = \frac{z+5}{2}.$$

Solution. The equation is not in symmetrical form. First equation can be written as

$$\frac{x+3}{3} = \frac{y+2}{-1} = \frac{z-3}{0}$$

$$a_1 = 3, b_1 = -1, c_1 = 0$$

The second equation of line can be written as

$$\frac{x+1}{2} = \frac{y+1}{\frac{3}{2}} = \frac{z+5}{2}$$

$$\Rightarrow \quad a_2 = 2, b_2 = \frac{3}{2}, c_2 = 2$$

If θ is the angle between two lines then

$$\cos\theta = \frac{a_1a_2 + b_1b_2 + c_1c_2}{\sqrt{a_1^2 + b_1^2 + c_1^2}\sqrt{a_2^2 + b_2^2 + c_2^2}}$$

$$\cos\theta = \frac{3\times 2 + (-1)\times\left(\frac{3}{2}\right) + 0\times 2}{\sqrt{9+1+0}\sqrt{4+\frac{9}{9}+4}} = \frac{\frac{9}{2}}{\frac{\sqrt{10}\sqrt{41}}{2}} = \frac{9}{\sqrt{410}}$$

$$\therefore \quad \theta = \cos^{-1}\frac{9}{\sqrt{410}}$$

Example 5.15. Find the equation of line passing through (– 1, 2, 0) and parallel to line

$$\frac{x-1}{2} = \frac{y-1}{1} = \frac{z}{-2}.$$

Solution. Since the required is parallel to $\frac{x-1}{2} = \frac{y-1}{1} = \frac{z}{-2}$. Therefore the direction ratio of parallel line will be 2, 1, – 2.

The required equation of line will be

$$\frac{x+1}{2} = \frac{y-2}{1} = \frac{z}{-2}$$

Example 5.16. Find the equation of line which passes through (2, – 1, 3) and perpendicular to line $\frac{x-1}{2} = \frac{y-1}{-2} = \frac{z+1}{1}$ and $\frac{x-2}{2} = \frac{y+1}{2} = \frac{z+3}{2}$.

Solution. Let the direction ratio of the required line be a, b, c. Since it is perpendicular to two given lines therefore

$$2a - 2b + c = 0 \quad \ldots(i)$$

$$a + 2b + 2c = 0 \quad \ldots(ii)$$

Solving (*i*) and (*ii*) using cross multiplication we get

$$\frac{a}{-6} = \frac{b}{-3} = \frac{c}{6} \quad \Rightarrow \quad \frac{a}{2} = \frac{b}{1} = \frac{c}{-2}$$

Thus the required line passes through (2, – 1, 3) therefore the equation of line is

$$\frac{x-2}{2} = \frac{y+1}{1} = \frac{z-2}{-2}$$

Example 5.17. Find the angle between the lines

$$\vec{r} = (\hat{i} + 3\hat{j} - 4\hat{k}) + \lambda_1(\hat{i} + \hat{j} + \hat{k})$$

and

$$\vec{r} = (2\hat{i} + \hat{j} + 3\hat{k}) + \lambda_2(2\hat{i} + \hat{j} + 2\hat{k})$$

Solution. Let θ be the angle between two lines

then
$$\cos\theta = \frac{\vec{b_1} \cdot \vec{b_2}}{|\vec{b_1}||\vec{b_2}|}$$

where
$$\vec{b_1} = (i + \hat{j} + \hat{k}) \quad \text{and} \quad \vec{b_2} = 2\hat{i} + \hat{j} + 2\hat{k}$$

$$|\vec{b_1}| = \sqrt{1^2 + 1^2 + 1^2} = \sqrt{3}$$

$$|\vec{b_2}| = \sqrt{2^2 + 1^2 + 2^2} = \sqrt{9} = 3$$

∴
$$\cos\theta = \frac{(\hat{i} + \hat{j} + \hat{k}) \cdot (2\hat{i} + \hat{j} + 2\hat{k})}{\sqrt{3} \times 3} = \frac{2+1+2}{3\sqrt{3}} = \frac{5}{3\sqrt{3}}$$

$$\theta = \cos^{-1}\left(\frac{5}{3\sqrt{3}}\right)$$

Example 5.18. Find the equation of a line passing through a point (2, 3, 4) and parallel to line

$$\vec{r} = (2\hat{i} + \hat{j} + \hat{k}) + \lambda(2\hat{i} + 3j + 4\hat{k})$$

Solution. The given line is parallel to $2\hat{i} + 3\hat{j} + 4\hat{k}$ and required line is also parallel to this line and passes through (2, 3, 4). Therefore the vector equation of line is

$$\vec{r} = (2\hat{i} + 3\hat{j} + 4\hat{k}) + \lambda_1(2\hat{i} + 3j + 4\hat{k})$$

Example 5.19. A line passes through (3, – 2, 1) and is perpendicular to $\vec{r} = (2\hat{i} + \hat{k}) + \lambda_1(2\hat{i} - 2\hat{j} + \hat{k})$ and $\vec{r} = (3\hat{i} - \hat{j} + \hat{k}) + \lambda_2(\hat{i} + 2\hat{j} + 2\hat{k})$. Obtain the equation of line.

Solution. The required line is parallel to the line $\vec{b_1} \times \vec{b_2}$ where $\vec{b_1} = 2\hat{i} - 2\hat{j} + \hat{k}$ and $\vec{b_2} = \hat{i} + 2\hat{j} + 2\hat{k}$

$$\vec{b_1} \times \vec{b_2} = \begin{vmatrix} \hat{i} & \hat{j} & \hat{k} \\ 2 & -2 & 1 \\ 1 & 2 & 2 \end{vmatrix} = \hat{i}(-4-2) - \hat{j}(4-1) + \hat{k}(4+2)$$

$$= -6\hat{i} - 3\hat{j} + 6\hat{k}$$

Thus the required line passes through (3, – 2, 1) and is parallel to vector $-6\hat{i} - 3\hat{j} + 6\hat{k}$.

Therefore the equation of line is

$$\vec{r} = (3\hat{i} - 2\hat{j} + \hat{k}) - \lambda_3(-6\hat{i} - 3\hat{j} + 6\hat{k})$$

EXERCISE 5.4

1. Find the angle between following pair of lines

   (*a*) $\frac{x-4}{1} = \frac{y+1}{2} = \frac{z}{-2}$ and $\frac{x-1}{-2} = \frac{y+1}{-4} = \frac{z-2}{4}$ (*b*) $\frac{x-3}{1} = \frac{y-2}{2} = \frac{z+4}{2}$ and $\frac{x}{3} = \frac{y-5}{2} = \frac{z+2}{6}$

   (*c*) $\frac{x-5}{5} = \frac{y-3}{-12} = \frac{z-4}{13}$ and $\frac{x-1}{-3} = \frac{y-2}{4} = \frac{z-4}{5}$

   $\left[\textbf{Ans.}\ (a)\ 0°\ (b)\ \cos^{-1}\left(\frac{19}{21}\right)\ (c)\ \cos^{-1}\left(\frac{1}{65}\right)\right]$

2. Find the equation of line passing through and parallel to line $\frac{x-4}{9} - \frac{y-2}{6} = \frac{z-4}{5}$.

   $\left[\textbf{Ans.}\ \frac{x-2}{4} - \frac{z-1}{6} = \frac{y-1}{5}\right]$

3. Find the equation of line passing through (2, 1, 3) and perpendicular to $\frac{x-1}{1} = \frac{y+2}{2} = \frac{z-4}{3}$ and $\frac{x-4}{-3} = \frac{y}{2} = \frac{z-1}{5}$.

   $\left[\textbf{Ans.}\ \frac{x-2}{2} = \frac{y-1}{-7} = \frac{z-3}{4}\right]$

4. Show that lines $\frac{x-4}{7} = \frac{y-5}{-5} = \frac{z}{1}$ and $\frac{x}{1} = \frac{y}{2} = \frac{z-1}{3}$ are perpendicular to each other.

5. If the lines $\frac{x+1}{-3} = \frac{y+3}{2\lambda} = \frac{z+3}{2}$ and $\frac{x+3}{3\lambda} = \frac{y-1}{1} = \frac{z-6}{5}$ are perpendicular to each other then find the value of λ.

   $\left[\textbf{Ans.}\ \lambda = \frac{-10}{7}\right]$

6. Find the angle between the pair of lines

(a) $\vec{r} = (-4\hat{i} + \hat{j} - 3\hat{k}) + \lambda_1(3\hat{i} + 5\hat{j} + 4\hat{k})$ and $\vec{r} = (-\hat{i} + 4\hat{j} + 5\hat{k}) + \lambda_2(\hat{i} + \hat{j} + 2\hat{k})$

(b) $\vec{r} = (\hat{i} + 2\hat{j} + 3\hat{k}) + \lambda_1(2\hat{i} + 3\hat{j} - 3\hat{k})$ and $\vec{r} = (-3\hat{i} + 5\hat{j} + \hat{k}) + \lambda_2(-\hat{i} + 8\hat{j} + 4\hat{k})$

(c) $\vec{r} = (5\hat{i} - 3\hat{j} + 3\hat{k}) + \lambda_1(\hat{i} - \hat{j} + \hat{k})$ and $\vec{r} = (2\hat{i} - \hat{j} + 6\hat{k}) + \lambda_2(3\hat{i} + 4\hat{j} + 5\hat{k})$

$\left[\textbf{Ans. } (a) \cos^{-1}\left(\frac{8}{5\sqrt{3}}\right), (b) \cos^{-1}\left(\frac{10}{9\sqrt{22}}\right), (c) \frac{\pi}{2}\right]$

7. Find the vector equationof the line passing through the point (2, – 1, – 1) which is parallel to line $6x - 2 = 3y + 1 = 2z - 2$.

[**Ans.** $\vec{r} = (2\hat{i} - \hat{j} - \hat{k}) + \lambda(\hat{i} + 2\hat{j} + 3\hat{k})$]

5.8. THE PLANE

A plane is a surface such that if any two points are taken on it, the line segment joining them lies completely on the surface.

Every first degree equation in x, y and z represent a plane *i.e.*, $ax + by + cz + d = 0$ is the general equation of a plane.

5.9. EQUATION OF A PLANE PASSING THROUGH A GIVEN POINT

The general equation of plane passing through (x_1, y_1, z_1) is $a(x - x_1) + b(y - y_1) + c(z - z_1) = 0$ where a, b and c are constants.

5.10. INTERCEPT FORM OF PLANE

The equation of plane intercepting lengths a, b and c with x-axis, y-axis and z-axis respectively is $\frac{x}{a} + \frac{y}{b} + \frac{z}{c} = 1$.

The equation is called intercept form because a, b and c are x, y, z intercepts of plane.

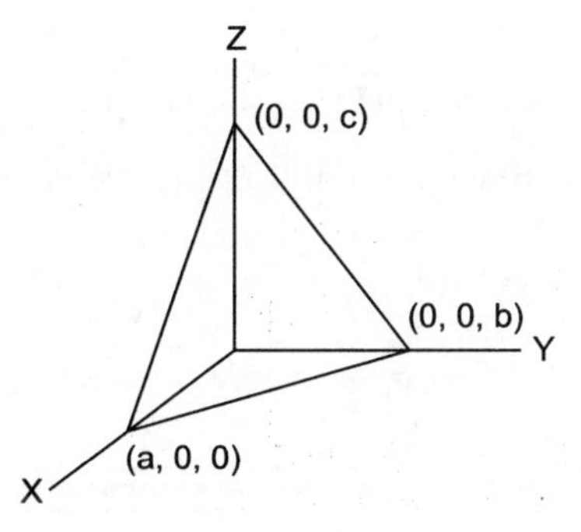

5.11. VECTOR EQUATION OF A PLANE PASSING THROUGH GIVEN POINT AND A NORMAL TO GIVEN VECTOR

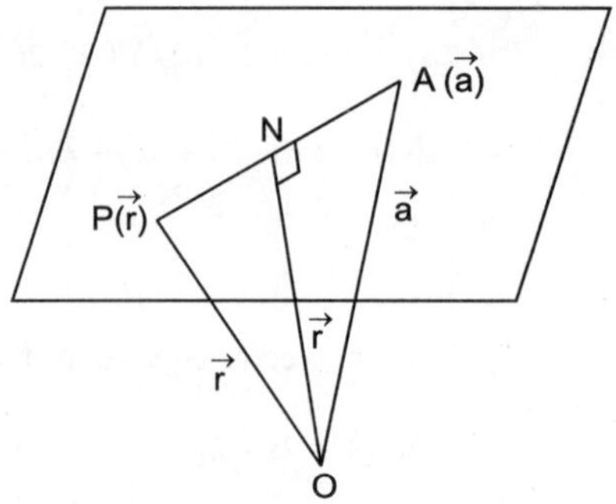

The vector equation of a plane passing through a point having position vector $\vec{a}$ and normal to vector $\vec{x}$ is $(\vec{r} - x\vec{a}) \cdot \vec{x} = 0$ or $\vec{r} \cdot \vec{x} = \vec{a} \cdot \vec{x}$.

The above equation can be also be written as $\vec{r} \cdot \vec{n} = \vec{d}$ where $\vec{d} = \vec{a} \cdot \vec{n}$.

5.12. EQUATION OF A PLANE IN NORMAL FORM

The vector equation of a plane normal to unit vector $\hat{n}$ and at a distance from the origin is $\vec{r} \cdot \hat{n} = d$.

In cartesian form, if l, m, n are direction cosines of the normal to a given plane which is at a distance p from the origin then the equation of plane is $lx + my + nz = p$.

Note: To reduce a given equation of plane *i.e.*, $ax + by + cz + d = 0$ to normal form, following steps will be followed:

Step 1. Keep terms containing x, y, z in left hand side and remaining terms in right hand side of equation.

Step 2. Make all the constant positive in right hand side of equation.

Step 3. Divide each term by $\sqrt{a^2 + b^2 + c^2}$.

The coefficient of x, y, z are *dc's* of the normal to the plane and constant on RHS gives the distance of plane from origin.

5.13. VECTOR EQUATION OF PLANE PASSING THROUGH THREE GIVEN POINTS

Let A, B and C be three points on a plane having position vector $\vec{a}, \vec{b}$ and $\vec{c}$ respectively.

Then the vector equation of plane passing through these three points is given by

$$(\vec{r} - \vec{a}) \cdot (\vec{a} \times \vec{b} + \vec{b} \times \vec{c} + \vec{c} \times \vec{a}) = 0$$

or
$$(\vec{r} - \vec{n}) = \vec{a} \cdot \vec{n} \text{ where } \vec{n} = \overrightarrow{AB} \times \overrightarrow{AC}$$

SOLVED EXAMPLES

Example 5.20. Find the equation of plane passing through the points (2, 1, 6), (1, 1, – 1) and (1, 2, 1).

Solution. The general expansion of plane passing through (2, 1, 6) is

$$a(x - 2) + b(y - 1) + c(z - 6) = 0 \qquad ...(i)$$

Since the given plane also passes through (1, 1, – 1) and (1, 2, 1)

$$\therefore \quad a(1-2)+b(1-1)+c(-1-6)=0 \quad \text{...(ii)}$$

$$a(1-2)+b(2-1)+c(1-6)=0 \quad \text{...(iii)}$$

$$\Rightarrow \quad -a-7c=0 \quad \text{...(iv)}$$

$$-a+b-5c=0 \quad \text{...(v)}$$

Solving (*iv*) and (*v*) we get

$$\frac{a}{7}=\frac{b}{2}=\frac{c}{-1}=\lambda$$

$$a=7\lambda,\ b=2\lambda,\ c=-\lambda$$

Substituting the values in (*i*), we get

$$5\lambda(x-2)+2\lambda(y-1)-\lambda(z-6)=0$$

$$\Rightarrow \quad 5(x-2)+2(y-1)-(z-6)=0$$

$$5x+2y-z-6=0$$

Example 5.21. Write the expansion of a plane whose intercept on the coordinate axis are – 8, 3 and 7.

Solution. The equation of plane with *a, b* and *c* intercept is

$$\frac{x}{a}+\frac{y}{b}+\frac{z}{c}=1$$

Here $a=-8,\ b=3,\ c=7$

$\therefore$ Equation of plane is

$$\frac{x}{-8}+\frac{y}{3}+\frac{z}{7}=1$$

$$\Rightarrow \quad -21x+56y+24z=168$$

Example 5.22. Reduce the equation 6x + 3y – 2z = 16 to intercept form. Also find the intercepts on the coordinates axis.

Solution. The equation of plane is $6x+3y-2z=16$

$$\frac{6x}{16}+\frac{3y}{16}-\frac{2z}{16}=0$$

$$\frac{x}{\frac{8}{3}}+\frac{y}{\frac{16}{3}}+\frac{z}{-8}=0$$

Thus equation is of $\frac{x}{a}+\frac{y}{b}+\frac{z}{c}=1$.

The intercepts are $a=\frac{8}{3}$, $b=\frac{16}{3}$ and $c=-8$.

Example 5.23. Find the vector equation of plane passing through $i+2j+\hat{k}$ and perpendicular to vector $2\hat{i}+\hat{j}+3\hat{k}$. Also reduce it to cartesian form.

Solution. We know that the vector equation of a plane passing through $\vec{a}$ and normal to $\vec{n}$ is

$$(\vec{r}-\vec{a})\cdot\vec{n}=0 \quad \Rightarrow \quad \vec{r}\cdot\vec{n}=\vec{a}\cdot\vec{n}$$

Here $\vec{a}=\hat{i}+2\hat{j}+\hat{k}$ and $\vec{n}=2\hat{i}+\hat{j}+3\hat{k}$

Therefore $$\vec{r}\cdot(\hat{i}+2\hat{j}+\hat{k}) = (\hat{i}+2\hat{j}+\hat{k})\cdot(2\hat{i}+2\hat{j}+3\hat{k})$$

$$\vec{r}(\hat{i}+2\hat{j}+\hat{k}) = 2+2+3$$

$$\vec{r}(\hat{i}+2\hat{j}+\hat{k}) = 7$$

The cartesian form can be obtained as

$$(x\hat{i}+y\hat{j}+2\hat{k})(\hat{i}+2\hat{j}+\hat{k}) = 7$$

$$x+2y+z=7$$

is required cartesian form.

Example 5.24. Find the vector equation of a plane which is at a distance 10 units from the origin and normal to vector is $2\hat{i}+\hat{j}+3\hat{k}$.

Solution. Here $d = 10$ and $\vec{n} = 2\hat{i}+\hat{j}+3\hat{k}$

$$\therefore \quad \hat{n} = \text{unit vector} = \frac{\vec{n}}{|\vec{n}|} = \frac{2}{\sqrt{14}}\hat{i} + \frac{\hat{i}}{\sqrt{14}} + \frac{3\hat{k}}{\sqrt{14}}$$

The required equation is

$$\vec{r}\cdot\left(\frac{2}{\sqrt{14}}\hat{i} + \frac{\hat{i}}{\sqrt{14}} + \frac{3\hat{k}}{\sqrt{14}}\right) = 10$$

The cartesian equation is given by

$$(x\hat{i}+y\hat{j}+z\hat{k})\left(\frac{2}{\sqrt{14}}\hat{i} + \frac{\hat{i}}{\sqrt{14}} + \frac{3\hat{k}}{\sqrt{14}}\right) = 10$$

$$2x+y+3z = 10\sqrt{14}$$

Example 5.25. Reduce the equations to normal form and hence find the length of perpendicular from origin to plane.

(*a*) $\vec{r}(4\hat{i}-\hat{j}+\hat{k}) = 6$ (*b*) $2x-4y+3z=18$

Solution. (*a*) The given equation is

$$\vec{n} = 4\hat{i}-\hat{j}+\hat{k}$$

$$\hat{n} = \frac{\vec{n}}{|\vec{n}|} = \frac{4\hat{i}-\hat{j}+\hat{k}}{\sqrt{16+1+1}} = \frac{4\hat{i}-j+\hat{k}}{\sqrt{18}}$$

$$\therefore \quad \vec{r}\cdot\frac{\vec{n}}{|\vec{n}|} = \frac{5}{|\vec{n}|}$$

$$\Rightarrow \quad \vec{r}\,\hat{n} = \frac{5}{|\vec{n}|} \quad \Rightarrow \quad \vec{r}\left(\frac{4\hat{i}-\hat{j}+\hat{k}}{\sqrt{18}}\right) = \frac{5}{\sqrt{18}}$$

The length of perpendicular is $\dfrac{5}{\sqrt{18}} = \dfrac{5}{3\sqrt{2}}$.

(*b*) The given equation of plane is $2x - 4y + 3z = 18$ divide throughout by $\sqrt{4+16+9} = \sqrt{29}$

$$\frac{x}{\sqrt{29}} - \frac{4y}{\sqrt{29}} + \frac{32}{\sqrt{29}} = \frac{18}{\sqrt{29}}$$

Length of perpendicular is $\frac{18}{\sqrt{29}}$ units.

EXERCISE 5.5

1. Find the equation of plane passing through
   (*a*) (2, 3, 4), (– 3, 5, 1) and (4, – 1, 2) [**Ans.** $x + y - 2 = 1$]
   (*b*) (1, 1, 1), (1, – 1, 2) and (– 2, – 2, 2) [**Ans.** $x - 3y - 6z + 8 = 0$]
   (*c*) (2, 1, 0), (3, – 2, – 2) and (3, 1, 7) [**Ans.** $7x + 3y - 2 = 17$]
2. Write the equation of plane whose intercepts are 2, 4 and 6. $\left[\textbf{Ans.} \frac{x}{1} + \frac{y}{2} + \frac{z}{3} = 2\right]$
3. Reduce the equation of following planes in intercept form and find its intercepts
   (*a*) $2x + 3y - z = 6$ (*b*) $3x + 6y + z = 18$ (*c*) $2x - y + z = 16$

   Ans. (*a*) $\frac{x}{3} + \frac{y}{2} - \frac{z}{6} = 1$; (3, 2, – 6)
   (*b*) $\frac{x}{6} + \frac{y}{23} + \frac{z}{18} = 1$; (6, 3, 18)
   (*c*) $\frac{x}{8} - \frac{y}{16} + \frac{z}{16} = 1$; (1, – 2, 2)
4. Find the cartesian form of equation of plane whose vector equation is
   (*a*) $\vec{r} \cdot (12\hat{i} + 3\hat{j} + 4\hat{k}) + 6 = 0$ (*b*) $\vec{r} \cdot (6\hat{i} - 2\hat{j} - \hat{k}) = 7$
   [**Ans.** (*a*) $12x + 3y + 4z = -6$, (*b*) $6x - 2y - z = 7$]
5. Find the vector as well as cartesian equation of a plane passing through the points (1, – 1, 1) and normal to the line joining the points (1, 2, 5) and (– 1, 3, 1). [**Ans.** $\vec{r}(2\hat{i} - \hat{j} + 4\hat{k}) = 7, 2x - y + 4z = 7$]
6. Find the equation of plane passing through the points (2, 3, 1) having 5, 3, 2 as direction ratios of normal to plane [**Ans.** $5x + 3y + 2z = 21$]
7. Reduce the equations to normal form and hence find the length of perpendicular from origin to plane
   (*a*) $2x - 3y - 6z = 14$ (*b*) $\vec{r} \cdot (\hat{i} - 2\hat{j} + 2\hat{k}) = -6$

   Ans. (*a*) $2, \frac{2}{7}, \frac{-3}{7}, \frac{-6}{7}$
   (*b*) $\vec{r} \cdot \left(\frac{-1}{3}\hat{i} + \frac{2}{3}\hat{j} - \frac{2}{3}\hat{k}\right) = 2$
8. Find the vector equation of a plane which is at a distance 7 from origin and which is normal to vector $\hat{i} - 2\hat{j} - 2\hat{k}$
   $\left[\textbf{Ans.} \vec{r}\left(\frac{1}{3}\hat{i} - \frac{2}{3}\hat{j} - \frac{2}{3}\hat{k}\right) = 7\right]$

❑❑❑

UNIT - III

DIFFERENTIAL CALCULUS

CHAPTER 6

Functions and Limits

INSIDE THIS CHAPTER

6.1. REAL NUMBERS

The set ***R***, of real number plays a vital role in the study of Calculus. R consists of all rational and irrational numbers. The real numbers possess a useful geometrical property viz., they can be represented as points on a straight line called the **Real line.**

If S is a subset of real numbers, we say that S is **bounded above** if there exists a real number K such that $x \leq K \ \forall \ x \in S$. The real number K is called a lower bound of S. The set is called **bounded** if it is bounded above as well as bounded below.

It may be observed that if K is an upper bound of S, then every real number greater than or equal to K is also an upper bound of S. The smallest of all the upper bounds of S is called the **least upper bound (l.u.b.)** or the **supermum (sup.)** of S. Similarly the greatest of all the lower bounds of S is called the **greatest lower bound (g.l.b.)** or the **infimum (Inf.)** of S.

Examples:

1. $S = \{1, 2, ..., 10\}$ is bounded. Inf. $S = 0$ and Sup. $S = 10$.

2. The set $N = \{1, 2, 3, ...\}$ of natural numbers is bounded below but not bounded above. 1 is a lower bound.

3. **R** is not bounded.

4. $S = \left\{1, \frac{1}{2}, \frac{1}{3}, ..., \frac{1}{n}, ...\right\}$ is bounded. Inf. $S = 0$ and Sup. $S = 1$.

5. $S = \{x : x \in R \text{ and } a < x < b\}$ is bounded. Inf. $S = a$ and Sup. $S = b$.

6.2. PROPERTIES OF R AS A COMPLETE ORDERED FIELD

The set R of real numbers satisfies the following properties

A.1. $a \in \mathbf{R}$ and $b \in \mathbf{R} \Rightarrow a + b \in \mathbf{R}$ (Closure law w.r.t. +)

A.2. $a + b = b + a \ \forall \ a, b \in \mathbf{R}$ (Commutative law w.r.t. +)

A.3. $(a + b) + c = a + (b + c) \ \forall \ a, b, c \in \mathbf{R}$ (Associative law w.r.t. +)

A.4. $a + 0 = a \ \forall \ a \in \mathbf{R}$ (Existence of additive identity)

A.5. To each a $\in$ **R**, $\alpha - a \in \mathbf{R}$ such that $a + (-a) = 0$ (Existence of additive identity)

M.1. $a \in \mathbf{R}$ and $b \in \mathbf{R} \Rightarrow ab \in \mathbf{R}$ (Closure law w.r.t. multiplication)

M.2. $ab = ba \ \forall \ a, b \in \mathbf{R}$ (Commutative law w.r.t multiplication)

M.3. $(ac)\ c = a\ (bc) \ \forall \ a, b, c \in \mathbf{R}$ (Associative law w.r.t multiplication)

M.4. $a \,.\, 1 = a \ \forall \ a \in \mathbf{R}$ (Existence of multiplicative identity)

M.5. To each $0 \neq a \in \mathbf{R}$, $\alpha\ a^{-1} \in \mathbf{R}$ such that $aa^{-1} = 1$ (Existence of multiplicative inverse)

A.M. $a(b + c) = ab + ac \ \forall \ a, b, c \in \mathbf{R}$ (Distributive law)

Q.1. For any $a, b \in \mathbf{R}$, exactly one of the following is true:

(*i*) $a > b$ (*ii*) $a = b$ (*iii*) $b > a$ (Trichotomy law)

Q.2. $a > b$ and $b > c \Rightarrow a > c$ (Transitive law)

Q.3. $a > b \Rightarrow a + c > b + c$

Q.4. $a > b$ and $c > 0 \Rightarrow ac > bc$

O.C.1. Every non-empty subset of **R** which is bounded above (bounded below) has the supermum (infimum) in **R**.

On account of **R** satisfying the above mentioned properties, **R** is called a **complete-ordered field.**

6.3. INTERVALS

If a and b are any two real numbers such that $a < b$, we can define the different types of **intervals** as follows:

The set $\{x : x \in \mathbf{R} \text{ and } a < x < b\}$ is called an **open interval** and is denoted by $]a, b[$.

The set $\{x : x \in \mathbf{R} \text{ and } a \leq x \leq b\}$ is called a **closed interval** and is denoted by $[a, b]$.

The sets
$$]a, b[= \{x : x \in \mathbf{R} \text{ and } a < x \leq b\}$$
$$[a, b] = \{x : x \in R \text{ and } a \leq x < b\}$$

are called **semi-closed** or **semi-open intervals.**

The open intervals $]a - \varepsilon, a + \varepsilon[$, $\varepsilon > 0$, is called a **neighbourhood (nbd.)** of a.

Example: Each of the following intervals is bounded.

$$]a, b[,\ [a, b],\ ]a, b[,\ [a, b]$$

6.4. FUNCTIONS

A special type of relation is called function. A function may be viewed as a rule which produces new elements out of some given elements.

Definition : Let, A and B be two non-empty sets. A relation f from set A to set B is said to be a function if every element of set A has one and only one image in set B.

In other words, a function from A to B is a rule which relates each element of A to a unique element of B.

The function from A to B is denoted by $f : A \to B$. If f is a function from A to B and $(a, b) \in f$, then 'b' is the image of the element 'a' under function f and 'a' is called the pre-image or inverse-image of 'b' under function f. For a function f from A to B, no element of A should have more than one image in B.

Note : Two or more elements of A may have same image in B.

6.5. DOMAIN AND RANGE OF A FUNCTION

Let f be a function from A to B *i.e.*, $f : A \to B$, then the set A is called the **domain** of function f and set B is called the **codomain** of function f. The set of images of all the elements of A under the function f is called the **range** of f.

$\therefore$ range of $f = \{(f(x) : \text{for all } x \in A\}$

For example : Let, $A = \{1, 2, 3, 4\}$, and $B = \{1, 4, 9, 16, 25\}$

Let, $f = \{(x, y) : x \in A \text{ and } y \in B \text{ and } y = x^2\}$

Then $f = (1, 1), (2, 4), (3, 9), (4, 16)$

Domain of $f = \{1, 2, 3, 4\} = A$

Range of $f = \{1, 4, 9, 16\}$.

6.6. REPRESENTATION OF A FUNCTION BY ARROW DIAGRAM

To represent a function of set A to B by arrow diagram, we represent the sets A and B by two closed regions and elements of the sets are represented by points inside the closed regions. The function $f : A \to B$ is shown by means of arrow drawn from the points in region A to the points in region B.

Let, $A = \{1, 2, 3, 4\}$,

$B = \{1, 4, 9, 16, 25\}$

Let, $f = \{(x, y) : x \in A$ and $y \in B$ and $y = x^2\}$

is shown by the figure.

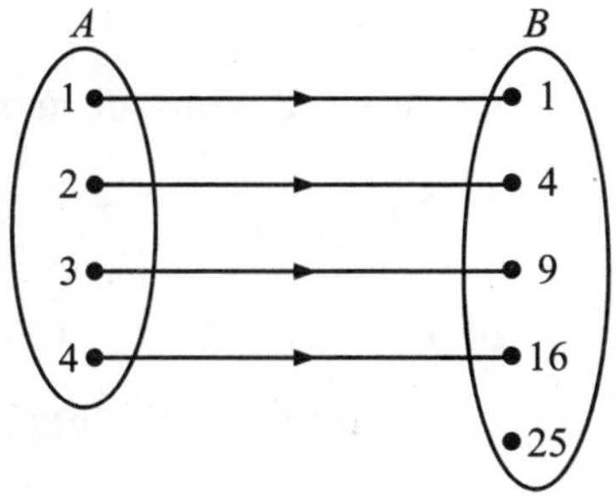

Fig. 6.1

6.7. REAL FUNCTION

Let, R be the set of all real numbers and A and B are two non-empty sets, then a function $f : A \to B$ is called a real valued function, if A and B are both subsets of R.

6.7.1. Domain and Range of a Real Function

Let f be a real function, then domain of function f is the set of all possible real numbers x for which $f(x)$ is a real number.

$\therefore$ Domain of $f(D_f) = \{x : x \in R, f(x) \in R\}$

and range of function f is the set of images $f(x)$ for all $x \in D_f$.

$\therefore$ Range of $f = \{f(x) : \text{for all } x \in D_f\}$

6.8. EQUAL FUNCTIONS

Two functions f and g are said to be equal, written as $f = g$, if and only if

(*i*) Domain of f = Domain of g

(*ii*) $f(x) = g(x)$ for all $x \in$ domain of f or (domain of g).

SOLVED EXAMPLES

Example 6.1. Let $A = \{9, 10, 11, 12, 13\}$ and let $f : A \to N$ be defined by $f(x)$ = the highest prime factor of n, $n \varepsilon A$. Write f as a set of ordered pairs and find the range of f.

Solution. $A = \{9, 10, 11, 12, 13\}$

Prime factor of $9 = 3 \times 3$; $10 = 2 \times 5$, $11 = 11 \times 1$; $12 = 3 \times 2 \times 2$; $13 = 1 \times 13$

$f(n)$ = highest prime factor of n

$\therefore$ $f(9) = 3,\ f(10) = 5, f(11) = 11,\ f(12) = 3,\ f(13) = 13$

$$f = \{(9, 3), (10, 5), (11, 11), (12, 3), (13, 13)\}$$

$$\text{Range of } f = \{3, 5, 11, 13\}$$

Example 6.2. Is $f = \{(1, 1), (2, 3), (3, 5), (4, 7)\}$ a function? If this is described by the formula $f(x) = ax + b$, then what values would be assigned to a and b.

Solution. For every value of x, we have unique value of $f(x)$

$\therefore$ f is a function

Now
$$f(1) = a \times 1 + b = 1 \Rightarrow a + b = 1 \quad \ldots(i)$$
$$f(2) = a \times 2 + b = 3 \Rightarrow 2a + b = 3 \quad \ldots(ii)$$

Solving (*i*) and (*ii*), we get

$$a = 2 \quad \text{and} \quad b = -1$$

Example 6.3. Find the domain of each of the following functions:

(*i*) $f(x) = \sqrt{x-1}$ (*ii*) $f(x) = \dfrac{1}{\sqrt{2-x}}$ (*iii*) $f(x) = \sqrt{9-x^2}$

Solution. (*i*) $f(x) = \sqrt{x-1}$

$f(x)$ is defined for all values of x satisfying $x - 1 \geq 0$ *i.e.*, $x \geq 1$

$\Rightarrow$ $x \,\varepsilon\, [1, \infty]$

$\therefore$ Domain of $f = [1, \infty]$

(*ii*) $f(x) = \dfrac{1}{\sqrt{2-x}}$

$f(x)$ is defined for all values of x satisfying $2 - x \geq 0$ *i.e.*, $x < 2$

$\Rightarrow$ $x \,\varepsilon\, [-\infty, 1]$

$\therefore$ Domain of $f = [-\infty, 1]$

(*iii*) $f(x) = \sqrt{9-x^2}$

$f(x)$ is defined for all values of x satisfying $9 - x^2 \geq 0$ *i.e.*, $x^2 - 9 \leq 0$

$\Rightarrow$ $(x-3)(x+3) \leq 0$

$\Rightarrow$ $x \,\varepsilon\, [-3, 3]$

$\therefore$ Domain of $f = [-3, 3]$

Example 6.4. Find the domain and range of $y = \dfrac{1}{2 - \sin 3x}$.

Solution.
$$y = \frac{1}{2 - \sin 3x}$$

$$-1 \leq \sin 3x \leq 1$$

$\Rightarrow$ $1 \geq -\sin 3x \geq -1$

$\Rightarrow$ $3 \geq 2 - \sin 3x \geq 1$

$\Rightarrow$ $1 \leq 2 - \sin 3x \leq 3$

$\Rightarrow$ $2 - \sin 3x \neq 0$

$\therefore$ y is defined for all real value of x.

$\therefore$ Domain $= R$

Now $$1 \le 2 - \sin 3x \le 3$$

$$\Rightarrow \quad \frac{1}{1} \ge \frac{1}{2 - \sin 3x} \ge \frac{1}{3}$$

$$\Rightarrow \quad 1 \ge y \ge \frac{1}{3} \quad \Rightarrow \quad \frac{1}{3} \le y \le 1$$

Hence, range $= \left[\frac{1}{3}, 1\right]$.

6.9. SOME IMPORTANT REAL FUNCTIONS AND THEIR GRAPHS

***(i)* Constant function :** If c is a fixed real number, then a function $f(x) = c$ for all $x \in R$ is called a constant function. Domain of $f(x) = R$ and range of $f(x) = \{c\}$

The graph of a constant function $f(x) = c$ is a straight line parallel to x-axis.

A constant function whose range is $\{0\}$, is called a zero function.

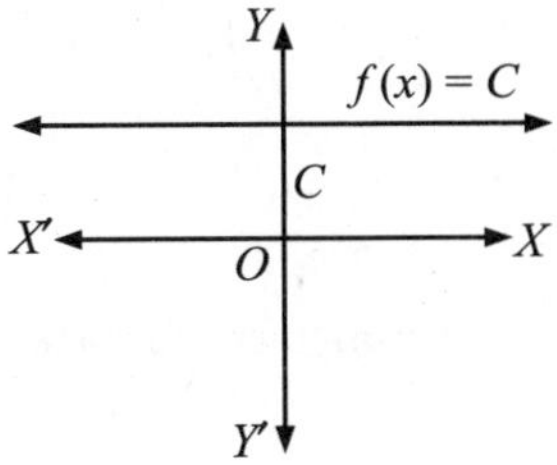

Fig. 6.2. Constant function

***(ii)* Identity function:** A function $f(x) : R \to R$ is said to be an identity function iff $f(x) = x$, for all $x \in R$.

The domain and range of identity function are both equal to R. The graph of an identity function is a straight line passing through origin and inclined at an angle of 45° with x-axis.

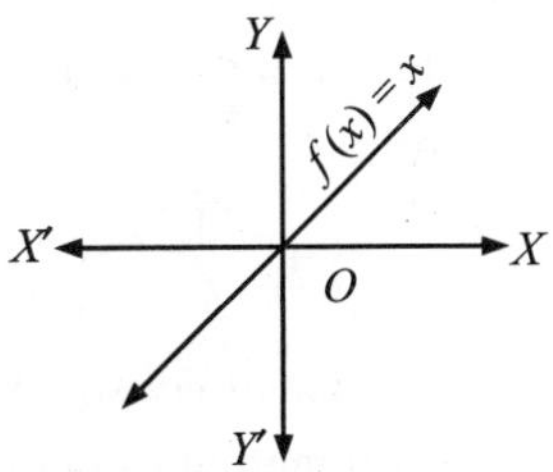

Fig. 6.3. Identity function

***(iii)* Modulus functions:** The function $f(x) : R \to R$ is said to be modulus function if

$$f(x) = \begin{cases} x, & \text{if } \; x > 0 \\ 0 & \text{if } \; x = 0 \\ -x & \text{if } \; x > 0 \end{cases}$$

It is denoted by $|x|$ and is also known as absolute value function.

The domain of modulus function $= R$ and range $= [0, \infty)$

The graph of modulus function is shown in Fig. 6.4.

(i) Line $y = x$ for $x \ge 0$

and *(ii)* Line $y = -x$ for $x \le 0$

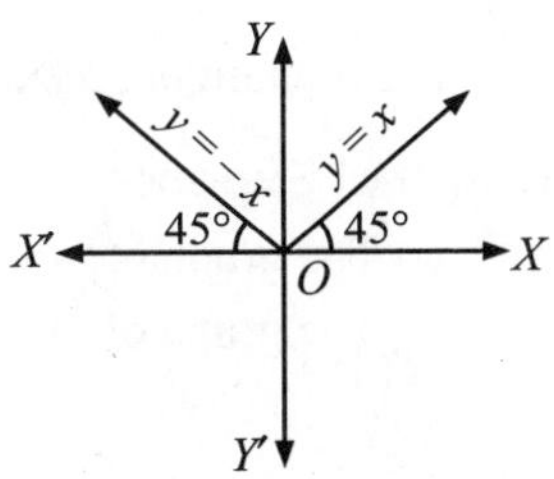

Fig. 6.4. Modulus function

***(iv)* Greatest integer function:** A function $f : R \to R$ defined by $f(x)$ = greatest integer less than or equal to x is called the greatest integer function. It is also known as step function or floor function and is denoted by $[x]$.

Thus, $[x] = x$, if x is an integer

= integer just less than x, if x is not an integer.

For example: $[2.6] = 2$, $[4] = 4$, $[-2.65] = -3$, $[0.8] = 0$

The domain of greatest integer function $= R$ and range $= Z$, set of all integers.

The graph of greatest integer function is shown in Fig. 6.5.

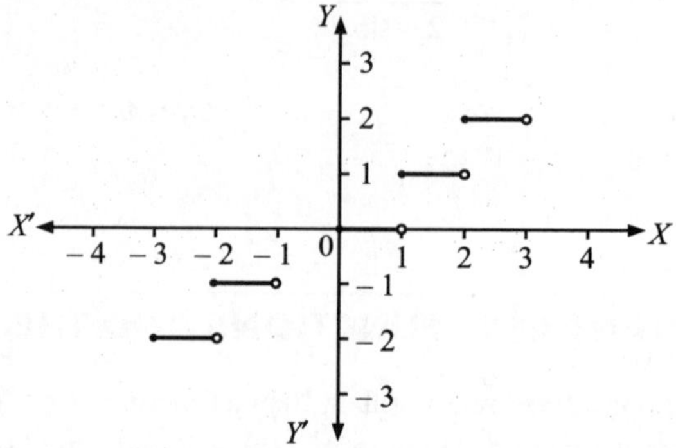

Fig. 6.5. Greatest integer function

(v) Signum function : The function f defined by

$$f(x) = \frac{|x|}{x}, \quad x \neq 0$$
$$= 0 \quad , \quad x = 0$$

is called the signum function and is denoted by (x).

clearly,
$$f(x) = 1, \text{ for } x > 0$$
$$= -1, \quad x < 0$$
$$= 0, \quad x = 0$$

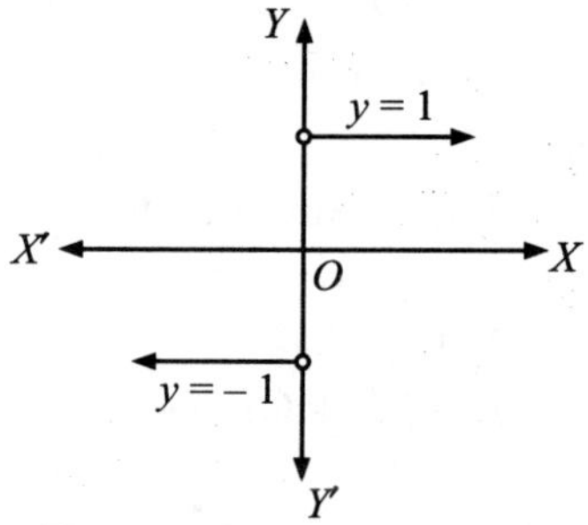

Fig. 6.6. Signum function

The domain of signum function = R and range = $\{-1, 0, 1\}$

The graph of signum function is shown in the Fig. 6.6.

(vi) Exponential function: If a is a +ve real number other than unity, then the function defined by

$$f(x) = a^x, x \in R$$

The domain of an exponential function = R and range = $(0, \infty)$

Graph of Exponential Function

(i) The graph of $f(x) = a^x$, when $a > 1$ is shown in Fig. 6.7 (i).

(ii) The graph of $f(x) = a^x$, when $0 < a < 1$ is shown in Fig. 6.7 (ii).

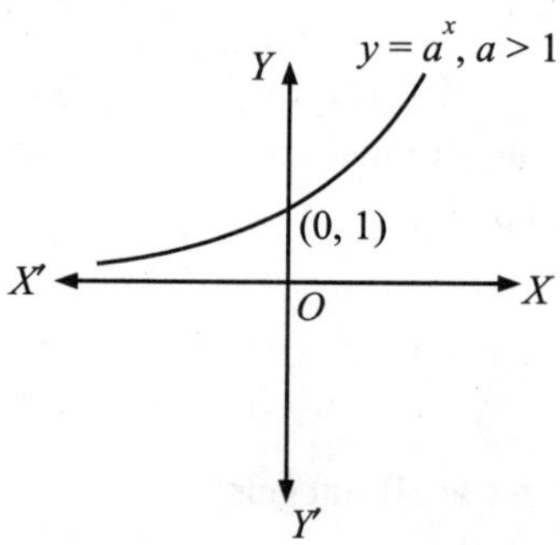

Fig. 6.7. (i) Exponential function $y = a^x$, $a > 1$

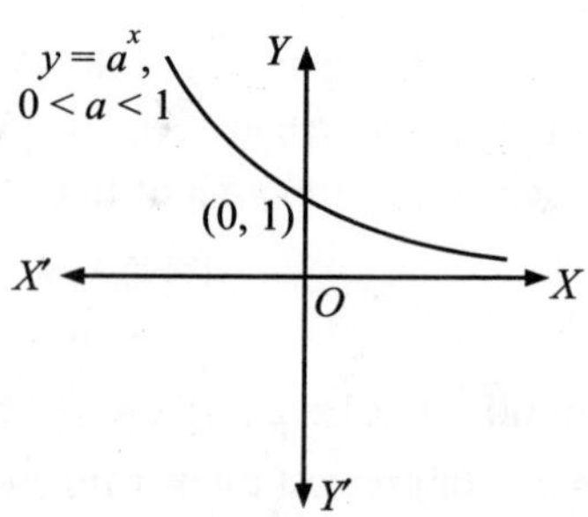

Fig. 6.7. (ii) Exponential function $y = a^x$, $0 < a < 1$

(*vii*) Logarithmic function : If a is a +ve real number other than unity, then the function $f(x) = \log a^x$, $x > 0$ is called the logarithmic function.

Domain of logarithmic function = $(0, \infty)$ *i.e.*, set of all +ve real number and range = R *i.e.*, set of all real numbers.

Graph of Logarithmic Function

(*i*) Graph of $f(x) = \log a^x$, $a > 1$ is shown in Fig. 6.8 (*i*).

(*ii*) Graph of $f(x) = \log a^x$, $0 < a < 1$ is shown in Fig. 6.8 (*ii*).

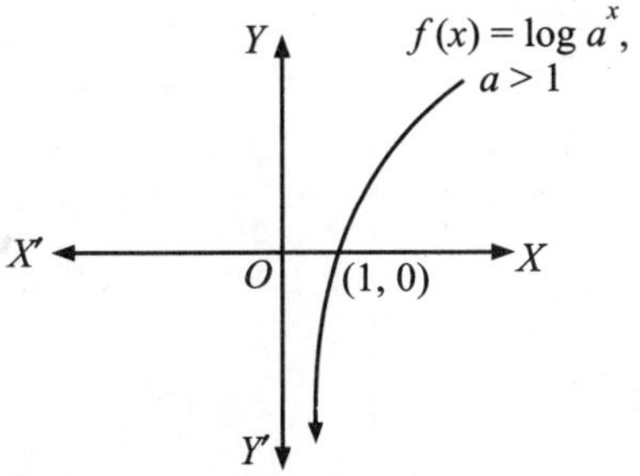

Fig. 6.8. (i) Logarithmic function, a > 1

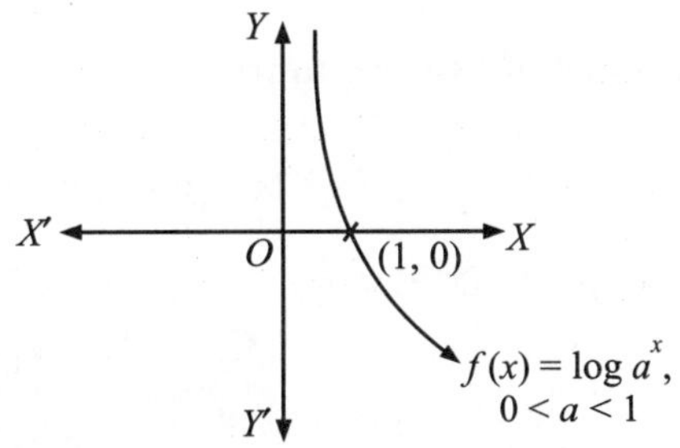

Fig. 6.8. (ii) Logarithmic function, 0 < a < 1

Note: The inverse of exponential function is logarithmic function.

(*viii*) Reciprocal function: Let, x be a real number, then a function $f(x) = \frac{1}{x}$ is called a reciprocal function. $\frac{1}{x}$ is not defined at $x = 0$, we define the recirpocal function as :

A reciprocal function is a function $f: R - \{0\} \to R$ defined by $f(x) = \frac{1}{x}$.

Domain of reciprocal function = $R - \{0\}$ and range = $R - \{0\}$.

(*ix*) Polynomial function : A polynomial function is a function $f: R \to R$ defined by $f(x) = a_0 + a_1x + a_2x^2 + ... + a_nx^n$ for all $x \in R$, where $a_0, a_1, a_2 .... a_n \in R$ and n is a non-negative integer. Domain of polynomial function = R and range = R^+ (set of all non-negative real numbers)

(*x*) Rational function : A rational function is a function of the form $f(x) = \frac{g(x)}{h(x)}$, where $g(x) = a_0 + a_1x + a_2x^2 + ... + a^nx^n$ and $h(x) = b_0 + b_1x + b_2 x^2 + ... + b_n x^n$ are two polynomial functions and $h(x) \neq 0$.

6.10. ALGEBRA OF REAL FUNCTIONS

(*i*) Addition of two real functions

Let, $f: x \to R$ and $g : x \to R$ be two real functions, then a function $(f + g)x = f(x) + g(x)$ for all $x \in D$.

where $D = D_f \cap D_g$

D_f = Domain of f and D_g = Domain of g.

For example : $f(x) = 3x + 1$ and $g(x) = x^2 + 3$, then

$(f + g)\, x = 3x + 1 + x^2 + 3 = x^2 + 3x + 4.$

***(ii)* Subtraction of two real functions**

Let, $f: x \to R$ and $g: x \to R$ be two real functions, then a function $(f-g)\,x = f(x) - g(x)$ for all $x \in D$.

For example : $f(x) = 3x + 1$ and $g(x) = x^2 + 3$, then

$(f-g)\,x = f(x) - g(x) = 3x + 1 - x^2 - 3 = -x^2 + 3x - 2.$

***(iii)* Multiplication of two real functions**

Let $f: x \to R$ and $g: x \to R$ be two real functions, then a function $fg(x) = f(x) \cdot g(x)$ for all $x \in D$.

For example : $f(x) = 3x + 1$ and $g(x) = x^2 + 3$, then

$fg(x) = (3x + 1)(x^2 + 3) = 3x^3 + 9x + x^2 + 3 = 3x^3 + x^2 + 9x + 3.$

***(iv)* Division of two real functions**

Let $f: x \to R$ and $g: x \to R$ be two real functions then a function $\left(\frac{f}{g}\right)x = \frac{f(x)}{g(x)}$ for all $x \in D$.

For example : $f(x) = x^2 + 3x + 2$, $g(x) = x + 2$, then

$$\left(\frac{f}{g}\right)x = \frac{f(x)}{g(x)} = \frac{x^2+3x+2}{x+2} = (x+1)$$

***(v)* Scalar multiplication of a function :** Let, $f: x \to R$ and C is any scalar, then a function $(Cf)x = C \cdot f(x)$ for all $x \in D$.

SOLVED EXAMPLES

Example 6.5. Let $f(x) = x^2$ and $g(x) = 2x + 1$ be two real functions. Find:

***(i)* $(f+g)\,x$** ***(ii)* $(f-g)\,x$,** ***(iii)* $(fg)\,x$,** ***(iv)* $\left(\frac{f}{g}\right)x$**

Solution:

$$f(x) = x^2$$

Domain $(f) = R$

$$g(x) = 2x + 1$$

$$\text{Domain } (g) = R$$

$\therefore$ Domain $(f) \cap$ Domain $(g) = R$

(i) $(f+g)\,x = f(x) + g(x) = x^2 + 2x + 1$

(ii) $(f-g)\,x = f(x) - g(x) = x^2 - (2x + 1) = x^2 - 2x - 1$

(iii) $(fg)\,x = f(x) \cdot g(x) = x^2(2x + 1) = 2x^3 - x^2$

(iv) $\left(\frac{f}{g}\right)x = \frac{f(x)}{g(x)} = \frac{x^2}{2x+1}$

Example 6.6. If $f(x) = \sqrt{x^2+1}$ and $g(x) = 2x^2 + 3$, then find $(f+g)\,x$.

Solution:

$$f(x) = \sqrt{x^2+1}, \quad \text{Domain } (f) = R$$

$$g(x) = 2x^2 + 3, \quad \text{Domain } (g) = R$$

$\therefore$ Domain $(f) \cap$ Domain $(g) = R$

$$(f+g)\,x = f(x) + g(x)$$

$$\therefore \quad (f+g)\,x = \sqrt{x^2+1} + 2x^2 + 3$$

Example 6.7. If $f(x) = |x|$, and $g(x) = x, x \in R$ then find:

(i) $(f+g)\,x$ **(ii)** $(f-g)\,x$ **(iii)** $(fg)x$ **(iv)** $\left(\frac{f}{g}\right)x$

Solution:

$$f(x) = |x|, \quad \text{Domain } (f) = R$$
$$f(x) = x \quad \text{when} \quad x \geq 0$$
$$= -x \quad \text{when} \quad x < 0$$
$$g(x) = x \quad \text{Domain } (g) = R$$
$$\text{Domain } (f) \cap \text{Domain } (g) = R$$

(i) $(f+g)\,x = f(x) + g(x) = x + x = 2x$ when $x \geq 0$

$= -x + x = 0$ when $x < 0$

(ii) $(f-g)\,x = f(x) - g(x) = x - x = 0$ when $x \geq 0$

$= -x - x = -2x$ when $x < 0$

(iii) $(fg)\,x = f(x) \cdot g(x) = x \cdot x = x^2$ when $x \geq 0$

$= -x \cdot x = -x^2$ when $x < 0$

(iv) $$\left(\frac{f}{g}\right)x = \frac{f(x)}{g(x)} = \frac{x}{x} = 1 \quad \text{when } x \geq 0$$

$$= \frac{-x}{x} = -1 \quad \text{when } x < 0$$

EXERCISE 6.1

1. Examine which of the following relations are functions ?
   (i) $f = \{(2, 1), (2, 3), (4, 3), (1, 2)\}$
   (ii) $g = \{(2, 1), (4, 2), (6, 3), (8, 4), (10, 5), (12, 6), (14, 7)\}$
   (iii) $h = \{(-4, 4), (4, 4), (3, 2)\}$ **[Ans.** (ii) and (iii)]
2. Let a relation R be defined on N such that $R = \{(x, y) : y = 2x, x, y \in N\}$. Write the domain, co-domain and range of R. Is this relation a function? **[Ans.** Dom = set of natural numbers N, co-domain = N
   Ran = set of even natural numbers.]
3. Express the following functions as sets of ordered pairs and determine their ranges
   (i) $f : A \to R, f(x) = x^2 + 1$, where $A = \{-1, 0, 2, 4\}$
   (ii) $g : A \to N, g(x) = 2x$, where $A = \{x : x \in N, x \leq 10\}$
   [Ans. (i) $f = \{(-1, 2), (0, 1), (2, 5), (4, 17)\}$, Range = $\{2, 1, 5, 17\}$
   (ii) $g = \{(1, 2), (2, 4), (3, 6) \ldots (10, 20)\}$, Range = $\{2, 4, 6, 8, 10, 12, 14, 16, 18, 20\}$]
4. A function f is defined by $f(x) = 2x - 3$. Find: (i) $f(0)$ (ii) $f(5)$, (iii) $f(-2)$. **[Ans.** (i) -3, (ii) 7, (iii) -7]
5. Let $f = \{(-1, -3), (0, -1), (1, 1), (2, 3)\}$ be a linear function from z into z. Find f. **[Ans.** $f(x) = 2x - 1$]
6. Let $X = \{1, 2, 3, 4\}$. Examine which of the following relations is a function from X to X :
   (i) $f = \{(2, 4), (2, 3), (1, 3), (3, 1), (4, 2)\}$
   (ii) $g = \{(1, 1), (2, 3), (3, 4), (4, 4)\}$ **[Ans.** (i)]
7. Given $f(x + 1) = 3x + 5$, find $f(x)$.
8. Let $f : R \to R$ be a function given by $f(x) = x^2 + 1$, find:
   (i) $f^{-1}\{-5\}$, (ii) $f^{-1}\{26\}$ (iii) $f^{-1}\{10, 37\}$
   [Ans. (i) ϕ, (ii) $\{-5, 5\}$, (iii) $\{3, -3, 6, -6\}$]

9. Find the domain of real function, $f(x) = \dfrac{x^2+2x+1}{x^2-7x+12}$. **[Ans.** Dom = $R - \{3, 4\}$]

10. Find the domain of:

(*i*) $y = \dfrac{x^2-1}{x-1}$ (*ii*) $y = \dfrac{5-x}{x-5}$ **[Ans.** (*i*) Dom = R – {1}, (*ii*) Dom = $R - \{5\}$]

11. Find the domain of $\dfrac{x+7}{x^2-8x+4}$. **[Ans.** Dom = $R - \{4 \pm 2\sqrt{3}\}$]

12. If $f(x) = x^2 - 3x + 1$, find $x \in R$ such that $f(2x) = f(x)$. **[Ans.** $x = 0, 1$]

13. If $f(x+1) = 3x + 7$, find (*i*) $f(-2)$ (*ii*) $f(3x)$. **[Ans.** (*i*) – 2, (*ii*) $9x + 4$]

14. Find the domain and range of the real function $f(x) = \dfrac{1}{x-3}$. **[Ans.** Dom = $R - \{3\}$, Range = $R - \{0\}$]

15. Find the domain and range of each of the following functions:

(*i*) $f(x) = \sqrt{x-1}$ (*ii*) $f(x) = \sqrt{x^2-1}$.

(*iii*) $f(x) = \dfrac{1}{1-x^2}$ (*iv*) $f(x) = \dfrac{1}{9-x^2}$.

[Ans. (*i*) Dom = $[1, \infty]$ and ran = R
(*ii*) Dom = $[x \in R; x \le -1 \text{ or } x \ge 1\}$, Ran = R
(*iii*) Dom = $R - \{-1, 1\}$, Ran = $[1, \infty]$
(*iv*) Dom = $R - \{3, -3\}$, Ran = $\left[\dfrac{1}{3}, \infty\right]$]

16. Find the domain and range of the function g defined by $g(x) = \dfrac{x+2}{|x+2|}$.
[Ans. Dom = $R - \{-2\}$, Ran = $\{-1, 1\}$]

17. If $f(x) = x^3$, find the value of $\dfrac{\{f(5) - f(1)\}}{5-1}$. **[Ans.** 31]

18. Find the domain and range of $f(x) = 1 - |x - 2|$. **[Ans.** Dom = R; Ran = $(-\infty, 1)$]

19. Let f and g be two functions over R and $f(x) = x^4 + 1$ and $g(x) = x + 1$, find: $(f + g)$, $(f - g)$, fg and $\dfrac{f}{g}$.

$\left[\textbf{Ans. } f + g = x^4 + x + 2,\ f - g = x^4 - x,\ fg = x^5 + x^4 + x + 1,\ \dfrac{f}{g} = \dfrac{x^4+1}{x+1}\right]$

20. Let $f(x)$ and $g(x)$ be two functions defined by $f(x) = |x|$, $x \in R$ and $g(x) = x$, $x \in R$.

Find: (*i*) $(f + g)\,x$ (*ii*) $(f - g)\,x$ (*iii*) $(fg)\,x$ (*iv*) $\left(\dfrac{f}{g}\right)x$.

[Ans. (*i*) $f + g = 2x$ when $x \ge 0$
$= 0$ when $x < 0$
(*ii*) $f - g = 0$ when $x \ge 0$
$= -2x$, when $x < 0$
(*iii*) $fg = x^2$, $x \ge 0$
$= -x^2$, $x < 0$
(*iv*) $\dfrac{f}{g} = 1$, $x \ge 0$
$= -1$, $x < 0$]

6.11. EVEN AND ODD FUNCTIONS

Even Function: A function $f(x)$ is said to be an even function if $f(-x) = f(x)$ for all x.

Consider the function $f(x) = x^2 + \sec x$

We observe that

$$f(-x) = (-x)^2 + \sec(-x) = x^2 + \sec x = f(x)$$

So, $f(x) = x^2 + \sec x$ is an even function

The graph of an even function is always symmetrical about y-axis

Odd Function: A function $f(x)$ is said to be an odd function

if $f(-x) = -f(x)$ for all x

Consider the function $f(x) = x^3 + \tan x$

We observe that

$$f(-x) = (-x)^3 + \tan(-x) = -x^3 - \sin x = -f(x)$$

$\therefore$ $f(x)$ is an odd function

The graph of an odd function is always symmetrical in opposite quadrants.

6.12. PERIODIC FUNCTIONS

A function $f(x)$ is said to be periodic, if there exists a positive real number T such that $f(x + T) = f(x)$ for all $x \in R$.

If T is the smallest positive real number such that $f(x + T) = f(x)$ for all $x \in R$, then T is called the fundamental or primitive period of $f(x)$.

Consider the function $f(x) = \sin x$. Let T be a positive real number such that

$$f(x + T) = f(x) \text{ for all } x \in R$$

$$\Rightarrow \quad \sin(x + T) = \sin x \text{ for all } x \in R$$

$$x + T = 2n\pi + (-1)^n x \text{ for all } x \in R$$

$$\Rightarrow \quad T = 2n\pi + (-1)^n x - x \text{ for all } x \in R$$

$$\Rightarrow \quad T = 2n\pi \text{ if } n \text{ is an even positive integer}$$

$$\Rightarrow \quad T = 2\pi, 4\pi, 6\pi, \ldots..$$

The smallest of these positive real numbers is 2π. Hence, $\sin x$ is a periodic function with period 2π.

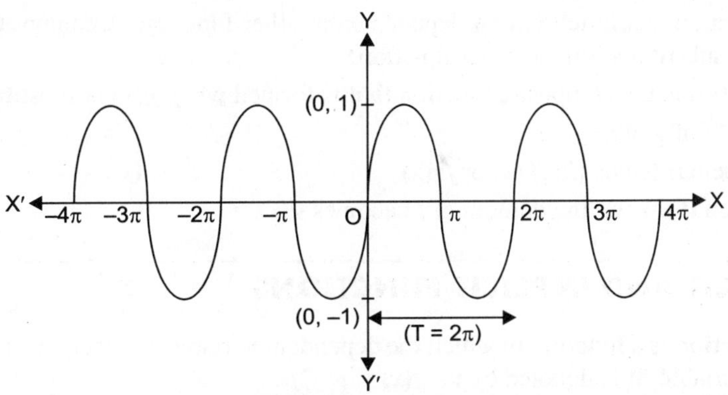

Fig. 6.9. Graph of f(x) = sin x

The functions cos x and tan x are also periodic with periods 2π and π respectively.

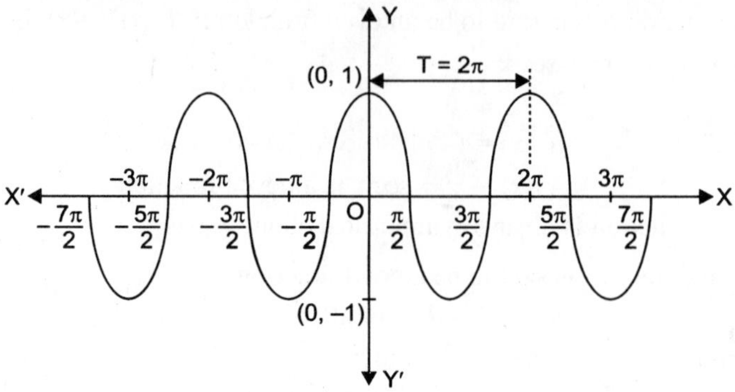

Fig. 6.10. Graph of f(x) = cos x

Following are some periodic functions with their periods:

Function	*Period*
sin x	2π
cos x	2π
tan x	π
cosec x	2π
sec x	2π
cot x	π
$\lvert \sin x \rvert$	π
$\lvert \cos x \rvert$	π

If $f(x)$ is a periodic function with period T, then the function $f(ax + b)$ is periodic with period $\dfrac{T}{|a|}$

It follows from the above result that sin $2x$, cos $(3x + 2)$ are periodic functions with periods π and $2\pi/3$ respectively.

6.13. COMPOSITE FUNCTIONS

A composite function is a function that depends on another function. A composite function is created when one function is substituted into another function.

For example, $f(g(x))$ is the composite function that is formed when $g(x)$ is substituted for x in $f(x)$

$f(g(x))$ is read as "f of g of x".

$f(g(x))$ can also be written as (fog) (x) or $fg(x)$

In the composition (fog) (x), the domain of f becomes $g(x)$.

6.14. EXPLICIT AND IMPLICIT FUNCTIONS

The **explicit function** is a function in which the dependent variable has been given "explicitly" in terms of the independent variable. It is denoted by $y = f(x)$

Examples of explicit function are

$$f(x) = ax^n + bx \text{ where } a, n, b \text{ are constants}$$

$$y = 5x^3 - 4$$

The **implicit function** is a function in which the dependent variable has not been given "explicity" in terms of the independent variable.

It is denoted by $f(x, y) = 0$. Examples of implicit functions are:

$$x^2 + y^2 - 1 = 0$$

$$y^4 + x^3 + 17 = 0$$

6.15. PARAMETRIC FUNCTIONS

Parametric functions are functions of a number of co-ordinates where each coordinate is expressed as another function of some parameter. In other words, sometimes x and y are given as function of a single variable *e.g.*, $x = f(t)$, $y = g(t)$ are two functions of a single variable. In such a case x and y are called *parametric functions.*

SOLVED EXAMPLES

Example 6.8. Determine whether the following functions are even or odd:

(*i*) $f(x) = x\left(\dfrac{a^x - 1}{a^x + 1}\right)$ **(*ii*)** $f(x) = \log\{x + \sqrt{x^2 + 1}\}$ **(*iii*)** $f(x) = \sin x - \cos x$

Solution. (*i*) We have,

$$f(x) = x\left(\frac{a^x - 1}{a^x + 1}\right)$$

$\therefore$

$$f(-x) = -x\left(\frac{a^{-x} - 1}{a^{-x} + 1}\right) = -x\left(\frac{\frac{1}{a^x} - 1}{\frac{1}{a^x} + 1}\right)$$

$$= -x\left(\frac{1 - a^x}{1 + a^x}\right) = x\left(\frac{a^x - 1}{a^x + 1}\right) = f(x)$$

So, $f(x)$ is an even function.

(*ii*) We have,

$$f(x) = \log\{x + \sqrt{x^2 + 1}\}$$

$\therefore$

$$f(-x) = \log\{-x + \sqrt{x^2 + 1}\}$$

$$= \log\left[\frac{-x + \sqrt{x^2 + 1}}{x + \sqrt{x^2 + 1}} \times (x + \sqrt{x^2 + 1})\right]$$

$$= \log\left[\frac{x^2 + 1 - x^2}{x + \sqrt{x^2 + 1}}\right] = \log\left[\frac{1}{x + \sqrt{x^2 + 1}}\right]$$

$$= -\log(x + \sqrt{x^2 + 1}) = -f(x)$$

So, $f(x)$ is an odd function

(*iii*) We have, $f(x) = \sin x - \cos x.$

Then, $f(-x) = \sin(-x) - \cos(-x) = -\sin x - \cos x$

Clearly, $f(-x)$ is neither equal to $f(x)$ nor $-f(x)$. So, $f(x)$ is neither an even function nor an odd function.

Example 6.9. Determine the period of each of the following functions:

(*i*) $\cos(2x + 5)$ (*ii*) $\sin^2 x$ (*iii*) $\tan \pi x$ (*iv*) $|\cos x|$

Solution. (*i*) We know that $\cos x$ is a periodic function with period 2π

Therefore, $f(x) = \cos(2x + 5)$ is a periodic function with period $\frac{2\pi}{2} = \pi.$

(*ii*) We have $f(x) = \sin^2 x = \frac{1 - \cos 2x}{2} = \frac{1}{2} - \frac{1}{2}\cos 2x$

As $\cos 2x$ is a periodic function with period $\frac{2\pi}{2} = \pi$. Therefore $f(x)$ is periodic with period π.

(*iii*) Since $\tan x$ is a periodic function with period equal to π. Therefore, $\tan \pi x$ is also periodic with period $\frac{\pi}{\pi} = 1.$

(*iv*) We have $f(x) = |\cos x| = \sqrt{\cos^2 x} = \sqrt{\frac{1 + \cos 2x}{2}}$

So, $f(x)$ is periodic with period $= \frac{2\pi}{2} = \pi$

Example 6.10. If $f, g: R \to R$ are defined respectively by $f(x) = x^2 + 3x + 1$, $g(x) = 2x - 3$, find:

(*i*) *fog* (*ii*) *gof* (*iii*) *fof* (*iv*) *gog*

Solution. Clearly, Range f = Domain g and Range g = Domain f. Therefore *fog*, *gof*, *fof* and *gog* all exist

(*i*) For any $x \in R$, we have

$$(fog)(x) = f(g(x)) = f(2x - 3) = (2x - 3)^2 + 3(2x - 3) + 1 = 4x^2 - 6x + 1$$

$\therefore$ $fog: R \to R$ is defined by $(fog)(x) = 4x^2 - 6x + 1$ for all $x \in R$

(*ii*) For any $x \in R$, we have

$$(gof)(x) = g(f(x)) = g(x^2 + 3x + 1) = 2(x^2 + 3x + 1) - 3 = 2x^2 + 6x - 1$$

$\therefore$ $gof: R \to R$ is defined by $(gof)(x) = 2x^2 + 6x - 1$ for all $x \in R$

(*iii*) For any $x \in R$, we have

$$(fof)(x) = f(f(x)) = f(x^2 + 3x + 1) = (x^2 + 3x + 1)^2 + 3(x^2 + 3x + 1) + 1$$
$$= x^4 + 6x^3 + 14x^2 + 15x + 5$$

$\therefore$ $fof: R \to R$ is defined by $(fof)(x) = x^4 + 6x^3 + 14x^2 + 15x + 5$ for all $x \in R$

(*iv*) For any $x \in R$, we have

$$(gog)(x) = g(g(x)) = g(2x - 3) = 2(2x - 3) - 3 = 4x - 9$$

$\therefore$ $gog: R \to R$ is defined by $(gog)(x) = 4x - 9$ for all $x \in R$

6.16. LIMITS

Consider the function $f(x) = x + 1$. Taking values of x near to 1. We write the corresponding values of $f(x)$ in the table given below:

Table 6.1

$x \to 1$ from the left

x	$f(x)$
0.9	1.9
0.99	1.99
0.999	1.999
0.9999	1.9999
....	....

Table 6.2

$x \to 1$ from the right

x	$f(x)$
1.1	2.1
1.01	2.01
1.001	2.001
....	....
....	....
1.00001	2.00001

From the above tables, we observe that as the value of x approaches 1, the value of $f(x)$ approaches 2 and we write as $x \to 1, f(x) \to 2$

We express this mathematically as

$$\lim_{x \to 2} (x+1) = 2$$

In general as $x \to a, f(x) \to b$, then b is called *limit of the function* $f(x)$ and written as

$$\lim_{x \to a} f(x) = b$$

When x approaches to 'a' from left hand side, the function $f(x)$ tends to the limit b.

We use the notation $x \to a^-$ to denote 'x' tends to 'a' from LHS' and is written as

$$\lim_{x \to a^-} f(x) = b$$

This is known as left hand limit $f(x)$

When x approaches to 'a' from right hand side, the function $f(x)$ tends to the limit b

We use the notation $x \to a^+$ to denote 'x tends to a from RHS' and is written as

$$\lim_{x \to a^+} f(x) = b$$

This is known as right hand limit of $f(x)$

If the left and right hand limits of a function $f(x)$ coincide we call that common value as the limit of $f(x)$ at $x = a$ and write it as $\lim_{x \to a} f(x)$.

And if the left hand limit and right hand limit different, then we say that the limit of $f(x)$ as $x \to a$ does not exist.

6.17. ALGEBRA OF LIMITS

Let f and g be two functions such that $\lim_{x \to a} f(x) = l$ and $\lim_{x \to a} g(x) = m$, then

(*i*) $\lim_{x \to a} kf(x) = k \lim_{x \to a} f(x)$ where k is a constant.

(ii) $\lim_{x\to a} (f \pm g)x = \lim_{x\to a} f(x) \pm \lim_{x\to a} g(x) = l \pm m$

(iii) $\lim_{x\to a} (f\,g)x = \lim_{x\to a} f(x) \cdot \lim_{x\to a} g(x) = lm$

(iv) $\lim_{x\to a} \left(\frac{f}{g}\right)x = \frac{\lim_{x\to a} f(x)}{\lim_{x\to a} g(x)} = \frac{l}{m}$, provided $m \neq 0$

(v) $\lim_{x\to a} \frac{1}{f(x)} = \frac{1}{\lim_{x\to a} f(x)} = \frac{1}{l}$, provided $l \neq 0$

(vi) $\lim_{x\to a} (f(x))^n = \left(\lim_{x\to a} f(x)\right)^n = l^n$ for all $n \in N$.

(vii) $\lim_{x\to a} |f(x)| = |\lim_{x\to a} f(x)| = |l|$

6.18. EVALUATION OF LIMITS OF ALGEBRAIC FUNCTIONS

The following methods are used:

(i) **Direct Substitution Method:** To evaluate $\lim_{x\to a} f(x)$, put $x = a$ in the given function. If $f(a)$ has definite value then

$$\lim_{x\to a} f(x) = f(a)$$

(ii) **Factorisation Method:** Consider $\lim_{x\to a} f(x) \frac{f(x)}{g(x)}$

At $x = a$ if $\frac{f(x)}{g(x)}$ reduced to $\frac{0}{0}$ form, then $(x - a)$ is a factor of both $f(x)$ and $g(x)$ which will cancel out.

(iii) **Rationalisation Method:** If the expression for which limit is to find out has square root either in numerator or denominator or both and when $x \to a$ takes the form $\frac{0}{0}, \frac{\infty}{\infty}$ etc., then denominator or numerator is rationalised. After rationalisation simplification is done and then direct substitution method is used to obtain limit.

(iv) **Using Standard Form:** Theorem: $\lim_{x\to a} \frac{x^n - a^n}{x - a} = na^{n-1}$, $a > 0$.

SOLVED EXAMPLES

Example 6.11. Evaluate $\lim_{x\to 0-} \frac{|x|}{x}$ **and** $\lim_{x\to 0+} \frac{|x|}{x}$.

Solution. By definition, $|x| = x$, if $x \geq 0$

and $|x| = -x$, if $x < 0$.

Now $\lim_{x\to 0-} f(x) = \lim_{\substack{x\to 0\\(x<0)}} \frac{|x|}{x} = \lim_{x\to 0} \frac{(-x)}{x} = -1,$

$$\lim_{x\to 0+} f(x) = \lim_{\substack{x\to 0\\(x>0)}} \frac{|x|}{x} = \lim_{x\to 0} \frac{x}{x} = 1$$

$\therefore$ $\lim_{x\to 0-} f(x) \neq \lim_{x\to 0+} f(x)$.

Hence, $\lim_{x\to 0} \frac{|x|}{x}$ does not exist.

Example 6.12. A function f is defined by

$$f(x) = \frac{1}{2} - x \text{ if } 0 < x < \frac{1}{2}$$
$$= \frac{3}{2} - x \text{ if } \frac{1}{2} < x < 1.$$

Evaluate $\lim_{x\to \frac{1}{2}} f(x)$.

Solution. We have $\lim_{x\to \frac{1}{2}} f(x) = \lim_{\substack{x\to \frac{1}{2}\\ \left(x<\frac{1}{2}\right)}} f(x) = \lim_{x\to \frac{1}{2}} f(x) = \lim_{x\to \frac{1}{2}} \left(\frac{1}{2} - x\right) = 0$

$$\lim_{x\to \frac{1}{2}+} f(x) = \lim_{\substack{x\to \frac{1}{2}\\ \left(x<\frac{1}{2}\right)}} f(x) = \lim_{x\to \frac{1}{2}} \left(\frac{3}{2} - x\right) = 1$$

Hence $\lim_{x\to \frac{1}{2}} f(x)$ does not exist.

Example 6.13. Show that $\lim_{x\to 1} (2x+3) = 5$.

Solution. Let $\varepsilon > 0$ be given. We have

$$|(2x+3) - 5| = |2x - 2| = 2|x - 1|$$

Now $|(2x+3) - 5| < \varepsilon$ when $2|x-1| < \varepsilon$

i.e., $|x - 1| < \frac{\varepsilon}{2}$

Choosing $\delta = \frac{\varepsilon}{2}$, $|(2x+3) - 5| < \varepsilon$ when $|x-1| < \delta$

Hence, $\lim_{x\to 1} (2x+3) = 5.$

Example 6.14. Find $\lim_{x\to 3} \frac{x^2-9}{x-3}$.

Solution. We have $\lim_{x\to 3} \frac{x^2-9}{x-3} = \lim_{x\to 3} \frac{(x-3)(x+3)}{(x-3)} = \lim_{x\to 3} (x+3)$, as $x \neq 3$.

$\therefore$ $\lim_{x\to 3} \frac{x^2-9}{x-3} = \lim_{x\to 3} (x+3) = 6.$

Example 6.15. **Show that** $\lim_{x \to 2}\left[\frac{1}{x-2} - \frac{1}{x^2-3x+2}\right] = 1.$

Solution. The required limit $= \lim_{x \to 2}\left[\frac{1}{x-2} - \frac{1}{x^2-3x+2}\right]$

$$= \lim_{x \to 2} \frac{(x-2)^2}{(x-2)(x-2)(x-1)} = \lim_{x \to 2} \frac{1}{(x-1)}, \text{ as } x \neq 2.$$

Hence, the required limit = 1.

Example 6.16. **Evaluate** $\lim_{x \to 0} \frac{\sqrt{1+x}-1}{x}$.

Solution. On rationalizing, we obtain

$$\lim_{x \to 0} \frac{\sqrt{1+x}-1}{x} = \frac{(\sqrt{1+x}-1)(\sqrt{1+x}+1)}{x(\sqrt{1+x}+1)} = \frac{(1+x)-1}{x(\sqrt{1+x}+1)} = \frac{1}{\sqrt{1+x}+1}$$

$$\therefore \quad \lim_{x \to 0} \frac{\sqrt{1+x}-1}{x} = \lim_{x \to 0} \frac{1}{\sqrt{1+x}+1} = \frac{1}{2}$$

Example 6.17. **Find** $\lim_{x \to \infty} \frac{x^2-4}{x^2+7x+12}$.

Solution. Dividing numerator and denominator by x^2

$$\lim_{x \to \infty} \frac{x^2-4}{x^2+7x+12} = \lim_{x \to \infty} \frac{1-\frac{4}{x^2}}{1+\frac{7}{x}+\frac{12}{x^2}} = \frac{1-4\lim_{x \to \infty}\left(\frac{1}{x^2}\right)}{1+7\lim_{x \to \infty}\left(\frac{1}{x}\right)+12\lim_{x \to \infty}\left(\frac{1}{x^2}\right)} = \frac{1-0}{1+0+0}$$

$$\therefore \quad \lim_{x \to \infty} \frac{x^2-4}{x^2+7x+12} = 1$$

Example 6.18. **Evaluate:**

(*i*) $\lim_{x \to \infty} \frac{\sqrt{x^2+1}}{x+1}$ (*ii*) $\lim_{x \to \infty} [\sqrt{x^2+3x} - x]$

Solution. (*i*) Dividing numerator and denominator by x

$$\lim_{x \to \infty} \frac{\sqrt{x^2+1}}{x+1} = \lim_{x \to \infty} \frac{\sqrt{1+\frac{1}{x^2}}}{1+\frac{1}{x}} = \frac{1+0}{1+0} = 1$$

(*ii*) We have $\sqrt{x^2+3x-x} = \frac{(x^2+3x)-x^2\,(\sqrt{x^2+3x}+x)}{(\sqrt{x^2+3x}+x)}$

$$= \frac{(x^2+3x)-x^2}{(\sqrt{x^2+3x}+x)} = \frac{3x}{\sqrt{x^2+3x}+x} = \frac{3}{\sqrt{1+\frac{3}{x}}+1}$$

$$\therefore \quad \lim_{x\to\infty} (\sqrt{x^2+3x}-x) = \frac{3}{\sqrt{1+0}+1} = \frac{3}{2}$$

Example 6.19. Evaluate $\lim_{x\to 9} \frac{x^{3/2}-27}{x-9}$

Solution. $\lim_{x\to 9} \frac{x^{3/2}-(9)^{3/2}}{x-9} = \frac{3}{2}\cdot(9)^{3/2-1} = \frac{3}{2}\times(9)^{1/2} = \frac{3}{2}\times 3 = \frac{9}{2}$.

Example 6.20. Evaluate $\lim_{x\to 1} \frac{x^{1/3}-1}{x^{1/6}-1}$

Solution.

$$\lim_{x\to 1} \frac{x^{1/3}-1}{x^{1/6}-1} = \lim_{x\to 1} \frac{x^{1/3}-1^{1/3}}{x-1}\cdot\frac{x-1}{x^{1/6}-1^{1/6}}$$

$$= \lim_{x\to 1} \frac{x^{1/3}-1^{1/3}}{x-1} \div \lim_{x\to 1} \frac{x^{1/6}-1^{1/6}}{x-1} \qquad (\because \quad 1 = 1^{1/3} = 1^{1/6})$$

$$= \frac{1}{3}\cdot(1)^{1/3-1} \div \frac{1}{6}\cdot(1)^{1/6-1} = \frac{6}{3} = 2$$

Example 6.21. If $\lim_{x\to 2} \frac{x^n-2^n}{x-2} = 80$ **and** $n \in N$**, find** n**.**

Solution.

$$\lim_{x\to 2} \frac{x^n-2^n}{x-2} = 80 \text{ (given)}$$

$$\Rightarrow \quad n\cdot 2^{n-1} = 80 = 5\cdot 2^{5-1}$$

$$\therefore \quad n = 5$$

6.19. EVALUATION OF LIMITS OF TRIGONOMETRIC FUNCTIONS

To find the limit of trigonometric functions, we will use the following standard results:

1. $\lim_{x\to 0} \sin x = 0$
2. $\lim_{x\to 0} \cos x = 1$
3. $\lim_{x\to 0} \frac{\sin x}{x} = 1$
4. $\lim_{x\to 0} \frac{\tan x}{x} = 1$
5. $\lim_{x\to a} \frac{\sin(x-a)}{x-a} = 1$
6. $\lim_{x\to a} \frac{\tan(x-a)}{x-a} = 1$

SOLVED EXAMPLES

Example 6.22. Evaluate:

(*i*) $\lim\limits_{x\to 0} \dfrac{1-\cos 2x}{x^2}$, (*ii*) $\lim\limits_{x\to 0} \dfrac{\tan x - \sin x}{\sin^3 x}$.

Solution. (*i*)

$$\lim_{x\to 0} \frac{1-\cos 2x}{x^2} = \lim_{x\to 0} \frac{\{1-(1-2\sin^2 x)\}}{x^2} = \lim_{x\to 0} \frac{2\sin^2 x}{x^2}$$

$$= 2\lim_{x\to 0} \frac{\sin x}{x} \lim_{x\to 0} \frac{\sin x}{x} = 2 \times 1 \times 1 = 2.$$

(*ii*)

$$\lim_{x\to 0} \frac{\tan x - \sin x}{\sin^3 x} = \lim_{x\to 0} \frac{\dfrac{\sin x}{\cos x} - \sin x}{\sin^3 x} = \lim_{x\to 0} \frac{\sin x(1-\cos x)}{\cos x \sin^3 x} = \lim_{x\to 0} \frac{1-\cos x}{\cos x \sin^2 x}$$

$$= \lim_{x\to 0} \frac{1-\cos x}{\cos x\,(1-\cos^2 x)} = \lim_{x\to 0} \frac{1}{\cos x(1+\cos x)} = \frac{1}{1(1+1)} = \frac{1}{2}.$$

Example 6.23. Evaluate:

(*i*) $\lim\limits_{x\to 0} \dfrac{\sin(a+x)-\sin(a-x)}{x}$, (*ii*) $\lim\limits_{x\to 0} \dfrac{\tan x - \sin x}{x^3}$.

Solution. (*i*) We have

$$\lim_{x\to 0} \frac{\sin(a+x)-\sin(a-x)}{x} = \lim_{x\to 0} \frac{2\cos\left(\dfrac{a+x+a-x}{2}\right)\sin\left(\dfrac{a+x-a+x}{2}\right)}{x}$$

$$= \lim_{x\to 0} \frac{2\cos a \cdot \sin x}{x} = 2\cos a \lim_{x\to 0}\left(\frac{\sin x}{x}\right) = 2\cos a$$

(*ii*)

$$\lim_{x\to 0}\left(\frac{\tan x - \sin x}{x^3}\right) = \lim_{x\to 0}\left(\frac{\dfrac{\sin x}{\cos x} - \sin x}{x^3}\right) = \lim_{x\to 0}\left[\frac{\sin x(1-\cos x)}{x^3 \cos x}\right]$$

$$= \lim_{x\to 0}\left[\frac{\sin x\left(2\sin^2 \dfrac{x}{2}\right)}{x^3 \cos x}\right] = 2\lim_{x\to 0}\left[\frac{\sin x}{x}\left(\frac{\sin \dfrac{x}{2}}{\dfrac{x}{2}}\right)^2 \cdot \frac{1}{4\cos x}\right]$$

$$= \frac{2}{4}\left[\lim_{x\to 0} \frac{\sin x}{x} \times \lim_{x\to 0}\left(\frac{\sin \dfrac{x}{2}}{\dfrac{x}{2}}\right)^2 \times \lim_{x\to 0}\left(\frac{1}{\cos x}\right)\right]$$

$$= \frac{1}{2}[1\times(1)^2 \times 1] = \frac{1}{2}.$$

6.20. EVALUATION OF EXPONENTIAL AND LOGARITHMIC LIMITS

To find the limits of exponential and logarithmic function. We will use the following standard results:

1. $\lim\limits_{x \to 0} \dfrac{e^x - 1}{x} = 1$ 2. $\lim\limits_{x \to 0} \dfrac{\log_e (1+x)}{x}$ 3. $\lim\limits_{x \to 0} \dfrac{a^x - 1}{x} = \log_e a$

SOLVED EXAMPLES

Example 6.24. Evaluate the following limits:

(i) $\lim\limits_{x \to 0} \dfrac{e^{4x} - 1}{x}$ **(ii)** $\lim\limits_{x \to 0} \dfrac{e^{3x} - e^{2x}}{x}$

Solution. (*i*)

$$\lim_{x \to 0} \frac{e^{4x} - 1}{x} = \lim_{x \to 0} \frac{4(e^{4x} - 1)}{4x} = 4 \lim_{x \to 0} \frac{e^{4x} - 1}{4x} = 4.1 = 4$$

(*ii*)

$$\lim_{x \to 0} \frac{e^{3x} - e^{2x}}{x} = \lim_{x \to 0} \frac{e^{3x} - 1 - e^{2x} + 1}{x}$$

$$= \lim_{x \to 0} \left[\frac{e^{3x} - 1}{x} - \frac{e^{2x} - 1}{x} \right] = \lim_{x \to 0} \frac{e^{3x} - 1}{x} - \lim_{x \to 0} \frac{e^{2x} - 1}{x}$$

$$= \lim_{x \to 0} \frac{3(e^{3x} - 1)}{3x} - \lim_{x \to 0} \frac{2(e^{2x} - 1)}{2x}$$

$$= 3 \lim_{x \to 0} \frac{e^{3x} - 1}{3x} - 2 \lim_{x \to 0} \frac{e^{2x} - 1}{2x}$$

$$= 3.1 - 2.1 = 3 - 2 = 1$$

Example 6.25. Evaluate the following limits:

(i) $\lim\limits_{x \to 0} \dfrac{3^x - 2^x}{x}$ **(ii)** $\lim\limits_{x \to 0} \dfrac{5^x - 2^x}{\sin x}$

Solution. (*i*)

$$\lim_{x \to 0} \frac{3^x - 2^x}{x} = \lim_{x \to 0} \frac{3^x - 1 - 2^x + 1}{x}$$

$$= \lim_{x \to 0} \left(\frac{3^x - 1}{x} - \frac{2^x - 1}{x} \right) = \lim_{x \to 0} \frac{3^x - 1}{x} - \lim_{x \to 0} \frac{2^x - 1}{x}$$

$$= \log e^3 - \log e^2 = \log_e \left(\frac{3}{2} \right)$$

(*ii*)

$$\lim_{x \to 0} \frac{5^x - 2^x}{\sin x} = \lim_{x \to 0} \frac{5^x - 1 - 2^x + 1}{\sin x}$$

$$= \lim_{x \to 0} \frac{\dfrac{5^x - 1 - 2^x + 1}{x}}{\dfrac{\sin x}{x}} = \lim_{x \to 0} \left(\frac{5^x - 1}{x} - \frac{2^x - 1}{x} \right) \times \frac{x}{\sin x}$$

$$= \left(\lim_{x\to 0} \frac{5^x - 1}{x} - \lim_{x\to 0} \frac{2^x - 1}{x} \right) \times \frac{1}{\lim_{x\to 0} \frac{\sin x}{x}}$$

$$= (\log e^5 - \log e^2) \times 1 = \log_e \left(\frac{5}{2}\right)$$

EXERCISE 6.2

Evaluate:

1. $\lim_{x\to 2} \frac{x^2 - 4}{x - 2}$ **[Ans.** 4]

2. $\lim_{x\to 3} \frac{x^3 - 27}{x - 3}$ **[Ans.** 27]

   [**Hint:** $x^3 - 27 = (x - 3)(x^2 + 3x + 9)$]

3. $\lim_{x\to 2} \frac{x^3 - 8}{x^2 - 4}$ **[Ans.** 3]

   [**Hint:** $x^3 - 8 = (x - 2)(x^2 + 2x + 4)$]

4. $\lim_{x\to 3} \frac{x^2 + 2x - 15}{x^2 - 9}$ $\left[\textbf{Ans. } \frac{4}{3}\right]$

5. Evaluate $\lim_{x\to 2} \frac{\sqrt{x+7} - 3}{x - 2}$ $\left[\textbf{Ans. } \frac{1}{6}\right]$

   $\left[\textbf{Hint. } \text{Rationalizing, } \lim_{x\to 2} \frac{\{(x+7) - 9\}}{(x-2)(\sqrt{x+7} + 3)} = \lim_{x\to 2} \frac{1}{\sqrt{x+7} + 3} = \frac{1}{6}\right]$

6. Evaluate $\lim_{x\to 0} \frac{\sin(2 + x) - \sin(2 - x)}{x}$ **[Ans.** 2 cos 2]

7. Show that $\lim_{x\to 0} \frac{1 - \cos x}{x} = 0.$

8. Show that $\lim_{x\to 0} \frac{3x + 2|x|}{5x - 3|x|}$ does not exist.

   $\left[\text{R.H.L.} = \lim_{x\to 0} \frac{3x + 2x}{5x - 3x} = \frac{5}{2}, \text{L.H.L.} = \lim_{x\to 0} \frac{3x - 2x}{5x + 3x} = \frac{1}{8}\right]$

9. Let $f(x) = \begin{cases} 2 + x, \text{ if } x \geq 0, \\ 2 - x, \text{ if } x < 0 \end{cases}$. Find $\lim_{x\to 0} f(x)$. **[Ans.** 2]

10. Show that $\lim_{x\to a} \frac{|x - a|}{x - a}$ does not exist.

11. $\lim_{x\to 0} \frac{\sin x - \sin a}{\sqrt{x} - \sqrt{a}}$ [**Ans.** $2\sqrt{a} \cos a$]

12. $\lim_{x\to \frac{\pi}{4}} \frac{\tan^3 x - \tan x}{\cos\left(x + \frac{\pi}{4}\right)}$ **[Ans.** – 4]

13. $\lim\limits_{x\to\frac{\pi}{3}} \dfrac{\sqrt{1-\cos 6x}}{\sqrt{2}\left(\dfrac{\pi}{3}-x\right)}$ **[Ans.** 3]

14. $\lim\limits_{x\to 0} \dfrac{e^{2x}-e^{x}}{\sin 2x}$ $\left[\textbf{Ans.}\left(\dfrac{1}{2}\right)\right]$

15. $\lim\limits_{x\to 0} \dfrac{a^{x}-a^{-x}}{x}$ **[Ans.** $2\log_e a$]

EXERCISE 6.3

Determine the nature of the following functions of even and odd.

1. $x^2 - |x|$ **[Ans.** even]
2. $\sin x - \cos x$ **[Ans.** neither even nor odd]
3. $\log[\sqrt{x^2+1}-x]$ **[Ans.** odd]
4. $\sin[\log(x+\sqrt{x^2+1})]$ **[Ans.** odd]

Determine the period for each of the following function.

5. $\cos 2\pi x$ **[Ans.** 1]
6. $\sin x - \cos x$ **[Ans.** 2π]
7. $\cos^2 \dfrac{x}{2}$ **[Ans.** 2π]
8. $4\cos^3 2x - 3\cos 2x$ $\left[\textbf{Ans.}\ \dfrac{\pi}{3}\right]$
9. If the function $f: R \to R$ be given by $f(x) = x^2 + 2$ and $g: R - [1] \to R$ be given by $g(x) = \dfrac{x}{x-1}$. Find fog and gof.

$$\left[\textbf{Ans.}\ \frac{x^2}{(x-1)^2}+2;\ \frac{x^2+2}{x^2+1}\right]$$

10. If $f: R \to \left\{\dfrac{7}{5}\right\} \to R - \left\{\dfrac{3}{5}\right\}$ be defined as $f(x) = \dfrac{3x+4}{5x-7}$ and $g: R - \left\{\dfrac{3}{5}\right\} \to R - \left\{\dfrac{7}{5}\right\}$ be defined as $g(x) = \dfrac{7x+5}{5x-3}$. Show that $gof = I_A$ and $fog = I_B$, where $B = R - \left\{\dfrac{3}{5}\right\}$ and $A = R - \left\{\dfrac{7}{5}\right\}$.

6.21. CONTINUITY

6.21.1. Continuity of a Real Function at a Point

Intuitively, a real function should be considered continuous at a point iff the graph of the function has no break at the point under consideration *i.e.*, iff the graph of the function can be drawn in the neighbourhood of the point (including the point) without lifting the pen from the plane of the paper, which is so iff the values of the function at the neighbouring points are '*close enough*' to the value at the given point and hence the continuity of a function at a point is closely related to the 'linking behaviour' of the function in the neighbourhood of that point. To understand this, we study some examples giving us an intuitive feeling for what is at issue.

(*i*) Consider the function f defined by

$$f(x) = \begin{cases} 1, & x \neq 1 \\ 2, & x = 1 \end{cases}$$

Clearly domain $f = \mathbf{R}$ so that the function values can be obtained for every real value of x.

$$\underset{x \to 1}{\text{Lt}}\ f(x) = \underset{x \to 1}{\text{Lt}}\ 1$$

$$[\because\ f(x) = 1, \text{ when } x \neq 1]$$

$$= 1$$

Also $\quad f(1) = 2$

Here, we observe that $\underset{x \to 1}{\text{Lt}}\ f(x)$ exists, but $\underset{x \to 1}{\text{Lt}}\ f(x) \neq f(1)$. A portion of the graph of the function is shown in Fig. 6.11.

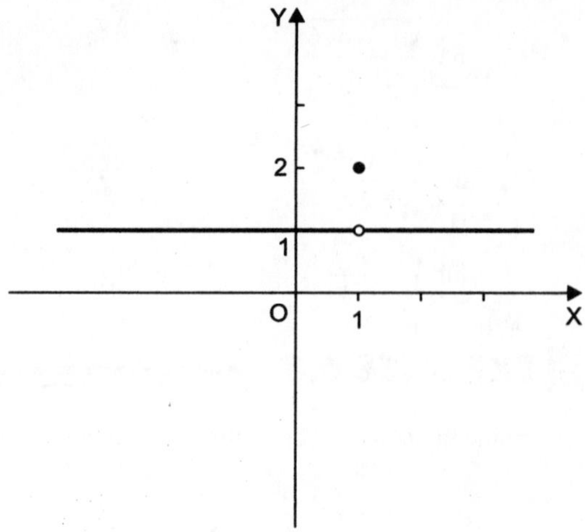

Fig. 6.11

Note that when we try to draw the graph near $x = 1$, we cannot draw it in one stroke *i.e.*, without lifting the pen from the plane of the paper. In fact, we have to lift the pen when we come to 1. Clearly, there is a break in the graph at $x = 1$ this is an example of a function not continuous at $x = 1$.

(*ii*) Consider the function f defined by

$$f(x) = \frac{2x^2 - 5x + 2}{x - 2}$$

Here $\quad$ domain of $f = R$ except 2

$$\underset{x \to 2}{\text{Lt}}\ f(x) = \underset{x \to 2}{\text{Lt}}\ \frac{2x^2 - 5x + 2}{x - 2} = \underset{x \to 2}{\text{Lt}}\ \frac{(2x-1)(x-2)}{x-2}$$

$$= \underset{x \to 2}{\text{Lt}}\ (2x - 1)$$

$$= 2.2 - 1 = 3$$

Here, we observe that $\underset{x \to 2}{\text{Lt}}\ f(x)$ exists, but $f(2)$ is not defined.

A portion of the graph of the function is shown in Fig. 6.12. The graph of f consists of all points on the straight line $y = 2x - 1$ except the point (2, 3). Note that when we try to draw the graph near $x = 2$, we cannot draw the graph of the function without lifting the pen. Clearly, there is a break in the graph at $x = 2$. This is another example of a function not continuous at $x = 2$.

Further, if we define $f(2) = 3$, then there is no break in the graph of f at $x = 2$ and $\underset{x \to 2}{\text{Lt}}\ f(x) = f(2)$, which we express by saying that f is continuous at $x = 2$.

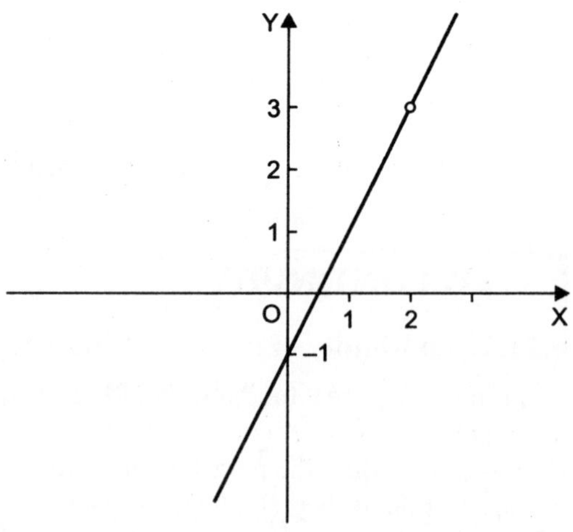

Fig. 6.12

(*iii*) Consider the function f defined by

$$f(x) = \begin{cases} x-1, & x<0 \\ 0, & x=0 \\ x+1, & x>0 \end{cases}$$

Clearly, domain of $f = \mathbf{R}$ so that the function values can be obtained for every real value of x.

When $x < 0$, the function values are dictated by $x - 1$,

$$\therefore \qquad \underset{x\to 1}{\text{Lt}}\ f(x) = \underset{x\to 0}{\text{Lt}}\ (x-1) = 0-1 = -1$$

When $x > 0$, the function values are dictated by $x + 1$,

$$\therefore \qquad \underset{x\to 0^+}{\text{Lt}}\ f(x) = \underset{x\to 1^+}{\text{Lt}}\ (x+1) = 0+1 = 1.$$

Thus $\qquad \underset{x\to 0^+}{\text{Lt}}\ f(x) \neq \underset{x\to 0^+}{\text{Lt}}\ f(x)$

$\Rightarrow \underset{x\to 0}{\text{Lt}}\ f(x)$ does not exist.

A portion of the graph of the function is shown in Fig. 6.13.

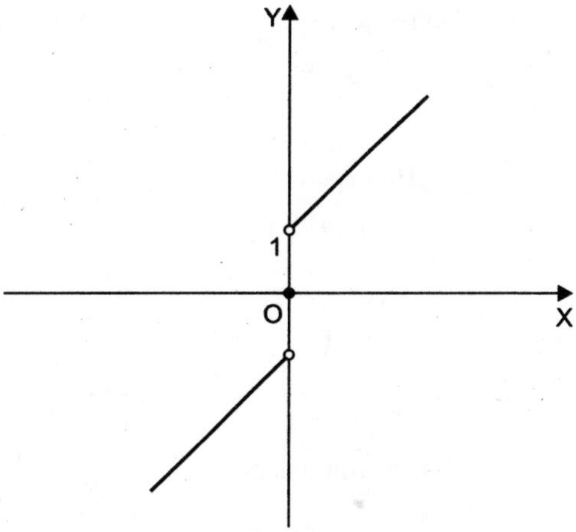

Fig. 6.13

Again we note that we cannot draw the graph of the function near 0 without lifting the pen. Clearly, there is a break in the graph of the function at $x = 0$. This is yet another example of a function not continuous at $x = 0$.

We know that on the real axis, point x can approach the given point c either from *left* or from the *right* of c. This means that we can talk about the continuity of a function at $x = c$ either from left or from right.

Definition. A function f is said to be left continuous or continuous from the left at $x = c$ iff

(*i*) $f(c)$ exists, (*ii*) $\underset{x\to c^-}{\text{Lt}}\ f(x)$ exists and (*iii*) $\underset{x\to c^-}{\text{Lt}}\ f(x) = f(c)$.

A function f is said to be right continuous or continuous from the right at $x = c$ iff

(*i*) $f(c)$ exists, (*ii*) $\underset{x\to c^+}{\text{Lt}}\ f(x)$ exists and (*iii*) $\underset{x\to c^+}{\text{Lt}}\ f(x) = f(c)$.

A function f is said to be continuous at $x = c$ iff

(*i*) $f(c)$ exists, (*ii*) $\underset{x\to c}{\text{Lt}}\ f(x)$ exists and (*iii*) $\underset{x\to c}{\text{Lt}}\ f(x) = f(c)$.

Hence, a function is continuous at $x = c$ iff it is both left as well as right continuous at $x = c$.

A function f fails to be continuous at $x = c$ for any of the following reasons:

1. f is not defined at $x = c$ *i.e.*, $f(c)$ does not exist.
2. $\underset{x\to c^-}{\text{Lt}}\ f(x)$ does not exist. This happens if either

   (*i*) $\underset{x\to c^-}{\text{Lt}}\ f(x)$ does not exist or (*ii*) $\underset{x\to c^-}{\text{Lt}}\ f(x)$ does not exist

   (*iii*) both $\underset{x\to c}{\text{Lt}}\ f(x)$ exist but are not equal.
3. $\underset{x\to c}{\text{Lt}}\ f(x)$ exists but $\underset{x\to c}{\text{Lt}}\ f(x) \neq f(c)$

When $\underset{x\to c}{\text{Lt}}\ f(x)$ exists but either $f(c)$ does not exist or $\underset{x\to c}{\text{Lt}}\ f(x) \neq f(c)$ we say that f has a removable discontinuity otherwise, we say that f has non-removable discontinuity.

SOLVED EXAMPLES

Example 6.26. Examine the following functions for continuity:

(*i*) $f(x) = \begin{cases} \dfrac{x^2-4}{x-2}, & x \neq 2 \\ 4, & x = 2 \end{cases}$ **at $x = 2$** (*ii*) $f(x) = \begin{cases} 1, & x < 0 \\ x-2, & x \geq 0 \end{cases}$ **at $x = 0$.**

Solution. (*i*)

$$\underset{x\to 2}{\text{Lt}}\ f(x) = \underset{x\to 2}{\text{Lt}}\ \frac{x^2-4}{x-2}$$

$$= \underset{x\to 1}{\text{Lt}}\ (x+2) \qquad [\because \ x \neq 2]$$

$$= 2 + 2 = 4$$

and $$f(2) = 4$$

$\Rightarrow$ $$\underset{x\to 2}{\text{Lt}}\ f(x) = f(2)$$

$\Rightarrow$ f is continuous at $x = 2$.

(*ii*) Here $$f(0) = 0 - 2 = -2,$$

$$\underset{x\to 0^-}{\text{Lt}}\ f(x) = \underset{x\to 0^-}{\text{Lt}}\ 1 = 1$$

and $$\underset{x\to 0^+}{\text{Lt}}\ f(x) = \underset{x\to 0^+}{\text{Lt}}\ (x-2)$$

$$= 0 - 2 = -2$$

$\Rightarrow$ $$\underset{x\to 0^-}{\text{Lt}}\ f(x) \neq \underset{x\to 0^+}{\text{Lt}}\ f(x)$$

$\Rightarrow$ f is discontinuous at $x = 0$ and it has non-removable discontinuity at $x = 0$.

However, $$\underset{x\to 0^+}{\text{Lt}}\ f(x) = -2 = f(0)$$

$\Rightarrow$ f is right continuous at $x = 0$

A portion of the graph of f is shown in Fig. 6.14.

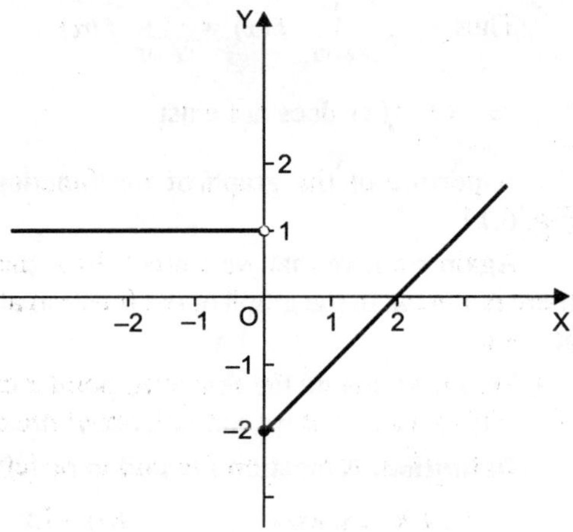

Fig. 6.14

Example 6.27. If the function $f(x) = \begin{cases} 3ax + b, & x > 1 \\ 11 & x = 1 \\ 5ax - 2b, & x < 1 \end{cases}$ **is continuous at $x = 1$, find the values of a and b.**

Solution. Here $f(x) = 11$ at $x = 1$ *i.e.*, $f(1) = 11$,

$$\underset{x\to 1^-}{\text{Lt}}\ f(x) = \underset{x\to 1^-}{\text{Lt}}\ (5ax - 2b)\ 5a \cdot 1 - 2b = 5a - 2b$$

and $$\underset{x\to 1^+}{\text{Lt}}\ f(x) = \underset{x\to 1^+}{\text{Lt}}\ (3ax + b) = 3a \cdot 1 + b = 3a + b$$

Since the function f is continuous at $x = 1$.

$$\underset{x\to 1^-}{\text{Lt}}\ f(x) = \underset{x\to 1^+}{\text{Lt}}\ f(x) = f(1)$$

$\Rightarrow$ $$5a - 2b = 3a + b = 11$$

$$\Rightarrow \quad \begin{cases} 5a - 2b = 11 & \dots (i) \\ 3a + b = 11 & \dots (ii) \end{cases}$$

Solving these equations, we get $a = 3$, $b = 2$

Example 6.28. Show that $f(x) = \begin{cases} |x-a| & x \neq a \\ 1 & x = a \end{cases}$ **is discontinuous at** $x = a$.

Solution.

$$\underset{x \to a^-}{\text{Lt}} f(x) = \underset{x \to a^-}{\text{Lt}} \frac{|x-a|}{x-a} = \underset{x \to a^-}{\text{Lt}} \frac{-(x-a)}{x-a}$$

$$[\because x \to a^- \Rightarrow x < a \Rightarrow x - a < 0 \Rightarrow |x-a| = -(x-a)]$$

$$= \underset{x \to a^-}{\text{Lt}} (-1) = -1$$

and

$$\underset{x \to a^+}{\text{Lt}} f(x) = \underset{x \to a^+}{\text{Lt}} \frac{|x-a|}{x-a} = \underset{x \to a^+}{\text{Lt}} \frac{x-a}{x-a}$$

$$[\because x \to a^+ \Rightarrow x > a \Rightarrow x - a > 0 \Rightarrow |x-a| = (x-a)]$$

$$= \underset{x \to a^+}{\text{Lt}} (1) = 1.$$

Thus, $\underset{x \to a^-}{\text{Lt}} f(x) \neq \underset{x \to a^+}{\text{Lt}} f(x) \Rightarrow \underset{x \to a}{\text{Lt}} f(x)$ does not exist.

$\Rightarrow$ f is discontinuous at $x = a$.

Remark. In this example, note that no choice of $f(a)$ can make the function continuous at $x = a$, therefore, f has a non-removable discontinuity at $x = a$.

Example 6.29. For what value of k is the function

$$f(x) = \begin{cases} \dfrac{k \cos x}{\pi - 2x}, & \text{if } x \neq \dfrac{\pi}{2} \\ 3, & \text{if } x = \dfrac{\pi}{2} \end{cases} \quad \textbf{continuous at } x = \frac{\pi}{2}\textbf{?}$$

Solution. Let $x = \frac{\pi}{2} + h$, so the when $x \to \frac{\pi}{2}$, $h \to 0$.

$$\therefore \quad \underset{x \to \frac{\pi}{2}}{\text{Lt}} f(x) = \underset{x \to \frac{\pi}{2}}{\text{Lt}} \frac{-k \cos x}{\pi - 2x} = \frac{k}{2} \underset{h \to 0}{\text{Lt}} \frac{k \cos\left(\frac{\pi}{2} + h\right)}{\pi - 2\left(\frac{h}{2} + h\right)}$$

$$= \underset{x \to 0}{\text{Lt}} \frac{-k \sin h}{-2h} = \frac{k}{2} \underset{x \to 0}{\text{Lt}} \frac{\sin h}{h} = \frac{k}{2} \times 1 = \frac{k}{2}$$

Also $f\left(\frac{\pi}{3}\right) = 3$. Since function f is continuous at $x = \frac{\pi}{2}$

$$\underset{x \to \frac{\pi}{2}}{\text{Lt}} f(x) = f\left(\frac{\pi}{2}\right) \Rightarrow \frac{k}{2} = 3 \Rightarrow k = 6$$

EXERCISE 6.4

1. Examine the following functions for continuity at the indicated points:

   (i) $f(x) = 2x + 3$ at $x = 1$ **[Ans.** Continuous]

   (ii) $f(x) = x^2$ at $x = 0$ **[Ans.** Continuous]

   (iii) $f(x) = \begin{cases} x^3 + 3, & x \neq 0 \\ 1, & x = 0 \end{cases}$ at $x = 0$ **[Ans.** Discontinuous]

   (iv) $f(x) = 2x^2 - 1$ at $x = 3$. **[Ans.** Continuous]

2. Prove that the function $f(x) = 5x - 3$ is continuous at $x = 0$, at $x = -3$ and at $x = 5$.

3. Show that the function $f(x) = \sin^2 x + x^2 - 2x$ is continuous at $x = 0$.

4. (i) If $f(x) = \dfrac{x^2 - 1}{x - 1}$ for $x \neq 1$ and $f(x) = 2$ when $x = 1$, show that the function is continuous at $x = 1$.

   (ii) A function f is defined as $f(x) = \begin{cases} \dfrac{x^2 - x - 6}{x - 3}, & x \neq 3 \\ 5, & x = 3 \end{cases}$, show that f is continuous at $x = 3$.

5. (i) Discuss the continuity of the function $f(x)$ at $x = 0$ if

$$f(x) = \begin{cases} 2x - 1, & x < 0 \\ 2x + 1, & x \geq 2 \end{cases}$$

[Ans. Continuous]

   (ii) Discuss the continuity of the function $f(x)$ at $x = 2$, where

$$f(x) = \begin{cases} 2 - x, & x < 2 \\ 2 + x, & x \geq 2 \end{cases}$$

[Ans. Discontinuous]

6. Examine the following functions for continuity:

   (i) $f(x) = \begin{cases} \dfrac{x}{\sin 3x}, & \text{when } x \neq 0 \\ 3, & \text{when } x = 0 \end{cases}$ at $x = 0$ **[Ans.** Discontinuous]

   (ii) $f(x) = \begin{cases} \dfrac{\sin 2x}{\sin 3x}, & \text{when } x \neq 0 \\ 2, & \text{when } x = 0 \end{cases}$ at $x = 0$ **[Ans.** Discontinuous]

   (iii) $f(x) = \begin{cases} \dfrac{|\sin x|}{x} & \text{if } x \neq 0 \\ 1 & \text{if } x = 0 \end{cases}$ at $x = 0$ **[Ans.** Discontinuous]

   (iv) $f(x) = \begin{cases} \dfrac{\sin x}{x}, & \text{if } x < 0 \\ x + 1, & \text{if } x \geq 0 \end{cases}$ at $x = 0$ **[Ans.** Continuous]

7. Find the value of the constant λ so that the function given below is continuous at x -1:

$$f(x) = \begin{cases} \dfrac{x^2 - 2x - 3}{x + 1}, & x \neq -1 \\ \lambda, & \text{for } x = -1 \end{cases}$$

[Ans. – 4]

8. Find the values of a and b so that the function f defined by

$$f(x) = \begin{cases} ax+1, & \text{if } x<3 \\ 4, & \text{if } x=3 \\ bx+3, & \text{if } x>3 \end{cases}$$ is continuous at $x = 3$.

$\left[\textbf{Ans. } a=1, b=\frac{1}{3}\right]$

9. For what value of λ (if it exists) is the function

$$f(x) = \begin{cases} \lambda(x^2-2x), & \text{if } x\le 0 \\ 4x+1, & \text{if } x>0 \end{cases}$$

continuous at $x = 0$? What about continuity at $x = \pm 1$?

[**Ans.** No value of λ can make function f continuous at $x = 0$; the function is continuous at $x = 1$ and at $x = -1$ for all values of λ]

❑❑❑

CHAPTER 7

Differentiation

INSIDE THIS CHAPTER

7.1. DERIVATIVE OF A FUNCTION

Let $y = f(x)$ be a given contionuous function. Then, the values of y depends upon the value of x and it changes with a change in the value of x. We use the word kincrement to denote a small change *i.e.*, increase or decrease in the value of x and y.

Let Δy be an increment in y corresponding to an increment to an increment Δx in x. Then

$$y = f(x) \quad \text{and} \quad y + \Delta y = f(x + \Delta x)$$

On subrtacting, we get $\quad \Delta y = f(x + \Delta x) - f(x)$

$$\therefore \quad \frac{\Delta y}{\Delta x} = \frac{f(x+\Delta x) - f(x)}{\Delta x}$$

So,
$$\lim_{\Delta x \to 0} \frac{\Delta y}{\Delta x} = \lim_{\Delta x \to 0} \frac{f(x+\Delta x) - f(x)}{\Delta x}$$

The above limit, if it exists finitely, is called the derivative or differential coefficient of $y = f(x)$ with respect to x and it is denoted by

$$\frac{dy}{dx} = \frac{d}{dx}[f(x)] \quad \text{or} \quad f'(x)$$

The process of finding the derivative is known as differentiation.

7.2. DERIVATIVE AT A POINT

The value of $f'(x)$ obtained by putting $x = a$, is called the derivative of $f(x)$ at $x = a$ and it is denoted by

$f'(a)$ or $\left[\dfrac{dy}{dx}\right]_{x=a}$.

If $f'(x)$ exists, we say that $f(x)$ is differentiable at $x = a$.

If $f'(x)$ exists for every value of x in the domain of the function, we say that $f(x)$ is differentiable.

7.3. GEOMETRICAL INTERPRETATION OF THE DERIVATIVE

Geometrically, the derivative of a function $f(x)$ at a point $x = a$ represents the slope of the tangent to the curve $y = f(x)$ at $(a, f(a))$.

Let $y = f(x)$ be differentiable on an open interval containing a. Let $P(a, f(a))$ and $Q(a + h, f(a + yh))$ be two neighbouring points on the graph of $y = f(x)$ (see figure). Then the slope of the secant line PQ is $\frac{f(a+h)-f(a)}{h}$. Since the tangnet at P is a limiting position of secant lines PQ, the slope of the tangent at P is the limit value of the slope of the secant lines as Q approaches P. But as Q approaches P, $h \to 0$.

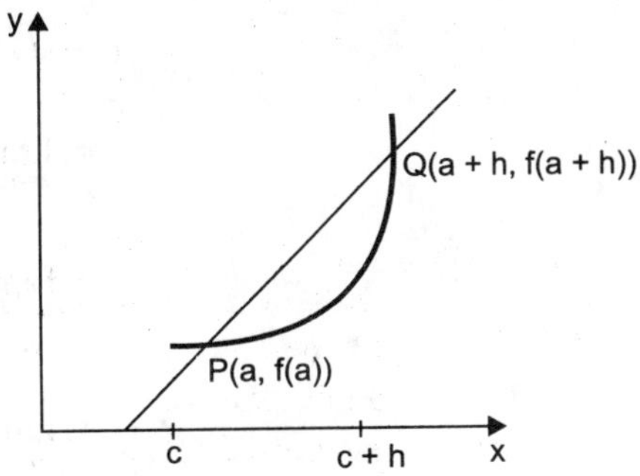

Fig. 7.1

This proves the result.

The provess of finding the dirivative of a function by using the definintion of the derivatives is called the differentiation from first principles or by *ab*-initio method or by delta method.

SOLVED EXAMPLES

Example 7.1. Show that the function $f(x) = x^2$ is differentiable at $x = 3$ and obtain is derivative at $x = 3$.

Solution. We have $f(x) = 3^2 = 9$

and $f(3 + h) = (3 + h)^2 = 9 + h^2 + 6h$

$$\therefore \quad \frac{f(3+h)-f(3)}{h} = \frac{9+h^2+6h-9}{h} = h+6$$

which approaches 6 as $h \to 0$.

Example 7.2. Show that the function $f(x) = |x|$ is not differentiable at $x = 0$.

Solution. We have $f(0) = |0| = 0$ and $f(0 + h) = |h|$

$$\therefore \quad \frac{f(0+h)-f(0)}{h} = \frac{|h|}{h} = \begin{cases} 1, & if\ h>0 \\ -1, & if\ h<0 \end{cases}$$

$$\text{Hence } \lim_{h\to 0^-} \frac{f(0+h)-f(0)}{h} = 1 \quad \text{and} \quad \lim_{h\to 0^-} \frac{f(0+h)-f(0)}{h} = -1$$

Thus, $\lim_{h\to 0} \frac{f(0+h)-f(0)}{h}$ does not exists and, therefore, if is not differentiable at $x = 0$.

Example 7.3. Find the derivative of $f(x) = x^2$.

Solution. By definition,

$$f'(x) = \lim_{h\to 0} \frac{f(0+h)-f(0)}{h}$$

$$= \lim_{h\to 0} \frac{(x+h)^2-x^2}{h} = \lim_{h\to 0} \frac{x^2+h^2+2xh-x^2}{h} = \lim_{h\to 0}(h+2x) = 2x.$$

Thus $2x$ is the derivative of the function x^2.

Example 7.4. Find the slope of the tangent to the curve $y = f(x) = \sqrt{x}$ at the point (9, 2).

Solution. By the geometrical interpretation, the slope of the tangent to the curve

$$y = \sqrt{x} \text{ at } (9, 2) = f'(9)$$

$$= \lim_{h\to 0} \frac{f(9+h)-f(9)}{h} = \lim_{h\to 0} \frac{\sqrt{9+h}-\sqrt{9}}{h}$$

$$= \lim_{h\to 0} \frac{\sqrt{9+h}-3}{h} = \lim_{h\to 0}\left[\frac{\sqrt{9+h}-3}{h} \cdot \frac{\sqrt{9+h}+3}{\sqrt{9+h}+3}\right]$$

$$= \lim_{h\to 0}\left[\frac{(9+h)-9}{h(\sqrt{9+h}+3)}\right] = \lim_{h\to 0}\left[\frac{1}{\sqrt{9+h}+3}\right] = \frac{1}{\sqrt{9+0}+3} = \frac{1}{6}.$$

Hence the tangent line to $y = \frac{1}{6}$ at (9, 2) has slope $\frac{1}{6}$.

7.4. DIFFERENTIATION OF SOME STANDARD FUNCTIONS FROM FIRST PRINCIPLES

In this article we will find find find derivatives or differentiation of differential coefficients of some standard function viz. x^n, e^x, a^x $\sin x$ $\cos x$, $\tan x$, cot, x cosec x, sec x, $\sin^{-1} x$, $\cos^{-1} x$, etc. Students are advised to learn these results these results as these results will be used will be used in the subsequent part of the chapter.

7.4.1. Differentiation of x^n From First Principles, where $n \in R$

We have
$$\frac{d(f(x))}{dx} = \lim_{h\to 0}\frac{f(x+h)-f(x)}{h}$$

Here $f(x) = x^n$. Therefore $f(x+h) = (x+h)^n$.

So,
$$\frac{d}{dx}(f(x)) = \lim_{h\to 0}\frac{(x+h)^n - x^n}{h} = \lim_{h\to 0}\frac{(x+h)^n - x^n}{(x+h)-x}$$

$$= \lim_{z\to x}\frac{z^n - x^n}{z-x} \quad \text{where } z = x+h \text{ and } z \to x \text{ as } h \to 0$$

$$= nx^{n-1}$$

Thus,
$$\frac{d}{dx}(f(x)) = nx^{n-1} \text{ i.e., } \frac{d}{dx}(x^n) = nx^{n-1}.$$

Example 7.5. Find the differentiation of the following functions w.r.t. x:

(i) x^5 ***(ii)*** $\frac{1}{x^3}$ ***(iii)*** $\sqrt{x}$ ***(iv)*** $\frac{1}{\sqrt{x}}$

Solution. We have $\frac{d}{dx}(x^n) = nx^{n-1}$. Therefore

(i)
$$\frac{d}{dx}(x^5) = 5x^{5-1} = 5x^4$$

(ii) $$\frac{d}{dx}\left(\frac{1}{x^3}\right) = \frac{d}{dx}(x^{-3}) = -3x^{-31-1} = \frac{-3}{x^4}$$

(iii) $$\frac{d}{dx}(\sqrt{x}) = \frac{d}{dx}(x^{1/2}) = \frac{1}{2}x^{1/2-1} = \frac{1}{2\sqrt{x}}$$

(iv) $$\frac{d}{dx}\left(\frac{1}{\sqrt{x}}\right) = \frac{d}{dx}(x^{-1/2}) = \frac{-1}{2}x^{-1/2-1} = \frac{-1}{2}x^{-3/2}$$

7.4.2. Differentiation of e^x From First Principles

We have $$\frac{d}{dx}(f(x)) = \lim_{h\to 0}\frac{f(x+h)-f(x)}{h}$$

Here $f(x) = e^x$. Therefore $f(x+h) = e^{x+h}$.

So, $$\frac{d}{dx}(f(x)) = \lim_{h\to 0}\frac{e^x \cdot e^h - e^x}{h} = \lim_{h\to 0} e^x\left(\frac{e^h-1}{h}\right) = a^x \lim_{h\to 0}\left(\frac{e^n-1}{h}\right) = e^x \cdot 1 = e^x$$

Thus, $$\frac{d}{dx}(f(x)) = a^x \log_e a \text{ i.e., } \frac{d}{dx}(a^x) = a^x \log_e a.$$

Example 7.6. Differentiate the following function w.r.t. x:

(i) 5^x **(ii) 10^x** **(iii) e^{2x}.**

Solution. We know that $\frac{d}{dx}(a^x) = a^x \log_e a, a > 0$. Therefore,

(i) $$\frac{d}{dx}(a^x) = (e^2)^x \log e^2 = e^2x.\ 2\log 3 = 2e^{2x}.$$

(ii) $$\frac{d}{dx}(10^x) = 10^x \log_e 10.$$

(iii) $$\frac{d}{dx}(e^{2x}) = \frac{d}{dx}((e^2)^x) = (e^2)^x \log e^2 = e^{2x}\ .\ 2\log e = 2e^{2x}.$$

7.4.3. Differentiation of $\text{Log}_e\, x$ From First Principles, $x > 0$

Let $f(x) = \log_e x$. Then $f(x+h) = \log_e (x+h)$

$$\therefore \quad \frac{d}{dx}(f(x)) = \lim_{h\to 0}\frac{f(x+h)-f(x)}{h} = \lim_{h\to 0}\frac{\log_e(x+h)-\log_e x}{h}$$

$$= \lim_{h\to 0}\frac{\log_e\left(1+\frac{h}{x}\right)}{h} = \lim_{h\to 0}\frac{\log_e\left(1+\frac{h}{x}\right)}{h/x}\cdot\frac{1}{x}$$

$$= \frac{1}{x} \qquad \left[\because \lim_{h\to 0}\frac{\log_e(1+x)}{x} = 1\right]$$

Thus, $$\frac{d}{dx}(f(x)) = \frac{1}{x} \text{ i.e., } \frac{d}{dx}(\log_e x) = \frac{1}{x}.$$

7.4.4. Differentiation of $\log_a x$ From First Principles, $x > 0$, $a > 0$ ($a \neq 1$)

Let $f(x) = \log_a x$. Then $f(x+h) = \log_a (x+h)$

$$\therefore \quad \frac{d}{dx}(f(x)) = \lim_{h\to 0} \frac{f(x+h) - f(x)}{h} = \lim_{h\to 0} \frac{\log_a (x+h) - \log_a x}{h}$$

$$= \lim_{h\to 0} \frac{\log_a \left(\frac{x+h}{x}\right)}{h}$$

$$= \lim_{h\to 0} \frac{\log_a \left(1 + \frac{h}{x}\right)}{h} = \lim_{h\to 0} \frac{\log_e \left(1 + \frac{h}{x}\right)}{\log_e a \cdot h} \qquad \left[\because \log_a \lambda = \frac{\log_e \lambda}{\log_e a}\right]$$

$$= \frac{1}{\log_e a} \lim_{h\to 0} \frac{\log_e \left(1 + \frac{h}{x}\right)}{x\left(\frac{h}{x}\right)} = \frac{1}{x \log_e a} \qquad \left[\because \lim_{h\to 0} \frac{\log\left(1 + \frac{h}{x}\right)}{\frac{h}{x}} = 1\right]$$

Thus, $$\frac{d}{dx}(f(x)) = \frac{1}{x \log_e a} \text{ i.e., } \frac{d}{dx}(\log_a x) = \frac{1}{x \log_e a}$$

Example 7.7. Differentiate the following function w.r.t. x:

(i) $\log_3 x$ **(ii)** $\dfrac{1}{\log_x 5}$

Solution. We know that $\frac{d}{dx}(\log_a x) = \frac{1}{x \log_e a}$ · Therefore,

(i) $$\frac{d}{dx}(\log_e x) = \frac{1}{x \log_e 3}$$

(ii) $$\frac{d}{dx}\left(\frac{1}{\log_x 5}\right) = \frac{d}{dx}(\log_5 x) = \frac{1}{x \log_e a}.$$

7.4.5. Differentiation of $\sin x$ From First Principles

Let $f(x) = \sin x$. Then $f(x+h) = \sin(x+h)$

$$\frac{d}{dx}(f(x)) = \lim_{h\to 0} \frac{f(x+h) - f(x)}{h} \lim_{h\to 0} \frac{\sin(x+h) - \sin x}{h}$$

$$= \lim_{h\to 0} \frac{2 \sin\left(\frac{h}{2}\right) \cos\left(\frac{2x+h}{2}\right)}{h} \qquad \left[\because \sin C - \sin D = 2 \sin \frac{C-D}{2} \cos \frac{C+D}{2}\right]$$

$$= \lim_{h\to 0} \frac{\left(2 \sin \frac{h}{2}\right) \cos\left(\frac{2x+h}{2}\right)}{2\left(\frac{h}{2}\right)}$$

$$= \lim_{h\to 0} \cos\left(x + \frac{h}{2}\right) \lim_{h\to 0} \frac{\sin\left(\frac{h}{2}\right)}{\left(\frac{h}{2}\right)} = (\cos x) \cdot 1 = \cos x \qquad \left[\because \lim_{h\to 0} \frac{\sin\left(\frac{h}{2}\right)}{\left(\frac{h}{2}\right)} = 1\right]$$

Thus, $$\frac{d}{dx}(f(x)) = \cos \text{ i.e., } \frac{d}{dx}(\sin x) = \cos x.$$

7.4.6. Differentiation of cos x Form First Principles

Let $f(x) = \cos x$. Then $f(x + h) = \cos (x + h)$

$$\therefore \quad \frac{d}{dx}(f(x)) = \lim_{h \to 0} \frac{f(x+h) - f(x)}{h}$$

$$= \lim_{h \to 0} \frac{\cos (x+h) - \cos x}{h}$$

$$= \lim_{h \to 0} \frac{-2 \sin\left(\frac{2x+h}{2}\right) \sin\left(\frac{h}{2}\right)}{h}$$

$$\left[\because \cos C - \cos D = -2 \sin\left(\frac{C+D}{2}\right) \cdot \sin\left(\frac{C-D}{2}\right)\right]$$

$$= -\lim_{h \to 0} \sin\left(x + \frac{h}{2}\right) \cdot \lim_{h \to 0} \frac{\sin\left(\frac{h}{2}\right)}{\left(\frac{h}{2}\right)} = (-\sin x) \cdot 1 \qquad \left[\because \lim_{h \to 0} \frac{\sin\left(\frac{h}{2}\right)}{\left(\frac{h}{2}\right)}\right]$$

$$= -\sin x.$$

Thus, $\frac{d}{dy}(f(x)) = -\sin x$ *i.e.*, $\frac{d}{dx}(\cos x) = -\sin x.$

7.4.7. Differentiation of tan x From First Principles

Let $f(x) = \tan x$. Then $f(x + h) = \tan (x + h)$

$$\therefore \quad \frac{d}{dx}(f(x)) = \lim_{h \to 0} \frac{f(x+h) - f(x)}{h} = \lim_{h \to 0} \frac{\tan (x+h) - \tan x}{h}$$

$$= \frac{\frac{\sin(x+h)}{\cos(x+h)}}{h} = \lim_{h \to 0} \frac{\sin(x+h) \cos x - \cos(x+h) \sin x}{h \cos x \cos (x+h)}$$

$$= \lim_{h \to 0} \frac{\sin h}{h} \cdot \frac{1}{\cos x \cos (x+h)} \qquad [\because \sin A \cos B - \cos A \sin B = \sin (A - B)]$$

$$= \lim_{h \to 0} \frac{\sin h}{h} \cdot \lim_{h \to 0} \frac{1}{\cos x \cos (x+h)}$$

$$= 1 \cdot \frac{1}{\cos x \cos x} = \sec^2 x. \qquad \left[\because \lim_{h \to 0} \frac{\sin h}{h} = 1\right]$$

Thus, $\frac{d}{dx}(f(x)) = \sec^2 x$ *i.e.*, $\frac{d}{dx}(\tan x) = \sec^2 x.$

7.4.8. Differentiation of cot x From First Principles

Let $f(x) = \cot x$. Then $f(x + h) = \cot (x + h)$

$$\therefore \quad \frac{d}{dx}(f(x)) = \lim_{h \to 0} \frac{f(x+h) - f(x)}{h} = \lim_{h \to 0} \frac{\cot (x+h) - \cot x}{h}$$

$$= \frac{\frac{\cos(x+h)}{\sin(x+h)} - \frac{\cos x}{\sin x}}{h}$$

$$= \lim_{h\to 0} \frac{\sin x \cos(x+h) - \cos x \sin(x+h)}{h \sin x \sin(x+h)}$$

$$= \lim_{h\to 0} \frac{\sin(x-(x+h))}{h \sin x \sin(x+h)} \qquad [\because \ \sin A \cos B - \cos A \sin B = \sin(A-B)]$$

$$= -\lim_{h\to 0} \frac{\sin h}{h} \cdot \lim_{h\to 0} \frac{1}{\sin x \sin(x+h)} \qquad [\because \ \sin(-h) = \sin h]$$

$$= (-1)\frac{1}{\sin x \sin x} = -\operatorname{cosec}^2 x.$$

Thus, $\dfrac{d}{dx}(f(x)) = -\operatorname{cosec}^2 x$ *i.e.,* $\dfrac{d}{dx}(\cot x) = -\operatorname{cosec}^2 x.$

7.4.9. Differentiation of sec *x* From Principles

Let $f(x) = \sec x$. Then $f(x+h) = \sec(x+h)$

$$\therefore \qquad \frac{d}{dx}(f(x)) = \lim_{h\to 0} \frac{f(x+h)-f(x)}{h} = \lim_{h\to 0} \frac{\sec(x+h) - \sec x}{h}$$

$$= \lim_{h\to 0} \frac{\dfrac{1}{\cos(x+h)} - \dfrac{1}{\cos x}}{h} = \lim_{h\to 0} \frac{\cos x - \cos(x+h)}{h \cos x \cos(x+h)}$$

$$= \lim_{h\to 0} \frac{2\sin\left(\dfrac{x+x+h}{2}\right)\sin\left(\dfrac{x+h+-x}{2}\right)}{h \cos x \cos(x+h)}$$

$$\left[\because \ \cos C - \cos D = 2\sin\left(\frac{C+D}{2}\right)\sin\left(\frac{D-C}{2}\right)\right]$$

$$= \lim_{h\to 0} \frac{2\sin\left(\dfrac{2x+h}{2}\right)\sin\left(\dfrac{h}{2}\right)}{h \cos x \cos(x+h)}$$

$$= \lim_{h\to 0} \frac{\sin\left(\dfrac{2x+h}{2}\right)}{\cos x \cos(x+h)} \cdot \lim_{h\to 0} \frac{\sin\left(\dfrac{h}{2}\right)}{\left(\dfrac{h}{2}\right)}$$

$$= \frac{\sin x}{\cos x \cos x}(1) = \tan x \sec x. \qquad \left[\because \ \frac{\sin\dfrac{h}{2}}{\dfrac{h}{2}} = 1\right]$$

Thus, $\dfrac{d}{dx}(f(x)) = \sec x \tan x$ *i.e.,* $\dfrac{d}{dx}(\sec x) = \sec x \tan x.$

7.4.10. Differentiation of cosec *x* From First Principles

Let $f(x) = \text{cosec } x$. Then $f(x+h) = \text{cosec }(x+h)$

$$\therefore \quad \frac{d}{dx}(f(x)) = \lim_{h\to 0}\frac{f(x+h)-f(x)}{h} = \lim_{h\to 0}\frac{\text{cosec }(x+h)-\text{cosec } x}{h}$$

$$= \lim_{h\to 0}\frac{\frac{1}{\sin(x+h)}-\frac{1}{\sin x}}{h} = \lim_{h\to 0}\frac{\sin x-\sin(x+h)}{h\sin x\sin(x+h)}$$

$$= \lim_{h\to 0}\frac{2\sin\left(\frac{x-x-h}{2}\right)\cos\left(\frac{x+x+h}{2}\right)}{h\sin x\sin(x+h)}$$

$$= \lim_{h\to 0}\frac{2\sin\left(-\frac{h}{2}\right)\cos\left(x+\frac{h}{2}\right)}{h\sin x\sin(x+h)}$$

$$= -\lim_{h\to 0}\frac{\sin\left(\frac{h}{2}\right)}{\frac{h}{2}}\cdot\lim_{h\to 0}\frac{\cos\left(x+\frac{h}{2}\right)}{\sin x\sin(x+h)} \qquad \left[\because \sin\left(-\frac{h}{2}\right) = -\sin\frac{h}{2}\right]$$

$$= (-1)\cdot\frac{\cos x}{\sin x\sin x} = -\cos x \text{ cosec } x.$$

Thus, $\frac{d}{dx}(f(x)) = -\text{cosec } x\cot x$ *i.e.*, $\frac{d}{dx}(\text{cosec } x) = -\text{cosec } x\cot x$.

7.4.11. Differentiation of $\sin^{-1} x$ From First Principles, where $-1 < x < 1$

Let $y = \sin^{-1} x$. Then

$-1 < x < 1 \Rightarrow$ the principle value of $y = \sin^{-1} x$ lies between $-\frac{\pi}{2}$ and $\frac{\pi}{2}$

$$\Rightarrow \quad y \in \left(-\frac{\pi}{2}, \frac{\pi}{2}\right)$$

Now, $y = \sin^{-1} x \quad \Rightarrow \quad x = \sin y$ [By def.]

Let Δx be a small change in x and let Δy be the corresponding change in y. then

$$x + \Delta x = \sin(y+\Delta y)$$

$$\Rightarrow \quad \Delta x = \sin(y+\Delta y) - x$$

$$\Rightarrow \quad \Delta x = \sin(y+\Delta y) - \sin y \qquad [\because \; x = \sin y]$$

$$\Rightarrow \quad \frac{\Delta x}{\Delta y} = \frac{\sin(y+\Delta y)-\sin y}{\Delta y}$$

$$\Rightarrow \quad \frac{\Delta y}{\Delta x} = \frac{\Delta y}{\sin(y+\Delta y)-\sin y}$$

$$\Rightarrow \quad \lim_{\Delta x\to 0}\frac{\Delta y}{\Delta x} = \lim_{h\to 0}\frac{\Delta y}{\sin(y+\Delta y)-\sin y} \qquad [\because \; \Delta x\to 0 \Rightarrow \Delta y\to 0]$$

$$\Rightarrow \quad \frac{dy}{dx} = \lim_{\Delta y \to 0} \frac{\Delta y}{s \sin\left(\frac{\Delta y}{2}\right) \cos\left(y + \frac{\Delta y}{2}\right)} \qquad \left[\because \lim_{\Delta x \to 0} \frac{\Delta y}{\Delta x} - \frac{dy}{dx}\right]$$

$$\Rightarrow \quad \frac{dy}{dx} = \lim_{\Delta y \to 0} \frac{\left(\frac{\Delta y}{2}\right)}{\sin\left(\frac{\Delta y}{2}\right)} \cdot \lim_{\Delta y \to 0} \frac{1}{\cos\left(y + \frac{\Delta y}{2}\right)}$$

$$\Rightarrow \quad \frac{dy}{dx} = \frac{1}{\cos y} \text{ which exists} \qquad \left[\because \; y \neq \pm \frac{y}{2} \therefore \; \cos y \neq 0\right]$$

Now, $\quad y \in \left(-\frac{\pi}{2}, \frac{\pi}{2}\right) \quad \Rightarrow \quad \cos y > 0$

$$\cos y = +\sqrt{1 - \sin^2 y}$$

$$\therefore \quad \frac{dy}{dx} = \frac{1}{\cos y} = \frac{1}{\sqrt{1 - \sin^2 y}} = \frac{1}{\sqrt{1 - \sin^2 y}} \qquad [\because \; x = \sin y]$$

Thus, $\quad \frac{d}{dx}(\sin^{-1} x) = \frac{1}{\sqrt{1 - x^2}}$, where $-1 < x < 1$.

7.4.12. Differentiation of $\tan^{-1} x$ From First Principles, where $x \in R$

Let $y = \tan^{-1} x$. Then $x = \tan y$.

let Δx be a small change in x and let Δy be the corresponding change in y. Then

$$x + \Delta x = \tan(y + \Delta y) \Rightarrow \Delta x = \tan(y + \Delta y) - x$$

$$\Rightarrow \quad \Delta x = \tan(y + \Delta y) - \text{tant } y \qquad [\because \; x = \tan y]$$

$$\Rightarrow \quad \frac{\Delta x}{\Delta y} = \frac{\tan(y + \Delta y) - \tan y}{\Delta y}$$

$$\Rightarrow \quad \frac{\Delta y}{\Delta x} = \frac{\Delta y}{\tan(y + \Delta y) - \tan y}$$

$$\Rightarrow \quad \lim_{\Delta x \to 0} \frac{\Delta y}{\Delta x} = \lim_{\Delta y \to 0} \frac{\Delta y}{\tan(y + \Delta y) - \tan y} \qquad [\because \; \Delta x \to 0 \; \Rightarrow \Delta y \to 0]$$

$$\Rightarrow \quad \frac{dy}{dx} = \lim_{\Delta y \to 0} \frac{\Delta y}{\frac{\sin(y + \Delta y)}{\cos(y + \Delta y)} - \frac{\sin y}{\cos y}} \qquad \left[\because \; \frac{dy}{dx} = \lim_{\Delta x \to 0} \frac{\Delta y}{\Delta x}\right]$$

$$\Rightarrow \quad \frac{dy}{dx} = \lim_{\Delta y \to 0} \frac{\Delta y \cos y \cos(y + \Delta y)}{\sin(y + \Delta y)\cos y - \cos(y + \Delta y)\sin y}$$

$$= \lim_{h \to 0} \frac{\Delta y}{\sin(y + \Delta y - y)} \cdot \cos y \cos(y + \Delta y)$$

$$= \lim_{\Delta y \to 0} \frac{\Delta y}{\sin \Delta y} \cdot \lim_{\Delta y \to 0} [\cos y \cos(y + \Delta y)]$$

$$= \cos^2 y = \frac{1}{\sec^2 y} = \frac{1}{1+\tan^2 y} = \frac{1}{1+x^2} \qquad [\because \tan y = x]$$

Thus, $$\frac{d}{dx}(\tan^{-1} x) = \frac{1}{1+x^2}.$$

7.4.13. Differentiation of $\cot^{-1} x$ From First Principles, where $x \in R$

Let $y = \cot^{-1} x$. Then $x = \cot y$

Let Δx be a small change in x and let Δy be the corresponding change in y.

Then, $$x + \Delta x = \cot(y + \Delta y)$$

$$\Rightarrow \quad \Delta x = \cot(y + \Delta y) - x$$

$$\Rightarrow \quad \Delta x = \cot(y + \Delta y) - \cot y \qquad [\because x = \cot y]$$

$$\Rightarrow \quad \frac{\Delta x}{\Delta y} = \frac{\cot(y+\Delta y) - \cot y}{\Delta y}$$

$$\Rightarrow \quad \frac{\Delta y}{\Delta x} = \frac{\Delta y}{\cot(y+\Delta y) - \cot y}$$

$$\Rightarrow \quad \lim_{\Delta y \to 0} \frac{\Delta y}{\Delta x} = \lim_{\Delta y \to 0} \frac{\Delta y}{\cot(y + Dy) - \cot y}$$

$$\Rightarrow \quad \lim_{\Delta y \to 0} \frac{\Delta y}{\Delta x} = \lim_{\Delta y \to 0} \frac{\Delta y}{\frac{\cos(y+\Delta y)}{\sin(y+\Delta y)} - \frac{\cos y}{\sin y}}$$

$$\Rightarrow \quad \frac{dy}{dx} = \lim_{\Delta y \to 0} \frac{\Delta y \sin y \sin(y+\Delta y)}{\sin y \cos(y+\Delta y) - \cos y \sin(y+\Delta y)}$$

$$\Rightarrow \quad \frac{dy}{dx} = \lim_{\Delta y \to 0} \frac{\Delta y \sin y \sin(y+\Delta y)}{-\sin(y+\Delta y - y)}$$

$$\Rightarrow \quad \frac{dy}{dx} = -\lim_{\Delta y \to 0} \frac{\Delta y}{\sin \Delta y} \cdot \lim_{\Delta y \to 0} [\sin y \sin(y+\Delta y)]$$

$$\Rightarrow \quad \frac{dy}{dx} = \sin^2 y = \frac{-1}{\operatorname{cosec}^2 y} = -\frac{1}{1+\cot^2 y} = -\frac{1}{1+x^2} \qquad [\because x = \cot y]$$

Thus, $$\frac{d}{dx}(\cot^{-1} x) = -\frac{1}{1+x^2}.$$

7.4.14. Differentiation of $\sec^{-1} x$ From First Principles, where $x \in R - (-1. 1)$

Let $y = \sec^{-1} x$. Then.

$$x \le -1 \text{ or } x \ge 1$$

$\Rightarrow$ The principle value of $y - \sec^{-1} x$ lies between 0 and π but $y \neq \frac{\pi}{2}$

$$\Rightarrow \quad y \in \left(0, \frac{\pi}{2}\right) \cup \left(\frac{\pi}{2}, \pi\right)$$

Now, $$y = \sec^{-1} x \quad \Rightarrow \quad x = \sec y.$$

Let Δx be a small change in x and let Δy be the corresponding change in y. Then

$$x + \Delta x = \sec (y\, \Delta y)$$

$$\Rightarrow \quad \Delta x = \sec (y + \Delta y) - x$$

$$\Rightarrow \quad \Delta x = \sec (y + \Delta y) - \sec y \qquad [\because \quad x = \sec y]$$

$$\Rightarrow \quad \frac{\Delta x}{\Delta y} = \frac{\sec (y + \Delta y) - \sec y}{\Delta y}$$

$$\Rightarrow \quad \frac{\Delta y}{\Delta x} = \frac{\Delta y}{\sec (y + \Delta y) - \sec y}$$

$$\Rightarrow \quad \lim_{\Delta y \to 0} \frac{\Delta y}{\Delta x} = \lim_{\Delta y \to 0} \frac{\Delta y}{\sec (y + \Delta y) - \sec y} \qquad [\because \quad \Delta x \to 0 \Rightarrow \Delta y \to 0]$$

$$\Rightarrow \quad \frac{dy}{dx} = \frac{\Delta y}{\dfrac{1}{\cos (y + \Delta y)} - \dfrac{1}{\cos y}} \qquad \left[\because \quad \frac{dy}{dx} = \lim_{\Delta y \to 0} \frac{\Delta y}{\Delta x}\right]$$

$$\Rightarrow \quad \frac{dy}{dx} = \frac{\cos y \cos (y + \Delta y) \cdot \Delta y}{\cos y - \cos (y + \Delta y)}$$

$$\Rightarrow \quad \frac{dy}{dx} = \lim_{\Delta y \to 0} \frac{\Delta y}{2 \sin \left(y + \dfrac{\Delta y}{2} \right) \sin \left(\dfrac{\Delta y}{2} \right)} \cos y \cos (y + \Delta y)$$

$$\Rightarrow \quad \frac{dy}{dx} = \lim_{\Delta y \to 0} \frac{\left(\dfrac{\Delta y}{2} \right)}{\sin \left(\dfrac{\Delta y}{2} \right) \sin \left(y + \dfrac{\Delta y}{2} \right)} \cos y \cos (y + \Delta y)$$

$$= \lim_{\Delta y \to 0} \frac{\left(\dfrac{\Delta y}{2} \right)}{\sin \left(\dfrac{\Delta y}{2} \right)} \lim_{\Delta y \to 0} \frac{\cos y \cos (y + \Delta y)}{\sin \left(y + \dfrac{\Delta y}{2} \right)}$$

$$= \frac{\cos y \cos y}{\sin y} = \frac{1}{\tan y \sec y}, \text{ which exists}$$

$$[\because \quad y \neq 0 \text{ and } y \neq \pi \Rightarrow \sec y \tan y \neq 0]$$

$$= \frac{1}{\sec y \sqrt{\sec^2 y - 1}}$$

Now, $\quad y \in \left(0, \dfrac{\pi}{2} \right) \cup \left(\dfrac{\pi}{2}, \pi \right)$

$$\Rightarrow \quad \sec y \tan y = 0$$

$$\Rightarrow \quad \frac{1}{\sec y \tan y} > 0$$

$$\Rightarrow \quad \frac{dy}{dx} = \frac{1}{\sec y \tan y} > 0 \quad \Rightarrow \quad \frac{dy}{dx} = \frac{1}{\sec y \sqrt{\sec^2 y - 1}} > 0$$

$$\Rightarrow \qquad \frac{dy}{dx} = \frac{1}{|x|\sqrt{x^2-1}}$$

Thus, $$\frac{d}{dx}(\sec^{-1} x) = \frac{d}{dx}(\sec^{-1} x)$$

7.4.15. Differentiation of cosec^{-1} *x* from First Principles, where $x \in R-(-1, 1)$

Let $y = \text{cosec}^{-1} x$. Then $x \le -1$ or $x \ge 1$

$\Rightarrow$ The principle value $y = \text{cosec}^{-1} x$ lies between $-\frac{\pi}{2}$ and $\frac{\pi}{2}$ but $y \neq 0$.

$$\Rightarrow \qquad y \in \left(-\frac{\pi}{2}, 0\right) \cup \left(0, \frac{\pi}{2}\right)$$

Now, $y = \text{cosec}^{-1} x \quad \Rightarrow \quad x = \text{cosec}\, y.$

Let Δx be a small change in x and let Δy be the cirresponding change in y. Then

$$x + \Delta x = \text{cosec}\,(y + \Delta y)$$

$$\Rightarrow \qquad \Delta x = \text{cosec}\,(y + \Delta y) - x$$

$$\Rightarrow \qquad \Delta x = \text{cosec}\,(y + \Delta y) - \text{cosec}\, y \qquad [\because \ x = \text{cosec}\, y]$$

$$\Rightarrow \qquad \frac{\Delta x}{\Delta y} = \frac{\text{cosec}\,(y+\Delta y) - \text{cosec}\, y}{\Delta y}$$

$$\Rightarrow \qquad \frac{\Delta y}{\Delta x} = \frac{\Delta y}{\text{cosec}\,(y+\Delta y) - \text{cosec}\, y}$$

$$\Rightarrow \qquad \lim_{\Delta y \to 0} \frac{\Delta y}{\Delta x} = \lim_{\Delta y \to 0} \frac{\Delta y}{\frac{1}{\sin(y+\Delta y)} - \frac{1}{\sin y}} \qquad [\because \ \Delta x \to 0 \Rightarrow \Delta y \to 0]$$

$$\Rightarrow \qquad \frac{dy}{dx} = \lim_{\Delta y \to 0} \frac{\Delta y \sin y \sin(y+\Delta y)}{\sin y - \sin(y+\Delta y)}$$

$$\Rightarrow \qquad \frac{dy}{dx} = \lim_{\Delta y \to 0} \frac{\Delta y}{2\cos\left(y + \frac{\Delta y}{2}\right)\sin\left(-\frac{\Delta y}{2}\right)} \sin y \sin(y+\Delta y)$$

$$\Rightarrow \qquad \frac{dy}{dx} = \lim_{\Delta y \to 0} \frac{\left(\frac{\Delta y}{2}\right)}{\sin\left(\frac{\Delta y}{2}\right)} \cdot \lim_{\Delta y \to 0} \frac{\sin y \sin(y+\Delta y)}{\cos\left(y + \frac{\Delta y}{2}\right)} \qquad \left[\because \left(-\frac{\Delta y}{2}\right) = -\sin\frac{\Delta y}{2}\right]$$

$$\Rightarrow \qquad \frac{dy}{dx} = \frac{\sin y \sin y}{\cos y} = -\frac{1}{\cos y \,\text{cosec}\, y}, \text{ which exists}$$

$$\left[\because \ y \neq 0, y \neq \frac{\pi}{2} \quad \because \ \cot y \,\text{cosec}\, y \text{ exists}\right]$$

Now, $$y \in \left(-\frac{\pi}{2}, 0\right) \cup \left(0, \frac{\pi}{2}\right)$$

$$\Rightarrow \qquad \frac{\sin^2 y}{\cos y} = \frac{1}{\cot y \,\text{cosec}\, y} > 0 \quad \Rightarrow \quad -\frac{\sin^2 y}{\cos y} = -\frac{1}{\text{cosec}\, y \cot y} < 0$$

$$\Rightarrow \qquad \frac{dy}{dx} = \frac{1}{\text{cosec } y \cot y} < 0 = -\frac{1}{\text{cosec } y \sqrt{\text{cosec}^2 y - 1}}$$

$$= -\frac{1}{\text{cosec } y \sqrt{\text{cosec}^2 y - 1}}$$

Thus, $$\frac{d}{dx}(\text{coces}^{-1} x) = \frac{1}{|x|\sqrt{x^2 - 1}}$$

7.5. DIFFERENTIATION OF SOME OTHER FUNCTIONS FROM FIRST PRINCIPLES

Example 7.8. Differentiation the following function with respect to x from first principles

(*i*) $\sin 2x$ **(*ii*) $\sin^2 x$** **(*iii*) $\sin \sqrt{x}$**

Solution. (*i*) Let $f(x) = \sin 2x$. Then

$$\frac{d}{dx}(f(x)) = \lim_{h\to 0} \frac{f(x+h) - f(x)}{h} = \lim_{h\to 0} \frac{\sin 2(x+h) - \sin 2x}{h}$$

$$= \lim_{h\to 0} \frac{2\,k \sin h \cos(2x+h)}{h} \qquad \left[\because \sin C - \sin D = 2\left(\frac{C-D}{2}\right)\cos\left(\frac{C+D}{2}\right)\right]$$

$$= 2\lim_{h\to 0} \frac{\sin h}{h} \cdot \lim_{h\to 0} \cos(2x+h) = 2(\cos 2x)\,(1) = 2\cos 2x$$

$$\therefore \qquad \frac{d}{dx}(\sin 2x) = 2\cos 2x$$

(*ii*) Let $f(x) = \sin^2 x$. Then

$$\frac{d}{dx}(f(x)) = \lim_{h\to 0} \frac{f(x+h) - f(x)}{h} = \lim_{h\to 0} \frac{\sin^2(x+h) - \sin^2 x}{h}$$

$$= \lim_{h\to 0} \frac{\sin(x+h+x)\sin(x+h-x)}{h}$$

$$[\because \sin^2 A - \sin^2 B = \sin(A+B)\sin(\sin A - B)]$$

$$= \lim_{h\to 0} \frac{\sin h}{h} \cdot \lim_{h\to 0} \sin(2x+h)(\sin 2x) = \sin 2x$$

$$\therefore \qquad \frac{d}{dx}(\sin^2 x) = \sin 2x$$

(*iii*) Let $f(x) = \sin \sqrt{x}$. Then

$$\frac{d}{dx} = \lim_{h\to 0} \frac{f(x+h) - f(x)}{h} = \lim_{h\to 0} \frac{\sin\sqrt{x+h} - \sin\sqrt{x}}{h}$$

$$= \lim_{h\to 0} \frac{2\sin\left(\frac{\sqrt{x+h} - \sqrt{x}}{2}\right)\cos\left(\frac{\sqrt{x+h} + \sqrt{x}}{2}\right)}{h} \qquad \left[\because \sin C - \sin D = 2\frac{C-D}{2}\cos\frac{C+D}{2}\right]$$

$$= \lim_{h\to 0} \frac{\sin\left(\frac{\sqrt{x+h}-\sqrt{x}}{2}\right)}{\left(\frac{\sqrt{x+h}-\sqrt{x}}{2}\right)} \cdot \frac{(\sqrt{x+h}-\sqrt{x})(\sqrt{x+h}+\sqrt{x})}{(\sqrt{x+h}+\sqrt{x})\,h} \cos\left(\frac{\sqrt{x+h}+\sqrt{x}}{2}\right)$$

$$= \lim_{h\to 0} \frac{\sin\left(\frac{\sqrt{x+h}-\sqrt{x}}{2}\right)}{\left(\frac{\sqrt{x+h}-\sqrt{x}}{2}\right)} \cdot \lim_{h\to 0} \frac{x+h-x}{(\sqrt{x+h}+\sqrt{x})\,h} \cdot \lim_{h\to 0} \cos\left(\frac{\sqrt{x+h}+\sqrt{x}}{2}\right)$$

$$= \lim_{h\to 0} \frac{\sin\theta}{\theta} \cdot \lim_{h\to 0} \frac{1}{\sqrt{x+h}+\sqrt{x}} \cdot \lim_{h\to 0} \cos\left(\frac{\sqrt{x+h}+\sqrt{x}}{2}\right)$$

where $\theta = \frac{\sqrt{x+h}-\sqrt{x}}{2}$ [when $h \to 0, \theta \to 0$] $= 1\left(\frac{1}{2\sqrt{x}}\right)(\cos\sqrt{x})$

$$\therefore \quad \frac{d}{dx}(\sin\sqrt{x}) = \frac{\cos\sqrt{x}}{2\sqrt{x}} = \frac{\cos\sqrt{x}}{2\sqrt{x}}$$

Example 7.9. Differentiate the following function w.r.t. x from first principles

(i) e^{2x} ***(ii)*** $e^{\sqrt{x}}$

Solution. (*i*) Let $f(x) = e^{2x}$. Then

$$\frac{d}{dx}(f(x)) = \lim_{h\to 0} \frac{f(x+h)-f(x)}{h} = \lim_{h\to 0} \frac{e^{2(x+h)}-e^{2x}}{h}$$

$$= \lim_{h\to 0} \frac{e^{2x}\cdot e^{2h}-e^{2x}}{h}$$

$$= 2e^{2x} \lim_{h\to 0}\left(\frac{e^{2h}-1}{2h}\right) = 2e^{2x} \lim_{h\to 0}\left(\frac{e^{y}-1}{y}\right) \quad \text{where } y = 2h$$

$$= 2e^{2x} \,.\, 1 = 2e^{2x} \qquad \left[\because \lim_{h\to 0} \frac{e^{y}-1}{y} = 1\right]$$

$$\therefore \quad \frac{d}{dx}(e^{2x}) = 2e^{2x}$$

(*ii*) Let $f(x) = e^{\sqrt{x}}$. Then

$$\frac{d}{dx} f(x) = \lim_{h\to 0} \frac{f(x+h)-f(x)}{h} = \lim_{h\to 0} \frac{e^{\sqrt{x+h}}-e^{\sqrt{x}}}{h}$$

$$= e^{\sqrt{x}} \lim_{h\to 0}\left(\frac{e^{\sqrt{x+h}-\sqrt{x}}-1}{h}\right)$$

$$= e^{\sqrt{x}} \lim_{h\to 0}\left(\frac{e^{\sqrt{x+h}-\sqrt{x}}-1}{\sqrt{x+h}-\sqrt{x}}\right) \cdot \left(\frac{\sqrt{x+h}-\sqrt{x}}{h}\right)$$

$$= e^{\sqrt{x}} \lim_{h\to 0}\left(\frac{e^{\sqrt{x+h}-\sqrt{x}}-1}{\sqrt{x+h}-\sqrt{x}}\right) \cdot \frac{(\sqrt{x+h}-\sqrt{x})(\sqrt{x+h}+\sqrt{x})}{h(\sqrt{x+h}+\sqrt{x})}$$

$$= e^{\sqrt{x}} \lim_{h\to 0}\left(\frac{e^y-1}{y}\right)\cdot \lim_{h\to 0}\frac{x+h-x}{(\sqrt{x+h}+\sqrt{x})}$$

$$y = \sqrt{x+h}-\sqrt{x} \qquad [\because \text{ when } h\to 0, y\to 0]$$

$$= e^{\sqrt{x}}\cdot 1\cdot\left(\frac{1}{\sqrt{x}+\sqrt{x}}\right) = \frac{e^{\sqrt{x}}}{2\sqrt{x}} \qquad \left[\because \lim_{h\to 0}\frac{e^y-1}{y}=1\right]$$

Example 7.10. Differentiate the following functions w.r.t. *x* from first principles

(*i*) $\sqrt{2x+3}$ (*ii*) $\frac{2x+3}{3x+2}$ (*iii*) $x^{-3/2}$

Solution. (*i*) Let $f(x) = \sqrt{2x+3}$. Then

$$\frac{d}{dx}(f(x)) = \lim_{h\to 0}\frac{f(x+h)-f(x)}{h} = \lim_{h\to 0}\frac{\sqrt{2(x+h)+3}-\sqrt{2x+3}}{h}$$

$$= \lim_{h\to 0}\frac{(\sqrt{2(x+h)+3}-\sqrt{2x+3})(\sqrt{2(x+h)+3}+\sqrt{2x+3})}{h(\sqrt{2(x+h)+3}+\sqrt{2x+3}}$$

$$= \lim_{h\to 0}\frac{(2x+2h+3-2x-3)}{h}\frac{1}{(\sqrt{2x+2h+3}+\sqrt{2x+3})}$$

$$= \lim_{h\to 0}\frac{2h}{h}\cdot\lim_{h\to 0}\frac{1}{(\sqrt{2x+2h+3}+\sqrt{2x+3})}$$

$$= 2\frac{1}{\sqrt{2x+3}+\sqrt{2x+3}} = \frac{2}{2(\sqrt{2x+3})} = \frac{1}{\sqrt{2x+3}}$$

(*ii*) Let $f(x) = \frac{2x+3}{3x+2}$. Then

$$\frac{d}{dx}(f(x)) = \lim_{h\to 0}\frac{f(x+h)-f(x)}{h} = \lim_{h\to 0}\frac{\frac{2(x+h)+3}{3(x+h)+2}-\frac{2x+3}{3x+2}}{h}$$

$$= \lim_{h\to 0}\frac{(2x+3+2h)(3x+2)-(2x+3)(3x+2+3h)}{h(3x+2)(3x+2+3h)}$$

$$= \lim_{h\to 0}\frac{(2x+3)(3x+2)+2h(3x+2)-(2x+3)(3x+2)-3h(2x+3)}{h(3x+2)(3x+2+3h)}$$

$$= \lim_{h\to 0}\frac{h[6x+4-6x-9]}{h(3x+2)(3x+2+3h)}$$

$$= \lim_{h\to 0}\frac{-5}{(3x+2)(3x+2+3h)} = -\frac{5}{(3x+2)^2}$$

(*iii*) Let $f(x) = x^{-3/2}$. Then $f(x+h) = (x+h)^{-3/2}$

$$\therefore \quad \frac{d}{dx}(f(x)) = \lim_{h\to 0}\frac{f(x+h)-f(x)}{h} = \lim_{h\to 0}\frac{(x+h)^{-3/2}-x^{-3/2}}{h}$$

$$= \lim_{h\to 0}\frac{(x+h)^{-3/2}-x^{-3/2}}{(x+h)-x}$$

$$= \lim_{h\to 0} \frac{z^{-3/2} - x^{-3/2}}{z - x}, \text{ where } z = x + h \text{ and } z \to x \text{ as } h \to 0$$

$$= \left(-\frac{3}{2}\right) x^{-3/2-1} = -\frac{3}{2} x^{-5/3} \qquad \left[\text{Using } \lim_{h\to 0} \frac{x^n - a^n}{x - a} = na^{n-1}\right]$$

Example 7.11. Differentiate the following functions w.r.t. x from first principles

(*i*) $\tan \sqrt{x}$ **(*ii*) $\cot \sqrt{x}$**

Solution. (*i*) Let $f(x) = \tan \sqrt{x}$.

Then $f(x + h) = \tan \sqrt{x + h}$

$$\therefore \quad \frac{d}{dx}(f(x)) = \lim_{h\to 0} \frac{f(x+h) - f(x)}{h} = \lim_{h\to 0} \frac{\tan \sqrt{x+h} - \tan \sqrt{x}}{h}$$

$$= \lim_{h\to 0} \frac{\sin(\sqrt{x+h} - \sqrt{x})}{\cos \sqrt{x+h} \cos \sqrt{xh}} \qquad \left[\because \tan A - \tan B = \frac{\sin(A - B)}{\cos A \cos B}\right]$$

$$= \lim_{h\to 0} \frac{\sin(\sqrt{x+h} - \sqrt{x})}{(x + h - x) \cos \sqrt{x} \cos \sqrt{x+h}}$$

$$= \lim_{h\to 0} \frac{\sin(\sqrt{x+h} - \sqrt{x})}{(\sqrt{x+h} - \sqrt{x})(\sqrt{x+h} + \sqrt{x}) \cos \sqrt{x} \cos \sqrt{x+h}}$$

$$= \lim_{h\to 0} \frac{\sin(\sqrt{x+h} - \sqrt{x})}{(\sqrt{x+h} - \sqrt{x})} \cdot \lim_{h\to 0} \frac{1}{(\sqrt{x+h} + \sqrt{x}) \cos \sqrt{x+h} \cos \sqrt{x}}$$

$$= 1 \cdot \frac{1}{2\sqrt{x} \cos \sqrt{x} \cos \sqrt{x}}$$

$$= 1 \cdot \frac{1}{2\sqrt{x} \cos \sqrt{x} \cos \sqrt{x}} \qquad \left[\because \lim_{h\to 0} \frac{\sin(\sqrt{x+h} - \sqrt{x}}{(\sqrt{x+h} - \sqrt{x})}\right]$$

(*ii*) Let $f(x) - \cot \sqrt{x}$. Then $f(x + h) = \cot \sqrt{x + h}$

$$\therefore \quad \frac{d}{dx}(f(x)) = \lim_{h\to 0} \frac{f(x+h) - f(x)}{h} \lim_{h\to 0} \frac{\cot \sqrt{x+h} - \cot \sqrt{x}}{h}$$

$$= \lim_{h\to 0} \frac{-\sin(\sqrt{x+h} - \sqrt{x})}{h \sin \sqrt{x+h} \sin \sqrt{x}} \qquad \left[\because \cot A - \cot B = \frac{-\sin(A - B)}{\sin A \sin B}\right]$$

$$= \lim_{h\to 0} \frac{-\sin(\sqrt{x+h} - \sqrt{x})}{[(x + h) - x] \sin \sqrt{x+h} \sin \sqrt{x}}$$

$$= \lim_{h\to 0} \frac{\sin(\sqrt{x+h} - \sqrt{x})}{(\sqrt{x+h} - \sqrt{x})(\sqrt{x+h} + \sqrt{x}) \sin \sqrt{x+h} \sin \sqrt{x}}$$

$$= \lim_{h\to 0} \frac{\sin(\sqrt{x+h} - \sqrt{x})}{\sqrt{x+h} - \sqrt{x}} \cdot \lim_{h\to 0} \frac{1}{(\sqrt{x+h} + \sqrt{x}) \sin \sqrt{x+h} \sin \sqrt{x}}$$

$$= \frac{-1}{2\sqrt{x} \sin \sqrt{x} \sin \sqrt{x}} = \frac{-\operatorname{cosec}^2 \sqrt{x}}{2\sqrt{x}}$$

EXERCISE 7.1

1. e^{3x} [**Ans.** $3e^{3x}$]

2. $\dfrac{1}{x^{3/2}}$ $\left[\textbf{Ans. } \dfrac{-1}{2x^{5/2}}\right]$

3. $\sin(2x+3)$ [**Ans.** $2\cos(2x+3)$]

4. $\cos\sqrt{x}$ $\left[\textbf{Ans. } -\dfrac{\sin\sqrt{x}}{2\sqrt{x}}\right]$

5. $\sqrt{\sin 2x}$ $\left[\textbf{Ans. } \dfrac{\cos 2x}{\sqrt{\sin 2x}}\right]$

6. $\sqrt{\tan x}$ $\left[\textbf{Ans. } \dfrac{\sec^2 x}{2\sqrt{\tan x}}\right]$

7. $\sin(\log x)$ $\left[\textbf{Ans. } \dfrac{\cos(\log x)}{x}\right]$

8. $\sqrt{\sin 3x}$ $\left[\textbf{Ans. } \dfrac{3\cos 3x}{2\sqrt{\sin 3x}}\right]$

9. $\cos x^2$ [**Ans.** $-2x\sin x^2$]

10. $\tan 2x$ [**Ans.** $2\sec^2 2x$]

11. $e^{\sqrt{2x}}$ $\left[\textbf{Ans. } \dfrac{e^{\sqrt{2x}}}{2x}\right]$

12. If $ax^2 + 2hxy + by^2 + 2gx + 2fy + c = 0$, then find $\dfrac{dy}{dx}$ fom first principles. $\left[\textbf{Ans. } \dfrac{dy}{dx} = -\left(\dfrac{ax+hy+g}{hx+by+f}\right)\right]$

13. If $f(x) = \dfrac{2}{x}$, find $f'(x)$ from first principles.

14. Find from first principles the derivative of the following functions w.r.t. x.

    (*i*) $\dfrac{2^2+1}{x}$ (*ii*) $\dfrac{2^2+1}{x}$ (*iii*) $\sqrt{\operatorname{cosec} x}$.

7.6. FUNDAMENTAL RULES FOR DIFFERENTIATION

In the previous articles, it became apparent that to differentiate a function by the direct use of the definition of a derivative after involves a long and tedious process. However, there are rules or formulas that allow us to evaluate derivatives quickly without explicitly going though the limit process. We shall look at some rules in this section.

Rule 1. Differentiation of a constant function is zero *i.e.*, $\dfrac{d}{dx}(c) = 0$.

Proof. Let $f(x) = c$ be a constant function. Applying the definition of derivative, we obtain

$$f'(x) = \operatorname*{Lim}_{h\to 0} \frac{f(x+h) - f(x)}{h}$$

$$f'(x) = \lim_{h \to 0} \frac{c-c}{h} = \lim_{h \to 0} \frac{0}{h} = 0.$$

For example, if $y = 9$, then $\frac{dy}{dx} = \frac{d(9)}{dx} = 0.$

Rule 2. Let $f(x)$ be a differentiable function and let c be a constant. Then $c.\ f(x)$ is also differentiable such that

$$\frac{d}{dx}[c \cdot f(x)] = c \cdot x \frac{d}{dx}[f(x)]$$

Proof. Since $f(x)$ is differentiable, therefore

$\lim_{h \to 0} \frac{f(x+h) - f(x)}{h}$ exists finitely and is equal to $\frac{d}{dx}[f(x)]$

Let $g(x) = c \cdot f(x)$. Then

$$\lim_{h \to 0} \frac{g(x+h) - g(x)}{h} = \lim_{h \to 0} \frac{c \cdot f(x+h) - c \cdot f(x)}{h}$$

$$= c \lim_{h \to 0} \frac{f(x+h) - f(x)}{h} = c \cdot \frac{d}{dx}[f(x)] \qquad \text{[From } (i)\text{]}$$

So, $g(x) = c \cdot f(x)$ is differentiable such that $\frac{d}{dx}[c \cdot f(x)] = \frac{d}{dx}[c \cdot f(x)]$

Rule 3. If $f(x)$ and $g(x)$ are differentiable functions, then $f(x) \pm g(x)$ are also differentiable such that

$$\frac{d}{dx}[f(x) \pm g(x)] = \frac{d}{dx}[f(x)] \pm \frac{d}{dx}[g(x)]$$

That is, the derivative of the sum or difference of two functions is the sum or difference of their derivatives.

Proof. We shall prove the result only for the case of a sum. To prove this, let

$$F(x) = f(x) + g(x)$$

Applying the definition of the derivative given

$$f'(x) = \lim_{h \to 0} \frac{F(x+h) - F(x)}{h} = \lim_{h \to 0} \frac{[f(x+h) + g(x+h)] - [f(x) + g(x)]}{h}$$

$$= \lim_{h \to 0} \frac{[f(x+h) - f(x)] - [g(x+h) - g(x)]}{h}$$

$$= \lim_{h \to 0} \left[\frac{f(x+h) - f(x)}{h} + \frac{g(x+h) - g(x)}{h} \right]$$

$$= \lim_{h \to 0} \frac{f(x+h) - f(x)}{h} + \lim_{h \to 0} \frac{g(x+h) - g(x)}{h}$$

The proof for the case of a difference is similar.

Note. Rule 3 can be extended to the derivative of a finite number of sums and differences of functions.

SOLVED EXAMPLES

Example 7.12. Differentiate the following functions

(i) $x + \frac{1}{2} - \frac{2}{3} \tan x + 4 \log_e x.$ ***(ii)*** $\left(\sqrt{x} + \frac{1}{\sqrt{x}}\right)^2$ ***(iii)*** $e^x + 2 \cos x + \log_e x^3 x.$

Solution. (*i*) Let $y = x + \frac{1}{x} - \frac{2}{3}\tan x + 4\log_e x$

$$\therefore \quad \frac{dy}{dx} = \frac{d}{dx}\left[x + \frac{1}{x} - \frac{2}{3}\tan x + 4\log_e x\right]$$

$$= \frac{d}{dx}(x) + \frac{d}{dx}\left(\frac{1}{x}\right) - \frac{2}{3}\frac{d}{dx}(\tan x) + 4\frac{d}{dx}(\log_e x) = 1 - \frac{1}{x^2} - \frac{2}{3}\sec^2 x + 4\frac{1}{x}$$

(*ii*) Let $y = \left(\sqrt{x} + \frac{1}{\sqrt{x}}\right)^2 = x + 2 + \frac{1}{x} = x + 2 + x^{-1}$

$$\therefore \quad \frac{dy}{dx} = \frac{d}{dx}[e^x + 2\cos x + \log_e x^3]$$

(*iii*) Let $e = e^x + 2\cos x + \log_e x^3$

$$\therefore \quad \frac{dy}{dx} = \frac{d}{dx}\left[e^x + 2\cos x + \log_e x^3\right]$$

$$= \frac{e}{dx}(e^x) + \frac{d}{dx}(\cos x) + \frac{d}{dx}(3\log_e x) \qquad [\because \log_e a^b = b\log_e a]$$

$$= e^x + 2(-\sin x) + 3\frac{d}{dx}(\log_3 x) = e^x - 2\sin x + 3\cdot\frac{1}{x}$$

Example 7.13. Find thederivative of

(*i*) $\frac{(x+1)(x-2)}{\sqrt{x}}$. **(*ii*)** $3^x + 3^2 + \frac{1}{3}e^x - 3e$

(*iii*) $2\sin x + 3\cos x - \frac{1}{2}\log x - e^x + 5\tan^{-1} x$ **(*iv*)** $(x^3 - 5x + 6)(x - 3)$

Solution. (*i*) Let $y = \frac{(x+1)(x-2)}{\sqrt{x}} = \frac{x^2 - x - 2}{\sqrt{x}}$

$$= \frac{x^2}{x^{1/2}} - \frac{x}{x^{1/2}} - \frac{2}{x^{1/2}} = x^{3/2} - x^{1/2} - 2x^{-1/2}$$

$$\therefore \quad \frac{dy}{dx} = \frac{3}{2}x^{1/2} - \frac{1}{2}x^{-1/2} - 2\left(-\frac{1}{2}x^{-3/2}\right) = \frac{3}{2}\sqrt{x} - \frac{1}{2\sqrt{x}} + \frac{1}{x^{3/2}}$$

(*ii*) Let $y = 3^x + 3^2 + \frac{1}{3}e^x - 3e$

$$\therefore \quad \frac{dy}{dx} = \frac{d}{dx}\left[3^x + 3^2\,\frac{1}{3}e^x - 3e\right] = \frac{d}{dx}(3^x) + \frac{d}{dx}(3^2) + \frac{1}{3}\frac{d}{dx}(e^x) - 3\frac{d}{dx}(e)$$

$$= 3^x\log 3 + 0 + \frac{1}{3}e^x - 0 = 3^x\log 3 + \frac{1}{3}e^x.$$

(*iii*) Let $$y = 2\sin x + 3\cos x - \frac{1}{2}\log x - e^x + 5\tan^{-1} x$$

$\therefore$ $$\frac{dy}{dx} = \frac{d}{dx}\left[2 + \sin x + 3\cos x - \frac{1}{2}\log x - e^x + 5\tan^{-1} x\right]$$

$$= 2\frac{d}{dx}(\sin x) + 3\frac{d}{dx}(\cos x) - \frac{1}{2}\frac{d}{dx}(\log x) - \frac{d}{dx}(e^x) + 5\frac{d}{dx}(\tan^{-1} x)$$

$$= 2\cos x + 3(-\sin x) - \frac{1}{2}\cdot\frac{1}{x} - e^x + 5\cdot\frac{1}{1+x^2}$$

$$= 2\cos x - 3\sin x - \frac{1}{2x} - e^x + \frac{5}{1+x^2}$$

(*iv*) Let $$y = (x^2 - 5x + 6)(x - 3)$$

$\therefore$ $$\frac{dy}{dx} = \frac{d}{dx}[(x^2 - 5x + 6)(x - 3)] = \frac{d}{dx}[x^3 - 8x^2 + 21x - 18]$$

$$= \frac{d}{dx}[x^3 - 8x^2 + 21x - 18]$$

$$= 3x^2 - 8 \times 2x + 21 \times 1 - 0 = 3x^2 - 16x + 21$$

Rule 4. (Product Rule). If $f(x)$ and $g(x)$ are two differentiable functions, then $f(x) \cdot g(x)$ is also differentiable such that

$$\frac{d}{dx}[f(x)\cdot g(x)] = f(x)\frac{d}{dx}[g(x)] + g(x)\frac{d}{dx}[f(x)]$$

Thut is, the dwerivative of the product of two functions is the first function times the derivatively of the second, plus the second function times the derivative of the first.

Proof. Let $F(x) = f(x).\ g(x)$, then be the definition of the derivative, we have

$$F'(x) = \lim_{h\to 0}\frac{F(x+h) - F(x)}{h}$$

$$= \lim_{h\to 0}\frac{f(x+h)\cdot g(x+h) - g(x)}{h}$$

$$= \lim_{h\to 0}\frac{f(x+h)\cdot g(x+h) - f(x)\cdot g(x) + [f(x+h)\cdot g(x) - f(x+h)\cdot g(x)]}{h}$$

(Adding and subtracting $f(x + h)\ g(x)$ in the numerator)

$$= \lim_{h\to 0}\frac{f(x+h)\cdot g(x+h) - f(x+h)\cdot g(x) + [f(x+h) - g(x+h) - f(x)]}{h}$$

$$= \lim_{h\to 0}\frac{f(x+h)[g(x+h) - g(x)] + g(x)[f(x+h) - f(x)]}{h}$$

$$= \lim_{h\to 0}\frac{f(x+h)[g(x+h) - g(x)]}{h} + \lim_{h\to 0}\frac{g(x)[f(x+h) - f(x)}{h}$$

$$= \lim_{h\to 0} f(x+h)\cdot\lim_{h\to 0}\frac{g(x+h) - g(x)}{h} + \lim_{h\to 0} g(x)\cdot\lim_{h\to 0}\frac{f(x+h) - f(x)}{h}$$

Since $f(x)$ is differentiable, it is continuous, and hence

$$\lim_{h\to 0} f(x+h) = f(x)$$

Thus $$F'(x) = f(x)\cdot g'(x) + g(x)\cdot f'(x)$$

Rule 5. (Quotient Rule). If $f(x)$ and $g(x)$ are two differentiable functions and $g(x) \neq 0$, then $\dfrac{f(x)}{g(x)}$ is also differentiable such that

$$\frac{d}{dx}\left[\frac{f(x)}{g(x)}\right] = \frac{g(x)\, f'(x) - f(x)\, g'(x)}{[g(x)]^2}.$$

That is, the derivative of the eqotient of two functions is the denominator times the derivative of the numerator, minus the numerator times the derivative of the denominator, all divided by the square of the denominator.

Remark. It is advisable to remember the result in the following form:

$$\frac{d}{dx}\left[\frac{N^r}{D^r}\right] = \frac{D^r \cdot \frac{d}{dx}(N^r) - N^r \cdot \frac{d}{dx}(D^r)}{(D^r)^2}$$ where N^r = Numerator, D^r = Denominator.

Proof. Its proof is left to the reader.

Rule 6. (Chain Rule). If $f(x)$ and $g(x)$ are differentiable function, then fog is also differentiable and

$$(fog)'(x) = f'(g(x)) \,.g'(x)$$

or
$$\frac{d}{dx}[(fog)(x)] = \frac{1}{dg(x)}[(fog)(x)] \cdot \frac{d}{dx}(g(x))$$

Proof. Since $f(x)$ and $g(x)$ are differentiable functions, therefore

$$\lim_{h\to 0} \frac{f(x+h) - f(x)}{h} = \frac{d}{dx}[f(x)]$$

and
$$\lim_{h\to 0} \frac{g(x+h) - g(x)}{h} = \frac{d}{dx}[g(x)] \text{ both exist}$$

Now
$$\frac{d}{dx}[(fog)(x)] = \lim_{h\to 0} \frac{fog\,(x+h) - fog\,(x)}{h} = \lim_{h\to 0} \frac{f[g(x+h)] - f[g(x)]}{h}$$

$$= \lim_{h\to 0} \frac{f[g(x+h] - f[g(x)]}{g(x+h) - g(x)} \cdot \frac{g(x+h) - g(x)}{h}$$

$$= \lim_{h\to 0} \frac{f[g(x+h) - f[g(x)]}{g(x+h) - g(x)} \cdot \lim_{h\to 0} \frac{g(x+h) - g(x)}{h}$$

$$= \lim_{g(h\to 0)\to g(x)} \frac{f[g(x+h)] - f[g(x)]}{g(x+h) - g(x)} \cdot \lim_{h\to 0} \frac{g(x+h) - g(x)}{h}$$

$\left[\because\ g(x) \text{ is differentiable } g(x) \text{ is continuous and hence } \lim_{h\to 0} g(x+h) = g(x)\right]$

$$= \frac{d}{dg(x)}[fog(x)] \cdot \frac{d}{dx}[g(x)]$$

Remark 1. The above rule can also be restated as follows:

If $z = f(y)$ and $y = g(x)$, then $\dfrac{dz}{dx} = \dfrac{dz}{dy} \cdot \dfrac{dy}{dx}$

or Derivative of z w.r.t. x = (Derivative of z w.r.t. y) × (Derivative of y w.r.t. x).

Remark 2. This rule can be extended.

It z is a function of u, u is a function of v and v is a function of x, then

$$\frac{dz}{dx} = \frac{dz}{du} \times \frac{du}{dv} \times \frac{dv}{dx}$$

SOLVED EXAMPLES

Example 7.14. Differentiate the following functions w.r.t. x

(*i*) $x^{-1/3} e^x$ (*ii*) $(x^2 + 3x) \cdot \log x$

(*iii*) $(x^2 - 5x + 6)(x^3 + 2)$ (*iv*) $x \sin x \log x$.

Solution.

(*i*) Let $y = x^{-1/3} e^x$

$$\therefore \quad \frac{dy}{dx} = x^{-1/3} \frac{d}{dx} + (e^x) + e^x \cdot \frac{d}{dx}(x^{-1/3})$$

$$= x^{-1/3} e^x + e^x + e^x\left(-\frac{1}{3}x^{-4/3}\right) = e^x\left(x^{-1/3} - \frac{1}{3}x^{-4/3}\right).$$

(*ii*) Let $y = (x^2 + 3x) . \log x$

$$\therefore \quad \frac{dy}{dx} = (x^2+3x)\frac{d}{dx}(\log x) + \log x \frac{d}{dx}(x^2+3x) = (x^2+3x)\cdot\frac{1}{x} + \log x \cdot (2x+3)$$

$$= x \ 3 + (2x + 3) \lg x.$$

(*iii*) Let $y = (x^2 - 5x + 6)(x^2 + 2)$

$$\therefore \quad \frac{dy}{dx} = (x^2-5x+6)\frac{d}{dx}(x^2+2) + (x^3+2)\frac{d}{dx}(x^2-5x+6)$$

$$= (x^2 - 5x + 6)(3x^2) + (x^3 + 2) \cdot (2x - 5)$$

$$= 3x^4 - 15x^3 + 18x^2 + 2x^4 - 5x^3 + 4x - 10$$

$$= 5x^4 - 20x^3 + 18x^2 + 4x - 10$$

(*iv*) Let $y = x \sin x \log x$

$$\therefore \quad \frac{dy}{dx} = \frac{d}{dx}[(x\sin x)\log x] = (x \sin x)\cdot\frac{d}{dx}(\log x) + \log x \cdot \frac{d}{dx}(x \sin x)$$

$$= x \sin x \cdot \frac{1}{x} + \log x \cdot \left[x \cdot \frac{d}{dx}(\sin x) + \sin x \frac{d}{dx}(x)\right]$$

$$= \sin x + x \log x \cos x + \sin x \log x.$$

Example 7.15. Differentiate are following functions w.r.t. x

(*i*) $\sqrt{1+\tan x}$ (*ii*) $\log \tan\left(\frac{\pi}{4}+\frac{\pi}{2}\right)$

Solution.

(*i*) Let $y = \sqrt{1+\tan x} = (1 + \tan x)^{1/2}$

$$\therefore \quad \frac{dy}{dx} = \frac{1}{2}(1+\tan x)^{\frac{1}{2}-1} \cdot \frac{d}{dx}(1+\tan x)$$

$$= \frac{1}{2\sqrt{1+\tan x}} \cdot (0 + \sec^2 x) = \frac{\sec^2 x}{2\sqrt{1+\tan x}}$$

(*ii*) Let $$y = \log\tan\left(\frac{\pi}{4}+\frac{x}{2}\right)$$

$\therefore$ $$\frac{dy}{dx} = \frac{d}{dx}\left[\tan\tan\left(\frac{\pi}{4}+\frac{x}{2}\right)\right] = \frac{1}{\tan\left(\frac{\pi}{4}+\frac{x}{2}\right)}\cdot\frac{d}{dx}\left[\tan\left(\frac{\pi}{4}+\frac{x}{2}\right)\right]$$

$$= \frac{1}{\tan\left(\frac{\pi}{4}+\frac{x}{2}\right)}\cdot\sec^2\left(\frac{\pi}{4}+\frac{x}{2}\right)\cdot\frac{d}{dx}\left(\tan\frac{\pi}{4}+\frac{x}{2}\right)$$

$$= \frac{1}{\sin\left(\frac{\pi}{4}+\frac{x}{2}\right)\cdot\cos\left(\frac{\pi}{4}+\frac{x}{2}\right)}\times\frac{1}{2}$$

$$= \frac{1}{\sin 2\left(\frac{\pi}{4}+\frac{x}{2}\right)} \qquad [\because\ 2\sin A\cos A = \sin 2A]$$

$$= \frac{1}{\sin\left(\frac{\pi}{2}+x\right)} = \frac{1}{\cos x} = \sec x$$

Example 7.16. Differentiate the following functions w.r.t. x:

(*i*) $\dfrac{e^x}{1+\sin x}$ **(*ii*)** $\dfrac{x+\sin x}{x+\cos x}$ **(*iii*)** $\dfrac{\sin x+\cos x}{\sin x-\cos x}$ **(*iv*)** $\dfrac{\sec x-1}{\sec x+1}$

Solution. (*i*) Using quotient rule, we have

$$\frac{d}{dx}\left(\frac{e^x}{1+\sin x}\right) = \frac{(1+\sin x)\cdot\frac{d}{dx}(e^x)\cdot\frac{d}{dx}(1+\sin x)}{(1+\sin x)^2}$$

(*ii*) Using quotient rule, we have

$$\frac{d}{dx}\left(\frac{x+\sin x}{x+\cos x}\right) = \frac{x+\cos x)\cdot\frac{d}{dx}(x+\sin x)-(x+\sin x)\cdot\frac{d}{dx}(x+\cos x)}{(x+\cos x)^2}$$

$$= \frac{(x+\cos x)\cdot(1+\cos x)-(x+\sin x)\cdot(1-\sin x)}{(x+\cos x)^2}$$

$$= \frac{x+\cos + x + x\cos x + x\sin x+\cos^2 x+\sin^2 x}{(x+\cos x)^2}$$

$$= \frac{\cos x-\sin x+x\cos x+x\sin x+\cos^2 x+\sin^2 x}{(x+\cos x)^2}$$

$$= \frac{\cos x-\sin x+x(\cos x+\sin x)+1}{(x+\cos x)^2}$$

(*iii*) Using quotient rule, we have

$$\frac{d}{dx}\left(\frac{\sin x+\cos x}{\sin x-\cos x}\right) = \frac{(\sin x-\cos x)\frac{d}{dx}(\sin x+\cos x)\,(-\sin x+\cos x)\frac{d}{dx}(\sin x-\cos x)}{(\sin x-\cos x)^2}$$

$$= \frac{(\sin x-\cos x)(\cos x-\sin x)-(\sin x+\cos x)(\cos x+\sin x)}{(\sin x-\cos x)^2}$$

$$= \frac{-(\sin x - \cos x)^2 - (\sin x + \cos x)^2}{(\sin x - \cos x)^2}$$

$$= \frac{-[(\sin x - \cos x)^2 + (\sin x + \cos x)^2]}{(\sin x - \cos x)^2}$$

$$= \frac{-2[(\sin^2 x + \cos^2 x]}{\sin^2 x + \cos^2 x - 2\sin x \, x \cos x} = \frac{-2}{1 - \sin 2x}$$

(*iv*) $$\frac{d}{dx}\left(\frac{\sec x - 1}{\sec x + 1}\right) = \frac{(\sec x + 1)\cdot \frac{d}{dx}(\sec x - 1) - (\sec x - 1)\cdot \frac{d}{dx}(\sec x + 1)}{(\sec x + 1)^2}$$

$$= \frac{(\sec x + 1)(\sec x \tan x - 0) - (\sec x - 1)(\sec x \tan x + 0)}{(\sec x + 1)^2}$$

$$= \frac{\sec^2 x \tan x + \sec x \tan x - \sec^2 x \tan x + \sec x \tan x}{(\sec x + 1)^2}$$

$$= \frac{2 \sec x \tan x}{(\sec x + 1)^2} = \frac{2 \sin x}{(1 + \cos x)^2}$$

EXERCISE 7.2

1. $e^{\sin\sqrt{x}}$ $\left[\textbf{Ans.} \dfrac{\cos\sqrt{x}\, e^{\sin\sqrt{x}}}{2\sqrt{x}}\right]$

2. $3^{x \log x}$ [**Ans.** $3^{x \log x} . \log 3. (1 + \log x)$]

3. $\sqrt{\dfrac{1+\sin x}{1-\sin x}}$ [**Ans.** $\sec x(\tan x + \sec x)$]

4. $\sqrt{\dfrac{1+x}{1-x}}$ $\left[\textbf{Ans.} \dfrac{-x}{\sqrt{1+x}\,(1-x)^{3/2}}\right]$

5. $\sin(\log \sin x)$ [**Ans.** $\cos(\log \sin x)\cdot \cot x$]

6. $\log\sqrt{\dfrac{1-\cos x}{1+\cos x}}$ [**Ans.** $\operatorname{cosec} x$]

7. $\dfrac{e^{2x}+e^{-2x}}{e^{2x}-e^{-2x}}$ $\left[\textbf{Ans.} \dfrac{-8}{(e^{2x}-e^{-2x})^2}\right]$

8. $\sqrt{\tan^{-1}\left(\dfrac{x}{2}\right)}$ $\left[\textbf{Ans.} \dfrac{1}{(4+x^2)\sqrt{\tan^{-1}\left(\dfrac{x}{2}\right)}}\right]$

9. $\sin^2\{\log(2x+3)]$ $\left[\textbf{Ans.} \sin\{2\log(2x+3)\}\left(\dfrac{2}{2x+3}\right)\right]$

10. $\dfrac{\sqrt{x^2+1}+\sqrt{x^2-1}}{\sqrt{x^2+1}-\sqrt{x^2-1}}$ $\left[\textbf{Ans.} 2x + \dfrac{2x^3}{\sqrt{x^4-1}}\right]$

11. $\log[x+2+\sqrt{x^2 \cdot\cdot 4x+1}]$ $\left[\textbf{Ans.} \frac{1}{\sqrt{x^2+4x+1}}\right]$

12. $\log(\cos x^2)$ [**Ans.** $-2x\tan x^2$]

13. $\cos(\log x)^2$ $\left[\textbf{Ans.} \frac{-2\log x \sin(\log x)^2}{x}\right]$

7.7. DIFFERENTIATION BY USING TRIGONOMETRICAL TRANSFORMATIONS

Sometimes, It becomes very easy to differentiate a function by using trigonometrical transformations. Usually this is done in case of inverse trigonmetric functions. Some important results on trigonometrical and inverse trigonometrical functions are given below for ready reference.

(*i*) $\sin 2 = 2\sin x\cos x$

(*ii*) $1+\cos 2x = 2\cos^2 x$ or $\cos 2x = 2\cos^2 x - 1$

(*iii*) $1-\cos 2x = 2\sin^2 x$ or $\cos 2x = 1-2\sin^2 x$

(*iv*) $\sin 2x = \frac{2\tan x}{1+\tan^2 x}$

(*v*) $\cos 2x = \frac{1-\tan^2 x}{1+\tan^2 x}$

(*vi*) $\tan 2x = \frac{2\tan x}{1-\tan^2 x}$

(*vii*) $\sin 3x = 3\sin x - 4\sin^3 x$

(*viii*) $\cos 3x = 4\cos^3 x - 3\cos x$

(*ix*) $\tan 3x = \frac{3\tan x - \tan^3 x}{1-3\tan^2 x}$

(*x*) $\sin^1 x \pm \sin^{-1} y = \sin^{-1}[x\sqrt{1-y^2} \pm y\sqrt{1-x^2}]$

(*xi*) $\cos^{-1} x \pm \cos^{-1} y = \cos^{-1}[xy \mp \sqrt{1-x^2}\sqrt{1-y^2}]$

(*xii*) $\tan^{-1} x + \tan^{-1} y = \tan^{-1}\left(\frac{x+y}{1-xy}\right)$

(*xiii*) $\tan^{-1} x + \tan^{-1} y = \tan^{-1}\left(\frac{x+y}{1-xy}\right)$

(*xiv*) $\sin^{-1} x + \cos^{-1} x = \frac{\pi}{2}$

(*xv*) $\tan^{-1} x + \cot^{-1} x = \frac{\pi}{2}$

(*xvi*) $\sec^{-1} x + \text{cosec}^{-1} x = \frac{\pi}{2}$

(*xvii*) $\cos^{-1}(-x) = \pi - \cos^{-1} x$

(*xviii*) $\tan^{-1}(-x) = -\tan^{-1} x$

(*xix*) $\sin^{-1}(-x) = -\sin^{-1} x$

(*xx*) $\sin^{-1} x = \text{cosec}^{-1}\left(\frac{1}{x}\right)$

(*xxi*) $\cos^{-1} x = \sec^{-1}\left(\frac{1}{x}\right)$

(*xxii*) $\tan^{-1} x = \cot^{-1}\left(\frac{1}{x}\right)$

Some Important Substitutions

Following are some substititions useful in finding derivatives:

Expression	*Substitution*
$a^2 + x^2$	$x = a \tan \theta$ or $a \cot \theta$
$a^2 - x^2$	$x = a \sin \theta$ or or $a \cos \theta$
$x^2 - a^2$	$x = a \sec \theta$ or $a \operatorname{cosec} \theta$
$\sqrt{\dfrac{a-x}{a+x}}$ or $\sqrt{\dfrac{a+x}{a-x}}$	$x = a \sec \theta$ or $a \operatorname{cosec} \theta$
$\sqrt{\dfrac{a^2-x^2}{a^2+x^2}}$ or $\sqrt{\dfrac{a^2+x^2}{a^2-x^2}}$	$x^2 = a^2 \cos 2\theta$

7.8. DIFFERENTIATION OF IMPLICIT FUNCTIONS

An equation of the form $y = f(x)$ is said to define y as an explicit function of x. In a relation of the form $f(x, y) = 0$, it is not always possible to express y in terms of x. For example, $x^3 + y^3 = 3axy$ is not easily solvable for y in terms of x. If this is the case, $f(x, y) = 0$ is called an implicit function $f(x, y) = 0$ defines implicit y as a function of x.

To find the value of $\frac{dy}{dx}$ when y is given as an implicit function of x, we differentiate both sides of the equation w.r.t. x regarding y as a function of x and ten solve the resulting equation for $\frac{dy}{dx}$, keeping in mind that

$$\frac{d}{dx}(y^2) = 2y \cdot \frac{dy}{dx};\ \frac{d}{dx}(y^3) = 3y^2 \cdot \frac{dy}{dx} \text{ and so on.}$$

Method to find $\frac{dy}{dx}$ from implicit function

(*i*) Differentiate both sides of the given equation w.r.t. x

(*ii*) Take all terms involving $\frac{dy}{dx}$ on L.H.S. and transfer the remaining on R.H.S. so as to get an equation in the form

$$\frac{dy}{dx}\phi(x, y) = \psi(x, y)$$

(*iii*) Then $\dfrac{dy}{dx} = \dfrac{\psi(x, y)}{\phi(x, y)}$

7.9. DIFFERENTIATION OF PARAMETRIC FORMS

Sometimes x and y are given as functions of a single variable, *e.g.*, $x = \phi(t)$, $y = \psi(t)$ are two functions and t is a variable. In such a case x and y are called parametric functions or parametric equation and t is called the parameter.

To find $\frac{dy}{dx}$ in case of parametric of parametric functions, we first obtain the relationship between x and y by eliminating the parameter t and then we differentials it with respect to x. But every time it is not convenient to eliminate the parameter. Therefore $\frac{dy}{dx}$ can also be obtained by the following formula

$$\frac{dy}{dx} = \frac{dy}{dx} \cdot \frac{dt}{dx} = \frac{dy / dt}{dx / dt}$$

7.10. LOGARITHMIC DIFFERENTIATION

The process of taking logarithms before differentiation is called logarithmic differentiation. The method of logarithmic differentiation is used to differentiate functions of the form $y = f(x)^{g(x)}$. With this method, we first take the natural logarithm of both sides of $y = f(x)^{g(x)}$ to obtain $\log y = \log [f(x)^{g(x)}]$ by using properties of logarithms, we differentiate both sides with respect to x and then solve for $\frac{dy}{dx}$. This method can also be used to differentiate functions which which are product of several functions. The following examples illustrate the method of logarithmic differentiation.

SOLVED EXAMPLES

Example 7.17. Differentiate the following functions with respect to x:

(*i*) $\sin^{-1}(3x - 4x)$ **(*ii*)** $\cos^{-1}(4x^3 - 3x)$

(*iii*) $\tan^{-1}\left(\frac{3x - x^3}{1 - 3x^2}\right)$ **(*iv*)** $\sin^{-1}\left(\frac{2x}{1 + x^2}\right)$

Solution. (*i*) Let $y = \sin^{-1}(3x - 4x)$

Putting $x = \sin\theta$, we get $y = \sin^{-1}(3\sin\theta - 4\sin^3\theta) = \sin^{-1}(\sin 3\theta) = 3\theta$

or $y = 3\sin^{-1} x$ $[\because\ x = \sin\theta \Rightarrow \theta = \sin^{-1} x]$

$$\therefore \quad \frac{dy}{dx} = \frac{d}{dx}(3\sin^{-1} x) = 3\frac{d}{dx}(\sin^{-1} x)$$

$$= 3 \cdot \frac{1}{\sqrt{1 - x^2}} = 3 \cdot \frac{1}{\sqrt{1 - x^2}}$$

(*ii*) Let $y = \cos^{-1}(4x^3 - 3x)$

Putting $x = \cos\theta$, we get $y = \cos^{-1}(4\cos^4\theta - 3\cos\theta) = \cos^{-1}(\cos 3\theta) = 3\theta$

or $y = 3\cos^{-1} x$ $[\because\ x = \cos\theta \Rightarrow \theta = \cos^{-1} x]$

$$\therefore \quad \frac{dy}{dx} = \frac{d}{dx}(3\cos^{-1} x) = 3\frac{d}{dx}(\cos^{-1} x) = 3 \cdot \left(\frac{1}{\sqrt{\sqrt{1 - x^2}}}\right) = -\frac{1}{\sqrt{1 - x^2}}$$

(*iii*) let $y = \tan^{-1}\left(\frac{3x - x^3}{1 - 3x^2}\right)$

Putting $x = \tan\theta$, we get $y = \tan^{-1}\left(\frac{3\tan\theta - \tan^3\theta}{1 - 3\tan^2\theta}\right) = \tan^{-1}(\tan 3\theta) = 3\theta$

or $\quad y = 3\tan^{-1} x \qquad [\because \; x \tan\theta \Rightarrow \theta = \tan^{-1} x]$

$$\therefore \quad \frac{dy}{dx} = \frac{d}{dx}(3\tan^{-1} x) = 3\cdot\frac{d}{dx}9\tan^{-1} x) = 3\cdot\left(\frac{1}{1+x^2}\right)$$

(*iv*) Let $\quad y = \sin^{-1}\left(\frac{2x}{1+x^2}\right)$

Putting $x = \tan\theta$, we get $\quad y = \sin^{-1}\left(\frac{2\tan\theta}{1+\tan^2\theta}\right) = \sin^{-1}(\sin 2\theta) = 2\theta$

or $\quad y = 2\tan^{-1} x \qquad [\because \; x = \tan\theta \Rightarrow \theta = \tan^{-1} x]$

$$\therefore \quad \frac{dy}{dx} = 2\cdot\frac{1}{1+x^2} = \frac{2}{1+x^2}$$

Example 7.18. If $y = \sin^{-1}[x\sqrt{1-x} - \sqrt{x}\sqrt{1-x^2}]$, find $\frac{dy}{dx}$.

Solution. We have $\quad y = \sin^{-1}[x\sqrt{1-x} - \sqrt{x}\sqrt{1-x^2}]$

$$= \sin^{-1}[x\sqrt{1-x} - \sqrt{x}\sqrt{1-x^2}]$$

or $\quad y = \sin^{-1} x - \sin^{-1}\sqrt{x}$

[Using $\sin^{-1} x - \sin^{-1} y = \sin^{-1}[x\sqrt{1-y^2} - y\sqrt{1-x^2}]$]

$$\frac{dy}{dx} = \frac{1}{\sqrt{1-x^2}} - \frac{1}{\sqrt{1-(\sqrt{x})^2}}\frac{d}{dx}(\sqrt{x})$$

$$= \frac{1}{\sqrt{1-x^2}} - \frac{1}{1-x} - \frac{1}{2\sqrt{x}}$$

Example 7.19. If $\log\sqrt{x^2+y^2} = \tan^{-1}\frac{y}{x}$, find $\frac{dy}{dx}$.

Solution. We have $\quad \log\sqrt{x^2+y^2} = \tan^{-1}\frac{y}{x}$

$$\Rightarrow \quad \frac{1}{2}\log(x^2+y^2) = \tan^{-1}\frac{y}{x} \qquad [\because \; \log_e m^n = n\log_e m]$$

$$\frac{1}{2}\cdot\frac{1}{x^2+y^2}\left[2x + 2y\frac{dy}{dx}\right] = \frac{1}{1+\frac{y^2}{x^2}}\cdot\frac{x\cdot\frac{dy}{dx} - y\cdot 1}{x^2}$$

$$\Rightarrow \quad \frac{x}{x^2+y^2} + \frac{y}{x^2+y^2}\frac{dy}{dx} = \frac{1}{x^2+y^2}\left[x\frac{dy}{dx} - y\right]$$

$$\Rightarrow \quad x + y\frac{dy}{dx} = x\frac{dy}{dx} - y$$

$$\Rightarrow \quad (x-y)\frac{dy}{dx} = x + y \qquad \text{[Regrouping]}$$

$$\Rightarrow \quad \frac{dy}{dx} = \frac{x+y}{x-y}$$

Example 7.20. If $x^3 + y^2 = 3axy$, find $\frac{dy}{dx}$.

Solution. We have $x^3 + y^3 = 3axy$

Differentiation both sides w.r.t. x, we get

$$3x^2\ 3y^2 \frac{dy}{dx} = 3a\left[x \cdot \frac{dy}{dx} + y \cdot 1\right]$$

$$\Rightarrow \quad 3(y^2 - ax) \cdot \frac{dy}{dx} = 3(ay - x^2) \quad \text{[Regrouping]}$$

$$\therefore \quad \frac{dy}{dx} = \frac{ay - x^2}{y^2 - ax}$$

Example 7.21. If $x\sqrt{1+y} + y\sqrt{1+x} = 0$, prove that $\frac{dy}{dx} = -\frac{1}{(1+x)^2}$

Solution. We have $x\sqrt{1+y} + y\sqrt{1+x} = 0$

$$\Rightarrow \quad x\sqrt{1+y} = -y\sqrt{1+x}$$

$$\Rightarrow \quad x^2(1+y) = y^2(1+x)$$

$$\Rightarrow \quad x^2 + x^2y = y^2 + y^2x$$

$$\Rightarrow \quad x^2 - y^2 = y^2 + y^2x$$

i.e., $(x - y)(x + y) = -xy(x - y)$

or $x + y = -xy(x - y)$ [$\because$ $x = y$ does not satisfy the given equation]

Solving this for y given $y = -\frac{x}{1+x}$

Differentiating both sides w.r.t. x, we get

$$\frac{dy}{dx} = -\left[\frac{(1+x).1 - x \cdot 1}{(1+x)^2}\right] = -\left[\frac{1+x-x}{(1+x)^2}\right] = -\frac{1}{(1+x)^2}$$

Hence $\frac{dy}{dx} = -\frac{1}{(1+x)^2}$

Example 7.22. If $x = a(\cos\theta + \theta\sin\theta)$ and $y = a(\sin\theta - \theta\cos\theta)$, find $\frac{dy}{dx}$.

Solution. We have $x = a(\cos\theta + \theta\sin\theta)$, $y = a(\sin\theta - \theta\cos\theta)$

$$\therefore \quad \frac{dx}{d\theta} = a[-\sin\theta + \theta . \cos\theta + \theta . 1] = a\theta\cos\theta$$

and

$$\frac{dy}{dx} = a[\cos\theta - (-\theta\sin\theta + \cos\theta \cdot 1] = a(\theta\sin\theta) = a\theta\sin\theta$$

Now

$$\frac{dy}{dx} = \frac{dy/d\theta}{dx/d\theta} = \frac{a\theta\sin\theta}{a\theta\cos\theta} = \tan\theta$$

Example 7.23. $a\left(\cos t + \log \tan \frac{t}{2}\right)$ **and** $y = a \sin t$**, find** $\frac{dy}{dx}$

Solution. We have, $x = a\left(\cos t + \log \tan \frac{t}{2}\right) \cdot y = a \sin t$

$$\therefore \quad \frac{dx}{dt} = a\left[-\sin t + \frac{1}{\tan \frac{t}{2}} \sec^2 \frac{t}{2} \cdot \frac{1}{2}\right]$$

$$= a\left[-\sin t + \frac{1}{2 \sin \frac{t}{2} \cos \frac{t}{2}}\right] = a\left[-\sin t + \frac{1}{\sin t}\right]$$

$$= a\left[\frac{-\sin^2 t + 1}{\sin t}\right] = \frac{a \cos^2 t}{\sin t}$$

and $$\frac{dy}{dx} = a \cos t$$

Now, $$\frac{dy}{dx} = \frac{dy/dt}{dx/dt} = \frac{a \cos t}{a \cos^2 t / \sin t} = \frac{\sin t}{\cos t} = \tan t$$

Example 7.24. Differentiation the following functions w.r.t. x**:**

(*i*) x^x (*ii*) $(\sin x)^{\log x}$

(*iii*) $\frac{e^{x^2} \cdot \tan^{-1} x}{\sqrt{1+x^2}}$ (*iv*) $\sqrt{(x-1)(x-2)(x-3)(x-4)}$

Solution.

(*i*) Let $y = x^x$

Taking logaerithms of both sides

$$\log y = x \log x$$

Differentiating both sides w.r.t. x, we get

$$\frac{1}{y} \cdot \frac{dy}{dx} = x \cdot \frac{1}{x} + \log x \cdot 1 = 1 + \log x$$

or $$\frac{dy}{dx} = y(1 + \log x) = x^x(1 + \log x)$$

(*ii*) Let $y = (\sin x)^{\log x}$

Taking logarithms of both sides

$$\log y = (\log x) \log (\sin x)$$

Differentiating both sides w.r.t x, we get

$$\frac{1}{y} \cdot \frac{dy}{dx} = (\log x) \cdot \left[\frac{1}{\sin x2} \cdot \cos x\right] + \log (\sin x) \frac{1}{x}$$

or $$\frac{dy}{dx} = y \cdot \left[(\log x)(\cot x) + \frac{\log (\sin x)}{x}\right]$$

$$= (\sin x)^{\log x}\left[(\log x)(\cot x) + \frac{\log (\sin x)}{x}\right]$$

(*iii*) $$y = \frac{e^{x^x} \tan^{-1} x}{\sqrt{1+x^2}}$$

Taking logarithms of both sides

$$\log y = x^2 + \log(\tan^{-1} x) - \frac{1}{2}\log(1+x^2)$$

Differentiating both sides w.r.t x, we get

$$\frac{1}{y}\cdot\frac{dy}{dx} = 2x + \frac{1}{\tan^{-1} x}\cdot\frac{1}{(1+x^2)} - \frac{1}{2}\cdot\frac{2x}{1+x^2}$$

$$\therefore \quad \frac{dy}{dx} = y\left[2x + \frac{1}{(1+x^2)\tan^{-1} x} - \frac{x}{1+x^2}\right]$$

$$= \frac{e^{x^2} \tan^{-1} x}{\sqrt{1+x^2}}\left[2x + \frac{1}{(1+x^2)\tan^{-1} x} - \frac{x}{1+x^2}\right]$$

(*iv*) Let $y = \sqrt{(x-1)(x-2)(x-3)(x-4)} = [(x-1)(x-2)(x-3)(x-4)]^{1/2}$

Taking logarithms of both sides

$$\log y = \frac{1}{2}[\log(x-1) + \log(x-2) + \log(x-3) + \log(x-4)$$

Differentiating both sides w.r.t., x, we get

$$\frac{1}{y}\frac{dy}{dx} = \frac{1}{2}\left[\frac{1}{x-1} + \frac{1}{x-2} + \frac{1}{x-3}\frac{1}{x-4}\right]$$

$$\therefore \quad \frac{dy}{dx} = \frac{1}{2}y\left[\frac{1}{x-1} + \frac{1}{x-2} + \frac{1}{x-3} + \frac{1}{x-4}\right]$$

$$= \frac{1}{2}\sqrt{(x-1)(x-2)(x-3)(x-4)}\left[\frac{1}{x-1} + \frac{1}{x-2} + \frac{1}{x-3} + \frac{1}{x-4}\right]$$

Example 7.25. If $x^m \cdot y^n = (x+y)^{m+n}$, prove that $\frac{dy}{dx} = \frac{y}{x}$.

Solution. We have $x^m \cdot y^n = (x+y)^{m+1}$

Taking logarithm of both sides given

$$m\log x + n\log y = (m+n)\log(x+y)$$

Differentiating both sides w.r.t. x, we get

$$m\cdot\frac{1}{x} + n\cdot\frac{1}{y}\frac{dy}{dx} = \frac{m+n}{x+y}\cdot\frac{dy}{dx}(x+y)$$

$$\Rightarrow \quad \frac{m}{x} + \frac{n}{y}\frac{dy}{dx} = \frac{m+n}{x+y}\left[1 + \frac{dy}{dx}\right]$$

$$\Rightarrow \quad \left[\frac{n}{y} - \frac{m+n}{x+y}\right] = \frac{m+n}{x+y} - \frac{m}{x}$$

$$\Rightarrow \quad \frac{nx-my}{y(x+y)}\cdot\frac{dy}{dx} = \frac{nx-my}{(x+y)x}$$

$$\Rightarrow \qquad \frac{dy}{dx} = \frac{y}{x}.$$

Example 7.26. If $y = \sqrt{\cos x + \sqrt{\cos x + \sqrt{\cos x + \ldots \text{to } \infty}}}$ = pove that $\dfrac{dy}{dx} = \dfrac{\sin x}{1-2y}$.

Solution. We have $\qquad y = \sqrt{\cos x + \sqrt{\cos x + \sqrt{\cos x + \ldots}}}$

$$y = \sqrt{\cos x + y} \qquad \text{or} \qquad y^2 = \cos x + y$$

Differentiating both sides w.r.t. x, we get

$$2y\frac{dy}{dx} = -\sin x + \frac{dy}{dx}$$

$$\Rightarrow \qquad (2y-1)\frac{dy}{dx} = -\sin x \qquad \Rightarrow \qquad \frac{dy}{dx} = \frac{-\sin x}{2y-1}$$

$$\Rightarrow \qquad \frac{dy}{dx} = \frac{\sin x}{1-2y}$$

EXERCISE 7.3

1. $\tan^{-1}\left\{\dfrac{x}{\sqrt{a^2-x^2}}\right\}$ $\qquad \left[\textbf{Ans. } \dfrac{1}{\sqrt{a^2-x^2}}\right]$

2. $\sin^{-1}\left\{\dfrac{x}{a+\sqrt{a^2-x^2}}\right\}$ $\qquad \left[\textbf{Ans. } \dfrac{1}{\sqrt{1-x^2}}\right]$

3. $\tan^{-1}\left(\dfrac{4x}{1-4x^2}\right)$ $\qquad \left[\textbf{Ans. } \dfrac{4}{1+4x^2}\right]$

4. $\tan^{-1}\left(\dfrac{\sin x}{1+\cos x}\right)$ $\qquad \left[\textbf{Ans. } \dfrac{1}{2}\right]$

5. $\tan^{-1}\left(\dfrac{\sqrt{x}+\sqrt{a}}{1-\sqrt{xa}}\right)$ $\qquad \left[\textbf{Ans. } \dfrac{1}{2\sqrt{x}(1+x)}\right]$

6. $\tan^{-1}\left(\dfrac{a+bx}{b-ax}\right)$ $\qquad$ [**Ans.** 1]

7. $\tan^{-1}\left\{\dfrac{x^{1/3}+a^{1/3}}{1-l(ax)^{1/3}}\right\}$ $\qquad \left[\textbf{Ans. } \dfrac{1}{3}\cdot\dfrac{x^{-2/3}}{1+x^{2/3}}\right]$

8. $\sin^{-1}\left(\dfrac{1}{\sqrt{1+x^2}}\right)$ $\qquad \left[\textbf{Ans. } \dfrac{-1}{1+x^2}\right]$

9. If $y = \sin^{-1}\left(\dfrac{2x}{1+x^2}\right) + \sec^{-1}\left(\dfrac{1+x^2}{1-x^2}\right)$ prove that $\dfrac{dy}{dx} = \dfrac{4}{1+x^2}$.

10. If $y = \tan^{-1}\left\{\dfrac{\sqrt{1+x^2}+\sqrt{1+x^2}}{\sqrt{1+x^2}-\sqrt{1-x^2}}\right\}$, find $\dfrac{dy}{dx}$.

11. If $y = \cot^{-1}\left\{\frac{\sqrt{1+\sin x}+\sqrt{1-\sin x}}{\sqrt{1+\sin x}-\sqrt{1-\sin x}}\right\}$, show that $\frac{dy}{dx} = \frac{4}{1+x^2}$.

12. If $y = \sec^{-1}\left(\frac{x+1}{x-1}\right) + \sin^{-1}\left(\frac{x-1}{x+1}\right)$. Find $\frac{dy}{dx}$. [**Ans.** 0]

13. $y = \cos^{-1}(2x) + 2\cos^{-1}\sqrt{1-4x^2}$ find $\frac{dy}{dx}$. $\left[\textbf{Ans.} \frac{2}{1-4x^3}\right]$

14. If $y = [x+\sqrt{1+x^2}]^m$ .Prove that $\frac{dy}{dx} = \frac{my}{\sqrt{1+x^2}}$.

15. If $x = \frac{dy}{dx} = \frac{my}{\sqrt{1+x^2}}$ and $y = \frac{\cos^3 t}{\sqrt{\cos 2t}}$, find $\frac{dy}{dx}$. [**Ans.** $-\cot 3t$]

16. If $ax^2 + 2hxy + by^2 + 2gx + 2fy + c = 0$, find $\frac{dy}{dx}$. $\left[\textbf{Ans.} -\left(\frac{ax-hy+g}{hx+by+f}\right)\right]$

17. If $\sin y = x\sin(a+y)$, prove that $\frac{dy}{dx} = \frac{\sin^2(a+y)}{\sin a}$.

18. If $\sqrt{1+x^2}+\sqrt{1-y^2} = a(x-y)$, prove that $\frac{dy}{dx} = \sqrt{\frac{1-y^2}{1-x^2}}$.

19. If $\frac{x}{x-y} = \log\left(\frac{a}{x-y}\right)$, Prove that $\frac{dy}{dx} = 2 - \frac{x}{y}$.

20. Find $\frac{dy}{dx}$, when

(*i*) $\cos(x+y) = y\sin x$ $\left[\textbf{Ans.} -\left[\frac{\sin(x+y)+y\cos x}{\sin(x+y)+\sin x}\right]\right]$

(*ii*) $\sin(xy) + \frac{x}{y} = x^2 - y$. $\left[\textbf{Ans.} \left[\frac{2xy^2 - y - y^3\cos(xy)}{xy^2\cos(xy) - x + y^2}\right]\right]$

21. If $\sqrt{1-x^6}+\sqrt{1-y^6} = a(x-y)$, prove that $\frac{dy}{dx} = \sqrt{\frac{1-y^2}{1-x^2}}$.

22. Find $\frac{dy}{dx}$ if $x = \frac{1-t}{1+t}$ = and $y = 2t^3 + 4^t$. $\left[\textbf{Ans.} \frac{(6t^2+4^t\log 4)(1+t)^2}{-2}\right]$

23. If $x = a\log t$ and $y = bt^2$, find $\frac{dy}{dx}$. $\left[\textbf{Ans.} \frac{2bt^2}{a}\right]$

24. Find $\frac{dy}{dx}$, if $x = a(t + \sin t)$ and $y = a(1 - \cos t)$. $\left[\textbf{Ans.} \tan\frac{t}{2}\right]$

25. Differentiate $(\sin x)^x$. [**Ans.** $(\sin x)^x [x\cot x + \log\sin x]$]

26. If $y = x^{x^x}$, find $\frac{dy}{dx}$. [**Ans.** $x^{x^x}[x^{x-1} + (\log x)\, x^x(1+\log x)]$]

27. If $y = (\sin x)^{\tan x} + (\cot x)^{\sec x}$, find $\frac{dy}{dx}$.

[**Ans.** $(\sin x)^{\tan x}[1 + \log(\sin x)\cdot\sec^2 x] + (\cos x)^{\sec x}\cdot\sec x\tan x[\log(\cos x) - 1]$]

28. Find the derivative of $\dfrac{\sqrt{x}\,(x+4)^{3/2}}{(4x-3)^{4/3}}$. $\left[\textbf{Ans.} \dfrac{\sqrt{x}\,(x+4)^{3/2}}{(4x-3)^{4/3}}\left[\dfrac{1}{2x}+\dfrac{3}{2(x+4)}-\dfrac{16}{3(4x-3)}\right]\right]$

29. If $x^y = y^x$, find $\dfrac{dy}{dx}$. $\left[\textbf{Ans.} \dfrac{y(x\log y - y)}{x(y\log x - x)}\right]$

30. Differentiate $(x+1)^2\ (x+2)^3\ (x+3)^4$. $\left[\textbf{Ans.} (x+1)^2\ (x+2)^3\ (x+3)^4\left[\dfrac{2}{x+1}+\dfrac{3}{x+2}+\dfrac{4}{x+3}\right]\right]$

31. If $y = \dfrac{(1+x)^{1/2}(3+x^2)^{3/2}}{(2-x^2)^{5/2}\,(4+2x^3)^{2/5}}$, find $\dfrac{dy}{dx}$.

$$\left[\textbf{Ans.} \frac{(1+x)^{1/2}\,(3+x^2)^{3/2}}{(2-x^2)^{5/2}\,(4+2x^3)^{2/5}}\left[\frac{1}{2(1+x)}+\frac{3x}{(3+x^2)}+\frac{5x}{(2-x^2)}-\frac{12x^2}{5(4+2x^3)}\right]\right]$$

32. If $e^{x+y} = xy$, show that $\dfrac{dy}{dx} = \dfrac{y(x-1)}{x(1-y)}$.

7.11. SUCCESSIVE DIFFERENTIATION

If $y = f(x)$ be a differentiable function of x, then $\dfrac{dy}{dx}$ is called the first differential coefficient of y w.r.t. x. If this derivative is again a differentiable function, then its derivative *i.e.*, $\dfrac{d}{dx}\left(\dfrac{dy}{dx}\right)$ is called the second differential coefficient of y w.r.t. x and is denoted by $\dfrac{d^2y}{dx^3}$. Similarly, the differential coefficient of $\dfrac{d^2y}{dx^2}$ *i.e.*, $\dfrac{d}{dx}\left(\dfrac{d^2y}{dx^2}\right)$ is called the third differential coefficient of y w.r.t. x and is denoted by $\dfrac{d^3y}{dx^3}$ In general, the nth differential coefficient of y w.r.t. x is denoted by $\dfrac{d^ny}{dx^n}$ and is given by $\dfrac{d^ny}{dx^n} = \dfrac{d}{dx}\left(\dfrac{d^{n-1}}{dx^{n-1}}\right)$.

If $y = f(x)$, then the successive differential coefficients are denoted by $y_1, y_2, \ldots, y_n, \ldots$ or by is $y', y'', y''', \ldots, y^n, \ldots$

The nth differential coefficient of y w.r.t. x is represented by any of the following ways: $\dfrac{d^ny}{dx^n}$ $D^n y$, y_n, $f^{(n)}(x)$, $D^n f(x)$, $f_n(x)$ etc.

SOLVED EXAMPLES

Example 7.27. If $x = a(\theta - \sin\theta)$ and $y = a(1 - \cos\theta)$, find $\dfrac{d^2y}{dx^2}$ at $\theta = \pi$.

Solution. Here, $x = a(\theta - \sin\theta)$ and $y = a(1-\cos\theta)$

$$\frac{dx}{d\theta} = a(1-\cos\theta) \quad \text{and} \quad \frac{dy}{d\theta} = a\sin\theta$$

$$\therefore \quad \frac{dy}{d\theta} = \frac{dy}{dx} + \frac{dx}{d\theta} = \frac{\sin\theta}{1-\cos\theta}$$

Differentiating w.r.t. x,

$$\frac{d^2y}{dx^2} = \frac{d}{d\theta}\left(\frac{\sin\theta}{1-\cos\theta}\right)\cdot\frac{d\theta}{dx}$$

$$= \frac{d}{d\theta}\left(\frac{\sin\theta}{1-\cos\theta}\right)\times\frac{1}{a(1-\cos\theta)}$$

$$= \frac{(1-\cos\theta)\cdot\cos\theta - \sin\theta\cdot\sin\theta}{(1-\cos\theta)^2}\times\frac{1}{a(1-\cos\theta)}$$

Hence $$\frac{d^2y}{dx^2} = \frac{\cos\theta - 1}{a(1-\cos\theta)^3} = \frac{-1}{a(1-\cos\theta)^2}$$

$\therefore$ $$\left.\frac{d^2y}{dx^2}\right]_{\theta=\pi} = -\frac{1}{a(1+1)^2} = -\frac{1}{4a}$$

Example 7.28. If $y = \log\,(x + \sqrt{1+x^2})$, prove that $(1+x^2)\dfrac{d^2y}{dx^2} + x\dfrac{dy}{dx} = 0.$

Solution. Here, $$y = \log\,(x + \sqrt{1+x^2})$$

$$\frac{dy}{dx} = \frac{1}{x+\sqrt{1+x^2}}\left[1 + \frac{1}{2}(1+x^2)^{-1/2}\cdot 2x\right]$$

$$= \frac{1}{x+\sqrt{1+x^2}}\left[1 + \frac{x}{\sqrt{1+x^2}}\right]$$

$$\frac{dy}{dx} = \frac{1}{x+\sqrt{1+x^2}}\cdot\frac{\sqrt{1+x^2}+x}{\sqrt{1+x^2}} = \frac{1}{\sqrt{1+x^2}} = \sqrt{1+x^2}\,\frac{dy}{dx} = 1$$

Squaring both sides,

$$(1+x^2)\left(\frac{dy}{dx}\right)^2 = 1$$

Again differentiating w.r.t x, we have

$$(1+x^2)\cdot 2\frac{dy}{dx}\cdot\frac{d^2y}{dx^2} + 2x\left(\frac{dy}{dx}\right)^2 = 0$$

Dividing throughout be $2\cdot\dfrac{dy}{dx}$, we get

$\Rightarrow$ $$(1+x^2)\frac{d^2y}{dx^2} + x\frac{dy}{dx} = 0$$

Example 7.29. If $y = \sin\,(a\sin^{-1}x)$, prove that $(1-x^2)\dfrac{d^2y}{dx^2} = x\dfrac{dy}{dx} - a^2y$.

Solution. Here, $$y = \sin\,(a\sin^{-1}x) \quad \ldots(i)$$

$\therefore$ $$\frac{dy}{dx} = \cos\,(a\sin^{-1}x)\cdot\frac{a}{\sqrt{1+x^2}} = \frac{a\cdot\cos\,(a\sin^{-1}x)}{\sqrt{1-x^2}}$$

$$\sqrt{1-x^2}\,\frac{dy}{dx} = a\cos(a\sin^{-1}x)$$

Squaring both sides,

$$(1-x^2)\left(\frac{dy}{dx}\right)^2 = a^2\cos^2(a\sin^{-1}x)$$

or $$(1-x^2)\left(\frac{dy}{dx}\right)^2 = a^2[1-\sin^2(a\sin^{-1}x)] \qquad [\because \cos^2\theta = 1-\sin^2\theta]$$

or $$(1-x^2)\left(\frac{dy}{dx}\right)^2 = a^2(1-y^2) \qquad \text{[From } (i)]$$

Again differentiating w.r.t. x, we have

$$(1-x^2)\frac{d}{dx}\left(\frac{dy}{dx}\right)^2 + \left(\frac{dy}{dx}\right)^2\cdot\frac{d}{dx}(1-x^2) = a^2\left[0-\frac{d}{dx}(y^2)\right]$$

or $$(1-x^2)\cdot 2\cdot\frac{dy}{dx}\cdot\frac{d^2y}{dx^2} - 2x\left(\frac{dy}{dx}\right)^2 = -2a^2y\frac{dy}{dx}$$

Dividing throughout by $2\frac{dy}{dx}$, we get

$$(1-x^2)\frac{d^2y}{dx} - x\frac{dy}{dx} = -a^2y$$

or $$(1-x^2)\frac{d^2y}{dx^2} = x\frac{dy}{dx} - a^2y.$$

Example 7.30. If $y = a\cos(\log x) + b\sin(\log x)$, prove that $x^2y_2 + xy_1 + y = 0$.

Solution. Here $y = a\cos(\log x) + b\sin(\log x)$

Differentating w.r.t x, we get

$$y_1 = -a\sin(\log x)\cdot\frac{1}{x} + b\cos(\log x)\cdot\frac{1}{x}$$

Multiplying every term by x.

$$xy_1 = -a\sin(\log x) + b\cos(\log x)$$

Again differentating both sides w.r.t. x, we get

$$\frac{d}{dx}(xy_1) = -a\cos(\log x)\cdot\frac{1}{x} - b\sin(\log x)\cdot\frac{1}{x}$$

or $$x\cdot y_2 + y_1\cdot 1 = -a\cos(\log x)\cdot\frac{1}{x} - b\sin(\log x)\cdot\frac{1}{x}$$

Cross multiplying $x^2y_2 + xy_1 = -y$

or $x^2y_2 + xy_1 + y = 0$

EXERCISE 7.4

1. If $y = \tan^{-1} x$, prove that $(1+x^2)\dfrac{d^2y}{dx^2} + 2x\dfrac{dy}{dx} = 0$

2. If $x = \sin t$, $y = \sin pt$; prove that $(1-x^2)\dfrac{d^2y}{dx^2} - x\dfrac{dy}{dx} + p^2y = 0$

3. Prove that the function $y = A \cos wx + B \sin wx$ satisfies the differential equation $\dfrac{d^2y}{dx^2} + w^2y = 0$.

4. If $y = \left(\dfrac{1}{x}\right)^x$, show that $y_2(1) = 0$.

5. If $x = a(\cos\theta + \theta\sin\theta)$, $y = a(\sin\theta - \theta\cos\theta)$; find $\dfrac{d^2y}{dx^2}$ $\left[\textbf{Ans.} \dfrac{1}{a\theta}\sec^3\theta\right]$

6. If $y = \sin(\sin x)$, prove that $\dfrac{d^2y}{dx^2} + \dfrac{dy}{dx}\tan x + y\cos^2 x = 0$.

7. If $y^3 + 3ax^2 + x^3$, show that $\dfrac{d^2y}{dx^2} + \dfrac{2a^2x^2}{y^5} = 0$.

8. If $y = e^{\tan^{-1}x}$, show that $(1+x^2)\dfrac{d^2y}{dx^2} + (2x-1)\dfrac{dy}{dx} = 0$.

9. If $x = 2\cos t - \cos 2t$, $y = 2\sin t - \sin 2t$, find $\dfrac{d^2y}{dx^2}$ at $t = \dfrac{\pi}{2}$. $\left[\textbf{Ans.} -\dfrac{3}{2}\right]$

10. If $y \dfrac{\log x}{x}$, show that $\dfrac{d^2y}{dx^2} = \dfrac{2\log x - 3}{x^3}$.

❑❑❑

CHAPTER 8

Applications of Derivatives

INSIDE THIS CHAPTER

8.1. APPROXIMATION OF ERRORS

Suppose a quantity u is dependent on two independent quantities x and y or in other words u is a function of x and y. Now if there is a change in the value of x from x to $x + \delta x$ (δx is amll and may be positive or negative), and a change in the value of y from y to $y + \delta y$ (δy is small and may be positive or negative), then there will be a change in the value of u say from u to $u + \delta u$. We may call the change in the value of x, *i.e.*, δx as 'increment in x' or 'error in x'. Similarly δy may be called as 'increment of y' or 'error in the value y' and so δu is the 'increment in the value of u' or 'error in the value u'.

Now we have $u = f(x, y)$

$$u + \delta u = f(x + \delta x, y + \delta y)$$

$$\therefore \quad \delta u = f[(x + \delta x), (y + \delta y)] - f(x, y) \qquad ...(i)$$

Expanding $f[(x + \delta x), (y + \delta y)]$ by Taylor's Theorem on two variables, we have

$$f(x + \delta x, y + \delta y) = f(x, y) + \left(\frac{\partial f}{\partial x}\delta x + \frac{\partial f}{\partial y}\delta y\right) + \frac{1}{2}\left(\frac{\partial^2 f}{\partial x^2}\delta x^2 + 2\delta x\delta y\frac{\partial^2 f}{\partial x\partial f} + \delta y^2\frac{\partial^2 f}{\partial y^2} + ...\right)$$

As δx and δy are supposed to be small, therefore their squares and higher powers can be neglected.

$$\therefore \quad f(x + \delta x, y + \delta y) = f(x, y) + \left(\frac{\partial f}{\partial x}\partial x + \frac{\partial f}{\partial y}\partial y\right) \qquad ...(ii)$$

From Eqs. (*i*) and (*ii*), we get

$$\delta u = \frac{\partial f}{\partial x}\delta x + \frac{\partial f}{\partial y}\delta y$$

or

$$\boldsymbol{\delta u = \frac{\partial u}{\partial x}\delta x + \frac{\partial u}{\partial y}\delta y} \qquad \left[\because du = \frac{\partial u}{\partial x}dx + \frac{\partial u}{\partial y}dy\right]$$

This formula is used in calculating the effect of small errors or increments in measured quantities and is useful in correcting the effect of small errors.

As above if u is a function of three or more variables

i.e., $u = f(x, y, z, ...)$

We have accordingly, $\delta u = \frac{\partial u}{\partial x}\delta x + \frac{\partial u}{\partial y}\delta y + \frac{\partial u}{\partial z}\delta z + ...$

$\frac{\delta x}{x} \times 100$ is called **percentage error** in the value of x, whereas δx is the change or **actual error** in the value of x.

Similarly $\frac{\delta y}{y} \times 100 = \%$ error in y and $\frac{\delta u}{u} \times 100 = \%$ error in u

where δy and δu are 'actual errors' in y and u respectively.

Note. $\frac{dx}{x}$ is called Relative Errors.

SOLVED EXAMPLES

Example 8.1. The edge of a cube is measured with a possible error of 0.05 cm. Find the relative error in the computed volume when the edge is found to be 7.5 cm. Also find percentage error in the computed volume.

Solution. Let x be the edge of the cube.

$\therefore$ Volume V of the cube $= x^3$

$$V = x^3$$

Taking log of both sides, we get

$$\log V = \log x^3 = 3 \log x.$$

Taking differentials, we get

$$\frac{1}{V} dV = \frac{1}{3} dx$$

$$\text{Error in computed volume} = dV = \frac{3V}{x} dx = 3x^2\, dx = 3 \times 7.5 \times 7.5 \times 0.05$$

$$\therefore \quad dV = 8.44 \text{ cubic cm.} \quad \textbf{Ans.}$$

Again % error in the computed volume $= \frac{dV}{V} \times 100 = 3\left(\frac{dx}{x} \times 100\right) = 3\left(\frac{0.05}{7.5} \times 100\right) = 2.$ **Ans.**

Example 8.2. The time of oscillation of a simple pendulum is given by the equation $T = 2\pi\sqrt{\frac{1}{g}}$. In an experiment carried out to find the value of 'g', error of 1.5% and 0.5% are possible in the values of l and T respectively. Show that the error in the calculated value of 'g' is 0.5%.

Solution. Given $\quad T = 2\pi\sqrt{\frac{1}{g}}$

$$\Rightarrow \quad T^2 = \frac{4\pi^2 l}{g} \qquad \text{(Squaring both sides)}$$

$$\therefore \quad g = \frac{4\pi^2 l}{T^2}$$

Taking log on both sides, we get

$$\log (g) = \log (4\pi^2) + \log l - 2 \log T$$

Differentiating, we get $\quad \frac{1}{g} dg = 0 + \frac{1}{l} dl - \frac{2}{T} dT$

$$\Rightarrow \qquad \frac{dg}{g} = \frac{dl}{l} - \frac{2dT}{T} \qquad ...(i)$$

Given $\qquad$ % error in $l = \frac{dl}{l} \times 100 = 1.5$

$$\therefore \qquad \frac{dl}{l} = \frac{1.5}{100}$$

% error in $T = \frac{dT}{T} \times 100 = 0.5$

$$\therefore \qquad \frac{dT}{T} = \frac{0.5}{100}$$

Putting the value of $\frac{dl}{l}$ and $\frac{dT}{T}$ in (*i*), we get

$$\frac{dg}{g} = \frac{1.5}{100} - \frac{2 \times 0.5}{100} = \frac{0.5}{100}$$

$\therefore \qquad$ % Error in $g = 0.5\%$

Example 8.3. Find the possible percentage error in computing the resistance r from the formula $\frac{1}{r} = \frac{1}{r_1} + \frac{1}{r_2}$, if r_1, r_2 are both in error by 2%.

Solution. Given $\qquad \frac{1}{r} = \frac{1}{r_1} + \frac{1}{r_2}$

$$-\frac{1}{r^2} dr = -\frac{1}{r_1^2} dr_1 - \frac{1}{r_2^2} dr_2$$

$$\Rightarrow \qquad \frac{dr}{r^2} = \frac{dr_1}{r_1^2} + \frac{dr_2}{r_2^2} \qquad ...(i)$$

% Error in $r_1 = \frac{dr_1}{r_1} \times 100 = 2$

$$\therefore \qquad dr_1 = \frac{r_1}{50} \qquad ...(ii)$$

% Error in $r_2 = \frac{dr_2}{r_2} \times 100 = 2$

$$\therefore \qquad dr_2 = \frac{r_2}{50} \qquad ...(iii)$$

From Eqs. (*i*), (*ii*) and (*iii*), we get

$$\frac{dr}{r^2} = \frac{r_1}{r_1^2} \times \frac{1}{50} + \frac{r_2}{r_2^2} \times \frac{1}{50}$$

$$\frac{dr}{r^2} = \frac{1}{50}\left(\frac{1}{r_1} + \frac{1}{r_2}\right) = \frac{1}{50} \times \frac{1}{r} \qquad \left[\because \frac{1}{r} = \frac{1}{r_1} + \frac{1}{r_2}\right]$$

$$\therefore \qquad \frac{dr}{r} = \frac{1}{50}$$

$\therefore \qquad$ % Error in $r = \frac{dr}{r} \times 100 = \frac{1}{50} \times 100 = 2.$ **Ans.**

EXERCISE 8.1

1. Find the percentage error in the area of an ellipse, when an error of 1% is made in measuring the semi-major and semi-minor axes. **[Ans.** 2]
2. Let $T = 2\pi\sqrt{l/g}$. Find the maximum percentage error in T due to possible errors of 1% and 2% in l and g respectively. **[Ans.** ± 1.5%]
3. Compute an approximate value $(1.04)^{3.01}$. **[Ans.** 0.131 approximately]
4. If the sides and angles of a triangle ABC vary in such a way that its circumradius remains constant, prove that
$$\frac{da}{\cos A}+\frac{db}{\cos B}+\frac{dc}{\cos C}=0.$$
5. If an error committed in measuring the side of square be 2%. Find the error in calculating the area. **[Ans.** 4%]
6. Use differentials to compute $f(0.9, -1.2)$ approximately where $f(x, y) = \tan^{-1}(xy)$. **[Ans.** 2.3062]

8.2. TANGENTS AND NORMALS

Let $y = f(x)$ be a continuous function of x. Let $P(x, y)$ be a point on the arc AB of the curve $y = f(x)$ and $Q(x + \Delta x, y + \Delta y)$ be another point on the curve continuous to P.

Joint QP and produce it to meet the axis of x in S.

Draw PM, QN perpendiculars to OX and $PR \perp$ from P on QN. Let PS' be the tangent to the curve at P.

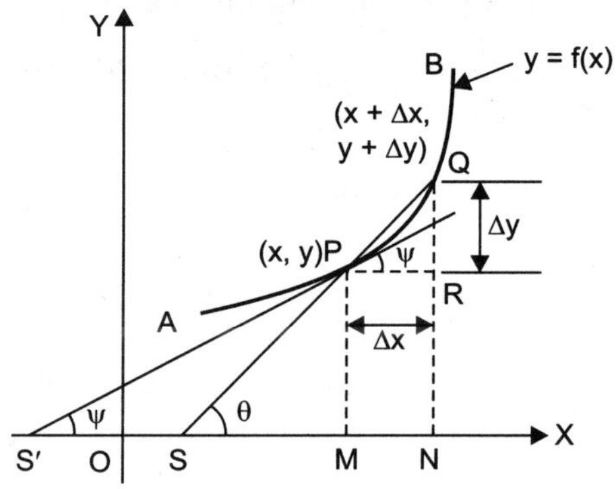

Let $\angle XSP = \theta$

$\therefore$ $\angle QPR = \angle XSP$

Now $$\tan \angle QPR = \frac{QR}{PR} = \frac{QN - RN}{MN} = \frac{(y+\Delta y)-y}{(x+\Delta x)-x} = \frac{\Delta y}{\Delta x}$$

$\therefore$ $$\tan\theta = \frac{\Delta y}{\Delta x} \quad ...(i)$$

Now as $Q \to P$, $\Delta x \to 0$, the secant QPS becomes the tangent PS' and the angle Q becomes the angle ψ.

$\therefore$ $$\tan\psi = \lim_{Q\to P}\tan\theta = \lim_{\Delta x\to 0}\frac{\Delta y}{\Delta x} \quad \text{[From } (i)]$$

$\therefore$ $$\tan\psi = \frac{dy}{dx}$$

$$\frac{dy}{dx} = \tan\psi = \text{Slope of the tangent at } P(x, y).$$

Remark 1. If the tangent is parallel to x-axis, then $\psi = 90°$.

$\therefore$ $$\frac{dy}{dx} = \tan\psi = \tan 0 = 0$$

2. If the tangent is perpendicular to x-axis, then $\psi = 90°$.

$\therefore$ $$\frac{dy}{dx} = \tan\psi = \tan 90° = \infty \quad \text{or} \quad \frac{dy}{dx} = \frac{1}{\frac{dy}{dx}} = 0$$

8.2.1. Equation of the Tangent at Any Point of the Curve

Let $P(x_1, y_1)$ be any point on the curve $y = f(x)$. Slope of tangent at $P(x_1, y_1)$ is the value of $\frac{dy}{dx}$ at the point (x_1, y_1). Let the value be denoted by m.

Now the tangent at P is a line through $P(x_1, y_1)$ having slope m.

∴ Equation of the tangent at P is

$$y - y_1 = m(x - x_1)$$

8.2.2. Equation of the Normal at Any Point of the Curve y = *f* (x)

The normal to a curve at any point $P(x_1, y_1)$ is the straight line through the point (x_1, y_1) perpendicular to the tangent to the curve at that point.

Let the slope of the tangent to the curve at $P(x_1, y_1)$ be m. Then slope of the normal to the curve at

$$P(x_1, y_1) = -\frac{1}{m} \qquad \text{(–ve reciprocal)}$$

∴ Equation of the normal to the curve at $P(x_1, y_1)$ is

$$y - y_1 = -\frac{1}{m}(x - x_1)$$

$$\Rightarrow \quad (x - x_1) + m(y - y_1) = 0$$

8.2.3. Algorithm for Finding the Equations of Tangent and Normal to the Curve *y* = *f* (*x*) at the Given Point *P*(x_1, y_1)

Step I: Find $\frac{dy}{dx}$ from the given equation $y = f(x)$.

Step II: Find the value of $\frac{dy}{dx}$ at the given point $P(x_1, y_1)$. This gives the slope of the tangent at $P(x_1, y_1)$.

Step III: Find the negative reciprocal of the slope of the tangent at $P(x_1, y_1)$. This gives the slope of the normal at $P(x_1, y_1)$.

Step IV:

(*a*) The equation of the tangent is $y - y_1 = m(x - x_1)$, where m denotes the slope found in step II.

(*b*) The equation of the normal at $P(x_1, y_1)$ is

$$y - y_1 = -\frac{1}{m}(x - x_1)$$

where m denotes the slope found in step II.

SOLVED EXAMPLES

Example 8.4. Find the equation of the tangent and the normal to the curve

$$y = 2x^2 - 3x - 1 \text{ at } (1, -2)$$

Solution. Equation of curve is $y = 2x^2 - 3x - 1$

$$\therefore \quad \frac{dy}{dx} = 4x - 3$$

This is the slope of tangent at (x, y).

Slope of tangent at $(1, -2) = 4 \times 1 - 3 = 1$.

Equation of tangent at (1, –2) is

$$y - (-2) = 1(x - 1)$$

$$\Rightarrow \quad y + 2 = x - 1 \Rightarrow x - y - 3 = 0$$

Slope of the normal at (1, –2) is –1 (–ve reciprocal of tangent)

$\therefore$ Equation of normal at (1, –2) is

$$y - (-2) = -1(x - 1)$$

$$\Rightarrow \quad x + y + 1 = 0.$$

Example 8.5. Find the equations of the tangent and the normal at the point 'θ' on the curve $x = a \sin^3 \theta, y = b \cos^3 \theta$.

Solution. We have $\dfrac{dx}{d\theta} = 3a \sin^2 \theta \cos \theta$ and $\dfrac{dy}{d\theta} = -3b \cos^2 \theta \sin \theta$

$$\therefore \quad \frac{dy}{dx} = \frac{\frac{dy}{d\theta}}{\frac{dx}{d\theta}} = \frac{-3b\cos^2\theta \sin\theta}{3a \sin^2\theta \cos\theta} = \frac{-b\cos\theta}{a\sin\theta}$$

$\therefore$ Equation of the tangent at the point 'θ' is

$$y - b\cos^3\theta = \frac{-b\cos\theta}{a\sin\theta}(x - a\sin^3\theta)$$

or
$$ay \sin\theta + bx\cos\theta = ab\sin\theta\cos\theta$$

or
$$\frac{x}{a\sin\theta} + \frac{y}{b\cos\theta} = 1 \qquad \text{[On dividing throughout by } ab\sin\theta\cos\theta]$$

Also, the equation of the normal at the point 'θ' is

$$y - b\cos^3\theta = \frac{a\sin\theta}{b\cos\theta}(x - a\sin^3\theta)$$

or
$$ax\sin\theta - by\cos\theta = (a^2\sin^4\theta - b^2\cos^4\theta)$$

Example 8.6. Find the equation of the tangent line to the curve $y = \sqrt{5x-3} - 2$ which is

(i) parallel to the line $2x - y + 9 = 0$.

(ii) perpendicular to the line $5y + 2\sqrt{2}x = 13$.

Solution. Equation of the curve is

$$y = \sqrt{5x-3} - 2 \qquad \text{...}(i)$$

$$\Rightarrow \quad y = (5x-3)^{\frac{1}{2}} - 2$$

$$\therefore \quad \frac{dy}{dx} = \frac{1}{2}(5x-3)^{-\frac{1}{2}} \cdot \frac{d}{dx}(5x-3) = \frac{5}{2\sqrt{5x-3}}$$

(i) Slope of the line $2x - y + 9 = 0$ is 2

If the tangent is parallel to $2x - y + 9 = 0$

Then slope of tangent = 2

$$\therefore \quad \frac{5}{2\sqrt{5x-3}} = 2$$

Squaring both sides, we get

$$25 = 16(5x - 3) \Rightarrow 80x = 25 + 48$$

$$\Rightarrow \qquad 80x = 73 \quad \text{or} \quad x = \frac{73}{80}.$$

Putting $x = \frac{73}{80}$ in (*i*), we get

$$y = \sqrt{5 \times \frac{73}{80} - 3} - 2 = -\frac{3}{4}$$

$\therefore$ The point of contact is $\left(\frac{73}{80}, \frac{-3}{4}\right)$.

Hence the required equation of the tangent is

$$y + \frac{3}{4} = 2\left(x - \frac{73}{80}\right) \quad \text{or} \quad 80x - 40y - 103 = 0$$

(*ii*) Slope of the line $5y + 2\sqrt{2}x = 13$ is $\frac{-2\sqrt{2}}{5}$

If the tangent is perpendicular to the line $5y + 2\sqrt{2}x = 13$ is $\frac{-252}{5}$. If the tangent is perpendicular to the line $5y + 2\sqrt{2}x = 13$, then slope of tangent is $\frac{5}{2\sqrt{2}}$.

$$\therefore \qquad \frac{5}{2\sqrt{5x-3}} = \frac{5}{2\sqrt{2}} \quad \text{or} \quad \sqrt{5x-3} = \sqrt{2}$$

$$\Rightarrow \qquad 5x - 3 = 2 \Rightarrow 5x = 5$$

$$\therefore \qquad x = 1$$

Substituting $x = 1$ in Eq. (*i*), we get

$$y = \sqrt{5 \times 1 - 3} - 2 = \sqrt{2} - 2$$

$\therefore$ At the point $(1, \sqrt{2} - 2)$, the tangent is perpendicular to the line

$$5y + 2\sqrt{2}x = 13$$

Hence the equation of the tangent is

$$y - (\sqrt{2} - 2) = \frac{5}{2\sqrt{2}}(x - 1)$$

$$\text{or} \qquad 2\sqrt{2}y - 4 + 4\sqrt{2} = 5x - 5 \quad \text{or} \quad 2\sqrt{2}y - 5x + 4\sqrt{2} + 1 = 0$$

8.3. ANGLE OF INTERSECTION OF TWO CURVES

Definition: The angle of intersection of two curves at a point of intersection is the angle between the tangents to the two curves at that point.

Thus if two curves intersect at P and PT_1, PT_2 are the tangents to the curve at P, the angle of intersection of the two curves is the angle $\angle T_1PT_2 = \theta$ between the tangents PT_1 and PT_2.

To find the angle of intersection of the curves $y = f(x)$ and $y = \phi(x)$.

Let the two curves intersect in a point $P(x_1, y_1)$ and let PT_1 and PT_2 be the tangents to the two curves respectively at the point $P(x_1, y_1)$.

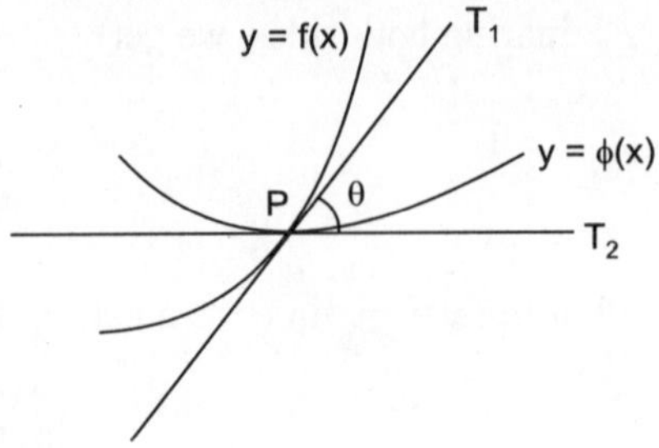

$\therefore$ Slope of the tangent PT_1 at

$$(x_1, y_1) = f'(x_1) = m_1 \text{ (say)}$$

and slope of the tangent PT_2 at

$$(x_1, y_1) = f'(x_1) = m_2 \text{ (say)}$$

$\therefore$ If θ is the angle between two tangents, then

$$\tan\theta = \left|\frac{m_1 - m_2}{1 + m_1 m_2}\right|$$

where m_1, m_2 are the values of $\frac{dy}{dx}$ for the two curves at the point $P(x_1, y_1)$.

Corollary 1. If the two curves touch each other, then the two tangents coincide at the point $P(x_1, y_1)$.

$\therefore$ The angle between the tangents *i.e.*, $\theta = 0$

$$\therefore \qquad \frac{m_1 - m_2}{1 + m_1 m_2} = \tan 0 = 0 \quad \Rightarrow \quad m_1 = m_2$$

Hence, when the two curve touch each other, the slopes of the tangents to the two curves at a point are equal.

Corollary 2. Condition of orthogonality: Two curves are said to be orthogonal if they intersect at right angles.

The curves will be orthogonal if $\theta = \frac{\pi}{2}$.

$$\therefore \qquad \frac{m_1 - m_1}{1 + m_1 m_2} = \tan 90° = \infty$$

$$\Rightarrow \qquad 1 + m_1 + m_2 = 0 \quad i.e., \quad m_1 m_2 = -1$$

Hence, if two curves cut orthogonally, the product of slopes of the tangents to the curves at the point of intersection = –1.

Rules for Finding the Angle of Intersection between Two Curves

Step I: Find the point or points of intersection of the two curves by solving their equations simultaneously.

Step II: Differentiate the equations of both the curves w.r.t. x and find $\frac{dy}{dx}$ at the point or points of intersection and name them m_1 and m_2.

Step III: If θ is the required angle between the curves, then

$$\tan\theta = \left|\frac{m_1 - m_2}{1 + m_1 m_2}\right|$$

SOLVED EXAMPLES

Example 8.7. Find the angle of intersection of the following curves : $xy = 6$ and $x^2y = 12$.

Solution. Equation of curves are

$$xy = 6 \qquad \text{...}(i)$$

$$x^2y = 12 \qquad \text{...}(ii)$$

From (*i*), we get
$$y = \frac{6}{x}$$

Putting this value of y in Eq. (*ii*), we get

$$x^2 \times \frac{6}{x} = 12 \quad \Rightarrow \quad 6x = 12; \quad x = 2$$

∴ From (*ii*), we get

∴ Given two curves intersect at $P(2, 3)$

Now differentiating (*i*) w.r.t. x, we get

$$x\frac{dy}{dx} + y = 0$$

$$\Rightarrow \quad \frac{dy}{dx} = \frac{-y}{x} \quad \text{...(iii)}$$

Differentiating (*ii*) w.r.t. x, we get

$$x^2\frac{dy}{dx} + 2xy = 0 \quad \Rightarrow \quad \frac{dy}{dx} = \frac{-2y}{x} \quad \text{...(iv)}$$

$$\therefore \quad m_1 = \left(\frac{dy}{dx}\right)_{(2,3)} = \frac{-3}{2} \quad \text{[From (iii)]}$$

and

$$m_2 = \left(\frac{dy}{dx}\right)_{(2,3)} = \frac{-2\times 3}{2} = -3 \quad \text{[From (iv)]}$$

If θ is the angle between the curves at $P(2, 3)$, then

$$\tan\theta = \left|\frac{m_1 - m_2}{1 + m_1 m_2}\right| = \left|\frac{-\frac{3}{2}+3}{1+\left(\frac{-3}{2}\right)(-3)}\right| = \frac{3}{11}$$

Example 8.8. Show that the curves $x = y^2$ and $xy = K$ cut at right angles, if $8K^2 = 1$.

Solution. The given curves are

$$x = y^2 \quad \text{...(i)}$$

and

$$xy = K \quad \text{...(ii)}$$

From (*i*), we obtain $x = y^2$. Putting this value of x in (*ii*), we obtain

$$y^3 = K \quad \Rightarrow \quad y = K^{1/3}$$

Putting $y = K^{1/3}$ in (*i*), we get $x = K^{2/3}$. So, the two curves intersect at the point $P(K^{2/3}, K^{1/3})$.

Differentiating (*i*) w.r.t. x, we get

$$1 = 2y\frac{dy}{dx} \quad \Rightarrow \quad \frac{dy}{dx} = \frac{1}{2y} \quad \Rightarrow \quad m_1 = \left(\frac{dy}{dx}\right)_P = \frac{1}{2K^{1/3}}$$

Differentiating (*ii*) w.r.t. x, we get

$$1.\,y + x\frac{dy}{dx} = 0 \Rightarrow \frac{dy}{dx} = \frac{-y}{x} \Rightarrow m_2 = \left(\frac{dy}{dx}\right)_P = \frac{-K^{1/3}}{K^{2/3}} = -\frac{1}{K^{1/3}}$$

For the curves (*i*) and (*ii*) to cut at right angles at P, we must have

$$m_1 m_2 = -1$$

$$\Rightarrow \quad \frac{1}{2K^{1/3}} \times \left(-\frac{1}{K^{1/3}}\right) = -1 \quad \Rightarrow \quad 2K^{2/3} = 1$$

$$\Rightarrow \quad (2K^{2/3})^3 = 13 \quad \Rightarrow \quad 8K^2 = 1$$

EXERCISE 8.2

1. Show that the tangents to the curve $y = 2x^3 - 3$ at the points where $x = 2$ and $x = -2$ are parallel.
2. Find the points on the curve $y = x^3 - 2x^2 - x$ at which the tangent lines are parallel to the line $y = 3x - 2$.

   $\left[\textbf{Ans. } (2, -2) \text{ and } \left(\frac{-2}{3}, \frac{-14}{27}\right)\right]$
3. Find a point on the curve $y = (x - 3)^2$, where the tangent is parallel to the line joining (4, 1) and (3, 0).

   $\left[\textbf{Ans. } \left(\frac{7}{2}, \frac{1}{4}\right)\right]$
4. At what point on the curve $y = x^2 - 4x + 5$ is the tangent perpendicular to the line $2y + x = 7$. [**Ans.** (3, 2)]
5. Find the equations of the tangent and normal to the curve $y(x - 2)(x - 3) - x + 7 = 0$ at the point where it meets x-axis. [**Ans.** $x - 20y = 7$ and $20x + y = 140$]
6. Find the equation of the tangent line to the curve $x = 1 - \cos\theta$, $y = \theta - \sin\theta$ at $\theta = \frac{\pi}{4}$.

   $\left[\textbf{Ans. } (\sqrt{2} - 1)x - y = 2(\sqrt{2} - 1) - \frac{\pi}{4}\right]$
7. Find the equation of the tangent line to the curve $y = x^2 + 4x - 16$ which is parallel to the line $3x - y + 1 = 0$.

   [**Ans.** $12x - 4y = 65$]
8. Determine the equations(s) of tangent(s) line to the curve $y = 4x^3 - 3x + 5$ which are perpendicular to the line $9x + x + 3$. [**Ans.** $9x - y - 3 = 0$; $9x - y + 13 = 0$]
9. Show that the normal at any point of the curve

   $$x = a\cos\theta + a\theta\sin\theta, \quad y = a\sin\theta - a\theta\cos\theta$$

   is at a constant distance from the origin.
10. Find the equation of the normal to the curve $3x^2 - y^2 = 8$ parallel to the line $x + 3y = 4$.

    [**Ans.** $x + 3y = 8$; $x + 3y + 8 = 0$]
11. Find the angle of intersection of the curves

    $$x^2 - y^2 = a^2 \quad \text{and} \quad x^2 + y^2 = a^2\sqrt{2}.$$

    $\left[\textbf{Ans. } \frac{\pi}{4}\right]$
12. Find the angle of intersection of the curves

    $$y^2 = 2x \quad \text{and} \quad x^2 + y^2 = 8$$

    [**Ans.** $\tan^{-1}(3)$]
13. Find the angle of intersection of the circle $x^2 + y^2 = 2a^2$ and the rectangular hyperbola $x^2 - y^2 = a^2$. $\left[\textbf{Ans. } \frac{\pi}{6}\right]$
14. Show that the curve $4x = y^2$ and $4xy = K$ cut at right angles, if $K^2 = 512$.

8.4. MAXIMA AND MINIMA

8.4.1. Introduction

A function $f(x)$ is maximum at $x = a$ if $f(a)$ is greater than any other value that the function can assume for value of x in the interval $(a - h, a + h)$ where h is a sufficiently small positive number.

That is $f(a) > f(x)$ or $f(x) < f(a)$ or $\{f(x) - f(a)\}$ is negative for every value of x in the immediate neighbourhood of a.

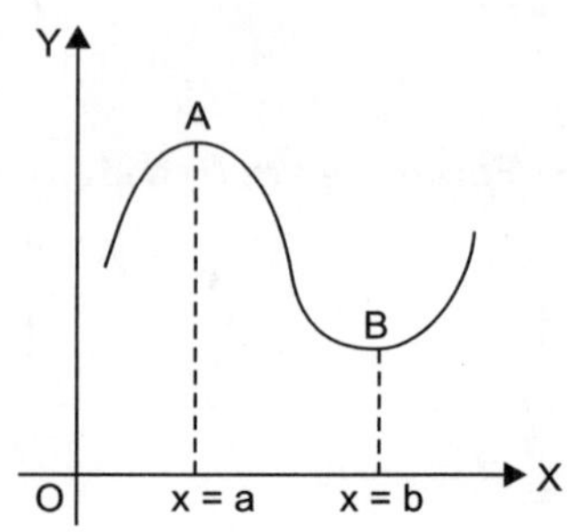

A function $f(x)$ is minimum at $x = b$ if $f(a)$ is less than any other value that function can assume in the interval $(b - h, b + h)$, where h is a sufficiently small positive number.

That is $f(b) < f(x)$ or $f(x) > f(b)$ or $\{f(x) - f(b)\}$ is positive for every value of x in the immediate neighbourhood of b.

The maximum or minimum value of a function is called extreme or stationary value of $f(x)$. A point where a maximum or a minimum occurs is called turning point. In the given figure, we have maximum and minimum at the turning points A and B respectively.

8.4.2. Local Maxima and Local Minima

Definition 1. Let f be a real function and let x_0 be an interior point in the domain of f. We say that x_0 is a local maximum of f, if there is an open interval containing x_0 such that $f(x_0) > f(x)$ for every x in that open interval.

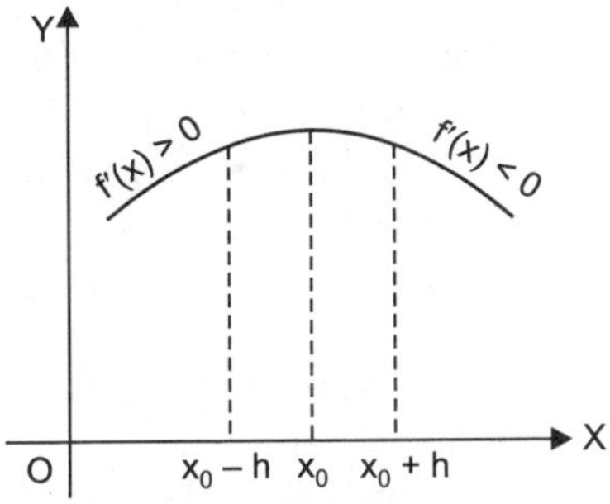

If x_0 is a point of local maximum of $f(x)$, then the graph of $f(x)$ around x_0 will be as shown in figure.

Here $f(x)$ is increasing in the interval $(x_0 - h, x_0)$ and decreasing in the interval $(x_0, x_0 + h)$.

$\therefore$ In $(x_0 - h, x_0)$, $f'(x) > 0$ and in $(x_0, x_0 + h)$, $f'(x) < 0$. This suggests that $f'(x)$ must be zero.

Definition 2. Let f be a real function and x_0 be an interior point in the domain of f. We see that x_0 is a local minimum of f if there is open interval containing x_0 such that $f(x_0) < f(x)$ for every x in that open interval.

If x_0 is a point of local minimum of $f(x)$, then the graph of $f(x)$ around x_0 will be as shown in figure.

Here $f(x)$ is decreasing in $(x - h, x_0)$ and increasing in $(x_0, x_0 + h)$. So $f'(x) < 0$ in $(x_0 - h, x_0)$ and $f'(x) > 0$ in $(x_0, x_0 + h)$ suggesting that $f'(x)$ must be zero.

8.4.3. Conditions for Maxima and Minima

Expanding $f(x)$ by Taylor's theorem, we have

$$f(x) = f(a) + (x-a)\,f'(a) + \frac{(x-a)^2}{2!} f''(a) + \frac{(x-a)^3}{3!} f'''(a) + \ldots$$

or
$$f(x) - f(a) = (x-a)\,f'(a) + \frac{1}{2}(x-a)^2\, f''(a) + \frac{1}{6}(x-a)^3 \times f'''(a) + \ldots$$

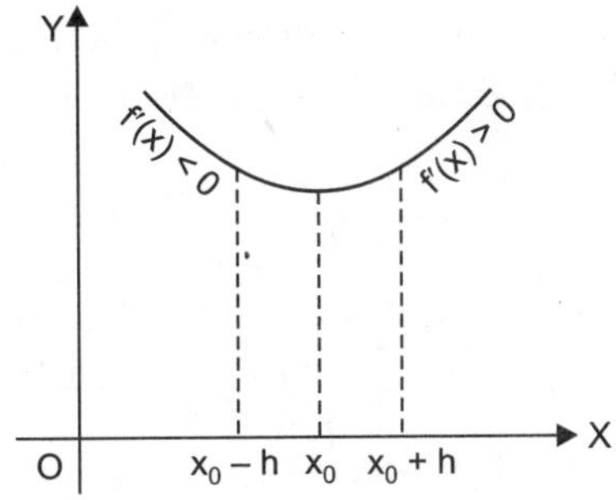

For sufficiently small values of $(x - a)$, $f'(a)$ is numerically much larger than the terms which follow. Thus the sign of $f(x) - f(a)$ depends on $(x - a)\,f'(a)$. It will have one sign when $x > a$ and another when $x < 0$. But $f(x) - f(a)$ must have one definite sign (–ve for max. and +ve for min.) whether $(x - a)$ is +ve or –ve. As such x_0 max. or min. is possible at $x = a$ unless

$$f'(a) = 0$$

When $f'(a) = 0$, we have

$$f(x) - f(a) = (x-a)^2\left[\frac{1}{2} f''(a) + \frac{1}{6}(x-a) f'''(a) + \ldots\right]$$

Again for sufficiently small values of $(x - a)$, $\frac{1}{2} f''(a)$ is much larger numerically than the terms which follow. Thus the sign of $f(x) - f(a)$ depends on $\frac{1}{2}(x - a)^2 f''(a)$ or $f''(a)$, since $(x - a)^2$ is always positive. Therefore for max. value of $f(x)$ at $x = a$, $f''(a)$ must be –ve and for min. value of $f(x)$ at $x = a$, $f''(a)$ must be +ve.

In case $f'(a) = 0$ and also $f''(a) = 0$, then following the above argument, we should have $f'''(a) = 0$ for max. or min. value of $f(x)$ at $x = a$. Then +ve sign of $f^{iv}(a)$ will decide whether $f(x)$ is max. or min. at $x = a$.

Working Rule for Finding Maxima and Minima

Step I: Denote the given function by $f(x)$.

Step II: Find $f'(x)$ and put it equal to zero.

Step III: Solve $f'(x) = 0$. Let its roots be $x = a, b, c, \ldots$

Step IV: Now find $f''(x)$ at $x = a, b, c, \ldots$ If $f''(a)$ is –ve, $f(x)$ is max. at $x = a$. If $f''(a)$ is +ve, $f(x)$ is min. at $x = a$. Similarly test at $x = b, c, \ldots$..

Step V: In case $f''(a) = 0$, find $f'''(x)$. If $f'''(a) \neq 0$ there is neither max. or min. at $x = a$. If $f'''(a) = 0$, find $f^{iv}(x)$. In case $f^{iv}(a)$ is –ve, $f(x)$ is max. at $x = a$ and if it is +ve, $f(x)$ is min. at $x = a$.

Step VI: This procedure is continued till we arrive at some conclusion.

SOLVED EXAMPLES

Example 8.9. Find all the points of local maxima and minima and corresponding maximum and minimum values of the function

$$f(x) = 2x^3 - 21x^2 + 36x - 20$$

Solution. We have $\quad f(x) = 2x^3 - 21x^2 + 36x - 20$

$\therefore \quad f'(x) = 6x^2 - 42x + 36$

For local maximum or local minimum,

$$f'(x) = 0 \Rightarrow 6x^2 - 42x + 36 = 0 \Rightarrow (x - 1)(x - 6) \Rightarrow x = 1, 6$$

Thus, $x = 1$ and $x = 6$ are the possible points of local maxima or minima.

Now test the function at each of these points.

We have $\quad f''(x) = 12x - 42$

When $x = 1$: We have $f''(1) = 12 - 42 = -30 < 0$. So, $x = 1$ is a point of local maximum.

The local maximum value is $f(1) = 2 - 21 + 36 - 20 = -3$

When $x = 6$: We have $f''(6) = 12(6) - 42 = 30 > 0$. So $x = 6$ is a point of local minimum.

The local minimum value is $f(6) = 2(6)^3 - 21(6)^2 + 36 \times 6 - 20 = -128$.

Example 8.10. Investigate for maximum and minium values of the function

$$f(x) = \sin x + \frac{1}{2}\cos 2x, \text{ where } 0 \leq x \leq \frac{\pi}{2}.$$

Solution. We have $\quad f(x) = \sin x + \frac{1}{2}\cos 2x$, where $0 \leq x \leq \frac{\pi}{2}$

$\therefore \quad f'(x) = \cos x - \sin 2x$

For local maximum or minimum,

$$f'(x) = 0$$

$$\Rightarrow \quad \cos x - \sin 2x = 0 \quad \Rightarrow \quad \cos x - 2\sin x \cos x = 0$$

$$\Rightarrow \quad \cos x(1 - 2\sin x) = 0 \quad \Rightarrow \quad \cos x = 0 \text{ or } 1 - 2\sin x = 0$$

$$\Rightarrow \quad \cos x = 0 \text{ or } \sin x = \frac{1}{2} \Rightarrow x = \frac{\pi}{2} \text{ or } x = \frac{\pi}{6} \qquad \left(\because 0 \le x \le \frac{\pi}{2}\right).$$

Thus, $x = \frac{\pi}{6}$ and $x = \frac{\pi}{2}$ are possible points of local maxima or local minima.

Now we test the function at each of these points.

We have $\quad f''(x) = -\sin x - 2\cos 2x$

When $x = \frac{\pi}{6}$: We have $f''\left(\frac{\pi}{6}\right) = \sin\frac{\pi}{6} + \frac{1}{2}\cos\frac{\pi}{3} = \frac{1}{2} + \frac{1}{2}\left(\frac{1}{2}\right) = \frac{3}{4}$

So, $x = \frac{\pi}{6}$ is the point of local maximum.

The local maximum value is

$$f\left(\frac{\pi}{6}\right) = \sin\frac{\pi}{6} + \frac{1}{2}\cos\frac{\pi}{3} = \frac{1}{2} + \frac{1}{2}\left(\frac{1}{2}\right) = \frac{3}{4}$$

When $x = \frac{\pi}{2}$: We have $f''\left(\frac{\pi}{2}\right) = -\sin\frac{\pi}{2} - 2\cos\pi = -1 + 2 = 1 > 0.$

So, $x = \frac{\pi}{2}$ is the point of local minimum.

The local minimum value is

$$f\left(\frac{\pi}{2}\right) = \sin\frac{\pi}{2} + \frac{1}{2}\cos\pi = 1 - \frac{1}{2} = \frac{1}{2}.$$

Example 8.11. If $y = \frac{ax - b}{(x-1)(x-4)}$ has a turning point $P(2, -1)$, find the values of a and b and show that y is maximum at P.

Solution. We have
$$y = \frac{ax - b}{(x-1)(x-4)} = \frac{ax - b}{x^2 - 5x + 4} \qquad \ldots(i)$$

$$\therefore \quad \frac{dy}{dx} = \frac{(x^2 - 5x + 4).a - (ax - b)(2x - 5)}{(x^2 - 5x + 4)^2} \qquad \ldots(ii)$$

$$\Rightarrow \quad \left(\frac{dy}{dx}\right)_P = \frac{(4 - 10 + 4)a - (2a - b)(4 - 5)}{(4 - 10 + 4)^2} = \frac{-b}{4}$$

Since P is a turning point of the curve (*i*), therefore

$$\left(\frac{dy}{dx}\right)_P = 0 \quad \Rightarrow \quad \frac{-b}{4} = 0 \quad \Rightarrow \quad b = 0 \qquad \ldots(iii)$$

Since, $P(2, -1)$ lies on $\quad y = \frac{ax - b}{(x-1)(x-4)}$. Therefore,

$$-1 = \frac{2a - b}{(2-1)(2-4)} \quad \Rightarrow \quad -1 = \frac{2a - b}{-2} \Rightarrow 2a - b = 2 \qquad \ldots(iv)$$

From (*iii*) and (*iv*), we get $a = 1, b = 0$.

Substituting the values of a and b in (ii), we get

$$\frac{dy}{dx} = \frac{(x^2-5x+4)-x(2x-5)}{(x^2-5x+4)^2} = \frac{-x^2+4}{(x^2-5x+4)^2}$$

$$\Rightarrow \quad \frac{d^2y}{dx^2} = \frac{(x^2-5x+4)^2(-2x)-(-x^2+4)\,2(x^2+5x+4)(2x-5)}{(x^2-5x+4)^4}$$

or

$$\frac{d^2y}{dx^2} = \frac{-2x(x^2-5x+4)+2(x^2-4)(2x-5)}{(x^2-5x+4)^{3-}}$$

Now $\quad \left(\frac{dy}{dx}\right)_{(2,1)} = 0$ and $\left[\frac{d^2y}{dx^2}\right]_{(1,2)} = \frac{(-2)(-4)}{(-2)^3} = -1 < 0$

So, y is maximum at P.

Example 8.12. Show that all the rectangles with a given perimeter, the square has the largest area.

Solution. Let x and y be the lengths of two sides of the rectangle of fixed parameter P, and let A be its area. Then

$$P = 2(x+y) \qquad \text{[Given condition]} \qquad ...(i)$$

and

$$A = xy \qquad \text{[to be maximized]} \qquad ...(ii)$$

Now, $\quad P = 2(x+y) \Rightarrow y = \frac{P}{2} - x$

$$\therefore \quad A = xy = x\left(\frac{P}{2}-x\right) = \frac{Px}{2} - x^2$$

$$\Rightarrow \quad \frac{dA}{dx} = \frac{P}{2} - 2x \quad \text{and} \quad \frac{d^2A}{dx^2} = -2$$

For maximum or minimum, we have $\frac{dA}{dx} = 0$

$$\therefore \quad \frac{dA}{dx} = 0 \Rightarrow \frac{P}{2} - 2x - 2x = 0 \Rightarrow P = 4x \Rightarrow 2x + 2y = 4x$$

$$\Rightarrow \quad 2x = 2y \Rightarrow x = y$$

Now, $\quad \left(\frac{d^2y}{dx^2}\right)_{x=y} = -2 < 0.$

Hence A is maximum when $x = y$ *i.e.* the rectangle is a square.

Example 8.13. A window is in the form of a rectangle surmounted by a semi-circular opening. The total perimeter of the window is 10 metres. Find the dimensions of the window to admit maximum light through it.

Solution. Let $ABCD$ be a rectangle and let the semi-circle be described on side AB as diameter. Let $AB = 2x$ and $AD = 2y$. Let A be the area of the figure. Then,

$$10 = 2x + 4y + \pi x \qquad \text{[Given condition]} \qquad ...(i)$$

and

$$A = (2x)(2y) + \frac{\pi x^2}{2} \qquad \text{[To be maximized]} \qquad ...(ii)$$

$$\Rightarrow \quad A = 4xy + \frac{\pi x^2}{2}$$

$$\Rightarrow \quad A = x[10 - 2x - \pi x] + \frac{\pi x^2}{2} \qquad \text{[Using } (i)\text{]}$$

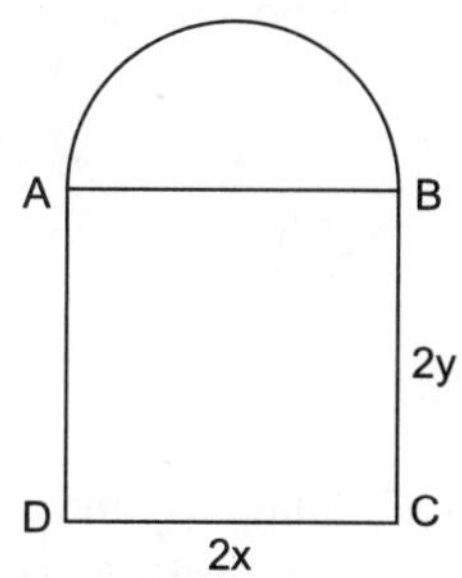

$$\Rightarrow \quad A = 10x - 2x^2 - \pi x^2 + \frac{\pi x^2}{2} \Rightarrow A = 10x - 2x^2 - \frac{\pi x^2}{2}$$

$$\Rightarrow \quad \frac{dA}{dx} = 10 - 4x - \pi x \text{ and } \frac{d^2A}{dx^2} = -4 - \pi$$

For maximum or minimum A, $\frac{dA}{dx} = 0$.

$$\therefore \quad \frac{dA}{dx} = 0$$

$$\Rightarrow \quad 10 - 4x - \pi x = 0 \Rightarrow x = \frac{10}{\pi + 4}$$

Now, $$\frac{d^2A}{dx^2} = -4 - \pi < 0 \text{ for all value of } x.$$

Thus, A is maximum when $x = \frac{10}{\pi+4}$. Putting $x = \frac{10}{\pi+4}$ in (*i*), we get $y = \frac{5}{\pi+4}$.

So, dimensions of the figure are $2x = \frac{20}{\pi+4}$ and $2y = \frac{10}{\pi+4}$.

Example 8.14. Show that the semi-vertical angle of the cone of maximum volume and of given slant height is $\tan^{-1}\sqrt{2}$.

Solution. Let θ be the semi-vertical angle of a cone of given slant height l. Let r be the radius and h be the height of the cone. Let V denotes the volume of the conc. Then

$$l^2 = r^2 + h^2 \qquad \text{[Given condition]} \qquad ...(i)$$

and $$V = \frac{1}{3}\pi r^2 h \qquad \text{[To be maximized]} \qquad ...(ii)$$

Now $$r^2 = l^2 - h^2$$

$$\therefore \quad V = \frac{1}{3}\pi(l^2 - h^2)h$$

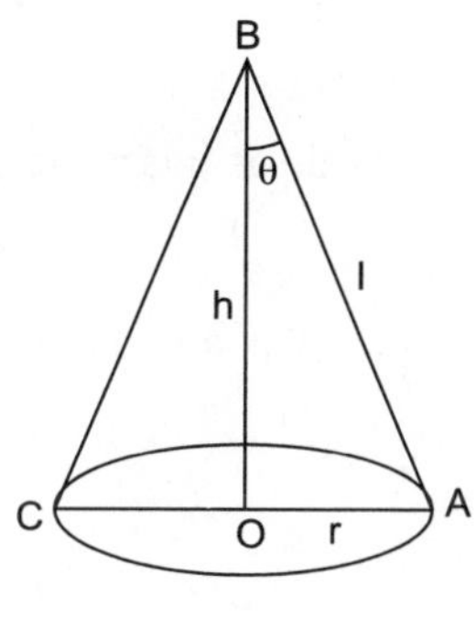

$$V = \frac{1}{3}\pi(l^2h - h^3)$$

$$\Rightarrow \quad \frac{dV}{dh} = \frac{1}{3}\pi(l^2 - 3h^2)$$

and $$\frac{d^2V}{dh^2} = \frac{1}{3}\pi(-6h) = -2\pi h$$

For maxima or minima, we have $\frac{dV}{dh} = 0$

$$\therefore \quad \frac{dV}{dh} = 0 \Rightarrow \frac{1}{3}\pi(l^2 - 3h^2) = 0 \Rightarrow l^2 - 3h^2 \Rightarrow l^2 = 3h^2 \qquad ...(iii)$$

Now, $$\left(\frac{d^2V}{dh^2}\right)_{h=\frac{1}{\sqrt{3}}} = -2\pi\times\frac{1}{\sqrt{3}} = \frac{-2\pi}{\sqrt{3l}} < 0$$

Hence V is maximum when $l^2 = 3h^2$.

From (*i*) and (*iii*), we have

$$r^2 + h^2 = 3h^2 \quad \Rightarrow \quad r^2 = 2h^2$$

$$\Rightarrow \quad \left(\frac{r}{h}\right)^2 = 2 \quad \Rightarrow \quad \frac{r}{h} = \sqrt{2}$$

$$\Rightarrow \quad \tan\theta = \sqrt{2} \quad \Rightarrow \quad \theta = \tan^{-1}\sqrt{2}$$

Thus, V is maximum when $\theta = \tan^{-1}\sqrt{2}$ *i.e.* when the semi-vertical angle of the cone is $\tan^{-1}\sqrt{2}$.

Example 8.15. A point in the hypotenuse of a right angled triangle is at distance *a* and *b* from the sides, show that the length of the hypotenuse have the minimum value $\left(a^{2/3}+b^{2/3}\right)^{3/2}$.

Solution. Let P be the point in the hypotenuse AB, whose distance from the sides are a and b thus

$PM = a$ and $PN = b$. Again, let y be the length of the hypotenuse.

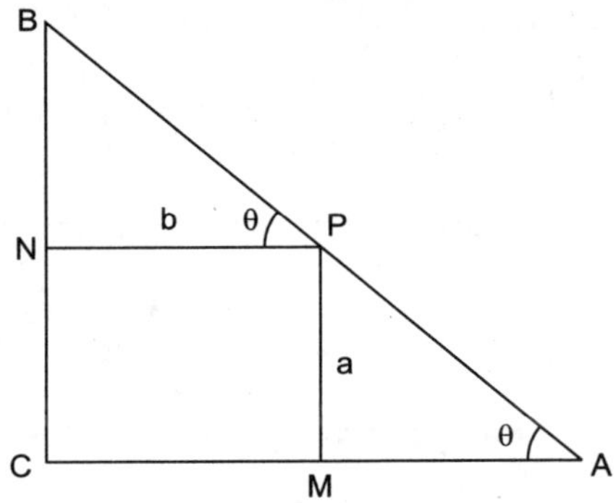

$\therefore$ $$y = AP + PB$$

or $$y = a \operatorname{cosec}\theta + b\sec\theta \quad ..(i)$$

$\therefore$ $$\frac{dy}{d\theta} = -a\operatorname{cosec}\theta\cot\theta + b\sec\theta\tan\theta$$

For maxima or minima, $\frac{dy}{d\theta} = 0$

$\therefore$ $b\sec\theta\tan\theta - a\operatorname{cosec}\theta\cot\theta = 0$

or $$b\sec\theta\tan\theta = a\operatorname{cosec}\theta\cot\theta$$

or $$\tan^3\theta = \frac{a}{b} \quad ...(ii)$$

Now $$\frac{d^2y}{d\theta^2} = a(\operatorname{cosec}^3\theta + \operatorname{cosec}\theta\cot^2\theta) + (\sec^3\theta + \sec\theta\tan^2\theta)$$

Again as θ is an acute angle, $\frac{d^2y}{d\theta^2}$ is +ve.

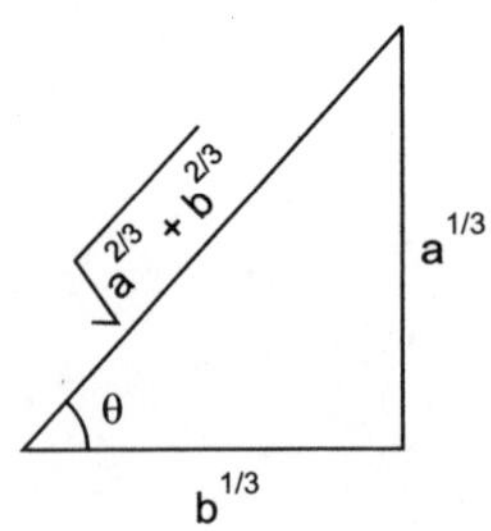

$\therefore$ y is minimum for $\tan\theta = \left(\frac{a}{b}\right)^{1/3}$ [From (*iii*)]

or $$\tan\theta = \frac{a^{1/3}}{b^{1/3}}$$

$\therefore$ $$\sec\theta = \frac{\sqrt{a^{2/3}+b^{2/3}}}{b^{1/3}} \quad \text{and} \quad \operatorname{cosec}\theta = \frac{\sqrt{a^{2/3}+b^{2/3}}}{a^{1/3}}$$

$\therefore$ The minimum value of the length of the hypotenuse

$$= a \cdot \frac{\sqrt{a^{2/3} + b^{2/3}}}{a^{1/3}} + b \cdot \frac{\sqrt{a^{2/3} + b^{2/3}}}{b^{1/3}}$$

$= (a^{2/3} + b^{2/3})^{3/2}$ which is the required result.

EXERCISE 8.3

1. Find the points of local maxima or local minima and corresponding local maximum and local minimum values of each of the following functions.

   (i) $f(x) = x^3 - 6x^2 + 9x + 15$ [**Ans.** Local max. at $x = 1$; Local max. value = 19; Local min at $x = 3$; Local min. value = 15]

   (ii) $f(x) = (x - 1)(x + 2)^2$ [**Ans.** Local max. at $x = -2$; Local max. value = 0; Local min at $x = 0$; Local min. value = –4]

2. Find the maximum profit that a company can make, if the profit function is given by

   $P(x) = 41 + 24x - 18x^2$. [**Ans.** Max. profit = 49]

3. Show that the surface area of a closed cuboid with square base and given volume is minimum, where it is a cube.

4. Find the radius of a closed right circular cylinder of volume 100 cubic centimeters which has the minimum total surface area. $\left[\textbf{Ans. } r = \left(\frac{50}{\pi}\right)^{1/3}\right]$

5. Find the area of the greatest isosceles triangle that can be inscribed in a given ellipse having its vertex coincident with one extremity of the major axis. $\left[\textbf{Ans. } \text{Area} = \frac{3\sqrt{3}}{4} ab\right]$

❑❑❑

CHAPTER 9

Partial Differentiation

INSIDE THIS CHAPTER

9.1. FUNCTION OF TWO VARIABLES

If to each point (x, y) in a certain subject of the two-dimensional plane there corresponds one and only value of z, then the correspondence is called a function of two variables. Just as in the one independent variable case, this correspondence is often represented by an equation. However, the equation is now of the form

$$z = f(x, y)$$

where z is the dependent variable, and x and y are the independent variable. The set of the allowable values for the independent variables constitutes the domain of the function. For example, the equation

$$z = f(x, y) = \frac{x^3 + y^3}{x - 2}$$

defines z as a function of two variables z and y. Because the denominator is zero when $x = 2$, the domain of f is all ordered pairs (x, y) such that $x \neq 2$.

We may extend the concept from two variables to three or more variables. Thus $u = f(x, y, z)$ denotes the value of a function at (x, y, z) etc.

9.2. PARTIAL DERIVATIVES

Consider a function if two independent variables written in the form $z = f(x, y)$. We might which to study the rate at which z changes as x changes if y is held constant. By holding y as a constant, $z = f(x, y)$ essentially becomes a function of x alone and we can calculate its derivative with respect to x. This derivative is called the partial derivative of z (or f) with respect to x and is denoted by

$$\frac{\partial z}{\partial x}, \frac{\partial}{\partial x}[f(x, y)], f_x(x, y) \text{ or } z_x(x, y)$$

Similarly, the partial derivatives of z (or f) with respect to y is denoted by

$$\frac{\partial z}{\partial y}, \frac{\partial}{\partial y}[f(x, y)], f_y(x, y) \text{ or } z_y(x, y)$$

and gives the rate of change of z with respect to y, with x held constant. In terms of limits, we have the following:

Definition: *If* $z = f(x, y)$, *the partial derivative of* z *with respect to* x, *denoted by* $\frac{\partial z}{\partial x}$, *is the function given by*

$$\frac{\partial z}{\partial x} = \lim_{k \to 0} \frac{f(x, y + k) - f(x, y)}{k}$$

provided this limit exists.

The partial derivative of z *with respect to* y, *denoted by* $\frac{\partial z}{\partial y}$, *is the function given by*

$$\frac{\partial z}{\partial x} = \lim_{k \to 0} \frac{f(x, y, + k) - f(x, y)}{k}$$

provided this limit exists.

Remarks:

1. From the definition, we see that to find $\frac{\partial z}{\partial x}$, we treat y as a constant and differentiate z with respect to x in the usual way. Similarly, to find $\frac{\partial z}{\partial y}$, we treat x as a constant and differentiate z with respect to y.
2. We use the letter ∂ (delta) rather than '*d*' to show that the other variables in the function have been held constant.
3. $\frac{\partial z}{\partial x}$ and $\frac{\partial z}{\partial y}$ are also called the first order partial derivatives of z.
4. In the one-variable case, the derivative measures the instantaneous rate of change of the function. In the two-variable case $\frac{\partial z}{\partial x}$ measures the instantaneous rate of change of z with respect to x(y is held constant) and $\frac{\partial z}{\partial y}$ measures the instantaneous rate of change of z with respect to y (x is held constant).

9.3. SECOND ORDER PARTIAL DERIVATIVES

If $z = f(x, y)$, then not only is z a function of x and y, but also $\frac{\partial z}{\partial x}$ and $\frac{\partial z}{\partial y}$ are each function of x and y.

This we can form partial derivatives of $\frac{\partial z}{\partial x}$ and $\frac{\partial z}{\partial y}$ just as we formed those of z to obtain the following second-order partial dervishes of z:

(*i*) $\frac{\partial}{\partial x}\left(\frac{\partial z}{\partial x}\right)$, denoted by $\frac{\partial^2 z}{\partial x^2}$ or f_{xx}.

(*ii*) $\frac{\partial}{\partial y}\left(\frac{\partial z}{\partial x}\right)$, denoted by $\frac{\partial^2 z}{\partial y\, \partial x}$ or f_{xy}.

(*iii*) $\frac{\partial}{\partial x}\left(\frac{\partial z}{\partial y}\right)$, denoted by $\frac{\partial^2 z}{\partial x \partial y}$ or f_{xy}.

(*iv*) $\frac{\partial}{\partial y}\left(\frac{\partial z}{\partial y}\right)$, denoted by $\frac{\partial^2 z}{\partial y^2}$ or f_{yy}.

Note 1. To find $\frac{\partial^2 z}{\partial x^2}\left(\text{respectively } \frac{\partial^2 z}{\partial y^2}\right)$, we take two successive derivative, each time treating y (respectively x) as a constant. Similarly, to find $\frac{\partial^2 z}{\partial y \partial x}\left(\text{respectively } \frac{\partial^2 z}{\partial x\, \partial y}\right)$, we take two successive derivatives, whereas, in the first differentiation y (respectively x) is treated as a constant and in the second differentiation x (respectively y) is treated as a constant.

Note 2. The derivatives $\frac{\partial^2 z}{\partial x \partial y}$ and $\frac{\partial^2 z}{\partial y \partial x}$ are called **mixed or cross partial derivatives.**

Usually, but not always $\frac{\partial^2 z}{\partial x \partial y} = \frac{\partial^2 z}{\partial y \partial x}$. That is, the order of partial differentiation is immaterial. We assume that this is the case for all functions that we consider.

SOLVED EXAMPLES

Example 9.1. Find the first order partial derivatives of

(*i*) $\log (x^2 + y^2)$ (*ii*) $x^2 \sin \frac{y}{x}$ (*iii*) $\tan^{-1}\left(\frac{x}{y}\right)$ (*iv*) $\tan^{-1}\left(\frac{x^2 + y^2}{x + y}\right)$.

Solution. (*i*) Let $z = \log (x^2 + y^2)$

Then $$\frac{\partial z}{\partial x} = \frac{1}{x^2 + y^2} \cdot \frac{\partial z}{\partial x}(x^2 + y^2)$$

or $$\frac{\partial z}{\partial x} = \frac{1}{x^2 + y^2} \cdot (2x) = \frac{2x}{x^2 + y^2}$$

Similarly $$\frac{\partial z}{\partial y} = \frac{2y}{x^2 + y^2}$$

(*ii*) Let $$z = x^2 \sin \frac{y}{x}$$

Then $$\frac{\partial z}{\partial x} = \frac{\partial}{\partial x}\left[x^2 \sin \frac{y}{x}\right] = \left[2x \,.\sin \frac{y}{x} - x^2 \,.\cos \frac{y}{x} \cdot \left(\frac{y}{x^2}\right)\right]$$

$$= \left(2x \,.\sin \frac{y}{x} - \cos \frac{y}{x}\right)$$

Similarly $$\frac{\partial z}{\partial y} = \frac{\partial}{\partial y}\left[x^2 \sin \frac{y}{x}\right] = x^2 \,.\cos \frac{y}{x} \cdot \frac{1}{x} = x \,.\cos \frac{y}{x}$$

(*iii*) Let $$z = \tan^{-1}\left(\frac{x}{y}\right)$$

Then $$\frac{\partial z}{\partial x} = \frac{1}{1 + \left(\frac{x}{y}\right)^2} \cdot \frac{\partial}{\partial x}\left(\frac{x}{y}\right) = \frac{y^2}{y^2 + x^2} \cdot \frac{1}{y} = \frac{y}{x^2 + y^2}$$

and $$\frac{\partial z}{\partial y} = \frac{1}{1+\left(\frac{x}{y}\right)^2}\cdot\frac{\partial}{\partial y}\left(\frac{x}{y}\right) = \frac{y^2}{y^2+x^2}\cdot\left(-\frac{x}{y^2}\right) = \frac{-x}{x^2+y^2}$$

(*iv*) Let $$z = \tan^{-1}\left(\frac{x^2+y^2}{x+y}\right)$$

Then $$\frac{\partial z}{\partial x} = \frac{1}{1+\left(\frac{x^2+y^2}{x+y}\right)^2}\cdot\frac{\partial}{\partial x}\left(\frac{x^2+y^2}{x+y}\right)$$

$$= \frac{(x+y)^2}{(x+y)^2+(x^2+y^2)^2}\cdot\frac{(x+y)\cdot\frac{\partial}{\partial x}(x^2+y^2)-(x^2+y^2)\cdot\frac{\partial}{\partial x}(x+y)}{(x+y)^2}$$

$$= \frac{(x+y)^2}{(x+y)^2+(x^2+y^2)^2}\cdot\frac{(x+y)(2x)-(x^2+y^2)(1)}{(x+y)^2}$$

$$\therefore \quad \frac{\partial z}{\partial x} = \frac{x^2+2xy-y^2}{(x+y)^2+(x^2+y^2)^2}$$

Since z is symmetrical w.r.t. x and y,

Similarly, $$\frac{\partial z}{\partial y} = \frac{y^2+2xy-x^2}{(x+y)^2+(x^2+y^2)^2}$$

Example 9.2. If $u = \log(x^3+y^2+z^3-3xyz)$, show that $\left(\frac{\partial}{\partial x}+\frac{\partial}{\partial y}+\frac{\partial}{\partial z}\right)^2 u = -\frac{9}{(x+y+z)^2}$.

Solution. We have $u = \log(x^3+y^3+z^3-3xyz)$

$$\frac{\partial u}{\partial x} = \frac{3x^2-3yz}{x^3+y^3+z^3-3xyz};\quad \frac{\partial u}{\partial y} = \frac{3y^2-3xz}{x^3+y^3+z^3-3xyz};\quad \frac{\partial u}{\partial z} = \frac{3z^2-3xy}{x^3+y^3+z^3-3xy}$$

Adding, $$\frac{\partial u}{\partial x}+\frac{\partial u}{\partial y}+\frac{\partial u}{\partial z} = \frac{3(x^2+y^2+z^2-xy-yz-zx)}{x^3+y^3+z^3-3xyz} = \frac{3}{x+y+z}$$

$[\because\ x^3+y^3+z^3-3xyz = (x+y+z)(x^2+y^2+z^2-xy-yz-zx)]$

Now, $$\left(\frac{\partial}{\partial x}+\frac{\partial}{\partial y}+\frac{\partial}{\partial z}\right)^2 u = \left(\frac{\partial}{\partial x}+\frac{\partial}{\partial y}+\frac{\partial}{\partial z}\right)\left(\frac{\partial}{\partial x}+\frac{\partial}{\partial y}+\frac{\partial}{\partial z}\right)u$$

$$= \left(\frac{\partial}{\partial x}+\frac{\partial}{\partial y}+\frac{\partial}{\partial z}\right)\left(\frac{\partial u}{\partial x}+\frac{\partial u}{\partial y}+\frac{\partial u}{\partial z}\right) = \left(\frac{\partial}{\partial x}+\frac{\partial}{\partial y}+\frac{\partial}{\partial z}\right)\left(\frac{3}{x+y+z}\right)$$

$$= -\frac{3}{(x+y+z)^2}-\frac{3}{(x+y+z)^2}-\frac{3}{(x+y+z)^2}$$

$$\therefore \quad \left(\frac{\partial}{\partial x}+\frac{\partial}{\partial y}+\frac{\partial}{\partial z}\right)^2 u = -\frac{9}{(x+y+z)^2}$$

Hence the result.

Example 9.3. If $x^x y^y z^z = c$, show that at $x = y = z$, $\dfrac{\partial^2}{\partial x\,\partial y} = -(x \log ex)^{-1}$.

Solution. We have $x^x y^y z^z = c \Rightarrow z^z = cx^{-x}y^{-y}$

Taking logarithms on both the sides, we obtain

$$z \log z = \log c - x \log x - y \log y$$

Differentiating partially w.r.t. y, we obtain

$$\left(z.\frac{1}{z} + 1.\log z\right)\frac{\partial z}{\partial y} = 0 - 0 - \left(y.\frac{1}{y} + 1.\log y\right)$$

or
$$\frac{\partial z}{\partial y} = -\frac{1+\log y}{1+\log z} = -\frac{\log e + \log y}{\log e + \log y}$$

∴
$$\frac{\partial z}{\partial y} = -\frac{\log ey}{\log ez}.$$

Similarly,
$$\frac{\partial z}{\partial x} = -\frac{\log ex}{\log ez} \qquad ...(i)$$

Now
$$\frac{\partial^2 z}{\partial x\,\partial y} = \frac{\partial}{\partial x}\left(-\frac{\log ey}{\log ez}\right) = -\log ey.\frac{\partial}{\partial x}(\log ez)^{-1}$$

$$= \log ey\,(\log ez)^{-2}.\frac{e}{ez}.\frac{\partial z}{\partial x}$$

$$= -\frac{\log ey}{(\log ez)}.\frac{1}{z}.\frac{\log ex}{\log ez} \qquad \text{[Using } (i)\text{]}$$

At $z = x = y$, we get
$$\frac{\partial^2 z}{\partial x\,\partial y} = -\frac{\log ex}{(\log ex)^2}.\frac{1}{x}.\frac{\log ex}{\log ex} = -\frac{1}{x \log ex}$$

Hence
$$\frac{\partial^2 z}{\partial x\,\partial y} = (x \log ex)^{-1}.$$

Example 9.4. If $\theta = t^n e^{-\frac{r^2}{4t}}$, find the value of n which makes $\dfrac{1}{r^2}\dfrac{\partial}{\partial r}\left(r^2 \dfrac{\partial\theta}{\partial r}\right) = \dfrac{\partial\theta}{\partial t}$.

Solution.
$$\theta = t^n.e^{-\frac{r^2}{4t}}$$

$$\frac{\partial\theta}{dt} = n.t^{n-1}.e^{-\frac{r^2}{4t}} + t^n.e^{-\frac{r^2}{dt}}\frac{\partial}{\partial t}\left(\frac{-r^2}{4t}\right)$$

$$= nt^{n-1}e^{-\frac{r^2}{4t}} + t^n.e^{-\frac{r^2}{4t}}.\frac{r^2}{4t^2}$$

$$= nt^{n-1}e^{-\frac{r^2}{4t}} + \frac{r^2}{4}.t^{n-2}e^{-\frac{r^2}{4t}} \qquad ...(i)$$

Again $$\frac{\partial \theta}{\partial r} = t^n . e^{-\frac{r^2}{4t}} . \frac{\partial}{\partial r}\left(-\frac{r^2}{4t}\right) = t^n . e^{-\frac{r^2}{4t}} . \left(-\frac{2r}{4t}\right)$$

$$= -\frac{r}{2} t^{n-1} . e^{-\frac{r^2}{4t}} \qquad ...(ii)$$

$$r^2 \frac{\partial \theta}{\partial r} = -\frac{r^3}{2} t^{n-1} . e^{-\frac{r^2}{4t}} \qquad \text{[Multiplying } (ii) \text{ by } r^2]$$

$$\frac{\partial}{\partial r}\left(r^2 \frac{\partial \theta}{\partial r}\right) = \frac{\partial}{\partial r}\left(\frac{r^3}{2} t^{n-1} . e^{-\frac{r^2}{4t}}\right) = -\frac{t^{n-1}}{2} \frac{\partial}{\partial r}\left(r^3 . e^{-\frac{r^2}{4t}}\right)$$

$$= -\frac{t^{n-1}}{2}\left\{3r^2 . e^{-\frac{r^2}{4t}} + r^3 . e^{-\frac{r^2}{4t}}\left(\frac{-2r}{4t}\right)\right\}$$

$$= -\frac{t^{n-1}}{2}\left\{3r^2 . e^{-\frac{r^2}{4t}} - \frac{r^4}{2t} e^{-\frac{r^2}{4t}}\right\} \qquad ...(iii)$$

$$\frac{1}{r^2}\frac{\partial}{\partial r}\left(r^2 \frac{\partial \theta}{\partial r}\right) = -\frac{t^{n-1}}{2}\left\{3e^{-\frac{r^2}{4t}} - \frac{r^2}{2t} e^{-\frac{r^2}{4t}}\right\} \qquad \text{[dividing } (iii) \text{ by } r^2]$$

Given $$\frac{1}{r^2}\frac{\partial}{\partial r}\left(r^2 \frac{\partial \theta}{\partial r}\right) = \frac{\partial \theta}{\partial t}$$

$$\therefore \quad -\frac{3}{2} t^{n-1} e^{-\frac{r^2}{4t}} + \frac{r^2}{4} t^{n-2} . e^{-\frac{r^2}{4t}} = n t^{n-1} e^{-\frac{r^2}{4t}} + \frac{r^2}{4} t^{n-2} . e^{-\frac{r^2}{4t}}$$

$$\Rightarrow \quad t^{n-2} e^{-\frac{r^2}{4t}}\left(-\frac{3}{2}t + \frac{r^2}{4}\right) = t^{n-2} e^{-\frac{r^2}{4t}}\left(nt + \frac{r^2}{4}\right)$$

$$\Rightarrow \quad -\frac{3}{2}t + \frac{r^2}{4} = nt + \frac{r^2}{4}$$

$$\Rightarrow \quad -\frac{3}{2}t = nt \quad \therefore \quad n = -\frac{3}{2}.$$

EXERCISE 9.1

1. If $z = x^3 + x^2y + y^3$, prove that $x\frac{\partial z}{\partial x} + y\frac{\partial z}{\partial y} = 3z$.

2. If $u = \log (x^2 + y^2)$, then prove that $\frac{\partial^2 u}{\partial x^2} + \frac{\partial^2 u}{\partial y^2} = 0$.

3. If $u = \sin^{-1}\frac{x}{y} + \tan^{-1}\frac{y}{x}$, prove that $x\frac{\partial u}{\partial x} + y\frac{\partial u}{\partial y} = 0$.

4. If $u = \sin (y + ax) - (y + ax)^2$, prove that $\frac{\partial^2 u}{\partial x^2} + \frac{\partial^2 u}{\partial y^2}$.

5. If $u = \tan^{-1}\left(\frac{y}{x}\right) + \log (x^2 + y^2)$, prove that $\frac{\partial^2 u}{\partial x^2} + \frac{\partial^2 u}{\partial y^2} = 0$.

6. If $u = e^{xyz}$, prove that $\dfrac{\partial^3 u}{\partial x\, \partial y\, \partial z} = (1 + 3xyz + x^2 y^2 z^3)e^{xyz}$.

7. If $V = (x^2 + y^2 + z^2)^{-1/2}$, prove that $\dfrac{\partial^2 V}{\partial x^2} + \dfrac{\partial^2 V}{\partial y^2} + \dfrac{\partial^2 V}{\partial z^2} = 0$.

8. If $z = xf(x + y) + yg(x + y)$, show that $\dfrac{\partial^2 z}{\partial x^2} - 2\dfrac{\partial^2 z}{\partial x\, \partial y} + \dfrac{\partial^2 z}{\partial y^2} = 0$.

9. If $x = e^{r\cos\theta} \cos(r \sin \theta)$ and $y = e^{r\cos\theta} \sin(r \sin\theta)$, prove that $\dfrac{\partial x}{\partial r} = \dfrac{1}{r}\dfrac{\partial y}{\partial \theta}, \dfrac{\partial y}{\partial r} = -\dfrac{1}{r}\dfrac{\partial x}{\partial \theta}$ and hence deduce that $\dfrac{\partial^2 x}{\partial r^2} + \dfrac{1}{r}\dfrac{\partial x}{\partial r} + \dfrac{1}{r^2}\dfrac{\partial^2 x}{\partial \theta^2} = 0$.

10. If $x = r \cos \theta, y = r \sin \theta$, prove that

(*i*) $\dfrac{\partial r}{\partial x} = \dfrac{\partial x}{\partial r}$ (*ii*) $\dfrac{1}{r}\dfrac{\partial x}{\partial \theta} = r\dfrac{\partial \theta}{\partial x}$ (*iii*) $\dfrac{\partial^2 \theta}{\partial x^2} + \dfrac{\partial^2 \theta}{\partial y^2} = 0$.

9.4. HOMOGENEOUS FUNCTION

An expression of the form $a_0 x^n + a_1 x^{n-1} y + a_2 x^{n-2} y^2 + ... + a_n y^n$ in which every term is of the nth degree is called a homogeneous function of degree n. This can be rewritten as

$$x^n \left[a_0 + a_1\left(\frac{y}{x}\right) + a_2\left(\frac{y}{x}\right)^2 + ... + a_n\left(\frac{y}{x}\right)^n \right].$$

Thus any function $f(x, y)$ which can be expressed in the form $x^n \phi\left(\dfrac{y}{x}\right)$ is called a homogeneous function of degree n in x and y.

For instance, $ax^2 + 2hxy + b^2 = x^2 \left[a + 2h\left(\dfrac{y}{x}\right) + b\left(\dfrac{y}{x}\right)^2 \right]$ is a homogeneous function of degree 2, in x and y.

In general, a function $f(x, y, z, t, ...)$ is said to be a homogeneous function of degree n in $x, y, z, t, ...$ if it can be expressed in the form $x^n\, f\left(\dfrac{y}{x}, \dfrac{z}{x}, \dfrac{t}{x}, ...\right)$.

9.5. EULER'S THEOREM ON HOMOGENEOUS FUNCTIONS

Theorem: If $z = f(x, y)$ is homogeneous function of x and y of degree n, then

$$x\frac{\partial z}{\partial x} + y\frac{\partial z}{\partial y} = nz.$$

Proof: Since $z = f(x, y)$ is a homogeneous function of degree n.

$$z = x^n\, \phi\left(\frac{y}{x}\right) \qquad ...(i)$$

Now, $$\frac{\partial z}{\partial x} = nx^{n-1}\,\phi\left(\frac{y}{x}\right) + x^n\,.\,\phi'\left(\frac{y}{x}\right).\left(-\frac{y}{x^2}\right)$$

and $$\frac{\partial z}{\partial y} = x^n\,\phi'\left(\frac{y}{x}\right).\frac{1}{x}$$

$\therefore$ $$x\frac{\partial z}{\partial x} + y\frac{\partial z}{\partial y} = nx^n\,\phi\left(\frac{y}{x}\right) - yx^{n-1}\,\phi'\left(\frac{y}{x}\right) + yx^{n-1}\,\phi'\left(\frac{y}{x}\right) = nx^n\,\phi\left(\frac{y}{x}\right) = nz \qquad \text{[Using } (i)]$$

Hence, $x\dfrac{\partial z}{\partial x} + y\dfrac{\partial z}{\partial y} = nz.$

SOLVED EXAMPLES

Example 9.5. Verify Euler's theorem for the function $z = \sin^{-1}\dfrac{x}{y} + \tan^{-1}\dfrac{y}{x}$.

Solution. $$z = x^0\sin^{-1}\frac{x}{y} + x^0\tan^{-1}\frac{y}{x} \qquad ...(i)$$

z is a homogeneous function of degree zero so by Euler's theorem

$$x\frac{\partial z}{\partial x} + y\frac{\partial z}{\partial y} = 0 \qquad ...(ii)$$

Let us find out the value of L.H.S. of (*ii*) by actual differentiation

Differentiating (*i*) w.r.t. x, we get

$$\frac{\partial z}{\partial x} = \frac{1}{\sqrt{1-\dfrac{x^2}{y^2}}}.\left(\frac{1}{y}\right) + \frac{1}{1+\dfrac{y^2}{x^2}}\left(-\frac{y}{x^2}\right) = \frac{1}{\sqrt{y^2-x^2}} - \frac{y}{\sqrt{x^2+y^2}}$$

$$x\frac{\partial z}{\partial x} = \frac{x}{\sqrt{y^2-x^2}} - \frac{xy}{x^2+y^2} \qquad ...(iii)$$

Differentiating (*i*) w.r.t. y, we get

$$\frac{\partial z}{\partial y} = \frac{1}{\sqrt{1-\dfrac{x^2}{y^2}}}.\left(-\frac{x}{y^2}\right) + \frac{1}{1+\dfrac{y^2}{x^2}}.\frac{1}{x} = \frac{-x}{y\sqrt{y^2-x^2}} + \frac{x}{x^2+y^2}$$

$$y\frac{\partial z}{\partial y} = -\frac{x}{\sqrt{y^2-x^2}} + \frac{xy}{x^2+y^2} \qquad ...(iv)$$

Adding (*iii*) and (*iv*), we get

$$x\frac{\partial z}{\partial x} + y\frac{\partial z}{\partial y} = 0$$ which is same as (*ii*), Hence the verification.

Example 9.6. If $z = \tan^{-1}\dfrac{x^3+y^3}{x-y}$, prove that $x\dfrac{\partial z}{\partial x} + y\dfrac{\partial z}{\partial y} = \sin 2z$.

Solution. We have $$z = \tan^{-1}\frac{x^3+y^3}{x-y}$$

$$\Rightarrow \qquad \tan z = \frac{x^3 + y^3}{x - y}.$$

Let $\qquad u = \tan z \qquad ...(i)$

$$\therefore \qquad u = \frac{x^3 + y^3}{x - y} = \frac{x^3\left[1+\left(\frac{y}{x}\right)^3\right]}{x\left[1-\frac{y}{x}\right]} = x^3\phi\left(\frac{y}{x}\right)$$

Thus, u is a homogeneous function of degree 2.

Hence by Euler's theorem,

$$x\frac{\partial u}{\partial x} + y\frac{\partial u}{\partial y} = 2u \qquad ...(ii)$$

From (*i*), $\qquad \frac{\partial u}{\partial x} = \sec^2 z\frac{\partial z}{\partial x}, \frac{\partial u}{\partial y} = \sec^2 z\frac{\partial z}{\partial y} \qquad ...(iii)$

Putting (*iii*) and (*i*) in (*ii*), we get

$$\sec^2 z\left(x\frac{\partial z}{\partial x} + y\frac{\partial z}{\partial y}\right) = 2\tan z$$

or

$$x\frac{\partial z}{\partial x} + y\frac{\partial z}{\partial y} = 2\frac{\sin z}{\cos z}.\cos^2 z = 2\sin z\cos z = \sin 2z$$

Example 9.7. If $u = \log\left(\frac{x^4 + y^4}{x + y}\right)$, show that $x\frac{\partial u}{\partial x} + y\frac{\partial u}{\partial y} = 3$.

Solution. We have $\qquad u = \log\left(\frac{x^4 + y^4}{x + y}\right)$

$$\Rightarrow \qquad e^u = \frac{x^4 + y^4}{x + y} = \frac{x^4\left[1+\left(\frac{y}{x}\right)^4\right]}{x\left[1+\frac{y}{x}\right]} = x^3\phi\left(\frac{y}{x}\right)$$

Let $z = e^u$

$$\therefore \qquad z = x^3\phi\left(\frac{y}{x}\right) \qquad ...(i)$$

Thus, z is a homogeneous function of degree 3.

Hence by Euler's Theorem,

$$x\frac{\partial z}{\partial x} + y\frac{\partial z}{\partial y} = 3z \qquad ...(ii)$$

From (*i*), $\qquad \frac{\partial z}{\partial x} = e^u\frac{\partial u}{\partial x}, \frac{\partial z}{\partial y} = e^u\frac{\partial u}{\partial y} \qquad ...(iii)$

Putting (*iii*) and (*i*) in (*ii*), we get

$$x\,e^u\frac{\partial u}{\partial x} + ye^u\frac{\partial u}{\partial y} = 3.e^u \qquad \text{or} \qquad x\frac{\partial u}{\partial x} + y\frac{\partial u}{\partial y} = 3$$

Example 9.8. If $z = f(x, y)$ is a homogeneous function of x and y of degree n, then

$$x^2 \frac{\partial^2 z}{\partial x^2} + 2xy \frac{\partial^2 z}{\partial x\, \partial y} + y^2 \frac{\partial^2 z}{\partial y^2} = n(n-1)z.$$

Solution. By Euler's theorem,

$$x \frac{\partial z}{\partial u} + y \frac{\partial z}{\partial y} = nz \qquad ...(i)$$

Differentiating (*i*) partially w.r.t. x and y, we obtain

$$\frac{\partial z}{\partial x} + x \frac{\partial^2 z}{\partial x^2} + y \frac{\partial^2 z}{\partial x \partial y} = n \frac{\partial z}{\partial x}, \qquad ...(ii)$$

$$x \frac{\partial^2 z}{\partial y \partial x} + \frac{\partial z}{\partial y} + y \,.\, \frac{\partial^2 z}{\partial y^2} = n \frac{\partial z}{\partial y} \qquad ...(iii)$$

Multiplying (*ii*) by x, (*iii*) by y and adding, we get

$$\left(x \frac{\partial z}{\partial x} + y \frac{\partial z}{\partial y} \right) + \left(x^2 \frac{\partial^2 z}{\partial x^2} + 2xy \frac{\partial^2 z}{\partial x \partial y} + y^2 \frac{\partial^2 z}{\partial y^2} \right) = n \left(x \frac{\partial z}{\partial u} + y \frac{\partial z}{\partial y} \right) \qquad \left(\because \frac{\partial^2 z}{\partial x \partial y} = \frac{\partial^2 z}{\partial y \partial x} \right)$$

Using (*i*), we get

$$nz + x^2 \frac{\partial^2 z}{\partial x^2} + 2xy \frac{\partial^2 z}{\partial x \partial y} + y^2 \frac{\partial^2 z}{\partial y^2} = n^2 z$$

Hence
$$x^2 \frac{\partial^2 z}{\partial x^2} + 2xy \frac{\partial^2 z}{\partial x \partial y} + y^2 \frac{\partial^2 z}{\partial y^2} = n^2 z - nz = n(n-1)z$$

Example 9.9. If $u = \tan^{-1} \left(\dfrac{x^3 + y^3}{x - y} \right)$, show that

$$x^2 \frac{\partial^2 u}{\partial x^2} + 2xy \frac{\partial^2 u}{\partial x\, \partial y} + y^2 \frac{\partial^2 u}{\partial y^2} = 2 \cos 3u \sin u.$$

Solution. We have
$$u = \tan^{-1} \left(\frac{x^3 + y^3}{x - y} \right)$$

$$\Rightarrow \qquad \tan u = \frac{x^3 + y^3}{x - y} = \frac{x^3 \left[1 + \left(\frac{y}{x} \right)^3 \right]}{x \left[1 - \frac{y}{x} \right]} = x^3 \phi \left(\frac{y}{x} \right)$$

Let $z = \tan u$ so that z is a homogeneous function of x and y of degree 2. By Euler's theorem,

$$x \frac{\partial z}{\partial x} + y \frac{\partial z}{\partial x} = 2z$$

or
$$x \sec^2 u \frac{\partial u}{\partial x} + y \sec^2 u \frac{\partial u}{\partial y} = 2 \tan u$$

or
$$x \frac{\partial u}{\partial x} + y \frac{\partial u}{\partial y} = \frac{2 \tan u}{\sec^2 u} = \sin 3u \qquad ...(i)$$

Partially differentiating (*i*) w.r.t. x and multiplying by x on both sides, we get

$$\frac{\partial u}{\partial x} + x\frac{\partial^2 u}{\partial x \partial y} + y\frac{\partial^2 u}{\partial x \partial y} = 2\cos 2u\frac{\partial u}{\partial x}$$

or $$x\frac{\partial u}{\partial x} + x^2\frac{\partial^2 u}{\partial x^2} + xy\frac{\partial^2 u}{\partial x \partial y} = 2x\cos 2u\frac{\partial u}{\partial x} \quad ...(ii)$$

Similarly, differentiating (*i*) partially w.r.t y and multiplying by y on both sides, we get

$$\frac{\partial u}{\partial y} + y\frac{\partial^2 u}{\partial y^2} + x\frac{\partial^2 u}{\partial x \partial y} = 2\cos 2u\frac{\partial u}{\partial y}$$

or $$y\frac{\partial u}{\partial y} + y^2\frac{\partial^2 u}{\partial y^2} + xy\frac{\partial^2 u}{\partial x \partial y} = 2y\cos 2u\frac{\partial u}{\partial y} \quad ...(iii)$$

Adding (*ii*) and (*iii*), we obtain

$$x^2\frac{\partial^2 u}{\partial x^2} + 2xy\frac{\partial^2 u}{\partial x \partial y} + y^2\frac{\partial^2 u}{\partial y^2} + x\frac{\partial u}{\partial x} + y\frac{\partial u}{\partial y} = 2\cos 2u\left(x\frac{\partial u}{\partial x} + y\frac{\partial u}{\partial y}\right)$$

Using (*i*), we obtain

$$x^2\frac{\partial^2 u}{\partial x^2} + 2xy\frac{\partial^2 u}{\partial x \partial y} + y^2\frac{\partial^2 u}{\partial y^2} + \sin 2u = 2\cos 2u\sin 2u$$

or $$x^2\frac{\partial^2 u}{\partial x^2} + 2xy\frac{\partial^2 u}{\partial x \partial y} + y^2\frac{\partial^2 u}{\partial y^2} = \sin 2u(2\cos 2u - 1)$$

$$= 2\sin 2u\cos 2u - \sin 2u \qquad [\because \ 2\sin A\cos A = \sin 2A]$$

$$= \sin 4u - \sin 2u = 2\cos 3u\sin u$$

$$\left[\because \ \sin A - \sin B - 2\cos\left(\frac{A+B}{2}\right).\cos\left(\frac{A-B}{2}\right)\right]$$

Hence, $x^2\dfrac{\partial^2 u}{\partial x^2} + 2xy\dfrac{\partial^2 u}{\partial x \partial y} + y^2\dfrac{\partial^2 u}{\partial y^2} = 2\cos 3u\sin u.$

Example 9.10. If $u = \sin^{-1}\dfrac{x+y}{\sqrt{x}+\sqrt{y}}$, prove that

(*i*) $x\dfrac{\partial u}{\partial x} + y\dfrac{\partial u}{\partial y} = \dfrac{1}{2}\tan u$ **(*ii*) $x^2\dfrac{\partial^2 u}{\partial x^2} + 2xy\dfrac{\partial^2 u}{\partial x \partial y} + y^2\dfrac{\partial^2 u}{\partial y^2} = -\dfrac{\sin u\cos 2u}{4\cos^3 u}$**

Solution. We have $u = \sin^{-1}\dfrac{x+y}{\sqrt{x}+\sqrt{y}} \Rightarrow \sin u = \dfrac{x+y}{\sqrt{x}+\sqrt{y}}$

Let $$z = \sin u \quad ...(i)$$

$\therefore$ $$z = \frac{x+y}{\sqrt{x}+\sqrt{y}} = \frac{x\left[1+\frac{y}{x}\right]}{\sqrt{x}\left[1+\sqrt{\frac{y}{x}}\right]} = x^{1/2}\,\phi\left(\frac{y}{x}\right)$$

Thus, z is a homogeneous function of degree $\dfrac{1}{2}$.

Hence by Euler's theorem

$$x\frac{\partial z}{\partial x}+y\frac{\partial z}{\partial y} = \frac{1}{2}z \quad ...(ii)$$

From (*i*),

$$\frac{\partial z}{\partial x} = \cos u\frac{\partial u}{\partial x}, \frac{\partial z}{\partial y} = \cos u\frac{\partial u}{\partial y} \quad ...(iii)$$

Substituting (*i*) and (*iii*) in (*ii*), we obtain

$$\cos u\left(x\frac{\partial u}{\partial x}+y\frac{\partial u}{\partial y}\right) = \frac{1}{2}\sin u$$

Hence

$$x\frac{\partial u}{\partial x}+y\frac{\partial u}{\partial y} = \frac{1}{2}\tan u \quad ...(iv)$$

Partially differentiating (*iv*) w.r.t. x and multiplying by x on both sides, we get

$$\frac{\partial u}{\partial x}+x\frac{\partial^2 u}{\partial x^2}+y\frac{\partial^2 u}{\partial x\partial y} = \frac{1}{2}\sec^2 u\frac{\partial u}{\partial x}$$

or

$$x\frac{\partial u}{\partial x}+x^2\frac{\partial^2 u}{\partial x^2}+xy\frac{\partial^2 u}{\partial x\partial y} = \frac{1}{2}x\sec^2 u\frac{\partial u}{\partial x} \quad ...(v)$$

Similarly differentiating (*iv*) w.r.t. y and multiplying by y on both sides, we get

$$\frac{\partial u}{\partial y}+y\frac{\partial^2 u}{\partial y^2}+x\frac{\partial^2 u}{\partial x\partial y} = \frac{1}{2}\sec^2 u\frac{\partial u}{\partial y}$$

or

$$y\frac{\partial u}{\partial y}+y^2\frac{\partial^2 u}{\partial y^2}+xy\frac{\partial^2 u}{\partial x\partial y} = \frac{1}{2}y\sec^2 u\frac{\partial u}{\partial y} \quad ...(vi)$$

Adding (*v*) and (*vi*), we obtain

$$x^2\frac{\partial^2 u}{\partial x^2}+2xy\frac{\partial^2 u}{\partial x\partial y}+y^2\frac{\partial^2 u}{\partial y^2}+x\frac{\partial u}{\partial x}+y\frac{\partial u}{\partial y} = \frac{1}{2}\sec^2 u\left(x\frac{\partial u}{\partial x}+y\frac{\partial u}{\partial y}\right)$$

Using (*iv*), we obtain

$$x^2\frac{\partial^2 u}{\partial x^2}+2xy\frac{\partial^2 u}{\partial x\partial y}+y^2\frac{\partial^2 u}{\partial y^2}+\frac{1}{2}\tan u = \frac{1}{2}\sec^2 u\left(\frac{1}{2}\tan u\right)$$

or

$$x^2\frac{\partial^2 u}{\partial x^2}+2xy\frac{\partial^2 u}{\partial x\partial y}+y^2\frac{\partial^2 u}{\partial y^2} = \frac{1}{2}\tan u\left[\frac{1}{2}\sec^2 u-1\right]$$

$$= \frac{1}{2}\frac{\sin u}{\cos u}\left[\frac{1}{2\cos^2 u}-1\right]$$

$$= \frac{1}{2}\frac{\sin u}{\cos u}\left[\frac{-(2\cos^2 u-1)}{2\cos^2 u}\right] = -\frac{\sin u\cos 2u}{4\cos^3 u}$$

EXERCISE 9.2

1. Verify Euler's theorem for the functions:

(i) $f(x, y) = ax^2 + 2hxy + by^2$

(ii) $u = \dfrac{x^{1/4} + y^{1/4}}{x^{1/5} + y^{1/5}}$

(iii) $z = \tan^{-1}\left(\dfrac{y}{x}\right)$

(iv) $f(x, y, z) = 3x^2yz + 5xy^2z + 4z^4$

2. If $u = \sin^{-1}\dfrac{x}{y} + \tan^{-1}\dfrac{y}{x}$, then prove that $x\dfrac{\partial u}{\partial x} + y\dfrac{\partial u}{\partial y} = 0.$

3. If $u = \log\dfrac{x^3 + y^3}{x^2 + y^2}$, prove that $x\dfrac{\partial u}{\partial x} + y\dfrac{\partial u}{\partial y} = 1.$

4. If $u = (x^1 + y^2 + z^2)^{-1/2}$, show that $x\dfrac{\partial u}{\partial x} + y\dfrac{\partial u}{\partial y} + z\dfrac{\partial u}{\partial z} = -u.$

5. If $z = \log(x^3 + y^3 - x^2y - xy^2)$, prove that $\dfrac{\partial z}{\partial x} + \dfrac{\partial z}{\partial y} = 2(x + y)^{-1}.$

6. If $z = \tan^{-1}\left(\dfrac{x + y}{\sqrt{x} + \sqrt{y}}\right)$, then prove that $x\dfrac{\partial z}{\partial x} + y\dfrac{\partial z}{\partial y} = \dfrac{1}{4}\sin 2z.$

7. If $u = \tan^{-1}\left(\dfrac{y}{x}\right)$, show that

(i) $x\dfrac{\partial u}{\partial x} + y\dfrac{\partial u}{\partial y} = 0$

(ii) $x^2\dfrac{\partial^2 u}{\partial x^2} + 2xy\dfrac{\partial^2 u}{\partial x \partial y} + y^2\dfrac{\partial^2 u}{\partial y^2} = 0.$

8. If $f(x, y) = \dfrac{1}{x^2} + \dfrac{1}{xy} + \dfrac{\log x - \log y}{x^2 + y^2}$, show that $x\dfrac{\partial f}{\partial x} + y\dfrac{\partial f}{\partial y} + 2f(x, y) = 0.$

9. If $u = (x^2 + y^2)^{1/3}$, show that $x^2\dfrac{\partial^2 u}{\partial x^2} + 2xy\dfrac{\partial^2 u}{\partial x \partial y} + y^2\dfrac{\partial^2 u}{\partial y^2} = -\dfrac{2u}{9}.$

10. If $u = \tan^{-1}\left(\dfrac{x + y}{\sqrt{x} + \sqrt{y}}\right)$, prove that $x^2\dfrac{\partial^2 u}{\partial x^2} + 2xy\dfrac{\partial^2 u}{\partial x \partial y} + y^2\dfrac{\partial^2 u}{\partial y^2} = \dfrac{\sin 2u}{8}(2\cos^2 u - 3)$

11. If $u = \operatorname{cosec}^{-1}\left[\dfrac{x^{1/2} + y^{1/2}}{x^{1/3} + y^{1/3}}\right]^{1/2}$, prove that $x^2\dfrac{\partial^2 u}{\partial x^2} + 2xy\dfrac{\partial^2 u}{\partial x\, \partial y} + y^2\dfrac{\partial^2 u}{\partial y^2} = \dfrac{1}{144}\tan u(\tan^2 u + 13).$

12. If $z = \log\left(\dfrac{x^3 + y^3}{x + y}\right)$, prove that

(i) $x\dfrac{\partial z}{\partial x} + y\dfrac{\partial z}{\partial y} = 2$

(ii) $x^2\dfrac{\partial^2 z}{\partial x^2} + 2xy\dfrac{\partial^2 z}{\partial x \partial y} + y^2\dfrac{\partial^2 z}{\partial y^2} = -2$

13. If $u = \sin^{-1}\left(\dfrac{x^2 + y^2}{x + y}\right)$, show that $x^2\dfrac{\partial^2 u}{\partial x^2} + 2xy\dfrac{\partial^2 u}{\partial x \partial y} + y^2\dfrac{\partial^2 u}{\partial y^2} = \tan^3 u.$

14. If u $= \sin^{-1}\left\{\dfrac{x^{1/3} + y^{1/3}}{x^{1/2} + y^{1/2}}\right\}^{1/2}$, then show that $x^2\dfrac{\partial^2 u}{\partial x^2} + 2xy\dfrac{\partial^2 u}{\partial x \partial y} + y^2\dfrac{\partial^2 u}{\partial y^2} = \dfrac{\tan u}{144}(13 + \tan^2 u).$

15. If $z = x^4y^2 \sin^{-1}\left(\frac{x}{y}\right) + \log x - \log y$, show that

$$x\frac{\partial z}{\partial x} + y\frac{\partial z}{\partial y} = 6x^4y^2 \sin^{-1}\left(\frac{x}{y}\right)$$

9.6. TOTAL DIFFERENTIAL

It may be recalled that if $y = f(x)$ is a differentiable function x, than the derivative, $\frac{dy}{dx}$, of y with respect to x is given by

$$\frac{dy}{dx} = \lim_{\Delta x \to 0} \frac{\Delta y}{\Delta x}$$

Where Δy denotes the increment in y due to an arbitrary small increment Δx in the variable x. Hence

$$\frac{\Delta y}{\Delta x} = \frac{dy}{dx} \text{ approximately when } \Delta x \text{ is small.}$$

i.e.,
$$\Delta y = \frac{dy}{dx}\Delta x \text{ approximately when } \Delta x \text{ is small.}$$

In other words, if x change by Δx, then the change in y, Δy, is approximately $\frac{dy}{dx}$ times the change in x. Thus the expression

$$\Delta y = \frac{dy}{dx}\Delta x$$

given the approximate increment in y for an arbitrary small increment Δx in x.

We now generalize this notion to the case of a function $z = f(x, y)$ of two variables. The partial derivative, $\frac{\partial z}{\partial x}$, gives the rate at which z changes when x changes (y remaining constant). Hence, if $\Delta_x z$ denotes the increment in z due to an arbitrary small increment Δx in the variable x from a point (x, y), then $\Delta_x z = \frac{\partial z}{\partial x}\Delta x$ approximately.

In the same way, if $\Delta_y z$ denote the increment in z due to an arbitrary small increment Δy in the variable from the point (x, y), then

$$\Delta_y z = \frac{\partial z}{\partial y}\Delta y \text{ approximately.}$$

Thus the increment in z when x varies alone is represented approximately by $\frac{\partial z}{\partial x}\Delta x$ and the increment when y varies alone approximately by $\frac{\partial z}{\partial y}\Delta y$. There remains the important problem of expressing the increment in z wen x and y vary together. It has been the function $z = f(x, y)$, corresponding to arbitrary small increments Δx and Δy in x and y is approximately

$$\Delta z = \frac{\partial z}{\partial x}\Delta x + \frac{\partial z}{\partial y}\Delta y$$

This is called the differential of z and is denoted by dz. For convenience, we denote the arbitrary increment Δx by dx and call it the differential of the independent variable x. Similarly, Δy is written by dy and called the differential of y. The differential of the dependent variable z is then defined in terms of the independent differentials by the formula

$$dz = \frac{\partial z}{\partial x}dx + \frac{\partial z}{\partial y}dy$$

It must always be remembered, in using this formula, that dx and dy are no more than arbitrary increments in the independent variables. An alternative notation for the differential of z is

$$df = f_x\, dx + f_y dy$$

The expression $dz = \frac{\partial z}{\partial x}dx \frac{\partial z}{\partial y}dy$ is often called the **total or complete** differential of the function $z = f(x, y)$.

9.7. TOTAL DERIVATIVE OF A COMPOSITE FUNCTION

Theorem: Let $z = f(x, y)$ possess continuous first order partial derivatives.
Let $x = f(t)$, $y = \psi(t)$ possess continuous first order partial derivatives.
Then the composite function $z = f[f(t), \psi(t)]$ has derivative given by

$$\frac{dz}{dt} = \frac{\partial z}{\partial x}\cdot\frac{dx}{dt} + \frac{\partial z}{\partial y}\cdot\frac{dy}{dt} \quad ...(i) \quad \left[\frac{dz}{dt} \text{ is called the total derivative of } z\right]$$

Proof: We have $z = f(x, y)$

Giving increment Δt to t, let the corresponding increments of x, y and z be Δx, Δy and Δz respectively.

$$z + \Delta z = f(x + \Delta x, y + \Delta y)$$

Subtracting,

$$\Delta z = r(x + \Delta x, y + \Delta y) - f(x, y)$$

$$= [f(x + \Delta x, y + \Delta y) - f(x, y + \Delta y)] + [f(x, y + \Delta y) - f(x, y)]$$

$$\therefore \quad \frac{\Delta z}{\Delta t} = \frac{f(x + \Delta x, y + \Delta y) - f(x, y + \Delta y)}{\Delta x}\cdot\frac{\Delta x}{\Delta t} + \frac{f(x, y + \Delta y) - f(x, y)}{\Delta y}\cdot\frac{\Delta y}{\Delta t}$$

Taking limits as $\Delta t \to 0$, Δx and Δy also $\to 0$, we have

$$\frac{dz}{dt} = \underset{\Delta y \to 0}{\text{Lt}}\left[\underset{\Delta x \to 0}{\text{Lt}}\left\{\frac{f(x + \Delta x, y + \Delta y) - f(x, y + \Delta y)}{\Delta x}\right\}\right]\frac{dx}{dt}$$

$$+ \underset{\Delta y \to 0}{\text{Lt}}\left\{\frac{f(x, y + \Delta y) - f(x, y)}{\Delta y}\right\}\frac{dy}{dt}$$

$$= \underset{\Delta y \to 0}{\text{Lt}}\left\{\frac{\partial f(x, y + \Delta y)}{\partial x}\right\}\cdot\frac{dx}{dt} + \frac{\partial f(x, y)}{\partial}\cdot\frac{dy}{dt}$$

$$\left[\text{Supposing } \frac{\partial f(x, y)}{\partial x} \text{ to be a continuous function of } y\right]$$

$$= \frac{\partial f(x, y)}{\partial x}\cdot\frac{dx}{dt} + \frac{\partial f(x, y)}{\partial y}\cdot\frac{dy}{dt}$$

$$\therefore \quad \frac{dz}{dt} = \frac{\partial z}{\partial x}.\frac{dx}{dt} + \frac{\partial z}{\partial y}.\frac{dy}{dt}$$

which is the desired formula.

Corollary 1. If $u = f(x, y, z)$ and x, y, z are function of t, then u is a composite function of t and

$$\frac{du}{dt} = \frac{\partial u}{\partial x}.\frac{dx}{dt} + \frac{\partial u}{\partial y}.\frac{dy}{dt} + \frac{\partial u}{\partial z}.\frac{dz}{dt}.$$

Corollary 2. If $z = f(x, y)$ and x and y are functions of u and v, then

$$\frac{\partial z}{\partial u} = \frac{\partial z}{\partial x}.\frac{\partial x}{\partial u} + \frac{\partial z}{\partial y}.\frac{\partial y}{\partial u}$$

$$\frac{\partial z}{\partial v} = \frac{\partial z}{\partial x}.\frac{\partial x}{\partial v} + \frac{\partial z}{\partial y}.\frac{\partial y}{\partial v}$$

9.8. DERIVATIVE OF AN IMPLICIT FUNCTION

Suppose that two variables x and y are related by a given implicit relation $f(x, y) = 0$. In general, This defines y as a multi-valued function of x. However, the simplex way of dealing with such a function is to introduce a third viable z by

$$z = f(x, y)$$

Of course, $z = 0$ for all x and y related by $f(x, y) = 0$. Thus the given implicit function can be studied by relating those values of x and y which make z equal to zero in the explicit function $z = f(x, y)$.

As x and y very in any way, independently or not, the complete differential of z is given by

$$dz = f_x\,dx + f_y\,dy$$

If (x, y) are values making $z = 0$ and if dx and dy are variations from these values keeping $z = 0$, then $dz = 0$. That is,

$$f_x\,dx + f_y\,dy = 0$$

Hence
$$\frac{dy}{dx} = -\frac{f_x}{f_y}$$

SOLVED EXAMPLES

Example 9.11. Find $\frac{dz}{dt}$ when

(i) $z = x^2 - y^2, x = e^t \cos t, y = e^t \sin t$ **(ii) $z = xy^2 + x^2y, x = at^2, y = 2at$**

(iii) If $u = x^2 - y^2 + \sin yz$ where $y = e^x$ and $z = \log x$; find $\frac{du}{dx}$.

Verify by direct substitution.

Solution. (i) Given that
$$z = x^2 - y^2$$

$$\therefore \quad \frac{\partial z}{\partial x} = 2x, \frac{\partial z}{\partial y} = -2y$$

and
$$x = e^t \cos t, y = e^t \sin t$$

$$\therefore \quad \frac{dx}{dt} = e^t (\cos t - \sin t), \frac{dy}{dt} = e^t (\sin t + \cos t)$$

Hence $$\frac{dz}{dt} = \frac{\partial z}{\partial x}\cdot\frac{dx}{dt} + \frac{\partial z}{\partial y}\cdot\frac{dy}{dt} \quad ...(i)$$

$$= 2xe^t(\cos t - \sin t) - 2y\,e^t(\sin t + \cos t)$$

$$= 2e^{2t}\cos t(\cos t - \sin t) - 2e^{2t}\sin t\,(\sin t + \cos t)$$

Again $$z = e^{2t}\cos^2 t - e^{2t}\sin^2 t$$

$\therefore$ $$\frac{dz}{dt} = 2e^{2t}(\cos^2 t - \cos t\sin t) - 2e^{2t}(\sin^2 t + \sin t\cos t)$$

Hence the verification.

(*ii*) $$z = xy^2 + x^2y$$

$\therefore$ $$\frac{\partial z}{\partial x} = y^2 + 2xy,\ \frac{\partial z}{\partial y} = 2xy + x^2$$

and $$x = at^2,\ y = 2at$$

$\therefore$ $$\frac{dx}{dt} = 2at,\ \frac{dy}{dt} = 2a$$

Putting in (*i*), we get $$\frac{dz}{dt} = (y^2 + 2xy)(2at) + (2xy + x^2)(2a)$$

$$= (4a^2t^2 + 4a^2t^3)(2at) + (4a^2t^3 + a^2t^4)(2a)$$

$$= a^3(16t^3 + 10t^4)$$

Again, $$z = xy^2 + x^2y = at^2\,.\,4a^2t^2 + a^2t^4\,.\,2at = 4a^3t^4 + 2a^3t^5$$

$\therefore$ $$\frac{dz}{dt} = a^3(16t^3 + 10t^4)$$

Hence the verification.

(*iii*) $$\frac{du}{dx} = \frac{\partial u}{\partial y}\cdot\frac{dy}{dx} + \frac{\partial u}{\partial z}\cdot\frac{dz}{dx} + \frac{\partial u}{\partial x}$$

$$= (-2y + z\cos yz)e^x + (y\cos yz)\,.\,\frac{1}{x} + 2x$$

$$= [-2e^x + \log x\cos(e^x\log x)]e^x + \left[e^x\cos(e^x\log x):\frac{1}{x}\right] + 2x$$

$$= 2(x - e^{2x}) + e^x\cos(e^x\log x)\left(\log x + \frac{1}{x}\right)$$

Example 9.12. If $u = x^2 - y^2 + \sin yz$ where $y = e^x$ and $z = \log x$; find $\frac{du}{dx}$.

Solution. $$u = f(x, y, z)$$

$$\frac{du}{dx} = \frac{\partial u}{\partial y}\cdot\frac{dy}{dx} + \frac{\partial u}{\partial z}\cdot\frac{dz}{dx} + \frac{\partial u}{\partial x}$$

$$= (-2y + z\cos yz)e^x + (y\cos yz)\,.\,\frac{1}{x} + 2x$$

$$= [-2e^x + \log x\cos(e^x\log x)]e^x + \left[e^x\cos(e^x\log x)\,.\,\frac{1}{x}\right] + 2x$$

$$= 2(x - e^{2x}) + e^x\cos(e^x\log x)\left[\log x + \frac{1}{x}\right].$$

Example 9.13. If $z = f(x, y)$, where $x = e^u + e^{-v}$, $y = e^{-u} - e^v$, show that

$$\frac{\partial z}{\partial u} - \frac{\partial z}{\partial v} = x\frac{\partial z}{\partial x} - y\frac{\partial z}{\partial y}$$

Solution. We have
$$\frac{\partial x}{\partial u} = e^u, \frac{\partial y}{\partial u} = -e^{-u}, \frac{\partial x}{\partial v} = -e^{-v}, \frac{\partial y}{\partial v} = -e^v$$

We know that,
$$\frac{\partial z}{\partial u} = \frac{\partial z}{\partial x}\cdot\frac{\partial x}{\partial u} + \frac{\partial z}{\partial y}\cdot\frac{\partial y}{\partial u}$$

$$= e^u\frac{\partial z}{\partial x} - e^{-u}\frac{\partial z}{\partial y} \qquad ...(i)$$

and
$$\frac{\partial z}{\partial v} = \frac{\partial z}{\partial x}\cdot\frac{\partial x}{\partial v} + \frac{\partial z}{\partial y}\cdot\frac{\partial y}{\partial v} = -e^{-v}\frac{\partial z}{\partial x} - e^v\frac{\partial z}{\partial y} \qquad ...(ii)$$

From eqns. (*i*) and (*ii*), we get

$$\frac{\partial z}{\partial u} - \frac{\partial z}{\partial v} = (e^u + e^{-v})\frac{\partial z}{\partial u} - (e^{-u} - e^v)\frac{\partial z}{\partial y} = x\frac{\partial z}{\partial u} - y\frac{\partial z}{\partial y}$$

Example 9.14. If $u = f(y - z, z - x, x - y)$, show that $\frac{\partial u}{\partial x} + \frac{\partial u}{\partial y} + \frac{\partial u}{\partial z} = 0$.

Solution. Let $X = y - z$, $Y = z - x$, $Z = x - y$

Thus $u = f(X, Y, Z)$ is a composite function of x, y, z.

We have
$$\frac{\partial u}{\partial x} = \frac{\partial u}{\partial X}\cdot\frac{\partial X}{\partial x} + \frac{\partial u}{\partial Y}\cdot\frac{\partial Y}{\partial x} + \frac{\partial u}{\partial Z}\cdot\frac{\partial Z}{\partial x} = \frac{\partial u}{\partial X}.0 + \frac{\partial u}{\partial Y}(-1) + \frac{\partial u}{\partial Z}.1$$

$$= -\frac{\partial u}{\partial Y} + \frac{\partial u}{\partial Z} \qquad ...(i)$$

Similarly,
$$\frac{\partial u}{\partial y} = -\frac{\partial u}{\partial Z} + \frac{\partial u}{\partial X}; \qquad ...(ii)$$

and
$$\frac{\partial u}{\partial z} = -\frac{\partial u}{\partial X} + \frac{\partial u}{\partial Y} \qquad ...(iii)$$

Adding (*i*), (*ii*) and (*iii*), we get

$$\frac{\partial u}{\partial x} + \frac{\partial u}{\partial y} + \frac{\partial u}{\partial z} = 0.$$

Example 9.15. If $u = f(x^2 + 2yz, y^2 + 2zx)$, prove that

$$(y^2 - zx)\frac{\partial u}{\partial x} + (x^2 - yz)\frac{\partial u}{\partial y} + (z^2 - xy)\frac{\partial u}{\partial z} = 0.$$

Solution. Let $X = x^2 + 2yz$ and $Y = y^2 + 2zx$

Thus
$$u = f(X, Y) \Rightarrow \frac{\partial u}{\partial x} = \frac{\partial u}{\partial X}\cdot\frac{\partial X}{\partial x} + \frac{\partial u}{\partial Y}\cdot\frac{\partial Y}{\partial x}$$

$$= \frac{\partial u}{\partial X}(2x) + \frac{\partial u}{\partial Y}(2z)$$

and
$$\frac{\partial u}{\partial y} = \frac{\partial u}{\partial X}\cdot\frac{\partial X}{\partial y} + \frac{\partial u}{\partial Y}\cdot\frac{\partial Y}{\partial y} = \frac{\partial u}{\partial X}(2z) + \frac{\partial u}{\partial Y}(2y)$$

Further $\qquad \dfrac{\partial u}{\partial z} = \dfrac{\partial u}{\partial X}(2y) + \dfrac{\partial u}{\partial Y}(2x)$

$$(y^2 - zx)\frac{\partial u}{\partial x} + (x^2 - yz)\frac{\partial u}{\partial y} + (z^2 - xy)\frac{\partial u}{\partial z}$$

$$= (y^2 - zx)\left[\frac{\partial u}{\partial X}(2x) + \frac{\partial u}{\partial y}(2z)\right] + (x^2 - yz)\left[\frac{\partial u}{\partial X}(2z) + \frac{\partial u}{\partial Y}(2y)\right]$$

$$+ (z^2 - xy)\left[\frac{\partial u}{\partial X}(2y) + \frac{\partial u}{\partial Y}(2x)\right]$$

$$= \frac{\partial u}{\partial X}[2x(y^2 - zx) + 2z(x^2 - yz) + 2y(z^2 - xy)]$$

$$\frac{\partial u}{\partial Y}[2z(y^2 - zx) + 2y(x^2 - yz) + 2x(z^2 - xy)]$$

$$= \frac{\partial u}{\partial X}.(0) + \frac{\partial u}{\partial Y}.(0) = 0$$

EXERCISE 9.3

1. If $u = f\left(\dfrac{y}{x}\right)$, show that $x\dfrac{\partial u}{\partial x} + y\dfrac{\partial u}{\partial y} = 0$.

2. If $u = \sin^{-1}(x - y)$, $x = 3t$, $y = 4t^3$, show that $\dfrac{du}{dt} = \dfrac{3}{\sqrt{1-t^2}}$.

3. If $u = x^3 + y^3$, where $x = a\cos t$, $y = b\sin t$, find $\dfrac{du}{dt}$ and verify the result.

4. If $z = \tan^{-1}\left(\dfrac{x}{y}\right)$, $x = u + v$, $y = u - v$, show that $\dfrac{\partial z}{\partial u} + \dfrac{\partial z}{\partial v} = \dfrac{u - v}{u^2 + v^2}$.

5. If $z = f(x, y)$ and $u = xy$, $v = \dfrac{y}{x}$, prove that $x\dfrac{\partial z}{\partial x} + y\dfrac{\partial z}{\partial y} = 2xy\dfrac{\partial}{\partial u}$.

6. Find the differential coefficient of $u = \sin\left(\dfrac{x}{y}\right)$, where $x = y = t^2$ w.r.t. 't'.

$$\left[\textbf{Ans. } \frac{du}{dt} = \frac{e^t}{t^2}\cos\left(\frac{e^t}{t^2}\right).\left(1 - \frac{2}{t}\right)\right]$$

7. Find the total differential coefficient of $u = x^2y$ with respect to x, when x and y are connected by the relation $x^2 + xy + y^2 = 1$. $\quad \left[\textbf{Ans. } \dfrac{du}{dx} = 2xy - \dfrac{x^2(2x + y)}{x + 2y}\right]$

8. If $x^2 - y^2 + u^2 + 2v^2 = 1$, $x^2 + y^2 - u^2 - v^2 = 2$, find $\dfrac{\partial u}{\partial x}$ and $\dfrac{\partial v}{\partial x}$. $\quad \left[\textbf{Ans. } \dfrac{\partial u}{\partial x} + \dfrac{3x}{u}, \dfrac{\partial v}{\partial x} = \dfrac{-2x}{v}\right]$

9. If $u = f(r, s, t)$ where $r = \dfrac{x}{y}$, $s = \dfrac{y}{z}$, $t = \dfrac{z}{x}$. Show that $x\dfrac{\partial u}{\partial x} + y\dfrac{\partial u}{\partial y} + z\dfrac{\partial u}{\partial z} = 0$.

10. Find $\dfrac{dy}{dx}$, if $x^y + y^x = a^b$. $\quad \left[\textbf{Ans. } \dfrac{-(yx^{y-1} + y^x \log y)}{(xy^{x-1} + x^y \log x)}\right]$

11. If $x = r\cos\theta, y = r\sin\theta$, transform the equation $xy\left(\dfrac{\partial^2 v}{\partial x^2} - \dfrac{\partial^2 v}{\partial y^2}\right) - (x^2 - y^2)\dfrac{\partial^2 v}{\partial x \partial y}$.

12. If $x = r\cos\theta, y = r\sin\theta$; show that $\dfrac{\partial x}{\partial r} = \dfrac{\partial r}{\partial x}, \dfrac{\partial x}{r\partial\theta} = r\dfrac{\partial\theta}{\partial x}$.

13. If $u = f(2x - 3y, 3y - 4z, 4z - 2x)$ prove that $\dfrac{1}{2}\dfrac{\partial u}{\partial x} + \dfrac{1}{3}\dfrac{\partial u}{\partial y} + \dfrac{1}{4}\dfrac{\partial u}{\partial z} = 0$.

14. If z is a function of x and y, and u and v be two other variables such that $u = lx + my$, $v = ly - mx$, show that

$$\frac{\partial^2 z}{\partial x^2} + \frac{\partial^2 z}{\partial y^2} = (l^2 + m^2)\left(\frac{\partial^2 z}{\partial u^2} + \frac{\partial^2 z}{\partial v^2}\right).$$

15. If $z = xyf\left(\dfrac{y}{x}\right)$ and z is constant, show that $\dfrac{f'\left(\dfrac{y}{x}\right)}{f\left(\dfrac{y}{x}\right)} = \dfrac{x\left(y + x\dfrac{dy}{dx}\right)}{y\left(y - x\dfrac{dy}{dx}\right)}$.

❑❑❑

UNIT - IV

INTEGRAL CALCULUS

CHAPTER 10

Integration

INSIDE THIS CHAPTER

10.1. INTRODUCTION

Integration is the inverse (*i.e.*, reverse) process of differentiation. The process of finding $f(x)$, when its derivative $f'(x)$ is given, is known as integration.

Let $$\frac{d}{dx}[\phi(x)] = f(x), \text{ then}$$

$\phi(x)$ is called the integral or antiderivative or primitive of $f(x)$ with respect to x.

Symbolically, it is written as

$$\int f(x)\,dx = \phi(x)$$

For example, since $\frac{d}{dx}(\sin x) = \cos x$, we have $\int \cos x\,dx = \sin x$. Moreover, if c is any constant, then $\frac{d}{dx}(\sin x + c) = \cos x$.

So, in general, $$\int \cos x\,dx = \sin x + c.$$

Thus, $$\frac{d}{dx}[\phi(x)] = f(x) \Rightarrow \int f(xc)\,dx = \phi(x) + c.$$

where c is a constant of integration.

Clearly, different values of c will give different integrals.

Thus, a given function may have an indefinite numbers of integrals. Because of this property, we call it indefinite integral.

Remarks:

1. The function whose integral is to be found out is called the integrand.
2. The symbol $\int$ (elongated S) denote the process of integration.
3. $\int dx$ is said to be "integral w.r.t. x".

10.2. FUNDAMENTAL INTEGRATION FORMULAE

Since $\frac{d}{dx}[\phi(x)] = f(x) \Leftrightarrow \int f(x)\,dx = \phi(x) + c,$

therefore based upon this definition and various standard differentiation formulae, we obtain the following integration formulae:

1. $\frac{d}{dx}\left(\frac{x^{n+1}}{n+1}\right) = x^n, n \neq -1 \quad \Rightarrow \quad \int x^n\,dx = \frac{x^{n+1}}{n+1} + c$

2. $\frac{d}{dx}(\log x) = \frac{1}{x} \quad \Rightarrow \quad \int \frac{1}{x}\,dx = \log|x| + c$

3. $\frac{d}{dx}(e^x) = e^x \quad \Rightarrow \quad \int e^x\,dx = e^x + c$

4. $\frac{d}{dx}\left(\frac{a^x}{\log a}\right) = a^x \quad \Rightarrow \quad \int a^x\,dx = \frac{a^x}{\log a} + c$

5. $\frac{d}{dx}(\sin x) = \cos x \quad \Rightarrow \quad \int \cos x\,dx = \sin x + c$

6. $\frac{d}{dx}(\cos x) = -\sin x \quad \Rightarrow \quad \int \sin x\,dx = -\cos x + c$

7. $\frac{d}{dx}(\tan x) = \sec^2 x \quad \Rightarrow \quad \int \sec^2 x\,dx = \tan x + c$

8. $\frac{d}{dx}(\cot x) = -\text{cosec}^2 x \quad \Rightarrow \quad \int \text{cosec}^2 x\,dx = -\cot x + c$

9. $\frac{d}{dx}(\sec x) = \sec x \tan x \quad \Rightarrow \quad \int \sec x \tan x\,dx = \sec x + c$

10. $\frac{d}{dx}(\text{cosec } x) = -\text{cosec } x \cot x \quad \Rightarrow \quad \int \text{cosec } x \cot x\,dx = -\text{cosec } x + c$

11. $\frac{d}{dx}(\log \sin x) = \cot x \quad \Rightarrow \quad \int \cot x\,dx = \log|\sin x| + c$

12. $\frac{d}{dx}(\log \cos x) = -\tan x \quad \Rightarrow \quad \int \tan x\,dx = -\log|\cos x| + c$

13. $\frac{d}{dx}[\log(\sec x + \tan x)] = \sec x \quad \Rightarrow \quad \int \sec x\,dx = \log|\sec x + \tan x| + c$

14. $\frac{d}{dx}[\log(\text{cosec } x - \cot x)] = \text{cosec } x \quad \Rightarrow \quad \int \text{cosec } x\,dx = \log|\text{cosec } x - \cot x| + c$

15. $\frac{d}{dx}(\sin^{-1} x) = \frac{1}{\sqrt{1-x^2}} \quad \Rightarrow \quad \int \frac{1}{\sqrt{1-x^2}}\,dx = \sin^{-1} x + c$

16. $\frac{d}{dx}(\cos^{-1} x) = \frac{-1}{\sqrt{1-x^2}} \quad \Rightarrow \quad \int \frac{1}{\sqrt{1-x^2}}\,dx = -\cos^{-1} x + c$

17. $\frac{d}{dx}(\tan^{-1} x) = \frac{1}{1+x^2}$ $\Rightarrow$ $\int \frac{1}{1+x^2}\,dx = \tan^{-1} x + c$

18. $\frac{d}{dx}(\cot^{-1} x) = \frac{-1}{1+x^2}$ $\Rightarrow$ $\int \frac{1}{1+x^2}\,dx = -\cot^{-1} x + c$

19. $\frac{d}{dx}(\sec^{-1} x) = \frac{1}{x\sqrt{x^2-1}}$ $\Rightarrow$ $\int \frac{1}{x\sqrt{x^2-1}}\,dx = \sec^{-1} x + c$

20. $\frac{d}{dx}(\text{cosec}^{-1} x) = \frac{-1}{x\sqrt{x^2-1}}$ $\Rightarrow$ $\int \frac{1}{x\sqrt{x^2-1}}\,dx = \text{cosec}^{-1} x + c$

10.3. SOME STANDARD RESULTS ON INTEGRATION

1. $\int K\, f(x)\, dx = K \int f(x)\, dx$, where K is a constant

i.e., the integral of the product of a constant and a function = constant × integral of the function.

2. $\frac{d}{dx}\left[\int f(x)\, dx\right] = f(x)$

i.e., the differentiation of an integral is the integrand itself or differentiation and integration are inverse operations.

3. $\int [f(x) \pm g(x)]\, dx = \int f(x)\, dx \pm \int g(x)\, dx$

i.e., the integral of the sum or difference of a finite number of functions is equal to the sum or difference of the integral of the various functions.

We can solve many problems with the help of the results given in this section.

SOLVED EXAMPLES

Example 10.1. Evaluate

(*i*) $\int \left(5x^3 + 2\cos x + \frac{3}{x} - 9e^x\right) dx$ (*ii*) $\int \sqrt{1+\sin 2x}\; dx$

(*iii*) $\int (1+x)(3-2x)(4+5x)\, dx$ (*iv*) $\int \left(\frac{3x^4 - 5x^3 + 4x^2 - x + 2}{x^3}\right) dx$

(*v*) $\left(\sqrt{x} - \frac{1}{\sqrt{x}}\right)^3$.

Solution. (*i*) We have

$$\int \left(5x^3 + 2\cos x + \frac{3}{x} - 9e^x\right) dx = 5\int x^3\, dx + 2\int \cos x\, dx + 3\int \frac{1}{x}\, dx - 9\int e^x\, dx$$

$$= \frac{5x^4}{4} + 2\sin x + 3\log|x| - 9e^x + c.$$

(*ii*) We have
$$\int \sqrt{1+\sin 2x}\, dx = \int \sqrt{\sin^2 x + \cos^2 x + 2\sin x \cos x}\, dx$$
$$= \int \sqrt{(\sin x + \cos x)^2}\, dx$$
$$= \int (\sin x + \cos x)\, dx = \int \sin x\, dx + \int \cos x\, dx = -\cos x + \sin x + c.$$

(*iii*) We have $\int (1+x)(3-2x)(4+5x)\, dx$

On multiplication, we get
$$\int (12 + 19x - 3x^2 - 10x^3)\, dx = 12\int 1\, dx + 19\int x\, dx - 3\int x^2\, dx - 10\int x^3\, dx$$
$$= 12x - \frac{19}{2}x^2 - x^3 - \frac{5}{2}x^4 + c.$$

(*iv*) We have
$$\int \left(\frac{3x^4 - 5x^3 + 4x^2 - x + 2}{x^3}\right) dx = \int \left(3x - 5 + \frac{4}{x} - \frac{1}{x^2} + \frac{2}{x^3}\right) dx \qquad \text{(dividing each term by } x^3\text{)}$$
$$= 3\int x\, dx - 5\int dx + 4\int \frac{1}{x}\, dx - \int x^{-2}\, dx + 2\int x^{-3}\, dx$$
$$= 3.\frac{x^2}{2} - 5x + 4\log|x| - \left(-\frac{1}{x}\right) + 2\left(\frac{x^{-2}}{-2}\right) + c$$
$$= \frac{3x^2}{2} - 5x + 4\log|x| + \frac{1}{x} - \frac{1}{x^2} + c.$$

(*v*) We have
$$\int \left(\sqrt{x} - \frac{1}{\sqrt{x}}\right)^3 dx = \int \left[(\sqrt{x})^3 - \left(\frac{1}{\sqrt{x}}\right)^3 - 3.\sqrt{x}.\frac{1}{\sqrt{x}}\left(\sqrt{x} - \frac{1}{\sqrt{x}}\right)\right] dx$$
$$= \int \left(x^{3/2} - x^{-3/2} - 3\sqrt{x} + \frac{3}{\sqrt{x}}\right) dx$$
$$= \int x^{3/2}\, dx - \int x^{-3/2}\, dx - 3\int x^{1/2}\, dx + 3\int x^{-1/2}\, dx$$
$$= \frac{x^{5/2}}{5/2} - \frac{x^{-1/2}}{(-1/2)} - 3.\frac{x^{3/2}}{3/2} + 3.\frac{x^{1/2}}{1/2} + c$$
$$= \frac{2}{5}x^{5/2} + 2x^{-1/2} - 2x^{3/2} + 6x^{1/2} + c$$

Example 10.2. Evaluate

(*i*) $\int \left(\frac{x^4+1}{x^2+1}\right) dx$ (*ii*) $\int \left(\frac{1-\cos 2x}{1+\cos 2x}\right) dx$ (*iii*) $\int \left(\frac{4 - 5\sin x}{\cos^2 x}\right) dx$ (*iv*) $\int \frac{\sin x}{1+\sin x}\, dx$

Solution. (*i*) We have $\int \frac{x^4+1}{x^2+1}\, dx$

On dividing numerator by the denominator, we get
$$\frac{x^4+1}{x^2+1} = x^2 - 1 + \frac{2}{x^2+1}$$

$$\therefore \quad \int\left(\frac{x^4+1}{x^2+1}\right)dx = \int\left(x^2-1+\frac{2}{x^2+1}\right)dx = \int x^2\,dx - \int 1.\,dx + 2\int\frac{1}{1+x^2}\,dx$$

$$= \frac{x^3}{3} - x + 2\ .\tan^{-1}x + c$$

(*ii*) We have $\int\left(\frac{1-\cos 2x}{1+\cos 2x}\right)dx = \int\left(\frac{2+\sin^2 x}{2\cos^2 x}\right)dx = \int \tan^2 x\,dx$

$$= \int(\sec^2 x - 1)\,dx = \int \sec^2 x\,dx - \int 1.\,dx = \tan x - x + c$$

(*iii*) We have $\int\left(\frac{4-5\sin x}{\cos^2 x}\right)dx = \int\frac{4}{\cos^2 x}\,dx - \int\frac{5\sin x}{\cos^2 x}\,dx$

$$= 4\int\sec^2 x\,dx - 5\int\frac{\sin x}{\cos x}\cdot\frac{1}{\cos x}\,dx$$

$$= 4\int\sec^2 x\,dx - 5\int\tan x\sec x\,dx$$

$$= 4\tan x - 5\sec x + c.$$

(*iv*) We have $\int\frac{\sin x}{1+\sin x}\,dx = \int\frac{\sin x\,(1-\sin x)}{(1+\sin x)\,(1-\sin x)}\,dx = \int\frac{\sin x - \sin^2 x}{1-\sin^2 x}\,dx$

$$= \int\frac{\sin x - \sin^2 x}{\cos^2 x}\,dx = \int\left(\frac{\sin x}{\cos^2 x}\,dx - \frac{\sin^2 x}{\cos^2 x}\right)dx$$

$$= \int\tan x\sec x\,dx - \int\tan^2 x\,dx$$

$$= \int\tan x\sec x\,dx - \int(\sec^2 x - 1)\,dx$$

$$= \int\sec x\tan x\,dx - \int\sec^2 x\,dx + \int 1.\,dx = \sec x - \tan x + x + c.$$

10.4. METHODS OF INTEGRATION

So far we have considered functions which can be put into standard forms easily by algebraic simplification or by trigonometric formulae. Many times the function to be integrated looks quite simple, but does not seems to fit in the standard form. It such cases we have to adopt other technique in finding the integrals:

We have the following methods of integration:

1. Integration by substitution.
2. Integration by parts.
3. Integration by resolution into partial fractions.

Integration by Substitution

Here, by a suitable substitution, the variable x in $\int f(x)\,dx$ is changed into another variable t so that the integrand $f(x)$, if changed into $F(t)$ is easily integrable. That is, we put $t = \phi(x)$ so that $\int f(x)\,dx$ reduces to $\int F(t)\,dt$ which is readily integrable.

For example consider the integral $\int \frac{\sqrt{2+\log x}}{x}\, dx$

Let $2 + \log x = t$. Then

$$\frac{1}{x} = \frac{dt}{dx} \quad \text{or} \quad \frac{dx}{x} = dt$$

$$\therefore \quad \int \sqrt{2+\log x}\,.\frac{dx}{x} = \int \sqrt{t}\, dt = \int t^{1/2}\, dt = \frac{t^{3/2}}{3/2} = \frac{2}{3}t^{3/2} + c = \frac{2}{3}(2+\log x)^{3/2} + c$$

Some practice will enable the student to arive at a suitable substitution.

We know that $\frac{d}{dx}\left[\frac{f^{n+1}(x)}{n+1}\right] = f^n(x)\, f'(x)$

Therefore all integrands which can be reduced to the form

$$f^n(x)\, f'(x)$$

can be integrated by putting $f(x) = t$ or $f'(x)\, dx = dt$

$$\therefore \quad \int f^n(x)\, f'(x)\, dx = \int t^n\, dt = \frac{t^{n+1}}{n+1} + c = \frac{f^{n+1}(x)}{n+1} + c.$$

SOLVED EXAMPLES

Example 10.3. Evaluate

(*i*) $\int \frac{1}{(2-3x)^4}\, dx$ (*ii*) $\int \cos(ax+b)\, dx$ (*iii*) $\int \frac{x+1}{x^2+2x-3}\, dx$ (*iv*) $\int \frac{\cos x - \sin x}{1+\sin 2x}\, dx$.

Solution. (*i*) We have $\int \frac{1}{(2-3x)}\, dx$

Put $2 - 3x = r$ so that $-3\, dx = dt$ or $dx = -\frac{1}{3}\, dt$

$$\therefore \quad \int \frac{1}{(2-3x)^4}\, dx = -\frac{1}{3}\int \frac{1}{t^4}\, dt = -\frac{1}{3}\cdot\frac{1}{(-3t^3)} + c = \frac{1}{9(2-3x)^3} + c$$

(*ii*) We have $\int \cos(ax+b)\, dx$

Put $ax + b = t$ so that $a\, dx = dt$ or $dx = \frac{dt}{a}$

$$\therefore \quad \int \cos(ax+b)\, dx = \int \cos t\, \frac{dt}{a} = \frac{1}{a}\int \cos t\, dt = \frac{1}{a}\sin t + c = \frac{1}{a}\sin(ax+b) + c.$$

(*iii*) We have $\int \frac{x+1}{x^2+2x-3}\, dx$

Put $x^2 + 2x - 3 = t$ so that $(2x + 2)\, dx = dt$

or $$2(x+1)\, dx = dt \quad \Rightarrow \quad (x+1)\, dx = \frac{dt}{2}$$

$$\therefore \quad \int \frac{x+1}{x^2+2x-3}\,dx = \int \frac{1}{t}\cdot\frac{dt}{2} = \int \frac{1}{t}\,dt = \frac{1}{2}\log|t| + c = \frac{1}{2}\log|x^2+2x-3| + c.$$

(*iv*) We have $\displaystyle\int \frac{\cos x - \sin x}{1+\sin 2x}\,dx = \int \frac{\cos x - \sin x}{\sin^2 x + \cos^2 x + 2\sin x\cos x}\,dx = \int \frac{\cos x - \sin x}{(\sin x + \cos x)^2} + c.$

Put $\sin x + \cos x = t$ so that $(\cos x - \sin x)\,dx = dt$

$$\therefore \quad \int \frac{\cos x - \sin x}{(\sin x + \cos x)^2}\,dx = \int \frac{dt}{t^2} = \int t^{-2}\,dt = \frac{t^{-1}}{-1} + c = -\frac{1}{t} + c = \frac{-1}{\sin x + \cos x} + c.$$

Example 10.4. Evaluate

(*i*) $\displaystyle\int \frac{e^x(1+x)}{\cos^2(xe^x)}\,dx$ (*ii*) $\displaystyle\int \frac{1}{(\sqrt{x}+x)}\,dx$ (*iii*) $\displaystyle\int \frac{\sqrt{\tan x}}{\sin x\cos x}\,dx$

(*iv*) $\displaystyle\int \frac{dx}{(\sqrt{2x+3}+\sqrt{2x-3})}$ (*v*) $\displaystyle\int \frac{1}{\sqrt{\sin^3 x \sin(x+\alpha)}}\,dx$ (*vi*) $\displaystyle\int \frac{1}{3\sin x + 4\cos x}\,dx$

Solution. (*i*) We have $\displaystyle\int \frac{e^x(1+x)}{\cos^2(xe^x)}\,dx$

Put $xe^x = t$ so that $(1.e^x + xe^x)\,dx = dt$ or $e^x(1+x)\,dx = dt$

$$\therefore \quad \int \frac{e^x(1+x)}{\cos^2(xe^x)}\,dx = \int \frac{dt}{\cos^2 t} = \int \sec^2 t\,dt = \tan t + c = \tan(xe^x) + c.$$

(*ii*) We have $\displaystyle\int \frac{1}{(\sqrt{x}+x)}\,dx = \int \frac{1}{\sqrt{x}(1+\sqrt{x})}\,dx$

Put $1+\sqrt{x} = t$ so that $\dfrac{1}{\sqrt{x}}\,dx = 2\,dt$

$$\therefore \quad \int \frac{e^x(1+x)}{\cos^2(xe^x)}\,dx = \int \frac{1}{\sqrt{x}(1+\sqrt{x})}\,dx = 2\int \frac{1}{t}\,dt$$

$$= 2\log|t| + c = 2\log|(1+\sqrt{x})| + c$$

(*iii*) We have $\displaystyle\int \frac{\sqrt{\tan x}}{\sin x\cos x} = \int \frac{\tan x}{(\sqrt{\tan x}).\sin x\cos x}\,dx = \int \frac{\sec^2 x}{\sqrt{\tan x}}\,dx$

Put $\tan x = t$ so that $\sec^2 x\,dx = dt$.

$$\therefore \quad \int \frac{\sec^2 x}{\sqrt{\tan x}}\,dx = \int \frac{1}{\sqrt{t}}\,dt = 2\sqrt{t} + c = 2\sqrt{\tan x} + c$$

(*iv*) We have

$$\int \frac{1}{\sqrt{2x+3}+\sqrt{2x-3}}\,dx = \int \frac{\sqrt{2x+3}-\sqrt{2x-3}}{(\sqrt{2x+3}+\sqrt{2x-3})(\sqrt{2x+3}-\sqrt{2x-3})}$$

$$= \int \frac{\sqrt{2x+3}}{(\sqrt{2x+3})^2 - (\sqrt{2x-3})^2}\,dx = \int \frac{\sqrt{2x+3}-\sqrt{2x-3}}{(2x+3)-(2x-3)}\,dx$$

$$= \int \frac{\sqrt{2x+3}-\sqrt{2x-3}}{6}\,dx = \frac{1}{6}\int\sqrt{2x+3}\,dx + \frac{1}{6}\int\sqrt{2x-3}\,dx$$

$$= \frac{1}{6}\frac{(2x+3)^{3/2}}{2.\frac{3}{2}} + \frac{1}{6}.\frac{(2x-3)^{3/2}}{2.\frac{3}{2}} + c$$

$$= \frac{1}{18}(2x+3)^{3/2} + \frac{1}{18}(2x-3)^{3/2} + c.$$

(*v*) We have $\sin^3 x \sin(x+\alpha) = \sin^3 x(\sin x\cos\alpha + \cos x\sin\alpha)$

$$= \sin^4 x(\cos\alpha + \cot x\sin\alpha)$$

$$\therefore \quad \int \frac{1}{\sqrt{\sin^3 x\sin(x+\alpha)}}\,dx = \int \frac{1}{\sin^2 x\sqrt{\cos\alpha+\cot x\sin\alpha}}\,dx = \int\frac{\operatorname{cosec}^2 x\,dx}{\sqrt{\cos\alpha+\cot x\sin\alpha}}$$

Put $\cos\alpha + \cot x\sin\alpha = t$ so that $-\operatorname{cosec}^2 x\sin\alpha\,dx = dt$

or $$\operatorname{cosec}^2 x\,dx = -\frac{dt}{\sin\alpha}$$

$$\therefore \quad \int\frac{\operatorname{cosec}^2 x\,dx}{\sqrt{\cos\alpha+\cot x\sin\alpha}} = \int\frac{1}{\sqrt{t}}\left(-\frac{dt}{\sin\alpha}\right)dx = -\frac{1}{\sin\alpha}\int t^{-1/2}\,dt$$

$$= -\frac{1}{\sin\alpha}\frac{t^{1/2}}{1/2} + c = -2\operatorname{cosec} a.\sqrt{t} + c$$

$$= -2\operatorname{cosec}\alpha(\cos\alpha + \cot x\sin\alpha)^{1/2} + c.$$

(*vi*) We have $\int \frac{dx}{3\sin x + 4\cos x}$

Put $3 = r\cos\theta$ and $4 = r\sin\theta$ so that $r = \sqrt{9+16} = 5$, $\tan\theta = \frac{4}{3}$.

$$\therefore \quad 3\sin x + 4\cos x = r(\sin x\cos\theta + \cos x\sin\theta) = 5\sin(x+\theta)$$

$$\Rightarrow \quad \int\frac{dx}{3\sin x+4\cos x} = \frac{1}{5}\int\frac{dx}{\sin(x+\theta)} = \frac{1}{5}\int\operatorname{cosec}(x+\theta)\,dx$$

$$= \frac{1}{5}\log\tan\frac{(x+\theta)}{2}, \text{ where } \theta = \tan^{-1}\frac{4}{3}.$$

10.5. SOME SPECIAL INTEGRALS

Remember the following formulae:

1. $\int\frac{dx}{\sqrt{a^2-x^2}} = \sin^{-1}\frac{x}{a} + c$

2. $\int\sqrt{a^2-x^2}\,dx = \frac{x}{2}\sqrt{a^2-x^2} + \frac{a^2}{2}\sin^{-1}\frac{x}{a} + c$

3. $\int\frac{dx}{\sqrt{a^2+x^2}} = \log.(x+\sqrt{a^2+x^2}) + c$

4. $\int \sqrt{a^2 + x^2}\, dx = \frac{x}{2}\sqrt{a^2 + x^2} + \frac{a^2}{2}\log\left(x + \sqrt{a^2 + x^2}\right) + c$

5. $\int \frac{dx}{\sqrt{x^2 - a^2}} = \log\left(x + \sqrt{x^2 - a^2}\right) + c$

6. $\int \sqrt{x^2 - a^2}\, dx = \frac{x}{2}\sqrt{x^2 - a^2} - \frac{a^2}{2}\log\left(x + \sqrt{x^2 - a^2}\right) + c$

7. $\int \frac{dx}{x\sqrt{x^2 - a^2}} = \frac{1}{a}\sec^{-1}\frac{x}{a} + c$

8. $\int \frac{dx}{x^2 - a^2} = \frac{1}{2a}\log\left(\frac{x-a}{x+a}\right) + c$

9. $\int \frac{dx}{a^2 - x^2} = \frac{1}{2a}\log\left(\frac{a+x}{a-x}\right) + c$

10. $\int \frac{dx}{x^2 - a^2} = \frac{1}{a}\tan^{-1}\frac{x}{a} + c$

In order to prove the above-mentioned formuale, remember the following useful substitutions:

1. If $a^2 - x^2$ occurs in the integrand, put $x = a \sin\theta$ or $a \cos\theta$.
2. If $a^2 + x^2$ occurs in the integrand, put $x = a\tan\theta$ or $a \cot\theta$.
3. If $x^2 - a^2$ occurs in the integrand, put $x = a\sec\theta$ or $a\,\text{cosec}\,\theta$.

10.5.1. To Evaluate the Integrals

(*i*) $\int \sqrt{ax^2 + bx + c}\, dx$; (*ii*) $\int \frac{1}{\sqrt{ax^2 + bx + c}}\, dx$

If a be positive, we have

$$\sqrt{ax^2 + bx + c} = \sqrt{a\left(x^2 + \frac{b}{a}x + \frac{c}{a}\right)}$$

$$= \sqrt{a}\sqrt{\left(x^2 + \frac{b}{a}\right)^2 + \frac{4ac - b^2}{4a^2}}$$

(*i*) Let a and $4ac - b^2$ be both positive. We write

$$\frac{(4ac - b^2)}{4a^2} = \sqrt{a}\int \sqrt{y^2 + k^2}\, dy$$

$$= \sqrt{a}\left[\frac{y\sqrt{y^2 + k^2}}{2} + \frac{k^2}{2}\sinh^{-1}\frac{y}{k}\right]$$

and
$$\int \frac{1}{\sqrt{(ax^2 + bx + c)}}\, dx = \frac{1}{\sqrt{a}}\int \frac{dy}{\sqrt{y^2 + k^2}} = \frac{1}{\sqrt{a}}\sinh^{-1}\frac{y}{k}$$

(*ii*) Let a be positive and $4ac - b^2$ be both negative. We write

$$\frac{(4ac - b^2)}{4a^2} = -t \quad \text{and put} \quad \frac{x + b}{2a} = y$$

so that we have, in this case.

$$\int \sqrt{ax^2+bx+c}\,dx = \sqrt{a}\int \sqrt{y^2-t^2}\,dy$$

$$= \sqrt{a}\left[\frac{\sqrt{y^2-t^2}}{2} - \frac{t^2}{2}\cosh^{-1}\frac{y}{t}\right]$$

and $$\int \frac{1}{\sqrt{ax^2+bx+c}}\,dx = \frac{1}{\sqrt{a}}\int \frac{dy}{\sqrt{y^2-t^2}} = \frac{1}{\sqrt{a}}\cosh^{-1}\frac{y}{t}$$

(*iii*) Let a be negative. We have

$$\sqrt{ax^2+bx+c} = \sqrt{-a}\sqrt{-x^2-\frac{bx}{a}-\frac{c}{a}}$$

$$= \sqrt{-a}\sqrt{\frac{b^2-4ac}{4a^2}-\left(x+\frac{b}{2a}\right)^2}$$

We have assumed that $b^2 > 4ac$, for if $b^2 < 4ac$, the expression under radical sign will always be negative.

As before, we write $(b^2-4ac)/4a^2 = t^2$ and $x + b/2a = y$, so that we have, in this case,

$$\sqrt{ax^2+bx+c} = \sqrt{-a}\sqrt{t^2-y^2}$$

$$= \sqrt{-a}\left[\frac{y\sqrt{t^2-y^2}}{2} + \frac{t^2}{2}\sin^{-1}\frac{y}{t}\right]$$

and $$\int \frac{1}{\sqrt{ax^2+bx+c}}\,dx = \frac{1}{\sqrt{-a}}\int \frac{dy}{\sqrt{t^2-y^2}} = \frac{1}{\sqrt{a}}\sin^{-1}\frac{y}{t}$$

10.5.2. To Evaluate the Integrals

(*i*) $\dfrac{Ax+B}{\sqrt{ax^2+bx+c}}$ (*ii*) $(Ax+B)\sqrt{ax^2+bx+c}$

We determine two constants λ and μ, such that

$$Ax + B = \lambda(2ax+b) + \mu,$$

where $2ax + b$ is the differential coefficient of $ax^2 + bx + c$. The constants λ, μ are given by

$$A = 2a\lambda,\ B = b\lambda + \lambda$$

$$\Rightarrow \qquad \lambda = \frac{A}{2a},\ \mu = \frac{2aB - Ab}{2a}$$

(*i*) $$\int \frac{Ax+B}{\sqrt{ax^2+bx+c}}\,dx = \int \frac{\lambda(2ax+b)+\mu}{\sqrt{ax^2+bx+c}}\,dx$$

$$= \lambda\int (ax^2+bx+c)^{-1/2}(2ax+b)dx + \mu\int \frac{dx}{\sqrt{ax^2+bx+c}}$$

$$= 2\lambda(ax^2+bx+c)^{1/2} + \mu\int \frac{dx}{\sqrt{ax^2+bx+c}}$$

and the integral on the right has already been considered previously.

(*ii*) Again, we have $\int (Ax + B)\sqrt{ax^2 + bx + c}\, dx$

$$= \lambda \int (2ax + b)\sqrt{ax^2 + bx + c}\, dx + \mu \int \sqrt{ax^2 + bx + c}\, dx$$

$$= \frac{2\lambda}{3}(ax^2 + bx + c)^{3/2} + \mu \int \sqrt{ax^2 + bx + c}\, dx$$

and $\int \sqrt{ax^2 + bx + c}\, dx$ has alread been considered previously.

SOLVED EXAMPLES

Example 10.5. Evaluate

(*i*) $\int \frac{dx}{50 + 2x^2}$ (*ii*) $\int \sqrt{\frac{x + a}{x - a}}\, dx$ (*iii*) $\int \frac{(2x + 3)}{\sqrt{x^2 + 1}}\, dx$

(*iv*) $\int \frac{dx}{\sqrt{7 - 3x - 2x^2}}$ (*v*) $\int \frac{(2x + 5)}{\sqrt{x^2 - 2x + 2}}\, dx$ (*vi*) $\int \frac{\sec^2 x}{5\tan^2 x - 12\tan x + 14}\, dx$.

Solution. (*i*) We have

$$\int \frac{dx}{50 + 2x^2} = \frac{1}{2}\int \frac{dx}{25 + x^2} = \frac{1}{2}\int \frac{dx}{5^2 + x^2} = \frac{1}{2} \times \frac{1}{5}\tan^{-1}\frac{x}{5} + c = \frac{1}{10}\tan^{-1}\frac{x}{5} + c$$

(*ii*) We have $\int \sqrt{\frac{x + a}{x - a}}\, dx = \int \sqrt{\frac{x + a}{x - a} \times \frac{x + a}{x + a}}\, dx$

$$= \int \frac{x + a}{\sqrt{x^2 - a^2}}\, dx = \int \frac{x}{\sqrt{x^2 - a^2}}\, dx + a\int \frac{dx}{\sqrt{x^2 - a^2}} = I_1 + aI_2$$

Now $I_1 = \int \frac{x}{\sqrt{x^2 - a^2}}\, dx$

Put $x^2 - a^2 = t^2$ so that $2x\, dx = 2t\, dt$ or $x\, dx = t\, dt$.

$$\therefore \quad I_1 = \int \frac{t\, dt}{t} = \int dt = t = \sqrt{x^2 + a^2}$$

Again $I_2 = \int \frac{dx}{\sqrt{x^2 - a^2}} = \log(x + \sqrt{x^2 - a^2})$

Hence $\int \sqrt{\frac{x + a}{x - a}}\, dx = I_1 + aI_2 = \sqrt{x^2 - a^2} + a\log(x + \sqrt{x^2 - a^2}) + c$.

(*iii*) We have $\int \frac{(2x + 3)}{\sqrt{x^2 + 1}}\, dx = \int \frac{2x}{\sqrt{x^2 + 1}}\, dx + 3\int \frac{dx}{x^2 + 1}$ ...(*i*)

Now $\int \frac{2x}{\sqrt{x^2 + 1}}\, dx = \int \frac{dt}{\sqrt{t}}$, where $x^2 + 1 = t$ so that $2x\, dx = dt$.

$$= 2\sqrt{t} = 2\sqrt{x^2 + 1}$$

and $\int \frac{dx}{\sqrt{x^2 + 1}} = \log(x + \sqrt{x^2 + 1})$

Substituting in (*i*), we obtain

$$\int \frac{2x+3}{\sqrt{x^2+1}}\,dx = 2\sqrt{x^2+1}+3\log(x+\sqrt{x^2+1})+c.$$

(*iv*) We have $\displaystyle\int \frac{dx}{\sqrt{7-3x-2x^2}} = \int \frac{dx}{\sqrt{2}\sqrt{\frac{7}{2}-\left(x^3+\frac{3}{2}x\right)}}$

$$= \frac{1}{\sqrt{2}}\int \frac{dx}{\sqrt{\left(\frac{7}{2}+\frac{9}{16}\right)-\left(x^2+2.x.\frac{3}{4}+\frac{9}{16}\right)}}$$

$$= \frac{1}{\sqrt{2}}\int \frac{dx}{\sqrt{\frac{65}{16}-\left(x+\frac{3}{4}\right)^2}} = \frac{1}{\sqrt{2}}\int \frac{1}{\sqrt{\left(\frac{\sqrt{65}}{4}\right)^2-\left(x+\frac{3}{4}\right)^2}}$$

$$= \frac{1}{\sqrt{2}}\sin^{-1}\left(\frac{x+3/4}{\frac{\sqrt{65}}{4}}\right)+c = \frac{1}{\sqrt{2}}\sin^{-1}\left(\frac{4x+3}{\sqrt{65}}\right)+c.$$

(*v*) We have $\displaystyle\int \frac{(2x+5)}{\sqrt{x^2-2x+2}}\,dx$

Put $x^2-2x+2=t$ so that $(2x-2)\,dx = dt$

$$\therefore \quad \int \frac{(2x+5)}{\sqrt{x^2-2x+2}}\,dx = \int \frac{[(2x-2)+7]}{\sqrt{x^2-2x+2}}\,dx = \int \frac{(2x-2)\,dx}{\sqrt{x^2-2x+2}}+7\int \frac{dx}{\sqrt{x^2-2x+2}}$$

$$= 2\sqrt{t}+7\frac{dx}{\sqrt{(x-1)^2+1}} = 2\sqrt{t}+7\log[x+\sqrt{(x-1)^2+1}]+c$$

$$= 2\sqrt{x^2-2x+2}+7\log[x+\sqrt{x^2-2x+2}]+c.$$

(*vi*) We have $\displaystyle\int \frac{\sec^2 x}{5\tan^2 x-12\tan x+14}\,dx$

Put $\tan x = r$ so that $\sec^2 x\,dx = dt$

$$= \int \frac{dt}{5t^2-12t+4} = \frac{1}{5}\int \frac{dt}{\left(t^2-\frac{12}{5}t+\frac{14}{5}\right)}$$

$$= \frac{1}{5}\int \frac{dt}{\left(t^2-\frac{12}{5}t+\frac{36}{25}\right)-\frac{36}{25}+\frac{14}{5}} = \frac{1}{5}\int \frac{dt}{\left(t-\frac{6}{5}\right)^2+\frac{34}{25}}$$

$$= \frac{1}{5}\int \frac{dt}{\left(t-\frac{6}{5}\right)^2+\left(\frac{\sqrt{34}}{5}\right)^2} = \frac{1}{5}\cdot\frac{1}{\sqrt{34}/5}\tan^{-1}\frac{(t-6/5)}{\sqrt{34}/5}+c$$

$$= \frac{1}{\sqrt{34}}\tan^{-1}\left(\frac{5\tan x-6}{\sqrt{34}}\right)+c$$

Example 10.6. Evaluate

(*i*) $\int (2x-5)\sqrt{2+3x-x^2}\,dx$ (*ii*) $\int \sqrt{1+2x-3x^2}\,dx$

(*iii*) $\int \frac{6x+7}{\sqrt{(x-5)(x-4)}}\,dx$ (*iv*) $\int \frac{3x+1}{2x^2+x+1}\,dx$.

Solution. (*i*) We have $\int (2x-5)\sqrt{2+3x-x^2}\,dx$

Put $2+3x-x^2 = t$ so that $(3-2x)\,dx = dt$

$$\therefore \int (2x-5)\sqrt{2+3x-x^2}\,dx = -\int [(3-2x)+2]\sqrt{2+3x-x^2}\,dx$$

$$= -\int (3-2x)\sqrt{2+3x-x^2}\,dx - 2\int \sqrt{2+3x-x^2}\,dx$$

$$= -I_1 - 2I_2 \qquad ...(i)$$

Now, $I_1 = \int (3-2x)\sqrt{2+3x-x^2}\,dx$

$\therefore$ $I_1 = \int \sqrt{t}\,dt = \frac{2}{5}t^{3/2} = \frac{2}{3}(2+3x-x^2)^{3/2}$

Again $I_2 = \int \sqrt{2+3x-x^2}\,dx$

$$= \int \sqrt{-(x^2-3x-2)}\,dx = \int \sqrt{-\left[\left(x-\frac{3}{2}\right)^2 - 2 - \frac{9}{4}\right]}\,dx$$

$$= \int \sqrt{-\left[\left(x-\frac{3}{2}\right)^2 - \left(\frac{\sqrt{17}}{2}\right)^2\right]}\,dx = \int \sqrt{\left(\frac{\sqrt{17}}{2}\right)^2 - \left(x-\frac{3}{2}\right)^2}\,dx$$

$$= \frac{\left(x-\frac{3}{2}\right)\sqrt{\left(\frac{\sqrt{17}}{2}\right)^2 - \left(x-\frac{3}{2}\right)^2}}{2} + \frac{\left(\frac{\sqrt{17}}{2}\right)^2}{2}\sin^{-1}\frac{\left(x-\frac{3}{2}\right)}{\frac{\sqrt{17}}{2}}$$

$\therefore$ $$I_2 = \frac{(2x-3)\sqrt{2+3x-x^2}}{4} + \frac{17}{8}\sin^{-1}\frac{(2x-3)}{\sqrt{17}}$$

Substituting in (*i*), we obtain

$$\int (2x-5)\sqrt{2+3x-x^2}\,dx = -\frac{2}{3}(2+3x-x^2)^{3/2}$$

$$= -\frac{(2x-3)\sqrt{2+3x-x^2}}{2} - \frac{17}{4}\sin^{-1}\frac{2x-3}{\sqrt{17}} + c$$

$$= \frac{1}{6}(4x^2-18x+1)\sqrt{2+3x-x^2} - \frac{17}{4}\sin^{-1}\frac{2x-3}{\sqrt{17}} + c.$$

(*ii*) We have $\int \sqrt{1+2x-3x^2}\,dx$

Now,
$$1+2x-3x^2 = -3\left[x^2-\frac{2}{3}x-\frac{1}{3}\right]$$
$$= -3\left[\left(x-\frac{1}{3}\right)^2-\frac{1}{9}-\frac{1}{3}\right] = 3\left[\left(\frac{2}{3}\right)^2-\left(x-\frac{1}{3}\right)^2\right]$$

$\therefore$
$$\int \sqrt{1+2x-3x^2}\,dx = \sqrt{3}\int \sqrt{\left(\frac{2}{3}\right)^2-\left(x-\frac{1}{3}\right)^2}\,dx$$
$$= \sqrt{3}\left[\frac{1}{2}\left(x-\frac{1}{3}\right)\sqrt{\left(\frac{2}{3}\right)^2-\left(x-\frac{1}{3}\right)^2}+\frac{2}{9}\sin^{-1}\frac{\left(x-\frac{1}{3}\right)}{2/3}\right]+c$$
$$= \frac{1}{6}(3x-1)\sqrt{1+2x-3x^2}+\frac{2}{9}\sqrt{3}\sin^{-1}\left(\frac{3x-1}{2}\right)+c.$$

(*iii*) We have $\int \frac{6x+7}{\sqrt{(x-5)(x-4)}}\,dx$

As,
$$(x-5)(x-4) = x^2-9x+20$$
Put $x^2-9x+20 = t$ so that $(2x-9)\,dx = dt$.

$\therefore$
$$\int \frac{6x+7}{\sqrt{(x-5)(x-4)}}\,dx = \int \frac{3(2x-9)+34}{\sqrt{x^2-9x+20}}\,dx$$
$$= 3\int \frac{2x-9}{\sqrt{x^2-9x+20}}\,dx+34\int \frac{dx}{\sqrt{x^2-9x+20}}$$
$$= 3I_1+34I_2 \qquad ...(i)$$

Let
$$I_1 = \int \frac{2x-9}{\sqrt{x^2-9x+20}}\,dx = \int \frac{dt}{\sqrt{t}} = \int t^{-1/2}\,dt = \frac{t^{1/2}}{1/2} = 2\sqrt{t}$$
$$= 2\sqrt{x^2-9x+20}$$

Again
$$I_2 = \int \frac{dx}{\sqrt{x^2-9x+20}} = \int \frac{dx}{\sqrt{\left(x^2-9x+\frac{81}{4}\right)-\frac{81}{4}+20}}$$
$$= \int \frac{dx}{\sqrt{\left(x-\frac{9}{2}\right)^2-\left(\frac{1}{2}\right)^2}} = \log\left[\left(x-\frac{9}{2}\right)+\sqrt{\left(x-\frac{9}{2}\right)^2-\left(\frac{1}{2}\right)^2}\right]$$
$$= \log\left[\left(x-\frac{9}{2}\right)+\sqrt{x^2-9x+20}\right]$$

Hence
$$\int \frac{6x+7}{\sqrt{(x-5)(x-4)}}\,dx = 6\sqrt{x^2-9x+20}+34\log\left[\left(x-\frac{9}{2}\right)+\sqrt{x^2-9x+20}\right]+c.$$

(*iv*) We have $\int \frac{3x+1}{2x^2+x+1}\,dx$

Let $\quad 3x + 1 = A(4x + 1) + B$

$\Rightarrow \quad 3 = 4A \quad \text{and} \quad 1 = A + B$

or $\quad A = \frac{3}{4} \quad \text{and} \quad B = \frac{1}{4}$

Therefore,

$$\int \frac{3x+1}{2x^2+x+1}\,dx = \frac{3}{4}\int \frac{(4x+1)\,dx}{2x^2+x+1} + \frac{1}{4}\int \frac{dx}{2x^2+x+1}$$

$$= \frac{3}{4}\int \frac{dt}{t} + \frac{1}{8}\int \frac{dx}{\left(x+\frac{1}{4}\right)^2 + \left(\frac{\sqrt{7}}{4}\right)^2}$$

$$= \frac{3}{4}\log t + \frac{1}{8}\cdot\frac{1}{\sqrt{7}/4}\tan^{-1}\left(\frac{x+1/4}{\sqrt{7}/4}\right) + c$$

$$= \frac{3}{4}\log(2x^2+x+1) + \frac{1}{2\sqrt{7}}\tan^{-1}\left(\frac{4x+1}{\sqrt{7}}\right) + c.$$

Integration by Parts

It u and v are any two differentiable functions of x, then

$$\int u\,.\,v\,dx = u\,.\int v\,dx - \int\left[\frac{du}{dx}\,.\int v\,dx\right]dx$$

In other words, this rule of integration by parts can be stated as:

Integral of the product of two functions

= First function × Integral of the second – Integral of [Diff. coeff. of the first × Integral of the second].

In finding integrals by this method proper choice of Ist and IInd function is essential. Although there is no fixed law for taking Ist and IInd function, yet following rule is helpful in the choice of functions Ist and IInd.

(*i*) If the two functions are of different types take that function as Ist which comes first in the word ILATE, where

I, stands for Inverse trigonometric function.

L, stands for Logarithmic function.

A, stands for Algebraic function.

T, stands for Trigonometric function.

and *E*, stands for Exponential function.

(*ii*) If both the function are trigonometric take that function as 2nd whose integral is simpler.

(*iii*) If both the functions are algebraic take that function as Ist whose differential coefficient is simplex.

(*iv*) Unity may be taken as one of the functions.

(*v*) The formula of integration by parts can be applied more than once if necessary.

SOLVED EXAMPLES

Example 10.7. Evaluate

(i) $\int x^3 e^x dx$ **(ii)** $\int x \tan^{-1} x\, dx$ **(iii)** $\int x^2 \cos^2 x\, dx$ **(iv)** $\int \log x\, dx$

Solution. (*i*) We have $\int x^3 e^x dx$

We choose: first function = x^3, second function = e^x

$$\therefore \quad \int x^3 e^x dx = x^3 e^x - \int 3x^2 . e^x dx \quad ...(i)$$

Integrating by parts again, we obtain

$$\int x^2 e^x dx = x^2 . e^x - \int 2xe^x dx \quad ...(ii)$$

and

$$\int xe^x dx = x . e^x - \int 1 . e^x dx = x e^x - e^x \quad ...(iii)$$

Substituting (*iii*) in (*ii*), we get

$$\int x^2 e^x dx = x^2 e^x - 2x e^x + 2e^x \quad ...(iv)$$

Substituting (*iv*) in (*i*), we get

$$\int x^3 e^x dx = x^3 e^x - 3x^2 e^x + 6x e^x - 6e^x + c.$$

(*ii*) We have $\int x \tan^{-1} x\, dx$

We choose: first function = $\tan^{-1} x$, second function = x

$$\therefore \quad \int x \tan^{-1} x\, dx = \tan^{-1} x . \frac{x^2}{2} - \int \frac{1}{1+x^2} . \frac{x^2}{2} dx$$

$$= \frac{x^2 \tan^{-1} x}{2} - \frac{1}{2}\int \frac{1+x^2-1}{1+x^2} dx$$

$$= \frac{x^2 \tan^{-1} x}{2} - \frac{1}{2}\int \left(1 - \frac{1}{1+x^2}\right) dx$$

$$= \frac{x^2 \tan^{-1} x}{2} - \frac{1}{2}(x - \tan^{-} x) + c$$

$$= \frac{1}{2}(x^2+1) \tan^{-1} x - \frac{1}{2} x + c .$$

(*iii*) We have $\int x^2 \cos^2 x\, dx = \int x^2 \left(\frac{1+\cos 2x}{2}\right) dx$ $\left[\because \cos^2 A = \frac{1+\cos 2A}{2}\right]$

$$= \frac{1}{2}\int x^2 dx + \frac{1}{2}\int x^2 \cos 2x\, dx$$

$$= \frac{1}{2} . \frac{x^3}{3} + \frac{1}{2}\left[x^2 . \frac{\sin 2x}{2} - \int 2x . \frac{\sin 2x}{2} dx\right]$$

[We choose: first function = x^2 ; second function = $\cos 2x$]

$$= \frac{1}{6} x^2 + \frac{1}{4} x^2 \sin 2x - \frac{1}{2}\int x \sin 2x\, dx$$

$$= \frac{1}{6}x^2 + \frac{1}{4}x^2 \sin 2x - \frac{1}{2}\left[x.\left(\frac{-\cos 2x}{2}\right) - \int 1.\left(\frac{-\cos 2x}{2}\right)dx\right]$$

$$= \frac{1}{6}x^2 + \frac{1}{4}x^2 \sin 2x + \frac{1}{4}x\cos 2x - \frac{1}{4}\int \cos 2x\, dx$$

$$= \frac{1}{6}x^2 + \frac{1}{4}x^2 \sin 2x + \frac{1}{4}x\cos 2x - \frac{1}{8}\sin 2x.$$

(*iv*) We have $\int \log x\, dx = \int \log x.1\, dx$

Integrating by parts, taking log x as the first function and 1 as the second function, we get

$$\int \log x.1\, dx = (\log x).\int 1 \log - \int\left(\frac{1}{x}.\int 1\, dx\right)dx = (\log x).x - \int \frac{1}{x}.x\, dx$$

$$= x\log x - \int dx = x\log x - x + c = x(\log x - 1) + c.$$

Note: To evaluate several integrals, we have to use the method of substitution as well as integration by parts in succession.

Example 10.8. Evaluate:

(*i*) $\int \sin\sqrt{x}\, dx$ (*ii*) $\int \frac{\log(\sec^{-1} x)}{x\sqrt{x^2-1}}dx$

Solution. (*i*) We have $\int \sin\sqrt{x}\, dx$

Put $\sqrt{x} = t$ so that $\frac{1}{2\sqrt{x}}dx = dt$ or $dx = 2\sqrt{x}\, dt$

or $dx = 2t\, dt.$

$\therefore$ $\int \sin\sqrt{x}\, dx = 2\int t.\sin t\, dt$

Integrating by parts, taking t as first function and sin t as the second function, we get

$$= 2\left[t(-\cos t) - \int 1.(-\cos t)\, dt\right] = -2t\cos t + 2\int \cos t\, dt$$

$$= -2t\cos t + 2\sin t + c = -2\sqrt{x}\cos\sqrt{x} + 2\sin\sqrt{x} + c.$$

(*ii*) We have $\int \frac{\log(\sec^{-1} x)}{x\sqrt{x^2-1}}dx$

Put $\sec^{-1} x = t$ so that $\frac{1}{x\sqrt{x^2-1}}dx = dt$

$\therefore$ $\int \frac{\log(\sec^{-1} x)}{x\sqrt{x^2-1}}dx = \int \log t\, dt = \int \log t.1\, dt$

$$= (\log t).t - \int \frac{1}{t}.t\, dt = t\log t - \int 1\, dt = (\log t - 1)t + c.$$

$$= [\log(\sec^{-1} x) - 1]\sec^{-1} x + c.$$

Sometimes we split an integral of the form

$$I = \int e^x [f(x) + f'(x)]\, dx$$

into sum of two integrals and integrals and integrate one of them by parts to cancel out the other. Thus

$$I = \int e^x f(x)\, dx + \int e^x f'(x)\, dx$$

$$= \int e^x f(x)\, dx + \left[e^x f(x) - \int e^x f(x)\, dx\right] + c = e^x f(x) + c$$

$$\therefore \quad \int e^x [f(x) + f'(x)]\, dx = e^x f(x) + c.$$

Example 10.9. Evaluate

(*i*) $\int e^x \left(\frac{1}{x} - \frac{1}{x^2}\right) dx$ (*ii*) $\int \frac{\log x}{(1 + \log x)^2} dx$

Solution. (*i*) We have

$$\int e^x \left(\frac{1}{x} - \frac{1}{x^2}\right) dx = \int e^x . \frac{1}{x} dx - \int e^x . \frac{1}{x^2} dx$$

$$= \left[\frac{1}{x}(e^x) - \int -\frac{1}{x^2} . e^x\, dx\right] - \int e^x . \frac{1}{x^2} dx$$

$$= \frac{1}{x} e^x + \int e^x . \frac{1}{x^2} dx - \int e^x . \frac{1}{x^2} dx = \frac{1}{x} e^x + c.$$

(*ii*) We have $\int \frac{\log x}{(1 + \log x^2)} dx$

Putting $\log x = t$ *i.e.*, $x = e^t$ so that $dx = e^t\, dt$

$$\therefore \quad \int \frac{\log x}{(1 + \log x)^2} = \int \frac{t}{(1+t)^2} e^t\, dt = \int \frac{(t+1) - 1}{(1+t)} e^t\, dt$$

$$= \int \left[\frac{1}{1+t} - \frac{1}{(1+t)^2}\right] e^t\, dt = \int \frac{1}{1+t} e^t\, dt + \int \frac{-1}{(1+t)} e^t\, dt$$

$$= \frac{1}{1+t} e^t - \int -\frac{1}{(1+t)^2} e^t\, dt + \int -\frac{1}{(1+t)^2} e^t\, dt$$

$$= \frac{e^t}{1+t} + c = \frac{x}{1 + \log x} + c.$$

Integration by Partial Fractions

This method is used to integrate function of the type $\frac{N(x)}{D(x)}$, where both $N(x)$ and $D(x)$ are polynomials.

Such functions are called rational functions. We may assume, without any loss of generality, that the numerator $N(x)$ and denominator $D(x)$ have no common polynomial factor and that the degree of $N(x)$ is less than degree of $D(x)$. For, if the degree of numerator were greater than or equal to the degree of denominator, we could use long division to divide $N(x)$ by $D(x)$ so that

$$\frac{N(x)}{D(x)} = Q(x) + \frac{R(x)}{D(x)}.$$

where $Q(x)$ is the quotient (a polynomial easily integrable) and $R(x)$ is the remainder whose degree is less than degree of $D(x)$. The method of expressing $\frac{N(x)}{D(x)}$ into partial fractions depends on factors of the denominator. It is a fact that every polynomial $D(x)$ can be expressed as a product of linear and irreducible quadratic factors with real coefficients. As such we have the following four cases to consider.

Case I: All factors in the denominator are linear and non-repeated.

Let $D(x) = (x - a_1)(x - a_2) \ldots (x - a_n)$. Then we assume that

$$\frac{N(x)}{D(x)} = \frac{A_1}{x - a_1} + \frac{A_2}{x - a_2} + \ldots + \frac{A_n}{x - a_n}$$

where $A_1, A_2, \ldots, A_n$ are constants are can be determined by equating the numerator on RHS to the numerator on LHS and then substituting $x = a_1, a_2, \ldots a_n$.

SOLVED EXAMPLES

Example 10.10. Evaluate

(*i*) $\int \frac{2x-1}{(x-1)(x+2)(x-3)} dx$ (*ii*) $\int \frac{2x}{(x^2+1)(x^2+2)} dx$

Solution. (*i*) Let

$$\frac{2x-1}{(x-1)(x+2)(x-3)} = \frac{A}{x-1} + \frac{B}{x+2} + \frac{C}{x-3} \quad \ldots(i)$$

$$\Rightarrow \quad 2x - 1 = A(x+2)(x-3) + B(x-1)(x-3) + C(x-1)(x-2) \quad \ldots(ii)$$

Putting $x + 2 = 0$ or $x = -2$ in (*ii*), we get $-5 = B(-3)(-5) \Rightarrow B = -\frac{1}{3}$

Putting $x - 3 = 0$ or $x = 3$ in (*ii*), we get $5 = C(2)(5) \Rightarrow C = \frac{1}{2}$

Putting $x - 1 = 0$ or $x = 1$ in (*ii*), we get $1 = A(3)(-2) \Rightarrow A = -\frac{1}{6}$

Substituting the values of A, B and C in (*i*), we obtain

$$\frac{2x-1}{(x-1)(x+2)(x-3)} = -\frac{1}{6}\cdot\frac{1}{x-1} - \frac{1}{3}\cdot\frac{1}{x+2} + \frac{1}{2}\cdot\frac{1}{x-3}$$

So,
$$\int \frac{2x-1}{(x-1)(x+2)(x-3)} dx = -\frac{1}{6}\int \frac{1}{x-1} dx - \frac{1}{3}\int \frac{1}{x+2} dx + \frac{1}{2}\int \frac{1}{x-3} dx$$

$$= -\frac{1}{6}\log(x-1) - \frac{1}{3}\log(x+2) + \frac{1}{2}\log(x-3) + c.$$

(*ii*) Put $x^2 = t$ so that $2x\,dx = dt$

$$\therefore \quad \int \frac{2x}{(x^2+1)(x^2+2)} dx = \int \frac{1}{(t+1)(t+2)} dt$$

Let $$\frac{1}{(t+1)(t+2)} = \frac{A}{t+1} + \frac{B}{t+2} \qquad ...(i)$$

Then, $$1 = A(t+2) + B(t+1) \qquad ...(ii)$$

Putting $t + 1 = 0$ or $t = -1$ in (*ii*), we get $A = 1$.

Putting $t + 2 = 0$ or $t = -2$ in (*ii*), we get $B = -1$.

Substituting the values of A and B in (*i*), we obtain

$$\frac{1}{(t+1)(t+2)} = \frac{1}{t+1} - \frac{1}{t+2}$$

So, $$\int \frac{2x\,dx}{(x^2+1)(x^2+2)} = \int \frac{1}{(t+1)(t+2)}\,dx = \int \frac{1}{t+1}\,dt - \int \frac{1}{t+2}\,dt$$

$$= \log|\,t+1\,|\,t+2\,| + c \qquad [\because \quad t = x^2]$$

Case II. All factors in the denominator are linear but some of which are repeated.

To each linear factor $(ax + b)r$ repeating r times, there will correspond the sum of r partial fractions as follows:

$$\frac{A_1}{ax+b} + \frac{A_2}{(ax+b)^2} + \ldots + \frac{A_r}{(ax+b)^r},$$

where $A_1, A_2, \ldots, A_r$ are constant.

Example 10.11. Evaluate $\displaystyle\int \frac{3x+1}{(x-2)^2(x+2)}\,dx$

Solution.

Let $$\frac{3x+1}{(x-2)^2(x+2)} = \frac{A}{x-2} + \frac{B}{(x-2)^2} + \frac{C}{x+2} \qquad ...(i)$$

Then, $$3x + 1 = A(x-2)(x+2) + B(x+2) + C(x-2)^2 \qquad ...(ii)$$

Putting $x - 2 = 0$ or $x = 2$ in (*ii*), we get $7 = 4B \Rightarrow B = \frac{7}{4}$.

Putting $x + 2 = 0$ or $x = -2$ in (*ii*), we get $-5 = 16C \Rightarrow C = -\frac{5}{16}$.

Comparing coefficient of x^2 on both sides of the identity (*ii*), we get

$$A + C = 0 \;\Rightarrow\; A = -C \;\Rightarrow\; A = \frac{5}{16}$$

Substituting the values of A, B and C in (*ii*), we obtain

$$\frac{3x+1}{(x-2)^2(x+2)} = \frac{5}{16}\cdot\frac{1}{x-2} + \frac{7}{4}\cdot\frac{1}{(x-2)^2} - \frac{5}{16}\cdot\frac{1}{x+2}$$

So, $$\int \frac{3x+1}{(x-2)^2(x+2)}\,dx = \frac{5}{16}\int \frac{1}{x-2}\,dx + \frac{7}{4}\int \frac{1}{(x-2)^2}\,dx - \frac{5}{16}\int \frac{1}{x+2}\,dx$$

$$= \frac{5}{16}\log|x-2| - \frac{7}{4(x-2)} - \frac{5}{16}\log|x+2| + c.$$

Case III. The factors in the denominator are linear and irreducible quadratic but quadratic factors are non-repeated.

To each irreducible quadratic factor $ax^2 + bx + c$ that occurs exactly once in the denominator, there will correspond a partial fraction of the type $\dfrac{Ax + B}{ax^2 + bx + c}$, where A and B are constant.

Example 10.12. Evaluate $\displaystyle\int \frac{x^3 - 5x}{(x^2 - 9)(x^2 + 1)} dx$

Solution.

Let $$\frac{x^3 - 5x}{(x^2 - 9)(x^2 + 1)} = \frac{A}{x-3} + \frac{B}{x+3} + \frac{Cx + D}{x^2 + 1} \quad ...(i)$$

$$\Rightarrow \quad x^3 - 5x = A(x + 3)(x^2 + 1) + B(x - 3)(x^2 + 1) + (Cx + D)(x^2 - 9) \quad ...(ii)$$

Putting $x - 3 = 0$ or $x = 3$ in (*ii*), we get $12 = 60A \Rightarrow A = \dfrac{1}{5}$

Putting $x + 3 = 0$ or $x = -3$ in (*ii*), we get $-12 = -60B \Rightarrow B = \dfrac{1}{5}$.

To find the constant C and D, let us choose $x = 0$ and $x = 1$.

If $x = 0$, then $0 = 3A - 3B - 9D = -9D$ or $D = 0$

If $x = 1$, then $-4 = 8A - 4B + (C + D)(-8) \Rightarrow C = \dfrac{3}{5}$.

Substituting the values of A, B, C and D in (*i*), we obtain

$$\frac{x^3 - 5x}{(x^2 - 9)(x^2 + 1)} = \frac{1}{5}\cdot\frac{1}{x-3} + \frac{1}{5}\cdot\frac{1}{x+3} + \frac{3}{5}\cdot\frac{x}{x^2 + 1}$$

$$\therefore \quad \int \frac{x^3 - 5x}{(x^2 - 9)(x^2 + 1)} dx = \frac{1}{5}\int \frac{1}{x-3} dx + \frac{1}{5}\int \frac{1}{x+3} dx + \frac{3}{5}\int \frac{x}{x^2+1} dx$$

$$= \frac{1}{5}\log|x - 3| + \frac{1}{5}\log|x + 3| + \frac{3}{10}\log|x^2 + 1| + c$$

$$= \frac{1}{5}\log|x^2 - 9| + \frac{3}{10}\log|x^2 + 1| + c.$$

Case IV: The factors in the denominator are linear and irreducible quadratic and factors are repeated.

To each repeated quadratic factor $(ax^2 + bx + c)^r$ repeated r times, there will correspond a sum of r partial fractions of the type

$$\frac{A_1 + B_1 x}{ax^2 + bx + C} + \frac{A_2 + B_2 x}{(ax^2 + bx + C)^2} + \ldots + \frac{A_r + B_r x}{(ax^2 + bx + C)^r}$$

Example 10.13. Evaluate $\displaystyle\int \frac{2x^4 + 9x^2 + 8}{x(x^2 + 2)} dx$.

Solution. The denominator of the given integral contains a linear factor which occurs exactly once and an irreducible quadratic factor which occurs twice. Thus we can write

$$\frac{2x^4+9x^2+8}{x(x^2+2)^2} = \frac{A}{x}+\frac{Bx+C}{x^2+2}+\frac{Dx+E}{(x^2+2)^2}$$

$$\Rightarrow \quad 2x^4+9x^2+8 = A(x^2+2)^2+(Bx+C)(x)(x^2+2)+(Dx+E)x$$
$$= (A+B)x^4+Cx^3+(4A+2B+D)x^2+(2C+E)x+4A$$

Equating like powers of x, we obtain

$$2 = A+B, \quad 0 = C, \quad 9 = 4A+2B+D, \quad 0 = 2C+E, \quad 8 = 8A$$

Solving gives

$$A = 2, \quad B = 0, \quad C = 0, \quad D = 1 \quad \text{and} \quad E = 0$$

$$\therefore \quad \frac{2x^4+9x^2+8}{x(x^2+2)^2} = \frac{2}{x}+\frac{x}{(x^2+2)^2}$$

$$\Rightarrow \quad \int \frac{2x^4+9x^2+8}{x(x^2+2)^2}\,dx = 2\int \frac{1}{x}\,dx+\int \frac{x}{(x^2+2)^2}\,dx = 2\log|x|+\frac{1}{2}\int (x^2+2)^{-2}\,2x\,dx$$

$$= 2\log|x|+\frac{1}{2}\frac{(x^2+2)^{-2+1}}{(-2+1)}+c = 2\log|x|-\frac{1}{2}\cdot\frac{1}{x^2+2}+c.$$

Case V: When only even powers of x occurs both in numerator and denominator. To resolve into partial fractions, put $x^2 = y$ in the given rational function.

Example 10.14. Evaluate

(*i*) $\displaystyle\int \frac{x^2}{(x^2+4)(x^2+9)}\,dx$ (*ii*) $\displaystyle\int \frac{\sin x}{\sin 4x}\,dx$

Solution. (*i*) Let $x^2 = y$.

Then $$\frac{x^2}{(x^2+4)(x^2+9)} = \frac{y}{(y+4)(y+9)}$$

Let $$\frac{y}{(y+4)(y+9)} = \frac{A}{y+4}+\frac{B}{y+9} \quad \text{...(i)}$$

$$\Rightarrow \quad y = A(y+9)+B(y+4) \quad \text{...(ii)}$$

Putting $y+9=0$ or $y=-9$ in (*ii*), we get $-9 = -5B \Rightarrow B = \frac{9}{5}$.

Putting $y+4=0$ or $y=-4$ in (*ii*), we get $-4 = 5A \Rightarrow A = -\frac{4}{5}$.

Substituting the values of A and B in (*i*), we obtain

$$\frac{y}{(y+4)(y+9)} = \frac{-4}{5}\cdot\frac{1}{y+4}+\frac{9}{5}\cdot\frac{1}{y+9}$$

Replacing y by x^2, we get

$$\frac{x^2}{(x^2+4)(x^2+9)} = -\frac{4}{5}\cdot\frac{1}{x^2+4}+\frac{9}{5}\cdot\frac{1}{x^2+9}$$

$$\therefore \quad \int \frac{x^2}{(x^2+4)(x^2+9)}\,dx = -\frac{4}{5}\int \frac{1}{x^2+4}\,dx + \frac{9}{5}\int \frac{1}{x^2+9}\,dx$$

$$= -\frac{4}{5}\cdot\frac{1}{2}\tan^{-1}\frac{x}{2} + \frac{9}{5}\cdot\frac{1}{3}\tan^{-1}\frac{x}{3} + c$$

$$= -\frac{2}{5}\tan^{-1}\frac{x}{2} + \frac{3}{5}\tan^{-1}\frac{x}{3} + c.$$

(*ii*) Let $\quad I = \int \frac{\sin x}{\sin 4x}\,dx = \int \frac{\sin x}{2\sin 2x\cos 2x}\,dx = \int \frac{\sin x}{4\sin x\cos x\cos 2x}\,dx$

$$= \frac{1}{4}\int \frac{1}{\cos x\cos 2x}\,dx = \frac{1}{4}\int \frac{\cos x}{\cos^2 x\cos 2x}\,dx$$

$$= \frac{1}{4}\int \frac{\cos x}{(1-\sin^2 x)(1-2\sin^2 x)}\,dx$$

Put $\sin x = t$ so that $\cos x\,dx = dt$

$$\therefore \quad I = \frac{1}{4}\int \frac{dt}{(1-t^2)(1-2t^2)}$$

Let $t^2 = y$.

Then $\quad \frac{1}{(1-t^2)(1-2t^2)} = \frac{1}{(1-y)(1-2y)}.$

Let $\quad \frac{1}{(1-y)(1-2y)} = \frac{A}{1-y} + \frac{B}{1-2y}$ ...(*i*)

Then $\quad 1 = A(1-2y) + B(1-y)$ ...(*ii*)

Putting $1 - y = 0$ or $y = 1$ in (*ii*), we get $A = -1$.

Putting $1 - 2y = 0$ or $y = \frac{1}{2}$ in (*ii*), we get $B = 2$.

$$\therefore \quad \frac{1}{(1-y)(1-2y)} = -\frac{1}{1-y} + \frac{2}{1-2y} \Rightarrow \frac{1}{(1-t^2)(1-2t^2)} = \frac{1}{1-t^2} + \frac{2}{1-2t^2}$$

$$\Rightarrow \quad I = \frac{1}{4}\int \frac{dt}{(1-t^2)(1-2t^2)} = \frac{1}{4}\int \left[-\frac{1}{1-t^2} + \frac{2}{1-2t^2}\right]dt$$

$$= -\frac{1}{4}\int \frac{1}{1-t^2}\,dt + \frac{2}{4}\int \frac{1}{1-(\sqrt{2}t)^2}\,dt$$

$$= -\frac{1}{4}\cdot\frac{1}{2}\log\left|\frac{1+t}{1-t}\right| + \frac{1}{2}\cdot\frac{1}{2\sqrt{2}}\log\left|\frac{1+\sqrt{2}t}{1-\sqrt{2}t}\right| + c$$

$$= -\frac{1}{8}\log\left|\frac{1+t}{1-t}\right| + \frac{1}{4\sqrt{2}}\log\left|\frac{1+\sqrt{2}t}{1-\sqrt{2}t}\right| + c$$

$$= -\frac{1}{8}\log\left|\frac{1+\sin x}{1-\sin x}\right| + \frac{1}{4\sqrt{2}}\log\left|\frac{1+\sqrt{2}\sin x}{1-\sqrt{2}\sin x}\right| + c.$$

EXERCISE 10.1

Evaluate the following integrals:

1. $2\sqrt{x} - \frac{3}{\sqrt{x}} + \frac{5}{x} + 7\tan^2 x - 9\cot^2 x$. $\left[\textbf{Ans. } \frac{4}{3}x^{3/2} - 6\sqrt{x} + 5\log x + 7\tan x + 9\cot x + 2x + c\right]$

2. $\frac{\cos 2x - \cos 2\alpha}{\cos x - \cos \alpha}$ $[\textbf{Ans. } 2(\sin x + x\cos\alpha) + c]$

3. $\frac{1}{1+\sin x}$. $[\textbf{Ans. } \tan x - \sec x + c]$

4. $\sqrt{1+\cos 2x}$. $\left[\textbf{Ans. } \sqrt{2}\sin x + c\right]$

5. $\frac{1}{\sqrt{3x+4} - \sqrt{3x+1}}$. $\left[\textbf{Ans. } \frac{2}{27}[(3x+4)^{3/2} + (3x+1)^{3/2}] + c\right]$

6. $\sin^3 x$. $\left[\textbf{Ans. } -\frac{3}{4}\cos x + \frac{1}{2}\cos 3x + c\right]$

7. $\frac{1+\tan x}{x + \log\sec x}$. $[\textbf{Ans. } \log|x + \log\sec x| + c]$

8. $\frac{x^2}{\sqrt{1-x^6}}$. $\left[\textbf{Ans. } \frac{1}{3}\sin^{-1} x^3 + c\right]$

9. $\frac{1}{1+\cot x}$. $\left[\textbf{Ans. } \frac{1}{2}x - \frac{1}{2}\log|\sin x + \cos x| + c\right]$

10. $\frac{1}{\sin x\cos^3 x}$. $\left[\textbf{Ans. } \log|\tan x| + \frac{1}{2}\tan^2 x + c\right]$

11. $\frac{(x+1)e^x}{\sin^2(xe^x)}$. $[\textbf{Ans. } -\cot(x^{ex}) + c]$

12. $\frac{x^2\tan^{-1}x^3}{1+x^6}$. $\left[\textbf{Ans. } \frac{1}{6}(\tan^{-1}x^3)^2 + c\right]$

13. $\frac{\sin x}{\sin(x-a)}$ $[\textbf{Ans. } \sin a\log\sin(x-a) + (x-a)\cos a + c]$

14. $\frac{\sin x\cos x}{a^2\sin^2 x + b^2\cos^2 x}$. $\left[\textbf{Ans. } \frac{1}{2(a^2-b^2)}\log|a^2\sin^2 x + b^2\cos^2 x| + c\right]$

15. $\frac{1}{\sin(x-a)\sin(x-b)}$. $\left[\textbf{Ans. } \frac{1}{\sin(a-b)}\log\frac{\sin(x-a)}{\sin(x-b)} + c\right]$

16. $\frac{x}{\cos^2 x}$. $[\textbf{Ans. } x\tan x - \log\sec x + c]$

17. $\sec^3 x$. $\left[\textbf{Ans. } \frac{1}{2}\left[\sec x\tan x + \log\tan\left(\frac{\pi}{4} + \frac{x}{2}\right)\right] + c\right]$

18. $e^{ax}\cos(bx + C)$. $\left[\textbf{Ans.}\ \dfrac{e^{ax}}{\sqrt{a^2+b^2}}\cos\left(bx + C - \tan^{-1}\dfrac{b}{a}\right) + c\right]$

19. $e^x\left[\sin^{-1}x + \dfrac{1}{\sqrt{1-x^2}}\right]$. $[\textbf{Ans.}\ e^x\sin^{-1}x + c]$

20. $\dfrac{x^2+1}{(x+1)}e^x$ $\left[\textbf{Ans.}\ e^x\left(\dfrac{x-1}{x+1}\right) + c\right]$

21. $\dfrac{xe^{2x}}{(2x+1)}$. $\left[\textbf{Ans.}\ \dfrac{e^{2x}}{4(2x+1)} + c\right]$

22. $\log(\log x) + \dfrac{1}{(\log x)^2}$. $\left[\textbf{Ans.}\ x\left[\log(\log x) - \dfrac{1}{\log x}\right] + c\right]$

23. $\dfrac{1}{1+3\sin^2 x}$. $\left[\textbf{Ans.}\ \dfrac{1}{2}\tan^{-1}(2\tan x) + c\right]$

24. $\dfrac{1}{9x^2+6x+10}$. $\left[\textbf{Ans.}\ \dfrac{1}{9}\tan^{-1}\left(\dfrac{3x+1}{3}\right) + c\right]$

25. $\dfrac{1}{\sqrt{2+2x-x^2}}$. $\left[\textbf{Ans.}\ \sin^{1}\left(\dfrac{x-1}{\sqrt{3}}\right) + c\right]$

26. $\dfrac{\cos x}{\sqrt{\sin^2 x - 2\sin x - 3}}$. $\left[\textbf{Ans.}\ \log\left[(\sin x - 1) + \sqrt{\sin^2 x - 2\sin x - 3}\right] + c\right]$

27. $\dfrac{x+2}{\sqrt{x^2+2x-1}}$. $\left[\textbf{Ans.}\ \sqrt{x^2+2x-1} + \log\left[(x+1) - \sqrt{x^2+2x-1}\right] + c\right]$

28. $\dfrac{3-2x}{\sqrt{(2-x)(1+x)}}$. $\left[\textbf{Ans.}\ 2\sqrt{2+x-x^2} + 2\sin^{-1}\left(\dfrac{2x-1}{3}\right) + c\right]$

29. $\sqrt{4-3x-2x^2}$. $\left[\textbf{Ans.}\ \dfrac{4x+3}{8}\sqrt{4-3x-2x^2} + \dfrac{41}{16\sqrt{2}}\sin^{-1}\dfrac{4x+3}{\sqrt{41}} + c\right]$

30. $(x-5)\sqrt{x^2+x}$. $\left[\textbf{Ans.}\ \dfrac{1}{3}(x^2+x)^{3/2} - \dfrac{11}{2}\left[\dfrac{x+\frac{1}{2}}{2}\sqrt{x^2+x} - \dfrac{1}{8}\log\left|x + \dfrac{1}{2} + \sqrt{x^2+x}\right|\right] + c\right]$

31. $(2x-5)\sqrt{x^2-4x+3}$. $\left[\textbf{Ans.}\ \dfrac{2}{3}(x^2-4x+3)^{3/2}\ \dfrac{1}{2}\left[(x-2)\sqrt{x^2-4x+3} - \log\left(\sqrt{x^2-4x+3} + x - 2\right) + c\right]\right]$

32. $\dfrac{x^3-x-2}{1-x^2}$. $\left[\textbf{Ans.}\ \log\left|\dfrac{x-1}{x+1}\right| - \dfrac{x^2}{2} + c\right]$

33. $\dfrac{x^2+8x+4}{x^3-4x}$. $[\textbf{Ans.}\ -\log|x| + 3\log|x-2| - \log|x+2| + c]$

34. $\dfrac{1}{x(x^n+1)}$. $\left[\textbf{Ans.} \dfrac{1}{n}\log\left|\dfrac{x^n}{x^n+1}\right|+c\right]$

35. $\dfrac{1}{(e^x-1)^2}$. $\left[\textbf{Ans.} \log\left|\dfrac{e^x}{e^x-1}\right|-\dfrac{1}{e^x-1}+c\right]$

36. $\dfrac{1}{x[6(\log x)^2+7\log x+2]}$. $\left[\textbf{Ans.} \log\left|\dfrac{2\log x+1}{3\log x+2}\right|+c\right]$

37. $\dfrac{1}{\sin^2 x+\sin 2x}$. $\left[\textbf{Ans.} \dfrac{1}{2}\log\left|\dfrac{\tan x}{\tan x+2}\right|+c\right]$

38. $\dfrac{3x+5}{x^3-x^2-x+1}$. $\left[\textbf{Ans.} \dfrac{1}{2}\log\left|\dfrac{x+1}{x-1}\right|-\dfrac{4}{x-1}+c\right]$

39. $\dfrac{x}{x^3+x^2+x+1}$ $\left[\textbf{Ans.} \dfrac{1}{4}\log|x^2+1|+\dfrac{1}{2}\tan^{-1}x-\dfrac{1}{2}\log|x+1|+c\right]$

40. $\dfrac{x^2+2}{(x^2+1)(x^2+4)}$. $\left[\textbf{Ans.} \dfrac{1}{6}\log|1-\cos x|+\dfrac{1}{2}\log|1+\cos x|+\dfrac{2}{3}\log|1+2\cos x|+c\right]$

10.6. THE DEFINITE INTEGRAL

Let $F(x)$ be any antiderivative of $f(x)$, then for any two values of the independent variable x, say a and b, the difference $F(b) - F(a)$ is called the definite integral of $f(x)$ from a to b and is denoted by $\int_a^b f(x)\,dx$. Thus

$$\int_a^b f(x)\,dx = F(b) - F(a) \qquad ...(i)$$

where $F(x)$ is any antiderivative of $f(x)$. The number a and b appearing with the integral sign $\int$ in (*i*) are called the limits of integration; a is the lower limit and b is the upper limit. Usually, $F(b) - F(a)$ is abbreviated by writing $[F(x)]_a^b$.

The reason for using the term 'definite integral' follows from the fact that the value of the definite integral is independent of the particular choice of the antiderivative of $f(x)$, then

$$\int_a^b f(x)\,dx = [F(x)+c]_a^b = [F(b)+c] - [F(a)+c] = F(b) - F(a)$$

which is same as before.

Thus, to find $\int_a^b f(x)\,dx$, if suffices to find an antiderivative of $f(x)$, say $F(x)$ and subtract the value of $F(x)$ at the lower unit a from its value at the upper limit b.

SOLVED EXAMPLES

Example 10.15. Evaluate

(*i*) $\int_0^1 (2x^3 + 3)^2\, dx$ (*ii*) $\int_0^1 \frac{1 - x^2}{1 + x^2}$ (*iii*) $\int_0^{\pi/2} \sqrt{\cos x}\, \sin^3 x\, dx$ (*iv*) $\int_0^{\pi/2} x^2 \sin x\, dx$

Solution. (*i*) We have

$$\int_0^1 (2x^3 + 3)^2\, dx = \int_0^1 (4x^6 + 12x^3 + 9)\, dx = 4\int_0^1 x^6\, dx + 12\int_0^1 x^3\, dx + 9\int_0^1 dx$$

$$= 4\left[\frac{x^7}{7}\right]_0^1 + 12\left[\frac{x^4}{4}\right]_0^1 + 9[x]_0^1$$

$$= 4\left(\frac{1}{7} - 0\right) + 12\left(\frac{1}{4} - 0\right) + 9\,(1 - 0) = \frac{4}{7} + 3 + 9 = \frac{4 + 21 + 63}{7} = \frac{88}{7}$$

(*ii*) We have $\int_0^1 \frac{1 - x^2}{1 + x^2}\, dx$

Dividing the numerator by the denominator, we get

$$\frac{1 - x^2}{1 + x^2} = -1 + \frac{2}{1 + x^2}$$

$$\therefore \quad \int_0^1 \frac{1 - x^2}{1 + x^2}\, dx = \int_0^1 \left(-1 + \frac{2}{1 + x^2}\right) dx = [-x + 2\tan^{-1} x]_0^1 = (-1 + 2\tan^{-1} 1) - (0 + 0)$$

$$= -1 + 2\left(\frac{\pi}{4}\right) - 0 = \frac{\pi}{2} - 1$$

(*iii*) We have $\int_0^{\pi/2} \sqrt{\cos x}\, \sin^3 x\, dx = \int_0^{\pi/2} (\cos x)^{1/2} . (1 - \cos^2 x) \sin x\, dx$

Put $\cos x = t$ so that $-\sin x\, dx = dt$

or $\sin x\, dx = -dt$

When $x = 0$, $t = \cos 0 = 1$

When $x = \frac{\pi}{2}$, $t = \cos \frac{\pi}{2} = 0$

$$\therefore \quad \int_0^{\pi/2} \sqrt{\cos x}\, \sin^3 x\, dx = -\int_1^0 t^{1/2}\, (1 - t^2)\, dt$$

$$= \int_0^1 t^{1/2}\, dt - \int_0^1 t^{5/2}\, dt = \frac{2}{3}[t^{1/2}]_0^1 - \frac{2}{7}[t^{7/2}]_0^1$$

$$= \frac{2}{3}(1 - 0) - \frac{2}{7}(1 - 0) = \frac{2}{3} - \frac{2}{7} = \frac{14 - 6}{21} = \frac{8}{21}$$

(*iv*) We have $\int_0^{\pi/2} x^2 \sin x \, dx$

Integrating by parts, we get

$$= x^2(-\cos x) - \int 2x(-\cos x)\,dx = -x^2 \cos x + 2\int x \cos x \, dx$$

Again integrating by parts, we obtain

$$= -x^2 \cos x + 2[x \sin x - (-\cos x)]$$
$$= -x^2 \cos x + 2x \sin x + 2 \cos x$$

$$\therefore \qquad \int_0^{\pi/2} x^2 \sin x \, dx = -[x^2 \cos x]_0^{\pi/2} + 2[x \sin x]_0^{\pi/2} + 2[\cos x]_0^{\pi/2}$$

$$= -\left(\frac{\pi^2}{4}\cos\frac{\pi}{2} - 0\right) + 2\left(\frac{\pi}{2}\sin\frac{\pi}{2} - 0\right) + 2\left(\cos\frac{\pi}{2} - \cos 0\right)$$

$$= 0 + 2 \times \frac{\pi}{2} + 2(0 - 1) = \pi - 2.$$

Example 10.16. Evaluate $\int_0^{\infty} \frac{dx}{x^2 + a^2}$

Solution. We have, $\int_0^{\infty} \frac{dx}{x^2+a^2} = \frac{1}{a}\left(\tan^{-1}\frac{x}{a}\right)_0^{\infty} = \frac{1}{a}[\tan^{-1}\infty - \tan^{-1} 0] = \frac{1}{a}\left[\frac{\pi}{2} - 0\right] = \frac{\pi}{2a}.$

Definite Integral as the Limit of a Sum

Let $f(x)$ be a continuous real-valued function, defines in the closed interval $[a, b]$ then we define

$$\int_a^b f(x)\,dx = \lim_{h \to 0} h[f(a) + f(a+h) + f(a+2h) + \ldots + f(a+(n-1)h)]$$

where $nh = b - a$.

The method of evaluating $\int_a^b f(x)\,dx$ by using the above definition is called integration a from first principles.

Some Useful Results

(*i*) $1 + 2 + 3 + \ldots + (n-1) = \frac{n(n-1)}{2}$

(*ii*) $1^2 + 2^2 + 3^2 + \ldots + (n-1)^2 = \frac{(n-1)n(2n-1)}{6}$

(*iii*) $1^3 + 2^3 + 3^3 + \ldots + (n-1)^3 = \left[\frac{n(n-1)}{2}\right]^2$

(*iv*) $a + ar + ar^2 + \ldots + ar^{n-1} = \dfrac{a(r^n - 1)}{r - 1}; r > 1$

(*v*) $\sin a + \sin (a + h) + \sin (a + 2h) + \ldots + \sin [a + (n - 1)h] = \dfrac{\sin\left\{a + \dfrac{(n-1)}{2}h\right\}\cdot \sin \dfrac{nh}{2}}{\sin\left(\dfrac{h}{2}\right)}$

(*vi*) $\cos a + \cos (a + h) + \cos (a + 2h) + \ldots + \cos [a + (n - 1)h] = \dfrac{\cos\left\{a + \dfrac{(n-1)}{2}h\right\}\cdot \sin\left(\dfrac{nh}{2}\right)}{\sin\left(\dfrac{h}{2}\right)}$

SOLVED EXAMPLES

Example 10.17. Evaluate

(*a*) $\int_0^1 (2x + 3)\, dx$ **(*b*)** $\int_0^1 (x^2 + 1)\, dx$

Solution. (*a*) Let $f(x) = 2x + 3$; $a = 1$, $b = 3$ and $nh = 3 - 1 = 2$

Then,
$$\int_1^3 (2x+3)\, dx = \lim_{h\to 0} h[f(1) + f(1+h) + \ldots + f(1 + (n-1)h]$$

$$= \lim_{h\to 0} h[5 + (5 + 2h) + (5 + 4h) + \ldots + (5 + 2(n-1)h)]$$

[$\because$ $f(1) = 5, f(1 + h) = 5 + 2h$, etc.]

$$= \lim_{h\to 0} h[5n + \{2h + 4h + \ldots + 2(n-1)\, h\}]$$

$$= \lim_{h\to 0} h[5n + 2\{(1 + 2 + 3 + \ldots + (n-1)\}h]$$

$$= \lim_{h\to 0} h\left[5n + 2\cdot\frac{n(n-1)}{2}.h\right] = \lim_{h\to 0} [5nh + nh(nh - h)]$$

$$= \lim_{h\to 0} [10 + 2(2 - h)] = 14$$ [$\because$ $nh = 2$]

(*b*) Let $f(x) = x^2 + 1$; $a = 0$, $b = 2$ and $nh = 2 - 0 = 2$

$\therefore$
$$\int_0^2 (x^2 + 1)\, dx = \lim_{h\to 0} h[f(0) + f(0+h) + \ldots + f(0 + (n-1)h]$$

$$= \lim_{h\to 0} h[1 + (h^2 + 1) + (4h^2 + 1) + (9h^2 + 1) + \ldots + ((n-1)^2 h^2 + 1)]$$

[$\because$ $f(a) = 1, f(a + h) = h^2 + 1, f(0 + 2h) = 4h^2 + 1$ etc.]

$$= \lim_{h\to 0} h\left[n + \frac{(n-1)n(2n-1)}{6}h^2\right] = \lim_{h\to 0}\left[nh + \frac{(nh-n)nh(2nh-h)}{6}\right]$$

$$= \lim_{h\to 0}\left[2 + \frac{(2-h)2(4-h)}{6}\right] = \frac{14}{3} \qquad [\because \quad nh = 2]$$

Example 10.18. Evaluate $\int_{-1}^{1} e^x\, dx$ as limit of sums.

Solution. Let $f(x) = e^x$; $a = -1$, $b = 1$ and $nh = [1 - (-1)] = 2$.

Then
$$\int_{-1}^{1} e^x\, dx = \lim_{h\to 0} h[f(-1) + f(-1+h) + ... + f(-1+(n-1)h)]$$

$$= \lim_{h\to 0} h[e^{-1} + e^{(-1+h)} + e^{(-1+2h)} + ... + e^{[-1+(n-1)h]}]$$

$$[\because \quad f(-1) = e^{-1}, f(-1+h) = e^{(-1+h)} \text{ etc.}]$$

$$= \lim_{h\to 0} e^{-1}h[1 + e^h + e^{2h} + ... + e^{(n-1)h}]$$

$$= \frac{1}{e}\lim_{h\to 0} h\left[\frac{(e^h)^n - 1}{e^h - 1}\right] = \frac{1}{e}\lim_{h\to 0}\frac{h(e^{nh}-1)}{e^h - 1}$$

$$= \frac{1}{e}\cdot\lim_{h\to 0}\frac{h(e^2-1)}{e^h-1} \qquad (\because \quad nh = 2)$$

$$= \frac{1}{e}\cdot\lim_{h\to 0}\frac{h(e^2-1)}{\left\{1 + h + \frac{h^2}{2!} + \frac{h^3}{3!} + ...\right\} - 1} = \frac{1}{e}(e^2 - 1) = \left(e - \frac{1}{e}\right)$$

Example 10.19. Evaluate $\int_{a}^{b} \sin x\, dx$ from first principles.

Solution. Let $f(x) = \sin x$ and let $nh = b - a$. Then

$$\int_a^b \sin x\, dx = \lim_{h\to 0} h[f(a) + f(a+h) + ... + \{a + (n-1)\}]$$

$$= \lim_{h\to 0} h[\sin a + \sin(a+h) + ... + \sin\{a + (n-1)\}]$$

$$= \lim_{h\to 0} h\cdot\frac{\sin\left\{a + \left(\frac{n-1}{2}\right)h\right\}\sin\left(\frac{nh}{2}\right)}{\sin\left(\frac{nh}{2}\right)}$$

$$= \lim_{h\to 0}\left[2\sin\left\{a + \frac{nh}{2} - \frac{1}{2}h\right\}\sin\left(\frac{nh}{2}\right)\times\frac{\left(\frac{h}{2}\right)}{\sin\left(\frac{h}{2}\right)}\right]$$

$$= \lim_{h \to 0} \left[2 \sin \left\{ a + \frac{1}{2}(b-a) - \frac{1}{2}h \right\} \sin\left(\frac{b-a}{2}\right) \times \frac{\left(\frac{h}{2}\right)}{\sin\left(\frac{h}{2}\right)} \right]$$

$$= \lim_{h \to 0} 2 \sin \left[\left(\frac{b+a}{2}\right) - \frac{1}{2}h \right] \sin\left(\frac{b-a}{2}\right) \cdot \lim_{h \to 0} \frac{\left(\frac{h}{2}\right)}{\sin\left(\frac{h}{2}\right)}$$

$$= 2 \sin\left(\frac{b+a}{2}\right) \cdot \sin\left(\frac{b-a}{2}\right)$$

$$= \cos a - \cos b \qquad [\because \; 2 \sin A \sin B = \cos(A - B) - \cos(A + B)]$$

EXERCISE 10.2

Evaluate the following integrals as the limit of sums:

1. $\int_a^b x\,dx$ $\left[\textbf{Ans.} \frac{1}{2}(b^2 - a^2)\right]$

2. $\int_0^2 (x+4)\,dx$ [**Ans.** 10]

3. $\int_2^3 x^2\,dx$ $\left[\textbf{Ans.} \frac{19}{3}\right]$

4. $\int_1^4 (x^2 - 4)\,dx$ $\left[\textbf{Ans.} \frac{27}{2}\right]$

5. $\int_1^3 (2x^2 + 5x)\,dx$ $\left[\textbf{Ans.} \frac{112}{3}\right]$

6. $\int_1^3 e^{-x}\,dx$ $\left[\textbf{Ans.} \frac{e^2 - 1}{e^3}\right]$

7. $\int_0^4 (x + e^{2x})\,dx$ $\left[\textbf{Ans.} \frac{15 + e^8}{2}\right]$

8. $\int_2^4 (x^2 - 3x + 2)\,dx$ $\left[\textbf{Ans.} \frac{14}{3}\right]$

9. $\int_0^1 |3x - 1|\,dx$ $\left[\textbf{Ans.} \frac{5}{6}\right]$

10. $\int_a^b \cos x\,dx$ [**Ans.** $\sin b - \sin a$]

Properties of Definite Integrals

These properties are very useful in evaluating the definite integrals. There are as under:

Property I: $$\int_a^b f(x)\,dx = \int_a^b f(t)\,dt$$

i.e., integration is independent of the change of variable.

Property II: $$\int_a^b f(x)\,dx = -\int_b^a f(x)\,dx$$

i.e., if the limits of a definite integral are interchanged then its value changes by minus sign only.

Property III: $$\int_a^b f(x)\,dx = \int_a^c f(x)\,dx + \int_b^c f(x)\,dx \qquad \text{where } a < c < b.$$

Property IV: $$\int_0^a f(x)\,dx = \int_0^a f(a-x)\,dx.$$

This property is useful to evaluate a definite integral without first finding the corresponding indefinite integrals which may be difficult or some time even impossible to find.

Property V: $$\int_{-a}^{a} f(x)\,dx = \begin{cases} 2\int_0^a f(x)\,dx, & \text{if } f(2a-x) = f(x) \\ 0, & \text{if } f(2a-x) = -f(x) \end{cases}$$

Property VI: $$\int_0^{2a} f(x)\,dx = \begin{cases} 2\int_0^a f(x)\,dx, & \text{if } f(2a-x) = f(x) \\ 0, & \text{if } f(2a-x) = -f(x) \end{cases}$$

Property VII: $$\int_a^b f(x)\,dx = \int_a^b f(a+b-x)\,dx.$$

SOLVED EXAMPLES

Example 10.20. Evaluate $\int_0^4 f(x)\,dx$ **where** $f(x) = \begin{cases} x^2, & 0 \le x \le 2 \\ x+2, & 2 \le x \le 4 \end{cases}$.

Solution. We have $$\int_0^4 f(x)\,dx = \int_0^2 f(x)\,dx + \int_0^4 f(x)\,dx$$

$$= \int_0^2 x^2\,dx + \int_2^4 (x+2)\,dx \qquad \text{[By definition of } f(x)]$$

$$= \left[\frac{x^3}{3}\right]_0^2 + \left[\frac{x^2}{2} + 2x\right]_2^4 = \left(\frac{8}{3} - 0\right) + \left(\frac{16}{2} + 8 - \frac{4}{2} - 4\right)$$

$$= \frac{8}{3} + (8 + 8 - 2 - 4) = \frac{8}{3} + 10 = \frac{38}{3}.$$

Example 10.21. Prove that $\int_0^{\pi/2} \frac{\sin x}{\sin x + \cos x}\, dx = \frac{\pi}{4}$.

Solution. Let $$I = \int_0^{\pi/2} \frac{\sin x}{\sin x + \cos x}\, dx \quad \text{...(i)}$$

Then $$I = \int_0^{\pi/2} \frac{\sin\left(\frac{\pi}{2} - x\right)}{\sin\left(\frac{\pi}{2} - x\right) + \cos\left(\frac{\pi}{2} - x\right)}\, dx \quad \left[\because \int_0^a f(x)\,dx = \int_0^a f(a-x)\,dx\right]$$

or $$I = \int_0^{\pi/2} \frac{\cos x}{\cos x + \sin x} \quad \text{...(ii)}$$

Adding (*i*) and (*ii*), we get

$$2I = \int_0^{\pi/2} \frac{\sin x}{\sin x + \cos x}\, dx + \int_0^{\pi/2} \frac{\cos x}{\sin x + \cos x}\, dx$$

$$= \int_0^{\pi/2} \frac{\sin x + \cos x}{\sin x + \cos x}\, dx = \int_0^{\pi/2} 1 . dx = [x]_0^{\pi/2} = \frac{\pi}{2} - 0 = \frac{\pi}{2}.$$

$$\therefore \quad I = \frac{\pi}{4} \text{ i.e., } \int_0^{\pi/2} \frac{\sin x}{\sin x + \cos x}\, dx = \frac{\pi}{4}.$$

Example 10.22. Evaluate $\int_0^{\pi/2} \frac{\sqrt{\sin x}}{\sqrt{\sin x} + \sqrt{\cos x}}$.

Solution. Let $$I = \int_0^{\pi/2} \frac{\sqrt{\sin x}}{\sqrt{\sin x} + \sqrt{\cos x}} \quad \text{...(i)}$$

Then $$I = \int_0^{\pi/2} \frac{\sqrt{\sin\left(\frac{\pi}{2} - x\right)}}{\sqrt{\sin\left(\frac{\pi}{2} - x\right)} + \sqrt{\cos\left(\frac{\pi}{2} - x\right)}} \quad \left[\because \int_0^a f(x)\,dx = \int_0^a f(a-x)\,dx\right]$$

or $$I = \int_0^{\pi/2} \frac{\sqrt{\cos x}}{\sqrt{\cos x} + \sqrt{\sin x}} \quad \text{...(ii)}$$

Adding (*i*) and (*ii*), we get

$$2I = \int_0^{\pi/2} \frac{\sqrt{\sin x}}{\sqrt{\sin x} + \sqrt{\cos x}}\, dx + \int_0^{\pi/2} \frac{\sqrt{\cos x}}{\sqrt{\cos x} + \sqrt{\sin x}}\, dx$$

$$= \int_0^{\pi/2} \frac{\sqrt{\sin x} + \sqrt{\cos x}}{\sqrt{\sin x} + \sqrt{\cos x}}\, dx = \int_0^{\pi/2} 1 . dx = [x]_0^{\pi/2} = \frac{\pi}{2} - 0 = \frac{\pi}{2}$$

$$\therefore \quad I = \frac{\pi}{4} \quad \text{i.e., } \int_0^{\pi/2} \frac{\sqrt{\sin x}}{\sqrt{\sin x} + \sqrt{\cos x}}\, dx = \frac{\pi}{4}.$$

Example 10.23. **Evaluate** $\int_0^{\pi} \frac{x}{1+\sin x}\, dx$.

Solution. Let $$I = \int_0^{\pi} \frac{x\, dx}{1+\sin x} \qquad ...(i)$$

Then $$I = \int_0^{\pi} \frac{(\pi - x)\, dx}{1+\sin(\pi - x)} \qquad \left[\because \int_0^a f(x)\, dx = \int_0^a f(a-x)\, dx\right]$$

$$I = \int_0^{\pi} \frac{\pi - x}{1+\sin x} \qquad ...(ii)$$

Adding (*i*) and (*ii*), we get

$$2I = \pi \int_0^{\pi} \frac{dx}{1+\sin x} = \pi \int_0^{\pi} \frac{1-\sin x}{1-\sin^2 x}\, dx$$

$$= \pi \int_0^{p} (\sec^2 x - \tan x \sec^2 x)\, dx = \pi[\tan x - \sec x]_0^{\pi/2}$$

$$= \pi[(\tan \pi - \sec \pi) - (\tan 0 - \sec 0)] = \pi[0 - (-1) - (0 - 1)] = 2\pi$$

$\therefore$ $$I = \pi.$$

Example 10.24. **Evaluate** $\int_{-\pi/2}^{\pi/2} \sin^7 x\, dx$.

Solution. Let $f(x) = \sin^7 x$

Then $f(-x) = \sin^7(-x) = [\sin(-x)]^7 = [-\sin x]^7 = -\sin^7 x = -f(x)$

So $f(x)$ is an odd function.

Hence $$\int_{-\pi/2}^{\pi/2} \sin^7 x\, dx = 0 \qquad \text{[Using Property V]}$$

Example 10.25. **Evaluate** $\int_0^{\pi/2} \log \sin x\, dx$.

Solution. Let $$I = \int_0^{\pi/2} \log \sin x\, dx \qquad ...(i)$$

Then $$I = \int_0^{\pi/2} \log \sin\left(\frac{\pi}{2} - x\right) dx \qquad \left[\because \int_0^a f(x)\, dx = \int_0^a f(a-x)\, dx\right]$$

or $$I = \int_0^{\pi/2} \log \cos x\, dx \qquad ...(ii)$$

Adding (*i*) and (*ii*), we get

$$2I = \int_0^{\pi/2} \log \sin x\, dx + \int_0^{\pi/2} \log \cos x\, dx$$

$$= \int_0^{\pi/2} (\log \sin x + \log \cos x)\, dx = \int_0^{\pi/2} \log(\sin x \cos x)\, dx$$

$$= \int_0^{\pi/2} \log\left(\frac{2\sin x\cos x}{2}\right) dx = \int_0^{\pi/2} \log\left(\frac{\sin 2x}{2}\right) dx$$

$$= \int_0^{\pi/2} \log \sin 2x \, dx - \log 2 . \int_0^{\pi/2} dx$$

$$= \int_0^{\pi/2} \log \sin 2x \, dx - \log 2 \int_0^{\pi/2} 1 . dx = \int_0^{\pi/2} \log \sin 2x \, dx - \log 2 . (x)_0^{\pi/2}$$

$$2I = \int_0^{\pi/2} \log \sin 2x - \frac{\pi}{2} \log 2 = I_1 - \frac{\pi}{2} \log 2 \quad ...(iii)$$

$$\therefore \quad I_1 = \int_0^{\pi/2} \log \sin 2x \, dx$$

Put $2x = t$, so that $2dx = dt$. When $x = 0$, $t = 0$ and when $x = \frac{\pi}{2}, t = \pi$.

$$\therefore \quad I_1 = \int_0^{p} \log \sin t \frac{dt}{2} = \frac{1}{2}\int_0^{\pi} \log \sin t \, dt$$

$$= \frac{1}{2} . 2 \int_0^{p/2} \log \sin t \, dt \quad [\because \ \log \sin (\pi - t) = \log \sin t \text{ and using prop. V}]$$

$$= \int_0^{\pi/2} \log \sin x \, dx = I$$

$\therefore$ From (*iii*), we get $\quad 2I = I - \frac{\pi}{2} \log 2 \quad \Rightarrow \quad I = -\frac{\pi}{2} \log 2$

Hence $\quad \int_0^{\pi/2} \log \sin x \, dx = -\frac{\pi}{2} \log 2.$

Example 10.26. Evaluate $\int_0^{\pi/2} \frac{x \sin x \cos x}{\cos^4 x + \sin^4 x} dx$ **.**

Solution. Let
$$I = \int_0^{\pi/2} \frac{x \sin x \cos x}{\cos^4 x + \sin^4 x} dx \quad ...(i)$$

Then
$$I = \int_0^{\pi/2} \frac{\left(\frac{\pi}{2} - x\right) \sin\left(\frac{\pi}{2} - x\right) \cos\left(\frac{\pi}{2} - x\right)}{\cos^4\left(\frac{\pi}{2} - x\right) + \sin^4\left(\frac{\pi}{2} - x\right)} \quad \left[\because \int_0^a f(x)\, dx = \int_0^a f(a-x)\, dx\right]$$

or
$$I = \int_0^{\pi/2} \frac{\left(\frac{\pi}{2} - x\right) \cos x \sin x}{\sin^4 x + \cos^4} dx \quad ...(ii)$$

Adding (*i*) and (*ii*), we get

$$2I = \int_0^{\pi/2} \frac{x \sin x \cos x}{\cos^4 x + \sin^4 x} dx + \int_0^{\pi/2} \frac{\left(\frac{\pi}{2} - x\right) \cos x \sin x}{\sin^4 x + \cos^4 x} dx$$

$$= \frac{\pi}{2} \int_0^{\pi/2} \frac{\sin x \cos x}{\sin^4 x + \cos^4 x} dx$$

Dividing numerator and denominator by $\cos^4 x$, we get

$$= \frac{\pi}{2} \int_0^{\pi/2} \frac{\tan x \sec^2 x}{\tan^4 x + 1} dx = \frac{\pi}{2} \int_0^{\pi/2} \frac{\tan x \sec^2 x}{(\tan^2 x)^2 + 1} dx$$

Put $\tan^2 x = t \Rightarrow 2 \tan x \sec^2 x \, dx = dt \Rightarrow \tan x \sec^2 x \, dx = \frac{1}{2} dt$

When $x = 0$, $t = \tan 0 = 0$ and when $x = \frac{\pi}{2}, t = \tan \frac{\pi}{2} = \infty$.

$$\therefore \qquad 2I = \frac{\pi}{2} \int \frac{\frac{1}{2} dt}{t^2 + 1} = \frac{\pi}{4} [\tan^{-1}]_0^{\infty}$$

$$\Rightarrow \qquad I = \frac{\pi}{8}\left(\frac{\pi}{2} - 0\right) = \frac{\pi^2}{16}.$$

Example 10.27. Evaluate $\int_0^1 \frac{\log (1+x)}{1+x^2} dx$.

Solution. Let $x = \tan \theta \quad \Rightarrow \quad dx = \sec^2 \theta \, d\theta$

$$\therefore \qquad I = \int_0^{\pi/4} \frac{\log (1 + \tan \theta)}{1 + \tan^2 \theta} \sec^2 \theta \, d\theta = \int_0^{\pi/4} \log (1 + \tan \theta) \, d\theta$$

$$= \int_0^{\pi/4} \log \left[1 + \tan\left(\frac{\pi}{4} - \theta\right)\right] \qquad \left[\because \int_0^a f(x) \, dx = \int_0^a f(a - x) \, dx\right]$$

$$= \int_0^{\pi/4} \log \left[1 + \frac{1 - \tan \theta}{1 + \tan \theta}\right] d\theta = \int_0^{\pi/4} \log \left(\frac{2}{1 + \tan \theta}\right) d\theta$$

$$= \int_0^{\pi/4} \log 2 \, d\theta - \int_0^{\pi/4} \log (1 + \tan \theta) \, d\theta = \frac{\pi}{4} \log 2 - I$$

$$\Rightarrow \qquad 2I = \frac{\pi}{4} . \log 2 \qquad \Rightarrow \qquad I = \frac{\pi}{8} \log 2$$

i.e., $$\int_0^1 \frac{\log (1 + x)}{1 + x^2} dx = \frac{\pi}{8} \log 2 .$$

Example 10.28. **Evaluate** $\int_0^{\pi} \frac{x\,dx}{a^2\cos^2 x + b^2\sin^2 x}$.

Solution. Let $I = \int_0^{\pi} \frac{x\,dx}{a^2\cos^2 x + b^2\sin^2 x}$ ...(*i*)

Then, $I = \int_0^{\pi} \frac{(\pi - x)}{a^2\cos^2(\pi - x) + b^2\sin^2(\pi - x)}\,dx$ $\left[\because \int_0^a f(x)\,dx = \int_0^a f(a-x)\,dx\right]$

or $I = \int_0^{\pi} \frac{\pi - x}{a^2\cos^2 x + b^2\sin^2 x}$ ...(*ii*)

Adding (*i*) and (*ii*), we get

$$2I = \int_0^{\pi} \frac{x + \pi - x}{a^2\cos^2 x + b^2\sin^2 x}\,dx = \pi\int_0^{\pi} \frac{1}{a^2\cos^2 x + b^2\sin^2 x}\,dx$$

$$= 2\pi\int_0^{\pi/2} \frac{1}{a^2\cos^2 x + b^2\sin^2 x}\,dx \quad \left[\int_0^{2a} f(x)\,dx = 2\int_0^a f(x)\,dx \text{ if } f(2a-x) = f(x)\right]$$

Dividing numerator and denominator by $\cos^2 x$, we get

$$2I = 2\pi\int_0^{\pi/2} \frac{\sec^2 x}{a^2 + b^2\tan^2 x}\,dx$$

Put $\tan x = t$ so that $\sec^2 x\,dx = dt$

When $x = 0$, $t = \tan 0 = 0$;

When $x = \frac{\pi}{2}$, $t = \tan\frac{\pi}{2} = \infty$

$$\therefore \quad I = \pi\int_0^{\infty} \frac{dt}{a^2 + b^2t^2} = \frac{\pi}{b^2}\cdot\frac{1}{\left(\frac{a}{b}\right)}\cdot\left[\tan^{-1}\left(\frac{t}{\left(\frac{a}{b}\right)}\right)\right]_0^{\infty}$$

$$= \frac{\pi}{ab}[\tan^{-1}\infty - \tan^{-1}0] = \frac{\pi}{ab}\left(\frac{\pi}{2} - 0\right) = \frac{\pi^2}{2ab}$$

i.e., $\int_0^{\pi} \frac{x\,dx}{a^2\cos^2 x + b^2\sin^2 x} = \frac{\pi^2}{2ab}$.

EXERCISE 10.3

1. $\int_0^{\pi/2} \frac{dx}{1 + \cot x}$. $\left[\textbf{Ans. } \frac{\pi}{4}\right]$

2. $\int_0^{\pi/2} \frac{dx}{1 + \tan^3 x}$. $\left[\textbf{Ans. } \frac{\pi}{4}\right]$

3. $\int_0^{\pi/2} \frac{\cos^5 x\,dx}{\sin^5 x + \cos^5 x}\,dx$ $\left[\textbf{Ans. } \frac{\pi}{4}\right]$

4. $\int_0^{\pi/2} (2 \log \sin x - \log \sin 2x)\, dx$. $\left[\textbf{Ans. } \frac{\pi}{2} \log \frac{1}{2}\right]$

5. $\int_0^{2\pi} \cos^5 x\, dx$. [**Ans.** 0]

6. $\int_0^{\pi/4} \log (1 + \tan x)\, dx$. $\left[\textbf{Ans. } \frac{\pi}{8} \log 2\right]$

7. $\int_0^{\pi} \log (1 + \cos x)\, dx$. [**Ans.** $-\pi \log 2$]

8. $\int_0^{a} \frac{\sqrt{x}}{\sqrt{x} + \sqrt{a - x}}\, dx$. $\left[\textbf{Ans. } \frac{a}{2}\right]$

9. $\int_{-4}^{4} |x + 2|\, dx$ [**Ans.** 20]

10. $\int_{-1}^{3/2} |x \sin \pi x|\, dx$. $\left[\textbf{Ans. } \frac{3\pi + 1}{\pi^2}\right]$

11. $\int_{-\pi/2}^{\pi/2} |\sin x|\, dx$. [**Ans.** 2]

12. $\int_{\pi/6}^{\pi/3} \frac{1}{1 + \sqrt{\cot x}}\, dx$. $\left[\textbf{Ans. } \frac{\pi}{12}\right]$

13. $\int_0^{\pi} \frac{x \sin x}{1 + \cos^2 x}\, dx$. $\left[\textbf{Ans. } \frac{\pi^2}{4}\right]$

14. $\int_0^{1} \log\left(\frac{1}{x} - 1\right) dx$ [**Ans.** 0]

15. $\int_0^{1} x(1 - x)^n\, dx$. $\left[\textbf{Ans. } \frac{1}{(n + 1)(n + 2)}\right]$

16. $\int_0^{\pi} \frac{x \tan x}{\sec x + \cos x}\, dx$. $\left[\textbf{Ans. } \frac{\pi}{2} (\pi - 2)\right]$

17. $\int_0^{\infty} \frac{x\, dx}{(1 + x)(1 + x^2)}$. $\left[\textbf{Ans. } \frac{\pi}{4}\right]$

18. $\int_0^{\pi/2} \log \tan x\, dx$. [**Ans.** 0]

19. $\int_0^{\pi} \frac{x\, dx}{1 + \cos \alpha \sin x}$ $\left[\textbf{Ans. } \frac{\pi\alpha}{\sin \alpha}\right]$

20. $\int_0^{\infty} \log\left(x + \frac{1}{x}\right) \cdot \frac{dx}{1 + x^2}$. [**Ans.** $\pi \log 2$]

10.7. REDUCTION FORMULAE

A formula by means of which a given integral can be reduced to some known integral is called a reduction formula. The basic technique involved in obtaining a reduction formula is integration by parts.

1. Reduction Formula for $\int \sin^n x\, dx$ and $\int \cos^n x\, dx$, n being a positive integer.

(*i*) Let $I_n = \int \sin^n x\, dx = \int \sin^{n-1} x . \sin x\, dx$

Integration by parts, we get

$$= \sin^{n-1} x(-\cos x) - \int (n-1)\sin^{n-2} x \cos x(-\cos x)\, dx$$

$$= -\sin^{n-1} x \cos x + (n-1)\int \sin^{n-2} x \cos^2 x\, dx$$

$$= -\sin^{n-1} x \cos x + (n-1)\int \sin^{n-2} x . (1-\sin^2 x)\, dx$$

$$= -\sin^{n-1} x \cos x + (n-1)\left[\int \sin^{n-2} x\, dx - \int \sin^n x\, dx\right]$$

$$= -\sin^{n-1} x \cos x + (n-1)[I_{n-2} - I_n]$$

or $$I_n + (n-1) I_n = -\sin^{n-1} x \cos x + (n-1) I_{n-2}$$

or $$nI_n = -\sin^{n-1} x \cos x + (n-1) I_{n-2}$$

or $$\boxed{I_n = -\frac{\sin^{n-1} x \cos x}{n} + \left(\frac{n-1}{n}\right) I_{n-2}} \quad \text{...}(i)$$

(*ii*) Let $I_n = \int \cos^n x\, dx = \int \cos^{n-1} x . \cos x\, dx$

Now integrating by parts and proceeding as above, we get

$$\boxed{I_n = \frac{\sin x \cos^{n-1} x}{n} + \left(\frac{n-1}{n}\right) I_{n-2}} \quad \text{...}(ii)$$

2. Walli's Formula: Walli's formula is applicable only when the limits are from 0 to $\frac{\pi}{2}$.

From (*i*), we have $$I_n = -\frac{\sin^{n-1} x \cos x}{n} + \left(\frac{n-1}{n}\right) I_{n-2}$$

In case limits are from $x = 0$ to $x = \frac{\pi}{2}$, we have

$$I_n = \left[-\frac{\sin^{n-1} x \cos x}{n}\right]_0^{\pi/2} + \left(\frac{n-1}{n}\right) I_{n-2}$$

or $$I_n = \left(\frac{n-1}{n}\right) I_{n-2} \quad \text{...}(iii)$$

Similarly applying limits from $x = 0$ to $x = \frac{\pi}{2}$ in (*ii*), we have

$$I_n = \left(\frac{n-1}{n}\right) I_{n-2}$$

Now with the help of this reduction formula, we will connect I_{n-2} with I_{n-4}, I_{n-4} with I_{n-6}, ..., I_3 with I_1 or I_2 with I_0 according as n is odd or even.

$$\therefore \quad I_n = \frac{n-1}{n} I_{n-2}$$

$$I_{n-2} = \frac{n-3}{n-2} I_{n-4}$$

$$\vdots \quad \vdots \quad \vdots$$

$$I_3 = \frac{2}{3} I_1; \quad \text{if } n \text{ is odd}$$

$$I_2 = \frac{1}{2} I_0; \quad \text{if } n \text{ is odd}$$

Now,
$$I_1 = \int_0^{\pi/2} \sin x\, dx = \int_0^{\pi/2} \cos x\, dx = 1$$

and
$$I_0 = \int_0^{\pi/2} \sin^0 x\, dx = \int_0^{\pi/2} \cos^0 x\, dx = \frac{\pi}{2}$$

Therefore, whether
$$I_n = \int_0^{\pi/2} \sin^n x\, dx \text{ or } \int_0^{\pi/2} \cos^n x\, dx$$

We have
$$I_n = \begin{cases} \dfrac{n-1}{n} \cdot \dfrac{n-3}{n-2} \cdot \dfrac{n-5}{n-4} \cdots\cdots \dfrac{2}{3}, & \text{when } n \text{ is odd} \\[2ex] \dfrac{n-1}{n} \cdot \dfrac{n-3}{n-2} \cdot \dfrac{n-5}{n-4} \cdots\cdots \dfrac{1}{2} \cdot \dfrac{\pi}{2}, & \text{when } n \text{ is even} \end{cases}$$

SOLVED EXAMPLES

Example 10.29. Evaluate $\int \sin^6 x\, dx$.

Solution. We know that

$$\int \sin^n x\, dx = -\frac{\sin^{n-1} x \cos x}{n} + \left(\frac{n-1}{n}\right) \int \sin^{n-2} x\, dx$$

Taking $n = 6$ and applying the above formula successively, we have

$$\int \sin^6 x\, dx = -\frac{\sin^{6-1} x \cos x}{6} + \frac{6-1}{6} \int \sin^{6-2} x\, dx$$

$$= -\frac{1}{6} \sin^5 x \cos x + \frac{5}{6} \int \sin^4 x\, dx$$

$$= -\frac{1}{6} \sin^5 x \cos x + \frac{5}{6} \left[-\frac{1}{4} \sin^3 x \cos x + \frac{3}{4} \int \sin^2 x\, dx \right]$$

$$= -\frac{1}{6} \sin^5 x \cos x - \frac{5}{24} \sin^3 x \cos x + \frac{5}{8} \int \sin^2 x\, dx$$

$$= -\frac{1}{6}\sin^5 x\cos x - \frac{5}{24}\sin^3 x\cos x + \frac{5}{8}\int \frac{1}{2}(1-\cos 2x)\,dx$$

$$= -\frac{1}{6}\sin^5 x\cos x - \frac{5}{24}\sin^3 x\cos x + \frac{5}{8}\left[\frac{1}{2}x - \frac{1}{4}\sin 2x\right]$$

$$= -\frac{1}{6}\sin^5 x\cos x - \frac{5}{24}\sin^3 x\cos x + \frac{5}{16}\sin x\cos x + \frac{5}{16}x + c.$$

Example 10.30. Evaluate $\int_0^{\pi/2} \cos^9 x\,dx$.

Solution. Here $n = 9$ (odd)

$$\therefore \quad \int_0^{\pi/2} \cos^9 x\,dx = \frac{8}{9}\cdot\frac{6}{7}\cdot\frac{4}{5}\cdot\frac{2}{3} = \frac{128}{315}.$$ [Using reduction formula]

Example 10.31. Evaluate $\int_0^{\pi/2} \sin^{10} x\,dx$ or $\int_0^{\pi/2} \cos^{10} x\,dx$.

Solution. Here $n = 10$ (even) in both the cases

$$\therefore \quad \int_0^{\pi/2} \sin^{10} x\,dx = \int_0^{\pi/2} \cos^{10} x\,dx$$

$$= \frac{9}{10}\cdot\frac{7}{8}\cdot\frac{5}{6}\cdot\frac{3}{4}\cdot\frac{1}{2}\cdot\frac{\pi}{2} = \frac{63\pi}{512}$$ [Using reduction formula]

Example 10.32. Show that $\int_0^a \frac{x^4}{\sqrt{a^2-x^2}} = \frac{3a^4\pi}{16}$.

Solution. Put $x = a\sin\theta$, so that $dx = a\cos\theta\,d\theta$.

Also when $x = 0$, $\sin\theta = 0$ *i.e.*, $\theta = 0$

and when $x = a$, $\sin\theta = 1$ *i.e.*, $\theta = \frac{\pi}{2}$

Then, $$\int_0^a \frac{x^4}{\sqrt{a^2-x^2}}\,dx = \int_0^{\pi/2} \frac{a^4\sin^4\theta\, a\cos\theta\,d\theta}{\sqrt{a^2-a^2\sin^2\theta}}$$

$$= a^4\int_0^{\pi/2} \sin^4\theta\,d\theta = a^4.\frac{3}{4}\cdot\frac{1}{2}\cdot\frac{\pi}{2}$$

$$= \frac{3\pi a^4}{16}$$ [Using reduction formula]

Example 10.33. Evaluate $\int_0^{2a} x^{3/2}(2a-x)^{1/2}\,dx$.

Solution. Let $x = 2a\sin^2\theta$, $dx = 4a\sin\theta\cos\theta\,d\theta$.

Also when $x = 0$, $\theta = 0$

and when $x = 2a$, $\theta = \frac{\pi}{2}$

Therefore,

$$\int_0^{2a} x^{3/2}(2a-x)^{1/2}\,dx = 16a^3\int_0^{\pi/2}\sin^4\theta\cos^2\theta\,d\theta = 16a^3\int_0^{\pi/2}\sin^4\theta(1-\sin^2\theta)\,d\theta$$

$$= 16a^3\left[\int_0^{\pi/2}\sin^4\theta\,d\theta - \int_0^{\pi/2}\sin^6\theta\,d\theta\right] = 16a^3\left[\left(\frac{3}{4}.\frac{1}{2}.\frac{\pi}{2}\right)-\left(\frac{5}{6}.\frac{3}{4}.\frac{1}{2}.\frac{\pi}{2}\right)\right]$$

$$= 16a^3\left(\frac{1}{6}\times\frac{3}{4}\times\frac{1}{2}\times\frac{\pi}{2}\right) = \frac{\pi a^2}{2}$$

Example 10.34. Evaluate $\displaystyle\int_0^{2a}\frac{x^{9/2}\,dx}{\sqrt{2a-x}}$.

Solution. Put $x = 2a\sin^2\theta$, so that $dx = 2a\,.\,2\sin\theta\cos\theta\,d\theta$.

Also when $x = 0$, $\sin^2\theta = 0$ *i.e.*, $\theta = 0$

and when $x = 2a$, $\sin^2\theta = 1$ *i.e.*, $\theta = \dfrac{\pi}{2}$

$$\therefore \quad \int_0^{2a}\frac{x^{9/2}\,dx}{\sqrt{2a-x}} = \int_0^{\pi/2}\frac{(2a\sin^2\theta)^{9/2}\,.\,4a\sin\theta\cos\theta\,d\theta}{\sqrt{(2a-2a\sin^2\theta)}}$$

$$= \int_0^{\pi/2}\frac{(2a)^{9/2}\,.\,4a\sin^{10}\theta\cos\theta\,d\theta}{(2a)^{1/2}\,.\cos\theta} = (2a)^4\,.\,4a\,.\int_0^{\pi/2}\sin^{10}\theta\,d\theta$$

$$= 64a^5\,.\,\frac{9}{10}.\frac{7}{8}.\frac{5}{6}.\frac{3}{4}.\frac{1}{2}.\frac{\pi}{2} = \frac{63a^5\pi}{8} \qquad \text{[Using reduction formula]}$$

Note. If the integrand is of the form

(*i*) $\sqrt{a^2-x^2}$, put $x = a\sin\theta$, or $x = a\cos\theta$. (*ii*) $\sqrt{a^2+x^2}$, put $x = a\tan\theta$.

(*iii*) $\sqrt{2a-x}$, put $x = 2a\sin^2\theta$. (*iv*) $\sqrt{a-x}$, put $x = a\sin^2\theta$.

3. Reduction Formula for $\int \sin^m x\cos^n x\,dx$

Let $\quad I_{m,n} = \int\sin^m x\cos^n x\,dx = \int\sin^{m-1}x(\cos^n x\,.\sin x)\,dx$

Integrating by parts, we obtain

$$I_{m,n} = -\frac{\cos^{n+1}x}{n+1}\sin^{m-1}x + \frac{m-1}{n+1}\int\cos^{n+1}x\sin^{m-2}x\cos x\,dx$$

$$= -\frac{\cos^{n+1}x\sin^{m-1}x}{n+1} + \frac{m-1}{n+1}\int\sin^{m-2}x\cos^n x(1-\sin^2 x)\,dx$$

$$= -\frac{\cos^{n+1}x\sin^{m-1}x}{n+1} + \frac{m-1}{n+1}\left[\int\sin^{m-2}x\cos^n x\,dx - \int\sin^m x\cos^n x\,dx\right]$$

$$= -\frac{\cos^{n+1}x\sin^{m-1}x}{n+1} + \frac{m-1}{n+1}[I_{m-2,n} - I_{m,n}]$$

or $$\left[1+\frac{m-1}{n+1}\right]I_{m,n} = -\frac{\cos^{n+1}x\sin^{m-1}x}{n+1}+\frac{m-1}{n+1}I_{m-2,n}$$

or $$\left[\frac{m+n}{n+1}\right]I_{m,n} = -\frac{\cos^{n+1}x\sin^{m-1}x}{n+1}+\frac{m-1}{n+1}I_{m-2,n}$$

or $$\boxed{I_{m.n} = -\frac{\cos^{n+1}x\sin^{m-1}x}{m+n}+\frac{m-1}{m+n}I_{m-2,m}}$$

4. Reduction formula for $\int_0^{\pi/2}\sin^m x\cos^n x\,dx$ where m and n are positive integers. (Walli's Formula)

Using the above reduction formula, we have

$$\int_0^{\pi/2}\sin^m x\cos^n x\,dx = \left[-\frac{\cos^{n+1}x\sin^{m-1}x}{m+n}\right]_0^{\pi/2}+\frac{m-1}{m+n}\int_0^{\pi/2}\sin^{m-2}x\cos^n x\,dx$$

or $$\int_0^{\pi/2}\sin^m x\cos^n x\,dx = \frac{m-1}{m+n}\int_0^{\pi/2}\sin^{m-2}x\cos^n x\,dx$$

or $$I_{m,n} = \frac{m-1}{m+n}I_{m-2,n} \qquad ...(i)$$

Case I: If m is even.

Replacing m by $m-2, m-4, m-6, ..., 4, 2$ in (*i*), we obtain

$$I_{m-2,n} = \frac{m-3}{m+n-2}I_{m-4,n}$$

$$I_{m-4,n} = \frac{m-5}{m+n-4}I_{m-6,n}$$

....

....

$$I_{4,n} = \frac{3}{4+n}I_{2,n}$$

$$I_{2,n} = \frac{1}{2+n}I_{0,n}$$

Therefore $$I_{m,n} = \frac{m-1}{m+n}.\frac{m-3}{m+n-2}.I_{m-4,n}$$

$$= \frac{m-1}{m+n}.\frac{m-3}{m+n-2}.\frac{m-5}{m+n-4}I_{m-6,m}$$

$$= \frac{m-1}{m+n}.\frac{m-3}{m+n-2}.\frac{m-5}{m+n-4}....\frac{1}{2+n}I_{0,n}$$

But $$I_{0,n} = \int_0^{\pi/2}\sin^0 x\cos^n x\,dx = \int_0^{\pi/2}\cos^n x\,dx = I_n$$

Thus,
$$I_{m,n} = \left(\frac{m-1}{m+n}\cdot\frac{m-3}{m+n-2}\cdot\frac{m-5}{m+n-4}\cdots\frac{1}{2+n}\right)I_n$$
$$= \begin{cases}\left(\frac{m-1}{m+n}\cdot\frac{m-3}{m+n-2}\cdot\frac{m-5}{m+n-4}\cdots\frac{1}{2+n}\right)\left(\frac{n-1}{n}\cdot\frac{n-3}{n-2}\cdots\frac{1}{2}\cdot\frac{\pi}{2}\right); \text{if } n \text{ is even}\\ \left(\frac{m-1}{m+n}\cdot\frac{m-3}{m+n-2}\cdot\frac{m-5}{m+n-4}\cdots\frac{1}{2+n}\right)\left(\frac{n-1}{n}\cdot\frac{n-3}{n-2}\cdots\frac{2}{3}\right); \text{if } n \text{ is odd}\end{cases}$$

Case II: If *m* is odd.

Replacing m by $m-2$, $m-6$, ... 5, 3 successively in (*i*) and proceeding as above, we have

$$I_{m,n} = \frac{m-1}{m+n}\cdot\frac{m-3}{m+n-2}\cdot\frac{m-5}{m+n-4}\cdots\frac{2}{3+n}\cdot I_{1,n}$$

Now,
$$I_{1,n} = \int_0^{\pi/2} \sin x \cos^n x\, dx = \left[-\frac{\cos^{n+1} x}{n+1}\right]_0^{\pi/2} = \frac{1}{n+1}$$

Therefore,
$$I_{m,n} = \left(\frac{m-1}{m+n}\cdot\frac{m-3}{m+n-2}\cdot\frac{m-5}{m+n-4}\cdots\frac{2}{3+n}\right)\left(\frac{1}{n+1}\right)$$
when n may be odd or even.

SOLVED EXAMPLES

Example 10.35. Evaluate $\int_0^{\pi/4} \sin^3 2x \cos^4 2x\, dx$.

Solution. Let
$$I = \int_0^{\pi/4} \sin^3 2x \cos^4 2x\, dx$$

Put $2x = \theta$, so that $dx = \frac{d\theta}{2}$

Also when $x = 0$, $\theta = 0$

and when $x = \frac{\pi}{4}, \theta = \frac{\pi}{2}$

Then
$$I = \frac{1}{2}\int_0^{\pi/2} \sin^3\theta \cos^4\theta\, d\theta = \frac{1}{2}\left(\frac{2}{7}\right)\left(\frac{1}{5}\right) = \frac{1}{35}.$$
[by Walli's formula]

Example 10.36. Evaluate $\int_0^1 x^4 (1-x^2)^{5/2}\, dx$.

Solution. Put $x = \sin\theta$, so that $dx = \cos\theta\, d\theta$

The new limits are $\theta = 0$ to $\theta = \frac{\pi}{2}$

Thus the given integral

$$\int_0^{\pi/2} \sin^4\theta\,(1-\sin^2\theta)^{5/2}\cos\theta\,d\theta = \int_0^{\pi/2} \sin^4\theta\,.\cos^5\theta\cos\theta\,d\theta = \int_0^{\pi/2} \sin^4\theta\cos^6\theta\,d\theta$$

$$= \left(\frac{3}{10}.\frac{1}{8}\right)\left(\frac{5}{6}.\frac{3}{4}.\frac{1}{2}.\frac{\pi}{2}\right) = \frac{3\pi}{512} \qquad \text{[by Walli's formula]}$$

Example 10.37. Show that $\int_0^1 x^{3/2}(1-x)^{3/2}\,dx = \frac{3\pi}{128}$**.**

Solution. Let $x = \sin^2\theta$ so that $dx = 2\sin\theta\cos\theta\,dx$

Also when $x = 0$, $\theta = 0$ and when $x = 1$, $\theta = \frac{\pi}{2}$

∴ The given integral

$$= \int_0^{\pi/2} \sin^3\theta\cos^3\theta;\,2\sin\theta\cos\theta\,d\theta = 2\int_0^{\pi/2} \sin^4\theta\cos^4\theta\,d\theta$$

$$= 2.\left(\frac{3}{8}.\frac{1}{6}\right)\left(\frac{3}{4}.\frac{1}{2}.\frac{\pi}{2}\right) = \frac{3\pi}{128} \qquad \text{[by Walli's formula]}$$

EXERCISE 10.4

1. Establish a reduction formula for $\int \sin^n 2x\,dx$. $\left[\textbf{Ans. } I_n = -\frac{\sin^{n-1}2x\cos 2x}{2n} + \left(\frac{n-1}{n}\right)I_{n-2}\right]$

2. Evaluate $\int_0^{\pi/2} \sin^7 x\,dx$. $\left[\textbf{Ans. } \frac{16}{35}\right]$

3. Evaluate $\int_0^{\infty} \frac{dx}{(1+x^2)^5}$. $\left[\textbf{Ans. } \frac{35\pi}{256}\right]$

4. Evaluate $\int_0^4 x^3\sqrt{4x-x^2}\,dx$. [**Ans.** 28 π]

5. Prove that $\int_0^2 x^3\sqrt{2x-x^2}\,dx = \frac{7\pi}{8}$.

6. Prove that $\int_0^1 \frac{x^3\sin^{-1}x}{\sqrt{1-x^2}}\,dx = \frac{7}{9}$.

7. Find the reduction formula of $I_n = \int_0^{\pi/4} \tan x\,dx$. Hence evaluate $\int_0^a x^5(2a^2-x^2)^{-3}\,dx$. $\left[\textbf{Ans. } \frac{1}{2}\log_2 - \frac{1}{4}\right]$

8. Show that $\int x^m(1-x)^{n-1}\,dx = \frac{x^{m+1}(1-x)^{n-1}}{m+n} + \frac{n-1}{m+n}\int x^m(1-n)^{n-2}\,dx$, m and n being positive integers.

9. Evaluate $\int_0^{\pi/2} \sin^3 x\cos^4 x\cos 2x\,dx$. $\left[\textbf{Ans. } \frac{2}{315}\right]$

10. Evaluate $\int x\sin^4 x\,dx$. $\left[\textbf{Ans. } -\frac{x\sin^3 x\cos x}{4} + \frac{\sin^4 x}{16} + \frac{3}{16}x^2 - \frac{3}{16}x\sin 2x - \frac{3}{32}\cos 2x\right]$

❑❑❑

CHAPTER 11

Applications of Integration

INSIDE THIS CHAPTER

11.1. QUADRATURE

The process of finding the area bounded by a given arc of the curve is called **quadrature.**

11.2. AREA BOUNDED BY THE CURVE y = f(x), THE x-AXIS, THE ORDINATES x = a AND x = b

In this section we evaluate the area of plane regions, length of plane curves, volume and surface of a solid of revolution.

11.2.1. Area Under a Cartesian Curves

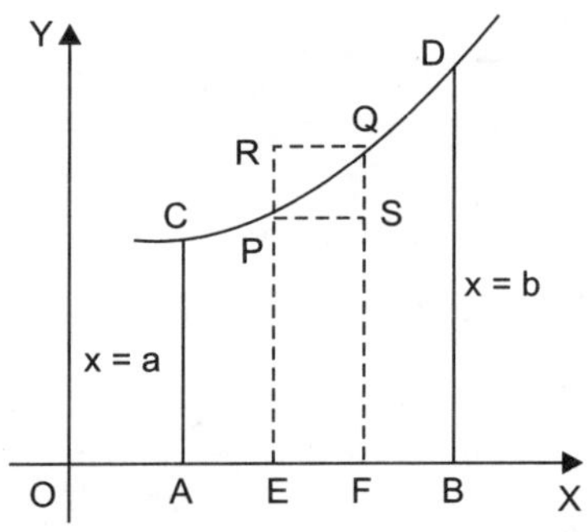

Theorem 1. Let $f(x)$ be continuous and finite in $[a, b]$. Then, the area bounded by, the curve $y = f(x)$, the x-axis and the ordinate $x = a$ and $x = b$ is equal to $\int_a^b y\, dx$.

Proof. Consider an arc of the given curve $y = f(x)$.

Let AC and BD be the ordinates drawn at $x = a$ and $x = b$ respectively. Now, it is required to find the area $ABDC$.

Consider a point $P(x, y)$ and a neighbouring point $Q(x + \Delta x, y + \Delta y)$ on the given curve. Draw PE and QF perpendicular on x-axis. Complete the rectangle $EFQR$. Also draw $PS \perp QF$.

Let us denote the areas $AEPC$ and $AFQC$ by $A(x)$ and $A(x + \Delta x)$ respectively. Then, area $(EFQP)$ $= A(x + \Delta x) - A(x)$.

Now, area ($\square$ $PEFS$) < area ($\square$ $EFQP$) < area ($\square EFQR$)

$$\Rightarrow \qquad y.\Delta x < A(x + \Delta x) - A(x) < (y + \Delta y) \,.\, \Delta x$$

$$\Rightarrow \qquad y < \frac{A(x + \Delta x) - A(x)}{\Delta x} < (y + \Delta y)$$

$$\Rightarrow \qquad \lim_{\Delta x\to 0} y < \lim_{\Delta x\to 0} \frac{A(x+\Delta x)-A(x)}{\Delta x} < \lim_{\Delta x\to 0}(y+\Delta y)$$

$$\Rightarrow \qquad = \frac{\pi}{ab}\left[\tan^{-1}\infty - \tan^{-1} 0\right] = \frac{\pi}{ab}\left(\frac{\pi}{2}-0\right) = \frac{\pi^2}{2ab}$$

i.e.,
$$\int_0^{\pi} \frac{x\,dx}{a^2\cos^2 x + b^2 \sin^2 x} = \frac{\pi^2}{2ab}.$$

$$\Rightarrow \qquad y < \frac{d}{dx}[A(x)] < y \quad \Rightarrow \quad \frac{d}{dx}A(x) = y$$

$$\Rightarrow \qquad \Big[A(x)\Big]_a^b = \int_a^b y\,dx \qquad \text{[Integrating between } x = a \text{ and } x = b]$$

$$\Rightarrow \qquad A(b) - A(a) = \int_a^b y\,dx$$

$$\Rightarrow \qquad [\text{Area when } x = b] - [\text{Area when } x = a] = \int_a^b y\,dx$$

$$\Rightarrow \qquad [\text{Area } ABDC] - 0 = \int_a^b y\,dx$$

$$\Rightarrow \qquad \text{Area } ABDC = \int_a^b y\,dx = \int_a^b f(x)\,dx$$

Similarly it can be proved that the area of the region bounded by a curve $x = f(y)$ the axis of y and the two lines $y = c$ and $y = d$ is $\int_c^d x\,dy$.

11.2.2. Area Under a Parametric Curves

Theorem 2. The area of the region bounded by curves.

$x = \phi(t)$ and $y = f(t)$, the x-axis and the ordinates at the points when $t = t_1$, $t = t_2$ is $\int_{t_1}^{t_2} y \frac{dx}{dt}\,dt$.

Proof. We know that the required area $= \int y\,dx = \int_{t_1}^{t_2} y\frac{dx}{dt}\,dt.$

Similarly the area bounded by the above curve, the y-axis and the abscissa at the points where $t = t_1$, $t = t_2$ is $\int_{t_1}^{t_2} x\frac{dy}{dt}\,dt$.

11.2.3. Area of Polar Curves

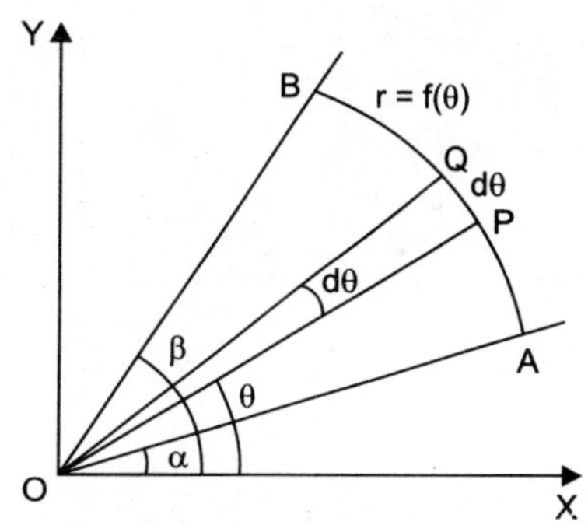

Let $r = f(\theta)$ be the equation of the curve in polar coordinates. Let OA, OB be the radii vectors of the points A and B on the curve, and let A be the area bounded by OA, OB and the curve. Let OA and OB make angle α and β respectively with y the initial line OX.

Let P and Q be the points (r, θ), $(r + \Delta r, \theta + \Delta\theta)$

On the curve very close to each other.

Consider the elementary area POQ which is equal to $\frac{1}{2}OP\,.\,OQ$ $\sin POQ = \frac{1}{2}r(r+\Delta r)\sin\Delta\theta$

As $\Delta\theta \to 0$, $\sin \Delta\theta \to \Delta\theta$; and

$$\text{Area } POQ \to \frac{1}{2} r^2 \,\Delta\theta \qquad [\because \ \Delta r,\ \Delta\theta \to 0]$$

$$\therefore \qquad \text{Area } AOB = \int_{\alpha}^{\beta} \frac{1}{2} r^2 \,d\theta, \quad \text{where } r = f(\theta)$$

$$\text{Total area} = \frac{1}{2}\int_{\alpha}^{\beta} r^2 \,d\theta$$

SOLVED EXAMPLES

Example 11.1. Find the area bounded by the curve $x^2 = 4y$ and the straight line $x = 4y - 2$.

Solution. Given curve is $x^2 = 4y$ ...(*i*)

which is an upward parabola with vertex at (0, 0) and is symmetrical about *y*-axis.

Equation of the line is $x = 4y - 2$ ...(*ii*)

Solving (*i*) and (*ii*) simultaneously, we get

$$(4y - 2)^2 = 4y$$

$$\Rightarrow \quad 16y^2 - 16y + 4 = 4y$$

$$\Rightarrow \quad 16y^2 - 20y + 4 = 0$$

$$\Rightarrow \quad 4y^2 - 5y + 1 = 0$$

$$\Rightarrow \quad (4y - 1)(y - 1) = 0$$

$$\Rightarrow \quad y = \frac{1}{4}, 1$$

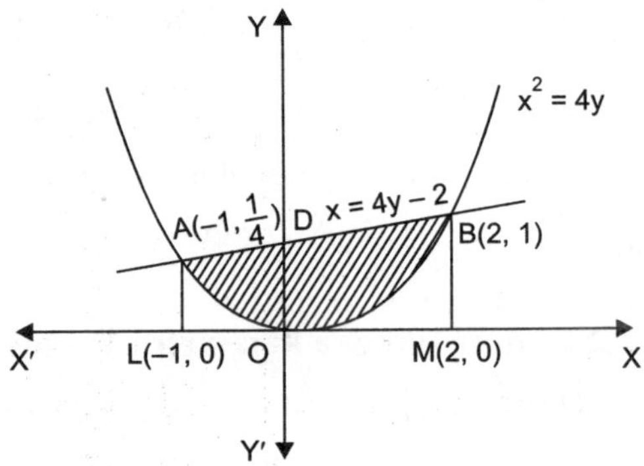

From Eq. (*ii*), when $y = \frac{1}{4}, x = 1 - 2 = -1$

When $y = 1, x = 4 - 2 = 2.$

$\therefore \quad A\left(-1, \frac{1}{4}\right)$ and B (2, – 1) are the point of intersection of (*i*) and (*ii*).

Clearly, from the figure shaded portion *OBDAO* is the region of the required area.

$$\text{Required Area} = \text{Area } LOMBDA - \text{Area } LMBOAL \qquad ...(iii)$$

Now, $\quad \text{Area } LMBDA = \int y\,dx \qquad \left[\text{where } x = 4y - 2 \Rightarrow y = \frac{1}{4}(x+2)\right]$

$$= \frac{1}{4}\int_{-1}^{2} (x + 2)\,dx = \frac{1}{4}\left[\frac{x^2}{2} + 2x\right]_{-1}^{2}$$

$$= \frac{1}{4}\left[(2 + 4) - \left(\frac{1}{2} + 2\right)\right] = \frac{1}{4}\left(6 + \frac{3}{2}\right) = \frac{15}{8} \qquad ...(iv)$$

and $\quad \text{Area } LMBOAL = \int_{-1}^{2} y\,dx, \quad \text{where } x^2 = 4y$

$$= \frac{1}{4}\int_{-1}^{2} x^2\,dx = \frac{1}{4}\left[\frac{x^3}{3}\right]_{-1}^{2}$$

$$= \frac{1}{2}[8 - (-1)] = \frac{9}{12} = \frac{3}{4} \text{ sq. units} \qquad ...(v)$$

From (*iii*), (*iv*) and (*v*), the required area

$$= \frac{15}{8} - \frac{3}{4} = \frac{9}{8} \text{ sq. units.}$$

Example 11.2. Find the area bounded by $y^2 = 9x$ and $x^2 = 9y$.

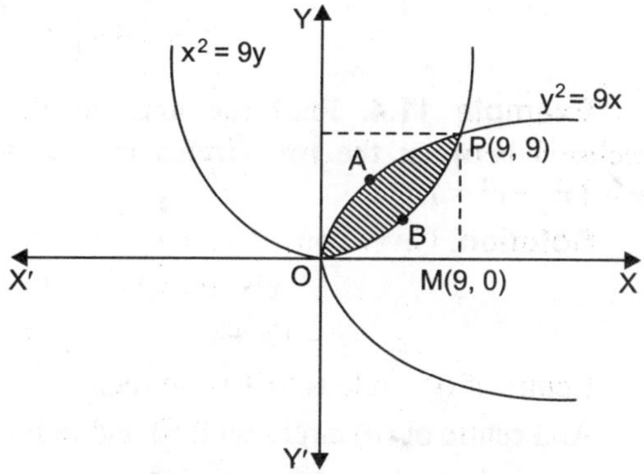

Solution. As shown in the figure, we have to find the area *OAPBO*.

Solving the given two equation simultaneously, we have

$$x^2 = 81y^2 = 81(9x) = 729x$$

$$\Rightarrow \quad x^4 - 729x = 0$$

$$\Rightarrow \quad x(x^3 - 729) = 0$$

$$\Rightarrow \quad x = 0, \quad x^3 = 729 = (9)^3$$

$$\Rightarrow \quad x = 9$$

∴ Print of intersection of two curves $y^2 = 9x$ and $x^2 = 9y$ are (0, 0) and (9, 9).

Now, Area *OAPBO* = Area *OAPMO* – Area *OBPMO*

$$= \int_0^9 y_1\, dx - \int_0^9 y_2\, dx = \int_0^9 3\sqrt{x}\, dx - \int_0^9 \frac{x^2}{9}\, dx$$

$$= 3\left[\frac{x^{3/2}}{3/2}\right]_0^9 - \frac{1}{9}\left[\frac{x^3}{3}\right]_0^9 = 2[(9)^{3/2}] - \frac{1}{27}(9)^3$$

$$= 2(3)^2 - \frac{1}{27} \times 729 = 2 \times 27 - 27 = 54 - 27 = 27 \text{ sq. units.}$$

Example 11.3. Find the area bounded by the circle $x^2 + y^2 = 16$ and the line $y = x$ in the first quadrant.

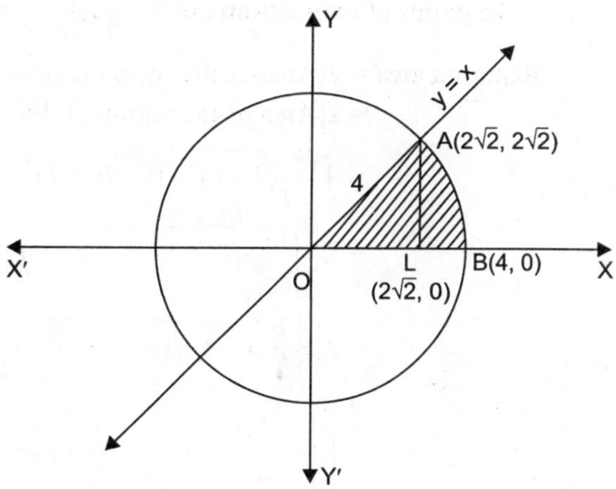

Solution. The given circle is

$$x^2 + y^2 = 16 \qquad \ldots(i)$$

The given line is $y = x$ ...(*ii*)

Putting $y = x$ from (*ii*) and (*i*), we get

$$2x^2 = 16 \Rightarrow x^2 = 8 \Rightarrow x = 2\sqrt{2}$$

[∵ *x* is +ve in the first quadrant]

Thus, the point of intersection of (*i*) and (*ii*) in the first quadrant is $A(2\sqrt{2}, 2\sqrt{2})$.

Draw *AL* perpendicular on the *x*-axis.

∴ Required area = area (*OLA*) + area (*LBAL*)

$$= \int_0^{2\sqrt{2}} x\, dx + \int_{2\sqrt{2}}^4 \sqrt{16 - x^2}\, dx$$

$$= \left[\frac{x^2}{2}\right]_0^{2\sqrt{2}} + \left[\frac{x\sqrt{16 - x^2}}{2} + \frac{16}{2}\sin^{-1}\frac{x}{4}\right]_{2\sqrt{2}}^4$$

$$= \frac{1}{2}[(2\sqrt{2})^2 - 0] + \left[(0 + 8\sin^{-1} 1) - \left(4 + 8\sin^{-1}\frac{1}{\sqrt{2}}\right)\right]$$

$$= \left[4 + \left(8 \times \frac{\pi}{2}\right) - 4 - \left(8 \times \frac{\pi}{2}\right)\right] = 2\pi \text{ sq. units.}$$

Example 11.4. Find the area of the region enclosed between the two circles $x^2 + y^2 = 1$ and $(x-1)^2 + y^2 = 1$.

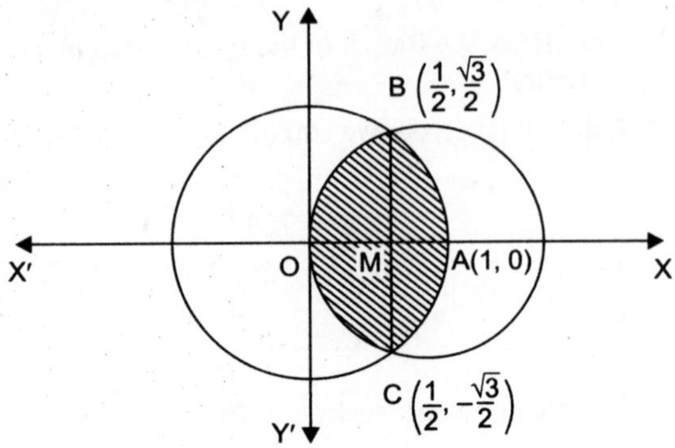

Solution. Given equation are

$$x^2 + y^2 = 1 \quad \text{...(i)}$$

$$(x-1)^2 + y^2 = 1 \quad \text{...(ii)}$$

Centre of (*i*) circle is (0, 0) and radius = 1

And centre of (*ii*) circle is (1, 0) and radius = 1

Now solving (*i*) and (*ii*), we get

$$-2x + 1 = 0$$

$$\Rightarrow \quad x = \frac{1}{2}$$

$$\therefore \quad y^2 = 1 - \left(\frac{1}{2}\right)^2 = 1 - \frac{1}{4} = \frac{3}{4}$$

$$\therefore \quad y = \frac{\sqrt{3}}{2}$$

$\therefore$ The points of intersection are $B\left(\frac{1}{2}, \frac{\sqrt{3}}{2}\right)$ and $C\left(\frac{1}{2}, -\frac{\sqrt{3}}{2}\right)$.

Required area = 2[Area of the region *OMAB*]

= 2[Area of the region *OMBO*] + 2[Area of the region *MABM*]

$$= \int_0^{1/2} \sqrt{1-(x-1)^2}\, dx + 2\int_{1/2}^{1} \sqrt{1-x^2}\, dx$$

$$= 2\left[\int_0^{1/2} \sqrt{1-(x-1)^2}\, dx + \int_{1/2}^{1} \sqrt{1-x^2}\, dx\right]$$

$$= 2\left[\left\{\frac{1}{2}(x-1)\sqrt{1-(x-1)^2} + \frac{1}{2}\sin^{-1}\left(\frac{x-1}{1}\right)\right\}_0^{1/2} + \left\{\frac{1}{2}\sqrt{1-x^2} + \frac{1}{2}\sin^{-1}\left(\frac{x}{1}\right)\right\}_{1/2}^{1}\right]$$

$$= \left[\left\{-\frac{\sqrt{3}}{4} + \sin^{-1}\left(-\frac{1}{2}\right) - \sin^{-1}(-1)\right\} + \left\{\sin^{-1}(1) - \frac{\sqrt{3}}{4} - \sin^{-1}\left(\frac{1}{2}\right)\right\}\right]$$

$$= -\frac{\sqrt{3}}{4} - \frac{\pi}{6} + \frac{\pi}{2} + \frac{\pi}{2} - \frac{\sqrt{3}}{2} - \frac{\pi}{6} = \left(\frac{2\pi}{3} - \frac{\sqrt{3}}{2}\right) \text{ sq. units.}$$

Example 11.5. Find the area of the region bounded by the ellipse $\frac{x^2}{9} + \frac{y^2}{4} = 1$ and the straight line $\frac{x}{3} + \frac{y}{2} = 1$.

Solution. We have $$\frac{x^2}{9}+\frac{y^2}{4}=1 \quad ...(i)$$

and $$\frac{x}{3}+\frac{y}{2}=1 \quad ...(ii)$$

Clearly (*i*) represent an ellipse and (*ii*) represents a straight line meeting the *x*-axis and *y*-axis at the point *A*(3, 0) and *B*(0, 2) representatively.

Solving (*i*) and (*ii*), we have

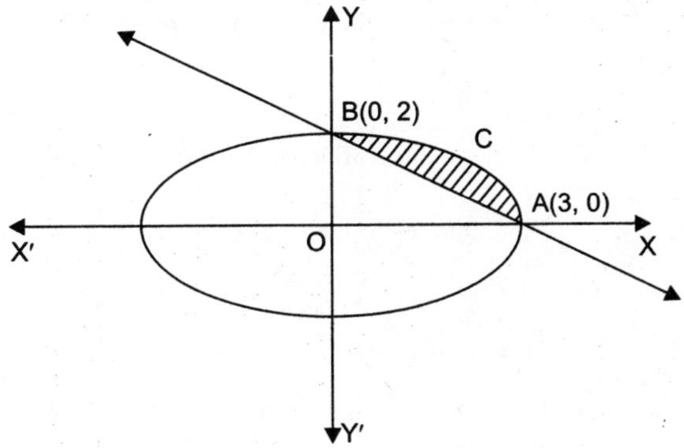

$$\frac{x^2}{9}+\left(1-\frac{x}{3}\right)^2=1$$

$$\Rightarrow \quad \frac{x^2}{9}+1+\frac{x^2}{9}-\frac{2x}{3}=1$$

$$\Rightarrow \quad \frac{2x^2}{9}-\frac{2x}{3}=0$$

$$\Rightarrow \quad \frac{2x}{3}\left(\frac{x}{3}-1\right)=0$$

$\Rightarrow$ Either $\frac{x}{3}=0$ or $\frac{x}{3}-1=0$

$\Rightarrow$ Either $x=0$ or $x=3$.

When $x = 0$, then from (*ii*), we get

$$\frac{0}{3}+\frac{y}{2}=1 \quad \Rightarrow \quad \frac{y}{2}=1 \quad \Rightarrow \quad y=2$$

When $x = 3$, then from (*ii*), we get

$$\frac{3}{3}+\frac{y}{2}=1 \quad \Rightarrow \quad 1+\frac{y}{2}=1$$

$$\Rightarrow \quad \frac{y}{2}=0 \quad \Rightarrow \quad y=0$$

Thus the point of intersection of an ellipse (*i*) and the straight line (*ii*) are *B*(0, 2) and *A*(3, 0). Join the points *A*(3, 0) and *B*(0, 2).

Hence the smaller portion bounded by these two curves is shown in the figure.

$\therefore$ Required area = Area of the shaded portion *ACBA*

= Area of the portion *OACBO* from $x = 0$ to $x = 3$

– Area of the portion *OABO* from $x = 0$ to $x = 3$

$$= \int_0^3 \frac{2}{3}\sqrt{9-x^2}\,dx - \int_0^3 \frac{2}{3}(3-x)dx \quad \left[\because \frac{x^2}{9}+\frac{y^2}{4}=1 \Rightarrow y=\frac{2}{3}\sqrt{9-x^2};\ \frac{x}{3}+\frac{y}{2}=1 \Rightarrow y=\frac{2}{3}(3-x)\right]$$

$$= \frac{2}{3}\times\frac{1}{2}\left[x\sqrt{9-x^2}+9\sin^{-1}\frac{x}{3}\right]_0^3 - \frac{2}{3}\left[3x-\frac{x^2}{2}\right]_0^3$$

$$= \frac{1}{3}\left[3(0)+9\sin\left(\frac{3}{3}\right)-\left\{0+9\sin^{-1}\left(\frac{0}{3}\right)\right\}\right]-\frac{2}{3}\left[\left(9-\frac{9}{2}\right)-(0-0)\right]$$

$$= \frac{1}{3}(9 \sin^{-1} 1) - \frac{2}{3}\left(\frac{9}{2}\right) = \frac{1}{3}\left(9 \times \frac{\pi}{2}\right) - \frac{2}{3} \times \frac{9}{2}$$

$$= \frac{3\pi}{2} - 3 = \frac{3}{2}(\pi - 2) \text{ sq. units.}$$

EXERCISE 11.1

1. Find the area of the region bounded by the line $y = 3x + 2$, the x-axis and the ordinates $x = -1$ and $x = 1$.

$$\left[\textbf{Ans. } \frac{13}{3} \text{ sq. units}\right]$$

2. By using integration, prove that the area of a circle of radius r units is πr^2 square units.

3. Find the area of the smaller region bounded by the ellipse $\frac{x^2}{a^2} + \frac{y^2}{b^2} = 1$ and the straight line $\frac{x}{a} + \frac{y}{b} = 1$.

$$\left[\textbf{Ans. } \frac{\pi ab}{4} - \frac{ab}{2}\right]$$

4. Find the area bounded by the curve $y^2 = 4a^2(x - 1)$ and the line $x = 1$ and $y = 4a$. $\left[\textbf{Ans. } \frac{16}{3}a \text{ sq. units}\right]$

5. Find the area enclosed between the parabola $y^2 = 4ax$ and the line $y = mx$. $\left[\textbf{Ans. } \frac{8a^2}{3m^3} \text{ sq. units}\right]$

6. Make a rough sketch of the region given below and find its area using integration

$[(x, y): 0 \le y \le x^2 + 3;\ 0 \le y \le 2x + 3;\ 0 \le x \le 3]$ $\left[\textbf{Ans. } \frac{50}{3} \text{ sq. units}\right]$

7. Find the area of smaller region bounded by the ellipse $\frac{x^2}{16} + \frac{y^2}{9} = 1$ and the straight line $\frac{x}{4} + \frac{y}{3} = 1$.

[**Ans.** $3(\pi - 2)$ sq. units]

8. Find the area of the region in the first quadrant enclosed by the x-axis, the line $y = x$ and the circle $x^2 + y^2 = 32$.

[**Ans.** 4π sq. units]

9. Find the area of the region enclosed between the two circles $x^2 + y^2 = 4$ and $(x - 2)^2 + y^2 = 4$.

$$\left[\textbf{Ans. } \frac{8\pi}{3} - 2\sqrt{3} \text{ sq. units}\right]$$

10. Find the area of circle $4x^2 + 4y^2 = 9$ which is interior to the parabola $x^2 = 4y$.

$$\left[\textbf{Ans. } \frac{\sqrt{2}}{6} + \frac{9}{4}\sin^{-1}\frac{2\sqrt{2}}{3} \text{ sq. units}\right]$$

11. Find the whole area of the tractrix $x = a \cos t + \frac{1}{2} a \log \tan^2 \frac{t}{2}, y = a \sin t$. [**Ans.** πa^2]

12. Find the whole area of the curve $a^2 y^2 = x^2(2a - x)$. [**Ans.** πa^2]

13. Show that the larger of the two areas into which the circle $x^2 + y^2 = 64a^2$ is divided by the parabola $y^2 = 12ax$ is $\frac{16}{3}a^2 (8\pi - \sqrt{3})$.

14. Show that the area common to the cardioid $r = a(1 - \cos \theta)$ and $r = a + (1 + \cos \theta)$ is $\frac{a^2}{2}(3\pi - 8)$.

15. Find the area of the cycloid $x = a(\theta - \sin \theta)$, $y = a(1 - \cos \theta)$. [**Ans.** $3\pi a^2$]

16. Find the area of the loop of the curve $x^3 + y^3 = 3axy$. $\left[\textbf{Ans. } \frac{3a^2}{2}\right]$

17. Find the area bounded by the curve $y = \sqrt{x}$ and the line $y = x$. $\left[\textbf{Ans. } \frac{1}{6}\right]$

18. Find the area bounded by the line $y = x$ and the curve $y = x^3$. $\left[\textbf{Ans. } \frac{1}{2}\right]$

19. Find the area of the region bounded by the curve $(y - 1)^2 = 4(x + 1)$ and the line $y = x - 1$. $\left[\textbf{Ans. } \frac{64}{3}\right]$

20. Using the method of integration to find the area of the region bounded by lines $2x + y = 4$, $3x - 2y = 6$ and $x - 3y + 5 = 0$. $\left[\textbf{Ans. } \frac{7}{2} \text{ sq. units}\right]$

Length of Curves (Rectification)

The process of finding the length of the arc of a curve is called **rectification.**

11.3. LENGTH OF ARC OF THE CURVE y = f(x)

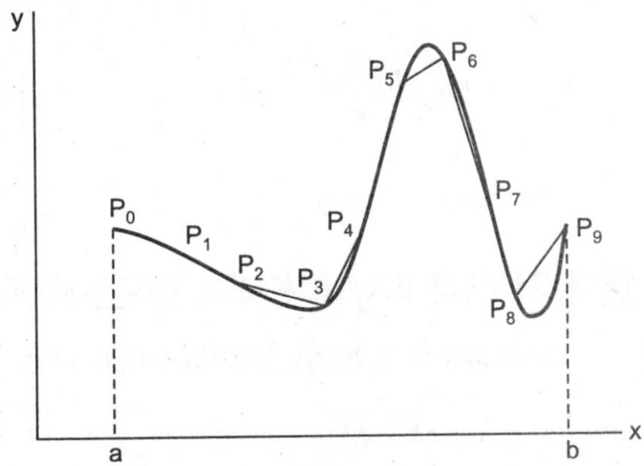

In this section we are going to look at computing the arc length of a function. Because it's easy enough to derive the formulas that we will use in this section, we will derive one of them and leave the other to you to derive.

We want to determine the length of the continuous function $y = f(x)$ on the interval $[a, b]$. We will also need to assume that the derivative is continuous on $[a, b]$.

Initially we will need to estimate the length of the curve. We will do this by dividing the interval up into n equal dividing the interval up into n equal subintervals each of width Dx and we will denote the point on the curve at each point by P_i. We can then approximate the curve by a series of straight lines connecting the points. Here is a sketch of this situation for $n = 9$.

Now denote the length of each of these line segments by $|P_{i-1} P_i|$ and the length of the curve will then be approximately.

$$L \approx \sum_{i=1}^{n} |P_{i-1} P_i|$$

and we get the exact length by taking n larger and larger. In other words, the exact length will be,

$$L = \lim_{n \to \infty} \sum_{i=1}^{n} |P_{i-1} P_i|$$

Now, let's get a better grasp on the length of each of these line segments. First, on each segment let's define $\Delta y_i = y_i - y_{i-1} = f(x_i) - f(x_{i-1})$. We can then compute directly the length of the line segments as follows.

$$|P_{i-1} P_i| = \sqrt{(x_i - x_{i-1})^2 + (y_i - y_{i-1})^2}$$

$$= \sqrt{\Delta x^2 + \Delta y_i^2}$$

By the **Mean Value Theorem,** we know that on the interval $| x_{i-1}, x_i |$ there is a point x_i^* so that,

$$f(x_i) - f(x_{i-1}) = f'(x_i^*)(x_i - x_{i-1})$$

$$Dy_i = f'(x_i^*)\,\Delta x$$

Therefore, the length can now be written as,

$$|P_{i-1}P_i| = \sqrt{(x_i - x_{i-1})^2 + (y_i - y_{i-1})^2}$$

$$= \sqrt{\Delta x^2 + [f'(\Delta_i^*)]^2\,\Delta x^2} = \sqrt{1 + [f'(\Delta_i^*)]^2}\,\Delta x$$

The exact length of the curve of then,

$$L = \lim_{n\to\infty} \sum_{i=1}^{n} |P_{i-1}P_i| = \lim_{n\to\infty} \sum_{i=1}^{n} \sqrt{1 + [f'(x_i^*)]^2}\,\Delta x$$

However, using the **definition of the definite integral,** this is nothing more than,

$$L = \int_a^b \sqrt{1 + [f'(x)]^2}\,dx$$

A slightly more convenient notation (in my opinion anyway) is the following:

$$L = \int_a^b \sqrt{1 + \left(\frac{dy}{dx}\right)^2}\,dx$$

In a similar fashion we can also derive a formula for $x = h(y)$ on $[c, d]$. This formula is,

$$L = \int_c^d \sqrt{1 + [h'(y)]^2}\,dy = \int_c^d \sqrt{1 + \left(\frac{dx}{dy}\right)^2}\,dy$$

SOLVED EXAMPLES

Example 11.6. Show that the length of the curve $y = \log \sec x$ between the points where $x = 0$ and $x = \frac{1}{3}\pi$ is $\log(2 + \sqrt{3})$.

Solution. The equation of the curve is $y = \log \sec x$

Differentiating, we get

$$\frac{dy}{dx} = \frac{1}{\sec x}\,.\,\sec x \tan x = \tan x$$

$\therefore$ Required length of the curve is

$$L = \int_0^{\pi/3} \sqrt{1 + \left(\frac{dy}{dx}\right)^2}\,dx = \int_0^{\pi/3} \sqrt{1 + \tan^2 x}\,dx$$

$$= \int_0^{\pi/3} \sec x\,dx = \Big[\log(\sec x + \tan x)\Big]_0^{\pi}$$

$$= \log\left(\sec\frac{x}{3} + \tan\frac{x}{3}\right) - 0 = \log(2 + \sqrt{3}).$$

Example 11.7. Find the perimeter of the circle $x^2 + y^2 = r^2$.

Solution. The equation of the curve is $x^2 + y^2 = r^2$ which is the standard equation of the circle with centre at origin and radius r. Also it is symmetrical to both the axes.

The circle meets x-axis *i.e.* $y = 0$ at $(r, 0)$ and $(-r, 0)$ and meets y-axis *i.e.* $x = 0$ at $(0, r)$ and $(0, -r)$.

$\therefore$ The required perimeter is $L = 4$ (Length of arc in the first quadrant)

$= 4$ [Length of arc from $(0, r)$ to $(r, 0)$]

$$= 4\int_{x=0}^{r} \sqrt{1+\left(\frac{dy}{dx}\right)^2}\, dx \qquad ...(i)$$

Now from $x^2 + y^2 = r^2$ we get $2x + 2y\left(\frac{dy}{dx}\right) = 0$.

$$\Rightarrow \qquad \frac{dy}{dx} = -\frac{x}{y} \quad \text{or} \quad \left(\frac{dy}{dx}\right)^2 = \frac{x^2}{y^2}$$

or
$$1+\left(\frac{dy}{dx}\right)^2 = 1+\frac{x^2}{y^2} = \frac{y^2+x^2}{y^2} = \frac{r^2}{y^2} \qquad [\because x^2+y^2=r^2]$$

$\therefore$ From (i), the required perimeter is

$$L = 4\int_0^r \sqrt{\frac{r^2}{y^2}}\, dx = 4r\int_0^r \frac{1}{y}\, dx$$

$$= 4r\int \frac{dx}{\sqrt{r^2-x^2}} = 4r\left[\sin^{-1}\left(\frac{x}{r}\right)\right]_0^r$$

$$= 4r\,[\sin^{-1}(1) - \sin^{-1}(0)] = 4r\left(\frac{\pi}{2}-0\right) = 2\pi r.$$

Example 11.8. Find the length of curve $y^2 = x^3$ from origin to the point (1, 1).

Solution. The equation of the curve is

$$y^2 = x^3 \qquad ...(i)$$

$$\therefore \qquad 2y\frac{dy}{dx} = 3x^2$$

$$\Rightarrow \qquad \frac{dy}{dx} = \frac{3x^2}{2y} = \frac{3x^2}{2x^{3/2}} = \frac{3}{2}x^{1/2} \qquad \text{[From } (i)]$$

Now
$$L = \int_0^1 \sqrt{1+\left(\frac{dy}{dx}\right)^2}\, dx$$

$$= \int_0^1 \sqrt{\left(1+\frac{9x}{4}\right)}\, dx = \frac{1}{2}\int_0^1 \sqrt{4+9x}\, dx = \frac{1}{2}\left[\frac{1}{9}\times\frac{2}{3}(4+9x)^{3/2}\right]_0^1$$

$$= \frac{1}{27}[(13)^{3/2} - (4)^{3/2}] = \frac{1}{27}(13\sqrt{13}-8).$$

Example 11.9. Find the length of the arc of the parabola $y^2 = 4ax$ extending from the vertex to an extremity of the latus rectum.

Solution. The coordinates of one extremity of the latus rectum of the parabola is $(a, 2a)$. Also differentiating $y^2 = 4ax$ w.r.t. y, we get

$$4a\frac{dx}{dy} = 2y \quad \text{or} \quad \frac{dx}{dy} = \frac{y}{2a}$$

∴ The required length of the curve

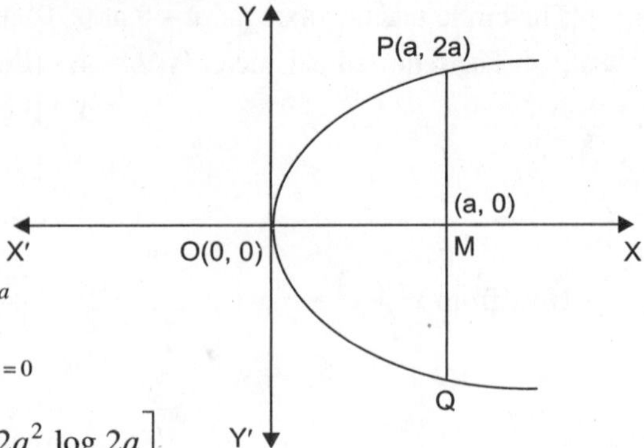

$$L = \int_0^{2a} \sqrt{1+\left(\frac{dx}{dy}\right)^2}\, dy$$

$$= \frac{1}{2a}\int_0^{2a} \sqrt{4a^2+y^2}\, dy$$

$$= \frac{1}{2a}\left[\frac{y}{2}\sqrt{4a^2+y^2} + \frac{4a^2}{2}\log\left(y+\sqrt{4a^2+y^2}\right)\right]_{y=0}^{2a}$$

$$= \frac{1}{2a}\left[a\sqrt{4a^2+4a^2} + 2a^2\log(2a+\sqrt{8a^2}) - 0 - 2a^2\log 2a\right]$$

$$= a\sqrt{2} + a\log\left[\frac{2a+2\sqrt{2a^2}}{2a}\right] = a\left[\sqrt{2}+\log(1+\sqrt{2})\right].$$

EXERCISE 11.2

1. Find the length of the arc of $f(x) = x^{3/2}$ on [0, 5]. $\left[\textbf{Ans. } \frac{335}{27}\right]$
2. Find the length of the arc of the parabola $x^2 = 8y$ from the vertex to an extremity of the latus rectum. $\left[\textbf{Ans. } 2\left[\sqrt{2}+\log(1+\sqrt{2})\right]\right]$
3. Find the length of the arc of the curve $y^2 = x^3$ from origin to the point (1, 1). $\left[\textbf{Ans. } \frac{1}{27}\left[13\sqrt{13}-8\right]\right]$
4. Find the length of the arc of the curve $y = \log \sec x$ from $x = 0$ to $x = \frac{\pi}{4}$. [**Ans.** log 3]
5. Find the length of arc of the parabola $y^2 = 12x$ cutoff by the latus rectum. $\left[\textbf{Ans. } 6\left[\sqrt{2}+\log(\sqrt{2}+1)\right]\right]$
6. Find the length of the curve $y^2 = (2x-1)^3$ cut-off by the line $x = 4$. $\left[\textbf{Ans. } \frac{1022}{27}\right]$
7. Find the actual length of $f(x) = \log(\sin x)$ on $\left[\frac{\pi}{4}, \frac{\pi}{2}\right]$. $\left[\textbf{Ans. } \log(\sqrt{2}+1)\right]$
8. Find the length of the curve $3y^2 = x(x-1)^2$, $y \geq 0$ between the points where the curve meets the x-axis. $\left[\textbf{Ans. } \frac{2}{\sqrt{3}}\right]$
9. Show that the length of the arc of the curve $y^2 = 4x$ from (0, 0) to (9, 6) is 11.3 approximately.

11.4. VOLUME OF A SOLID OF REVOLUTION

Theorem. The volume of the solid generated by the revolution of the area of bounded by the curve $y = f(x)$, x-axis and ordinates $x = a$, $x = b$ about the x-axis is $\int_a^b xy^2\, dx$.

Proof. Let AB be the curve given by the equation $y = f(x)$ and $AC = BD$ be the two given ordinates $x = a$ and $x = b$. Let $P(x, y)$, $Q(x+\Delta x, y+\Delta y)$ be two neighbouring points on the given curve. From P and Q drawn PM and QN perpendicular to the x-axis. Complete the rectangles NP and MQ, we have

$$MN = ON - OM$$

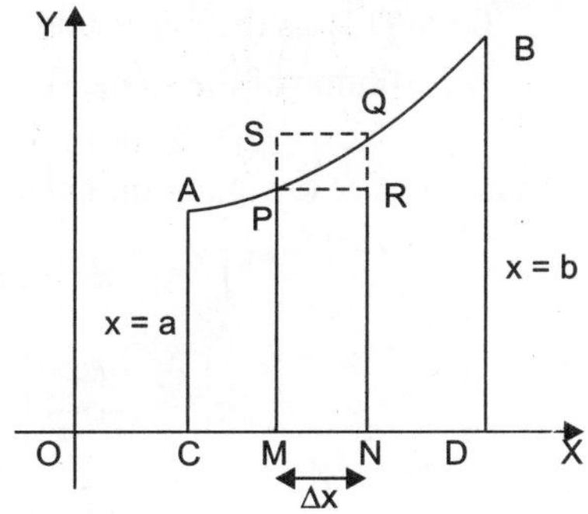

$\therefore \qquad MN = x + \Delta x - x = \Delta x$

Let the volume of the solids generated by the revolution of areas *ACMPA* and *ACNQPA* about *x*-axis be *V* and $V + \Delta V$ respectively. Then ΔV is the volume of the solid generated by the revolution of the area *PMNQP* about *x*-axis.

$$\therefore \qquad \Delta(PM)^2 \,.\, MN < \Delta V < P(QN)^2 \,.\, MN$$

$$\Rightarrow \qquad \pi y^2 \,\Delta x < \Delta V < \pi(y + \Delta y)^2 \,\Delta x$$

$$\Rightarrow \qquad \pi y^2 < \frac{\Delta V}{\Delta x} < \pi(y + \Delta y)^2$$

$$\Rightarrow \qquad \pi y^2 < \frac{dV}{dx} < \pi y^2, \quad \text{when } Q \to P\text{` [or } \Delta x \to 0, \Delta y \to 0]$$

$$\therefore \qquad \frac{dV}{dx} = \pi y^2 \quad \text{or} \quad dV = \pi y^2\, dx$$

Integrating $\displaystyle \int_{x=a}^{b} dV = \int_a^b xy^2$

or $\displaystyle \int_a^b xy^2\, dx = \Big[V\Big]_a^b$ = (Value of V at $x = b$) – (Value of V at $x = a$)

= Volume generated by the arc *ACDBA* – (0)

$\therefore$ At $x = a$, *PM* and *AC* coincide

= Volume generated by the arc *ACDBA*

Hence the required volume $= \displaystyle \int_a^b xy^2\, dx$

Similarly, the volume of the solid of revolution obtained by revolving about *y*-axis, the area bounded by a curve $x = g(y)$, two lines $y = c$, $y = d$ and *y*-axis is $\displaystyle \int_{y=c}^{y-d} \pi x^2\, dy$.

SOLVED EXAMPLES

Example 11.10. Find the volume of the solid generated by the revolution of the curve $y = \sin x$ from $x = 0$ to $x = 2\pi$ about x-axis.

Solution. Required volume $= \displaystyle \int_0^{2\pi} \pi y^2\, dx$

$$= \int_0^{2\pi} \pi \sin x\, dx = \frac{1}{2}\pi \int_0^{2\pi} \pi \sin^2 x\, dx$$

$$= \frac{1}{2}\pi \int_0^{2\pi} (1 - \cos 2x)\, dx = \frac{1}{2}\pi\left[x - \frac{1}{2}\sin^2 2x\right]_0^{2x}$$

$$= \frac{1}{2}\pi\,(2\pi) = \pi^2.$$

Example 11.11. Find the volume of a sphere of radius *r*.

Solution. Let the shape be generated by the revolution of the semi-circle *ABC* of radius *r* about its diameter *CA*.

Taking CA as the x-axis and its mid-point O as the origin, the equation of the circle ABC is $x^2 + y^2 = r^2$.

$\therefore$ Volume of the shape

= 2(Volume of the solid generated by the revolution about x-axis of the quadrant OAB).

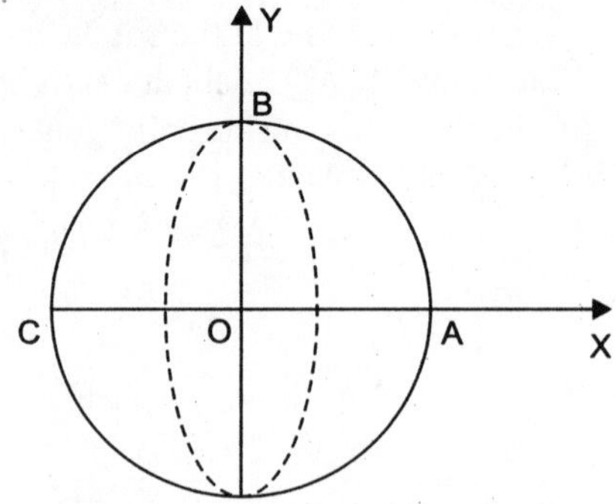

$$= 2\int_0^r \pi y^2\, dx = 2\pi\int_0^r (r^2 - x^2)\, dx$$

$$= 2\pi\left[r^2 x - \frac{x^3}{3}\right]_0^r$$

$$= 2\pi\left[r^2 - \frac{r^3}{3} - (0-0)\right] = \frac{4}{3}\pi r^3.$$

Example 11.12. The loop of the curve $2ay^2 = x(x-a)^2$ revolves about x-axis. Find the volume of the solid so generated.

Solution. Putting $y = 0$, we find that for the loop x varies from 0 to a. The solid is obtained when the upper half of the loop is revolved about x-axis.

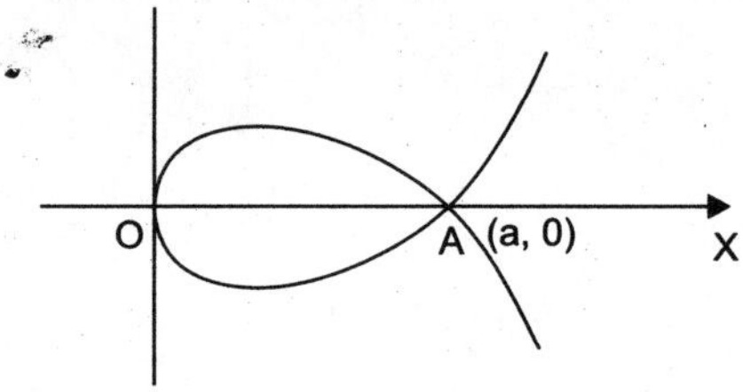

$\therefore$ Required volume $= \int_0^a \pi y^2\, dx$

$$= \frac{\pi}{2a}\int_0^a x(x-a)^2$$

$$= \frac{\pi}{2a}\int_0^a x(x^2 - 2ax + a^2)\, dx = \frac{\pi}{2a}\int_0^a (x^3 - 2ax^2 + a^2x)\, dx$$

$$= \frac{\pi}{2a}\left[\frac{1}{4}x^4 - \frac{2}{3}ax^3 + \frac{1}{2}a^2x^2\right]_0^a$$

$$= \frac{\pi}{2a}\left[\frac{1}{4}a^4 - \frac{2}{3}a^4 + \frac{1}{2}a^4\right] = \frac{\pi a^3}{2}\left[\frac{1}{4} - \frac{2}{3} + \frac{1}{2}\right]$$

$$= \frac{\pi a^3}{24}.$$

Example 11.13. Find the volume of the solid generated by revolving the curve $x^{2/3} + y^{2/3} = a^{2/3}$ about x-axis.

Solution. The required volume is generated by revolving the area $ABA'OA$ about x-axis.

= 2 × volume generated by revolving the area $ABOA$ about x-axis.

$$= 2\int_0^a \pi y^2\, dx = 2\pi\int_0^a (a^{2/3} - x^{2/3})^3\, dx$$

$$(\because\ y^{2/3} = a^{2/3} - x^{2/3})$$

Put $x = a\sin^3\theta \Rightarrow dx = 3a\sin^2\theta\cos\theta\, d\theta$

When $x = 0,\ \theta = 0$

and $x = a,\ \theta = \dfrac{\pi}{2}$

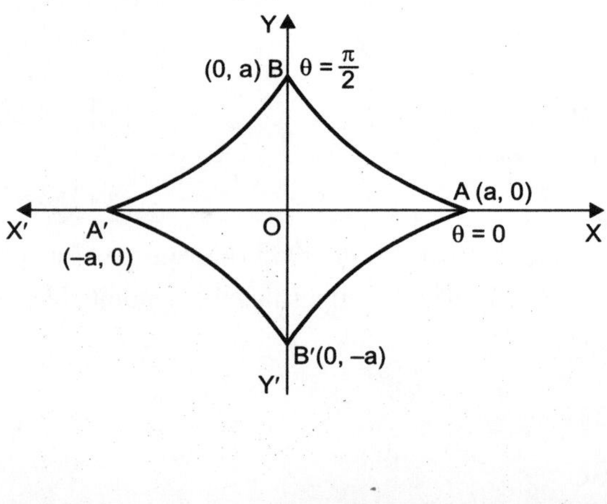

Thus, Required volume $= 2\pi\int_0^{\pi/2} a^2 \cos^6\theta . 3a\sin^2\theta\cos\theta\, d\theta$

$$= 6\pi a^3 \int_0^{\pi/2} \sin^2\theta\cos^7\theta\, d\theta$$

$$= 6\pi a^3 \left(\frac{1}{9}\right)\left(\frac{6}{7}\times\frac{4}{5}\times\frac{2}{3}\right) = \frac{32\pi a^3}{105}.$$

Example 11.14. The region enclosed by the curves $y = \sin x, y = \cos x$, and the axis between $x = 0$ to $x = \frac{\pi}{2}$ is revolved about the x-axis. Find the volume of the solid thus generated.

Solution. The given curves $y = \sin x$ and $y = \cos x$ intersect at $x = \frac{\pi}{4}$.

∴ The required volume

$$= \int_0^{\pi/4} \pi\sin^2 x\, dx + \int_{\pi/4}^{\pi/2} \pi\cos^2 x\, dx$$

$$= \frac{\pi}{2}\int_0^{\pi/4} (1-\cos 2x)\, dx + \frac{\pi}{2}\int_{\pi/4}^{\pi/2} (1+\cos 2x)\, dx$$

$$= \frac{\pi}{2}\left[x - \frac{\sin 2x}{2}\right]_0^{\pi/4} + \frac{\pi}{2}\left[x + \frac{\sin 2x}{2}\right]_{\pi/4}^{\pi/2}$$

$$= \frac{\pi}{2}\left[\frac{\pi}{4} - \frac{1}{2}\right] + \frac{\pi}{2}\left[\left(\frac{\pi}{2} - \frac{\pi}{4}\right) + \frac{1}{2}(0-1)\right]$$

$$= \frac{1}{4}\pi^2 - \frac{1}{2}\pi = \frac{\pi}{4}(\pi - 2).$$

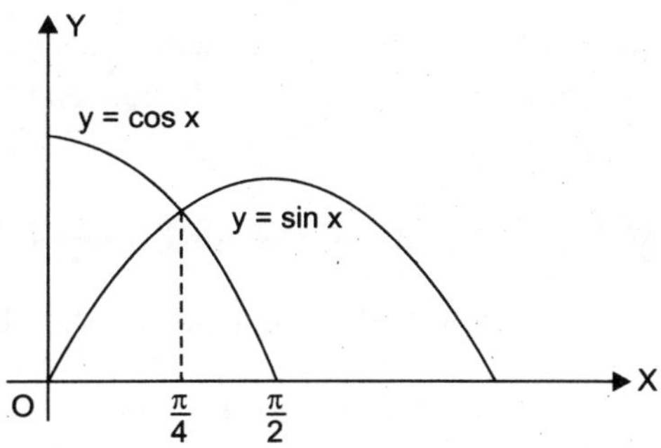

EXERCISE 11.3

1. Find the volume of the solid generated by rotating completely about the axis of x, the area enclosed between $y^2 = x^3 + 5x$ and the lines $x = 2$ and $x = 4$.
2. Find the volume of the ellipsoid formed by the revolution of the ellipse $\frac{x^2}{a^2} + \frac{y^2}{b^2} = 1$ about the major axis.
   $\left[\textbf{Ans. } \frac{4}{3}\pi ab^2\right]$
3. Find the volume formed by the revolution of loop of the curve $y^2(a + x) = x^2(3a - x)$, about the x-axis.
   [**Ans.** $\pi a^3(8 \log 2 - 3)$]
4. Prove that the volume of the reel formed by the revolution of the cycloid $x = a(q + \sin\theta), y = a(1 - \cos\theta)$ about the tangent at the vertex is $\pi^2 a^3$.
5. The part of the parabola $y^2 = 4ax$ cutoff by the latus-rectum revolves about the tangent at the vertex. Find the volume of the reel thus generated.
   $\left[\textbf{Ans. } \frac{4}{5}\pi a^3\right]$
6. Find the volume generated by the revolution of the astroid $x = a\cos^3\theta, y = a\sin^3\theta$, about the x-axis.
   $\left[\textbf{Ans. } \frac{35}{105}\pi a^3\right]$

7. Find the volume of the solid obtained by revolving the cardioid $r = (1 + \cos\theta)$ about the initial line.
$$\left[\textbf{Ans. } \frac{8}{5}\pi a^3\right]$$

8. Find the volume of the solid form by the revolution of the curve $xy^2 = 4(2 - x)$ through four right angles about the x-axis. **[Ans.** $4\pi^2$]

11.5. CENTRE OF GRAVITY

Centre of gravity of a plane lamina: Take two perpendicular lines in the plane lamina as the axes of reference. A small elementary area near P(x, y) would have its value $\delta x \delta y$. Its mass would be $\rho \delta x \delta y$, where ρ is density per unit area.

$$\therefore \qquad \bar{x} = \frac{\iint x\rho\, dx\, dy}{\iint \rho\, dx\, dy}; \quad \bar{y} = \frac{\iint y\rho\, dx\, dy}{\iint \rho\, dx\, dy}$$

Limits are to be chosen to cover the whole area.

The above results can also be converted into polar co-ordinates by putting $x = r\cos\theta$, $y = r\sin\theta$, $dx\, dy = r\, d\theta\, dr$.

SOLVED EXAMPLES

Example 11.15. Find co-ordinates of the centre of gravity of a quadrant of an ellipse, assuming the surface density at all points to be uniform and equal to 1.

Solution. Take the equation of the ellipse to be

$$\frac{x^2}{a^2} + \frac{y^2}{b^2} = 1$$

Then,
$$\bar{x} = \frac{\iint x\, dx\, dy}{\iint dx\, dy} = \frac{\int_0^b \left(\int_0^{\frac{a}{b}\sqrt{b^2 - y^2}} (x\, dx)\, dy\right)}{\int_0^a \int_0^{b\sqrt{a^2 - x^2}/a} dy\, dx} = \frac{\frac{1}{2}\int_0^b \frac{a^2}{b^2}(b^2 - y^2)\, dy}{\int_0^a \frac{b}{a}\sqrt{a^2 - x^2}\, dx}$$

$$= \frac{\frac{a^2}{b^2}\left|b^2 y - \frac{y^3}{3}\right|_0^b}{\int_0^a \frac{b}{a}\sqrt{a^2 - x^2}\, dx} = \frac{\frac{2}{3}a^2 b}{\int_0^{\frac{\pi}{2}} \frac{b}{a}.a^2\cos^2 t\, dt}$$

$= \dfrac{4a}{3\pi}$ on simple evaluation of integral by putting $x = a\sin t$ in the denominator integral.

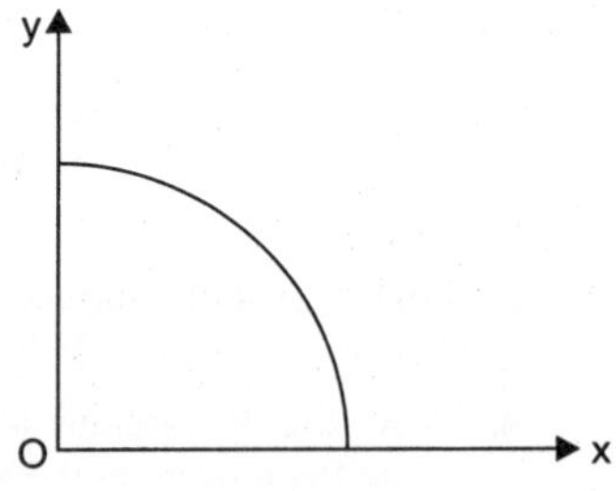

Similarly,
$$\bar{y} = \frac{\iint y\, dx\, dy}{\iint dx\, dy} = \frac{\int_0^a \left(\int_0^{b\sqrt{a^2 - x^2}/a}\right) dx}{\frac{1}{4}\pi ab} = \frac{4b}{3\pi}$$

Example 11.16. Using double integration, find the C.G. of the lamina in the shape of a quadrant of the curve $\left(\frac{x}{a}\right)^{2/3} + \left(\frac{y}{b}\right)^{2/3} = 1$, **the density being** $\rho = Kxy$, **where** K **is a constant.**

Solution. Let $G(\bar{x}, \bar{y})$ be the C.G of the lamina OAB, so that

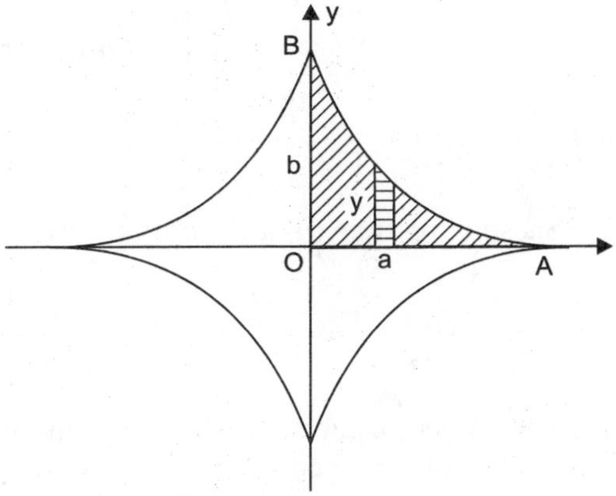

$$x = \frac{\iint Kxy\,x\,dx\,dy}{\iint Kxy\,dx\,dy} = \frac{\iint x^2y\,dx\,dy}{\iint xy\,dx\,dy}$$

where the integrals are taken over the area OAB so that the y varies from 0 to y (to be found from the equation of the curve in terms of x) and then x varies from 0 to a.

Then $$\bar{x} = \frac{\int_0^a \int_0^y x^2 y\,dy\,dx}{\int_0^a \int_0^y xy\,dy\,dx}$$

$$= \frac{\int_0^a x^2 \left|\frac{y^2}{2}\right|_0^y dx}{\int_0^a x\left|\frac{y^2}{2}\right|_0^y dx} = \frac{\int_0^a x^2y^2\,dx}{\int_0^a xy^2\,dx}$$

For any point on the curve, we have

$$x = a\cos^3\theta,\ y = b\sin^3\theta$$

$$dx = -3a\cos^2\theta\sin\theta\,d\theta,$$

when $x = 0$, $\theta = \dfrac{\pi}{2}$

$$x = a,\ \theta = 0$$

Hence $$\bar{x} = \frac{\int_{\frac{\pi}{2}}^0 a^2\cos^6\theta b^2\sin^6\theta(-3a\cos^2\theta\sin\theta)\,d\theta}{\int_{\frac{\pi}{2}}^0 a\cos^3\theta b^2\sin^6\theta(-3a\cos^2\theta\sin\theta)\,d\theta} = \frac{\int_0^{\frac{\pi}{2}}\sin^7\theta\cos^8\theta\,d\theta}{\int_0^{\frac{\pi}{2}}\sin^7\theta\cos^5\theta\,d\theta} = \frac{128a}{429}$$

Similarly $$y = \frac{\int_0^a\int_0^y \mathrm{K}xy\cdot y\,dx\,dy}{\int_0^a\int_0^y \mathrm{K}xy\cdot dx\,dy} = \frac{128b}{429}$$

Hence required C.G. is $\mathrm{G}\left(\dfrac{128a}{429}, \dfrac{128b}{429}\right)$. **Ans.**

11.6. MEAN AND R.M.S. VALUES

***(a)* Mean value:** The mean value of the ordinate y, of a function $y = \phi(x)$, over the range $x = a$ to $x = b$ is the limit of the mean value of the equidistant ordinates $y_1, y_2, ..., y_n$, as n tends to infinity.

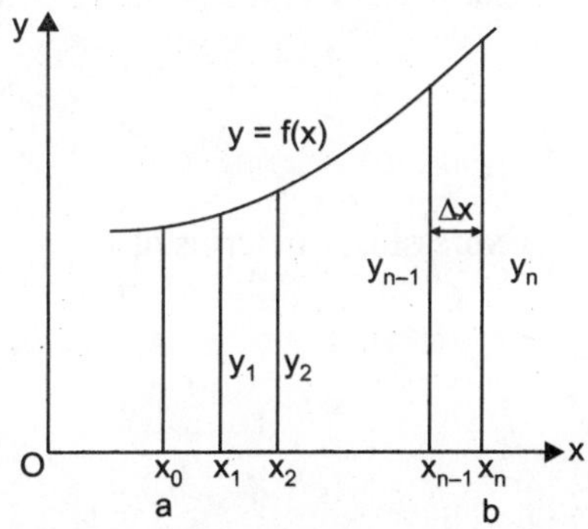

Thus if y_m is the mean value

$$y_m = \lim_{n\to\infty} \frac{y_1 + y_2 + ... + y_n}{n} \quad ...(i)$$

$$= \lim_{n\to\infty} \frac{y_1 + y_2 + ... + y_n}{n} \quad ...(ii)$$

Now, $$n\Delta x = (b-a) = \int_a^b dx \qquad \ldots(iii)$$

and $$\lim_{n\to\infty} [y_1\Delta x + y_2\Delta x + \ldots + y_n\Delta x] = \lim_{\substack{n\to\infty\\ \Delta x\to 0}} \sum_{x=a}^{x=b} y\Delta x = \int_a^b y\,dx$$

Substituting from (*ii*) and (*iii*) in (*i*), we have

$$y_m = \frac{\int_a^b y dx}{\int_a^b dx} \qquad \ldots(iv)$$

The analysis can be extended to the mean values of the functions of two or three variables. Thus the mean values of the function $z = \phi(x, y)$ over an area A and of $u = \phi(x, y, z)$ over the region of volume V are

$$z_m = \frac{\iint_A f(x, y)\, dA}{\iint_A dA},\ u_m = \frac{\iiint_V f(x, y, z)\, dv}{\iiint_V dv} \qquad \ldots(v)$$

The integrals being taken over the area A and throughout the volume V respectively.

***(b)* Root mean Square value:** This term is usuallly applied to periodic function only. If y isa periodic function of x, of period p, the root mean square value of y is the square root of the mean value of y^2 over the range $x = c$ to $x = c + p$, c being any constant. Thus,

$$\text{R.M.S. of } y = \sqrt{\frac{\int_c^{c+p} y^2\, dx}{p}} = \frac{\sqrt{\int_c^{c+p} y^2\, dx}}{\sqrt{\int_c^{c+p} dx}}$$

The root mean square values play an importatn role in the calculations of the A.C. electrical circuits.

SOLVED EXAMPLES

Example 11.17. A point moves from rest with uniform acceleration. Prove that in any interval form the starting point, the space-average of the velocity is $\frac{4}{3}$ of the time-average.

Solution. If v is the velocity at time t, at a distance s from the starting point, and f is the uniform acceleration, the equations of motion are

$$v = ft,\ v^2 = 2fs \quad \text{and} \quad s = \frac{1}{2}ft^2 \qquad \ldots(i)$$

The space-average of the velocity V_s over the distance travelled s is

$$V_s = \frac{\int_0^x v\, ds}{\int_0^x v\, ds}$$

Expressing v in terms of s,

$$V_s = \frac{1}{s}\int_0^s \sqrt{2fs}\, ds$$

$$= \frac{2}{3}\sqrt{2fs} \qquad \ldots(iii)$$

(*iii*) gives the space-average of the velocity

If T is the time required to cover the same distance s as above, then the time average of the velocity V_T is

$$V_T = \frac{\int_0^T v\,dt}{\int_0^T dt} \qquad \ldots(iv)$$

Expressing v in terms of t by (*i*), we get

$$V_t = \frac{1}{T}\int_0^T ft\,dt = \frac{1}{2}fT \qquad \ldots(v)$$

Since $s = \frac{1}{T}fT^2$ or $T = \sqrt{\frac{2s}{f}}$, we get the time average of velocity as

$$V_T = \frac{1}{2}f \cdot \sqrt{\frac{2s}{f}} = \frac{1}{2}\sqrt{2\,fs} \qquad \ldots(vi)$$

Comparing (*iii*) and (*vi*), we have

$$\frac{V_s}{V_T} = \frac{4}{3}$$

That is the space-average of velocity is $\frac{4}{3}$ the time-average.

Example 11.18. Find the mean and the r.m.s. value of the ordinate y of the cycloid $x = a(\theta + \sin\theta)$, $y = a(1 - \cos\theta)$ over the range $\theta = -\pi$ to $\theta = \pi$, Hence show that the value of the latter is $\sqrt{2}$ times that of the former.

Solution. The mean value of y_m, over the given range is

$$y_m = \frac{\int y\,dx}{\int dx} = \frac{\int_{-\pi}^{\pi} y\frac{dy}{d\theta}d\theta}{\int_{-\pi}^{\pi} y\frac{dx}{d\theta}d\theta}$$

Substituting for y and $\frac{dx}{d\theta}$ from the equations of the cycloids, we have for the integrals

$$\int_{-\pi}^{\pi} y\frac{dy}{dx}d\theta = \int_{-\pi}^{\pi} a(1-\cos\theta)\,a(1+\cos\theta)\,d\theta = a^2\int_{-\pi}^{\pi} \sin^2\theta\,d\theta$$

$$= 4a^2\int_{-\pi}^{\frac{\pi}{2}} (1+\cos\theta)\,d\theta \qquad \ldots(ii)$$

and

$$\int_{-\pi}^{\pi} \frac{dx}{d\theta}d\theta = a\int_{-\pi}^{\pi} (1+\cos\theta)d\theta$$

$$= a[\theta + \sin\theta]_{-\pi}^{\pi} = 2a\pi \qquad \ldots(iii)$$

Putting the values of the integrals from (*ii*) and (*iii*) in (*i*)

$$y_m = \frac{\pi a^2}{2a\pi} = \frac{a}{2}$$

Next the r.m.s. value of y is given by

$$y_{r.m.s.} = \sqrt{\frac{\int y^2\,dx}{\int dx}} \qquad \ldots(v)$$

over the range $\theta = -\pi$ to $\theta = \pi$

Now $\int y^2\, dx = \int_{-\pi}^{\pi} y^2 \frac{dx}{d\theta} d\theta = \int_{-\pi}^{\pi} a^2(1-\cos\theta)^2\, a(1+\cos\theta)\, d\theta$

$$= 8a^3 \int_{-\pi}^{\pi} \sin^4 \frac{\theta}{2} \cdot \cos^2 \frac{\theta}{2}\, d\theta \qquad \ldots(vi)$$

$$= 16a^3 \int_{-\frac{\pi}{2}}^{\frac{\pi}{2}} \sin^4 \phi \cdot \cos^2 \phi\, d\phi, \text{ where } \phi = \frac{\theta}{2}$$

$$= 32a^3 \int_0^{\frac{\pi}{2}} \sin^4 \phi \cos^2 \phi\, d\phi = \pi a^3$$

Using (*iii*), the expression under the radical sign in (*v*) is

$$\frac{\pi a^3}{2\pi a} = \frac{a^2}{2}$$

and hence $$y_{\text{r.m.s.}} = \frac{a}{\sqrt{2}} \qquad \ldots(vii)$$

The ratio of y_m to $y_{\text{r.m.s.}}$ is easily obtained from (*iv*) and (*vii*) which shows that $y_{\text{r.m.s.}} = \frac{\sqrt{2}}{y_m}$.

Example 11.19. Find the mean value of r² over the area of the cardiode *r* = *a*(1 + cos θ).

Solution. The mean value of r^2 is $\dfrac{\iint r^2\, dA}{\iint dA}$ ...(*i*)

Integrals being taken over the area of the cardiode

Now $\iint r^2\, dA = \int_0^{2\pi} \int_0^{a(1+\cos\theta)} r^2\, r dr\, d\theta = \frac{a^4}{4} \int_0^{2\pi} (1+\cos\theta)^4\, d\theta$

$$= 4a^4 \int_0^{2\pi} \cos^8 \frac{\theta}{2} d\theta = 8a^4 \int_0^{\pi} \cos^8 \phi\, d\phi, \text{ where } \frac{\theta}{2} = \phi$$

$$= 16a^4 \cdot \frac{7}{8} \cdot \frac{5}{6} \cdot \frac{3}{4} \cdot \frac{1}{2} \cdot \frac{\pi}{2}$$

$$= \frac{35}{16} \pi a^4 \qquad \ldots(ii)$$

and $\iint dA = \int_0^{2\pi} \int_0^{2a(1+\cos\theta)} r\, dr\, d\theta$

$$= \frac{a^2}{2} \int_0^{2\pi} (1+\cos\theta)^2\, d\theta = 2a^2 \int_0^{2\pi} \cos^4 \frac{\theta}{2} d\theta$$

$$= 4a^2 \int_0^{\pi} \cos^4 \phi\, d\theta, \text{ where } \frac{\theta}{2} = \phi = \frac{\theta}{2} = \phi = 8a^2 \cdot \frac{3}{4} \cdot \frac{1}{2} \cdot \frac{\pi}{2}$$

$$= \frac{3}{2} \pi a^2$$

From (*i*), (*ii*) and (*iii*), the mean value of r^2 is

$$\frac{\frac{35\pi a^4}{16}}{\frac{3\pi a^2}{2}} = \frac{35}{24} a^2$$

EXERCISE 11.4

1. Using double integration, find the C.G. of the lamina in the shape of a quadrant of the curve $\left(\frac{x}{a}\right)^{2/3}+\left(\frac{y}{b}\right)^{2/3}$ = 1, the destiny being $\rho = Kxy$, where K is a constant.

2. If the density at any point of the solid octant of the ellipsoid $\left(\frac{x}{a}\right)^2+\left(\frac{y}{b}\right)^2+\left(\frac{z}{c}\right)^2$ varies xy^2, find the coordinates of the C.G. of the solid. $\left[\textbf{Ans. } \frac{16}{35}a, \frac{16}{35}b, \frac{16}{35}c\right]$

3. A quantity of gas confined behind the piston expands adiabatically according to law pv^n = constant. Find the mean value of pressure p as volume v increases from v_1 to v_2. Given at $v = v_1, p = =1, v = v_2, p = p_2$.

$\left[\textbf{Ans. } \frac{p_1v_1 - p_2v_2}{(n-1)(v_2-v_1)}\right]$

4. A particle moves along x-axis according to the reqlation $2x = t^2 + 2t$. Find the mean velocity w.r.t. time 't' between $t = 2$ to $t = 2$. Also find the average velocity w.r.t. 'x' for values of x corresponding to $t = 1, t = 2$.

$\left[\textbf{Ans. } \frac{5}{2}, \frac{38}{15}\right]$

5. In the above problem, find r.m.s value of velocity w.r.t. time 't' and w.r.t. 'x', corresponding to the values $t = 1, t = 2$. $\left[\textbf{Ans. } (i)\sqrt{\frac{19}{3}}, (ii)\frac{\sqrt{13}}{2\sqrt{2}}\right]$

6. Find the r.m.s. value of the ordinate of one arch of cycloid $x = a(\theta - \sin\theta), y = a(1-\cos\theta)$ w.r.t. 'x' where $0 \le \theta \le 2\pi$. Further show that mean value of the ordinate of one arch of the above cycloid is $\sqrt{\frac{9}{10}}$ times the r.m.s. value. $\left[\textbf{Ans. } \sqrt{\frac{5}{2}}\cdot a\right]$

7. Find the average value and root mean square value of the cardioide $r = a(1-\cos\theta)$ w.r.t. area of the cardioide where $(0 \le \theta \le 2\pi)$.

8. Find the mean value of $f(t) = \sin^2(\pi t + \alpha)$. $\left[\textbf{Ans. } \frac{1}{2}\right]$

9. Find the average ordinate w.r.t. arc length of the semi-circle $y = \sqrt{4-x^2}$

$\left[\textbf{Hint: } \text{Average ordinate} = \frac{\int y\,ds}{\int ds} \text{ where } s \text{ is the arc length given by } \int_{-2}^{+2}\sqrt{1+\left(\frac{dy}{dx}\right)^2}\,dx\right]$

$\left[\textbf{Ans. } \frac{4}{\pi}\right]$

❑❑❑

UNIT - V

ORDINARY DIFFERENTIAL EQUATIONS

CHAPTER 12

First Order Ordinary Differential Equations

INSIDE THIS CHAPTER

12.1. DIFFERENTIAL EQUATIONS AND THEIR FORMATION

Differential equation is an important branch of mathematics which has wide applications in Engineering, Physical Sciences and Biological Sciences. Many practical problems like decay of a certain substance, change in temperature of a body, increase in population can be solved with the help of differential equations.

Definitions

Differential Equation: An equation containing an independent variable, dependent variable and differential coefficients of dependent variable with respect to independent variable is called a differential equation.

For instance

(*i*) $\frac{dy}{dx} = x$ (*ii*) $\frac{dy}{dx} = \cot x$ (*iii*) $\frac{d^4y}{dx^4} - 4\frac{dy}{dx} + 4y = 5\cos 3x$

(*iv*) $\frac{d^2y}{dx^2} + 16y = 0$ (*v*) $\sin x\, dx + \cos y\, dy = 0$

are examples of differential equations.

Order of a Differential Equation: The order of a differential equation is the order of the highest order derivative occurring in the differential equation.

For example

1. $\frac{d^2y}{dx^2} + 3\frac{dy}{dx} + 2y = e^x$ is a differential equation of order 2 because the order of highest derivative occurring in it is 2.

2. $\frac{d^3y}{dx^3} - 6\left(\frac{dy}{dx}\right)^2 - 4y = 0$ is a differential equation of order 3 because the order of highest derivative occurring in it is 3.

3. $\left(\frac{d^2y}{dx^2}\right)+\left(\frac{dy}{dx}\right)^4 = ex$ is a differential equation of order 2 because the order of highest derivative occurring in it is 2.

Degree of a Differential Equation: The degree of a differential equation is the degree of the highest order derivative, when differential coefficients are made free from radicals and fractions. In other words, the degree of a differential equations is the power of the highest order derivative occurring in a differential equation when it is written as a polynomial in differential coefficients.

For example

1. $x^2\frac{d^2y}{dx^2}+x\frac{dy}{dx}+y = 0$ is a differential equation of degree 1 as in this equation the power of the highest order derivative is 1.

2. $x\left(\frac{d^3y}{dx^3}\right)+\left(\frac{dy}{dx}\right)^4+y^2 = 0$ is a differential equation of degree 2 as in this equation the power of the highest order derivative is 2.

3. $\left(\frac{d^2y}{dx^2}\right)=\sqrt{1+\frac{dy}{dx}}$. In this equation derivatives are not free from radicals. Removing the radicals it takes the form

$$\left(\frac{d^2y}{dx^2}\right)^2 = 1+\left(\frac{dy}{dx}\right)^2$$

Clearly, the degree of this differential equation is 2.

SOLVED EXAMPLES

Example 12.1. Determine the the order and degree each of the following differential equations

(*i*) $x^2\left(\frac{dy}{dx}\right)+2x-6x^3=0$ (*ii*) $2y\left(\frac{d^2y}{dx^2}\right)+\left(\frac{dy}{dx}\right)^4=0$

(*iii*) $\sqrt{1-x^2}\,dx+\sqrt{1-y^2}\,dy=0$ (*iv*) $\left[1+\left(\frac{dy}{dx}\right)^2\right]^{3/2}=5\frac{d^2y}{dx^2}$

Solution. (*i*) $x^2\left(\frac{dy}{dx}\right)+2xy-6x^3=0$

The orderof the highest derivative is 1 and its degree is 1.

∵ Order is 1 and degree is 1

(*ii*) $$2y\left(\frac{d^2y}{dx^2}\right)+\left(\frac{dy}{dx}\right)^4=0$$

The order of the highest derivative is 2 and its degree is 1.

∴ Order is 2 and degree is 1.

(*iii*) $$\sqrt{1-x^2}\,dx + \sqrt{1-y^2}\,dy = 0$$

or $$\frac{dy}{dx} = -\frac{\sqrt{1-x^2}}{\sqrt{1-y^2}}$$

The order of the highest differential is1 and its degree is 1.

$\therefore$ Order is 1 and degree is 1

(*iv*) We have $$\left[1+\left(\frac{dy}{dx}\right)^2\right]^{3/2} = 5\frac{d^2y}{dx^2}$$

$\Rightarrow$ $$\left[1+\left(\frac{dy}{dx}\right)^2\right]^3 = 25\left(\frac{d^2y}{dx^2}\right)^2$$

It is free from radicals and fractions

The order of highest derivative $\left(\frac{d^2y}{dx^2}\right)$ is 2.

Hence order and degree of the given differential equation are 2 each.

EXERCISE 12.1

1. $\left(\frac{d^2y}{dx^2}\right)+x^2\left(\frac{dy}{dx}\right)^2=0$ **[Ans.** Order = 2, Degree = 1]

2. $\frac{d^2y}{dx^2}+4y=0$ **[Ans.** Order = 2, Degree = 1]

3. $x\sqrt{1-y^2}\,dx+y\sqrt{1-x^2}\,dy=0$ **[Ans.** Order = 1, Degree = 1]

4. $(2x-2y+5)dx+(x-y+3)dy$ **[Ans.** Order = 1, Degree = 1]

5. $y=px+\sqrt{a^2p^2+b^2}$, where $p=\frac{dy}{dx}$ **[Ans.** Order = 1, Degree = 2]

6. $x\frac{dy}{dx}+\frac{3}{dy/dx}=y^2$ **[Ans.** Order = 1, Degree = 2]

7. $\sqrt{1+\left(\frac{d^2y}{dx^2}\right)}=\frac{d^3y}{dx^3}$ **[Ans.** Order = 3, Degree = 2]

8. $\frac{d^3y}{dx^3}+\left(\frac{d^2y}{dx^2}\right)^3+\frac{dy}{dx}+4y=\sin x$ **[Ans.** Order = 3, Degree = 1]

9. $\left(\frac{d^2s}{dt^2}\right)^2+3\left(\frac{ds}{dt}\right)^3+4=0$ **[Ans.** Order = 2, Degree = 1]

10. $x\frac{d^2y}{dx^2}+2y+\left(\frac{dy}{dx}\right)^4=0$ **[Ans.** Order = 2, Degree = 2]

11. $y=x\frac{dy}{dx}+a\sqrt{1+\left(\frac{dy}{dx}\right)^2}$ **[Ans.** Order = 1, Degree = 1]

12.2. DIFFERENTIAL EQUATIONS OF FIRST ORDER AND FIRST DEGREE

A differential equation of first order and first degree involves x, y and $\frac{dy}{dx}$. So it can be put in any of the following forms:

$$\frac{dy}{dx} = f(x, y) \quad \text{or} \quad f\left(x, y, \frac{dy}{dx}\right) = 0 \quad \text{or} \quad M\,dx + N\,dy = 0.$$

Here the differential equation being of first order its general solution (or primitive) will contain only one arbitrary constant. All differential equations of the first order cannot be always solved. However, they can be solved by suitable methods if they belong to any one of the following standard forms:

Type I: Variable separable.

Type II: Homogeneous differential equations.

Type III: Linear equations.

Type IV: Bernoulli's equation

Type V: Exact differential equations.

12.3. TYPE I : VARIABLE SEPARABLE

If in an equation, it is possible to get all the functions of x and dx to one side, and all the functions of y and dy to the othder, the variavble are said to be separable.

Rule to solve an equation in which the variables are separable

Consider the equation $\frac{dy}{dx} = XY$, where X is a funcyion of x only and Y is a function of y only.

Step 1. Given differential equation is $\frac{dy}{dx} = XY$.

Step 2. $\frac{dy}{Y} = Xdx$ *i.e.*, variables have been separate

Step 3. Integrating both sides, $\int \frac{dy}{Y} = \int X\,dx + c$, where c is an arbitrary constant, is the required solution.

The arbitrary constant can be chosen in any form suitable for the answer, *i.e*, we can replace it by log c, $\tan^{-1} c$, $\sin c$, e^c etc.

SOLVED EXAMPLES

Example 12.2. Solve $xy(y + 1)dy = (x^2 + 1)dx$.

Solution. Separating the variables, we get

$$y(y + 1)dy = \frac{x^2 + 1}{x}dx$$

Integrating both sides, we get

$$\int y(y + 1)dy = \int \left(x + \frac{1}{x}\right)dx + c$$

$$\Rightarrow \quad \int y(y+1)dy = \int \left(x+\frac{1}{x}\right)dx + c$$

$$\Rightarrow \quad \frac{y^3}{3}+\frac{y^2}{2} = \frac{x^2}{2}+\log|x|+c$$, which is the required solution.

Example 12.3. **Solve** $\dfrac{dy}{dx} = e^{x-y} + x^2e^{-y}$.

Solution.
$$\frac{dy}{dx} = e^{x-y} + x^2e^{-y} = e^x \,.\, e^{-y} + x^2\, e^{-y}$$

or
$$\frac{dy}{dx} = e^{-y}(e^x + x^2)$$

Separating the variables, $\quad e^y\, dy = (e^x + x^2)\, dx$

Integrating both sides, we get

$$\int e^y\, dy = \int (e^x + x^2)\, dx + c$$

$$e^y = e^x + \frac{x^3}{3} + c$$

or
$$e^x - e^y + \frac{x^3}{3} + c = 0$$, which is the required solution.

Example 12.4. **Solve the following differential equation:**

(*i*) $\dfrac{dy}{dx} = 1 + x + y + xy$ **(*ii*)** $\sec^2 x \tan y\, dx + \sec^2 y \tan x\, dy = 0$.

Solution. (*i*) We have
$$\frac{dy}{dx} = 1 + x + y + xy$$

$$\Rightarrow \quad \frac{dy}{dx} = 1 + x + y\,(1 + x) = (1 + x)\,(1 + y)$$

$$\Rightarrow \quad \frac{dy}{1+y} = (1 + x)\, dx$$ [Variables are separately]

Integrating both sides, we get

$$\int \frac{dy}{1+y} = \int (1 + x)\, dx$$

$$\Rightarrow \quad \log|1 + y| = x + \frac{x^2}{2} + c$$, which is the required solution.

(*ii*) We have

$$\sec^2 x \tan y\, dx + \sec^2 y \tan x\, dy = 0$$

$$\Rightarrow \quad \sec^2 x \tan y\, dx = -\sec^2 y \tan x\, dy$$

$$\Rightarrow \quad \frac{\sec^2 x}{\tan x}\, dx = -\sec^2 y \tan x$$

$$\Rightarrow \quad \frac{\sec^2 x}{\tan x}\, dx = -\frac{\sec^2 y}{\tan y}\, dy$$ [Variablese are separable]

Integrating both sides, we get

$$\int \frac{\sec^2 x}{\tan x}\, dx = -\int \frac{\sec^2 y}{\tan y}\, dy$$

$\Rightarrow$ $$\log|\tan x| = -\log|\tan y| + \log c$$

$\Rightarrow$ $$\log|(\tan x)(\tan y)| = \log c$$

$\Rightarrow$ $\tan x \tan y = c$, which is the required solution.

Example 12.5. Find the particular solution of the differential equation $\log_e\left(\frac{dy}{dx}\right) = 3x + 4y$, given that $y = 0$ when $x = 0$.

Solution. We have $$\log_e\left(\frac{dy}{dx}\right) = 3x + 4y$$

$\Rightarrow$ $$e^{3x+4y} = \frac{dy}{dx} \qquad \text{[By def. of log]}$$

$\Rightarrow$ $$e^{3x} \cdot e^{4y} = \frac{dy}{dx} \Rightarrow e^{-4y}\, dy = e^{3x}\, dx.$$

Now integrating both sides, we get

$$\int e^{-4y}\, dy = \int e^{3x}\, dx$$

$\Rightarrow$ $$\frac{e^{-4y}}{4} = \frac{e^{3x}}{3} + c$$

$\Rightarrow$ $$3e^{-4y} + 4e^{3x} = -12c \qquad ...(i)$$

Now it is given that $y = 0$, when $x = 0$.

Putting $x = 0$ and $y = 0$ in (*i*), we get

$$3e^0 + 4e^0 = -12c \Rightarrow 3 + 4 = -12c$$

$\Rightarrow$ $$c = -\frac{7}{12}$$

Hence, the required solution is

$$3e^{-4y} + 4e^{3x} = -12\left(-\frac{7}{12}\right) = 7$$

$\Rightarrow$ $$3e^{-4y} + 4e^{3x} = 7.$$

EXERCISE 12.2

Solve the following differential equations:

1. $x\frac{dy}{dx} + y = y^2$. [**Ans.** $cxy = (y - 1)$]
2. $\frac{dy}{dx} + \left(\frac{xy + y}{xy + x}\right) = 0$. [**Ans.** $x + y + \log|xy| = c$]
3. $(1 - x^2)\, dy + xy\, dx = xy^2\, dx$. [**Ans.** $(y - 1)^2 |1 - x^2| = c^2 y^2$]
4. $\frac{dy}{dx} = \frac{xe^x \log x + e^x}{x \cos y}$. [**Ans.** $\sin y = e^x \log x + c$]

5. $y\sqrt{1+x^2} + x\sqrt{1+y^2}\,\dfrac{dy}{dx} = 0.$ $\left[\textbf{Ans. } \sqrt{1+y^2} + \sqrt{1+x^2} + \dfrac{1}{2}\log\left|\dfrac{\sqrt{1+x^2}-1}{\sqrt{1+x^2}+1}\right| + \dfrac{1}{2}\log\left|\dfrac{\sqrt{1+y^2}-1}{\sqrt{1+y^2}+1}\right| = c\right]$

6. $\dfrac{dy}{dx} = \dfrac{e^x(\sin^2 x + \sin 2x)}{y(2\log y + 1)}.$ [**Ans.** $y^2 \log y = e^x \sin^2 x + c$]

7. $y \log y\, dx + x \log x\, dy = 0.$ [**Ans.** $e^{c/\log x}$]

8. $xy\,\dfrac{dy}{dx} = 1 + x + y + xy.$ [**Ans.** $x - y + \log x\,(1 + y) + c = 0$]

9. $(1 + x)(1 + y^2)\,dx + (1 + y)(1 + x^2)dy = 0.$ $\left[\textbf{Ans. } \tan^{-1} x + \tan^{-1} y + \dfrac{1}{2}\log[(1+x^2)(1+y^2)] = c\right]$

10. $\dfrac{dy}{dx} = y \cot 2x,\ y\left(\dfrac{\pi}{4}\right) = 2.$ [**Ans.** $y^2 = 4 \sin 2x$]

11. $\sqrt{1 + x^2 + y^2 + x^2 + y^2} + xy\dfrac{dy}{dx} = 0.$ $\left[\textbf{Ans. } \sqrt{1+x^2} + \sqrt{1+y^2} = \log\dfrac{1+\sqrt{1+x^2}}{x} + c\right]$

12. $\sqrt{(a + x)}\, dy + x\, dx = 0.$ $\left[\textbf{Ans. } y + \dfrac{2}{3}(x - 2a)\sqrt{a + x} = c\right]$

13. $x \cos x \cos y + \sin y\,\dfrac{dy}{dx} = 0.$ [**Ans.** $x \sin x + \cos x = \log \cos y + c$]

14. Solve the initial value problem $e^x(\cos y\, dx - \sin y\, dy) = 0,\ y(0) = 0.$ [**Ans.** $y = \cos^{-1}(e^{-x})$]

15. Solve the differential equation $x\dfrac{dy}{dx} + \cot y = 0$, given that $y = \dfrac{\pi}{4}$ when $x = \sqrt{2}$. [**Ans.** $x = 2 \cos y$]

16. Find the particular solution of the differential equation $\cos y\, dy + \cos x \sin y\, dx = 0$ given that $y = \dfrac{\pi}{2}$ when $x = \dfrac{\pi}{2}$. [**Ans.** $\log \sin y + \sin x = 1$]

17. Solve the differential equation $\dfrac{dy}{dx} = \dfrac{2x(\log x + 1)}{\sin y + \cos y}$, given that $y = 0$, when $x = 1$.

[**Ans.** $2y \sin y = 2x^2 \log x + x^2 - 1$]

18. The slope of the tangent at a point $P(x, y)$ on a curve is $\left(-\dfrac{y+3}{x+2}\right)$. If the curve passes through the origin, find the equation of the curve. [**Ans.** $2y \sin y = 2x^2 \log x + x^2 - 1$]

19. Find the equation of a curve whose differential equation is $(1 + y^2)dx - xy\, dy = 0$ and which passes through $(1, 0)$. [**Ans.** $x^2 = (1 + y^2)$]

20. If $\dfrac{dV}{dt} = g \cos \alpha - KV$, g, α, K being contains, find V in terms of t if $V = 0$, when $t = 0$.

$\left[\textbf{Ans. } V = \dfrac{g \cos \alpha}{K}(1 - e^{-Kt})\right]$

21. A particle falls under gravity in a resisting medium whose resistance varies with velocity. Find the relation between distance and velocity if initially the particle starts from rest. $\left[\textbf{Ans. } s = \dfrac{g}{K^2}\log\left(\dfrac{g}{g - KV}\right) - \dfrac{V}{K}\right]$

12.4. TYPE II : HOMOGENEOUS DIFFERENTIAL EQUATIONS

Homogeneous Function: A homogeneous function of nth degree in x and y is that which can be put in the form:

$$x^n f\left(\frac{y}{x}\right).$$

An equation of the form $$\frac{dy}{dx} = \frac{f(x, y)}{g(x, y)}$$

where $f(x, y)$ and $g(x, y)$ are homogeneous function of x and y of the same degree is called a homogeneous differential equation.

Let $$f(x, y) = x^n \phi_1\left(\frac{y}{x}\right); \quad g(x, y) = x^n \phi_2\left(\frac{y}{x}\right)$$

$\therefore$ From (*i*), $$\frac{dy}{dx} = \frac{x^n \phi_1\left(\frac{y}{x}\right)}{x^n \phi_2\left(\frac{y}{x}\right)} = \frac{\phi_1\left(\frac{y}{x}\right)}{\phi_2\left(\frac{y}{x}\right)}$$

$\Rightarrow$ $$\frac{dy}{dx} = f_1\left(\frac{y}{x}\right)$$

Put $y = vx$ then $$\frac{dy}{dx} = v + x\frac{dv}{dx}$$

Equation (*i*) becomes, $$v + x\frac{dv}{dx} = f(v) \quad \Rightarrow \quad x\frac{dv}{dx} = f(v) - v$$

Separating the variables, $$\frac{dv}{f(v) - v} = \frac{dx}{x}$$ which can be integrated.

In the solution putting $v = \frac{y}{x}$, we get the required solution.

Working Rule:

Step 1. Put $y = vx$, then $\frac{dy}{dx} = v + x \cdot \frac{dv}{dx}$.

Step 2. Transfer the above equation in terms of v and x by using above substitution.

Step 3. Solve the resulting equation by 'separation of variable method'.

Step 4. Replace v by $\frac{y}{x}$ after integration.

Following examples illustrate the procedure.

SOLVED EXAMPLES

Example 12.6. Solve $x\,dy - y\,dx = \sqrt{x^2 + y^2}\,dx$.

Solution. The given equation can be written as

$$\frac{dy}{dx} = \frac{y + \sqrt{y^2 + x^2}}{x} \qquad \text{[Homogeneous eqution]} \quad ...(i)$$

Put $y = vx$ so that

Equation (i) becomes,

$$v + x\frac{dv}{dx} = v + \sqrt{1 + v^2} \quad \text{or} \quad x\frac{dv}{dx} = \sqrt{1+v^2}$$

Separating the variables, $\dfrac{dv}{\sqrt{1+v^2}} = \dfrac{dx}{x}$

Integrating both sides, we get

$$\int \frac{dv}{\sqrt{1+v^2}} = \int \frac{dx}{x} + c$$

$$\Rightarrow \quad \log|v + \sqrt{1+v^2}| = \log|x| + \log|c|$$

$$\Rightarrow \quad \log|v + \sqrt{1+v^2}| = \log|cx|$$

$$\Rightarrow \quad v + \sqrt{1+v^2} = cx$$

$$\Rightarrow \quad \frac{y}{x} + \sqrt{1 + \left(\frac{y}{x}\right)^2} = cx \qquad \left[\because v = \frac{y}{x}\right]$$

$$\Rightarrow \quad y + \sqrt{x^2 + y^2} = cx$$

which is the required solution.

Example 12.7. Solve $(x^2 - y^2)\,dx + 2xy\,dy = 0$ given that $y = 1$ when $x = 1$.

Solution. We are given that

$$(x^2 - y^2)\,dx + 2xy\,dy = 0$$

$$\Rightarrow \quad \frac{dy}{dx} = \frac{y^2 - x^2}{2xy} = \frac{x^2\left(\frac{y^2}{x^2} - 1\right)}{2xy}$$

$$\therefore \quad \frac{dy}{dx} = \frac{\left(\frac{y}{x}\right)^2 - 1}{2\left(\frac{y}{x}\right)} \qquad \text{[Homogeneous equation]} \quad ...(i)$$

Put $y = vx$, so that $\dfrac{dy}{dx} = v + x\dfrac{dv}{dx}$

Equation (i) becomes, $v + x\dfrac{dv}{dx} = \dfrac{v^2 - 1}{2v}$

$$\Rightarrow \quad x\frac{dv}{dx} = \frac{v^2-1}{2v} - v = \frac{v^2-1-2v^2}{2v} = -\frac{1+v^2}{2v}$$

$$\therefore \quad \frac{2v}{1+v^2}dv = -\frac{dx}{x}$$ [Variables are separable]

Integrating both sides, we get

$$\int \frac{2v}{1+v^2}dv = -\int \frac{dx}{x} + \log c$$

$$\Rightarrow \quad \log(1+v^2) = -\log|x| + \log c$$

$$\Rightarrow \quad \log\left(1+\frac{y^2}{x^2}\right) = -\log|x| + \log c$$

$$\Rightarrow \quad \log(x^2+y^2) - \log x^2 = \log|x| + \log c$$

$$\Rightarrow \quad \log(x^2+y^2) - 2\log|x| = -\log|x| + \log c$$

$$\Rightarrow \quad \log(x^2+y^2) = \log|x| + \log c$$

$$\Rightarrow \quad \log(x^2+y^2) = \log|cx| \Rightarrow (x^2+y^2) = \pm cx$$

$$\Rightarrow \quad x^2+y^2 = c_1 x \text{ where } c_1 = \pm c$$

For $x = 1, y = 1$, we get $(1)^2 + 1 = c_1(1)$

$$\Rightarrow \quad c_1 = 2$$

Hence $x^2 + y^2 = 2x$ is the required solution.

Example 12.8. Solve $(x^3 - 3xy^2)dx = (y^3 - 3x^2y)dy$.

Solution. Given equation is $\frac{dy}{dx} = \frac{x^3-3xy^2}{y^3-3x^2y}$ [Homogeneous equation] ...(*i*)

Put $y = vx$, then $\frac{dy}{dx} = v + x\frac{dv}{dx}$

Equation (*i*) becomes, $v + x\frac{dv}{dx} = \frac{x^3-3v^2x^3}{v^3x^3-3vx^3}$

$$\Rightarrow \quad x\frac{dv}{dx} = \frac{1-3v^2}{v^3-3v} - v = \frac{1-3v^2-v^4+3v^2}{v^3-3v} = \frac{1-v^4}{v^3-3v}$$

Separating the variables, we get

$$\frac{v^3-3v}{1-v^4}dv = \frac{dx}{x}$$

Integrating both sides, we get

$$\int \frac{v^3-3v}{1-v^4}dv = \int \frac{dx}{x} + c \quad \text{...(ii)}$$

Let us find $\int \frac{v^3-3v}{1-v^4}dv$

$$\int \frac{v^3-3v}{1-v^4}dv = -\frac{1}{4}\int \frac{-4v^3}{1-v^4}dv - \frac{3}{2}\int \frac{2v}{1-v^4}dv$$

$$= -\frac{1}{4}\log|1-v^4| - \frac{3}{2}\int \frac{dt}{1-t^2}$$ where $t = v^2$

$$= -\frac{1}{4}\log|1-v^4| - \frac{3}{2}\cdot\frac{1}{2}\log\left|\frac{1+t}{1-t}\right|$$

$$= -\frac{1}{4}\left[\log|1-v^4| + 3\log\left|\frac{1+v^2}{1-v^2}\right|\right] = -\frac{1}{4}\log\left[|1-v^4|\cdot\left|\frac{1+v^2}{1-v^2}\right|^3\right]$$

$$= -\frac{1}{4}\log\left[\frac{(1+v^2)^4}{(1-v^2)^2}\right] = -\frac{1}{2}\log\left|\frac{(1+v^2)^2}{1-v^2}\right|$$

$\therefore$ From (*ii*), we get

$$-\frac{1}{2}\log\left|\frac{(1+v^2)^2}{1-v^2}\right| = \log|x| + c$$

$\Rightarrow$ $$\log\left|\frac{(1+v^2)^2}{1-v^2}\right| = -2\log|x| + 2c = -\log x^2 + \log c_1 = \log\frac{c_1}{x^2}$$

$\Rightarrow$ $$\frac{(1+v^2)^2}{1-v^2} = \frac{c_1}{x^2} \Rightarrow \frac{\left[1+\frac{y^2}{x^2}\right]^2}{1-\frac{y^2}{x^2}} = \frac{c_1}{x^2}$$

$\Rightarrow$ $$\frac{(x^2+y^2)^2}{x^4}\cdot\frac{x^2}{x^2-y^2} = \frac{c_1}{x^2}$$

$\Rightarrow$ $(x^2+y^2)^2 = c_1(x^2-y^2)$ which is the required solution.

EXERCISE 12.3

Solve the following differential equations:

1. $x^2\frac{dy}{dx} = 2xy + y^2$. [**Ans.** $y = c_1x(x+y)$ where $c_1 = \pm c$]

2. $(x+y)dx + (y-x)dy = 0$. $\left[\textbf{Ans. } \log(x^2+y^2) = 2\tan^{-1}\frac{y}{x} + c\right]$

3. $x\frac{dy}{dx} = y(\log y - \log x + 1)$ $\left[\textbf{Ans. } \log\left(\frac{y}{x}\right) = cx\right]$

4. $x^2 - \frac{dy}{dx} = x^2 - 2y^2 + xy$ $\left[\textbf{Ans. } \frac{1}{2\sqrt{2}}\log\left|\frac{x+\sqrt{2}y}{x-\sqrt{2}y}\right| = \log cx\right]$

5. $\frac{dy}{dx} = \frac{y}{x} \pm \tan\left(\frac{y}{x}\right)$ $\left[\textbf{Ans. } \sin\left(\frac{y}{x}\right) = cx\right]$

6. $x(x-y)\frac{dy}{dx} = y(x+y)$ $\left[\textbf{Ans. } \frac{x}{y} + \log(xy) = c\right]$

7. (*i*) $\frac{dy}{dx} = \frac{x-y}{x+y}$ [**Ans.** $x^2 - 2xy - y^2 = c$]

   (*ii*) $\frac{dy}{dx} = \frac{x+y}{x-y}$ $\left[\textbf{Ans. } \tan^{-1}\left(\frac{y}{x}\right) = \frac{1}{2}\log(x^2+y^2) + c\right]$

8. $xdy - ydx = \sqrt{x^2 - y^2}\, dx$ $\left[\textbf{Ans. } \sin^{-1}\left(\frac{y}{x}\right) = \log x + c\right]$

9. $x^2dy + y(x + y)dx = 0$ given that $y = 1$ when $x = 1$ $\left[\textbf{Ans. } x^2 y = \frac{2x+y}{3}\right]$

10. (*i*) $(x^2y - 2xy^2)dx = x^3 - 3x^2y$ [**Ans.** $x^2 = cy^3 e^{x/y}$]

(*ii*) $\left[x\cos\frac{y}{x} + y\sin\frac{y}{x}\right]y - \left[y\sin\frac{y}{x} - x\cos\frac{y}{x}\right]x\frac{dy}{dx} = 0.$ $\left[\textbf{Ans. } xy\cos\frac{y}{x} = c\right]$

11. $\left(x\tan\frac{y}{x} - y\sin\frac{y}{x}\right)dx + x\sec^2\frac{y}{x}dy = 0$ $\left[\textbf{Ans. } x\tan\frac{y}{x} = c\right]$

12. $(x^2 + xy)dy = (x^2 + y^2)dx$ [**Ans.** $c(x - y)^2 = xe^{-y/x}$]

13. (*i*) $(x^3 - y^3)dx + xy^2\, dy = 0$ $\left[\textbf{Ans. } y^3 = 3x^3 \log\left(\frac{c}{x}\right)\right]$

(*ii*) $(x^3 - y^3)dx = (x^2y + xy^2)dx$ [**Ans.** $e^{-y/x} = c(x - y)$]

(*iii*) $(1 + e^{y/x})dy + e^{y/x}\left(1 - \frac{y}{x}\right)dx = 0$ [**Ans.** $y + xe^{y/x} = c$]

12.5. TYPE III : LINEAR DIFFERENTIAL EQUATIONS

A differential euqation is linear if the line dependent variable (y) and its derivative appear only in first degree. The general form of a linear differential equation of first order is

$$\frac{dy}{dx} = Py = Q \qquad ...(i)$$

where P and Q are functions of x (or constants)

For example,

1. $\frac{dy}{dx} + 5y = x^2$ is a linear differential equation of order 1

2. $x\frac{d^2y}{dx^2} + 5\frac{dy}{dx} = 8$ is a linear differential equation of order 2.

This type of differential equations are solved when they are multiplied by a factor, which is called integrating factors, because by multiplication of this factor the left hand side of the differential equation (*i*) becomes exact differential of some function.

Multiplying both sides of (*i*) by $e^{\int P\,dx}$, we get

$$e^{\int P\,dx}\left(\frac{dy}{dx} + Py\right) = Qe^{\int P\,dx} \Rightarrow \frac{d}{dx}\left(y\,.\,e^{\int P\,dx}\right) = Q\,.\,e^{\int P\,dx}$$

On integrating both sides w.r.t. x, we get

$$y\,.\,e^{\int P\,dx} = \int Q\,.\,e^{\int P\,dx}\,dx + c \qquad ...(ii)$$

which is the required solution, where c is the constant of integration.

$e^{\int P\,dx}$ is called the integrating factor (I.F.). The solution (*ii*) in short may also be written as

$$y\,.\,(\text{I.F.}) = \int Q\,.\,(\text{I.F.})\,dx + c$$

Alongwith for solving a linear differential equation of the form $\frac{dy}{dx} + Py = Q$.

Step I. Write the differential equation in the form $\frac{dy}{dx} + Py = 0$ and obtain P and Q.

Step II. Find integrating factor (I.F.) given by I.F. = $e^{\int P\,dx}$.

Step III. Multiply both side of equation in step I by I.F.

Step IV. Integrate both sides of the equation obtained in step III. w.r.t. x to obtain

$$y\,.\,(\text{I.F.}) = \int Q\,.\,(\text{I.F.})\,dx + c$$

This gives the required solution.

Note. Sometimes a linear differential equation can be put in the form $\frac{dx}{dy} + Px = Q$ where P and Q are functions of y or constants. Note that y is independent variable and x is a dependent variable. The following algorithm is used to solve these types of equations.

Algorithm for Solving Linear Differential Equation of the form $\frac{dx}{dy} + Px = Q$.

Step I. Sometimes a linear differential equation can be put in the form $\frac{dx}{dy} + Px = Q$ and obtain P and Q.

Step II. Find I.F. by using I.F. $= e^{\int P\,dy}$.

Step III. Multiply both sides of the differential equation in Step I by I.F.

Step IV. Integrate both sides of the equation obtained in Step III w.r.t. y to obtain the solution given by

$$x\,.\,(\text{I.F.}) = \int Q\,.\,(\text{I.F.})\,dy + c$$

where c is a constant of integration.

Following examples illustrate the produre.

SOLVED EXAMPLES

Example 12.9. Solve $\cos^2 x \frac{dx}{dy} + y = \tan x$.

Solution. We are given that

$$\cos^2 x \frac{dx}{dy} + y = \tan x$$

Dividing throughout by $\cos^2 x$ to make the coefficient of $\frac{dy}{dx}$ unity.

$$\frac{dy}{dx} + \sec^2 x\,.\,y = \sec^2 x \tan x \qquad \text{...}(i)$$

It is of the form $\frac{dy}{dx} + Py = Q$ where

$$P = \sec^2 x,\ Q = \sec^2 x \tan^2 x$$

$$\text{I.F.} = e^{\int P\,dx} = e^{\int \sec^2 x\,dx} = e^{\tan x}$$

Multiplying both in sides of (*i*) by I.F. $= e^{\tan x}$, we get

$$e^{\tan x}\frac{dy}{dx} + \sec^2 x \,.\, e^{\tan x} y = \sec^2 x \tan x\, e^{\tan x}$$

$$y\,.\,e^{\tan x} = \int \sec^2 x \tan x\,.\,e^{\tan x}\,dx + c$$

$$= \int te^t\,dt + c \quad \text{where } t = \tan x$$

$$= t\,e^t - e^t + c = e^t(t-1) + c = e^{\tan x}(\tan x - 1) + c$$

or $\quad y = \tan x - 1 + ce^{-\tan x}$, which is the required solution.

Example 12.10. Solve $\frac{dy}{dx} + \frac{2x}{x^2+1}\,.\,y = \frac{1}{(x^2+1)^2}$ given that $y = 0$, when $x = 1$.

Solution. Given equation is

$$\frac{dy}{dx} + \frac{2x}{x^2+1}\,.\,y = \frac{1}{(x^2+1)^2} \qquad ...(i)$$

It is of the form $\quad \frac{dy}{dx} + Py = Q$

where $\quad P = \frac{2x}{x^2+1},\quad Q = \frac{1}{(x^2+1)^2}$

$$\text{I.F.} = e^{\int P\,dx} = e^{\int \frac{2x}{x^2+1}dx} = e^{\log(x^2+1)}$$

Multiplying both the sides of (*i*) by I.F. $= x^2 + 1$, we get $(x^2+1)\frac{dy}{dx} + 2xy = \frac{1}{x^2+1}$.

$$y(x^2+1) = \int \frac{1}{(x^2+1)^2}\,.\,(x^2+1)\,dx + c = \int \frac{1}{x^2+1}\,dx + c$$

or $\quad y(x^2+1) = \tan^{-1} x + c$

when $x = 1, y = 0$

$$\therefore \quad 0 = \tan^{-1} + c \quad \text{or} \quad 0 = \frac{\pi}{4} + c \quad \therefore \quad c = \frac{\pi}{4}$$

$\therefore$ Required solution is $y(x^2+1) = \tan^{-1} x - \frac{\pi}{4}$.

Example 12.11. Solve $(x + 2y^3)\frac{dy}{dx} = y$.

Solution. The given equation is $(x + 2y^3)\frac{dy}{dx} = y$

or $$y\,.\,\frac{dx}{dy} = x + 2y^3 \quad \text{or} \quad \frac{dx}{dy} - \frac{1}{y}\,.\,x = 2y^2 \qquad ...(i)$$

It is of the form $\frac{dx}{dy} + Px = Q$ where $P = -\frac{1}{y}$. $Q = 2y^2$

$$\text{I.F.} = e^{\int -\frac{1}{y} dy} = e^{-\log y} = e^{\log y^{-1}} = y^{-1} = \frac{1}{y}$$

Multiplying both sides of (*i*) by I.F. $= \frac{1}{y}$, we get $\frac{1}{y}\frac{dx}{dy} - \frac{1}{y^2}x = 2y$

∴ The solution is $$x \cdot \frac{1}{y} = \int 2y^2 \cdot \frac{1}{y} dy + c = \int 2y\, dy + c$$

or $\frac{x}{y} = y^2 + c$ or $x = y^3 + cy$, which is the required solution.

Example 12.12. Solve $(1 + y^2)\, dx = (\tan^{-1} y - x)\, dy$.

Solution. The given equation is

$$(1 + y^2)\, dx = (\tan^{-1} y - x)\, dy$$

or $$(1 + y^2) \frac{dy}{dx} = \tan^{-1} y - x$$

Dividing throughout by $1 + y^2$ to make the coefficient of $\frac{dy}{dx}$ unity.

$$\Rightarrow \quad \frac{dx}{dy} + \frac{1}{1+y^2} x = \frac{\tan^{-1} y}{1+y^2}$$

It is of the form $\frac{dx}{dy} + Px = Q$ where $P = \frac{1}{1+y^2}$, $Q = \frac{\tan^{-1} y}{1+y^2}$

$$\text{I.F.} = e^{\int \frac{1}{1+y^2} dy} = e^{\tan^{-1} y}$$

Multiplying both sides of (*i*) by I.F. $= e^{\tan^{-1} y}$, we get

$$e^{\tan^{-1} y} \frac{dx}{dy} + e^{\tan^{-1} y} \cdot \frac{1}{1+y^2} x = \frac{\tan^{-1} y}{1+y^2} \cdot e^{\tan^{-1} y}$$

∴ The solution is $$x \cdot e^{\tan^{-1} y} = \int \frac{\tan^{-1} y}{1+y^2} \cdot e^{\tan^{-1} y} dy + c$$

$$= \int t e^t\, dt + c \quad \text{where} \quad t = \tan^{-1} y$$

$$= e^t (t - 1) + c = e^{\tan^{-1} y} (\tan^{-1} y - 1) + c$$

or $$x = \tan^{-1} y - 1 + ce^{-\tan^{-1} y}$$, which is the required solution.

EXERCISE 12.4

Solve the following differential equations:

1. $x\dfrac{dy}{dx} - y = (x-1)e^x.$ [Ans. $y = e^x + cx$]

2. $\dfrac{dy}{dx} = y\tan x - 2\sin x.$ [Ans. $2y\cos x = \cos 2x + c$]

3. $(1+x^2)\dfrac{dy}{dx} + y = \tan^{-1}x.$ [Ans. $y = \tan^{-1}x - 1 + ce^{-\tan^{-1}x}$]

4. $\sin 2x\dfrac{dy}{dx} = y + \tan x.$ [Ans. $y = \tan x + c\sqrt{\tan x}$]

5. $\dfrac{dy}{dx} + y\cos x = \sin 2x.$ [Ans. $y = 2(\sin x - 1) + ce^{-\sin x}$]

6. $\dfrac{dy}{dx} + \dfrac{y}{(1-x^2)^{3/2}} = \dfrac{x+\sqrt{1+x^2}}{(1-x^2)^2}.$ $\left[\textbf{Ans. } y = \dfrac{x}{\sqrt{1-x^2}} + ce^{\frac{-x}{\sqrt{1-x^2}}}\right]$

7. $x\dfrac{dy}{dx} - y = (x+1)e^{-x}.$ [Ans. $y = -e^{-x} + cx$]

8. $x\dfrac{dy}{dx} + y = \log x.$ $\left[\textbf{Ans. } y = \dfrac{x}{2}\log x - \dfrac{x}{4} + \dfrac{c}{x}\right]$

9. $\dfrac{dy}{dx} - \dfrac{2x}{1+x^2}y = x^2 + 2.$ [Ans. $y = (x^2+1)\{(x + \tan^{-1}x) + c\}$]

10. $(2x - 10y^3)\dfrac{dy}{dx} + y = 0.$ $\left[\textbf{Ans. } x = 2y^3 + \dfrac{c}{y^2}\right]$

11. $dx + x\,dy = e^{-y}\sec^2 y\,dy.$ [Ans. $xe^y = \tan y + c$]

12. $(1+y^2) + (x - e^{\tan^{-1}y})\dfrac{dy}{dx} = 0.$ $\left[\textbf{Ans. } 2xe^{-\tan^{-1}y} = e^{2\tan^{-1}y} + c\right]$

13. $(x^2-1)\dfrac{dy}{dx} + 2(x+2)y = 2(x+1).$ $\left[\textbf{Ans. } y\sin x = \dfrac{2}{3}\sin^3 x + c\right]$

14. $\dfrac{dy}{dx} + (\sec x)y = \tan x.$ [Ans. $y(\sec x + \tan x) = \sec x + \tan x - x + c$]

15. $(1+x^2)\dfrac{dy}{dx} + 2xy - 4x^2 = 0.$ [Ans. $3y(1+x^2) = 4x^3 + c$]

16. $x(1-x^2)\dfrac{dy}{dx} + (2x^2-1)y = x^3.$ $\left[\textbf{Ans. } y = x + cx\sqrt{1-x^2}\right]$

17. $x\,.\,(x-1)\dfrac{dy}{dx} - y = x^2(x-1)^2.$ $\left[\textbf{Ans. } \dfrac{xy}{x-1} = \dfrac{x^3}{3} + c\right]$

18. $\sin x\dfrac{dy}{dx} + 3y = \cos x.$ $\left[\textbf{Ans. } y\tan^3\dfrac{x}{2} = -\dfrac{1}{3}\tan^3\dfrac{x}{2} + 2\tan\dfrac{x}{2} - x + c\right]$

19. $\dfrac{dy}{dx} + y\cot x = 2x + x^2\cot x, y(0) = 0.$ [Ans. $y = x^2$]

20. $\dfrac{dy}{dx} + 2y\tan x = \sin x$, given that $y = 0$ when $x = \dfrac{\pi}{3}.$ [Ans. $y\sec^2 x = \sec x - 2$]

12.6. TYPE IV : EQUATIONS REDUCIBLE TO LINEAR FORM (BERNOULLI'S EQUATION)

The equation of the form $\frac{dy}{dx} + Py = Qy^n$ where P and Q are functions of x only or constants and n is a non-zero constant other than unity are known as Bernoulli's equations. These equations can be reduced to the linear form by dividing both sides by y^n and then putting $y^{1-n} = z$ (say) in the following manner.

$$y^{-n}\frac{dy}{dx} + Py^{1-n} = Q$$

Putting $y^{1-n} = z$ or $(1-n)\ y^{-n}\frac{dy}{dx} = \frac{dz}{dx}$, we get

$$\frac{1}{1-n}\frac{dz}{dx} + Pz = Q \qquad \text{or} \qquad \frac{dz}{dx} + (1-n)\,Pz = Q(1-n)$$

which is a linear differential equation in z and can be solved easily by the previous method discussed in Type III.

SOLVED EXAMPLES

Example 12.13. Solve: $\frac{dy}{dx} + \frac{y}{x} = \frac{y^2}{x^2}$.

Solution. The given equation can be written as

$$\frac{1}{y^2}\frac{dy}{dx} + \frac{1}{xy} = \frac{1}{x^2}$$

Put $-\frac{1}{y} = z$ so that $\frac{1}{y^2}\frac{dy}{dx} = \frac{dz}{dx}$

The given equation reduces to a linear differential equation

$$\frac{dz}{dx} - \frac{z}{x} = \frac{1}{x^2} \qquad \ldots(i)$$

$$\text{I.F.} = e^{-\int \frac{1}{x}dx} = e^{-\log x} = e^{\log x^{-1}} = \frac{1}{x}$$

Multiplying both sides of (*i*) by I.F. $= \frac{1}{x}$, we get

$$\frac{1}{x}\frac{dz}{dx} - \frac{z}{x^2} = \frac{1}{x^3}$$

Integrating this w.r.t. x, we get

$$z \cdot \frac{1}{x} = \int \frac{1}{x^3}dx + c = \frac{x^{-2}}{2} + c$$

or $$-\frac{1}{xy} = -\frac{1}{2x^2} + c$$

or $$\frac{1}{xy} = \frac{1}{2x^2} - c,$$ which is the required solution.

Example 12.14. Solve: $\dfrac{dy}{dx} = x^3y^3 - xy$.

Solution. The given differential equation can be written as $\dfrac{dy}{dx} + xy = x^3y^3$.

Dividing both sides by y^3, we get $\dfrac{1}{y^3}\dfrac{dy}{dx} + x \cdot \dfrac{1}{y^2} = x^3$

Put $\dfrac{1}{y^2} = z$ so that $\dfrac{-2}{y^3}\dfrac{dy}{dx} = \dfrac{dz}{dx}$

or $$\frac{1}{y^3}\frac{dy}{dx} = -\frac{1}{2}\frac{dz}{dx}, \text{ we get}$$

$$-\frac{1}{2}\frac{dz}{dx} + xz = x^3 \quad \text{or} \quad \frac{dz}{dx} - 2xz = -2x^3 \qquad \ldots(i)$$

This is a linear differential equation in z. The integrating factor (I.F.) is given by

$$\text{I.F.} = e^{\int -2x\,dx} = e^{-x^2}$$

Multiplying both sides of (*i*) by e^{-x^2}, we get

$$e^{-x^2}\frac{dz}{dx} - 2xze^{-x^2} = -2x^3e^{-x^2}$$

Integrating both sides w.r.t. x, we get

$$z \cdot e^{-x^2} = \int -2x^3e^{-x^2}\,dx + c$$

or $$ze^{-x^2} = \int -2x^3e^t\left(-\frac{dt}{2x}\right) + c \qquad \text{where } t = -x^2$$

or $$z \cdot e^{-x^2} = -\int te^t\,dt + c$$

or $$z \cdot e^{-x^2} = -e^t(t-1) + c \qquad \text{[Integrating by parts]}$$

or $$z \cdot e^{-x^2} = -e^{-x^2}(-x^2 - 1) + c$$

or $$\frac{1}{y^2} \cdot e^{-x^2} = e^{-x^2}(x^2 + 1) + c$$

or $$\frac{1}{y^2} - x^2 - 1 = ce^{x^2}, \text{ which is the required solution.}$$

Example 12.15. Solve: $xy(1 + xy^2)\dfrac{dy}{dx} = 1$.

Solution. Given equation is

$$xy(1 + xy^2)\frac{dy}{dx} = 1$$

or $$\frac{dx}{dy} = x^2y^3 + xy$$

or $$\frac{dx}{dy} - yx = y^3x^2$$

Dividing throughout by x^2, we have

$$x^{-2}\frac{dx}{dy}-\frac{y}{x}=y^3$$

Putting $\frac{1}{x}=z$, so that $\frac{-1}{x^2}\frac{dx}{dy}=\frac{dz}{dy}$

or $\frac{1}{x^2}\frac{dy}{dx}=\frac{-dz}{dy}$, we get

$$\frac{-dz}{dy}-yz=y^3$$

or $$\frac{dz}{dx}+yz=-y^3 \qquad \ldots(i)$$

This is a linear differential equation in z. The integrating factor (I.F.) is given by

$$\text{I.F.}=e^{\int y\,dy}=e^{y^2/2}$$

Multiplying both the sides of (i) by I.F. $=e^{y^2/2}$, and integrating w.r.t. y, we get

$$z\cdot e^{y^2/2}=-\int y^3\cdot e^{y^2/2}\,dy+c$$

$$=-2\int te^t\,dt+c \qquad \left[\text{Put } \frac{y^2}{2}=t \quad \therefore \quad y\,dy=dt\right]$$

$$=-2\left[t\cdot e^t-\int 1\cdot e^t\,dt+c\right]$$

or $$x^{-1}\cdot e^{y^2/2}=-2e^{y^2/2}\left(\frac{y^2}{2}-1\right)+c \qquad [\because \ z=x^{-1}]$$

or $$\frac{1}{x}=2-y+ce^{-y^2/2}.$$

EXERCISE 12.5

1. $\frac{dy}{dx}+\frac{1}{x}=\frac{e^y}{x^2}$ [**Ans.** $2x=2cx^2e^y+e^y$]

2. $\frac{dy}{dx}+\frac{y}{x}=x^3y^3$ $\left[\textbf{Ans.}\ (1-x^2)\sqrt{y}=\frac{1}{2}\left(x-\frac{x^3}{3}\right)+c\right]$

3. $\cos x\frac{dy}{dx}+y\sin x=y^4\sin 2x$ [**Ans.** $2\cos^3 x=(3\cos^4 x+c)y^3$]

4. $(x+1)\frac{dy}{dx}+1=2e^{-y}$ [**Ans.** $e^y\cdot(x+1)=2x+c$]

5. $\cos x\,dy=y(\sin x-y)\,dx$ $\left[\textbf{Ans.}\ \frac{1}{y}\sec x=\tan x+c\right]$

6. $\dfrac{dy}{dx} - \dfrac{\tan y}{1+x} = (1+x)e^x \sec y$ $\left[\textbf{Ans. } \dfrac{\sin y}{1+x} = e^x + c\right]$

7. $\dfrac{dy}{dx} + y = 3e^x y^3$ $\left[\textbf{Ans. } \dfrac{1}{y^2} = 6x^x + ce^{2x}\right]$

8. $x\dfrac{dy}{dx} + y = x^3 y^6$ $\left[\textbf{Ans. } \dfrac{1}{y^5} = \dfrac{5}{2}x^3 + cx^5\right]$

9. $3\dfrac{dy}{dx} + \dfrac{2}{x+1}y = \dfrac{x^3}{y^2}$ $\left[\textbf{Ans. } y^2(x-1)^2 = \dfrac{x^6}{6} + \dfrac{2x^5}{5} + \dfrac{x^4}{4} + c\right]$

10. $\dfrac{dy}{dx} = e^{x-y}(e^x - e^y)$ $[\textbf{Ans. } e^y = e^x - 1 + ce^{-e^x}]$

11. $x\dfrac{dy}{dx} + y \log y = xye^x$ $[\textbf{Ans. } x \log y = xe^x - e^x + c]$

12.7. TYPE V : EXACT DIFFERENTIAL EQUATIONS

A differential equation is said to be an exact differential equation if it is obtained from its primitive by differentiations, and without any further operation of elimination or reduction.

For example, the differential equation $x\dfrac{dy}{dx} + y = 0$ is an exact differential equation because it can be obtained by differentiation of $xy = c$. The differential equation $x + y\dfrac{dy}{dx} = 0$ is also an exact differential equation because it is obtained by differentiating $x^2 + y^2 = c^2$ w.r.t. x.

Now we will find out the condition under which any differential equation of first order and first degree will be an exact differential equation.

Theorem 1. The necessary and sufficient for the equation $Mdx + Ndy = 0$ to be exact differential equation is

$$\frac{\partial M}{\partial y} = \frac{\partial N}{\partial x}.$$

Proof. Necessary Condition: Let the differential equation $Mdx + Ndy = 0$ be an exact differential equation. Then, there exist a function $u = f(x, y)$ such that

$$M\,dx + N\,dy = du, \text{ where } u \text{ is a function of } x \text{ and } y$$

$$= \frac{\partial u}{\partial x}\,.\,dx + \frac{\partial u}{\partial y}\,.\,dy$$

Equating coefficient of dx, $M = \dfrac{\partial u}{\partial x}$

Equating coefficent of dy, $N = \dfrac{\partial u}{\partial y}$

So that $\dfrac{\partial M}{\partial y} = \dfrac{\partial^2 u}{\partial x \partial y}, \quad \dfrac{\partial N}{\partial x} = \dfrac{\partial^2 u}{\partial x \partial y}$

Hence $\qquad \dfrac{\partial M}{\partial y} = \dfrac{\partial N}{\partial x} \qquad \left[\because \dfrac{\partial^2 u}{\partial y \partial x} = \dfrac{\partial^2 u}{\partial x \partial y}\right]$

is the required necessary condition for exactness.

Sufficient Condition. Let $Mdx + Ndy = 0$ be a differential equation such that $\dfrac{\partial M}{\partial y} = \dfrac{\partial N}{\partial x}$. Then we have to show that the differential is an exact differential equation.

Let $\qquad P(x, y) = \int M\, dx$

$$\therefore \qquad \frac{\partial P}{\partial x} = M \quad \text{and} \quad \frac{\partial^2 P}{\partial y \partial x} = \frac{\partial M}{\partial y}$$

But $$\frac{\partial N}{\partial x} = \frac{\partial M}{\partial y} = \frac{\partial^2 P}{\partial y \partial x}$$

or $$\frac{\partial N}{\partial x} = \frac{\partial}{\partial x}\left(\frac{\partial P}{\partial y}\right) \qquad \left(\because \frac{\partial^2 P}{\partial y \partial x} = \frac{\partial^2 P}{\partial x \partial y}\right)$$

Integrating both sides w.r.t. x, we get

$$N = \frac{\partial P}{\partial y} + f(y)$$

$$\therefore \qquad Mdx + Ndy = \frac{\partial P}{\partial x} dx + \left[\frac{\partial P}{\partial y} + \phi(y)\right] dy \qquad \left[\because M = \frac{\partial P}{\partial x},\quad N = \frac{\partial P}{\partial y} + \phi(y)\right]$$

$$= \left[\frac{\partial P}{\partial x} . dx + \frac{\partial P}{\partial y} . dy\right] + \phi(y)\, dy$$

or $$Mdx + Ndy = dP + \phi(y)\, dy$$

Let $$d[f(y)] = \phi(y)dy$$

$$\therefore \qquad Mdx + Ndy = dP + d[f(y)] = d[P + f(y)]$$

which shows that $Mdx + Ndy$ is an exact differential.

Hence $Mdx + Ndy = 0$ is an exact differential equation.

Working Rule for Solving an Exact Differential Equation

Step 1. Compare the given equation with $Mdx + Ndy = 0$ and find out M and N. Then find out $\dfrac{\partial M}{\partial y}$ and $\dfrac{\partial N}{\partial x}$. If $\dfrac{\partial M}{\partial y} = \dfrac{\partial N}{\partial x}$ we conclude that the given equation is exact.

Step 2. Integrate M w.r.t. x keeping y constant.

Step 3. Integrating w.r.t. y only those terms of N which do not contain x.

Step 4. On adding the terms obtained in Step 2 and Step 3 and equating to an arbitrary constant, we get required solution.

The above rule implies that the solution of an exact differential equation $Mdx + Ndy = 0$ is

$$\int_{(y\text{-constant})} M\, dx + \int (\text{terms to } N \text{ that do not contain } x)\, dy = c.$$

SOLVED EXAMPLES

Example 12.16. Solve $(12x + 5y - 9)dx + (5x + 2y - 4)dy = 0$.

Solution. The given equation is

$$(12x + 5y - 9)dx + (5x + 2y - 4) = 0 \quad ...(i)$$

Comparing it with $Mdx + Ndy = 0$

Here $M = 12x + 5y - 9$ and $N = 5x + 2y - 4$

$$\therefore \quad \frac{\partial M}{\partial y} = 5, \ \frac{\partial N}{\partial x} = 5$$

Since $\frac{\partial M}{\partial y} = \frac{\partial N}{\partial x}$, the equation (*i*) is exact.

Hence the solution of (*i*) is

$$\int_{(y\text{-constant})} M\,dx + \int (\text{terms to } N \text{ that do not contain } x)\,dy = c$$

i.e., $$\int (12x + 5y - 9)\,dx + \int (2y - 4)\,dy = c$$

or $$6x^2 + 5xy - 9x + y^2 - 4y = c$$

or $$6x^2 + 5xy + y^2 - 9x - 4y = c.$$

Example 12.17. Solve $(e^y + 1)\cos x\,dx + e^y \sin x\,dy = 0$.

Solution. The given equation is

$$(e^y + 1)\cos x\,dx + e^y \sin x\,dy = 0 \quad ...(i)$$

Comparing it with $Mdx + Ndy = 0$

Here $M = (e^y + 1)\cos x; \quad N = e^y \sin x$

$$\frac{\partial M}{\partial y} = e^y \cos x, \quad \frac{\partial N}{\partial x} = e^y \cos x$$

Since $\frac{\partial M}{\partial y} = \frac{\partial N}{\partial x}$, the equation (*i*) is exact.

Hence the solution of (*i*) is

$$\int_{(y\text{-constant})} M\,dx + \int (\text{terms to } N \text{ that do not contain } x)\,dy = c$$

i.e. $$\int (e^y + 1)\cos x\,dx = c$$

or $$(e^y + 1)\int \cos x\,dx = c \quad \text{or} \quad (e^y + 1)\sin x = c.$$

Example 12.18. Solve $(5x^4 + 3x^2y^2 - 2xy^3)\,dx + (2x^3 y - 3x^2y^2 - 5y^4)\,dy = 0$.

Solution. The given equation is

$$(5x^4 + 3x^2 y^2 - 2xy^3)\,dx + (2x^3 y - 3x^2 y^2 - 5y^4)\,dy = 0 \quad ...(i)$$

Comparing it with $Mdx + Ndy = 0$

Here $M = 5x^4 + 3x^2y^2 - 2xy^3; N = 2x^3y - 3x^2y^2 - 5y^4$

$$\frac{\partial M}{\partial y} = 6x^2y - 6xy^2, \quad \frac{\partial N}{\partial x} = 6x^2y - 6xy^2$$

Since $\dfrac{\partial M}{\partial y} = \dfrac{\partial N}{\partial x}$, the equation (*i*) is exact.

Hence the solution of (*i*) is

$$\int_{(y\text{-constant})} M\,dx + \int (\text{terms to } N \text{ that do not contain } x)\,dy = c$$

i.e., $$\int_{(y\text{-constant})} (5x^4 + 3x^2y^2 - 2xy^3)\,dx + \int -5y^4\,dy = c$$

or $$x^5 + x^3y^2 - x^2y^3 - y^5 = c.$$

Example 12.19. Solve $\left[y\left(1+\dfrac{1}{x}\right)+\cos y\right]dx + [x + \log x - x\sin y]\,dy = 0.$

Solution. The given equation is

$$\left[y\left(1+\frac{1}{x}\right)+\cos y\right]dx + [x + \log x - x\sin y]\,dy = 0 \qquad \ldots(i)$$

Comparing it with $Mdx + Ndy = 0$

Here $$M = y\left(1+\frac{1}{x}\right) + \cos y, \quad N = x + \log x - x\sin y$$

$$\frac{\partial M}{\partial y} = 1 + \frac{1}{x} - \sin y, \qquad \frac{\partial N}{\partial x} = 1 + \frac{1}{x} - \sin y$$

Since, $\dfrac{\partial M}{\partial y} = \dfrac{\partial N}{\partial x}$, the equation (*i*) is exact.

Hence the solution of (*i*) is

$$\int_{(y\text{-constant})} M\,dx + \int (\text{terms to } N \text{ that do not contain } x)\,dy = c$$

i.e., $$\int_{(y\text{-constant})} \left[y\left(1+\frac{1}{x}\right)+\cos y\right]dx = c \quad \text{or} \quad y\int\left(1+\frac{1}{x}\right)dx + \cos y\int 1\,dx = c$$

$$y(x + \log x) + x\cos y = c.$$

Example 12.20. Find the value of λ for which the differential equation $(xy^2 + \lambda x^2y)\,dx + (x+y)x^2\,dy = 0$ is exact.

Solution. The given equation is

$$(xy^2 + \lambda x^2y)\,dx + (x+y)x^2\,dy = 0 \qquad \ldots(i)$$

Comparing it with $Mdx + Ndy = 0$

Here $$M = xy^2 + \lambda x^2y, \; N = x^3 + x^2y$$

$$\frac{\partial M}{\partial y} = 2xy + \lambda x^2, \qquad \frac{\partial N}{\partial x} = 3x^2 + 2xy$$

Since $$\frac{\partial M}{\partial y} = \frac{\partial N}{\partial x}$$

$\Rightarrow$ $$2xy + \lambda x^2 = 3x^2 + 2xy \quad \text{or} \quad \lambda x^2 = 3x^2 \quad \text{or} \quad \lambda = 3.$$

EXERCISE 12.6

Solve the following differential equations:

1. $y \sin 2x\, dx - (1 + y^2 + \cos^2 x)\, dy = 0$ $\left[\textbf{Ans. } y\cos 2x + \frac{2}{3}y^3 + 3y = 0\right]$
2. $(ax + hy + g)\, dx + (hx + by + f)\, dy = 0.$ [**Ans.** $ax^2 + 2hxy + by^2 + 2gx + 2fy = c$]
3. $[\cos x \tan y + \cos (x + y)]\, dx + [\sin x \sec^2 y + \cos (x + y)]\, dy = 0.$ [**Ans.** $\sin x \tan y + \sin (x + y) = c$]
4. $(x^2 - 4xy - 2y^2)\, dx + (y^2 - 4xy - 2x^2)\, dy = 0.$ [**Ans.** $x^3 - 6x^2y - 6xy^2 + y^3 = c$]
5. $(3x^2 + 6xy^2)\, dx + (6x^2y + 4y^3)\, dy = 0.$ [**Ans.** $x^3 + 3x^2y^2 + y^4 = c$]
6. $(a^2 - 2xy - y^2)\, dx - (x + y)^2\, dy = 0.$ $\left[\textbf{Ans. } a^2x - x^2y - xy^2 - \frac{y^3}{3} = c\right]$
7. $\sec^2 x \tan y\, dx + \sec^2 y \tan x\, dy = 0.$ [**Ans.** $\tan y \tan x = c$]
8. $(\sec x \tan x \tan y - e^x)\, dx + \sec x \sec^2 y\, dy = 0.$ [**Ans.** $\sec x \tan y = c$]
9. $(2xy \cos x^2 - 2xy + 1)\, dx + (\sin x^2 - x^2)\, dy = 0.$ [**Ans.** $y \sin x^2 - x^2y + x = c$]
10. $(y^2e^{xy^2} + 4x^3)\, dx + (2xy\, e^{xy^2} - 3y^2)\, dy = 0.$ $\left[\textbf{Ans. } e^{xy^2} + x^4 - y^3 = 0\right]$
11. $(x^4 - 2xy^2 + y^4)\, dx - (2x^2y - 4xy^3 + \sin y)\, dy = 0.$ $\left[\textbf{Ans. } \frac{x^5}{5} - x^2y^2 + xy^4 + \cos y = c\right]$
12. $(2xy + y - \tan x)\, dx + (x^2 - x \tan^2 y + \sec^2 y)\, dy = 0.$ [**Ans.** $x^2y + xy - x \tan y + \tan y = c$]
13. $\frac{dy}{dx} + \frac{y \cos x + \sin y + y}{\sin x + x \cos y + x} = 0.$ [**Ans.** $y \sin x + (\sin y + y)x = 0$]

❑❑❑

CHAPTER 13

Ordinary Differential Equations of Higher Order

INSIDE THIS CHAPTER

13.1. LINEAR DIFFERENTIAL EQUATIONS WITH CONSTANT COEFFICIENTS

The equation of the form

$$\frac{d^n y}{dx^n} + a_1 \frac{d^{n-1} y}{dx^{n-2}} + a_2 \frac{d^{n-2} y}{dx^{n-2}} + \ldots + a_{n-1} \frac{dy}{dx} + a_n y = X \qquad \ldots(i)$$

where $a_1, a_2 \ldots, a_n$ are all constants and Q is a function of x alone is called a linear differential equation of nth other with constant coefficients.

The general solution of a differential equation of nth order has n arbitrary constants.

We write $\qquad D = \frac{d}{dx}, D^2 = \frac{d^2}{dx^2}, \ldots D^n = \frac{d^n}{dx^n}$

Now (i) may be written as

$$D^n y + a_1 D^{n-1} y + \ldots + a_{n-1} Dy + a_n y = X$$

or $$(D^n + a_1 D^{n-1} + \ldots + a_{n-1} D + a_n) y = X \qquad \ldots(ii)$$

or $$f(D)y = X$$

where $f(D) = D^n + a_1 D^{n-1} + \ldots + a_{n-1} D + a_n$

The solution of $f(D)y = 0$

For the sake of convenience, we consider a second order linear differential equation

$$\frac{d^2 y}{dx^2} + a_1 \frac{dy}{dx} + a_2 y = 0. \qquad \ldots(iii)$$

Let $y = e^{mx}$ be a solution of (iii). Substituting in (iii), we get

$$m^2 e^{mx} + a_1 m e^{mx} + a_2 e^{mx} = 0 \quad \Rightarrow \quad e^{mx}(m^2 + a_1 m + a_2) = 0$$

Thus $y = e^{mx}$ is a solution of (iii) if and only if

$$m^2 + a_1 m + a_2 = 0 \qquad \ldots(iv)$$

The equation (iv) is called the auxiliary equation of (iii).

Following cases arise:

Case I. The roots of (*iv*) are real and district

Let m_1 and m_2 be the distinct roots of (*iv*). Then e^{m_1x}, e^{m_2x} are the solutions of (*iii*). Hence the general solution of (*iii*) is

$$y = c_1e^{m_1x}, c_2e^{m_2x} \qquad ...(v)$$

Case II. The roots of (*iv*) are real and equal.

Let m and m be the equation roots of (*iv*). From (*v*).

$$y = (c_1 + c_2)e^{mx} \quad \Rightarrow \quad ae^{mx}, a = c_1 + c_2$$

which cannot be the general solution of (*iii*) as it contains only one arbitrary constants. To obtain the general solution, we proceed as follows:

$$\Rightarrow \qquad m + m = -a_1 \quad \Rightarrow \quad 2m + a_1 = 0 \qquad ...(vi)$$

We now show that $y = xe^{mx}$, is also a solution of (*iii*),

We have $$\frac{dy}{dx} = 1.e^{mx} + mxe^{mx} = (1 + mx)\, e^{mx} \qquad ...(vii)$$

$$\frac{d^2y}{dx^2} = me^{mx} + m(1 + mx)\, e^{mx} = 2mx^{mx} + m^2\, xe^{mx} \qquad ...(viii)$$

Substituting (*vii*) and (*viii*) in (*iii*), we obtain

$$e^{mx}[(2m + m^2x) + a_1(1 + mx) + a_2x)] = 0$$

or $$(m^2 + a_1m + a_2)\, x + (2m + a_1) = 0$$

This equation is identically true in view of (*iv*) and (*vi*).

$\therefore \qquad y = ae^{mx}, xe^{mx}$ are two solutions of (*iii*).

Hence the general solution of (*iii*) is

$$y = ae^{mx} + bxe^{mx}, \text{ where a and b are constants.}$$

$$\therefore \qquad y = (a + bx)\, e^{mx}$$

Case III. The roots of (*iv*) are imaginary

Let $a + ib$ and $a - ib$ be the roots of (*iii*).

Using (*v*), the general solution of (*iii*) is

$$y = c_1e^{(a+ib)x} = e^{ax}[c_1e^{ibx} + c_2e^{-ibx}$$
$$= e^{ax}[c_1(\cos bx + i \sin bx) + c_2(\cos bx - i \sin bx)$$
$$= e^{ax}[(c_1 + c_2) \cos bx + (ic_1 - ic_2) \sin bx]$$

Hence $y = e^{ax}(A \cos bx + B \sin bx)$ is the general solution of (*iii*).

where $$A = c_1 + c_2, \quad B = ic_1 - ic_2$$

Based on above disecussion, we give the following rules of the solution of $f(D)y = 0$,

$$f(D) = D^n + a_1D^{n-1} + ... + a_{n-1}D + a_n$$

13.2. RULES FOR FINDING COMPLEMENTARY FUNCTION (C.F.)

1. The auxiliary equation is obtained by replacing D by m in $f(D)$:

   $$m^n + a_1m^{n-1} + ... + a_{n-1}m + a_n = 0$$

   Let its roots be $m_1, m_2, ..., m_n$.

2. If all the roots are district, then

$$y = c_1e^{m_1x} + c_2e^{m_2x} + \ldots + c_ne^{m_nx}$$

is the solution of $f(D)y = 0$

Here $c_1, c_2, \ldots, c_n$ are n arbitrary constants.

3. (*i*) If $m_1 = m_2 = m_3 = m$, then the solution of $f(D)y = 0$ is

$$y = (c_1 + c_2x)\,e^{mx} + c_3e^{m_3x} + \ldots + c_ne^{m_nx}$$

(*ii*) If $m_1 = m_2 = m_3 = m$, then the solution of $f(D)y = 0$ is

$$y = (c_1 + c_2x + c_3x^2)e^{mx} + c_4e^{m_4x} + \ldots + c_ne^{m_nx} \text{ and so on.}$$

4. If $m_1 = a + ib$, $m_2 = a - ib$, then the solution of $f(D)y = 0$ is

$$y = e^{ax}(c_1 \cos bx + c_2 \sin bx) + c_3e^{m_3x} + \ldots c_ne^{m_nx} \text{ 000000}$$

SOLVED EXAMPLES

Example 13.1. Solve $\frac{d^2y}{dx^2} - 3\frac{dy}{dx} - 4y = 0.$

Solution. Given equation in the symbolic form is

$$(D^2 - 3D - 4)\,y = 0$$

Its auxiliary equation is

$$m^2 - 3m - 4 = 0$$

$$(m - 4)(m + 1) = 0 \quad \Rightarrow \quad m = 4 - 1$$

Hence the solution is $y = c_1e^{4x} + c_2e^{-x}$.

Example 13.2. Solve $\frac{d^3y}{dx^3} - 7\frac{dy}{dx} - 6y = 0.$

Solution. Given equation in the symbolic form is

$$(D^3 - 7D - 6)\,y = 0$$

Its auxiliary equation is

$$m^3 - 7m - 6 = 0$$

Putting $m = -1$ L.H.S of auxiliary equation $= -1 + 7 - 6 = 0$

$\therefore$ $m + 1$ is a factor of L.H.S of A.E.

$\therefore$ A.E. may be written as

$$(m + 1)(m^2 - m - 6) = 0$$

or $$(m + 1)(m - 3)(m + 2) = 0$$

$\therefore$ $$m = -1, -2, 3$$

Hence the solution is $y = c_1e^{-x} + c_2e^{-2x} + c_3e^{3x}$.

Example 13.3. Solve $(D^3 + D^2 - D - 1)\,y = 0$

Solution. The auxiliary equation is $m^3 + m^2 - m - 1 = 0$

$\Rightarrow$ $$(m - 1)(m^2 + 2m + 1) = 0 \Rightarrow (m - 1)(m + 1)^2 = 0$$

$\Rightarrow$ $$m = 1, -1, -1.$$

Hence the solution is $y = c_1e^{x} + (c_2 + c_3x)\,e^{-x}$.

Example 13.4. Solve $(D^2 + D + 1)\, y = 0$.

Solution. The auxiliary equation is $m^2 + m + 1 = 0$

$$\therefore \qquad m = \frac{-1 \pm \sqrt{1-4}}{2} = \frac{1}{2} \pm \frac{\sqrt{3}}{2} i, \; i = \sqrt{-1}$$

Hence the solution is $$y = e^{-x/2}\left[c_1 \cos \frac{\sqrt{3}}{2} x + c_2 \sin \frac{\sqrt{3}}{2} x\right].$$

Example 13.5. Solve $(D^4 + a^4)\, y = 0$.

Solution. The auxiliary equation is $m^4 - a^4 = 0$

$$\Rightarrow \qquad (m^2 - a^2)(m^2 + a^2) = 0 \quad \Rightarrow \quad m = \pm a, \pm ai$$

Hence the solution is $$y = c_1 e^{ax} + c_2 e^{-ax} + c_3 \cos ax + c_4 \sin ax$$

Example 13.6. Solve $(D^3 + 3D^2 + 3D + 1)\, y = 0$.

Solution. The auxiliary equation is $m^3 + 3m^2 + 3m + 1 = 0$

$$\Rightarrow \qquad (m+1)^3 = 0$$

$$\therefore \qquad m = -1, -1, -1.$$

Hence the solution is $$y = (c_1 + c_2 x + c_3 x^2)\, e^{-x}.$$

Example 13.7. Solve $(D^4 - 2D^2 + 2D^2 - 2D + 1)\, y = 0$

Solution. The auxiliary equation is

$$m^4 - 2m^3 + 2m^2 - 2m + 1 = 0$$

$$\Rightarrow \qquad (m-1)(m^3 - m^2 + m - 1) = 0$$

$$\Rightarrow \qquad (m-1)[(m-1)(m^2+1)] = 0$$

$$\Rightarrow \qquad m = 1, 1, \pm i$$

Example 13.8. Solve $\dfrac{d^2y}{dx^2} - 3\dfrac{dy}{dx} + 2y = 0.$

Subject to condition $y(0) = 0$ and $y'(0) = 1$.

Solution. Given equation in symbolic form is

$$(D^2 - 3D + 2)y = 0$$

The auxiliary equation is

$$m^2 - 3m + 2 = 0 \quad \Rightarrow \quad (m-1)(m-2) = 0$$

$$\therefore \qquad m = 1, 2$$

$$\therefore \qquad y = c_1 e^x + c_2 e^{2x}. \qquad \text{...}(i)$$

It is given that $y(0) = 0$ *i.e.*, if $x = 0$, then $y = 0$.

Putting these values in (*i*), we have

$$0 = c_1 + c_2 \qquad \text{...}(ii)$$

On differentiating (*i*) w.r.t. x, we get

$$\frac{dy}{dx} = c_1 e^x + 2c_2 e^{2x} \qquad \text{...}(iii)$$

It is given that $y'(0) =$ *i.e.*, if $x = 0$, then $\dfrac{dy}{dx} = 0$.

Putting these values in (*iii*), we get

$$1 = c_1 + 2c_2$$

Solving (*ii*) and (*iv*), we get $c_1 = -1$, $c_2 = 1$. Putting the values of c_1 and c_2 in (*i*), we get

$$y = -e^x + e^{2x}$$

as the complete solution of the given differential equation.

EXERCISE 13.1

1. $(D^2 - 5D + 6)\,y = 0.$ [**Ans.** $y = c_1e^{2x} + c_2e^{3x}$]
2. $(9D^2 + 12D + 4)\,y = 0.$ [**Ans.** $y = (c_1 + c_2x)\,e^{-2x3}$]
3. $(D^3 - 2D^2 + D)\,y = 0.$ [**Ans.** $y = c_1 + (c_2 + c_3x)\,e^{-x}$]
4. $(D^4 - 2D^2 + 1)\,y = 0.$ [**Ans.** $y = (c_1 + c_2x)\,e^x + (c_3 + c_4x)\,e^{-x}$]
5. $(D^3 - 2D^2 + 4D - 8)\,y = 0.$ [**Ans.** $y = c_1\,e^{2x} + c_2\cos 2x + c_3\sin 2x$]
6. $(D^3 + 1)\,y = 0.$ $\left[\textbf{Ans. } y = c_1e^{-x} + e^{x/2}\left(c_2\cos\frac{\sqrt{3}}{2}x + c_3\sin\frac{\sqrt{3}}{2}x\right)\right]$
7. Solve the differential equation $\frac{d^2y}{dx^2} + 4\frac{dy}{dx} + 29y = 0$, given that at $x = 0, y = 0, \frac{dy}{dx} = 15.$

   [**Ans.** $y = 3e^{-2x}\sin 5x$]

13.3. THE GENERAL SOLUTION OF $f(D)y = X$

The general solution of $f(D)y = X$ is obtained in the following manner:

1. Find the solution, $y = u$ of the equation $f(D)y = 0$ as already discussed.
   This solution $y = u$ is called the complementary function (C.F.) of $f(D)y = X$.
2. Find a particular solution, $y = v$ of $f(D)y = X$. This solution $y = v$ is also called particular integral (P.I.)
3. Then $y = u + v$ or $y =$ C.F. + P.I. is the general solution of $f(D)y = X$.

13.4. PARTICULAR INTEGRAL (P.I.)

Let $f(D)y = X$ be a linear differential equation with constant coefficients. Then the particular integral or particular solution of this differential equation is that function of x, which when operated by $f(D)$ given X.

Now as $f(D)\frac{1}{f(D)}X = X$, therefore $\frac{1}{f(D)}$ can be regarded as the inverse operator of $f(D)$.

Similarly D and $\frac{1}{D}$ are inverse operations. If D stands for differentiation then $\frac{1}{D}$ will stand for integration.

Hence the particular integral of the equation $f(D)y = X$ will be $\frac{1}{f(D)}$ because it satisfies the given equation.

$$\text{P.I.} = \frac{1}{f(D)}X$$

13.5. RULES FOR FINDING PARTICULAR INTEGRAL (P.I.)

Rule I: When $X = e^{ax}$, then

$$\frac{1}{f(D)}e^{ax} = \frac{1}{f(a)}e^{ax}$$

If f $(a) = 0$ then $\frac{1}{f(D)}e^{ax} = x\cdot\frac{1}{f'(a)}e^{ax}$

If $f'(a) = 0$ then $\frac{1}{f(D)}e^{ax} = x^2\cdot\frac{1}{f''(a)}e^{ax}$ provided $f''(a) \neq 0$ and so on.

To find the P.I. $\frac{1}{f(D)}e^{ax}$, replace D by a, provided $f(a) \neq 0$ In case $f(a) = 0$, multiply the numerator by x and differentiate the denominator w.r.t D and apply the above rule provided $f'(a)$ ¹ 0 and so on.

SOLVED EXAMPLES

Example 13.9. Solve $(9D^2 + 12D + 4)\, y = 6e^{-2x/3}$.

Solution. The uxiliary equation is $9m^2 + 12m + 4 = 0$

$$\Rightarrow \quad (3m + 2)(3m + 2) = 0$$

$$\Rightarrow \quad m = \frac{-2}{3}, -\frac{2}{3}$$

$$\therefore \quad \text{C.F.} = (c_1 + c_2x)\, e^{-2x/3}$$

$$\text{P.I.} = \frac{1}{9D^2+12D+4}6e^{-2x/3} = 6x.\frac{1}{18D+12}e^{-2x/3}$$

$$= \frac{6x^2}{18}e^{-2x/3} = \frac{x^2e^{-2x/3}}{3}$$

Hence the complete solution is

$$y = \text{C.F.} + \text{P.I.}$$

i.e.,
$$y = (c_1 + c_2x)\, e^{-2x/3} + \frac{x^3}{3}e^{-2x/3}.$$

Example 13.10. Solve $(D^3 + 1)\, y = (e^x + 1)^2$.

Solution. The auxiliary equation is

$$m^3 + 1 = 0 \quad \text{or} \quad (m + 1)(m^2 - m + 1) = 0 \quad \text{or} \quad m = -1, \frac{2}{1} \pm \frac{\sqrt{3}}{2}i$$

$$\therefore \quad \text{C.F.} = c_1e^{-x} + e^{x/2}\left(c_2 \cos\frac{\sqrt{3}}{2}x + c_3 \sin\frac{\sqrt{3}}{2}x\right)$$

$$\text{P.I.} = \frac{1}{D^3+1}(e^x+1)^2 = \frac{1}{D^3+1}(e^{2x}+2e^x+1)$$

$$= \frac{1}{D^3+1}e^{2x} + 2\cdot\frac{1}{D^3+1}e^x + \frac{1}{D^3+1}e^{0x}$$

$$= \frac{1}{2^3+1}e^{2x} + 2\cdot\frac{1}{1^3+1}e^x + \frac{1}{0^3+1}e^{0x} = \frac{1}{9}e^{2x} + e^x + 1$$

Hence the complete solution is

$$y = \text{C.F.} + \text{P.I.}$$

i.e.,
$$y = c_1e^{-x} + e^{x/2}\left(c_2 \cos\frac{\sqrt{3}}{2}x + c_3 \sin\frac{\sqrt{3}}{2}\right) + \frac{1}{9}e^{2x} + e^x + 1.$$

Example 13.11. Solve $\frac{d^2y}{dx^2} - 7\frac{dy}{dx} + 6y = e^{2y}$ **=, given that $y = 0$ when $x = 0$.**

Solution. Given equation in symbolic form is

$$(D^2 - 7D + 6)\,y = e^{2x}$$

The auxiliary equation is

$$m^2 - 7m + 6 = 0 \quad \text{or} \quad (m-1)(m-6) = 0$$

$\Rightarrow$
$$m = 1, 6$$

$\therefore$
$$\text{C.F.} = c_1e^x + c_2e^{6x}$$

$$\text{P.I.} = \frac{1}{D^2 - 7D + 6}e^{2x} = \frac{1}{(2)^2 - 7(2) + 6}e^{2x} = -\frac{1}{4}e^{2x}$$

Hence the complete solution is $y = \text{C.F.} + \text{P.I.}$

i.e.,
$$y = c_1e^x + c_2e^{6x} - \frac{1}{4}e^{2x}$$

When $x = 0$, $y = 0$

$\therefore$ From (*i*)
$$0 = c_1e^0 + c_2e^0 - \frac{1}{4}e^0 = c_1 + c_2 - \frac{1}{4}$$

$\Rightarrow$
$$c_2 = \frac{1}{4} - c_1$$

$\therefore$ From (*i*), required solution is

$$y = c_1e^x + \left(\frac{1}{4} - c_1\right)e^{6x} - \frac{1}{4}e^{2x}$$

or
$$y = c_1(e^x - e^{6x}) + \frac{1}{4}(e^{6x} - e^{2x}).$$

EXERCISE 13.2

Solve the following differential equations:

1. $(D^2 - 3D + 2)\,y = e^{3x}$. $\left[\textbf{Ans. } y = c_1e^x + c_2e^{2x} + \frac{1}{2}e^{3x}\right]$

2. $(2D + 1)^2\,y = 4e^{-x/2}$ $\left[\textbf{Ans. } y = (c_1 + c_2x)\,e^{-x/2} + \frac{x^2}{2}e^{-x/2}\right]$

3. $(D^2 - 2D + 1)\,y = (1 + e^{-x})^2$ $\left[\textbf{Ans. } y = (c_1 + c_2x)\,e^x + 1 + \frac{1}{2}e^{-x} + \frac{1}{9}e^{-2x}\right]$

4. $[D^2 - (a + b)D + ab]\,y = e^{ax} + e^{bx}$. $\left[\textbf{Ans. } y = c_1e^{4x} + c_2e^{bx} + \frac{x}{(a-b)}(e^{ax} - e^{bx})\right]$

5. $(4D^2 - 1)\,y = e^{x/2}$. $\left[\textbf{Ans. } y = c_1e^{x/2} + c_2e^{-x/2} + \frac{x}{4}e^{x/2}\right]$

6. $(D^2 - 6D + 8)\,y = (e^{2x} - 1)^2.$ $\left[\textbf{Ans. } y = c_1e^{x/2} + c_2e^{4x} + \frac{x}{2}e^{4x} + xe^{2x} + \frac{1}{8}\right]$

7. $\frac{d^3y}{dx^3} + 3\frac{d^2y}{dx^2} + 3\frac{dy}{dx} + y = e^{-x}$ $\left[\textbf{Ans. } y = (c_1 + c_2x + c_3x^2)\,e^{-x} + \frac{1}{6}x^3 e^{-x}\right]$

8. $(D^2 - 3D + 2)\,y = e^x + e^{2x}$ $[\textbf{Ans. } y = c_1e^x + c_2e^{2x} - xe^x + xe^{2x}]$

9. $(D^2 - 6D + 9)\,y = 6e^{3x} + 7e^{-2x} - \log 2.$ $\left[\textbf{Ans. } y = (c_1 + c_2x)\,e^{3x} + 3x^2e^{3x} + \frac{7}{25}e^{-2x} - \frac{1}{9}\log 2\right]$

10. $\frac{d^2y}{dx^2} + 2p\frac{dy}{dx} + (p^2 + q^2)\,y = e^{ax}$ $\left[\textbf{Ans. } y = e^{-px}\,(c_1\,x\cos qx + c_2 \sin qx) + \frac{1}{(a+p)^2 + q^2}\,e^{ax}\right]$

Rule II. When $X = \sin(ax + b)$ or $\cos(ax + b)$, then

$$\frac{1}{f(D)^2}\sin(ax + b) = \frac{1}{f(-a)^2}\sin(ax + b) \text{ Provided } f(-a)^2 \neq 0$$

and
$$\frac{1}{f(D^2)}\cos(ax + b) = \frac{1}{f(-a)^2}\cos(ax + b) \text{ provided } f(-a)^2 \neq 0$$

If $f(-a^2) = 0$ then

$$\frac{1}{f(D)^2}\cos(ax + b) = x \cdot \frac{1}{f'(-a^2)}\sin(ax + b) \text{ provided } f'(-a^2) \neq 0$$

and
$$\frac{1}{f(D)^2}\cos(ax + b) = x \cdot \frac{1}{f'([-a^2)}\cos(ax + b) \text{ provided } f'(-a)^2 \neq 0$$

and so on.

To find the P.I. $\frac{1}{f(D^2)}\sin(ax + b)$ or $\frac{1}{f(D)^2}\cos(ax + b)$, replace D^2 by $-a^2$, provided $f(-a^2) \neq 0$. In case $f(-a^2) = 0$, multiply the numerator by x and differentiate the denominator w.r.t. D and apply the above rule, provided $f'(-a^2) \neq 0$ and so on.

SOLVED EXAMPLES

Example 13.12. Solve $\frac{d^2y}{dx^2} - 5\frac{dy}{dx} + 6y = \sin 3x$.

Solution. Given equation in symbolic form is

$$(D^2 - 5D + 6)\,y = \sin 3x$$

Its auxiliary equation is $m^2 - 5m + 6 = 0$

$$\Rightarrow \quad (m - 2)(m - 3) = 0$$

$$\Rightarrow \quad m = 2, 3$$

$$\therefore \quad \text{C.F.} = c_1e^{2x} + c_2e^{3x}$$

$$\text{P.I.} = \frac{1}{D^2 - 5D + 6}\sin 3x = \frac{1}{-3^2 - 5D + 6}\sin 3x \qquad [\because\ D^2 \to -a^2 = -2^2 = -4]$$

$$= -\frac{1}{5D + 3}\sin 3x = -\frac{5D - 3}{(5D + 3)(5D - 3)}\sin 3x$$

$$= -\frac{5D-3}{25D^2-9}\sin 3x = -\frac{5D-3}{25(-9)-9}\sin 3x \qquad [\because \; D^2 \to -a^2 = -3^2 = -a]$$

$$= \frac{1}{234}\left[5\frac{d}{dx}(\sin 3x) - 3\sin 3x\right]$$

$$= \frac{1}{234}(15\cos 3x - 3\sin 3x) = \frac{1}{78}(5\cos 3x - \sin 3x)$$

$\therefore$ The complete solution is $y =$ C.F. + P.I.

i.e.,
$$y = c_1 e^{2x} + c_2 e^{3x} + \frac{1}{78}(5\cos 3x - \sin 3x)$$

Example 13.13. Solve $(D^2 - 4)\, y = \cos^2 x$.

Solution. The auxiliary equation is $m^2 - 4 = 0 \;\Rightarrow\; m = \pm 2$.

$\therefore$ C.F. $= c_1 e^{2x} + c_2 e^{-2x}$

We know $\cos 2x = 2\cos^2 x - 1 \;\Rightarrow\; \cos^2 x = \frac{1}{2}(1 + \cos 2x)$

$$\text{P.I.} = \frac{1}{D^2-4}\cos^2 x = \frac{1}{2}\cdot\frac{1}{D^2-4}(e^{0x} + \cos 2x) \qquad [\because \; e^{0x} = 1]$$

$$= \frac{1}{2}\cdot\left[\frac{1}{(0-4)}e^{0x} + \frac{1}{(-4-4)}\cos 2x\right] \qquad [\because \; D^2 \to -2^2 = -4]$$

$$= -\frac{1}{8} - \frac{1}{16}\cos 2x$$

$\therefore$ The complete solution is $y =$ C.F. + P.I.

i.e.,
$$y = c_1 e^{2x} + c_2 e^{-2x} - \frac{1}{8} - \frac{1}{6}\cos 2x.$$

Example 13.14. Solve $\frac{d^3y}{dx^3} - 3\frac{d^2y}{dx^2} + 4\frac{dy}{dx} - 2y = e^x + \cos x$.

Solution. Given equation in symbolic form is

$$(D^3 - 3D^2 + 4D - 2)\, y = e^x + \cos x$$

Its auxiliary equation is $m^3 - 3m^2 + 4m - 2 = 0$

or $(m - 1)(m^2 - 2m + 2) = 0$, *i.e.*, $m = 1, 1 \pm i$

$\therefore$ C.F. $= c_1 e^x + e^x (c_2 \cos x + c_3 \sin x)$

$$\text{P.I.} = x\cdot\frac{1}{(D^3 - 3D^2 + 4D - 2}e^x + \frac{1}{(D^3 - 3D^2 + 4D - 2)}\cos x$$

$$= x\cdot\frac{1}{3D^2 - 3D^2 + 4D - 2)}e^x + \frac{1}{(D^3 - 3D^2 + 4D - 2)}\cos x$$

$$[\because \; D^2 \to -1^2]$$

$$= x\cdot\frac{1}{(7-6)}e^x + \frac{1}{3D+1}(\cos x)$$

$$= xe^x + \frac{3D-1}{9D^2-1}(\cos x) = xe^x - \frac{1}{10}(-3\sin x - \cos x).$$

∴ Complete solution is $y = \text{C.F.} + \text{P.I.}$

i.e.,
$$y = c_1e^x + e^x(c_2 \cos x + c_3 \sin x) + xe^x + \frac{1}{10}(3 \sin x + \cos x).$$

EXERCISE 13.3

Solve the following differential eauations:

1. $(D^2 - 3D + 2)y = \cos 2x.$ $\left[\textbf{Ans. } y = c_1e^x + c_2e^{2x} - \frac{1}{20}\cos 2x - \frac{3}{20}\sin 2x\right]$

2. $(D^3 + D^2 + D + 1)y = \sin 2x.$ $\left[\textbf{Ans. } y = c_1e^{-x} + c_2 \cos x + c_3 \sin x + \frac{1}{15}(2\cos 2x - \sin 2x\right]$

3. $\dfrac{d^2y}{dx^2} + 2\dfrac{dy}{dx} + y = e^{2x} - \cos^2 x.$ $\left[\textbf{Ans. } y = (c_1 + c_2x)e^{-x} + \frac{1}{9}e^{2x} + \frac{1}{50}(3\cos 2x - 4\sin 2x)\right]$

4. $(D^3 + 1)y = \cos 2x.$ $\left[\textbf{Ans. } y = c_1e^x + e^{x/2}\left(c_2 \cos\frac{\sqrt{3}}{2} + c_3 \sin\frac{\sqrt{3}}{2}\right) + \frac{1}{65}(\cos 2x - 8\sin 2x)\right]$

5. $(D^2 - D - 2) = \sin 2x + e^x.$ $\left[\textbf{Ans. } y - c_1e^{2x} + c_2e^{-x} - \frac{1}{20}(3\sin 2x - \cos 2x) - \frac{1}{2}e^x\right]$

6. $\dfrac{d^2y}{dx^2} + 3\dfrac{dy}{dx} + 2y = 4\cos^2 x.$ $\left[\textbf{Ans. } y = c_1e^{-x} + c_2e^{-x} + 1 + \frac{1}{10}(3\sin 2x - \cos 2x)\right]$

7. $(D^3 + D^2 - D - 1)y = \cos 2x.$ $\left[\textbf{Ans. } y = c_1e^x + (c_2 + c_3x)e^{-x} - \frac{2}{25}\sin 2x - \frac{1}{25}\cos 2x\right]$

8. $(D^2 + a^2)\text{y} = \sin ax.$ $\left[\textbf{Ans. } y = c_1 \cos ax + c_2 \sin ax - \frac{x}{2a}\cos ax\right]$

9. $\dfrac{d^2y}{dx^3} + y = \sin 3x - \cos^2\dfrac{x}{2}.$

$\left[\textbf{Ans. } y - c_1e^{-x} + e^{x/2}\left(c_2 \cos\frac{\sqrt{3}}{2}x + c_3 \sin\frac{\sqrt{3}}{2}x\right) + \frac{1}{130}(\sin 3x + 27\cos 3x) - \frac{1}{2} - \frac{1}{4}(\cos x - \sin x)\right]$

10. $(D^2 - 4D + 4)y = e^{-4x} + 5\cos 3x.$ $\left[\textbf{Ans. } y = (c_1 + c_2x)e^{2x} + \frac{1}{36}e^{-4x} - \frac{5}{169}(12\sin 3x + 5\cos 3x)\right]$

11. $(D^2 - 3D + 2)y = 6e^{-3x} + \sin 2x.$ $\left[\textbf{Ans. } y = c_1e^x + c_2e^{2x} + \frac{3}{10}e^{-3x} + \frac{1}{20}(3\cos 2x - \sin 2x)\right]$

12. $\dfrac{d^3y}{dx^3} + 2\dfrac{d^2y}{dx^2} + \dfrac{dy}{dx} = e^{-x} + \sin 2x.$ $\left[\textbf{Ans. } y = c_1 + (c_2 + c_3x)e^{-x} - \frac{x^2}{2}e^{-x} + \frac{1}{50}(3\cos 2x - 4\sin 2x)\right]$

13. $(D^2 - 1)y = 3\cos^2 x + 2\sin^3 2x.$ $\left[\textbf{Ans. } y = c_1 \cos x + c_2 \sin x + \frac{3}{2} - \frac{1}{2}(\cos 2x + \sin 2x) + \frac{1}{70}\sin 6x\right]$

14. $(D^2 - 3D + 2)y = \cos 3x \cos 2x.$ $\left[\textbf{Ans. } y = c_1e^x + c_2e^{2x} - \frac{1}{1580}(15\sin 5x + 23\cos 5x) + \frac{1}{20}(\cos x - 3\sin x)\right]$

15. $(D^2 - 4D + 3)y = \sin 3x \cos 2x.$ $\left[\textbf{Ans. } y = c_1e^x + c_2e^{3x} + \frac{1}{884}(10\cos 5x - 11\sin 5x) + \frac{1}{20}(\sin x + 2\cos x)\right]$

16. $(D^2 + 4)y = \sin 3x + \cos 2x.$ $\left[\textbf{Ans. } y = c_1 \cos 2x + c_2 \sin 2x - \frac{1}{5}\sin 3x + \frac{x}{4}\sin 2x\right]$

17. $(D^4 + 2D^3 - 3D^2)y = 3e^{2x} + 4\sin x.$ $\left[\textbf{Ans. } y = c_1 + c_2x + c_3e^x + c_4e^{3x} + c_4e^{-3x} + \frac{3}{20}e^{2x} + \frac{2}{5}(\cos x + 2\sin x)\right]$

Rule III. When $X = x^n$, then $\frac{1}{f(D)} x^x = [f(D)]^{-1} x^n$

In order to evaluate $\left\{\frac{1}{f(D)}\right\} x^n$. bring out common the lowest degree term in D from $f(D)$ so that resulting factor in the denominator is of the form $[1 + f(D)]$ or $[1 - f(D)]$ which is taken in the numerator with a negative index. next we expand $[1 \pm f(D)]^{-1}$ in powers of D by the binomial theorem and operate upto x^n with the expansion obtained.

This expansion should be done upto the term D^n, since $D^{n+1} x^n = 0$ and all higher differential coefficients of x^n are zero. Now several terms obtained in the expansion of $[f(D)]^{-1}$ are operated on x^n and the result obtained in such a way will be the particular integral corresponding to x^n.

The following expansion will be helpful in expanding $[f(D)]^{-1}$ in ascending powers of D.

(*i*) $(1 + x)^n = 1 + nx + \frac{n(n-1)}{2!} x^2 + \frac{n(n-1)(n-2)}{31} x^3 + \ldots$

(*ii*) $(1 + x)^n = 1 - nx + \frac{n(n-1)}{2!} x^2 - \frac{n(n-1)(n-2)}{31} x^3 + \ldots$

(*iii*) $(1 - x)^{-1} = 1 + x + x^2 + x^3 + x^4 + \ldots$

(*iv*) $(1 - x)^{-1} = 1 - x + x^2 - x^3 + x^4 + \ldots$

(*v*) $(1 - x)^{-2} = 1 + 2x + 3x^2 + 4x^3 + \ldots$

(*vi*) $(1 - x)^{-2} = 1 - 2x + 3x^2 - 4x^3 + \ldots$

SOLVED EXAMPLES

Example 13.15. Solve $(D^2 - 6D + 9)\, y = 2x^2 - x + 3$.

Solution. The auxiliary equation is

$$m^2 - 6m + 9 = 0 \Rightarrow (m-3)^2 = 0 \Rightarrow m = 3, 3$$

$$\therefore \quad \text{C.F.} = (c_1 + c_2 x)\, e^{3x}$$

$$\text{P.I.} = \frac{1}{D^2 - 6D + 9}(2x^2 - x + 3) = \frac{1}{(D-3)^2}(2x^2 - x + 3)$$

$$= \frac{1}{9}\left(1 - \frac{D}{3}\right)^{-2}(2x^2 - x + 3)$$

$$= \frac{1}{9}\left(1 + \frac{2D}{3} + \frac{3D^2}{9} + \ldots\right)(2x^2 - x + 3)$$

$$= \frac{1}{9}\left(1 + \frac{2D}{3} + \frac{D^2}{2}\right)(2x^2 - x + 3) \quad \text{[Leaving higher powers of } D\text{]}$$

$$= \frac{1}{9}\left[2x^2 - x + 3 + \frac{2}{3}(4x - 1) + \frac{1}{3}(4)\right] = \frac{1}{9}\left(2x^2 + \frac{5}{3}x + \frac{11}{3}\right)$$

Hence the complete solution is $y = \text{C.F.} + \text{P.I.}$

i.e., $$y = (c_1 + c_2 x)\, e^{3x} + \frac{1}{9}\left(2x^2 + \frac{5}{3}x + \frac{11}{3}\right)$$

Example 13.16. **Solve $(D^3 + 2D^2 + D)\, y = e^{2x} + x^2 + x$.**

Solution. The auxiliary equation is $m^3 + 2m^3\ m = 0$

$$\Rightarrow \quad m(m^2 + 2m + 1) = 0 \Rightarrow m = 0, -1, -1$$

$$\therefore \quad \text{C.F.} = \frac{1}{D^3 + 2D^2 + D}(e^{2x} + x^2 + x)$$

$$= \frac{1}{D^3 + 2D^2 + D} e^{2x} + \frac{1}{D^3 + 2D^2 + D} \cdot (x^2 + x)$$

$$= \frac{1}{2^3 + 2 \cdot 2^2 + 2} e^{2x} + \frac{1}{D[1 + (D^2 + 2D)]} \cdot (x^2 + x)$$

$$= \frac{1}{18} e^{2x} + \frac{1}{D}[1 - (D^2 + 2D)]^{-1} \cdot (x^2 + x)$$

$$= \frac{1}{18} e^{2x} + \frac{1}{D}[1 - (D^2 + 2D) + (D^2 + 2D)^2 - \ldots] \cdot (x^2 + x)$$

$$= \frac{1}{18} e^{2x} + \frac{1}{D}[x^2 + x - [2 + 2(2x + 1)] + 4.2] \qquad [\because \ 4D^2 (x^2 + x = 4.2\,]$$

$$= \frac{1}{18} e^{2x} + \frac{1}{D}(x^2 - 3x + 4)$$

$$= \frac{1}{18} e^{2x} + \int (x^2 - 3x + 4)\, dx \qquad \left[\because \ \frac{1}{D} X = \int X\, dx\right]$$

$$= \frac{1}{18} e^{2x} + \frac{x^2}{3} - \frac{3x^2}{2} + 4x$$

Hence the complete solution is y = C.F. + P.I.

i.e.,
$$y = c_1 + (c_2 + c_3 x)\, e^{-x} + \frac{1}{18} e^{2x} + \frac{x^3}{3} - \frac{3}{2} x^2 + 4x.$$

Example 13.17. **Solve $(D^2 - 2D + 3)\, y = x^2 + \cos x$.**

Solution. The auxiliary equation is $m^2 - 2m + 3 = 0$

$$\Rightarrow \quad m = \frac{2 \pm \sqrt{4 - 12}}{2} = 1 \pm \sqrt{2i}$$

$$\therefore \quad \text{C.F.} = e^x (c^x \cos \sqrt{2x} + c_2 \sin \sqrt{2x}) = \frac{1}{D^2 - 2D + 3}(x^2 + \cos x)$$

$$= \frac{1}{3} \cdot \left[1 + \left(\frac{D^2 - 2D}{3}\right)\right]^{-1} \cdot x^2 + \frac{1}{D^2 - 2D + 3} \cos x$$

$$= \frac{1}{3}\left[1 - \left(\frac{D^2 - 2D}{3}\right) + \left(\frac{D^2 - 2D}{3}\right)^2 - \ldots\right] x^2 + \frac{1}{(-1^2 - 2D + 3)} \cos x$$

$$= \frac{1}{3}\left[x^2 + \frac{1}{3}(2 - 2.2x) + \frac{4}{9} \cdot 2\right] + \frac{1}{2(1 - D)} \cos x$$

$$= \frac{1}{3}\left(x^2 + \frac{4}{3} x + \frac{2}{9}\right) + \frac{(1 + D)}{2(1 - D^2)(1 + D)} \cos x$$

$$= \frac{1}{3}\left(x^2 + \frac{4}{3}x + \frac{2}{9}\right) + \frac{(1+D)}{2(1-D^2)}\cos x$$

$$= \frac{1}{3}\left(x^2 + \frac{4}{3}x + \frac{2}{9}\right) + \frac{(1+D)}{2(1+1)} \qquad [\because \; D^2 \to -1^2 = -1]$$

$$= \frac{1}{3}\left(x^2 + \frac{4}{3}x + \frac{2}{9}\right) + \frac{1}{4}(\cos x - \sin x)$$

hence the required solution is y = C.F. + P.I.

i.e.,
$$y = e^x(c_1 \cos\sqrt{2}x + c_2 \sin\sqrt{2}x) + \frac{1}{3}\left(x^2 + \frac{4}{3}x + \frac{2}{9}\right) + \frac{1}{4}(\cos x - \sin x).$$

EXERCISE 13.4

Solve the following differential equations:

1. $\dfrac{d^2y}{dx^2} - y = 2 + 3x.$ $\quad$ [**Ans.** $y = c_1e^x + c_2e^{-x} - 2 - 3x$]

2. $(D^2 + 5D + 4)\,y = x^2 + 7x + 9.$ $\quad \left[\textbf{Ans. } y = c_1e^{-x} + c_2e^{-4x} + \frac{1}{4}\left(x^2 + \frac{9}{2}x + \frac{23}{8}\right)\right]$

3. $\dfrac{d^3y}{dx^3} - \dfrac{d^2y}{dx^2} - 6\dfrac{dy}{dx} = 1 + x^3.$ $\quad \left[\textbf{Ans. } y = c_1 + c_2e^{-2x} + c_3e^{3x} - \frac{1}{6}\left[\frac{25}{18}x - \frac{1}{6}x^2 + \frac{1}{3}x^3\right]\right]$

4. $\dfrac{d^2y}{dx^2} - 4y = x^2 + 2x.$ $\quad \left[\textbf{Ans. } y = c_1e^{2x} + c_2e^{-2x} - \frac{1}{4}\left(x^2 + \frac{1}{2}\right)\right]$

5. $(D^2 - 5D + 6)\,y = x + \sin 3x.$ $\quad \left[\textbf{Ans. } y = c_1e^{2x} + c^2e^{3x} + \frac{1}{6}\left(x + \frac{5}{6}\right) - \frac{1}{234}(3\sin 3x - 15\cos 3x)\right]$

6. $(D^2 + 3D + 2)\,y = e^{2x} + x^2 + \sin x.$
$\left[\textbf{Ans. } y = c_1e^{-x} + c_2e^{-2x} + \frac{1}{12}e^{2x} + \frac{1}{4}(2x^2 - 6x + 7) - \frac{1}{10}(3\cos x - \sin x)\right]$

7. $(D^2 + D + 1)\,y = \cos 2x + x^2.$
$\left[\textbf{Ans. } y = c_1e^{-x/2}\left(c_1\cos\frac{\sqrt{3}}{2}x + c_2\sin\frac{\sqrt{3}}{2}x\right) + (x^2 - 2x) + \frac{1}{13}(2\sin 2x - 3\cos 2x)\right]$

8. $\dfrac{d^3y}{dx^3} + 2\dfrac{d^2y}{dx^2} + \dfrac{dy}{dx} = e^{2x} + x^2 + x.$ $\quad \left[\textbf{Ans. } y = c_1 + (c_2 + c_3x)\,e^{-x} + \frac{1}{18}e^{2x} + \frac{x^3}{3} - \frac{3x^2}{2} + 4x\right]$

9. $\dfrac{d^3y}{dx^3} + \dfrac{d^2y}{dx^2} - \dfrac{dy}{dx} - y = \sin 2x = x^3.$ $\quad \left[\textbf{Ans. } y = c_1e^x + (c_2 + c_3x)e^{-x} + \frac{1}{25}(2\cos 2x - \sin 2x) - x^2 + 2x - 4\right]$

10. $\dfrac{d^2y}{dx^2} - 4y = x^2.$ $\quad \left[\textbf{Ans. } y = c_1e^{2x} + c_2e^{-2x} - \frac{1}{4}\left(x^2 + \frac{1}{2}\right)\right]$

11. $(D^4 - 2D^3 + D^2)\,y = x^3.$ $\quad \left[\textbf{Ans. } y = c_1 + c_2x + (c_3 + c_4x)\,e^x + \frac{x^5}{20} + \frac{x^4}{2} + 3x^3 + 12x^2\right]$

12. $(D-1)^2(D+1)^2\,y = \sin^2\dfrac{x}{2} + e^x + \cos 2x.$ $\quad \left[\textbf{Ans. } y = (c_1 + c_2x)\,e^x + (c_3 + c_4x)\,e^{-x} + \frac{1}{2} - \frac{1}{8}\cos x + \frac{x^2}{8}e^x + x\right]$

13. $(D^2 - 4D + 4)\,y = x^2 + e^x + \cos 2x.$ $\quad \left[\textbf{Ans. } y = (c_1 + c_2x)e^{2x} + e^x - \frac{1}{8}\sin 2x + \frac{1}{4}(x^2 + 2x + 3\_)\right]$

14. If $\frac{d^2y}{dt^2}+\frac{8}{b}(x-a)=0$; a, b and g are positive numbers and $x = a'$, $\frac{dx}{dt} = 0$ when $t = 0$ show that

$$x = a + (a' - a)\cos\sqrt{\frac{g}{b}}t.$$

Rule IV. When $X = e^{ax}V$, where V is any function of x, then

$$\frac{1}{f(D)}[e^{ax}V] = e^{ax}\frac{1}{f(D+a)}.V$$

To find P.I. $\frac{1}{f(D)}[e^{ax}V]$, repalce D by $(D + a)$ and write it as $e^{ax}\frac{1}{f(D+a)}V$, and then apply the rules stated above, depending upon the nature of the function V.

SOLVED EXAMPLES

Example 13.18. Solve $(D^2 - 4D + 3)y = e^x \cos 2x$.

Solution. The auxiliary equation is

$$m^2 - 4m + 3 = 0$$

or $$(m-1)(m-3) = 0$$

or $$m = 1, 3$$

$\therefore$ $$\text{C.F.} = c_1e^x + c_2e^{3x}$$

$$\text{P.I.} = \frac{1}{D^2-4D+3}e^x\cos 2x = e^x\frac{1}{(D+1)^2-4(D+1)+3}\cos 2x$$

$$= e^x\frac{1}{D^2-2D}\cos 2x = e^x\frac{1}{-2^2-2D}\cos 2x \quad [\because D^2 \to -2^2]$$

$$= -\frac{1}{2}e^x\frac{1}{2+D}\cos 2x = -\frac{1}{2}e^x\frac{(2-D)}{(2-D)(2+D)}\cos 2x$$

$$= -\frac{1}{2}e^x\frac{2-D}{4-D^2}\cos 2x = -\frac{1}{2}e^x.\frac{2-D}{4-(-2^2)}\cos 2x$$

$$= -\frac{1}{16}e^x(2\cos 2x - D\cos 2x) = -\frac{1}{16}e^x(2\cos 2x - 2\sin 2x)$$

$$= -\frac{1}{8}e^x(\cos 2x + \sin 2x)$$

Hence, the complete solution is y = C.F. + P.I.

i.e., $$y = c_1e^x + c_2e^{3x} - \frac{1}{8}e^x(\cos 2x + \sin 2x)$$

Example 13.19. Obtain the general solution of the differential equation

$$y'' - 2y' + 2y = x + e^x \cos x.$$

Solution. The given equation in symbolic form is

$$(D^2 - 2D + 2)y = x + e^x \cos x$$

Its auxiliary equation is $m^2 - 2m + 2 = 0$

$$m = \frac{2 \pm \sqrt{4-8}}{2} = 1 \pm i$$

$$\therefore \qquad \text{C.F.} = e^x (c_1 \cos x + c_2 \sin x)$$

$$\text{P.I.} = \frac{1}{D^2 - 2D + 2}(x + e^x \cos x)$$

$$= \frac{1}{D^2 - 2D + 2}(x) + \frac{1}{D^2 - 2D + 2}(e^x \cos x)$$

$$= \frac{1}{2 - 2D + D^2}(x) + e^x \frac{1}{(D+1)^2 - 2(D+1) + 2}(\cos x)$$

$$= \frac{1}{2}\left[1 - \left(\frac{2D - D^2}{2}\right)\right]^{-1}(x) + e^x \cdot \frac{1}{D^2 + 1} \cdot \cos x$$

$$= \frac{1}{2}\left[1 + \left(\frac{2D - D^2}{2}\right)\right](x) + e^x \cdot x \cdot \frac{1}{2D} \cos x \qquad \text{[Case of failure]}$$

$$= \frac{1}{2}[1 + D] \cdot x + e^x \cdot \frac{x}{2} \sin x \qquad \text{[Leaving higher powers]}$$

$$= \frac{1}{2}(x + 1) + \frac{xe^x}{2} \sin x$$

Hence, the complete solution is $y = \text{C.F.} + \text{P.I.}$

i.e.,

$$y = e^x(c_1 \cos x + c^2 \sin x) + \frac{1}{2}(x + 1) + \frac{xe^x}{2} \sin x.$$

Example 13.20. Solve $\dfrac{d^2y}{dx^2} + 2y = x^2 e^{3x} + e^x \cos 2x.$

Solution. Given equation in symbolic form is

$$(D^2 + 2)y = x^2 e^{3x} + e^x \cos 2x$$

Its auxiliary equation is $m^2 + 2 = 0 \Rightarrow m = \pm\sqrt{2}i$

$$\therefore \qquad \text{C.F.} = c_1 \cos \sqrt{2x} + c_2 \sin \sqrt{2x}$$

$$\text{P.I.} = \frac{1}{D^2 + 2}(x^2 e^{3x} + e^x \cos 2x)$$

$$= \frac{1}{D^2+2}(x^2e^{3x}) + \frac{1}{D^2+2}(e^x \cos 2x)$$

$$= e^{3x}\frac{1}{(D+3)^2+2}x^2 + e^x\frac{1}{(D+1)^2+2}\cos 2x$$

$$= e^{3x}\cdot\frac{1}{D^2+6D+11}x^2 + e^x\cdot\frac{1}{D^2+2D+3}\cdot\cos 2x$$

$$= e^{3x}\cdot\frac{1}{11\left(1+\frac{6D+D^2}{11}\right)}(x^2) + e^x\cdot\frac{1}{-4+2D+3}\cos 2x \qquad [\because\ D^2 \to 2^2 = -4]$$

$$= \frac{1}{11}e^{3x}\left[1+\left(\frac{6D+D^2}{11}\right)\right]^{-1}(x^2) + e^x\cdot\frac{1}{2D-1}\cos 2x$$

$$= \frac{1}{11}e^{3x}\left[1-\left(\frac{6D+D^2}{11}\right)+\left(\frac{6D+D^2}{11}\right)^2-\ldots\right](x^2) + e^x\cdot\frac{(2D+1)}{(2D-1)(2D+1)}\cos 2x$$

$$= \frac{1}{11}e^{3x}\left[1-\frac{6D}{11}-\frac{D^2}{11}-\frac{36}{121}D^2\right](x^2) + e^x\frac{(2D+1)}{4D^2-1}\cos 2x$$

$$= \frac{1}{11}e^{3x}\left[1-\frac{6D}{11}+\frac{25}{121}D^2\right]x^2 + e^x\cdot\frac{(2D+1)}{4(-4)-1}\cos 2x \qquad [\because\ D^2 \to -2^2 = -4]$$

$$= \frac{1}{11}e^{3x}\left[x^2-\frac{12}{11}x+\frac{50}{121}\right] - \frac{1}{17}e^x(2D+1)\cos 2x$$

$$= \frac{1}{11}e^{3x}\left(x^2-\frac{12}{11}x+\frac{50}{121}\right) - \frac{1}{17}e^x(4\sin 2x - \cos 2x)$$

Hence, the complete solution is y = C.F. + P.I.

i.e.,
$$y = c_1\cos\sqrt{2}x + c_2\sin\sqrt{2}x + \frac{1}{11}e^{3x}\left(x^2-\frac{12}{11}x+\frac{50}{121}\right) + \frac{1}{17}e^x(4\sin 2x - \cos 2x).$$

Example 13.21. Solve $\dfrac{d^2y}{dx^2} - 2\dfrac{dy}{dx} + y = xe^x \sin x.$

Solution. Given equation in symbolic form is $(D^2 - 2D + 1)y = xe^x \sin x$

The auxiliary equation is $m^2 - 2m + 1 = 0$ or $(m-1)^2 = 0 \Rightarrow m = 1, 1$

$\therefore$ C.F. $= (c_1 + c_2x)e^x$

$$\text{P.I.} = \frac{1}{(D-1)^2}e^x\cdot\sin x = e^x\cdot\frac{1}{(D+1-1)^2}x\sin x$$

$$= e^x\frac{1}{D^2}x\sin x = e^x\cdot\frac{1}{D}\int x\sin x\,dx$$

$$= e^x \frac{1}{D}[x(-\cos x) - \int 1(-\cos x)\,dx] \qquad \text{[Integrating by parts]}$$

$$= e^x \frac{1}{D}(-x\cos x + \sin x)$$

$$= e^x \int(-x\cos x + \sin x)\,dx = e^x[-[x\sin x - \int 1.\sin x\,dx] - \cos x]$$

$$= e^x[-x\sin x - \cos x - \cos x] = -e^x(x\sin x + 2\cos x)$$

Hence, the complete solution is y = C.F. + P.I.

i.e., $$y = (c_1 + c_2x)e^x - e^x(x\sin x + 2\cos x)$$

EXERCISE 13.5

Solve the following differential equations:

1. $(D^2 - 5D + 6)y = xe^{4x}$ $\left[\textbf{Ans. } y = c_1e^{2x} + c_2e^{3x} + \frac{1}{2}e^{4x}\left(x - \frac{3}{2}\right)\right]$

2. $(D^2 + 1)y = xe^{2x}$ $\left[\textbf{Ans. } y = c_1\cos x + c_2\sin x + \frac{1}{5}e^{2x}\left(x - \frac{4}{5}\right)\right]$

3. $(D^2 - 2D + 1)y = x^2e^{3x}$ $\left[\textbf{Ans. } y = (c_1 + c_2x)e^x + \frac{1}{4}e^{3x}\left(x^2 - 2x + \frac{3}{2}\right)\right]$

4. $(D^2 + 3D + 2)y = e^{2x}\sin x$ $\left[\textbf{Ans. } y = c_1e^{-x} + c_2e^{-2x} - \frac{1}{170}e^{cx}(7\cos x - 11\sin x)\right]$

5. $\frac{d^2y}{dx^2} - 3\frac{dy}{dx} + 2y = xe^{3x} + \sin 2x$ $\left[\textbf{Ans. } y = c_1e^x + c_2e^{2x} + \frac{1}{4}e^{3x}(2x - 3) + \frac{1}{20}(3\cos 2x - \sin 2x)\right]$

6. $(D^2 - 3D + 2)y = 2e^x\cos\frac{x}{2}$ $\left[\textbf{Ans. } y = c_1e^x + c_2e^{2x} - \frac{16}{5}e^x\left(\sin\frac{x}{2} + \frac{1}{2}\cos\frac{x}{2}\right)\right]$

7. $(D^4 + D^2 + 1)y = e^{-x/2}\cos\left(\frac{\sqrt{3}}{2}x\right)$

$$\left[\textbf{Ans. } y = e^{x/2}\left(c_1\cos\frac{\sqrt{3}}{2}x + c_2\sin\frac{\sqrt{3}}{2}x\right) + e^{-x/2}\left(c_3\cos\frac{\sqrt{3}}{2}x + c_4\sin\frac{\sqrt{3}}{2}x\right) + \frac{xe^{-x/2}}{12}\left[\sqrt{3}\sin\frac{\sqrt{3}}{2}x + 3\cos\frac{\sqrt{3}}{2}x\right]\right]$$

8. $(D^2 - 4D + 3)y = e^x\cos 2x + \cos 3x$ $\left[\textbf{Ans. } y = c_1e^{3x} + c_2e^x - \frac{1}{30}(2\sin 3x + \cos 3x) - \frac{e^x}{8}(\sin 2x + \cos 2x)\right]$

9. $(D^3 + 2D^2 + D)y = x^2e^{2x} + \sin^2 x$ $\left[\textbf{Ans. } y = c_1 + (c_2 + c_3x)e^{-x} + \frac{1}{2} - \frac{1}{8}\cos x + \frac{x^2}{8}e^x + x\right]$

10. $(D^2 - 6D + 13)y = 8e^{3x}\sin 4x + 2^x$

$$\left[\textbf{Ans. } y = e^{3x}(c_1\cos 2x + c_2\sin 2x) - \frac{2}{3}e^{3x}\sin 4x + \frac{2^x}{(\log 2)^2 - 6(\log 2) + 13}\right]$$

11. $(D^2 + 4D + 3)y = e^{-x}\sin x + xe^{3x}$ $\left[\textbf{Ans. } y = c_1e^{-x} + c_2e^{-3x} - \frac{1}{5}e^{-x}(\sin x + 2\cos x) + \frac{1}{3}\left(x - \frac{4}{3}\right)\right]$

12. $(D^2 - 4D + 3)y = 2xe^{3x}\, 3e^x \cos 2x$ $\left[\textbf{Ans. } y = c_1e^x + c_2e^{3x} + e^{3x}\left(\frac{x^2}{2} - \frac{x}{2}\right) - \frac{3}{8}e^x (\cos 2x + \sin 2x)\right]$

13. $(D^3 - 3D^2 + 3D - 1)y = xe^x + e^x$ $\left[\textbf{Ans. } y = (c_1 + c_2x + c_3x^2)e^x + \frac{1}{24}(4 + x)x^3e^x\right]$

14. $(D^2 - 2D)y = e^x \sin x$ $\left[\textbf{Ans. } y = c_1 + c_2e^{2x} - \frac{1}{2}e^x \sin x\right]$

15. $\frac{d^2y}{dx^2} + 4\frac{dy}{dx} - 12y = (x - 1)e^{2x}$ $\left[\textbf{Ans. } y = c_1e^{2x} + c_2e^{-6x} + e^{2x}\left(\frac{x^2}{16} - \frac{9x}{64}\right)\right]$

16. $(D^4 + 6D^3 + 11D^2 + 6D)y = 20e^{-2x} \sin x$ [**Ans.** $y = c_1 + c_2e^{-x} + c_3e^{-2x} + c_4e^{-3x} + 2e^{2x} (\sin x - 2 \cos x)$]

13.6. L-C CIRCUIT

1. Consider an electrical circuit containing an inductance L and capacitance C.

Let i be the current and q, the charge in the condenser plate at any time, so that

(*i*) The voltage drop across 'L'

$$= L\frac{di}{dt} = L\frac{d^2q}{dt^2} \qquad \left[\because\ i = \frac{dq}{dt}\right]$$

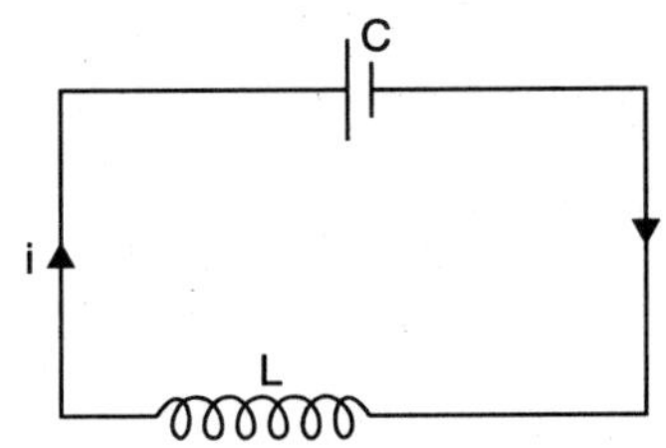

(*ii*) Voltage drop across $C = \frac{q}{C}$

As there is no applied e.m.f. in the circuit, therefore, by Kirchhoff's first law, we have

$$L\frac{d^2q}{dt^2} + \frac{q}{C} = 0$$

or

$$\frac{d^2q}{dt^2} + \frac{q}{LC} = 0$$

$$\frac{d^2q}{dt^2} + \mu^2 q = 0 \qquad \left[\text{where } \mu^2 = \frac{1}{LC}\right]$$

The equation represents free electrical oscillation of the current having period

$$\frac{2\pi}{\mu} = 2\pi\sqrt{LC}$$

2. L-R series Circuit: Let i be the current flowing in the circuit containing resistance R and I inductance L in series with voltage source E at any time 't'

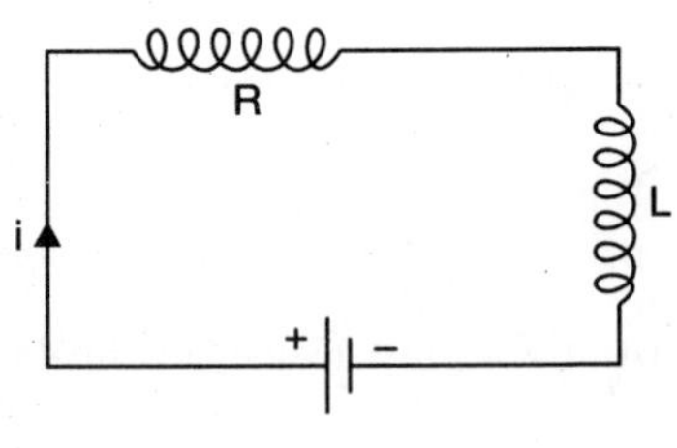

By voltage law $$Ri + L\frac{di}{dt} = E$$

$$\frac{di}{dt} + \frac{R}{L}i = \frac{E}{L}$$

This is a linear differential equation

$$\text{I.F.} = e^{\int \frac{R}{L}dt} = e^{\frac{R}{L}t}$$

In solution is $$ie^{\frac{R}{L}t} = \int \frac{E}{L}e^{\frac{R}{L}t}\,dt + A = \frac{E}{L}\cdot\frac{L}{R}\cdot e^{\frac{R}{L}t} + A = \frac{E}{L}e^{\frac{R}{L}t} + A$$

Initially at $t = 0$, $i = 0 \Rightarrow A = \frac{-E}{R}$

$\therefore$ Equation becomes $$i = \frac{E}{R}(1 - e^{\frac{-R}{L}t})$$

which shows that i increases with t and attains maximum value $\frac{E}{R}$.

3. L-R-C Series Circuit: Let i be the current in the circuit containing resistance R, inductance L, and capacitance C in the series with voltage source E at any time t.

By Voltage Law (Kirchhoff's Law)

$$Ri + L\frac{di}{dt} + \frac{q}{C} = E$$

$$L\frac{d^2q}{dt^2} + R\frac{dq}{dt} + \frac{q}{C} = E \qquad \left(\because i = \frac{dq}{dt}\right)$$

is the required differential equation which can be solved easily.

SOLVED EXAMPLES

Example 13.22. Solve the equation $L\frac{di}{dt} + Ri = E_0 \sin \omega t$, where L, R and E_0 are constants and discuss the case when t increases indefinitely.

Solution. We have

$$L\frac{di}{dt} + Ri = E_0 \sin \omega t$$

$$\frac{di}{dt} + \frac{R}{L}i = \frac{E_0}{L}\sin \omega t$$

Writing in symbolic form with $\frac{d}{dt} = D$

$$\therefore \quad \left(D + \frac{R}{L}\right)i = \frac{E_0}{L} = 0 \Rightarrow D = \frac{-R}{L}$$

Its A.E. is $D + \frac{R}{L}$

$$\text{C.F.} = Ae^{\frac{-R}{L}t}$$

$$\text{P.I.} = \frac{1}{D+\frac{R}{L}}\left(\frac{E_0 \sin \omega t}{L}\right) = \frac{E_0}{L}\frac{\left(D-\frac{R}{L}\right)\sin \omega t}{D^2 - \frac{R^2}{L^2}} = \frac{E_0}{L}\cdot\frac{1}{-\omega^2 - \frac{R^2}{L^2}}\left(D-\frac{R}{L}\right)\sin \omega t$$

$$= -\frac{E_0}{L}\cdot\frac{L^2}{\omega^2 L^2 + R^2}\left(\omega \cos \omega t - \frac{R}{L}\sin \omega t\right) = \frac{-E_0 L}{\omega^2 L^2 + R^2}\left(\frac{L\omega \cos \omega t - R \sin \omega t}{L}\right)$$

$$= \frac{-E_0}{\omega^2 L^2 + R^2}(L\omega \cos \omega t - R \sin \omega t) = -\frac{E_0}{\omega^2 L^2 + R^2}(r \sin \alpha \cos \omega t - r \cos \alpha \sin \omega t)$$

Put $\quad L\omega = r \sin \alpha$

$\quad R = r \cos \alpha$

$\therefore \quad \frac{L\omega}{R} = \tan \alpha$

and $\quad \sqrt{L^2\omega^2 + R^2} = r$

$\therefore$
$$\text{P.I.} = \frac{-E_0}{\omega^2 L^2 + R^2} r \sin (\alpha - \omega t)$$

$$= \frac{E_0}{(\omega^2 L^2 + R^2)}\sqrt{L^2\omega^2 + R^2}\sin\left(\omega t - \tan^{-1}\frac{L\omega}{R}\right)$$

$$= \frac{E_0}{(\omega^2 L^2 + R^2)}\sin\left(\omega t - \tan^{-1}\frac{L\omega}{R}\right)$$

$\therefore$ The solution is given by $i = Ae^{\frac{-R}{L}t} + \frac{E_0}{\sqrt{\omega^2 L^2 + R^2}}\sin\left(\omega t - \tan^{-1}\frac{L\omega}{R}\right)$ when t increases indefinitely,

then $Ae^{\frac{-R}{L}t}$ tends to zero.

So that
$$i = \frac{E_0}{\sqrt{\omega^2 L^2 + R^2}}\sin\left(\omega t - \tan^{-1}\frac{L\omega}{R}\right).$$

Example 13.23. Solve the differential equation $L\frac{di}{dt} + \frac{1}{C}\int i\, dt = 0$ which means that self-inductance and capacitance in a circuit neutralize each other.

Determine the constants in such a way that i_0 is the maximum current and $i = 0$ when $t = 0$.

Solution. We have $L\frac{di}{dt} + \frac{1}{C}\int i\, dt = 0$

Differentiating this equation

$$L\frac{d^2 i}{dt^2} + \frac{i}{C} = 0$$

or
$$\frac{d^2 i}{dt^2} + \frac{1}{LC}i = 0$$

Its solution is $$i = c_1 \cos \frac{t}{\sqrt{LC}} + c_2 \sin \frac{t}{\sqrt{LC}}$$

When $t = 0$, $i = 0 \Rightarrow c_1 = 0$

Thus, $$i = c_2 \sin \frac{t}{\sqrt{LC}}$$

Again, the maximum value of sine of any angle being 1, the maximum value of i in this case is
$$i_0 = c_2$$

Hence, its solution is $$i = i_0 \sin \frac{t}{\sqrt{LC}}.$$

Example 13.24. For an electric circuit (constants L, R, C), the charge q on a plate on condenser is given by

$$L\frac{d^2q}{dt^2} + R\frac{dq}{dt} + \frac{q}{C} = E \sin \omega t$$

and the current $i = \frac{dq}{dt}$, the circuit is tuned to resonance so that $\omega^2 = \frac{1}{LC}$.

If $R^2 < \frac{4L}{C}$ and if $q = i = 0$ at $t = 0$ show that

$$q = \frac{E}{\omega R}\left\{-\cos \omega t + e^{-\frac{Rt}{2L}}\left(\cos pt + \frac{R}{2L_p} \sin pt\right)\right\}$$

$$i = \frac{E}{R}\left\{\sin \omega t - \frac{1}{p\sqrt{LC}} e^{-\frac{Rt}{2L}} \sin pt\right\}$$

where $$p^2 = \frac{1}{LC} - \frac{R^2}{4L^2}.$$

Solution. The given equation in symbolic form is

$$\left(LD^2 + RD + \frac{1}{C}\right)q = E \sin \omega t, \quad \text{where } D = \frac{d}{dt}$$

$\therefore$ Its A.E. is $$LD^2 + RD + \frac{1}{C} = 0$$

$$D = \frac{-R \pm \sqrt{R^2 - \frac{4L}{C}}}{2L} = \frac{-R}{2L} \pm i\frac{\sqrt{\frac{4L}{C} - R^2}}{2L}$$

$$= \frac{-R}{2L} \pm i\sqrt{\frac{4L}{4L^2C} - \frac{R^2}{4L^2}} = \frac{-R}{2L} \pm i\sqrt{\frac{1}{LC} - \frac{R^2}{4L^2}}$$

$$= \frac{-R}{2L} \pm ip \qquad \left[\because \sqrt{\frac{1}{LC} - \frac{R^2}{4L^2}} = p\right]$$

$$\text{C.F.} = e^{\frac{-R}{2L}t}(c_1 \cos pt + c_2 \sin pt)$$

$$\text{P.I.} = \frac{1}{LD + RD + \frac{1}{C}}(E \sin \omega t) = \frac{E}{L(-\omega^2) + RD + \frac{1}{C}} \sin \omega t$$

$$= \frac{E}{RD}(\sin \omega t) \qquad \left[\because \ \omega^2 = \frac{1}{LC}\right]$$

$$= \frac{E}{R}\left(\frac{-\cos \omega t}{\omega}\right)$$

Hence, $$q = e^{\frac{-R}{2L}t}(c_1 \cos pt + c_2 \sin pt) - \frac{E}{R}\frac{\cos \omega t}{\omega}$$

Now applying the initial condition, *i.e.*, $q = 0$ when $t = 0$

$$\therefore \qquad c_1 - \frac{E}{R\omega} = 0 \ \Rightarrow \ c_1 = \frac{E}{R\omega}$$

Differentiating (*i*), we get

$$\frac{dq}{dt} = \left(\frac{-R}{2L}\right) e^{\frac{-R}{2L}t}(c_1 \cos pt + c_2 \sin pt) + e^{\frac{-R}{2L}t}(-c_1 p \sin pt + c_2 p \cos pt) - \frac{E}{R}\frac{\sin \omega t}{\omega}.\omega$$

or $$\frac{dq}{dt} = \left(\frac{-R}{2L}\right) e^{\frac{-R}{2L}t}(c_1 \cos pt + c_2 \sin pt) + e^{\frac{-R}{2L}t}(-c_1 p \sin pt + c_2 p \cos pt) - \frac{E}{R} \sin pt$$

Using $i = \frac{dq}{dt} = 0$ at $t = 0$

$$0 = \frac{-R}{2L}c_1 + c_2 p \qquad \text{or} \qquad c_2 = \frac{1}{p}\frac{Rc_1}{(2L)} = \left(\frac{R}{2pL}\right)\left(\frac{E}{R\omega}\right)$$

$$\therefore \qquad c_2 = \frac{E}{2\omega Lp} \text{ and } c_1 = \frac{E}{R\omega}$$

Thus, $$q = \frac{E}{R\omega}\left(-\cos \omega t + e^{\frac{-Rt}{2L}}(\cos pt + \frac{R}{2Lp} \sin pt\right)$$

Differentiating q, we get the value as required,

$$\frac{dq}{dt} = \frac{E}{R\omega}\left[\omega \sin \omega t + \left(\frac{-R}{2L}\right)e^{\frac{-Rt}{2L}}(\cos pt + \frac{R}{2Lp} \sin pt)\right] + e^{\frac{-Rt}{2Lp}}\left(-p \sin pt + \frac{Rp}{2Lp} \cos pt\right)$$

$$i = \frac{E}{R\omega}\left[\omega \sin \omega t - e^{\frac{-Rt}{2L}}\left(\frac{-R}{2L} \cos pt - \frac{R^2 \sin pt}{4L^2 p} - p \sin pt + \frac{R}{2L} \cos pt\right)\right]$$

$$= \frac{E}{R\omega}\left[\omega \sin \omega t - e^{\frac{-Rt}{2L}}\left(\frac{R^2}{4L^2 p} \sin pt + p \sin pt\right)\right]$$

$$= \frac{E}{R\omega}\left[\omega \sin \omega t - e^{\frac{-Rt}{2L}}\left(\frac{R^2}{4L^2 p} + p\right)\sin pt\right]$$

$$= \frac{E}{R\omega}\left[\omega \sin \omega t - \frac{e^{\frac{-Rt}{2L}}}{p}\left(\frac{R^2}{4L^2} + p^2\right)\sin pt\right]$$

$$= \frac{E}{R\omega}\left[\omega \sin \omega t - \frac{e^{\frac{-Rt}{2L}}}{p}\left(\frac{1}{LC}\right)\sin pt\right] \qquad \left[\because \frac{R^2}{4L^2} + p^2 = \frac{1}{LC}\right]$$

$$= \frac{E}{R}\left[\sin \omega t - \frac{e^{\frac{-Rt}{2L}}}{\omega p}\cdot\frac{1}{LC}\sin pt\right]$$

$$i = \frac{E}{R}\left[\sin \omega t - \frac{e^{\frac{-Rt}{2L}}}{p}\cdot\frac{1}{\sqrt{LC}}\sin pt\right] \qquad \left[\because \omega^2 = \frac{1}{LC}\right]$$

is the required solution.

EXERCISE 13.6

1. When a resistance R ohms is connected in series with an inductance L henries with an e.m.f. of E volts, the current i amperes at any time t is given by

$$L\frac{di}{dt} + Ri = E$$

If $E = 10 \sin t$ volts and $i = 0$ when $t = 0$, find i as a function of t.

$$\left[\textbf{Ans. } \frac{10}{L^2 + R^2}\left(R \sin t - L \cos t + Le^{\frac{-Rt}{L}}\right)\right]$$

2. An emf $E \sin pt$ is applied at $t = 0$ to a circuit containing a condenser C and inductance L in series. The current i satisfies the equation

$$L\frac{di}{dt} + \frac{1}{C}\int i\, dt = E \sin pt$$

If $p^2 = \frac{1}{LC}$ and initially the current i and charge q are zero, show that the current in the circuit at time t is given by $\frac{E}{2L} t \sin pt$.

3. A condenser of capacity C, resistance R and inductance L are connected in series. The equation giving the current at any time t is

$$\frac{d^2i}{dt^2} + \frac{R}{L}\cdot\frac{di}{dt} + \frac{i}{CL} = 0 \quad \text{where } CR^2 < 4L$$

Show that the amplitudes of successive vibration are in G.P.

4. The equation of e.m.f. in terms of current i for an electrical circuit having resistance R and condenser of capacity C, in series, is

$$E = Ri + \int \frac{i}{C}\, dt$$

Find the current i at any time 't' when

$$E = E_m \sin \omega t$$

$$\left[\textbf{Ans. } i = ke^{\frac{-t}{RC}} + \frac{\omega C E_m}{\sqrt{1 + R^2C^2\omega^2}} \sin(\omega t + \theta), \text{ where } \theta = \cot^{-1}(RC\omega)\right]$$

5. Show that $L\frac{d^2q}{dt^2} + \frac{1}{C}q = E \sin nt$, is the differential equation for electric ahrge q is an electric circuit consisting of an inductance L, capacity C and an alternating e.m.f. $E \sin nt$ applied in series. If the initial current and charge on the condenser are zero, prove that the current i is given by

$$i = \frac{nE}{L(n^2 - \omega^2)}(\cos \omega t - \cos nt) \text{ where } \omega^2 = \frac{1}{LC} \text{ if } n = \omega \text{ prove the current } i = \frac{Et}{2L} \sin \omega t.$$

6. An unchanged condenser of capacity C is charged by applying an e.m.f. of $E \sin\left(\frac{t}{\sqrt{LC}}\right)$ through loads of self-inductance L and negligible resistance. Prove that at time t, the change on one of the plates is

$$\frac{EC}{2}\left(\sin \frac{1}{\sqrt{LC}} - \frac{t}{\sqrt{LC}} \cos \frac{t}{\sqrt{LC}}\right)$$

❑❑❑

UNIT - VI

14. Numerical Methods

CHAPTER 14

Numerical Methods

INSIDE THIS CHAPTER

14.1. INTRODUCTION

In applied mathematics, the most frequently occurring problem is to find the roots of equations of the form.

$$f(x) = 0.$$

The equation $f(x) = 0$ is called *Algebraic* or *transcendental* according as $f(x)$ is purely a polynomial in x or contains some other functions such as logarithmic, exponential and trigonometric functions etc. For example, $1 + \cos x - 5x$, $x \tan x - \cosh x$, $e^{-x} - \sin x$ are transcendental functions. A polynomial in x of degree n is an expression of the form $f(x) = a_0 x^n + a_1 x^{n-1} + ... + a_n$, where a's are constant and n is a positive integer. The zeros or roots of the polynomial $f(x)$ are those values of x for which $f(x)$ is zero. Geometrically, if the graph of $f(x)$ crosses the x-axis at the point $x = a$, then $x = a$ is a root of the equation $f(x) = 0$. We conclude that 'a' is a root of the equation $f(x) = 0$ if and only if $f(a) = 0$.

By finding the solution of an equation $f(x) = 0$, we mean to find zeros of $f(x)$. In this chapter, we shall discuss graphical and some numerical methods for the solution of equations of the form $f(x) = 0$. Here $f(x)$ may be algebraic or transcendental or combination of both.

14.2. BISECTION (OR BOLZANO) METHOD

This method of solving an algebraic and transcendental equation consists in locating the root of the equation $f(x) = 0$, between two numbers say a and b such that $f(x)$ is continuous for $a \le x \le b$ and $f(a)$ and $f(b)$ are of opposite signs so that the product $f(a) . f(b) < 0$, i.e., the curve crosses the x-axis between $x = a$ and $x = b$. Then the first approximation to the root is

$$x_1 = \frac{a+b}{2}$$

If $f(x_1) = 0$, then x_1 is a root of $f(x) = 0$, otherwise root lies between a and x_1 or x_1 and b according as $f(x_1)$ is +ve or –ve.

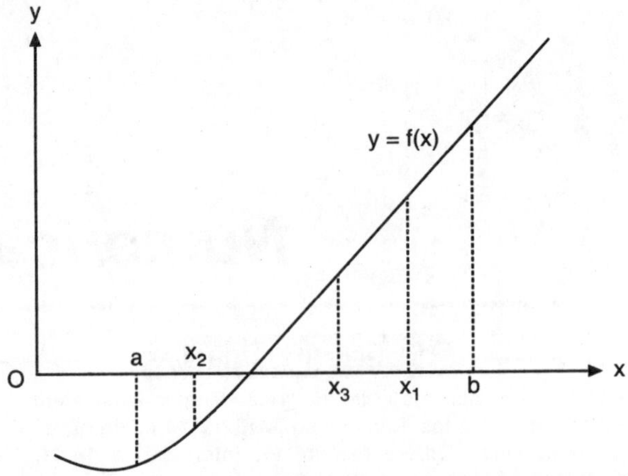

Then as before we bisect the interval and continue the process until the root is found to desired accuracy. In the figure above, $f(x_1)$ is +ve, therefore root lies between a and x_1. Then, the second approximation to the root is

$$x_2 = \frac{a + x_1}{2}$$

If $f(x_2)$ is –ve, the root lies between x_1 and x_2 and the third approximation to the root is

$$x_3 = \frac{x_1 + x_2}{2}$$

We repeat the process until the root is known to the desired accuracy.

SOLVED EXAMPLES

Example 14.1. Find a root of the equation $x^3 - 4x - 9 = 0$ using bisection method in four stages.

Solution.

Let $f(x) = x^3 - 4x - 9$

Since, $f(2.706) = -0.009488$ which is negative

and $f(2.707) = -0.008487$ which is positive

Hence root lies between 2.706 and 2.707.

First Approximation: $x_1 = \dfrac{2.706 + 2.707}{2} = 2.7065$

Now, $f(x_1) = -0.0005025$

$\therefore$ $f(2.7065)$ is negative.

Hence root lies between 2.7065 and 2.707.

Second Approximation: $x_2 = \dfrac{2.7065 + 2.707}{2} = 2.70675$

Now, $f(x_2) = 0.003992$

$\therefore$ $f(2.70675)$ is positive.

Hence root lies between 2.7065 and 2.70675.

Third Approximation: $x_3 = \dfrac{2.7065 + 2.70675}{2} = 2.706625$

Now, $f(x_3) = 0.001744$

$\therefore$ $f(2.706625)$ is positive.

Hence root lies between 2.7065 and 2.706625.

Fourth Approximation: $x_4 = \dfrac{2.7065 + 2.706625}{2} = 2.7065625$

Hence root the root is 2.7065625 correct to three decimal places.

Example 14.2. Find the real root of the equation $x \log_{10} x = 1.2$ by bisection method correct to four decimal places.

Solution.

Let $f(x) = x \log_{10} x - 1.2$

Since, $f(2.74) = -0.000563$ which is negative

and $f(2.75) = 0.0081649$ which is positive

Hence a root lies between 2.74 and 2.75.

First Approximation: $x_1 = \dfrac{2.74 + 2.75}{2} = 2.745$

Now, $f(x_1) = f(2.745) = 0.003798$

$\therefore$ $f(2.745)$ is positive.

Hence root lies between 2.74 and 2.745.

Second Approximation: $x_2 = \dfrac{2.74 + 2.745}{2} = 2.7425$

Now, $f(x_2) = f(2.7425) = 0.001617$

$\therefore$ $f(2.7425)$ is positive.

Hence root lies between 2.74 and 2.7425.

Third Approximation: $x_3 = \dfrac{2.74 + 2.7425}{2} = 2.74125$

Now, $f(x_3) = f(2.74125) = 0.0005267$

$\therefore$ $f(2.74125)$ is positive.

Hence root lies between 2.74 and 2.74125.

Fourth Approximation: $x_4 = \dfrac{2.74 + 2.74125}{2} = 2.740625$

Now, $f(x_4) = f(2.740625) = -0.00001839$

$\therefore$ $f(2.740625)$ is negative.

Hence root lies between 2.740625 and 2.74125.

Fifth Approximation: $x_5 = \dfrac{2.740625 + 2.74125}{2} = 2.7409375$

Now, $f(x_5) = f(2.7409375) = 0.000254$

$\therefore$ $f(2.7409375)$ is positive.

Hence root lies between 2.740625 and 2.749375.

Sixth Approximation: $$x_6 = \frac{2.740625 + 2.7409375}{2} = 2.74078125$$

Now, $$f(x_6) = f(2.74078125) = 0.0001178$$

$\therefore$ $f(2.74078125)$ is positive.

Hence root lies between 2.740625 and 2.74078125.

Seventh Approximation: $$x_7 = \frac{2.740625 + 2.74078125}{2} = 2.740703125$$

Now, $$f(x_7) = f(2.740703125) = 0.00004973$$

$\therefore$ $f(2.740703125)$ is positive.

Hence root lies between 2.740625 and 2.740703125.

Eighth Approximation: $$x_8 = \frac{2.740625 + 2.740703125}{2} = 2.740664063$$

Now, $$f(x_8) = f(2.740664063) = 0.00001567$$

$\therefore$ $f(2.740664063)$ is positive.

Hence root lies between 2.740625 and 2.740664063.

Ninth Approximation: $$x_9 = \frac{2.740625 + 2.740664063}{2} = 2.740644532$$

Since x_8 and x_9 are same upto four decimal places hence the approximate real root is 2.7406.

Example 14.3. Find the positive root of $x - \cos x = 0$ by bisection method.

Solution.

$$\begin{aligned} f(x) &= x - \cos x \\ f(0.05) &= 0.5 - \cos(0.5) = -0.37758 \text{ which is negative} \\ f(1) &= 1 - \cos(1) = 0.45970 \text{ which is positive.} \end{aligned}$$

Hence the root lies between 0.5 and 1.

First Approximation: $$x_1 = \frac{0.5+1}{2} = 0.75$$

Now, $$\begin{aligned} f(0.75) &= 0.75 - \cos(0.75) \\ &= 0.018311 \end{aligned}$$

$\therefore$ $f(0.75)$ is positive.

Hence the root lies between 0.5 and 0.75.

Second Approximation: $$x_2 = \frac{0.5+0.75}{2} = 0.625$$

Now, $$\begin{aligned} f(0.625) &= 0.625 - \cos(0.625) \\ &= -0.18596 \end{aligned}$$

$\therefore$ $f(0.625)$ is negative.

Hence the root lies between 0.625 and 0.750.

Third Approximation: $$x_3 = \frac{0.625+0.750}{2} = 0.6875$$

Now, $$\begin{aligned} f(0.6875) &= 0.6875 - \cos(0.6875) \\ &= -0.08533 \end{aligned}$$

$\therefore$ $f(0.6875)$ is negative.

Hence the root lies between 0.6875 and 0.75.

Fourth Approximation: $x_4 = \dfrac{0.6875 + 0.75}{2} = 0.71875$

Now, $f(0.71875) = 0.71875 - \cos(0.71875)$

$= -0.033879$

$\therefore$ $f(0.71875)$ is negative.

Hence the root lies between 0.71875 and 0.75.

Fifth Approximation: $x_5 = \dfrac{0.71875 + 0.75}{2} = 0.73438$

Now, $f(0.73438) = -0.0078664$

$\therefore$ $f(0.73438)$ is negative.

Hence the root lies between 0.73438 and 0.75.

Sixth Approximation: $x_6 = \dfrac{0.73438 + 0.75}{2} = 0.742190$

Now, $f(0.74219) = 0.0051999$

$\therefore$ $f(0.74219)$ is positive.

Hence the root lies between 0.73438 and 0.742190.

Seventh Approximation: $x_7 = \dfrac{0.73438 + 0.742190}{2} = 0.73829$

Now, $f(0.73829) = -0.0013305$

$\therefore$ $f(0.73829)$ is negative.

Hence the root lies between 0.73829 and 0.74219.

Eighth Approximation: $x_8 = \dfrac{0.73829 + 0.74219}{2} = 0.7402$

Now, $f(0.7402) = 0.7402 - \cos(0.7402) = 0.0018663$

$\therefore$ $f(0.3402)$ is positive.

Hence the root lies between 0.73829 and 0.7402.

Ninth Approximation: $x_9 = \dfrac{0.73829 + 0.7402}{2} = 0.73925$

Now, $f(0.73925) = 0.00027593$

$\therefore$ $f(0.73925)$ is positive.

Hence the root lies between 0.73829 and 0.73925.

Tenth Approximation: $x_{10} = \dfrac{0.73829 + 0.73925}{2} = 0.7388$

Hence the root is 0.7388 correct to three decimal places.

EXERCISE 14.1

1. Find a root of the equation $x^3 - x - 4$ between 1 and 2 to three places of decimal by bisection method.

[**Ans.** 1.796]

2. Using bisection method determine a real root of the equation
$$f(x) = 8x^3 - 2x - 1.$$ [Ans. 0.66]
3. Compute one positive root of $2x - 3\sin x - 5 = 0$ by bisection method correct to three significant figures. [Ans. 2.88]
4. Compute one root of $e^x - 3x = 0$ correct to two decimal places, using bisection method. [Ans. 1.51]
5. Find the root of $\tan x + x = 0$ upto two decimal places which lies between 2 and 2.1. [Ans. 2.03]
6. Find square root of 12 to four decimal places by bisection method upto five iterations. [Ans. 3.4687]
7. Compute one root of $x + \log x - 2 = 0$ correct to two decimal places which lies between 1 and 2. [Ans. 1.56]
8. Find the root of the equation $x^4 - x - 10 = 0$, using bisection method. [Ans. 1.8125]
9. Solve the equation $x - \exp\left(\frac{1}{x}\right) = 0$, by bisection method. [Ans. 1.718]
10. Find real root of $\cos x + xe^x$ correct to three decimal places by bisection method. [Ans. 0.517]

14.3. REGULA-FALSI METHOD (METHOD OF FALSE POSITION)

Consider the equation $f(x) = 0$ and let a and b be two numbers such that $f(a)$ and $f(b)$ are of opposite signs, i.e., the root lies between a and b (Discrete's Rule of Signs). Then the curve $y = f(x)$ crosses the x-axis, say at point P as shown in the figure. The abscissa of the point P gives the root of the equation $f(x) = 0$. This method is based upon the principle that any portion of a smooth curve is practically straight for a short distance. Hence we assume that the graph of $y = f(x)$ is a straight line between points $A[a, f(a)]$ and $B[b, f(b)]$. the points A and B are on opposite sides of the x-axis. This equation of the line AB is given by

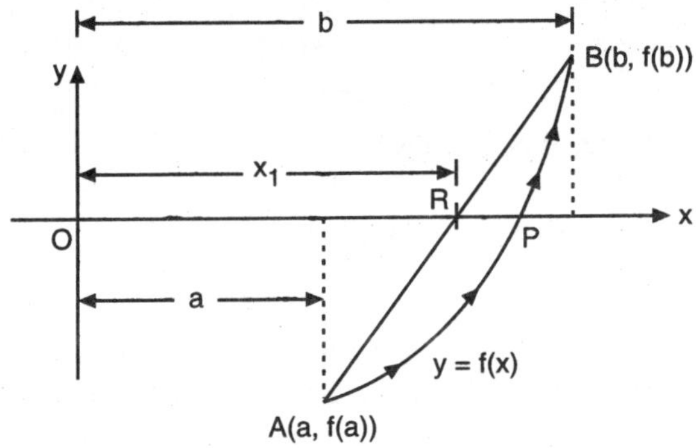

$$\frac{y - f(a)}{x - a} = \frac{f(b) - f(a)}{b - a}$$

The x-coordinate of the point of the intersection of line AB and the axis of x gives an approximate value of the desired root which is obtained by putting $y = 0$ in the equation of the straight line AB.

$$\therefore \qquad x_1 = OP_1 = \frac{af(b) - bf(a)}{f(b) - f(a)}$$

where $x = x_1$ is the first approximation to the root of the equation $f(x) = 0$.

Further we wish to know whether the root lies between a and x_1 and b. For this if signs of $f(a)$ and $f(x_1)$ are of opposite nature, the better root will lie between a and x_1 otherwise it lies between x_1 and b. In any case start the above procedure again and repeat it till the best approximate root is located.

Working rule to find out the root of an equation $f(x)$ by Regula-Falsi Method.

1. Obtain the equation $f(x) = 0$.
2. Choose the value of $x = a$ and $x = b$ such that $f(a)$ and $f(b)$ are of opposite signs.
3. The first approximation x_1 of the root is given by

$$x_1 = \frac{af(b) - bf(a)}{f(b) - f(a)}$$

4. If $f(a)$ and $f(x_1)$ are of opposite signs, then the root lies between 'a' and 'x_1', otherwise it lies between x_1 and b.
5. Find the next approximation by applying the formula in Step 2.

Repeat the method till the desired accuracy in obtained.

SOLVED EXAMPLES

Example 14.4. Find a real root of the equation $x^3 - 2x - 5 = 0$ by the method of false position correct to three decimal places.

Solution.

Let $f(x) = x^3 - 2x - 5 = 0;\ a = 2,\ b = 3$

Now $f(a) = f(2) = 2^3 - 2(2) - 5 = -1$

and $f(b) = f(3) = 3^3 - 2(3) - 5 = 16$

Therefore, the root of $f(x) = 0$ lies in between 2 and 3.

First Approximation:

$$x_1 = \frac{af(b) - bf(a)}{f(b) - f(a)} = \frac{2(16) - 3(1)}{16 - (-1)}$$

$$= \frac{35}{17} = 2.0588$$

Now, $f(x_1) = f(2.0588) = (2.0588)^3 - 2(2.0588) - 5$
$= -0.391$

$\therefore$ $f(2.0588)$ is negative.

Hence the root lies between 2.0588 and $b = 3$.

Second Approximation:

$$x_2 = \frac{x_1 f(b_1) - bf(x_1)}{f(b) - f(x_1)} = \frac{2.0588 \times 16 - 3(-0.391)}{16 - (-0.391)}$$

$$= \frac{34.1138}{16.391} = 2.08125$$

Now, $f(x_2) = f(2.08125) = (2.08125)^3 - 2(2.08125) - 5$
$= -0.147$

$\therefore$ $f(2.08125)$ is negative.

Hence the root lies between $x_2 = 2.08125$ and $b = 3$.

Third Approximation:

$$x_3 = \frac{x_2 f(b) - bf(x_2)}{f(b) - f(x_2)} = \frac{2.08125 \times 16 - 3(-0.147)}{16 - (-0.147)}$$

$$= \frac{33.741}{16.147} = 2.0896$$

Now, $f(x_3) = f(2.0896) = (2.0896)^3 - 2(2.0896) - 5$
$= -0.0581$

$\therefore$ $f(2.0896)$ is negative.

Hence the root lies between $x_3 = 2.0896$ and $b = 3$.

Fourth Approximation: $x_4 = \dfrac{x_3 f(b) - bf(x_3)}{f(b) - f(x_3)} = \dfrac{2.0896 \times 16 - 3(-0.051)}{16 - (-0.051)}$

$$= \frac{33.5989}{16.0551} = 2.0927$$

Now, $f(x_4) = f(2.0927) = (2.0927)^3 - 2(2.0.927) - 5 = -0.0206$

$\therefore$ $f(2.0927)$ is negative.

Hence the root lies between $x_4 = 2.0927$ and $b = 3$.

Fifth Approximation: $x_5 = \dfrac{x_4 f(b) - bf(x_4)}{f(b) - f(x_4)} = \dfrac{2.0927 \times 16 - 3(-0.0206)}{16 - (-0.0206)}$

$$= \frac{33.545}{16.0206} = 2.0939$$

Now, $f(x_5) = f(2.0939) = (2.0939)^3 - 2(2.0.939) - 5 = -0.00726$

$\therefore$ $f(2.0939)$ is negative.

Hence the root lies between $x_5 = 2.0939$ and $b = 3$.

Sixth Approximation: $x_6 = \dfrac{x_5 f(b) - bf(x_5)}{f(b) - f(x_5)} = \dfrac{2.0939 \times 16 - 3(-0.00726)}{16 - (-0.00726)}$

$$= \frac{33.52418}{16.00726} = 2.0943$$

Now, $f(x_6) = f(2.0943) = (2.0943)^3 - 2(2.0943) - 5$

$$= -0.0028$$

$\therefore$ $f(2.0943)$ is negative.

Hence the root lies between $x_6 = 2.0943$ and $b = 3$.

Seventh Approximation: $x_7 = \dfrac{x_6 f(b) - bf(x_6)}{f(b) - f(x_6)} = \dfrac{2.0943 \times 16 - 3(-0.0028)}{16 - (-0.0028)}$

$$= \frac{33.5172}{16.0028} = 2.0944$$

Since x_6 and x_7 agree upto 3 decimal places hence the required root is 2.094 correct upto three decimal places.

Example 14.5. Using Regula-Falsi Method, compute the real root of the equation.

$$\cos x - x e^x = 0.$$

Solution.

Let $f(x) = \cos x - x e^x$

$f(0) = 1, f(1) = \cos(1) - e = -2.17798$ [In cos (1), 1 is in radian]

$\therefore$ The root lies between 0 and 1.

First Approximation:

Let $a = 0$ and $b = 1$.

$\therefore$ $$x = \frac{af(b) - bf(a)}{f(b) - f(a)} = \frac{0 - 1 \times 1}{-2.17798 - 1} = \frac{1}{3.17798} = 0.31466.$$

Now $f(0.31466) = \cos(0.31466) - (0.31466)\, e^{0.31466}$

$$= 0.9509 - 0.4310 = 0.5199.$$

The root lies between 0.31466 and 1.

Second Approximation:

Let $a = 0.31466$ and $b = 1$.

$$f(0.31466) = \cos(0.31466) - 0.31466\, e^{0.31466}$$
$$= 0.9509 - 0.4310 = 0.5199$$
$$f(1) = -2.17798$$
$$x_2 = \frac{af(b) - bf(a)}{f(b) - f(a)} = \frac{0.31466 \times -2.17798 - 1 \times (0.5199)}{-2.17798 - 0.5199}$$
$$= \frac{-1.2052}{-2.6978} = 0.4467.$$

Now, $f(0.4467) = 0.2036.$

The root lies between 0.4467 and 1

Third Approximation: $a = 0.4467,\ b = 1$

$$x_3 = \frac{0.4467 \times (-2.17798) - 1 \times (0.2036)}{-2.17798 - 0.2036} = 0.4940.$$

Now $f(0.4940) = 0.0708.$

The root lies between 0.4940 and 1

Fourth Approximation: $a = 0.4940,\ b = 1$

$$x_4 = \frac{0.4940 \times (-2.17798) - 1 \times (0.0708)}{-2.17798 - 0.0708} = 0.5099$$

$f(0.5099) = 0.0237$

The root lies between 0.5099 and 1

Fifth Approximation: $a = 0.5099,\ b = 1$

$$x_5 = \frac{0.5099 \times (-2.17798) - 1 \times (0.0237)}{-2.17798 - 0.0237} = 0.5151.$$

Now $f(0.5151) = 0.0080.$

The root lies between 0.5151 and 1

Sixth Approximation: $a = 0.5151,\ b = 1$

$$x_6 = \frac{0.5151 \times (-2.17798) - 1 \times (0.0080)}{-2.17798 - 0.0080} = 0.5169.$$

$\therefore$ **Since x_5 and x_6 agree upto two decimal places, the required root correct to 2 decimal places** is 0.51.

EXERCISE 14.2

1. Find an approximate real root of the equation $x^3 - 5x + 3 = 0$, correct to three decimal places by the method of Regula-Falsi. **[Ans.** 0.6566]
2. Find the root of the equation $2x - \log_{10} x = 7$, which lies between 3.5 and by the method of false position. **[Ans.** 3.78916]
3. Solve $x^3 - 9x + 1 + 0$ for the root lying between 2 and 4 by the method of false position. **[Ans.** 2.7391]
4. Find the real approximation root of the equation $xe^x - 2 = 0$ by Regula-Falsi method. **[Ans.** 0.52605]
5. Solve $2x = 1 - 2 \log_{10} x$ by the method of false position. **[Ans.** 0.672]
6. Use Regula-Falsi method to solve $x = \cos x$. **[Ans.** 0.7391]
7. Find the cube root of 11 by Regula-Falsi method. **[Ans.** 2.224]

8. Apply false position method to find the smallest positive root of the equation $x - e^{-x} = 0$. **[Ans.** 0.567]
9. Find the real root of the equation $x^6 - x^4 - x^3 - 1 = 0$ correct to four decimal places using the method of false position (Regula-Falsi method). **[Ans.** 1.7365]

14.4. NEWTON-RAPHSON METHOD

Given an approximate value of a root of an equation, a better and closer approximation to the root can be found by using an iterative process called Newton's method or Newton-Raphson method.

Let x_0 be an approximate value of a root of the equation $f(x) = 0$

Let x_1 be the exact root nearer to x_0

Then $x_1 = x_0 + h$ where h is very small, positive or negative.

$$\therefore \quad f(x_1) = f(x_0 + h) = 0 \text{ since } x_1 \text{ is the exact root of } f(x) = 0$$

By Taylor's expansion

$$f(x_1) = f(x_0 + h) = f(x_0) + hf'(x_0) + \frac{h^2}{2!} f''(x_0) + \ldots = 0$$

i.e., if h is small neglecting $h^2, h^3, \ldots$, etc., we get

$$f(x_0) + hf'(x_0) = 0$$

$$\therefore \quad h = -\frac{f(x_0)}{f'(x_0)} \quad \text{if } f'(x_0) \neq 0$$

$$\therefore \quad x_1 = x_0 + h = x_0 - \frac{f(x_0)}{f'(x_0)} \text{ approximately}$$

x_1 is a better approximate root than x.

Starting with this x_1, we get

$$x_2 = x_1 - \frac{f(x_1)}{f'(x_1)} \text{ which is still better.}$$

Continuing like this, we iterate this process until $| x_{r+1} - x_r |$ is less than quantity desired.

$$\therefore \quad x_{n+1} = x_n - \frac{f(x_n)}{f'(x_n)}, n = 0, 1, 2, \ldots$$

This is the iterative formula of Newton-Raphson method.

SOLVED EXAMPLES

Example 14.6. Find the approximate value for the real root of $x \log_{10} x - 1.2 = 0$, correct to five decimal places by Newton-Raphson method.

Solution. $f(x) = x \log_{10} x - 1.2$

Therefore, $f'(x) = \log_{10} x + x \cdot \frac{1}{x} \cdot \log_{10} e = \log_{10} x + \log_{10} e$

$= \log_{10} x + 0.43429$ $[\because e = 2.71825]$

Now, $f(2) = 2 \log_{10} 2 - 1.2 = -0.59794$ (–ve)

and $f(3) = 3 \log_{10} 3 - 1.2 = 0.23136$ (+ve)

Hence the root of $f(x) = 0$ lies between 2 and 3. We notice that out of $| f(2) |$ and $| f(3) |$, $| f(3) |$ is minimum.

We take $x_0 = 3$ so as to minimize the number of iterations.

Now, putting $n = 0$ in the Newtons-Raphson formula.

$$x_{n+1} = x_n - \frac{f(x_n)}{f'(x_n)} = x_n - \frac{x_n \log_{10} x_n - 1.2}{\log_{10} x_n + 0.43429} = \frac{0.43429\, x_n + 1.2}{\log_{10} x_n + 0.43429}$$

The first approximation is $x_1 = \dfrac{0.43429\, x_0 + 1.2}{\log_{10} x_0 + 0.43429} = \dfrac{0.43429 \times 3 + 1.2}{\log_{10} 3 + 0.43429}$

or $x_1 = \dfrac{1.30287 + 1.2}{0.47712 + 0.43429} = \dfrac{2.50287}{0.91141} = 2.74615$.

The second approximation is $x_2 = \dfrac{0.43429\, x_1 + 1.2}{\log_{10} x_1 + 0.43429} = \dfrac{1.19262 + 1.2}{0.43872 + 0.43429}$

$$= \frac{2.39263}{0.87301} = 2.74066$$

The third approximation is $x_3 = \dfrac{0.43429\, x_2 + 1.2}{\log_{10} x_2 + 0.43429} = \dfrac{1.19023 + 1.2}{0.43785 + 0.43429}$

$$= \frac{2.39025}{0.87215} = 2.74065 \text{ (approx.)}$$

The fourth approximation is $x_4 = \dfrac{0.43429\, x_3 + 1.2}{\log_{10} x_3 + 0.43429} = \dfrac{1.19023 + 1.2}{0.43785 + 0.43429} = 2.74065$

So $x_4 = x_3$, we stop at this stage and hence the required root correct to five decimal places is 2.74065.

Remark. If we take $x_0 = 2$, five number of iterations gives the value of the root correct to the desirable accuracy.

Example 14.7. Use Newton's formula to prove that square root of N can be obtained by the recursion formula.

$$x_{n+1} = x_n \left(1 - \frac{x_n^2 - N}{2N}\right)$$

Hence find the square root of 26.

Solution.

Let us take $f(x) = 1 - \dfrac{N}{x^2}$, then $f(x) = 0$ gives $x = \sqrt{N}$, and $f'(x) = \dfrac{2N}{x^3}$.

Now, the Newton-Raphson formula

$$x_{n+1} = x_n - \frac{f(x_n)}{f'(x_n)}$$

gives

$$x_{n+1} = x_n - \frac{\left(\dfrac{x_n^2 - N}{x_n^2}\right)}{\left(\dfrac{2N}{x_n^2}\right)} = x_n - \frac{x_n^3}{2N}\left(\frac{x_n^2 - N}{x_n^2}\right)$$

or $$x_{n+1} = x_n - \left(1 - \frac{x_n^2 - N}{2N}\right)$$

To find the square root of 26, we put $n = 0$; $x_0 = 5.0$, $N = 26$, to get the first approximation x_1 as given by

$$x_1 = 5\left(1 - \frac{25-26}{52}\right) = 5.096.$$

The second approximation x_2 is given by

$$x_2 = 5.096\,(1 + 0.000592) = 5.099$$

Hence $\sqrt{26} \cong 5.099.$

Example 14.8. Find a positive value of $(17)^{1/3}$ correct to four decimal place by Newton-Raphson method.

Solution. The iterative formula is

$$x_{n+1} = \frac{1}{3}\left(2x_n + \frac{a}{x_n^2}\right) \quad \ldots (i)$$

Here $a = 17$

Take $x_0 = 2.5$ $\left[\because \sqrt[3]{8} = 2 \text{ and } \sqrt[3]{27} = 3\right]$

$$x_1 = \frac{1}{3}\left(2x_0 + \frac{17}{x_0^2}\right) = \frac{1}{3}\left(5 + \frac{17}{6.25}\right) = 2.5733$$

Putting $n = 1$ eqn. (i), we get

$$x_2 = \frac{1}{3}\left(2x_1 + \frac{17}{x_1^2}\right) = \frac{1}{3}\left(5.1466 + \frac{17}{6.6220}\right) = 2.5713$$

Again putting $n = 2$ in eqn. (ii), we get

$$x_3 = \frac{1}{3}\left(2x_2 + \frac{17}{x_2^2}\right) = \frac{1}{3}\left(5.1426 + \frac{17}{6.61158}\right) = 2.57128$$

Putting $n = 3$ in eqn. (iii), we get

$$x_4 = \frac{1}{3}\left(2x_3 + \frac{17}{x_3^2}\right) = \frac{1}{3}\left(5.14256 + \frac{17}{6.61148}\right) = 2.57128$$

Since x_3 and x_4 agree to four places of decimal hence the required root is 2.5713 correct to four decimal places.

EXERCISE 14.3

Use Newton-Raphson method to obtain a root, correct to three decimal place of the following equations:

1. $x^3 - 3x - 5 = 0$. **[Ans.** 2.279]
2. $\sin x = 1 - x$. **[Ans.** 0.511]
3. $x - \cos x = 0$. **[Ans.** 0.739]
4. $x + \log x = 2$. **[Ans.** 1.756]

5. $\sin x = \frac{1}{2}x$. [**Ans.** 1.896]

6. $\tan x - x = 0$. [**Ans.** 4.4934]

7. $\cos x - x^2 - x = 0$. [**Ans.** 0.55000]

8. Using Newton-Raphson method, find the real root of $x \sin x + \cos x = 0$ which is near $x = \pi$ correct to 3 decimal places. [**Ans.** 2.798]

9. Use Newton's formula to prove that square root of N can be obtained by the recursion formula.

$$x_{i+1} = x_i \left(1 - \frac{x_i^2 - N}{2N}\right)$$

10. Show that the square root of $N = AB$ is given by

$$\sqrt{N} = \frac{S}{4} + \frac{N}{S}, \text{ where } S = A + B.$$

11. Under what conditions, Newton-Raphson method becomes linearly convergent? Explain.

14.5. GAUSS ELIMINATION METHOD

In this method, the unknowns are elimianted successively by transforming the given system and the system is reduced to an upper triangle system from which the unknown are found by back substitution.

Consider the equations

$$a_1x + b_1y + c_1z = d_1 \quad \text{...(i)}$$
$$a_2x + b_2y + c_2z = d_2 \quad \text{...(ii)}$$
$$a_3x + b_3y + c_3z = d_3 \quad \text{...(iii)}$$

where x, y, z are unknowns.

Step 1. To elimiante x from (*ii*) and (*iii*)

Assuming $a_1 \neq 0$, x can be elimianted from (*ii*), by subtracting $\left(\frac{a_1}{a_2}\right)$ times (*i*) from (*ii*). Similarly x can be eliminated from (*iii*), by subtracting $\left(\frac{a_3}{a_1}\right)$ times (*i*) from (*iii*). Thus, we get the new system

$$a_1x + b_1y + c_1z = d_1 \quad \text{...(iv)}$$
$$b_2'y + c_2'z = d_2' \quad \text{...(v)}$$
$$b_3'y + c_3'z = d_3' \quad \text{...(vi)}$$

Here equation (*iv*) is called the pivotal equation and a_1 is called the first pivot.

Step 2. To elimiante y from (*vi*)

Assuming $b_2' \neq 0$, y can be eliminated from (*vi*), by subtracting (b_3'/b_2') times (*v*) from (*vi*). Thus we get the new system as

$$a_1x + b_1y + c_1z = d_1 \quad \text{...(vii)}$$
$$b_2'y + c_2'z = d_2' \quad \text{...(viii)}$$
$$c_3''z = d_3'' \quad \text{...(ix)}$$

Here (*viii*) equation is called the **pivotal equation** and b'_2 is the **new pivot.**

Step 3. To evaluate the unknowns

The values of x, y, z are found from the system containing equations (*vii*), (*viii*) and (*ix*) by back substitution.

Note. This method fails if any one of the pivots a_1, b'_2, or c''_3 becomes zero. In such cases we rewrite the equations in a different order so that the pivots are non-zero.

SOLVED EXAMPLES

Example 14.9. Solve the following system by Gauss elimination method

$$2x + y + z = 10$$
$$3x + 2y + 3z = 18$$
$$x + 4y + 9z = 16.$$

Solution. We have

$$2x + y + z = 10 \quad ...(i)$$
$$3x + 2y + 3z = 18 \quad ...(ii)$$
$$x + 4y + 9z = 16 \quad ...(iii)$$

Divide (*i*) by 2 and subtract multiplied by 3 from (*ii*) then subtract from (*iii*), we get

$$x + \frac{1}{2}y + \frac{1}{2}z = 5 \quad ...(iv)$$
$$+\frac{1}{2}y + \frac{3}{2}z = 3 \quad ...(v)$$
$$\frac{7}{2}y + \frac{17}{2}z = 11 \quad ...(vi)$$

Now divide (*v*) by $\frac{1}{2}$ and then subtract after multiplied by $\frac{7}{2}$ from (*vi*), we get,

$$x + \frac{1}{2}y + \frac{1}{2}z = 5 \quad ...(vii)$$
$$y + 3z = 6 \quad ...(viii)$$
$$-2z = -10 \quad ...(ix)$$

From back substitution from (*ix*), (*viii*) and (*vii*), we get

$$z = 5$$

and
$$y + 3z = 6$$
$$y + 3(5) = 6$$
$$y = 6 - 15$$
$$y = -9$$

and
$$x + \frac{1}{2}(-9) + \frac{1}{2}(5) = 5$$
$$x = 5 + \frac{9}{2} - \frac{5}{2} = 7.$$

Hence the solution is $x = 7, \quad y = -9, \quad z = 5.$

Example 14.10. Apply Gauss elimination method to solve the equations

$$2x + y + 4z = 12$$
$$8x - 3y + 2z = 23$$
$$4x + 11y - z = 33.$$

Solution. Dividing first equation by 2, we get

$$x + \frac{1}{2}y + 2z = 6 \quad ...(i)$$

Now subtract (i) after multiplied by 8 and 4 respectively from second and third equation, we get

$$-7y - 14x = -45 \quad ...(ii)$$

$$9y - 9z = 9 \quad ...(iii)$$

Now divide (ii) by –7, we get

$$y + 2z = \frac{45}{7} \quad ...(iv)$$

Now multiplying (iv) by 9 and subtract from (iii), we get

$$-2yz = 9 - \frac{405}{7}$$

or

$$-2yz = -\frac{342}{7} \quad ...(v)$$

Hence the system of equations reduces to upper triangular form as follows:

$$\left.\begin{aligned} x + \frac{1}{2}y + 2z &= 6 \\ y + 2z &= \frac{45}{7} \\ -27z &= \frac{342}{7} \end{aligned}\right\} \quad ...(vi)$$

By back substitution, we get

$$z = \frac{342}{189} = 1.81$$

and

$$y + 2(1.81) = \frac{45}{7}$$

$$y = \frac{45}{7} - 2(1.81) = 6.43 - 3.62 = 2.81$$

and

$$x + \frac{1}{2}(2.81) + 2(1.81) = 6$$

$\therefore$

$$x = 6 - \frac{1}{2}(2.81) - 2(1.81) = 0.975$$

Hence the solution is $x = 0.975, y = 2.81, z = 1.81.$

14.6. ITERATIVE METHODS

The previous methods for finding the solution of simultaneous linear equations are known as direct methods, because after certain amount of fixed calculations we obtain the solutions. Iterative methods are such that in which we start the process by assuming a solution which is nearer to the true solution and we obtain the solution by successive repeatation of the process. The number of iteration depends on the degree of accuracy required. Under the category of iteractive methods, we shall describe the following two methods:

(i) Jacobi's method, and

(ii) Gauss-Seidel method.

14.7. JACOBI METHOD

Let us consider a system of equations:

$$\left.\begin{aligned} a_{11}x + a_{12}y + a_{13}z &= b_1 \\ a_{21}x + a_{22}y + a_{23}z &= b_2 \\ a_{31}x + a_{32}y + a_{33}z &= b_3 \end{aligned}\right\} \quad ...(i)$$

Let us assume that

$$|a_{11}| > |a_{12}| + |a_{13}|,$$
$$|a_{22}| > |a_{21}| + |a_{23}|$$
$$|a_{33}| > |a_{31}| + |a_{32}|,$$

Then the above equations can be rearranged as:

$$\left.\begin{aligned} x &= \frac{1}{a_{11}}(b_1 - a_{12}y - a_{12}z) \\ y &= \frac{1}{a_{22}}(b_2 - a_{21}x - a_{23}z) \\ z &= \frac{1}{a_{33}}(b_3 - a_{31}x - a_{32}y) \end{aligned}\right\} \quad ...(ii)$$

If $x^{(0)}, y^{(0)}, z^{(0)}$ are the initial values of x, y and z respectively, then

$$\left.\begin{aligned} x^{(1)} &= \frac{1}{a_{11}}(b_1 - a_{12})y^{(0)} - a_{13}z^{(0)} \\ y^{(1)} &= \frac{1}{a_{22}}(b_2 - a_{21}x^{(0)}) - a_{23}z^{(0)} \\ z^{(1)} &= \frac{1}{a_{33}}(b_3 - a_{31}x^{(0)}) - a_{32}y^{(0)} \end{aligned}\right\} \quad ...(iii)$$

Again using these values $x^{(1)}, y^{(1)}, z^{(1)}$ in equation (*ii*), we get

$$\left.\begin{aligned} x^{(2)} &= \frac{1}{a_{11}}(b_1 - a_{12})y^{(1)} - a_{13}z^{(1)} \\ y^{(2)} &= \frac{1}{a_{22}}(b_2 - a_{21}x^{(1)}) - a_{23}z^{(1)} \\ z^{(3)} &= \frac{1}{a_{33}}(b_3 - a_{31}x^{(1)}) - a_{32}y^{(1)} \end{aligned}\right\}$$

Preceding in the same way, if the nth iterates are $x^{(n)}, y^{(n)}, z^{(n)}$, the iteractions are as follows:

$$\left.\begin{aligned} x^{(n+1)} &= \frac{1}{a_{11}}(b_1 - a_{12}y^{(n)} - a_{13}z^{(n)}) \\ y^{(n+1)} &= \frac{1}{a_{22}}(b_2 - a_{21}x^{(n)}) - a_{23}z^{(n)}) \\ z^{(n+1)} &= \frac{1}{a_{33}}(b_3 - a_{31}x^{(n)} - a_{32}y^{(n)}) \end{aligned}\right\}$$

The above procedure is continued till the desired accuracy in solutions is attained.

SOLVED EXAMPLES

Example 14.11. Solve the following equations by Jacobi method:

$$20x + y - 2z = 17$$
$$3x + 20y - z = -18$$
$$2x - 3y + 20z = 25.$$

Solution. Above equations can be written as

$$\left.\begin{aligned} x &= \frac{1}{20}(17 - y + 2z) \\ y &= \frac{1}{20}(-18 - 3x + z) \\ z &= \frac{1}{20}(25 - 2x + 3y) \end{aligned}\right\} \quad \ldots(i)$$

Let us take first approximation $x^{(1)} = 0, y^{(1)} = 0, z^{(1)} = 0$. Then the second approximation using (*i*), we get

$$x^{(2)} = \frac{1}{20}(17 - y^{(1)} + 2z^{(1)}) = \frac{17}{20} = 0.85$$

$$y^{(2)} = \frac{1}{20}(-18 - 3x^{(1)} + z^{(1)}) = \frac{1}{20}(-18 - 0 + 0) = -\frac{18}{20} = -0.9$$

$$z^{(2)} = \frac{1}{20}(25 - 2x^{(1)} + 3y^{(1)}) = \frac{1}{20}(25 - 0 + 0) = \frac{25}{20} = 1.25$$

Substituting these values in R.H.S. of (*i*), we get third approximation

$$x^{(3)} = \frac{1}{20}(17 - y^{(2)} + 2z^{(2)})$$

$$= \frac{1}{20}(17 + 0.9 + 2(1.25)) = \frac{20.4}{20} = 1.02$$

$$y^{(3)} = \frac{1}{20}(-18 - 3x^{(2)} + z^{(2)})$$

$$= \frac{1}{20}(-18 - 3(0.85) + 1.25) = -\frac{19.3}{20} = -0.965$$

$$z^{(3)} = \frac{1}{20}(25 - 2(0.85) + 3(-0.9)) = \frac{20.6}{20} = 1.03$$

Now substitute these values into (*i*), we get fourth approximation

$$x^{(4)} = \frac{1}{20}(17 - y(3) + 2z^{(3)}) = \frac{1}{20}(17 + 0.965 + 2(1.03))$$

$$= \frac{20.025}{20} = 1.0013$$

$$y^{(4)} = \frac{1}{20}(-18 - 3x^{(3)} + z^{(3)})$$

$$= \frac{1}{20}(-18 - 3(1.02) + 1.03) = -\frac{20.03}{20} = -1.0015$$

$$z^{(4)} = \frac{1}{20}(25 - 2x^{(3)} + 3y^{(3)})$$

$$= \frac{1}{20}(25 - 2(1.02) + 3(-0.965)) = \frac{20.065}{20} = 1.0032.$$

Substitute these values into (*i*), we get fifth approximation

$$x^{(5)} = \frac{1}{20}(17 - y^{(4)} + 2z^{(4)})$$

$$= \frac{1}{20}(17 + 1.0015 + 2(1.0032)) = \frac{20.0079}{20} = 1.0004$$

$$y^{(5)} = \frac{1}{20}(-18 - 3x^{(4)} + z^{(4)})$$

$$= \frac{1}{20}(-18 - 3(1.0013) + 1.0032) = \frac{20.0007}{20} = -1.00003$$

$$z^{(5)} = \frac{1}{20}(25 - 2x^{(4)} + 3y^{(4)})$$

$$= \frac{1}{20}(25 - 2(1.0013) + 3(-1.0015)) = \frac{19.9929}{20} = 0.9996.$$

Substituting these value into (*i*), we get sixth approximation

$$x^{(6)} = \frac{1}{20}(17 - y^{(5)} + 2z^{(5)})$$

$$= \frac{1}{20}(17 + 1.00003 + 2(0.9996)) = \frac{19.99923}{20} = 0.99996$$

$$y^{(6)} = \frac{1}{20}(-18 - 3x^{(5)} + z^{(5)})$$

$$= \frac{1}{20}(-18 - 3(1.0004) + 0.9996) = -\frac{20.0016}{20} = -1.00008$$

$$z^{(6)} = \frac{1}{20}(25 - 2x^{(5)} + 3y^{(5)})$$

$$= \frac{1}{20}(25 - 2(1.0004) + 3(-1.00003)) = \frac{19.99911}{20} = 0.99995$$

The values in the fifth and sixth approximation are nearly same. Hence the solution is

$$x = 1,\ y = -1,\ z = 1.$$

Example 14.12. Solve the system of equations by Jacobi iteractive method:

$$27x + 6y - z = 85$$

$$6x + 15y + 2z = 72$$

$$x + y + 54z = 110.$$

Solution. Here, we see that diagonal elements are dominant. Hence, the iteraction process can be applied. Rearranging the equations, we get

$$x = \frac{1}{27}[85 - 6y + z]$$

$$y = \frac{1}{15}[72 - 6x - 2z]$$

$$z = \frac{1}{54}[110 - x - y]$$

Let the initial values be the $x = 0 = y = z$.

Thus, $$x^{(1)} = \frac{1}{27}[85 - 0 - 0] = 3.1481$$

$$y^{(1)} = \frac{1}{15}[72 - 0 - 0] = 4.8000$$

$$z^{(1)} = \frac{1}{54}[110 - 0 - 0] = 2.0370$$

Now, the second approximation can be obtained by

$$x^{(2)} = \frac{1}{27}[85 - 6y^{(1)} + z^{(1)}]$$

$$y^{(2)} = \frac{1}{15}[72 - 6x^{(1)} - 2z^{(1)}]$$

$$z^{(2)} = \frac{1}{54}[110 - x^{(1)} - y^{(1)}]$$

i.e., $$x^{(2)} = \frac{1}{27}[85 - 6 \times 4.8 + 2.0370] = 2.1569$$

$$y^{(2)} = \frac{1}{15}[72 - 6 \times 3 \times 3.1481 - 2 \times 2.0370] = 3.2692$$

and $$z^{(2)} = \frac{1}{54}[110 - 3.1481 - 4.80] = 1.8899.$$

Similarly, the third approximation is given by

$$x^{(3)} = \frac{1}{27}[85 - 6y^{(2)} + z^{(2)}]$$

$$y^{(3)} = \frac{1}{15}[72 - 6x^{(2)} - 2z^{(2)}]$$

$$z^{(3)} = \frac{1}{54}[110 - x^{(2)} - y^{(2)}]$$

i.e., $$x^{(3)} = \frac{1}{27}[85 - 6 \times 3.2692 + 1.8899] = 2.4917$$

$$y^{(3)} = \frac{1}{15}[72 - 6 \times 2.1569 - 2 \times 1.8899] = 3.6853$$

$$z^{(3)} = \frac{1}{54}[110 - 2.1569 - 3.2692] = 1.9366$$

The fourth iteration is given by

$$x^{(4)} = \frac{1}{27}[85 - 6y^{(3)} + z^{(3)}]$$

$$y^{(4)} = \frac{1}{15}[72 - 6x^{(3)} - 2z^{(3)}]$$

$$z^{(4)} = \frac{1}{54}[110 - x^{(3)} - y^{(3)}]$$

i.e.,

$$x^{(4)} = \frac{1}{27}[85 - 6 \times 3.6853 + 1.9366] = 2.4009$$

$$y^{(4)} = \frac{1}{15}[72 - 6 \times 2.4917 - 2 \times 1.9366] = 3.5451$$

$$z^{(4)} = \frac{1}{54}[110 - 2.4917 - 3.6853] = 1.9226.$$

Again the next iteration is

$$x^{(5)} = \frac{1}{27}[85 - 6y^{(4)} + z^{(4)}]$$

$$= \frac{1}{27}[85 - 6 \times 3.5451 + 1.9226] = 2.4316$$

$$y^{(5)} = \frac{1}{15}[72 = 6x^{(4)} - 2z^{(4)}]$$

$$= \frac{1}{15}[72 - 6 \times 2.4009 - 2 \times 1.9226] = 3.5883$$

$$z^{(5)} = \frac{1}{54}[110 - x^{(4)} - y^{(4)}]$$

$$= \frac{1}{54}[110 - 2.4009 - 2.5451] = 1.9269.$$

Hence, the solution is $x = 2.4316$, $y = 3.5833$, $z = 1.9269$.

14.8. GAUSS-SEIDAL METHOD

This method is a modification of Jacobi's method. This method converges twice as fast as the Jacobi method. This method is also known as *method of successive displacement*. To understand this method let us consider a system of 'n' equations in which $a_{ii} \neq 0$.

$$\left.\begin{array}{l} a_{11}x_1 + a_{12}x_2 + a_{13}x_3 + \ldots + a_{1n}x_n = b_1 \\ a_{21}x_1 + a_{22}x_2 + a_{23}x_3 + \ldots + a_{2n}x_n = b_2 \\ a_{31}x_1 + a_{32}x_2 + a_{33}x_3 + \ldots + a_{3n}x_n = b_3 \\ \ldots \quad \ldots \quad \ldots \quad \ldots \quad \ldots \\ \ldots \quad \ldots \quad \ldots \quad \ldots \quad \ldots \quad \ldots \\ a_{n1}x_1 + a_{n2}x_2 + a_{n3}x_3 + \ldots + a_{nn}x_n = b_n \end{array}\right\} \quad \ldots(i)$$

Above equations can be written as

$$\left.\begin{aligned} x_1 &= \frac{1}{a_{11}}(b_1 - a_{12}x_2 - a_{13}x_3 - \ldots - a_{1n}x_n) \\ x_2 &= \frac{1}{a_{22}}(b_2 - a_{21}x_1 - a_{23}x_3 - \ldots - a_{2n}x_n) \\ x_3 &= \frac{1}{a_{33}}(b_3 - a_{31}x_1 - a_{32}x_2 - \ldots - a_{3n}x_n) \\ \ldots & \quad \ldots \quad \ldots \quad \ldots \quad \ldots \quad \ldots \\ x_n &= \frac{1}{a_{nn}}(b_n - a_{n1}x_1 - a_{n2}x_2 - \ldots - a_{n(n-1)}x_{n-1}) \end{aligned}\right\} \quad \ldots(ii)$$

Now let us assume first approximations $x_1^{(1)}, x_2^{(1)}, ..., x_n^{(1)}$. Substitute these values into the first equation of (*ii*), we get

$$x_1^{(2)} = \frac{1}{a_{11}}(b_1 - a_{12}x_2^{(1)} - a_{13}x_3^{(1)} - \ldots - a_{1n}x_n^{(1)}).$$

Now substitute $x_1^{(2)}, x_2^{(2)}, x_3^{(1)} ..., x_n^{(1)}$ into the R.H.S. of second equation of (*ii*), we get

$$x_2^{(2)} = \frac{1}{a_{22}}(b_2 - a_{21}x_1^{(2)} - a_{23}x_3^{(1)} - \ldots - a_{2n}x_n^{(1)}).$$

In the third equation of (*ii*) substitute $x_1^{(2)}, x_2^{(2)}, x_3^{(1)}, ..., x_n^{(1)}$; we get

$$x_3^{(2)} = \frac{1}{a_{33}}(b_3 - a_{31}x_1^{(2)} - a_{32}x_2^{(2)} - a_{3n}x_n^{(1)}).$$

Proceeding in this way, we get $x_n^{(2)}$ and this comletes the first stage of iteration. The whole process is repeated till we get the values of $x_1, x_2, x_3, ..., x_n$ to the desired degree of accuracy. Therefore, Gauss-Seidel Method is also known as a metod of successive displacement.

Both the methods Jacobi and Gauss-Seidel, converge for any type of first approximation if every equation of the system (*ii*) satisfies the condition that the sum of the absolute values of the coefficients $\frac{a_{ij}}{a_{ii}}$ is atmost equal to or in atleast one equation less than unity. That is,

$$\sum_{\substack{j=1 \\ (j \neq 1)}}^{n} \left| \frac{a_{ij}}{a_{ii}} \right| \leq 1, \; i = 1, 2, 3, ..., n$$

The sign < holds good in the case of atleast one equation.

SOLVED EXAMPLES

Example 14.13. Solve the following system by Gauss-Seidel method:

$$\begin{aligned} 10x + 2y + z &= 9 \\ 2x + 20y - 2z &= -44 \\ -2x + 3y + 10z &= 22. \end{aligned}$$

Solution. These equations can be written as

$$\left.\begin{aligned} x &= \frac{1}{10}(9 - 2y - z) \\ y &= \frac{1}{20}(-44 - 2x + 2z) \\ z &= \frac{1}{10}(22 + 2x - 3y) \end{aligned}\right\} \quad \ldots(i)$$

Let us start with first approximation $x^{(1)} = 0, y^{(1)} = 0, z^{(1)} = 0$; substitute these values into first equation of (*i*), we get

$$x^{(2)} = \frac{1}{10}(9 - 2y^{(1)} - z^{(1)}) = \frac{1}{10}(9 - 0 - 0) = 0.9.$$

Now substitute $x^{(2)} = 0.9, y^{(1)} = 0, z^{(1)} = 0$ into second equation of (*i*), we get

$$y = \frac{1}{20}(-44 - 2x^{(2)} + 2z^{(1)})$$

$$= \frac{1}{20}(-44 - 2(0.9) + 0) = -2.29.$$

and substitute the values $x^{(2)} = 0.9, y^{(2)} = -2.29, z^{(1)} = 0$ into third equation of (*i*), we get

$$z^{(2)} = \frac{1}{10}(22 + 2x^{(2)} + 3y^{(2)})$$

$$= \frac{1}{10}(22 + 2(0.9) - 3(-2.29)) = 3.067.$$

Here first stage of iteration is completed. For next iteration, substitute the values $x^{(2)} = 0.9, y^{(2)} = -2.29, z^{(2)} = 3.067$ into first equation of (*i*), we get

$$x^{(3)} = \frac{1}{10}(9 - 2y^{(2)} - z^{(2)})$$

$$= \frac{1}{10}(9 - 2(-2.29) - 3.067) = 1.051.$$

Substitute $x^{(3)} = 1.051, y^{(3)} = -2.99, z^{(2)} = 3.067$ into second equation of (*i*), we get

$$y^{(3)} = \frac{1}{20}(-44 - 2x^{(3)} + 2z^{(2)})$$

$$= \frac{1}{20}(-44 - 2(1.051) + 2(3.067)) = -1.99.$$

Now substitute $x^{(3)} = 1.051, y^{(3)} = -1.99, z^{(2)} = 3.067$ into third equation of (*i*), we get

$$z^{(3)} = \frac{1}{10}(22 + 2x^{(3)} - 3y^{(3)})$$

$$= \frac{1}{10}(22 + 2(1.051) - 3(-1.99)) = 3.007.$$

Second stage of iteration is now completed. For next iteration, substitute $x^{(3)} = 1.051, y^{(3)} = -1.99, z^{(3)} = 3.007$ into first equation of (1), we get

$$x^{(4)} = \frac{1}{10}(9 - 2y^{(3)} - z^{(3)})$$

$$= \frac{1}{10}(9 - 2(-1.99) - (3.007) = 0.997$$

Substitute $x^{(4)} = 0.997, y^{(3)} = -1.99, z^{(3)} = 3.007$ into second equation of (*i*), we get

$$y^{(4)} = \frac{1}{20}(-44 - 2x^{(4)} + 2z^{(3)})$$

$$= \frac{1}{20}(-44 - 2(0.997) + 2(3.007)) = -1.99$$

Now substitute x(4) = 0.997, y(4) = – 1.99, z(3) = 3.007 into third equation of (*i*), we get

$$z^{(4)} = \frac{1}{10}(22 + 2x^{(4)} - 3y^{(4)})$$

$$= \frac{1}{10}(22 + 2(0.997) - 3(-1.99)) = 2.99$$

Thus $x^{(4)}, y^{(4)}, z^{(4)}$ are very close to $x^{(3)}, y^{(3)}, z^{(3)}$ respectively. Hence the solution is

$$x = 1.0, y = -2.0, z = 3.0$$

Example 14.14. Find the solution of the system by Gauss-Siedel method:

$$83x + 11y - 4z = 95$$
$$7x + 52y + 13z = 104$$
$$3x + 8y + 29z = 71$$

Solution. Above equations can be written as

$$x = \frac{1}{83}(95 - 11y + 4z) \quad ...(i)$$

$$y = \frac{1}{52}(104 - 7x - 13z) \quad ...(ii)$$

$$z = \frac{1}{29}(71 - 3x - 8y) \quad ...(iii)$$

Taking first approxination $x^{(1)} = 0, y^{(1)} = 0, z^{(1)} = 0$ and substitute these values in (*i*), we get

$$x^{(2)} = \frac{1}{83}(95 - 11y^{(1)} + 4z^{(1)}) = \frac{1}{83}(95 - 0 + 0) = 1.14$$

Put $x^{(2)} = 1.14, y^{(1)} = 0, z^{(1)} = 0$ into (2), we get

$$y^{(2)} = \frac{1}{52}(104 - 7x^{(2)} - 13z^{(1)})$$

$$= \frac{1}{52}(104 - 7(1.14) - 0) = 1.85$$

Put $x^{(2)} = 1.14, y^{(2)} = 1.85, z^{(1)} = 0$ into (3), we get

$$z^{(2)} = \frac{1}{29}(71 - 3x^{(2)} - 8y^{(2)})$$

$$= \frac{1}{29}(71 - 3(1.14) - 8(1.85) = 1.82$$

Now, put $x^{(2)} = 1.14, y^{(2)} = 1.85, z^{(2)} = 1.82$ into (1), we get

$$x^{(3)} = \frac{1}{83}(95 - 11y^{(2)} + 4z^{(2)})$$

$$= \frac{1}{83}(95 - 11(1.85) + 4(1.82)) = 0.99$$

Put $x^{(3)} = 0.99, y^{(2)} = 1.85, z^{(2)} = 1.82$ into (2), we get

$$y^{(3)} = \frac{1}{52}(104 - 7x^{(3)} - 13z^{(2)})$$

$$= \frac{1}{52}(104 - 7(0.99) - 13(1.82)) = 1.41$$

Put $x^{(3)} = 0.99$, $y^{(3)} = 1.41$, into (3), we get

$$z^{(3)} = \frac{1}{29}(71 - 3x^{(3)} - 8y^{(3)}) = \frac{1}{29}(71 - 3(0.99) - 8y(1.41))$$

$$= \frac{56.75}{29} = 1.95$$

Again put $x^{(3)} = 0.99$, $y^{(3)} = 1.41$, $z^{(3)} = 1.95$ into (1), we get

$$x^{(4)} = \frac{1}{83}(95 - 11y^{(3)} + 4z^{(3)})$$

$$= \frac{1}{83}(95 - 11(1.41) + 4(1.95)) = \frac{87.29}{83} = 1.05$$

Put $x^{(4)} = 1.05$, $z^{(3)} = 1.95$, into (2), we get

$$y^{(4)} = \frac{1}{52}(104 - 7x^{(4)} - 13z^{(3)})$$

$$= \frac{1}{52}(104 - 7(1.05) - 13(1.95)) = \frac{71.3}{52} = 1.37$$

Put $x^{(4)} = 1.05$, $y^{(4)} = 1.37$ into (3), we get

$$z^{(4)} = \frac{1}{29}(71 - 3x^{(4)} - 8y^{(4)})$$

$$= \frac{1}{29}(71 - 3(1.05) - 8(1.37)) = \frac{56.89}{29} = 1.96$$

Thus $x^{(4)}$, $y^{(4)}$, $z^{(4)}$ are sufficiently close to $x^{(3)}$, $y^{(3)}$, $z^{(3)}$ respectively. Hence the solution is

$$x = 1.05, y = 1.37, z = 1.96.$$

EXERCISE 14.4

Solve the system of equations by Gauss elimination method

1. $$x + 2y + z = 3$$
$$2x + 3y + 3z = 10$$
$$3x - y + 2z = 13$$
[Ans. $x = 2, y = 1, z = 3$**]**

2. $$10x + y + z = 12$$
$$2x + 10y + z = 13$$
$$x + y + 5z = 7$$
[Ans. $x = 1, y = 1, z = 3$**]**

3. $$2x + 3y - z = 5$$
$$4x + 4y - 3z = 3$$
$$2x - 3y + 2z = 2$$
[Ans. $x = 3, y = 2, z = 1$**]**

4. $$5x_1 + x_2 + x_3 + x_4 = 4$$
$$x_1 + 7x_2 + x_3 + x_4 = 12$$
$$x_1 + x_2 + 6x_3 + x_4 = -5$$
$$x_1 + x_2 + x_3 + 4x_4 = -6$$
[Ans. $x_1 = 1, x_2 = 2, x_3 = -1, x_4 = -2$**]**

5. $$x + y + z + w = 2$$
$$2x - y + 2z - w = -5$$
$$3x + 2y + 3z + 4w = 7$$
$$x - 2y - 3z + 2s = 5$$
[Ans. $x = 0, y = 1, z = -1, w = 2$**]**

Solve the following system of equations by Gauss-Jacobi method:

6.
$$2x + y + z = 4$$
$$x + 2y + z = 4$$
$$x + 2y + 2z = 4$$
[**Ans.** 1, 1, 1]

7.
$$10x - 2y + z = 12$$
$$x + 9y - z = 10$$
$$2x - y + 11z = 20$$
[**Ans.** 1.2624, 1.1591, 1.6940]

8.
$$8x - y + z = 18$$
$$2x + 5y - 2z = 3$$
$$x + y - 3z = -16$$
[**Ans.** 2, 0.9998, 2.9999]

9.
$$8x + y + z = 8$$
$$2x + 4y + z = 4$$
$$x + 3y + 3z = 5$$
[**Ans.** 0.83, 0.32, 1.07]

10.
$$8x - 6x + z = 13.67$$
$$3x + 11y - 2z = 17.59$$
$$2x - 6y + 9z = 29.29$$
[**Ans.** 2.45, 1.62, 3.79]

11.
$$5x - 2y + z = -4$$
$$x + 6y - 2z = -1$$
$$3x + y + 5z = 13$$
[**Ans.** –1.001, 0.999, 3]

Solve the following system of equations using Gauss-Seidel method:

12.
$$10x + y + 2z = 44$$
$$2x + 10y + z = 51$$
$$x + 2y + 10z = 61$$
[**Ans.** $x = 3, y = 4, z = 5$]

13.
$$37x + 6y - z = 85$$
$$6x + 15y + 2z = 72$$
$$x + y + 54z = 110$$
[**Ans.** 2.43, $y = 3.57, z = 1.926$]

14.
$$10x + y + z = 12$$
$$2x + 10y + z = 13$$
$$2x + 2y + 10z = 14$$
[**Ans.** $x = 1, y = 1, z = 1$]

15.
$$2x - y + z = 5$$
$$x + 3y - 2z = 7$$
$$x + 2y + 3z = 10$$
[**Ans.** $x = 3, y = 2, z = 1$]

16.
$$5x + 2y + z = 12$$
$$x + 4y + 2z = 15$$
$$x + 2y + 5z = 20$$
[**Ans.** $x = 0.996, y = 1.97, z = 3.16$]

17.
$$54x + y + z = 110$$
$$2x + 15y + 6z = 72$$
$$-x + 6y + 27z = 85$$
[**Ans.** $x = 1.926, y = 3.573, z = 2.425$]

18.
$$5x - y + z = 10$$
$$2x + 4y = 12$$
$$x + y + 5z = -1$$
[**Ans.** $x = 2.556, y = 1.722, z = 1.055$]

19.
$$10x - 2y - z - u = 3$$
$$-2x + 10y - z - u = 15$$
$$-x - y + 10z - 2u = 27$$
$$-x - y - 2z + 10u = -9$$
[**Ans.** $x = 0.997, y = 1.998, z = 2.999, u = -0.00$]

14.9. FINITE DIFFERENCES

The calculus of finite differences, in its broad meaning deals with the changes that take place in the value of the function, the dependent variable, due to changes in the independent variable. It is a study of relations that exist between the values assumed by the functions say 'y' or $f(x)$ whenever the independent variable say 'x' changes by finite jumps or intervals whether equal or unequal. The calculus of finite differences plays a very important role in the theory of interpolation, numerical differentiation and integration which are further useful in solving a number of technical problems like measurement of heat radiation, path of a particle in a flow, numerical solution of a differential equation, determination of the potential and current distribution along an electrical network etc. Firstly, we study the variations in the function when the independent variable changes by *equal intervals*. Later on we will also consider the variations in the function when the independent variable changes by unequal intervals.

Consider the function $y = f(x) = x^2 - x + 3$

The value of y corresponding to any value of x can be obtained by substituting that value of x on the right hand side of the equation. For example:

x	y
10	93
20	383
30	873
40	1563
50	2453

Now suppose that instead of numerical values of x differing by 10, we have the following consecutive values of x differing by 'h' ;

$$x_0, x_1, x_2, x_3, x_4, ..., x_n ...$$

The values of $y = f(x)$ corresponding to these values of x are:

$$y_0, y_1, y_2, y_3, y_4, ..., y_n ...$$

where $x_1 = x_0 + h, x_2 = x_1 + h = x_0 = 2h$

$x_3 = x_2 + h = x_0 + 3h, x_4 = x_0 + 4h, ..., x_n = x_0 + nth$

and $y_0 = f(x_0), y_1 = f(x_1) = f(x_0 + h),$

$y_2 = f(x_2) = f(x_0 + 2h), y_3 = f(x_3) = f(x_0 + 3h),$

$y_4 = f(x_4) = f(x_0 + 4h), ..., y_n = f(x_0 + nh), ...$

The value of the independent variable 'x' is called the **Argument** and the value of the dependent variable 'y' or $f(x)$ corresponding to that argument is called the **Entry**. The equal difference 'h' between the consecutive values of the independent variable 'x' is called the **interval of differencing** or **step length.**

Interval of differencing or step length.

Forward Differences

The differences $y_1 - y_0, y_2 - y_1, y_3 - y_2$ etc., are called the **first forward differences** of the function $y = f(x)$.

Denoting these above differences by $\Delta y_0, \Delta y_1, \Delta y_2, \ldots,$ we have :

$$\Delta y_0 = y_1 - y_0 = f(x_0 + h) - f(x_0)$$

$$\Delta y_1 = y_2 - y_1 = f(x_0 + 2h) - f(x_0 + h)$$

$$\Delta y_2 = y_3 - y_2 = f(x_0 + 3h) - f(x_0 + 2h), \text{ etc.}$$

In general, $\Delta y_n = y_{n+1} - y_n$, where n is an integer, is called the first forward difference of y_n. The differences of these first differences are called **second forward differences**. Denoting these by $\Delta^2 y_0, \Delta^2 y_1$ etc., we have

$$\Delta^2 y_0 = \Delta y_1 - \Delta y_0 = y_2 - 2y_1 + y_0$$

$$\Delta^2 y_1 = \Delta y_2 - \Delta y_1 = y_3 - 2y_2 + y_1, \text{ etc.}$$

The differences of these second forward differences are called **third forward differences.**

Denoting these by $\Delta^3 y_0$ etc., we have

$$\Delta^3 y_0 = \Delta^2 y_1 - \Delta^2 y_0 = y_3 - 3y_2 + 3y_1 - y_0$$

and so on.

In general $$\Delta^m y_n = \Delta^{m-1} y_{n+1} - \Delta^{m-1} y_n$$

defines the ***m*th forward differences** of y_n.

These differences are shown in the following table called the **Forward Difference Table** and shows how the forward differences of all orders can be formed.

Forward Difference Table

Argument	*Entry y*	*Δy (first diff.)*	*Δ²y (second diff.)*	*Δ³y (third diff.)*	*Δ³y (fourth diff.)*	*Δ³y (fifth diff.)*	*Δ³y (sixth diff.)*
x_0	y_0						
		$y_1 - y_0 = \Delta y_0$					
x_1	y_1		$\Delta y_1 - \Delta y_0 = \Delta^2 y_0$				
		$y_2 - y_1 = \Delta y_1$		$\Delta^3 y_0$			
x_2	y_2		$\Delta^2 y_1$		$\Delta^4 y_0$		
		$y_3 - y_2 = \Delta y_2$		$\Delta^3 y_1$		$\Delta^5 y_0$	
x_3	y_3		$\Delta^2 y_2$		$\Delta^4 y_1$		$\Delta^6 y_0$
		$y_4 - y_3 = \Delta y_3$		$\Delta^3 y_2$		$\Delta^5 y_1$	
x_4	y_4		$\Delta^2 y_3$		$\Delta^4 y_2$		
		$y_5 - y_4 = \Delta y_4$		$\Delta^3 y_3$			
x_5	y_5		$\Delta^2 y_4$				
		$y_6 - y_5 = \Delta y_5$					
x_6	y_6						

Note. 1. y_0 the first entry is termed the **leading term** and the differences $\Delta y_0, \Delta^2 y_0, \Delta^3 y_0, \Delta^4 y_0, \Delta^5 y_0, \Delta^6 y_0 \ldots$ are known as the **leading differences.**

2. Δ is not a quantity but a symbol standing for 'operation'. Thus Δ^2 is not the square of a quantity but denotes the repetition of the operator Δ.

14.10. INTERPOLATION

Interpolation has been described as the art of reading between the line of a table and in elementary mathematics, it means the process of computing intermediate values of a function from a given set of tabular values of the function. Suppose the following table represents a set of corresponding values of x and y.

x	x_0	x_1	x_2	x_3	.....	x_n
y	y_0	y_1	y_2	y_3	.....	y_n

Now, we require the value of $y = y_i$ corresponding to a value $x = x_i$ where $x_0 < x_i < x_n$.

Extrapolation is used to denote the process of finding the values outside the interval (x_0, x_n). But in general the interpolation is used in both processes.

14.11. NEWTON'S FORWARD DIFFERENCE INTERPOLATION FORMULA

Let $y = f(x)$ denote a function which takes the values $y_0, y_1, y_2, ... y_n$ for the $(n + 1)$ equidistant values x_0, $x_1,, x_2, ... ,x_n$ of the independent variable x,

where $$x_n = x_0 + nh$$

and $$y_n = f(x_0 + nh)$$

Assume that it is required to find the value of y for some intermediate value of $x = x_0 + p_h$, where p is a constant.

Now $$y = f(x) = f(x_0 + ph) = E^p f(x_0) = (1 + \Delta)^p f(x_0)$$

$$= \left[1 + p\Delta + \frac{p(p-1)}{2!}\Delta^2 + \frac{p(p-1)(p-2)}{3!}\Delta^3 + ...\right] f(x_0)$$

[By Binomial Theorem for any index]

$$= f(x_0) + p\Delta f(x_0) + \frac{p(p-1)}{2!}\Delta^2 f(x_0) + \frac{p(p-1)(p-2)}{3!}\Delta^2 f(x_0) + ...$$

OR

$$y = y_0 + p\Delta y_0 + \frac{p(p-1)}{2!}\Delta^2 y_0 + \frac{p(p-1)(p-2)}{3!}\Delta^3 y_0 + ...$$

This is Newton's forward difference formula, which is used to find the value of y for value of x which lie near the starting of x, *i.e.*, x_0.

SOLVED EXAMPLES

Example 14.15. Given sin 45° = 0.771, sin 50° = 0.7660, sin 55° = 0.8192, sin 60° = 0.8660. Find sin 52°, using Newton's forward formula.

Solution. Let the angle is degree be x and $y = f(x) = \sin x$

Again let $$x = x_0 + ph \Rightarrow p = \frac{x - x_0}{h}$$

Here $$x_0 = 50°, h = 5°$$

$\therefore$ $$p = \frac{52 - 50}{5} = 0.4$$

The forward difference table is as under:

x	y	Δy	$\Delta^2 y$	$\Delta^3 y$
45°	0.7071			
		0.0589		
			– 0.0057	
50°	**0.7660**			– 0.0007
		0.0532		
55°	0.8192		**– 0.0064**	
		0.0468		
60°	0.8660			

$$y_0 = 0.7660,\ \Delta y_0 = 0.0532,\ \Delta y_0 = -\ 0.0064$$

Using Newton's forward difference formula

$$f(x_{52°}) = y_0 + p\Delta y_0 + \frac{p(p-1)\Delta^2 y_0}{2!}$$

$$\sin 52° = 0.7660 + 0.4 \times 0.532 + \frac{0.4(0.3)(-0.0064)}{2}$$

$$= 0.7660 + 0.0213 - 0.0038$$

$$\therefore \quad \sin 52° = 0.7835$$

Example 14.16. The population of a city in the decennial census were as under. Estimate the population for the year 1896.

Year	1891	1901	1911	1921	1931
Population (thousands)	46	66	81	93	101

Solution. Let $x = x_0 + ph$

Here $x_0 = 1891$

$h = 10$

$x = 1896$

$$\therefore \quad p = \frac{1896-1891}{10} = 0.5 \qquad \left[\because \quad p = \frac{x - x_0}{h}\right]$$

The forward difference table is as below:

x	y	Δ	Δ^2	Δ^3	Δ^4
1891	**46**				
		20			
1901	66		**– 5**		
		15		**2**	
1911	81		– 3		**– 3**
		12		–1	
1921	93		– 4		
		8			
1931	101				

By Newton's forward difference formula.

$$y = y_0 + p\Delta y_0 + \frac{p(p-1)}{2!}\Delta^2 y_0 + \frac{p(p-1)(p-2)}{3!}\Delta^3 y_0 + \frac{(p-1)(p-2)(p-3)}{4!}\Delta^4 Y_0 + ...$$

Here $y_0 = 46$

$\Delta y_0 = 20, \Delta^2 y_0 = -5, \Delta^3 y_0 = 2, \Delta^4 y_0 = -3$

$$\therefore \quad y_{x=1896} = 46 + (0.5)(20) + \frac{(0.5)(0.5-1)}{2!}(-5) + \frac{(0.5)(0.5-1)(0.5-2)}{3!}(2) + \frac{(0.5)(0.5-1)(0.5-2)(0.5-3)}{4!}(-3)$$

$$= 46 + 10 + 0.625 + 0.125 + 0.117$$

$$= 56.867 \text{ (thousands)}.$$

Note: In the above problem if it is find the population of the year 1890, we can solve the problem by putting the value of p as obtained below:

$$1890 = 1891 + p\,(10) \qquad i.e., \qquad p = -0.1$$

Example 14.17. From the following table of half-yearly premium for policies maturing at different ages, estimate the premium for policies maturing at age of 46.

Age	45	50	55	60	65
Premium (in rupees)	114.84	96.16	83.32	74.48	68.48

Solution. The difference table is:

Age (x)	*Premium (in rupees) (y)*	Δy	$\Delta^2 y$	$\Delta^3 y$	$\Delta^4 y$
45	114.84				
		– 18.68			
50	96.16		5.84		
		–12.84		– 1.84	
55	83.32		4		0.68
		– 8.84		– 1.16	
60	74.48		2.84		
		– 6			
65	68.48				

Here $h = 5, a = 45, a + ph = 46$

$\therefore \quad 45 + 5p = 46 \Rightarrow u = 0.2$

By Newton's forward difference formula,

$$y_{46} = y_{45} + p\Delta y_{45} + \frac{p(p-1)}{2!}\Delta^2 y_{45} + \frac{p(p-1)(p-2)}{3!}\Delta^3 y_{45} + \frac{p(p-1)(p-2)(p-3)}{4!}\Delta^4 y_{45}$$

$$= 114.84 + (0.2)(-18.68) + \frac{(0.2)(0.2-1)}{2!}(5.84) + \frac{(0.2)(0.2-1)(0.2-2)}{3!}(-1.84) + \frac{(0.2)(0.2-1)(0.2-2)(0.2-3)}{4!}(0.68)$$

$$= 110.525632$$

Hence the premium for policies maturing at the age of 46 is ₹ 110.52.

EXERCISE 14.5

1. Using Newton's forward formula find the value of $x = 1.6$, if

x	1.0	1.4	1.8	22
$f(x)$	3.49	4.82	5.96	6.5

[**Ans.** 5.54]

2. The value of $f(x)$ for $x = 0, 1, 2, \ldots\ldots 6$ are given by

x	0	1	2	3	4	5	6
$f(x)$	2	4	10	16	20	24	38

Estimate the value of (3.2) using only four of the given values. [**Ans.** 17.28]

3. Find the value of at $x = 45$, using the following data:

x	40	50	60	70	80
y	31	73	124	159	190

[**Ans.** 47.87]

4. Given log 100 = 2, log 101 = 2.0043
   log 103 = 2.0128, log 104 = 2.0170 [**Ans.** 2.0086]
   Find 102.

5. The following are data from the steam table:

Temperature °C	140	150	160	170	180
Pressure kg/ flcm²	3.685	4.854	6.302	8.076	10.225

Using the Newton's forward difference formula, find the pressure of the steam for temperature of 142°C.

6. Find the cubic polynomial which takes the following values:

x	0	1	2	3
y	1	0	1	10

Hence find y at $x = 4$ [**Ans.** (*i*) $x^3 - 2x^2 + 1$ (*ii*) 33]

7. A thrid degree polynomial passes through the points (0, – 1) (1, 1) (2, 1) and (3, – 2). Find the polynomial.

$$\left[\textbf{Ans. } -\frac{1}{6}x^3 - \frac{1}{x}x^2 + \frac{8}{3}x - 1\right]$$

8. Using Newton's forward difference interpolation formula to obtain the interpolating polynomial $f(x)$ satifying the following data:

x	1	2	3	4
$f(x)$	26	18	4	1

If another point $x = 5, f(x) = 26$ is added to the above data, will interpolation polynomial be the same as before or different? Explain why?

$$\left[\textbf{Ans. } -\frac{17}{6}x^3 - 20x^2 + \frac{143}{6}x^2 + 11, \text{same}\right]$$

9. Estimate the value of $f(22)$ from the following available data:

x	20	25	30	35	40	45
$f(x)$	354	332	291	260	231	204

[**Ans.** 352]

10. Following are the marks obtained by 492 conditions in a certain examination.

Marks	0-42	40-45	45-50	50-55	55-60	60-65
No. of Candidates	210	43	54	74	32	79

Find out the number of candidates.

(*a*) Who secured more than 48 but not more than 50 marks.

(*b*) Less than 48 but not more than 45 marks.

11. Find the polynomial of degree four which takes the following values:

x	2	4	6	8	10
y	0	0	1	0	0

$$\left[\textbf{Ans. } \frac{1}{64}(x^4 - 24x^3 + 196x^2 - 624x + 640)\right]$$

14.12. NEWTON-COTE'S QUADRATURE FORMULA

Let $I = \int_a^b f(x)dx$ where $f(x)$ takes the values $y_0, y_1, y_2, ... y_n$ for $x = x_0, x_1, x_2, ... x_n$.

Let the interval of integration (a, b) be divided into a equal sub-intervals, each of width $h = \frac{b-a}{n}$ so that $x_0 = a, x_1 = x_0 + h, x_2 = x_0 + 2h, ... x_n = x_0 + nh = b$.

$$\therefore \qquad I = \int_{x_0}^{x_0+nh} f(x)dx$$

Since $\qquad x = x_0 + rh$ or $\frac{x - x_0}{h} = r$

then $\qquad dx = h\,dr$

Then the limit of integration becomes 0 to n.

$$\therefore \qquad I = h\int_0^n f(x_0 + ph)dp$$

$$= h\int_0^n \left[y_0 + p\Delta y_0 + \frac{p(p-1)}{2!}\Delta^2 y_0 + \frac{p(p+1)(p-2)}{3!}\Delta^3 y_0 + ... \right]dp$$

[by Newton's forward interpolation formula]

$$= h\left[py_0 + \frac{p^2}{2}\Delta y_0 + \frac{1}{2}\left(\frac{p^3}{3} - \frac{p^2}{2}\right)\Delta^2 y_0 + \frac{1}{6}\left(\frac{p^4}{4} - p^3 + p^2\right)\Delta^3 y_0 + ...\right]_0^n$$

$$I = nh\left[y_0 + \frac{n}{2}\Delta y_0 + \frac{n(2n-3)}{12}\Delta^2 y_0 + \frac{n(n-2)^2}{24}\Delta^3 y_0 + ...\right] \qquad ...(i)$$

This is the required general quadrature formula. In this we have integrated function all the n-intervals at the same time. Thus we can say that one can find the area he the same time. This formula has no great importance from the point of view of applications the help of this formula we can find other important formulae by taking the particular values of n.

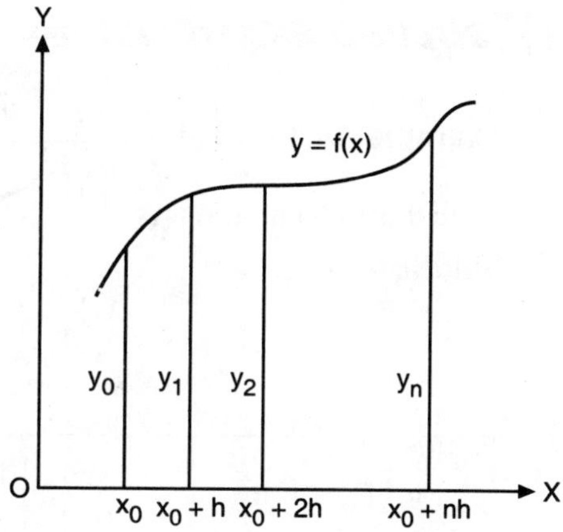

1. Trapezoidal Rule

The trapezoidal rule approximates the area under a curve by connecting successive points on the curve to form trapezoids of uniform width, and then summing the area under these trapezoidal to obtain the approximate area under the curve.

By the general quadrature formula, we know that

$$I = \int_{x_0}^{x_0+nh} f(x)dx = nh\left[y_0 + \frac{n}{2}\Delta y_0 + \frac{n(2n-3)}{12}\Delta^3 \frac{n(n-2)^2}{24}\Delta^3 y_0 \ldots\right]$$

Putting $n = 1$ in above equation and taking the curve $y = f(x)$ between the points (x_0, y_0) and (x_1, y_1) as a straight line i.e., as polynomial of degree one so that difference of order high than one vanish, we get

$$\int_{x_0}^{x_0+h} f(x)dx = 1.h\left[y_0 + \frac{1}{2}\Delta y_0\right] = h\left[y_0 + \frac{1}{2}(y_1 - y_0)\right]$$

$$= \frac{h}{2}[y_0 + y_1]$$

Similarly for next interval $(x_0 + h, x_0 + 2h)$, we get

$$\int_{x_0+h}^{x_0+2h} f(x)dx = \frac{h}{2}[y_1 + y_2]$$

$$\int_{x_0+2h}^{x_0+3h} f(x)dx = \frac{h}{2}[y_2 + y_3]$$

....

....

$$\int_{x_0+(n-1)h}^{x_0+nh} f(x)\,dx = \frac{h}{2}[y_{n-1} + y_n]$$

Now adding above n integrals we get

$$\int_{x_0}^{x_0+nh} f(x)\,dx = \frac{h}{2}\left[(y_0 + y_n) + 2(y_1 + y_2 + \ldots + y_{n-1})\right]$$

$$= h\left[\left(\frac{y_0 + y_n}{2}\right) + (y_1 + y_2 + \ldots + y_n)\right]$$

$= h$ [Mean of first and last ordinates + Sum of all the intermediate ordinates]

which is known as **trapezoidal rule**. By increasing the number of subintervals *i.e.* making h very small, we can improve the accuracy of the value of the given interval.

SOLVED EXAMPLES

Example 14.18. Find $\int_0^1 \frac{1}{1+x^2}\,dx$ by trapezoidal rule, where the interval of integration is subdivided into 6 equal parts.

Solution. Here,

$$h = \frac{1-0}{6} = \frac{1}{6}$$

x	0	$\frac{1}{6}$	$\frac{2}{6}$	$\frac{3}{6}$	$\frac{4}{6}$	$\frac{5}{6}$	1
$y=f(x)$	1	0.97297	0.9	0.8	0.69230	0.59016	0.5

$$\int_0^1 \frac{1}{1+x^2}\,dx = \frac{h}{2}[(y_0+y_n)+2(y_1+y_2+\ldots+y_{n-1})]$$

$$= \frac{1}{12}[(1+0.5)+2(0.97297+0.9+0.8+0.69230+0.59016)]$$

$$= \frac{9.41086}{12} = 0.78424$$

Exact value of the integral is $\left[\tan^{-1}x\right]_0^1 = \frac{\pi}{4} = 0.7853981$

Example 14.19. Find the integral $\int_4^{5.2} \log x\,dx$ by trapezoidal rule.

Solution. Divide the interval (4.5.2) into six parts. Then the values of log x for each points of success.

x	4	4.2	4.4	4.6	4.8	5.0	5.2
y log x	1.38629	1.43508	1.48160	1.52605	1.56861	1.60943	1.64865

Here $y_0 = 1.38629, y_3 = 1{,}52605\ y_4 = 1.56861, y_5\ 1.60943$ and $y_6 = 1.64865$.

By Trapezoidal rule

$$\int_{4.0}^{5.2} \log x dx = \frac{h}{2}[y_0 + 2(y_1+y_2+y_3+y_4+y_5)y_6]$$

$$= \frac{0.2}{2}[1.38629 + 2(1.43508 + 1.4160 + 1.526 - 5 + 1.600942) + 1.64865]$$

$$= 0.1\ [3.03494 + 2(1.62077) = 0.1\ [18.27648] = 1.827648$$

Example 14.20. Find the approximate value of integral $\int_0^{\pi/2} \sin x dx$.

Solution. Divide (0, π/2) into ten parts each of width $h = \frac{\pi}{20}$. Then the values of y sin x are given in table below:

x	$y = \sin x$	x	$y = \sin x$
0	0.00000	7π/20	0.89101
π/20	0.15643	8π/20	0.95106
2π/20	0.30902	9π/20	0.98769
3π/20	0.45399	10π/20	1.00000
4π/20	0.58779		
5π/20	0.70711		
6π/20	0.70711		

Here y_0 = 1.38629, y_1 = 0.15643, y_2 = 0.30902, y_3 = 0.45399, y_4 = 0.58779, y_5 = 0.70711, y_6 = 0.80902, y_7 = 0.89101, y_8 = 0.95106, y_9 = 0.98769, and y_{10} = 1.0000.

By Trapezoidal rule

$$\int_{4.0}^{5.2} \log x\, dx = \frac{h}{2}[y_0 + 2(y_1 + y_2 + y_3 + y_4 + y_5) + y_6 + y_7 + y_8 + y_9) + y_{10}]$$

$$= \frac{\pi/20}{2}[0 + 2(0.15643 + 0.30902 + 0.4399 + 0.58779 + 0.70711$$

$$+ 0.80902 + 0.89101 + 0.95106 + 0.98764 + 1.00000000]$$

$$= \frac{\pi}{40}(12.70624) = 0.99795$$

2. Simpson's One Third (1/3) Rule

We know that the general quadrature formula

$$I = \int_{x_0}^{x_0+nh} f(x)dx = nh\left[y_0 + \frac{n}{2}\Delta y_0 + \frac{n(2n-3)}{12}\Delta^2 y_0 + ...\right]$$

This rule can be obtained by putting $n = 2$ in general quadrature formula and taking the can through (x_0, y_0) (x_1, y_1) and (x_2, y_2) as polynomial of degree two and neglecting the third and higher order differences, we get

$$\int_{x_0}^{x_0+2h} f(x)dx = 2h\left[y_0 + \Delta y_0 + \frac{1}{6}\Delta^2 y_0\right]$$

$$= \frac{2h}{6}[6y_0 + 6(y_1 - y_0) + (y_2 - 2y_1 + y_0)] = \frac{h}{3}(y_0 + 4y_1 + y_2)$$

Similarly,

$$\int_{x_0+2h}^{x_0+4h} f(x)\,dx = \frac{h}{3}(y_2 + 4y_3 + y_4)....,\int_{x_0+(n-2)h}^{x_0+nh} f(x)dx = \frac{h}{3}(y_{n-2} + 4y_{n-1} + y_n)$$

Adding the above integrals, we get

$$I = \int_{x_0}^{x_0+nh} f(x)dx$$

$$= \frac{h}{3}[(y_0 + y_n) + 4(y_1 + y_3 + ... + y_{n-1}) + 2(y_2 + y_4 + ... + + y_{n-2})]$$

or, $$I = \frac{h}{3}[(y_0 + y_n) + y \text{ (sum of odd ordinates)} + 2 \text{ (sum of even ordinates)}]$$

which is known as Simpson's one-third rule.

While using this formula, the given interval of integration must be divided into an even number of sub-intervals, since we find the area over two sub-intervals of a time.

This rule gives the correct value of the integral only if $f(x)$ is a second order (quadratic) function. This rule i.e., Simpson's one third rule can be applied only if the number of known ordinates is odd (or even number of intervals are given). We require at least three pairs of values of (x, y) for the Simpson's 1/3 rule. This rule is known as parabolic rule because the interpolating polynomial is of degree 2.

SOLVED EXAMPLES

Example 14.21. **Evaluate $\int_0^6 \frac{dx}{1+x^2}$ by simpson's $\frac{1}{3}$ rule.**

Solution. Divide the interval (0,6) into 6 equal parts so that widcrth of each subinterval $h = \frac{6-0}{6} = 1$.

The value of $y = \frac{1}{1+x^2}$ at each point of subdivision are given in the table:

x	0	1	2	3	4	5	6
y	1.00000	0.50000	0.20000	0.10000	0.58824	0.03846	0.02702

From the table, $y_0 = 1.0000\, y_1 = 0.5000\, y_2 = 0.20000$, $y_3 = 0.1000\, y_4 = 0.58824$, $y_5 = 0.03846$ and $y_6 = 0.02702$. Then by Simpson's $\frac{1}{3}$ rule

$$\int_0^6 \frac{dx}{1+x^2} = \frac{h}{3}[y_0 + 4(y_1 + y_3 + y_5) + 2(y_2 + y_4) + y_6]$$

$$= \frac{1}{3}[1.0000 + 4(0.500000) + 0.10000 + 0.03846) + 2(0.20000 + 0.,58824) + 0.002702]$$

$$= \frac{1}{3}[1.02702 + 4(0.63846) + 2(0.78824)]$$

$$= \frac{1}{3}(5.15734) = 1.71911$$

Example 14.22. The speed *v* metres per second, of a car, *t* second after it starts, is known in the following table:

t	0	12	24	36	48	60	72	84	96	108	120
v	0	3.60	10.08	18.90	21.60	18.54	10.26	5.40	4.50	5.40	9.00

Using Simpson's rule, find the distance travelled by the car in 2 minutes or 120 sec.

Solution. If s metres is the distance covered in *t* second, then

$$\frac{ds}{dt} = v \quad \text{or} \quad ds = v\,dt$$

By integration

$$[s]_{t=0}^{120} = \int_0^{120} v dt$$

$$= \frac{h}{3}[(y_0 + y_{10}) + 4(y_1 + y_3 + y_5 + y_7 + y_9) + 2(y_2 + y_4 + y_6 + y_8)] \qquad \left[\text{By Simpson's } \frac{1}{3} \text{ rule}\right]$$

$$= \frac{12}{3}[(0 + 9.00) + 4(3.60 + 18.90 + 1854 + 5.40 + 5.40) + 2(10.08 + 21.60 + 10.26 + 4.50)]$$

$$= 123.96 \text{ metres}$$

Example 14.23. A river is 80 m wide. The depth '*y*' of the river at a distance '*x*' from one bank is given by the following table:

0	0	10	20	30	40	50	60	70	80
y	0	4	7	9	12	15	14	8	3

Find approximately the area of cross-section of the river using Simpson's one third rule.

Solution. We may use either Simpson's 1/3 rule of Trapezoidal rule to compute the area of the cross-section of the river.

By Simpson's $\frac{1}{3}$ rule.

$$I = \int_0^{80} y dx = \frac{h}{3}[(y_0 + y_8) + 4(y_1 + y_3 + y_5 + y_7) + 2(y_2 + y_4 + y_6)]$$

We obtain the required area *A* which is given as :

$$A = \int_0^{80} y dx = \frac{h}{3}[(y_0 + y_8) + 4(y_1 + y_3 + y_5 + y_7) + 2(y_2 + y_4 + y_6)]$$

We obtain the required area *A* which is given as :

$$A = \int_0^{80} y dx = \frac{10}{3}[(0 + 3) + 4(4 + 9 + 15 + 8) + 2(7 + 12 + 14)]$$

$$= \frac{2130}{3} = 710 \text{ m}^2$$

3. Simpson's 3/8 Rule

We known that the general quadrature formula,

$$I = \int_{x_0}^{x_0+nh} f(x)dx = nh\left[y_0 + \frac{n}{2}\Delta y_0 + \frac{n(2n-3)}{12}\Delta^2 y_0 + \frac{n(n-2)^2}{24}\Delta^3 y_0 + \ldots\right] \quad \ldots (i)$$

Putting $n = 3$ in eqn. (*i*) and taking the curve through (x_0, y_0). (x_1, y_1), (x_2, y_2) and (x_3, y_3) polynomial of thrid order so that difference above the order vanish, we get

$$\int_{x_0}^{x_0+3h} f(x)dx = 3h\left(y_0 + \frac{3}{2}\Delta y_0 + \frac{3}{4}\Delta^2 y_0 + \frac{1}{8}\Delta^3 y_0\right)$$

$$= 3h\left[y_0 + \frac{3}{2}(y_1 - y_0) + \frac{1}{4}(y_2 - 2y_1 + y_0) + \frac{1}{8}(y_3 - y_2 + 3y_1 - y_0)\right]$$

$$= \frac{3h}{8}(y_0 + 3y_1 + 3y_2 + y_3)$$

Similarly,

$$\int_{x_0+3h}^{x_0+5h} f(x)dx = \frac{3h}{8}(y_3 + 3y_4 + 3y_5 + y_6) \text{ and so on.}$$

Adding all such expression from x_0 to $x_0 + nh$, where n is a multiple of 3, we obtain

$$\int_{x_0}^{x_0+nh} f(x)dx = \frac{3h}{8}\cdot[(y_0 + y_n) + 3(y_1 + y_2 + y_4 + y_5 + \ldots y_{n-1}) + 2(y_3 - y_6 + \ldots y_{n-3}]$$

This is known as Simpson's 38 rule. (Simpson's three eight rule) while applying this rule it is to noted that n *i.e.*, the number of equal sub-divisions are the interval of integration, should be a multiple of 3. In this method the function $f(x)$ over any three consecutive subintervals is replaced by a polynomial of degree three $y = ax^3 + bx^2 + cx + d = 0$, because its difference of orders higher than three will vanish.

SOLVED EXAMPLES

Example 14.24. Evaluate the integral $\int_0^6 \frac{dx}{1+x^2}$ by using: (*i*) Trapezoidal rule, (*ii*) Simpson's 1/3 rule, (*iii*) Simpson 3/8 rule.

Solution. We have $f(x) = \frac{1}{1+x^2}$ as given in table

x	0	1	2	3	4	5	6
$f(x) = y$	y	0.5	0.2	0.1	0.0588	0.0385	0.027
	y_0	y_1	y_2	y_3	y_4	y_5	y

(*i*) By Trapezoidal rule.

$$\int_0^6 \frac{dx}{1+x^2} = \frac{h}{2}[(y_0 + y_6) + 2(y_1 + y_2 + y_3 + y_4 + y_5)]$$

$$= \frac{1}{2}[(1 + 0.0027) + 2(0.5 - 0.2 + 0.1 + 0.0588 + 0.0385] = 1.4108$$

(*ii*) By Simpson's 1/3 rule,

$$\int_0^6 \frac{dx}{1+x^2} = \frac{h}{3}\left[(y_0+y_6)+4(y_1+y_3+y_5+2(y_2+y_4)\right]$$

$$= \frac{1}{3}[(1+0.027)+4(0.5+0.1+0.0385)+2(0.2+0.0588)] = 1.3662$$

(*iii*) By Simpson's 3/8 rule.

$$\int_0^6 \frac{dx}{1+x^2} = \frac{3h}{8}[y_0+y_6+3(y_1+y_2+y_4+y_5)+2y_3]$$

Also $$\int_0^6 \frac{dx}{1+x^2} = \left|\tan^{-1} x\right|_0^6 = \tan^{-1} 6 = 1.4056$$

Example 14.25. Calculate $\int_0^{\pi/2} e^{\sin x}dx$, correct to four decimal places.

Solution. First we divide the given integral $\left(0, \frac{\pi}{2}\right)$ into 3 equal parts.

Here, $f(x) = e^{\sin x}$, $n = 3$ $a = 0$, $b = \frac{\pi}{2}$ $\therefore$ $h = \frac{\pi/2-0}{3} = \frac{\pi}{6}$

Now we contract the table of the values of the function is.

x	0	$\frac{\pi}{6}$	$\frac{\pi}{3}$	$\frac{\pi}{2}$
$y=f(x)=e^{\sin x}$	1	1.64872	2.36320	2.71828
	y_0	y_1	y_2	y_3

So by the Simpson's $\frac{3}{8}$ rule.

$$I = \int_0^{\pi/2} e^{\sin x}dx = \frac{3h}{8}[(y_0+y_3)+3(y_1+y_2)]$$

$$= \frac{3\times\pi}{8\times 6}[1+2.71828+3(1.64872+2.36320)]$$

$$= \int_0^{\pi} e^{\sin x}dx = 0.091111$$

Example 14.26. A train is moving at the speed of 30 m/sec. Suddenly brakes are applied. The speed of the train per second after *t* second is given by,

Time (t)	0	5	10	15	20	25	30	35	40	45
Speed (v)	30	24	19	16	13	11	10	8	7	5

Apply Simpson's three-eighth rule to determine the distance moved by the train in 45 sec.

Solution. If s metres is the distance covered in t seconds, then

$$\frac{ds}{dv} = v$$

$$\Rightarrow \qquad [s]_0^{t=45} = \int_0^{45} v\,dt$$

Since the no. of subintervals is 9 (a multiple of 3) hence by using Simpson's $\left(\frac{3}{8}\right)^{th}$ rule,

$$\int_0^{45} vdt = \frac{3h}{8}[(v_0+v_9)+3(v_1+v_2+v_4+v_5+v_7+v_8)+2(v_3+v_6)]$$

$$= \frac{15}{8}[(30+5)+3(24+19+13+11+8+7)+2(16+10)]$$

$$= 624.375 \text{ metres}$$

Hence the distance moved the train in 45 seconds is 624.375 metrs.

EXERCISE 14.6

1. Evaluate $\int_0^1 x^3 dx$ the by Trapezoidal rule. **[Ans.** 0.260]

2. Given the $e^0 = 1$, $e^1 = 2.72$, $e^2 = 7.39$ $e^3 = 20.09$, $e^4 = 54.60$ find an approximation value of by $\int_0^4 e^x$ Trapezoidal rule. **[Ans.** 58.00]

3. Evaluate $\int_0^1 (4x-3x^2)dx$ the taking 10 intervals by Trapezoidal rule. **[Ans.** 0.995]

4. Evaluate $\int_0^1 \sqrt{1-x^3}$ by (*i*) Simpson's rule and (*ii*) Trapezoidal rule, taking six interval correct to two decimal places. **[Ans.** 0.183]

5. Evaluate $\int_0^2 \frac{dx}{x} dx$ taking 4 subintervals, correct to five decimal places (*i*) Simpson's one third rule (*ii*) Trapezoidal rule. **[Ans.** 069326, 0.69702]

6. Compute by Simpson's one-thrid rule, the intergral $\int_0^1 x^2(1-x)dx$ correct to three places of decimal, taking step length equal to 0.1. **[Ans.** 0.083]

7. Evaluate the $\int_0^1 \sin x^2 dx$ by (*i*) Trapezoidal rule and (*ii*) Simspon's one thrid rule, correct to four decimals taking $x = 10$. **[Ans.** 0.3112, 03103]

8. Evaluate $\int_0^{\pi/2} \sqrt{\sin x^2}dx$ taking $x = 6$, correct to four significant figures by (*i*) Simpson's one third rule and (*ii*) Trapezoidal rule. **[Ans.** 1.187, 1,170]

9. Evaluate $\int_0^{1.5} \frac{x}{\cos x} dx$ taking $n = 10$, by (*i*) Trapezoidal rule and (*ii*) Simpson's one-thrid rule. **[Ans.** 0.13394, 0.133400]

10. Compute $\int_0^{1.5} e^x dx$ by (*i*) Trapezoidal rule and (*ii*) Simpson's one thrid rule taking $x = 10$. **[Ans.** 1.764,1.763]

11. Find the value of $\int_0^{\pi/2} \sqrt{\cos x}\, dx$ by (*i*) Trapezoidal rule and (*i*) Simpson's one-third rule taking $x = 6$. **[Ans.** 1.170, 1.187]

12. Calculate approximate value of $\int_3^{-3} \sin x^4 dx$ by using (*i*) Trapezoidal rule and (*ii*) Simspon's rule, taking $n = 6$. **[Ans.** 115, 98]

13. Evaluate $\int_0^{0.4} \cos x \, dx$ taking four equal intervals by (*i*) Trapezoidal rule and (*ii*) Simpson's one-third rule.

[**Ans.** 0.3891, 0.3894]

14. Using Simpson's rule evaluate $\int_0^{\pi/2} \sqrt{2 + \sin dx}$ with seven ordinates. [**Ans.** 2.545]

15. Using Simpson's rule evaluate $\int_2^6 \frac{1}{\log_e x} dx$ taking $n = 4$. [**Ans.** 3.1832]

16. Using Simpson's rule evaluate $\int_1^2 \sqrt{x - \frac{1}{x}} dx$ with five ordinates. [**Ans.** 1.007]

17. A river is 80 unit wide. The depth at distance x unit from one bank d is given by the following table.

x	0	10	20	30	40	50	60	70	80
d	0	4	7	9	12	15	14	8	3

[**Ans.** 710 sq. units]

18. Find the approximate value of $\int_0^{\pi 2} \sqrt{\cos \theta} d\theta$ using Simpson's rule with six intervals.

[**Ans.** 1.1872]

19. Use Simpson's three-eighths rule to obtain an approximate value of $\int_0^{0.3} \left(1 - 8x^3\right)^{1/2} dx$.

[**Ans.** 0.2899]

❑❑❑

UNIT - VII

STATISTICS AND PROBABILITY

CHAPTER 15

Basic Statistics

INSIDE THIS CHAPTER

15.1. INTRODUCTION

Statistics is as old as the human society iself. The word 'statistics' seems to have been drived from the Latin word 'Stratus'or the Italian word 'Statisca' or German word 'Statistik'. All these words mean a political state. In the olden days, stat mber of road accidents, etc.

Statistics as a set of numerical data should have the following characteristics:

(*i*) These should be, as far as possible, expressed quantitatively and not qualitatively.

(*ii*) There should be a collectin of observations and not a single observation. An individual observation cannot be called statistics.

(*iii*) Statistics should be collected with a definite purpose in view.

(*iv*) Statistics in an experiment should be comparable.

15.2. FREQUENCY DISTRIBUTION

Let us consider the marks obtained by 50 students of a class in Mathematics, maximum marks being 100, as follows:

25, 75, 34, 5, 46, 25, 15, 42, 29, 38
46, 12, 76, 36, 54, 58, 29, 9, 64, 50
16, 80, 62, 46,75, 46, 38, 64, 68, 62
45, 70, 68, 54, 56, 95, 54, 78, 46, 30
82, 18, 41, 44, 65, 24, 45, 54, 47, 62

Data is given above is called ungrouped data. This data may further be considered by expressing the marks in various groups. Suppose the marks obtained are divided into five groups: 0–20, 20–40, 40–60, 60–80, 80–100. These groups are also known as **classes** *e.g.*, the class 40–60 includes all marks between 40 and 60 (including 40 and excluding 60). A quantity which varies from one individual to another is known as a **variable** or variate *e.g.*, marks obtained by the students is a variate (x). The number of students against the marks is known as the **frequency** (f) of the variable. In other words, frequency (f) means how often the variate repeats *e.g.*, frequency of 46 marks is 5, which that there are 5 students who have obtained 46 marks etc. Similarly from the above data, we notice that there are 6 students in the class 0–20 and 10 students in class 20–40 etc. i.e., the frequencies of the classes 0–20 and 20–40 are 6 and 10 respectively. The above data can be considered as follows:

Marks obtained	*Mid-value (x)*	*Frequency (f)*	*Cumulative frequency*
0–20	10	6	6
20–40	30	10	16
40–60	50	18	34
60–80	70	13	47
80–100	90	3	50

This type of representation is called **grouped frequency distribution**. A frequency distribution is usually denoted as $x_i/f_i : i = 1, 2, \ldots. n$, where $x_1, x_2, \ldots, x_n$ are the values of the variate x with frequencies $f_1, f_2, \ldots. f_n$ respectively. We write

$$f_1 + f_2 + \ldots + f_n = \sum_{i=1}^{n} f_i = N$$

Some other definition associated with a frequency distribution are give below

(*i*) **Class Limits:** The boundary values of a class are called the class limits. The smaller of the two class limits is known as the lower limit and the greater value is known as the upper limit. In the class 20–40, 20 is the lower limit and 40 is the upper limit.

(*ii*) **Class Interval:** The difference of the upper and lower limits is called the class interval or width of the class.

(*iii*) **Cumulative frequency:** Cumulative frequency of a class is the total of the observations (frequencies) upto and including the observations of that class *e.g.*, cumulative frequency of the class 40–60 is 34.

15.3. MEASURES OF CENTRAL TENDENCY

The most important objective of our statistical analysis is to determine a single value for the entire mass of data so that it describes the overall level of the group of observation and can be considered a representative value of the whole set of data. In other words, it is difficult for anyone to understand or remember a large group of raw data. One would like to know the vertical value which represents all the items in a series. Such a value is called '*Central tendency*' or '*average value*'. For instance, it is very difficult to remember and understand the data concerning the income of millions of Indians. However, if it is said that in 2010 average income of the people in India was ₹ 19,000 per annum, it will be easy for us to guess the economic condition of most of the Indians. It is the average value which is called *central tendency* of the series. It is also called *measures of location*. Thus central tendency refers to a central value or a representative value of a statistical series.

Definitions

According to Croxton and Cowdon, "*An average is a single value with in the range of the data that is used to represent all of the values in the series. Since on average is somewhere within the range of data, it is sometimes called measure of central value.*"

According to Clark, "*An average is a figure that represents the whole group.*"

According to Kelloy and Smith, "*An average is sometimes called a measure of central tendency because individual values of the variable usually cluster around it.*"

15.4. TYPES OF STATISTICAL AVERAGES

Averages are broadly classified into two categories:

1. Mathematical averages 2. Positional averages

Flow chart reveals their further categorisations:

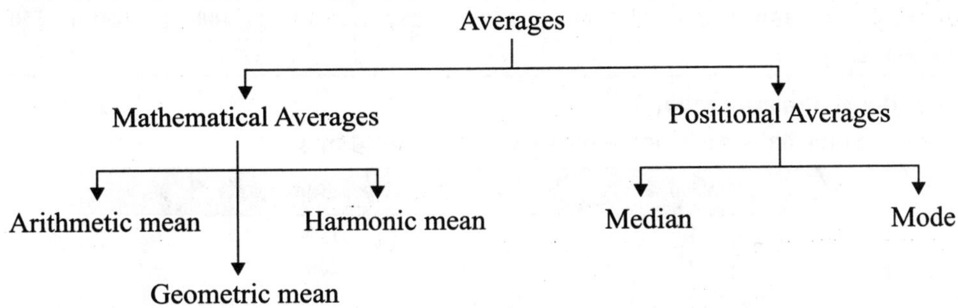

15.5. ARITHMETIC MEAN

The most popular and widely used measure for representing the entire data by one value is arithmetic mean. The arithmetic mean is the sum of the values in a group divided by the number of items in that group. It is (*a*) Simple arithmetic mean, and (*b*) weighted arithmetic mean.

15.5.1. Calculation of Simple Arithmetic Mean – Ungrouped Data

In this method, Arithmetic mean is calculated by adding the values of all observations and dividing the total by the number observations. Thus, if $x_1, x_2, x_3, ..., x_N$ are the N values of a variate x, the arithmetic mean ($\bar{x}$) is given by

1. Direct Method: $\bar{X} = \dfrac{x_1 + x_2 + x_3 + ... + x_n}{N}$ or $\dfrac{\Sigma X}{N}$

where, $\bar{X}$ is arithmetic mean,

ΣX is sum of all the values of variable.

N is the number of items or observations.

2. Short-cut Method. Where the number of items is very large, it is convenient to use the short-cut method. In the short-cut method, we make use of an arbitrary origin. This arbitrary origin is called A. It is also called *assumed mean.* We seek the difference between this A and the actual value of an item. This difference is called deviation and denoted as d. Symbolically,

$$\bar{X} = A + \frac{\Sigma d}{N}$$

where, $\bar{X}$ is arithmetic mean,

A is assumed mean,

d is deviation of items from assumed, and

N is the number of items or observations.

SOLVED EXAMPLES

Example 15.1. The following table contains the half-yearly bonus paid to 10 workers in a factory:

S. No.	1	2	3	4	5	6	7	8	9	10
Half-Yearly Bonus (₹)	150	200	300	650	250	180	400	500	550	220

Find out the arithmetic mean.

Solution. Let the half-yearly bonus of workers by denoted by *x*.

S.No.	*Half yearly Bonus x (in ₹)*
1.	150
2.	200
3.	300
4.	650
5.	250
6.	180
7.	400
8.	500
9.	550
10.	220
MN = 10	ΣX = 3400

$$\therefore \quad \overline{X} = \frac{x_1 + x_2 + x_3 + ... + x_n}{N} \text{ or } \frac{\Sigma X}{N} = \frac{3400}{10} = 340.$$

Thus the mean half-yearly bonus is ₹ 340.

(Assume mean or A = 400)

S.No.	*Half-yearly bonus x (in ₹)*	*d = x – 400*
1.	150	–250
2.	200	–200
3.	300	100
4.	650	250
5.	250	–150
6.	180	–220
7.	400	0
8.	500	100
9.	550	150
10.	220	–180
N = 10		Σ *d* = – 600

Here $\quad A = 400; \Sigma d = -600 ; N = 10$

$$\therefore \quad \overline{X} = 400 + \frac{(-600)}{10} = 400 - 60 = ₹\ 340.$$

Note: *The mean is the same under both the Methods. Moreover, the direct method involves more calculations when the values of items are large. This can be avoided by employing short-cut method.*

15.5.2. Calculation of Simple Arithmetic Mean – Discrete Series

1. Direct Method. In a discrete series the values of the variable are multiplied by their respective frequencies and the products so obtained are totaled. The total is divided by the number of items, which in a discrete, is equal to the total of the frequencies. The resulting quotient is a simple arithmetic average of the series. Symbolically,

$$\bar{X} = \frac{f_1x_1 + f_2x_2 + f_3x_3 + ... + f_nx_n}{f_1 + f_2 + f_3 + ... + f_n} = \frac{\Sigma fX}{N}$$

where,

f is frequency,

ΣfX is sum of product of x and f, and

N is the number of items or observations.

2. Short-cut Method. When short-cut method is used, arithmetic mean is computed by the following formula:

$$\bar{X} = A + \frac{\Sigma fd}{N}$$

where,

$\bar{X}$ is arithmetic mean,

A is assumed mean,

d is deviation of items from assumed, and

N is the number of items or observations.

3. Step Deviation Method. This method is a variant of shortcut method. It is adopted when the deviations from the assumed mean have some *common factor*. The following formula is used to calculated the arithmetic mean by step deviation method.

$$\bar{X} = A + \frac{\Sigma fd'}{N} \times h$$

where,

$\bar{X}$ is arithmetic mean,

A is assumed mean,

h is step deviation,

d' is $\frac{x - A}{h}$, and

N is the number of items or observations.

Note. *Symbol d′ has been used to distinguish between the step deviation and the absolute deviation.*

SOLVED EXAMPLES

Example 15.2. Calculate the mean of the following frequency distribution of marks in a test in statistics:

Marks:	10	20	30	40	50	60	70	80
No. of students:	3	6	10	12	9	6	2	2

Solution. Let the marks be denoted by X and the number of students by f.

Marks x	*Number of students* f	fx
10	3	30
20	6	120
30	10	300
40	12	480
50	9	450
60	6	360
70	2	140
80	2	160
	$N = \Sigma f = 50$	$\Sigma fx = 2040$

$$\bar{X} = \frac{\Sigma fX}{N} = \frac{2040}{50} = 40.8$$

Hence, average or mean marks in statistics = 40.8.

Marks x	*Number of students* f	$d = x - 40$	fd
10	3	– 30	– 90
20	6	– 20	– 120
30	10	– 10	– 100
40	12	0	0
50	9	10	90
60	6	20	120
70	2	30	60
80	2	40	80
	$N = \Sigma f = 50$		$\Sigma fd = 40$

$$\bar{X} = A + \frac{\Sigma fd}{N} = 40 + \frac{40}{50} = 40.8$$

Hence, average or mean marks in statistics = 40.8.

Marks x	*Number of students* f	$d' = \frac{x - 40}{10}$	fd'
10	3	–3	–9
20	6	–2	–12
30	10	–1	–10
40	12	0	0
50	9	1	9
60	6	2	12
70	2	3	6
80	2	4	8
	$N = \Sigma f = 50$		$\Sigma fd' = 4$

$$\bar{X} = A + \frac{\Sigma fd'}{N} \times h = 40 + \frac{4}{50} \times 10 = 40.8$$

Hence, average or mean marks in statistics = 40.8.

15.5.3. Calculation of Simple Arithmetic Mean – Continuous Series

The process of the calculation of arithmetic average in a continuous series is the same as in the case of discrete series. In a continuous series the midpoints of the various class intervals are written down to replace the class intervals. Once it is done, there is no difference between a continuous series and a discrete series. In continuous series, the arithmetic mean compuied by applying any one of the three methods given below:

1. Direct method $\bar{X} = \frac{\Sigma fm}{N}$
2. Short-cut method $\bar{X} = A + \frac{\Sigma fd}{N}$
3. Step deviation method $\bar{X} = A + \left(\frac{\Sigma fd'}{N} \times h\right)$

where,

$\bar{X}$ is arithmetic mean,

A is assumed mean,

m is the mid-point of various classes,

h is step deviation,

d' is $\frac{x - A}{h}$

N is the number of items or observations.

SOLVED EXAMPLES

Example 15.3. **Find the arithmetic mean for the following data:**

Marks	*Number of students*
0–10	**5**
10–20	**10**
20–30	**40**
30–40	**20**
40–50	**25**

Solution. (*a*) Direct method: $\bar{X} = \frac{\Sigma fm}{N}$

Marks	*Mid-values (m)*	*Number of students (f)*	$f \times m$
0-10	5	5	25
10-20	15	10	15
20-30	25	40	1000
30-40	35	20	700
40-50	45	25	1125
		$\Sigma f = 100$	$\Sigma fm = 3000$

$$\therefore \quad \bar{X} = \frac{\Sigma fm}{N} = \frac{3000}{100} = 300 \text{ marks}$$

(*b*) Short-cut method: $\bar{X} = A + \frac{\Sigma fd}{N}$

Mid-values (x)	*Number of students (f)*	*Deviation from Assumed mean (d = x – 25)*	*f × d*
5	5	–20	–100
15	10	–10	–100
25	40	0	0
35	20	10	200
45	25	20	500
	$\Sigma f = 100$		$\Sigma fd = 500$

$$\therefore \quad \bar{X} = A + \frac{\Sigma fd}{N} = 25 + \frac{500}{100} = 25 + 5 = 30 \text{ marks.}$$

(*c*) Step deviation method: $\bar{X} = A + \left(\frac{\Sigma fd'}{N} + h\right)$

Mid-values (x)	*Number of students (f)*	$d' = \frac{x-25}{10}$	*f × d′*
5	5	–2	–10
15	10	–1	–10
25	40	0	0
35	20	1	20
45	25	2	50
	$\Sigma f = 100$		$\Sigma fd' = 50$

$$\therefore \quad \bar{X} = A + \left(\frac{\Sigma fd'}{N} \times h\right) = 25 + \frac{50}{100} \times 10 = 25 + 5 = 30 \text{ marks.}$$

Example 15.4. For the following data, calculate arithmetic mean.

Marks	*No. of students*
Less than 10	**5**
Less than 20	**17**
Less than 30	**31**
Less than 40	**41**
Less than 50	**49**

Solution. A cumulative frequency distribution should first be converted into a simple frequency distribution as under:

Marks	*No. of students*
0–10	5
10–20	17 – 5 = 12
20–30	31 – 17 = 14
30–40	41 – 31 = 10
40–50	49 – 41 = 8

Now mean value of the data is obtained using direct method as under:

Marks	*Mid-values (m)*	*Number of students (f)*	$f \times m$
0–10	5	5	25
10–20	15	12	180
20–30	25	14	350
30–40	35	10	350
40–50	45	8	360
		$\Sigma f = 49$	$\Sigma fm = 1265$

$$\therefore \qquad \bar{X} = \frac{\Sigma fm}{N} = \frac{1265}{49} = 25.82 \text{ marks.}$$

Example 15.5. Find the average marks of students from the following table:

Marks	*No. of students*
Above 0	**80**
Above 10	**77**
Above 20	**72**
Above 30	**65**
Above 40	**55**
Above 50	**43**
Above 60	**28**
Above 70	**16**
Above 80	**10**
Above 90	**8**
Above 100	**0**

Solution. The given data is in the form of more than type cumulative frequency distribution to calculate mean we shall convert it into frequency distribution as given below:

Marks	*No. of students*
0–10	80 – 77 = 3
10–20	77 – 72 = 5
20–30	72 – 65 = 7
30–40	65 – 55 = 10
40–50	55 – 43 = 12
50–60	43 – 28 = 15
60–70	28 – 16 = 12
70–80	16 – 10 = 6
80–90	10 – 8 = 2
90–100	8 – 0 = 8

Now mean value of the data is obtained using step-deviation method as under:

Mid-values (x)	*Number of students (f)*	$d' = \frac{x-55}{10}$	$f \times d'$
5	3	–5	–15
15	5	–4	–20
25	7	–3	–21
35	10	–2	–20
45	12	–1	–12
55	15	0	0
65	12	1	12
75	6	2	12
85	2	3	6
95	8	4	32
	$\Sigma f = 80$		$\Sigma fd' = -26$

$$\therefore \quad \bar{X} = A + \left(\frac{\Sigma fd'}{N} \times C\right) = 55 + \frac{(-26)}{800} \times 10 = 55 - 3.25 = 51.75 \text{ marks.}$$

Example 15.6. The arithmetic mean of a series of 40 items was calculated by a student as ₹ 265. But while calculating it an item ₹ 115 was misread as ₹ 150. Find the correct arithmetic mean.

Solution. As we know that,

$$\bar{X} = \frac{\Sigma X}{N} \text{ or } \Sigma X = \bar{X}N$$

Given that:

$$\bar{X} = 265; N = 40$$

$$\therefore \quad 265 = \frac{\Sigma X}{40}, \text{ and}$$

$$\therefore \quad \Sigma X\,(wrong) = 265 \times 40 = 10600$$

Correct value = 115

Incorrect value = 150

$$\therefore \quad \text{Correct } (\bar{X}) = \frac{\Sigma X \text{ (wrong)} + \text{(correct value)} - \text{(Incorrect value)}}{N}$$

$$= \frac{10600 + 115 - 150}{40} = \frac{10565}{40} = ₹\ 264.12.$$

EXERCISE 15.1

1. What is meant by central tendency? Discuss the essentials of a measure of central tendency.
2. Calculate the arithmetic averages of the following information:
   Income of 7 families (in ₹) : Also show that

   $$\Sigma(X - \bar{X}) = 0$$

   550, 490, 670, 890, 435, 590, 575 **[Ans.** ₹ 600]
3. Calculate the mean of the following series:

Size:	4	5	6	7	8	9	10
Frequency:	6	12	15	28	20	14	5

4. Find the arithmetic mean of first n natural numbers. $\left[\textbf{Ans. } \frac{n+1}{2}\right]$
5. Compute the mean of the following data:

Height (in cm):	219	216	213	210	207	204	201	198	195
Frequency:	2	4	6	10	11	7	5	4	1

6. Calculate arithmetic mean of the following data:

Profits (in ₹):	0-10	10-20	20-30	30-40	40-50	50-60
Frequency:	12	18	27	20	17	16

7. Find the mean of the following distribution:

Marks	*No. of students*
Below 10	15
Below 20	35
Below 30	60
Below 40	84
Below 50	96
Below 60	127
Below 70	198
Below 80	250

[Ans. 50.4 Marks]

15.6. MEDIAN

Median is the value of the middle item of a series when it is arranged in ascending or descending order of magnitude. It is the value in the series which divides the series into two equal parts, one part consisting the values equal to median or smaller than it and the other part having the value equal to the median or larger than it. Unlike mean, median is the positional average. The position here means the place of value in the series. Median as such is the positional average of the data and has a position more or less at the centre of the values.

Definition of Median

In the words of **A.L. Bowley,** "*If the number of group are ranked in order according to the measurement under consideration then the measurement of the number most nearly one half is the median,*"

According to **Croxton** and **Cowden,** "*The median is usually defined as the value which divides a distribution so that an equal number of items is on either side of it.*"

15.6.1. Calculation of Median – Ungrouped Data

Let $x_1, x_2, x_3, ..., x_N$ be the N values of a varibale written in ascending (or descending) order of magnitude. The median denoted by M or Md is given by

$$\text{Median} = \text{Value of the middle item}$$

(*i*) When N is odd, then median $= \left(\frac{N+1}{2}\right)$th value

(*ii*) When N is even, the mediam $= \dfrac{\left(\frac{N}{2}\right)\text{th value} + \left(\frac{N}{2}+1\right)\text{th value}}{2}$

15.6.2. Calculation of Median – Discrete Series

In this case the values of the variable are arranged in ascending or descending order of magnitude. A table prepared showing the corresponding frequencies and cumulative frequencies. Then the median is calculated by the following formula:

$$\text{Median} = \left(\frac{N+1}{2}\right)\text{th value; where } N = \Sigma f$$

15.6.3. Calculation of Median – Continuous Series

In case of continuous frequency distribution, median can be calculated by using the following steps:

1. Compute cumulative frequency (*c.f.*).
2. Median item is located by finding out size of $\left(\frac{N}{2}\right)$th item.
3. Locate the median group in cumulative frequency column where the size of $\left(\frac{N}{2}\right)$th item falls.
4. Apply the formula to calculate the median from located group:

$$\text{Median} = L + \frac{\frac{N}{2} - c.f.}{f} \times C \text{ where,}$$

L = lower limit of the median class

N = sum of frequencies

f = frequency of the median class

$c.f.$ = cumulative frequency of the class, preceding the median class and

C = width of the class interval of median class.

SOLVED EXAMPLES

Example 15.7. The consumption of printing paper reams (in units) for the first 11 months of a computer operator is given as

20, 25, 30, 15, 17, 35, 26, 18, 40, 45, 50

Find the median.

Solution. By arranging the data in ascending order, we get the series

15, 17, 18, 20, 25, 26, 30, 35, 40, 45, 50

The number of terms in this series is 11 which is odd

Hence, the required median (Md) = value of the $\left(\frac{11+1}{2}\right)$th observation

= value of the 6th observation = 12.

Example 15.8. Calculate the median of the following data that relates to the monthly salaries of employees (in thousand rupees): 110, 115, 108, 112, 120, 116, 140, 135, 128, 132

Solution. By arranging the data in ascending order, we get the series

108, 110, 112, 115, 116, 120, 128, 132, 135, 140

The number of terms in this series is 10 which is even

Median will be the average of $\left(\frac{N}{2}\right)$th value and $\left(\frac{N}{2}+1\right)$th value, that is, average of $\frac{10}{2}$th value and $\left(\frac{10}{2}+1\right)$th value. Hence, median will be average of 5th and 6th term.

Hence, the required median (Md) = $\frac{116+120}{2}$ = 118

Thus, the median salary is ₹ 118,000.

Example 15.9. Obtain the median size of shoes sold from the following data:

Size:	5	$5\frac{1}{2}$	6	$6\frac{1}{2}$	7	$7\frac{1}{2}$	8	$8\frac{1}{2}$	9	$9\frac{1}{2}$	10	$10\frac{1}{2}$	11	$11\frac{1}{2}$
No. of pairs:	30	40	50	150	300	600	950	820	750	440	250	150	40	39

Solution. *Calculation of median*

Size (x)	*No. of pairs (f)*	*Cumulative frequency (c.f.)*
5	30	30
$5\frac{1}{2}$	40	70
6	50	120
$6\frac{1}{2}$	150	270
7	300	570
$7\frac{1}{2}$	600	1170
8	950	2120
$8\frac{1}{2}$	820	2940
9	750	3690
$9\frac{1}{2}$	440	4130
10	250	4380
$10\frac{1}{2}$	150	4530
11	40	4570
$11\frac{1}{2}$	39	4609
	$N = \Sigma f = 4609$	

$$\text{Median} = \left(\frac{N+1}{2}\right)\text{th value} = \left(\frac{4609+1}{2}\right)\text{th value} = 2305^{\text{th}} \text{ value.}$$

It shows that median value corresponds to 2305th value in the series. This value appears first of all in 2940th cumulative frequency of the series. Therefore, median shall be the value corresponding to the 2940th cumulative frequency, which is $8\frac{1}{2}$. Hence, the median size of shoes sold is $8\frac{1}{2}$.

Example 15.10. An insurance company obtained the following data for accident claims from a particular region. Obtain the median from this data.

Amount of claim (in thousand rupees)	*Frequency*
1–3	**6**
3–5	**53**
5–7	**85**
7–9	**56**
9–11	**21**
11–13	**16**
13–15	**4**
15–17	**4**

Solution. *Computation of median*

Amount of claim (in thousand rupees)	*Frequency (f)*	*Cumulative frequency (c.f.)*
1–3	6	6
3–5	53	59
5–7	85	144
7–9	56	200
9–11	21	221
11–13	16	237
13–15	4	241
15–17	4	245
	$N = \sum f = 245$	

Here $N = 245$, which is an odd number

$\frac{N}{2} = \frac{245}{2} = 122.5$, which falls in the class 5–7 (see the row of the cumulative frequency 144 which contains 122.5). Hence, the median class is 5–7

L = lower limit of the median class = 5

f = frequency of the median class = 85

$c.f.$ = cumulative frequency of the class, preceding the median class = 59

C = width of the class interval of median class = 2

$$\text{Median} = L + \frac{\frac{N}{2} - c.f.}{f} \times C = 5 + \frac{\frac{245}{2} - 59}{85} \times 2 = 5 + \frac{122.5 - 59}{85} \times 2$$

$$= 5 + \left(\frac{63.5}{85} \times 2\right) = 5 + \frac{127}{85} = 5 + 1.49 = 6.49.$$

Example 15.11. Calculate the median from the following data:

Age (in years)	*Number of persons (f)*
46–50	5
41–45	11
36–40	22
31–35	35
26–30	26
21–25	13
16–20	10
11–15	7

Solution. This is inclusive series given in the descending order. It should be first converted into an exclusive series and placed in the ascending order, as in the following table:

Class intervals	*No. of persons (f)*	*Cumulative frequency (c.f.)*
10.5–15.5	7	7
15.5–20.5	10	17
20.5–25.5	13	30
25.5–30.5	26	56
30.5–35.5	35	91
35.5–40.5	22	113
40.5–45.5	11	124
45.5–50.5	5	129
	$N = \Sigma f = 129$	

Here $N = 129$, which is an odd number

$\frac{N}{2} = \frac{129}{2} = 64.5$, which falls in the class 30.5–35.5 (see the row of the cumulative frequency 91 which contains 64.5). Hence the median class is 30.5–35.5

L = lower limit of the median class = 30.5

f = frequency of the median class = 35

$c.f.$ = cumulative frequency of the class, preceding the median class = 56

C = width of the class interval of median class = 5

$$\text{Median} = L + \frac{\frac{N}{2} - c.f.}{f} \times C = 30.5 + \frac{\frac{129}{2} - 56}{35} \times 5 = 30.5 + \frac{64.5 - 56}{35} \times 5$$

$$= 30.5 + \frac{8.5}{7} = 30.5 + 1.2 = 31.7 \text{ years.}$$

Example 15.12. In the frequency of 100 families given below, the number of families corresponding to expenditure groups 10–20 and 40–50 are missing from the table. However, the median is known to be 30. Find the missing frequencies.

Expenditure	0–10	10–20	20–30	30–40	40–50	50–60
No. of families	10	?	25	30	?	10

Solution. Let the missing frequency of the group 10-20 be f_1 and the missing frequency of 40-50 groups be f_2.

Here $$N = 100, \text{ i.e., } 100 = 10 + f_1 + 25 + 30 + f_2 + 10$$

or $$f_1 + f_2 = 25 \quad ...(i)$$

Median is given in this problem as 30.

Expenditure	*No. of families (f)*	*Cumulative frequency (c.f.)*
0–10	10	14
10–20	f_1	$10 + f_1$
20–30	25	$35 + f_1$
30–40	30	$65 + f_1$
40–50	f_2	$65 + f_1 + f_2$
50–60	10	$75 + f_1 + f_2$
	$N = \Sigma f = 100$	

Median item of the series is $\frac{100}{2}$ = 50th item, which means it lies in the class interval 30-40.

(Given Median = 30)

Now, $$\text{Median} = L + \frac{\frac{N}{2} - c.f.}{f} \times C$$

Here, L = Lower limit of the median class = 30

f = frequency of the median class = 30

$c.f.$ = cumulative frequency of the class, preceding the median class = $35 + f_1$

C = width of the class interval of median class = 10

$$\text{Median} = L + \frac{\frac{N}{2} - c.f.}{f} \times C$$

$$30 = 30 + \frac{50 - (35 + f_1)}{30} \times 10$$

$$\frac{30 - 30}{10} \times 30 = 50 - (35 + f_1) = 0 = 50 - 35 - f_1$$

$$f_1 = 50 - 35 = 15 \quad ...(ii)$$

Substituting $f_1 = 15$ from equation (*ii*), in (*i*)

$$\text{As } f_1 + f_2 = 25$$

$\Rightarrow$ $$15 + f_2 = 25 \Rightarrow f_2 = 25 - 15 = 10.$$

Thus the missing frequencies are 15 and 10.

EXERCISE 15.2

1. The number of runs scored by 11 players of a cricket team of a college are 5, 19, 42, 11, 50, 30, 21, 0, 52, 36, 27. Find the median. **[Ans.** 27 runs]

2. Calculate the median fro the following data:

Number of students	6	4	16	7	8	2
Marks	20	9	25	50	40	80

3. The followin gives the weekly expenditure of 100 families. Find he median weekly expenditure

Weekly expenditure (in ₹)	*Number of families*
0–10	14
10–20	23
20–30	27
30–40	21
40–50	15

[**Ans.** 24.81]

4. If the median of distribution given below is 28.5, find the values of x and y.

Class Interval	*Number of families*
0–10	5
10–20	x
20–30	20
30–40	15
40–50	y
50–60	5
Total	60

[**Ans.** $x = 8, y = 7$]

5. The percentage of marks obtained by 68 students in an examination are given below. Compute the median:

Marks	Below 20	20-40	40-60	60-80	Above 80
No. of students	0	5	22	25	16

[**Ans.** 65.6]

6. The following table gives the marks obtained by 65 students in statistics in a certain examination. Caclulate the median

Marks	*Number of students*
More than 70%	8
More than 60%	18
More than 50%	40
More than 40%	45
More than 30%	50
More than 20%	63
More than 10%	65

[**Ans.** 53.4 Marks]

7. Define Median. Discuss its merits and demerits.
8. If the arithmetic mean of the data is 28. Find the (a) missing frequency and (b) the median of the series:

Profit (in ₹)	0-10	10-20	20-30	30-40	40-50	50-60
Shops	12	18	27	–	17	6

9. Calculate the median from the following data:

Value	Frequency
Less than 10	4
Less than 20	16
Less than 30	40
Less than 40	76
Less than 50	96
Less than 60	112
Less than 70	120
Less than 80	125

[**Ans.** 36.25]

10. In a frequency distribution of 100 familites given below, the number of families corresponding to expenditure groups 20-40 and 60-80 are missing from the table. However,the median is known to be 50. Find the missing frequencies

Expenditure	0-20	20-40	40-60	60-80	80-100
No. of families	14	?	27	?	15

[**Ans.** 23 and 21]

15.7. MODE

Mode is defined to be the size of the variable which occurs most frequently or the point of maximum frequency or the point of greatest density. In other words, the mode or modal value of the distribution is that value for which the frequency is maximum. It is the value of the variable corresponding the maximum of the ideal curve which gives the closest fit to the actual distribution. The word '*mode*' is derived from the French word *la Mode* which means a thing in vogue or fashion. When one speaks of the "average student", the "most common wage", the "common man", or the "typical farm" and the like, he is unconsciously referring to mode. If it is said that the most common wage in a particular industry is ₹ 200 per month, what it mean is that the largest number of persons get this single figure of ₹ 200 as wage. Other figures of wages are not as popular as this one, and the number of persons getting them is less than the number getting ₹ 200 per month.

Definitions

According to **Kenny**, "*The distribution of the variable occurs most frequently in a distribution is called the Mode.*"

According to **Croxton** and **Cowden,** "*The mode may be regarded as the most typical of a series of values.*"

15.7.1. Calculation of Mode

Mode is calculated differently for different types of series.

1. Computation of Mode – Individual Series

(*a*) If the number of observations is small we can inspect mode at a glance by looking which one of the observation occurs most frequently.

(*b*) If the number of observations is large we convert the individual series into discrete series to identify quickly the most frequently repeated value.

2. Computation of Mode—Discrete Series

(*a*) If there is a single mode we can locate mode at a glance by looking into the frequency column for maximum frequency.

(*b*) In case of discrete frequency distribution, inspection method is possible only when there is regularity and homogeneity in the series. However, the series may sometimes not be regular or homogeneous. More than one value command the highest frequency in the series. In such cases grouping method of calculation of mode is used.

Preparation of grouping table. In a grouping table, normally there are six columns. If necessary more columns may be constructed. The details of these six columns are explained below:

Column 1: Original frequencies (given in the data).

Column 2: Given frequencies are added in twos and the highest total is marked.

Column 3: Leaving the first frequency, the remaining frequencies are added in twos and the highest total is marked.

Column 4: Given frequencies are added in threes and the highest total is marked.

Column 5: Leaving the first frequency, the given frequencies are added in threes.

Column 6: Leaving the first two frequencies, the given frequencies are added in threes.

The preparation of the analysis table and finding the mode are explained with the help of below example.

3. Computation of Mode – Continuous Series

(*a*) If there is a single class with maximum frequency, we call this class as model class and within this class mode is obtained by the formula:

$$\text{Mode } (M_0) = L + \frac{f_1 - f_0}{2f_1 - f_0 - f_2} \times C$$

where, L = Lower limit of the modal class

f_1 = Frequency of the modal class

f_0 = Frequency of the class preceding the modal class

f_2 = Frequency of the class succeeding the modal class, and

C = Width of the modal class.

Remark: If the mode lies in the first class interval, then f_0 is taken to be zero. If the mode lies in the last class interval, then f_2 is taken to be zero.

(*b*) However, if the values are cluster at more than one class interval, we decide the modal class by grouping method and then use the formula given above. If this formula fails then we use

$$\text{Mode } (M_0) = L + \frac{f_2}{f_0 - f_2} \times C$$

where symbols have their usual meaning.

(*c*) In dealings with a variety of data you may come across situations where these three measures of central tendency are very close to each other or at divergence. This largely depends upon the nature of the distribution. In a perfectly symmetrical unimodal distribution these three measures come very close to each other or even become identical. As the symmetry of the distribution varies the three measures (Mean, Median and Mode) register divergence. A very crude sort of relationship among them is shown by the equation:

Mode = 3 Median – 2 Mean

(*d*) We can also locate mode graphically by drawing a histogram. Steps for its location are:

Step 1: Draw a histogram of the series

Step 2: Locate the modal class bar – the one having greatest height is the modal class bar.

Step 3: Draw two diagonal lines from the top right corner and left corner of the tallest rectangle to the top right corner and left corner of the adjacent rectangles.

Step 4: Draw a perpendicular line from the point of intersection of the two diagonal lines on the *x*-axis. The value on the *x*-axis marked by the line will represent the modal value.

SOLVED EXAMPLES

Example 15.13. Find out the mode of the following marks obtained by 15 students in a class.

4 6 5 7 9 8 10 4 7 6 5 8 7 7 9

Solution. (*a*) By arranging data:

4 4 5 5 6 6 7 7 7 7 8 8 9 9 10

It will be observed from the array that 7 is repeated four times *i.e.*, more than any other item in the series, so 7 is the mode *i.e.*, modal marks.

(*b*) By converting into discrete series:

Marks	*Tally bars*	*Frequency*
4	||	2
5	||	2
6	||	2
7	||||	4
8	||	2
9	||	2
10	|	1
	Total 15	

∴ Mode = 7 marks.

Example 15.14. Calculate the mode from the following frequency distribution:

Size of items:	8	9	10	11	12	13	14	15
No. of items:	5	6	8	7	9	8	9	6

Solution. Here, the maximum frequency 9 belongs to two values of the items 12 and 14. However, Owing to the irregular distribution of frequencies, we use the grouping method to decide which one may be considered as the maximum frequency. Grouping table is constructed below to determine mode.

Grouping Table

Size of items	*Col. 1*	*Col. 2*	*Col. 3*	*Col. 4*	*Col. 5*	*Col. 6*
8	5					
		11				
9	6			19		
			14			
10	8				21	
		15				
11	7			**24**		**24**
			16			
12	9					
		17				
13	8				**26**	
			17			
14	9					23
		15				
15	6					

Analysis Table

Col. No.	*Size of items*							
	8	*9*	*10*	*11*	*12*	*13*	*14*	*15*
1					1		1	
2					1	1		
3						1	1	
4				1	1	1		
5					1	1	1	
6			1	1	1			
Total			1	2	5	4	3	

As size 12 occurs the maximum number of times, *i.e.,* 5 times, the mode is 12.

Example 15.15. Calculate mode from the following data:

Marks	*Number of students*	*Marks*	*Number of students*
Above 0	**80**	**Above 60**	**28**
Above 10	**77**	**Above 70**	**16**
Above 20	**72**	**Above 80**	**10**
Above 30	**65**	**Above 90**	**8**
Above 40	**55**	**Above 100**	**0**
Above 50	**43**		

Solution. Since this is cumulative frequency distribution, we first convert it into a simple frequency distribution.

Marks	*Number of students*	*Marks*	*Number of students*
0–10	3	50–60	15
10–20	5	60–70	12
20–30	7	70–80	6
30–40	10	80–90	2
40–50	12	90–100	8

By inspection the modal class is 50 – 60.

$$\text{Mode } (M_0) = L + \frac{f_1 - f_0}{2f_1 - f_0 - f_2} \times C$$

where, $L = 50; f_1 = 15; f_0 = 12; f_2 = 12;$ and $C = 10.$

$$\therefore \quad \text{Mode } (M_0) = 50 + \frac{15 - 12}{30 - 12 - 12} \times 10 = 50 + \frac{3}{6} \times 10 = 50 + 5 = 55.$$

Thus Mode = 55 marks.

Example 15.16. Calculate the mode of the following series:

Marks	200–220	220–240	240–280	280–300	300–320	320–340
No. of students	7	15	40	6	4	2

Solution. In order to calculate mode we first make class intervals equal. We have fixed 20 as the class interval. The adjusted distribution is as follows:

Marks	*Number of students*
200–220	7
220–240	15
240–260	20
260–280	6
280–300	6
300–320	4
320–340	2

Since concentration of items is around the class 240-260, hence 240-260 is the modal class. It can be verified with the help of the grouping method*. Applying the formula:

$$\text{Mode } (M_0) = L + \frac{f_1 - f_0}{2f_1 - f_0 - f_2} \times C$$

where, $L = 240; f_1 = 20; f_0 = 15; f_2 = 20$, and $C = 20$.

$$\therefore \quad \text{Mode } (M_0) = 240 + \frac{20 - 15}{40 - 15 - 20} \times 20$$

$$= 240 + \frac{5}{5} \times 20 = 240 + 20 = 260.$$

Thus Mode = 260 marks.

Example 15.17. In an asymmetrical distribution mean is 16 and the median is 20. Calculate mode.

Solution. Mode = 3 Median – 2 Mean

$= 3 \times 20 - 2 \times 16 = 60 - 32 = 28$

$\therefore$ Mode = 28.

Example 15.18. A student noted the number of cars passing through a spot on road for 100 periods each of 3 minutes and summarised it in the table given below. Find the mode of the data:

Number of cars	*Frequency*
0–10	**7**
10–20	**14**
20–30	**13**
30–40	**12**
40–50	**20**
50–60	**11**
60–70	**15**
70–80	**8**

Solution. From the above table it is clear that the maximum frequency is 20 and it lies in the class 40–50. Thus the modal class is 40–50.

Here $L = 40$, $h = 10$, $f_1 = 20$, $f_0 = 12$ and $f_2 = 11$

$$\therefore \quad \text{Mode} = L + \left(\frac{f_1 - f_0}{2f_1 - f_0 - f_2}\right) \times h = 40 + \left(\frac{20 - 12}{40 - 12 - 11}\right) \times 10$$

$$= 40 + \frac{8}{17} \times 10 = 40 + 4.7 = 44.7$$

Hence, Mode = 44.7 cars.

EXERCISE 15.3

1. A cricket player scored the following runs in 13 on a day matches:
65, 30, 07, 32, 60, 65, 30, 28, 30, 25, 06, 30
Find his modal score. [**Ans.** 30 runs]

2. Calculate mode for the following data:

No. of Persons	*Families*
1	26
2	113
3	120
4	95
5	60
6	42
7	21
8	14
9	5

[**Ans.** 3]

3. Find the mode from the following distribution:

x	30–40	40–50	50–60	60–70	70–80	80–90	90–100
f	6	10	16	14	10	5	2

[**Ans.** 57.5]

4. Convert mode from the following data:

Marks	*No. of Candidates*
0–9	6
10–19	29
20–29	87
30–39	181
40–49	247
50–59	263
60–69	113
70–79	49
80–89	9
90–99	2

[**Ans.** 51.097]

5. Determine the value of mode for the following data using the formula:

Mode = 3 Median – 2 Mean

Marks	No. of students
Less than 10	5
Less than 20	15
Less than 30	98
Less than 40	242
Less than 50	367
Less than 60	405
Less than 70	425
Less than 80	438
Less than 90	439

[**Ans.** $\overline{X}$ = 39.55, Median = 38.44, Mode = 36.22]

6. For what value of k for which the mode of the following data is 7:

$$3, 5, 5, 7, 3, 6, 7, 9, 6, 7, 3, 5, 7, 3, k$$

7. (*a*) In an asymmetrical distribution mean is 58 and the median 61. Calculate mode.

   (*b*) If mode in a usably asymmetrical distribution is 12 and median is 16, what would be the most probable mean?

   [**Ans.** (*a*) 67, (*b*) 18]

15.8. GRAPHICAL REPRESENTATION OF STATISTICAL DATA

Graphical representation is a way of analysing numerical data. It exhibits the relation between data, ideas, information and concepts in a diagram. It is easy to understand and it is one of the most important learning strategies. It always depends on the type of information in a particular domain. There are different types of graphical representation:

1. Histogram
2. Frequency Polygon
3. Cummulative frequency curve or on ogive

Histogram

It is a graphical representation of a frequency distribution in the form of rectangles with class-intervals as bases as the corresponding frequencies as heights, there being no gap between any two successive rectangles.

Note: If the given frequency table is in the inclusive form, we first convert it into an exclusive form and then draw the histogram.

Drawing of Histogram

Step 1. We represent along x-axis the class limits on a suitable scale.

Step 2. We construct rectangles with bases along x-axis and heightsalong y-axis.

Note: It must be assumed that scale for both the axis should be the same.

Drawing a Frequency Polygon

A frequency polygon is obtained by joining the middle points of the upper sides (tops) of the adjacent rectangles of the histogram by means of line segments.

To complete the pnlygon, the mid-point at each end is joined to the mid-point of the immediately lower or high imagined class interval as the case may be of zero frequency. Note that this ensures that toal area of a frequency polygon is the same as that of the corresponding histogram.

A frequency polygon can also be drawn independently by plotting the class marks along x-axis, frequencies along y-axis, and then joining the plotted points by segment lines.

Note: If both histogram and frequency polygon are to be drawn, then it is advisable first to draw histogram and then join the mid-points of the tops of the rectangles of the histogram to get frequency polygon. But if only frequency polygon is to be drawn then first represent the class marks along x-axis and frequencies along y-axis and then plot the point corresponding to the frequency at each class mark.

Drawing of Ogive

Just as histogram or a frequency polygon is a graph of frequency distribution, cummulative frequency curve or an o give is a graph of cumulative frequency distribution.

To draw the cumulative frequency curve we plot the points with the upper limits of the classes as abscissae and the corresponding cumulative frequencies as ordinates. We then join the points by a free hand smooth curve to get the ogive. It is a using curve.

Note: While drawing a cumulative frequency curve or an ogive we assume that there are class intervals before (and after) the given class intervals with zero frequency.

SOLVED EXAMPLES

Example 15.19. Following in the distribution of total household expenditure (in ₹) if manual workers in a city:

Class intervals (in rupees)	*Frequencies*
100–150	25
150–200	40
200–250	33
250–300	28
300–350	30
350–400	22
400–450	16
450–500	8
Total	202

Draw a histogram and a frequency polygon representing the above data.

Solution. Histogram and frequency polygon representing the distribution of total household expenditure (in ₹) of manual workers in a city.

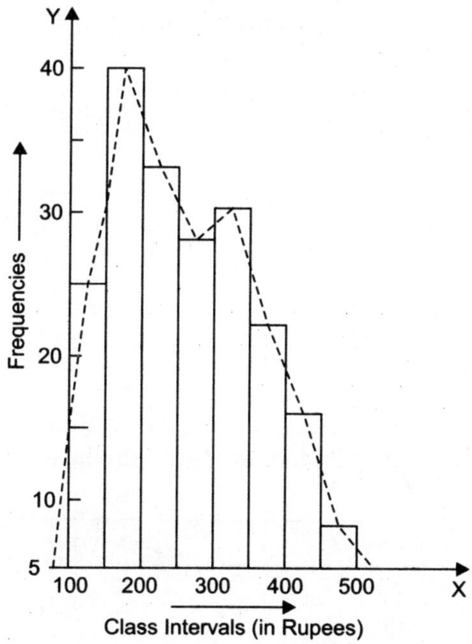

Example 15.20. Following in the distribution of total household expenditure (in ₹) if manual workers in a city

Class intervals	*Group A*	*Group B*
50–52	**4**	**2**
47–49	**10**	**3**
44–46	**15**	**4**
41–43	**18**	**8**
38–40	**20**	**12**
35–37	**12**	**17**
32–34	**13**	**22**
Total	**92**	**68**

Construct a frequency polygon for each of these groups on the same axes.

Solution. Frequency polygon for group A and group B representing the scoresof two groups of class IV students in a test of reading ability.

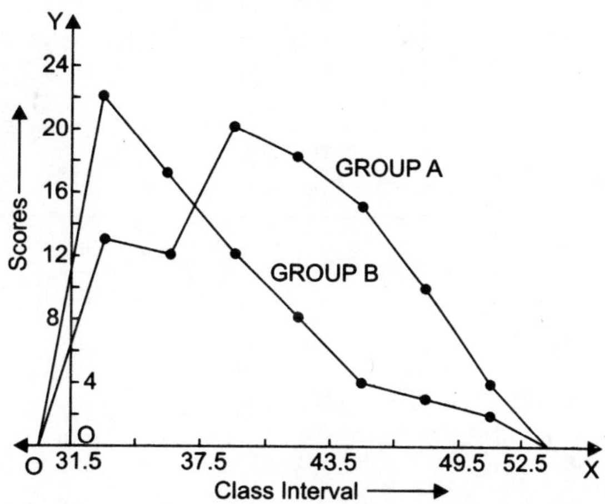

Let us convert the given distributions in such a manner that the intervals are contributions. It is shown below:

Class intervals	*Group A*	*Group B*
49.5–52.5	4	2
46.5–49.5	10	3
43.5–46.5	15	4
40.5–43.5	18	8
37.5–40.5	20	12
34.5–37.5	12	17
31.5–34.5	13	22
Total	92	68

Example 15.21. The frequency distribution of scores obtained by 230 candidates in a medical entrance test is as follows:

Intervals (scores)	*Frequency*
400–450	**20**
450–500	**35**
500–550	**40**
550–600	**32**
600–650	**24**
650–700	**27**
700–750	**18**
750–800	**34**
Total	**230**

Draw a cumulative frequency polygon (ogive) to represent the above data.

Solution. The cumulative frequency table for the given frequency table is as given as given below:

Intervals (scores)	*Frequency*	*Cumulative frequency*
400–450	20	20
450–500	35	55
500–550	40	95
550–600	32	127
600–650	24	151
650–700	27	178
700–750	18	196
750–800	34	230

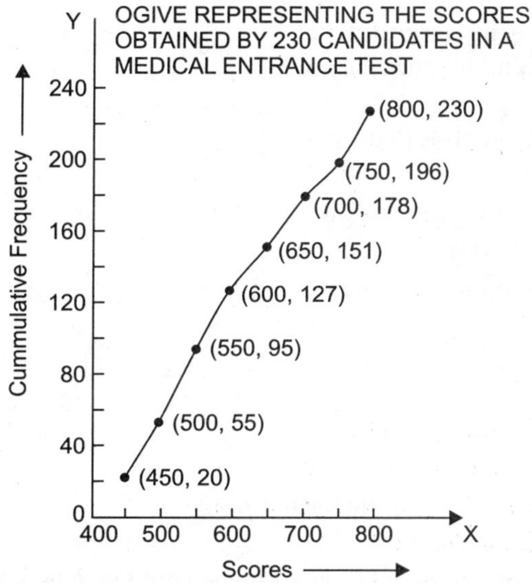

15.9. GRAPHICAL DETERMINATION OF MEDIAN

Median can be located graphically by ogive curves or cumulative frequency curves.

Step 1. Calculate ascending cumulative frequencies (less than) and descending cumulative frequencies (more than).

Step 2. Draw two ogive – one by less than and other by 'more than' methods.

Step 3. Draw perpendicular line from the point of intersection on x-axis.

Step 4. The point of intersectoin on x-axis gives the median.

SOLVED EXAMPLES

Example 15.22. Determine medan graphically from the following table:

Marks:	0–5	5–10	10–15	15–20	20–25	25–30	30–35	35–40
Students:	7	10	20	13	17	10	14	9

Solution.

Marks	*f*	*Marks less than*	*Less than cumulative frequencies (c.f.)*	*Marks more than*	*More than cumulative frequencies (c.f.)*
0–5	7	5	7	0	100
5–10	10	10	17	5	93
10–15	20	15	37	10	83
15–20	13	20	50	15	63
20–25	17	25	67	20	50
25–30	10	30	77	25	33
30–35	14	35	91	30	23
35–40	9	40	100	35	9
	$N = 100$				

Step 1. Calculate ascending cumulative frequencies (less than) and descending cumulative frequencies (more than).

Step 2. Draw two ogives–one by 'less than' and other by 'more than' methods.

Step 3. From the intersecting point of two ogives, draw a perpendicular of X-axis.

Step 4. The point where perpendicular touches X-axis, median value is determined.

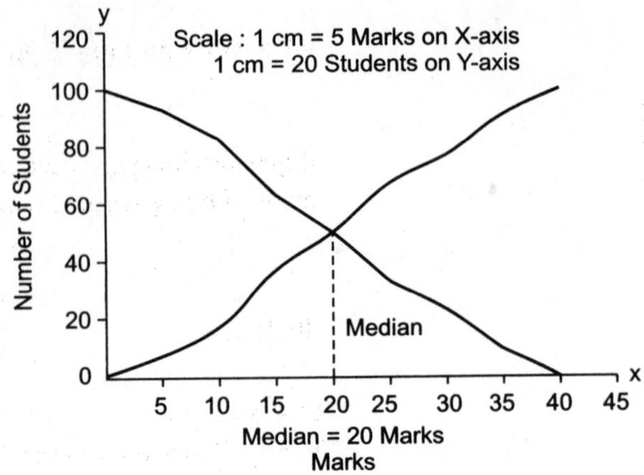

Merits

1. It is easy to calculate and understand
2. It indicates the value of middle item in the distribution. So it is well defined as an ideal average.
3. It is not affectedby extreme values.
4. It is useful in open end distribution because only position is to be known.
5. It is useful in the distribution of unequal class.
6. It can be determined graphically.
7. It is a proper average for qualitative data such as beauty, honesty etc.

Demerits

1. Data need to be arranged in ascending and descending order.
2. It is not based on all the observations of the series.
3. It is affected by fluctuations of sampling.
4. It is not accurate when data is huge
5. In continued series, interpolation by the formula is required to calculate median.

15.10. GRAPHICAL DETERMINATION OF MODE

The value of mode can be determined graphically in a frequency distribution. Following are the steps of locating:

1. Prepare a histogram of the given data.
2. The highest rectangle will be the modal class.
3. Draw two lines diagonally inside the modal class rectangle to the upper corner of the adjacent bar.
4. From the point of intersection of these lines, draw a perpendicular X-axis which gives the modal value.

SOLVED EXAMPLES

Example 15.23. Determine the value of mode of the following distribution graphically and verify the results:

Marks	0–10	10–20	20–30	30–40	40–50	50–60
No. of Students	5	12	14	10	8	6

Solution.

Scale: 2 cm = 10 Marks on X-axis
1 cm = 2 Students onY-axis

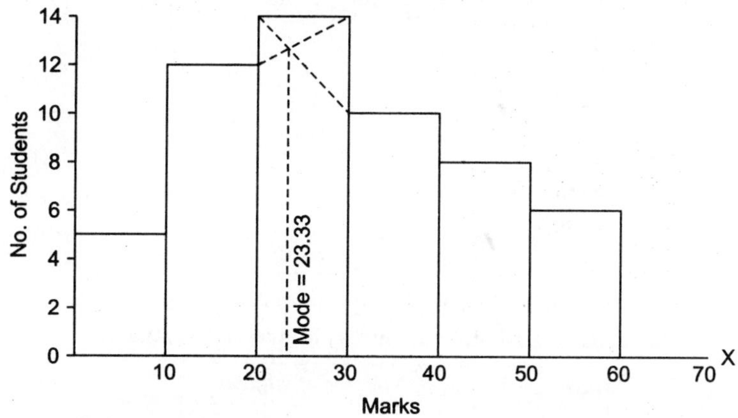

Verification: Mode lies in the class 20–30

$$\text{Mo} = l_1 + \frac{f_1 - f_0}{2f_1 - f_0 - f_2} \times i, \text{ where } l_1 = 20, f_1 = 14, f_0 = 12, f_2 = 10, i = 10$$

$$\text{Mo} = 20 + \frac{14 - 12}{2 \times 14 - 12 - 10} \times 10 = 20 + \frac{2}{6} \times 10 = 20 + 3.33$$

$\therefore$ Mode = 23.33 Marks

EXERCISE 15.4

1. In a study on certain disease, the following data were obtained:

Age at first detection (in years)	*Number of patients*
4–8	2
8–12	12
12–16	15
16–20	25
20–24	18
24–28	12
28–32	3
32–36	1

2. Represent the data given below by a histogram:

Age (in years)	*Number of workers*
11–13	3
13–15	4
15–17	5
17–19	6
19–21	5
21–23	4
23–25	3

3. Draw a histogram and frequency from the following data, giving the age of doctors working in C.G.H.S. in a city:

Age (in years)	*Number of doctors*
20–25	21
25–30	22
30–35	50
35–40	110
40–45	87
45–50	51
50–55	18
55–60	23

4. For the following data, draw a histogram and a frequency polygon:

Frequency distribution of marks obtained by 51 students

Marks	*Number of students*
0–10	5
10–20	10
20–30	4
30–40	6
40–50	7
50–60	3
60–70	2
70–80	2
80–90	3
90–100	9
Total	51

5. For the following data draw the corresponding ogive:

Marks	*Number of students*
0–10	6
10–20	4
20–30	8
30–40	2
40–50	4
Total	24

6. Plot a frequency curve and an ogive for the following data showing the parential income of 122 school of students of a class:

Income (in ₹)	*Number of students*
under 250	0
250 to less than 500	5
500 to less than 750	20
750 to less than 1000	43
1000 to less than 1200	37
1250 to less than 1500	10
1500 to less than 1750	5
1750 to less than 2000	2
Total	122

7. Represent the following data by an ogive:

Group	*Frequency*
0–10	4
10–20	4
20–30	7
30–40	10
40–50	12
50–60	8
60–70	5

8. The following table gives the frequency distribution of the marks obtained in a test give by 300 candidates. Construct a table showing the cumulative frequency and draw an ogive:

Marks obtained	*No. of candidates*
0–10	10
10–20	10
20–30	20
30–40	20
40–50	25
50–60	25
60–70	40
70–80	55
80–90	45
90–100	20
100–110	20
110–120	10

9. Draw a 'less than' ogive from the following data and hence find out the value of median

Class	*Frequency*
20–25	6
25–30	9
30–35	13
35–40	23
40–45	19
45–50	15
50–55	9
55–60	6

10. Draw the histogram and estimate the value of mode from the following data:

Marks	0–10	10–20	20–30	30–40	40–50	50–60	60–70	70–80	80–90
No. of students	0	2	3	7	13	11	9	2	1

15.11. MEASURES OF DISPERSION

The concept of dispersion is related to the extent of scatter or variability in observations. The variability, in an observation, is often measured as its derivation from a central value. A suitable average of all such dervations is called the measure of dispersion. Since most of the measures of dispersion. Since most of the measures of dispersion are based on the average of deviations of observations from an average, they are also known as the *average of second order.* As opposed to this, the measures of central tendency are known as the *averages of first order.* Some important definitions of dispersion are given below:

(*i*) "*Dispersion is the measure of variation of the items.*" **–A.L. Bowely**

(*ii*) "*Dispersion is the measure of extent to which individual items vary.*" **–L.R. Connor**

(*iii*) "*The measure of the scatteredness of the mass of figures in a series about an average is called the measure of variation or dispersion.*" **–Simpson and Kafka**

(*iv*) "*The degree to which numerical data tend to spread about an average value is called variation or dispersion of the data.*" **–Spiegel**

15.12. RANGE

The range is the most simple measure of dispersion and is based on the location of the largest and the smallest values in the data. Thus, the **range** *is defined to be the difference between the largest and smallest observed values in a data set.*

Symbolically, Range = $L - S$

where, L = Largest item,

S = Smallest item,

To compare the variability of two or more distributions given in different units of measurement, we cannot use absolute measure but we need a relative measure which is independent of the units of measurement.

The relative measure corresponding to range, called the coefficient of range, is obtained by applying the following formula:

$$\text{Coefficient of range} = \frac{L-S}{L+S}$$

SOLVED EXAMPLES

Example 15.24. The following are the sales figure of a firm for the last 12 months

Months	1	2	3	4	5	6	7	8	9	10	11	12
Sales (₹ '000)	80	82	82	84	84	86	86	88	88	90	90	92

Calculate the range and coefficient of range for sales.

Solution. Given that $L = 92$ and $S = 80$. Therefore,

$$\text{Range} = L - S = 92 - 80 = 12$$

and
$$\text{coefficient of range} = \frac{L-S}{L+S} = \frac{92-80}{92+80} = \frac{12}{172} = 0.069$$

Example 15.25. Calculate range and coefficient of range from the following data:

Marks	10–20	20–30	30–40	40–50	50–60	60–70
Number of students	6	10	12	8	4	3

Solution. We are given the data of class interval. In such a situation, the range can be calculated by finding out the difference between the upper limit of the highest class and the lower limit of the lowest class.

Given that, $L = 70$ and $S = 10$

$$\text{Range} = L - S = 70 - 10 = 60$$

and Coefficient of range $= \dfrac{L-S}{L+S} = \dfrac{70-10}{70+10} = \dfrac{60}{80} = 0.75$

EXERCISE 15.5

1. What is the meaning of dispersion and what are its objectives ?
2. What are the requisites of a good measure of dispersion ?
3. Distinguish between absolute and relative measure of dispersion.
4. The following are the per day earnings of semiagricultural labourers:

   200, 210, 208, 160, 220, 250

   Calculate the range and coefficient of range. [**Ans.** Range = 50, Coefficient of range = 0.42]
5. Calculate range and coefficient of range for the following data:

Income (₹)	50	70	80	90	100	120	130	150
Number of workers	2	8	12	7	4	3	8	6

[**Ans.** Range = 100; Coefficient of range = 0.5]

6. Calculate the range and coefficient of range for the following distribution:

Class	1–5	6–10	11–15	16–20	21–25	26–30
Frequency	2	8	15	35	20	10

[**Ans.** Range = 30; Coefficient of range = 0.97]

15.13. MEAN DEVIATION

Mean deviation is also known as average deviation. It is a measure of average of actual deviations of individual items from their averages (*i.e.* mean, median or mode of a distribution). Actual deviation mean ignoring the sign of deviations which render mean deviation in capable of further algebraic treatment. Mean deviation is also known as the *first movement of dispersion.*

(*i*) Mean Deviation for Unground Data or Individual Observations

If $x_1, x_2, ..., x_n$ are n values of a variable X, then the mean deviation from an average A (median or mean) is given by

$$\text{M.D.} = \frac{1}{n}\sum_{i=1}^{n} |x_i - A| = \frac{1}{n}\Sigma |d_i|, \text{ where } d_i = x_i - \text{A}$$

(*ii*) Mean Deviation of a Discrete Frequency Distribution

If $x_1, x_2, ..., x_n$ are n values of a variable and $f_1, f_2, ..., f_n$ be the corresponding frequencies, then mean deviation from an average A (median or AM) is given by

$$\text{M.D.} = \frac{1}{N}\Sigma f_i |x_i - A|, \text{ where } \sum_{i=1}^{n} f_i = N$$

(iii) Mean Deviation of a Grouped or Continuous Frequency Distribution

For calculating mean deviation of a continous frequency distribution the procedure is same as for a discrete frequency distribution. The only difference is that here we have to obtain the mid-points of the various classes and take the deviations of these mid points from the given central value (median or mean).

The relative measure of mean absolute deviation called the coefficient of mean absolute deviation is obtained by dividing the mean absolute deviation by a measure of central tendency (arithmetic mean or median) used for calculating mean deviation.

$$\text{Coefficient of M.D.} = \frac{\text{M.D.}}{\bar{x} \text{ or } Me}, \text{ where } \bar{x} \text{ is mean and } Me \text{ is median}$$

SOLVED EXAMPLES

Example 15.26. Find the mean deviation from the mean for the following data:

6, 7, 10, 12, 13, 4, 8, 20

Solution. Let $\bar{X}$ be the mean of the given data. Then,

$$\bar{X} = \frac{6+7+10+12+13+4+8+20}{8} = 10$$

Computation of Mean Deviation

x_i	$\lvert d_i \rvert = \lvert x_i - \bar{X} \rvert = \lvert x_i - 10 \rvert$
6	4
7	3
10	0
12	2
13	3
4	6
8	2
20	10
Total	$\Sigma d_i = 30$

We have, $\Sigma \lvert d_i \rvert = 30$ and $n = 8$

$$\therefore \quad \text{Mean Deviation (M.D.)} = \frac{1}{n}\Sigma \lvert d_i \rvert = \frac{30}{8} = 3.75$$

Thus, the mean deviation from the mean for the given data is 3.75.

Example 15.27. Calculate the mean deviation from the median for the following distribution:

x_i	10	15	20	25	30	35	40	45
f_i	7	3	8	5	6	8	4	9

Solution. We have to calculate mean deviation about median. So, first we calculate median.

x_i	f_i	*Cumulative frequency*	$\lvert d_i \rvert = \lvert x_i - 30 \rvert$	$f_i \lvert d_i \rvert$
10	7	7	20	140
15	3	10	15	45
20	8	18	10	80
25	5	23	5	25
30	6	29	0	0
35	8	37	5	40
40	4	41	10	40
45	9	50	15	135
	$N = \Sigma f_i = 50$			

We have, $N = 50 \Rightarrow \left(\frac{N+1}{2}\right) = 25.5$

The cumulative frequency just greater than $\left(\frac{N+1}{2}\right)$ is 29 and the corresponding value of x is 30.

Hence, median = 30.

Now, Mean deviation = $\frac{1}{N}\Sigma f_i \lvert d_i \rvert = \frac{505}{50} = 10.1$

Example 15.28. Calculate the mean deviation from the median of the following data:

Wages per week (in ₹)	10–20	20–30	30–40	40–50	50–60	60–70	70–80
Number of workers	4	6	10	20	10	6	4

Solution.

Calculation of Mean Deviation from Median

Wages per week (in ₹)	*Mid-Values* (x_i)	*Frequency* f_i	*Cumulative frequency*	$\lvert d_i \rvert = \lvert x_i - 45 \rvert$	$f \lvert d_i \rvert$
10–20	15	4	4	30	120
20–30	25	6	10	20	120
30–40	35	10	20	10	100
40–50	45	20	40	0	0
50–60	55	10	50	10	100
60–70	65	6	56	20	120
70–80	75	4	60	30	120
		$N = \Sigma f_i = 60$			

Here, $N = 60.$

So, $\frac{N}{2} = 30.$

The cumulative frequency just greater than $f = 30$ is 40 and the corresponding class is 40–50. So, 40–50 is the median class.

$\therefore$ $l = 40,\ f = 20,\ h = 10,\ F = 20.$

Now, $$\text{Median} = l + \frac{N/2 - F}{f} \times h = 40 + \frac{30 - 20}{20} \times 10 = 45$$

$$\text{Mean deviation from median} = \frac{\Sigma f_i |d_i|}{N} = \frac{680}{60} = 11.33$$

15.14. STANDARD DEVIATION

The variance of a variate X is the arithmetic mean of the squares of all deviations of X from the arithmetic mean of the observations and is denoted by Var (X) or σ^2.

The positive square root of the variance of a variate X is known as its standard deviation and is denoted by σ.

Thus, $$\text{Standard deviation} = +\sqrt{\text{Var}(X)}$$

Case I: For an individual series. If $x_1, x_2, ..., x_n$ are n values of a variable X, then by definition

$$\text{Var}(X) = \frac{1}{n}\sum_{i=1}^{n}(x_i - \overline{x})^2 \quad \text{or} \quad \sigma^2 = \frac{1}{n}\sum_{i=1}^{n}(x_i - \overline{x})^2$$

where $\overline{x}$ is the mean of these observations.

Alternatively $$\text{Var}(X) = \frac{1}{n}\sum_{i=1}^{n} x_i^2 - \left(\frac{1}{n}\Sigma x_i\right)^2 \quad \text{or} \quad \sigma = \sqrt{\frac{\sum_{i=1}^{n} x_i^2}{n} - \overline{x}^2}$$

The above formula is used when the items are very small, on the other hand the relevance of this method is particularly useful when computers are used where by the values, even when they are of high magnitude, can be used directly for calculating σ.

If the values of variable X are large, the calculation of variance from the above formula is quite tedious and time consuming. In this case, we take deviation from an arbitary point A (say). If $d_i = x_i - A : i = 1, 2, ..., n$, then the above formula reduces to

$$\text{Var}(X) = \frac{1}{n}\sum_{i=1}^{n} d_i^2 - \left(\frac{1}{n}\sum_{i=1}^{n} d_i\right)^2$$

or $$\text{Standard deviation} = \sqrt{\frac{1}{n}\sum_{i=1}^{n} d_i^2 - \left(\frac{1}{n}\sum_{i=1}^{n} d_i\right)^2}$$

Case II: For a frequency distribution

(*i*) *When the series is discrete*. If x_i/f_i; $i = 1, 2, ..., n$ is a discrete frequency distribution of a variate X, then

$$\text{Var}(X) = \frac{1}{N}\sum_{i=1}^{n} f_i (x_i - \overline{x})^2, \text{ where } N = \Sigma f_i$$

or $$\text{Standard deviation} = \sqrt{\frac{1}{N}\sum_{i=1}^{n} f_i (x_i - \overline{x})^2}$$

Alternatively, $$\text{Var}(X) = \frac{1}{N}\sum_{i=1}^{n} f_i x_i^2 - \left(\frac{1}{N}\Sigma f_i x_i\right)^2$$

or $$\text{Standard deviation} = \sqrt{\frac{1}{N}\sum_{i=1}^{n} f_i x_i^2 - \left(\frac{1}{N}\Sigma f_i x_i\right)^2}$$

If the values of X or (and) f are large the calculation of variance from the above formulae is quite tedious and time consuming. In such a case, we take deviation of the values of variable X from an arbitrary point A(say). If $d_i = x_i - A$; $i = 1, 2, ..., n$, then the above formula reduce to

$$\text{Var}(X) = \frac{1}{N}\sum_{i=1}^{n} f_i d_i^2 - \left(\frac{1}{N}\sum_{i=1}^{n} f_i d_i\right)^2$$

or $$\text{Standard deviation} = \sqrt{\frac{1}{N}\sum_{i=1}^{n} f_i d_i^2 - \left(\frac{1}{N}\sum_{i=1}^{n} f_i d_i\right)^2}$$

Sometimes $d_i = x_i - A$ are divisible by a common number h (say). If we define $u_i = \frac{x_i - A}{h}$, $i = 1, 2, ..., n$, then we obtain the following formula for variance.

$$\text{Var}(x) = h^2\left[\frac{1}{N}\sum_{i=1}^{n} f_i u_i^2 - \left(\frac{1}{N}\Sigma f_i u_i\right)^2\right]$$

or $$\text{Standard deviation} = \sqrt{h^2\left[\frac{1}{N}\sum_{i=1}^{n} f_i u_i^2 - \left(\frac{1}{N}\Sigma f_i u_i\right)^2\right]}$$

(*ii*) *When the series is continuous.* In a grouped or continuous frequency distribution, any of the methods discussed above for a discrete frequency distribution can be used.

15.15. COEFFICIENT OF VARIATION

This relative measurement is developed by Karl Pearson and is most popularly used to measure relative variation of two or more than two series. It shows the relationship between the standard deviation and the arithmetic mean expressed in terms of percentage. This measure is used to compare uniformity, consistency and variability in two different series. The coefficient of variation (C.V.) is defined as

$$\text{C.V.} = \frac{\sigma}{x}\times 100$$

where σ and $\bar{x}$ are the standard deviation and mean of the data.

For comparing the variability of two series, we calculate the coefficient of variation for each series. The series having greater C.V. is said to be more variable or conversely less consistant, less uniform, less table or less homogeneous that the other and the series having lesser C.V. is said to be more consistent (or homogeneous) than the other.

SOLVED EXAMPLES

Example 15.29. The score of a batsman in 10 matches were as follows:

25, 50, 45, 30, 70, 42, 36, 48, 34, 60

Compute the variance and standard deviation.

Solution. (*a*) **Actual Mean Method:**

Calculation of Variance

Score (x)	$X = x - \bar{x}$	X^2
25	–19	361
50	+6	36
45	+1	1
30	–14	196
70	–26	676
42	–2	4
36	–8	64
48	+4	16
34	–10	100
60	+6	256
$\sum x = 440$		$\sum X^2 = 1710$

Now, we get $\bar{x} = \frac{\sum x}{n} = \frac{440}{10} = 44$

Here $\sum X^2 = 1710,\ n = 0$

$\therefore$ $\text{Var}(X) = \frac{1}{n}\sum(x_i - \bar{x})^2 = \frac{1710}{10} = 171$

Hence, $\text{S.D.} = \sqrt{\text{Var}(x)} = \sqrt{171} = 13.076$

(*b*) **Direct Method:**

Calculation of Variance

Score (x)	x^2
25	625
50	2500
45	2025
30	900
70	4900
42	1764
36	1296
48	2304
34	1156
60	3600
$\sum x = 440$	$\sum x^2 = 21070$

Now $\bar{x} = \frac{\sum x}{n} = \frac{440}{10} = 44$

Here $\sum x^2 = 21070$, $n = 10$ and $\bar{x} = 44$

Apply the following formula:

$$\text{Var}(x) = \frac{\sum x^2}{n} - \bar{x}^2 = \frac{21070}{10} - (44)^2 = 2107 - 1936 = 171$$

Hence, $\text{S.D.} = \sqrt{\text{Var}(x)} = \sqrt{171} = 13.076$

(c) Assumed Mean Method:

Calculation of Variance

Score (x)	*d = (x – 45)*	d^2
25	–20	400
50	+5	25
45	0	0
30	–15	225
70	+25	625
42	–3	9
36	–9	81
48	+3	9
34	–11	121
60	+15	225
$n = 10$	$\Sigma d = -10$	$\Sigma d^2 = 1720$

Now, we get $\text{Var}(x) = \dfrac{\Sigma d^2}{n} - \left(\dfrac{\Sigma d}{n}\right)^2$

Here $\Sigma d^2 = 1720,\ n = 10,\ \Sigma d = -10$

$$\text{Var}(x) = \frac{1720}{10} - \left(\frac{-10}{10}\right)^2 = 172 - (-1)^2 = 171$$

Hence, $\text{S.D.} = \sqrt{\text{Var}(x)} = \sqrt{171} = 13.076$

Example 15.30. Calculate the standard deviation of the following data:

Size	4	5	6	7	8	9	10
Frequency	6	12	15	28	20	14	5

Solution. (a) Actual Mean Method:

Calculation of Standard Deviation

Size (x)	*Frequency (f)*	fx	$X = (x - \bar{x})$	X^2	fX^2
4	6	24	3.06	9.3636	56.1816
5	12	60	–2.06	4.2436	50.1232
6	15	90	–1.06	1.1236	16.8540
7	28	196	–0.06	0.0036	0.1008
8	20	160	+0.94	0.8836	17.6720
9	14	126	+1.94	3.7636	52.6904
10	5	50	+2.94	8.6436	43.2180
	$N = 100$	$\Sigma fx = 706$			$\Sigma fX^2 = 237.6400$

Here, $\Sigma fx = 706$ and $N = \Sigma f = 100$

$\therefore$ $$\bar{x} = \frac{706}{100} = 7.06$$

Applying formula, we get $\sigma = \sqrt{\dfrac{\Sigma fX^2}{N}}$

Here, $\Sigma fX^2 = 237.6400$ and $N = 100$

$$\therefore \quad \sigma = \sqrt{\frac{237.64}{100}} = \sqrt{2.3764} = 1.541$$

(*b*) Direct Method: **Calculation of Standard Devation**

Size (x)	*Frequency (f)*	*fx*	x^2	fx^2
4	6	24	16	96
5	12	60	25	300
6	15	90	36	540
7	28	196	49	1372
8	20	160	64	1280
9	14	126	81	1134
10	5	50	100	500
	$N = 100$	$\Sigma fx = 706$		$\Sigma fx^2 = 5222$

Here, $\Sigma fx = 706$ and $N = 100$

$$\therefore \quad \bar{x} = \frac{\Sigma fx}{\Sigma f} = \frac{706}{100} = 7.06$$

Applying formula, we get $\sigma = \sqrt{\frac{\Sigma fx^2}{N} - \bar{x}^2} = \sqrt{\frac{5222}{100} - (7.06)^2}$

$$= \sqrt{52.22 - 49.8436} = \sqrt{2.3764} = 1.541$$

(*c*) **Assumed Mean Method:**

Calculation of Standard Deviation

Size (x)	*Frequency (f)*	*d = (x – 7)*	*fd*	fd^2
4	6	–3	–18	54
5	12	–2	–24	48
6	15	–1	–15	15
7	28	0	0	0
8	20	+1	+20	20
9	14	+2	+28	56
10	5	+3	+15	45
	$N = 100$		$\Sigma fd = 6$	$\Sigma fd^2 = 238$

Here, $\Sigma fd^2 = 238$, $\Sigma fd = 6$, $N = 100$

Applying formula $\sigma = \sqrt{\frac{\Sigma fd^2}{N} - \left(\frac{\Sigma fd}{N}\right)^2} = \sqrt{\frac{238}{100} - \left(\frac{6}{100}\right)^2} = \sqrt{2.38 - (0.06)^2}$

$$= \sqrt{2.38 - 0.0036} = \sqrt{2.3764}$$

$$\sigma = 1.541$$

Example 15.31. Find out the standard deviation of the following frequency distribution:

Age in years	50–55	45–50	40–45	35–40	30–45	25–30
Number of labourers	22	29	31	47	51	70

Solution. (*a*) **Assumed Mean Method**

Calculation of Standard Deviation

Age in years (*x*)	*Number of labourers* (*f*)	*Mid-point* (*m*)	*m* – 42.5 *d*	*fd*	*fd*²
50–55	22	52.5	+10	220	2200
45–50	29	47.5	+5	145	725
40–45	31	42.5	0	0	0
35–40	47	37.5	–5	–235	1175
30–35	51	32.5	–10	–510	5100
25–30	70	27.5	–15	–1050	15750
	$N = 250$			$\Sigma fd = -1430$	$\Sigma fd^2 = 24950$

Here $\Sigma fd^2 = 24590$, $\Sigma fd = -1430$, $N = 250$

$$\therefore \qquad \sigma = \sqrt{\frac{\Sigma fd^2}{N} - \left(\frac{\Sigma fd}{N}\right)^2} = \sqrt{\frac{24590}{250} - \left(\frac{-1430}{250}\right)^2}$$

$$= \sqrt{998 - (-5.72)^2} = \sqrt{99.8 - 32.718} = \sqrt{67.082} = 8.19$$

∴ Standard deviation = 8.19 years

(*b*) **Step Deviation Method:** This method is mostly used in practice.

Calculation of Standard Deviation

Age in years (*x*)	*Number of labourers* (*f*)	*Mid-point* (*m*)	$d = m - 4.25$	$u = \frac{m - 42.5}{5}$	*fu*	*fu*²
50–55	22	52.5	+10	+2	+44	88
45–50	29	47.5	+5	+1	+29	29
40–45	41	42.5	0	0	0	0
35–40	47	37.5	–5	–1	–47	47
30–35	51	32.5	–10	–2	–102	204
25–30	70	27.5	–15	–3	–210	630
	$N = 250$			$\Sigma fu = -286$	$\Sigma fu^2 = 998$	

Here, $\Sigma fu^2 = 998$, $\Sigma fu = -286$, $N = 250$ and $h = 5$

$$\therefore \qquad \sigma = \sqrt{\frac{\Sigma fu^2}{N} - \left(\frac{\Sigma fu}{N}\right)^2} \times h$$

$$= \sqrt{\frac{998}{250} - \left(-\frac{286}{250}\right)^2} \times 5 = \sqrt{3.992 - (-1.144)^2} \times 5$$

$$= \sqrt{3.992 - 1.309} \times 5 = \sqrt{2.683} \times 5 = 1.638 \times 5 = 8.19$$

∴ Standard deviation = 8.19 years.

Example 15.32. The mean and variance of 7 observations are 8 and 16 respectively. If 5 of the observations are 2, 4, 10, 12, 14, find the remaining two observations.

Solution. Let x and y be the remaining two observations.

$$\text{Mean} = 8$$

$$\Rightarrow \quad \frac{2+4+10+12+14+x+y}{7} = 8$$

$$\Rightarrow \quad 4x + x + y = 56$$

$$\Rightarrow \quad x + y = 14 \quad \text{...(i)}$$

$$\text{Variance} = 16$$

$$\Rightarrow \quad \frac{1}{7}(2^2 + 4^2 + 10^2 + 12^2 + 14^2 + x^2 + y^2) - (\text{Mean})^2 = 16$$

$$\Rightarrow \quad \frac{1}{7}(4 + 16 + 100 + 144 + 196 + x^2 + y^2) - 64 = 16$$

$$\Rightarrow \quad x^2 + y^2 = 100$$

Now $$(x+y)^2 + (x-y)^2 = 2(x^2 + y^2)$$

$$\Rightarrow \quad 196 + (x-y)^2 = 2 \times 100$$

$$\Rightarrow \quad (x-y)^2 = 4 \quad \Rightarrow \quad x - y = +2$$

If $x - y = 2$, then $x + y = 14$ and $x - y = 2$

$\Rightarrow \quad x = 8, y = 6$

If $x - y = -2$, then $x + y = 14$ and $x - y = -2$

$$\Rightarrow \quad x = 6, y = 8$$

Hence, the remaining two observations are 6 and 8.

Example 15.33. Mean and S.D. of 100 items are found by a student are 50 and 0.1. If at the time of calculation, two items are wrongly taken as 40 and 50 instead of 60 and 30. Find correct mean and standard deviation.

Solution. Given: $\bar{x} = 50,\ n = 100,\ \sigma = 0.1$ Thus,

$$\bar{x} = \frac{\Sigma x}{n}$$

or $$\Sigma x = n\bar{x} = 100 \times 50 = 5000$$

$$\text{Correct } \Sigma x = 5000 - \text{wrong items} + \text{correct items}$$

$$= 5000 - 40 - 50 + 60 + 30 = 5000$$

Correct Mean, $$\bar{x} = \frac{5000}{100} = 50$$

Now, $$\sigma = \sqrt{\frac{\Sigma x^2}{n} - (\bar{x})^2} \quad \text{or} \quad \sigma^2 = \frac{\Sigma x^2}{n} - (\bar{x})^2$$

Substituting the values: $\sigma = 0.1$, $\bar{x} = 50$, $n = 100$, we get

$$(0.1)^2 = \frac{\Sigma x^2}{100} - (50)^2$$

$$\Rightarrow \quad 0.01 = \frac{\Sigma x^2}{100} - 2500 \quad \Rightarrow \quad \Sigma x^2 = 250001$$

Correct $\Sigma x^2 = 250001 - (40)^2 - (50)^2 + (60)^2 + (30)^2$

$= 250001 - 1600 - 2500 + 3600 + 900 = 250401$

$\therefore$ Correct standard deviation $= \sqrt{\frac{\Sigma x^2}{n} - (\bar{x})^2} = \sqrt{\frac{250401}{100} - (50)^2} = 2.0025$

Example 15.34. The following is the record of goals scored by team A in football session

Number of goal scored	0	1	2	3	4
Number of matches	1	9	7	5	3

For the team *B*, mean number of goal scored per match was 2 with a standard deviation 1.25 goals. Find which team may be considered more consistent.

Solution. In order to determine the consistency of teams, we will have to find the coefficients of variations of two teams.

Computation of Mean and Standard deviation of goals scored by team A

Number of goals scored x_i	*Number of matches* f_i	$f_i x_i$	$f_i x_i^2$
0	1	0	0
1	9	9	9
2	7	14	28
3	5	15	45
4	3	12	48
	$\Sigma f_i = 25$	$\Sigma f_i x_i = 50$	$\Sigma f_i x_i^2 = 130$

$\therefore \qquad \bar{x}_A = \frac{\Sigma f_i x_i}{\Sigma f_i} = \frac{50}{25} = 2$

and, $\qquad \sigma_A^2 = \frac{\Sigma f_i x_i}{\Sigma f_i} - \bar{x}_A^2 = \frac{130}{25} - 4 = 5.2 - 4 = 1.2$

$\therefore \qquad \sigma_A = \sqrt{1.2} = 1.095$

It is given that $\bar{x}_B = 2$ and $\sigma_B = 1.25$

Now, coefficient of variation in goals scored by team $A = \frac{\sigma_A}{\bar{x}_A} \times 100$

$= \frac{1.095}{2} \times 100 = 54.75\%$

Coefficient of variation of goals scored by team $B = \frac{\sigma_B}{\bar{x}_B} \times 100 = \frac{1.25}{2} \times 100 = 62.50\%$

Since coefficient of variation of goals scored by team *A* is lesser than that of team *B*, hence team *A* is more consistent.

Example 15.35. An analysis of monthly wages paid to workers in two firms A and B, belonging to the same industry, gives the following results:

	Firm A	*Firm B*
Number of wage earners	586	648
Mean of monthly wages (₹)	5253	5253
Variance of the distribution of wages	100	121

(*i*) which firm *A* or *B* pays larger amount as monthly wages?

(*ii*) which firm *A* and *B* shows greater variability in individual wages?

Solution. *Firm A:*

Number of wage earners (say), $n_1 = 586$

Mean of monthly wages (say) $\bar{x}_1 =$ ₹ 5253

$$\therefore \quad \text{Mean of monthly wages} = \frac{\text{Total monthly wage}}{\text{Number of workers}}$$

$$\Rightarrow \quad 5253 = \frac{\text{Total monthly wages}}{586}$$

$$\Rightarrow \quad \text{Total monthly wages} = 5253 \times 586 = ₹\ 3078258$$

and $$\text{C.V.} = \frac{\sigma_1}{x_1} \times 100 \quad \text{or} \quad \text{C.V.} = \frac{100}{5253} \times 100 = 1.90$$

Firm B:

Number of wage earners (say) $n_2 = 648$

Mean of monthly wages (say), $\bar{x}_2 =$ ₹ 5253

$$\therefore \quad \text{Mean of monthly wages} = \frac{\text{Total monthly wages}}{\text{Number of workers}}$$

$$5253 = \frac{\text{Total monthly wages}}{648}$$

$$\Rightarrow \quad \text{Total monthly wages} = 5253 \times 648 = ₹\ 3403944$$

and $$\text{C.V.} = \frac{\sigma_2}{\bar{x}_2} \times 100 \quad \text{or} \quad \text{C.V.} = \frac{121}{5253} \times 100 = 2.30\%$$

(*i*) Clearly, firm *B* pays out larger amount as monthly wages.

(*ii*) Since coefficient of variation is higher for factory *B*, there is greater variability in individual wages.

Merits and Demerits of Standard Deviation

Merits:

1. In measuring variation, the standard deviation (SD) is the best measure because of its mathematical characteristics. And it includes each and every item and is amenable to algebraic treatment.
2. Combined standard deviation of two or more groups can be computed.
3. Variability of two or more distributions can be compared by coefficient of variation which is considered to be most appropriate.
4. Standard deviation is prominently used in further statistical work. For example, in calculation correlation, skewness, etc.

Demerits:

1. It is difficult to compute standard deviation as compared to other measures.
2. It gives more weight to extreme items and less to those which are near to the mean. For instance, the deviation 4 and 8 are in the ratio of 1 : 2 but their squares, *i.e.*, 16 and 64 would be in the ratio of 1 : 4.

EXERCISE 15.6

1. The number of patients seen in the emergency ward of a hospital for a simple of 5 days in the last month were: 153, 147, 151, 156 and 153. Determine the mean deviation and interpret.

[**Ans.** 3 patients (approx.), number of patients deviate on the average by 3 patients]

2. Find out the average deviation from median for the following distribution:

Values	6	12	18	24	30	36	42
Frequency	4	7	9	18	15	10	5

[**Ans.** Median = 24, M.D. = 7.5]

3. The number of telephone calls received at an exchange in 245 successive one-minute intervals are shown in the following frequency distribution:

Number of calls	0	1	2	3	4	5	6	7
Frequency	14	21	25	43	51	40	39	12

Compute the mean deviation about median. [**Ans.** 1.49]

4. Find the mean deviation from the mean for the following data:

Values	5	7	9	10	12	15
Frequency	8	6	2	2	2	6

[**Ans.** 3.39]

5. Find the mean deviation about the mean for the following data:

Marks obtained	10–20	20–30	30–40	40–50	50–60	60–70	70–80
Number of students	2	3	8	14	8	3	2

[**Ans.** 10]

6. Find the mean deviation about the median of the following frequency distribution:

Class	0–6	6–12	12–18	18–24	24–30
Frequency	8	10	12	9	5

[**Ans.** 6.318]

7. Find the standard deviation of 10 persons whose income in rupees is given below:

312 292 227 235 269 255 333 348 321 299 [**Ans.** 39.24]

8. Find the standard deviation of the following data:

Size of item	10	11	12	13	14	15	16
Frequency	2	7	11	15	10	4	1

[**Ans.** 1.342]

9. In a survey on patients, the following data was obtained. Find the standard deviation of the data.

Age (in years)	10–19	20–29	30–39	40–49	50–59	60–69	70–79	80–89
Number of cases	1	0	1	10	17	38	9	3

[**Ans.** 11.28]

10. Find the standard deviation of the following data:

x	4.5	14.5	24.5	34.5	44.5	54.5	64.5
f	1	5	12	22	17	9	4

[**Ans.** 12.93]

11. You are provided with the following raw data of two variables x and y.

$$\Sigma x = 235, \Sigma y = 250, \Sigma x^2 = 6750, \Sigma y^2 = 6840$$

Ten pairs of values are included in the survey. Calculate their standard deviations.

[**Ans.** $\sigma_x = 11.08$, $\sigma_y = 7.68$]

12. Find the standard deviation for the following discrete distribution:

x	8	12	16	20	24
$P(x)$	$\frac{1}{8}$	$\frac{1}{6}$	$\frac{3}{8}$	$\frac{1}{4}$	$\frac{1}{12}$

[**Ans.** $2\sqrt{5}$]

13. Let $x_1, x_2, \ldots x_n$ be n values of a variable X. If these values are changed to $x_1 + a, x_2 + a, \ldots, x_n + a$, where $a \in R$, show that the variance remains unchanged.

14. If $x_1, x_2, \ldots, x_n$ be n values of a variable X and let 'a' be a non-zero real number. Then, prove that the variance of the observations $ax_1, ax_2, \ldots, ax_n$ is a^2 var (x).

15. In a series of 5 observations, the mean and variance are 4, 4 and 8.24. If three observations are 1, 2, 6, find other two. [**Ans.** 4 and 9]

16. For a group of 200 candidates the mean and standard deviation were found to be 40 and 15 respectively. Later on it was found that the score 43 was misread as 34. Find the correct mean and correct standard deviation.

[**Ans.** 40.045, 14.995]

17. The mean and standard deviation of 20 observations are found to be 10 and 2 respectively. On rechecking it was found that an observation 8 was incorrect. Calculate the correct mean and standard deviation in each of the following cases:

(*i*) If wrong item is omitted (*ii*) If it is replaced by 12. [**Ans.** (*i*) 10.10, 1.997, (*ii*) 10.2, 1.9899]

18. Two brands of tyres are tested with the following results:

Life ('000 miles)	20–25	25–30	30–35	35–40	40–45	45–50
Brand X	8	15	12	18	13	9
Brand Y	6	20	32	30	12	0

Which brand has a greater variation?

[**Ans.** C.V. of X = 21.82%, C.V. of Y = 16.101%. Brand X has greater variation]

19. Suppose that the samples of polythene bags from two manufacturers A and B are tested by a buyer for bursting pressures, giving the following results

Brusting Pressurse	***Number of bags***	
	A	***B***
5.0–9.9	2	9
10.0–14.9	9	1
15.0–19.9	29	18
20.0–24.9	54	32
25.0–29.9	11	27
30.0–34.9	5	13

(*a*) Which set of bags has the highest bursting pressure?

(*b*) Which has more uniform pressure? If prices are the same, which manufacturer's bags would be by the buyer? Why?

Ans. Manufacturer *A*: $\bar{x}_1 = 21, \sigma_1 = 4.875, C.V. = 23.32\%$

Manufacturer *B*: $\bar{x}_2 = 21.81, \sigma_2 = 7.074, C.V. = 32.44\%$

(*a*) Bags of manufacturer *B* have higher bursting pressure;

(*b*) Bags of manufacturer *A* have more uniform pressure

(*c*) Bags of manufacturer *A* should be preferred by buyer as they have uniform pressures

15.16. SPEARMAN'S RANK CORRELATION COEFFICIENT METHOD

Karl Pearson's coefficient of correlation discussed degree of covariability of liner relationship between two quantitative variables. But often we come across situations when definite measurements on the variables are not possible *i.e.*, the factor, under study cannot be measured in quantitative terms. For example, the evaluation of a group of students on the basis of leadership ability, the ordering of women in a beauty contest, the ranking of students by two or more judges in an interview and so on. In all such cases, objects or individuals may be ranked and arranged in order of merit or proficiency on two variable and when these two sets of ranks covery or have agreement between them, we measure the degree of covariation or relationship by **coefficient of rank correlation.**

In rank correlation we may have three types of problems:

(I) when ranks are given.

(II) when ranks are not given.

(III) when equal ranks are given to more than two attributes.

(I) When Ranks are Given

When actual ranks are given to us the steps required fro computing rank correlation are:

Step 1. Take the difference of two ranks, *i.e.*, $(R_1 - R_2)$ and denote these differences by D.

Step 2. Square these difference and obtain the total $\sum D^2$.

Step 3. Apply the formula: $R = 1 - \dfrac{6\sum D^2}{n(n^2 - 1)}$

when n is the total number of pairs of observations.

SOLVED EXAMPLES

Example 15.36. Two ladies were asked to rank 7 different types of lipsticks. The ranks given by them are given below:

Lipisticks	*A*	*B*	*C*	*D*	*E*	*F*	*G*
Anita	2	1	4	3	5	7	6
Sunita	1	3	2	4	5	6	7

Calculate Spearman's rank correlation coefficient.

Solution. Let ranks given by Anita be denoted by R_1 and Sunita by R_2.

Calculation of rank correlation coefficient

R_1	R_2	$D = R_1 - R_2$	D^2
2	1	+1	1
1	3	−2	4
4	2	+2	4
3	4	−1	1
5	5	0	0
7	6	+1	1
6	7	−1	1
			$\Sigma D^2 = 12$

Now $$R = 1 - \frac{6\sum D^2}{n(n^2-1)}$$

Here $\Sigma D^2 = 12,\ n = 7$

∴ $$R = 1 - \frac{6\times 12}{7(7^2-1)} = 1 - 0.214 = 0.786.$$

(II) When Ranks are not Given

In this case we are given only the data. We assign rank to both the series x and y by giving the rank 1 to highest values in both the series (or to the lowest values in both the series) and so on.

SOLVED EXAMPLES

Example 15.37. From the data given below, calculate the coefficient of rank correlation between *X* and *Y*.

X	78	89	97	69	59	79	68	57
Y	125	137	156	112	107	136	123	108

Solution. Here $n = 8$

Computation of rank correlation coefficient

X	Y	*Rank in X* (R_1)	*Rank in Y* (R_2)	*Rank difference* $D = R_1 - R_2$	D^2
78	125	4	4	0	0
89	137	2	2	0	0
97	156	1	1	0	0
69	112	5	6	−1	1
59	107	7	8	−1	1
79	136	3	3	0	0
68	123	6	5	1	1
57	108	8	7	1	1
				$\Sigma D = 0$	$\Sigma D^2 = 4$

Coefficients of rank correlation:

$$R = 1 - \frac{6\sum D^2}{n(n^2-1)} = 1 - \frac{6\times 4}{8(64-1)} = 1 - \frac{3}{63} = 0.952.$$

(III) When Equal Ranks are Given to More than Two Attributes

It two or more individuals are placed together in any classification with respect to an attribute, *i.e.*, if in case of variable data, their are more than one item with the same rank in either or both the series (*i.e.*, Tie rank), then the Spearman's Rank Correlation coefficient formula given above does not give the correlation coefficient for Tie Rank. The problem is solved by assigning average rank to each of these individuals who are put in tie. For example, suppose an item is repeated at rank 5th (*i.e.*, the 5th and 6th items are having same values), then the common rank assigned to 5th and 6th items is $(5 + 6)/2 = 5.5$, which is the average of 5 and 6, the ranks which these items would have been assigned if they were different. The next rank assigned will be 7. But if an item is repeated thrice at rank 2, then the common rank assigned to each value will be $(2 + 3 + 4)/3 = 3$, which is the arithmetic mean of 2, 3 and 4. The next rank to be assigned would be 5. In order to find the Rank Correlation coefficient of Repeated Ranks or Tie Ranks, an Adjustment or Correction Factor is added to the Spearman's Rank Correlation formula, *i.e.*, as

$$R = 1 - \frac{6\Sigma D^2}{n(n^2-1)} \qquad ...(i)$$

Correction Factor. "In the formula (*i*), add the factor $\frac{m(m^2-1)}{12}$ to $\sum D^2$, where m is the number of times an item is repeated. This correction factor is to be added for each repeated value in both the series."

The modified formula for Tie Rank Correlation Coefficient is given by

$$R = 1 - \frac{6\left[\sum D^2 + \frac{1}{12}(m_1^3 - m_1) + \frac{1}{12}(m_2^3 - m_3) + ...\right]}{n(n^2-1)}$$

where $m_1, m_2, m_3, ...,$ are the number of times a value is repeated.

SOLVED EXAMPLES

Example 15.38. From the following data of the makes obtained by 8 students in Mathematics and Physics paper compute rank coefficients of correlation.

Marks in Mathematics	15	20	28	12	40	60	20	80
Marks in Physics	40	30	50	30	20	10	30	60

Solution. *Computation of rank correlation*

Marks in Mathematics X	*Rank assigned R_1*	*Marks in Physics Y*	*Rank assigned R_2*	D^2
15	2	40	6	16.00
20	3.5	30	4	0.25
28	5	50	7	4.00
12	1	30	4	9.00
40	6	20	2	16.00
60	7	10	1	36.00
20	3.5	30	4	0.25
80	8	60	8	0.00
				$\Sigma D^2 = 81.5$

$$R = 1 - \frac{6\left[\sum D^2 + \frac{1}{12}(m_1^3 - m_1) + \frac{1}{12}(m_2^3 - m_2)\right]}{n^3 - n}$$

Here $\Sigma D^2 = 81.5, n = 8$

The item 20 is repeated 2 times is series X. So $m_1 = 2$. In series Y the item 30 occurs 3 times and so $m_2 = 3$.

Substituting these values in the above formula.

$$R = 1 - \frac{6\left[81.5 + \frac{1}{12}(2^3 - 2) + \frac{1}{12}(3^3 - 3)\right]}{8^3 - 8}$$

$$= 1 - \frac{6(81.5 + 0.5 + 2)}{504} = 1 - \frac{6 \times 84}{504} = 1 - \frac{504}{504} = 0.$$

Example 15.39. From the following table, calculate the rank correlation coefficient:

X	48	33	40	9	16	16	65	24	16	57
Y	13	13	24	6	15	4	20	9	6	19

Solution. We prepare the following table.

Computation of rank correlation coefficient

X	Y	Rank X R_1	Rank Y R_2	$D = R_1 - R_2$	D^2
48	13	8	5.5	2.5	6.25
33	13	6	5.5	0.5	0.25
40	24	7	10	– 3	9
9	6	1	2.5	– 1.5	2.25
16	15	3	7	– 4	16
16	4	3	1	2	4
65	20	10	9	1	1
24	9	5	4	1	1
16	6	3	2.5	0.5	0.25
57	19	9	8	1	1
					$\Sigma D^2 = 41$

In the X-series, we notice that the value 16 is repeated thrice. The common rank given to these values is 3, which is the average of 2, 3 and 4 $\left[i.e., \frac{(2+3+4)}{3} = 3\right]$, the ranks which these values would has assumed if they were different. The next rank given to the next value is 5. Here $m_1 = 3$. The correction factor for X-series is.

$$\frac{m_1(m_1^2 - 1)}{12} = \frac{3(3^2 - 1)}{12} = 2.$$

In the Y-series, the value 6 and 13 are repeated twice. The value of 6 occurs twice and its average rank is $2.5 = \frac{(2+3)}{2}$. The next value 9 is assigned the rank 4. Here $m_2 = 2$. Again the value 13 is repeated twice at different places. The average rank is $5.5 = \frac{5+6}{2}$. The next value 5 is assigned the rank 7. Here $m_3 = 2$. The total correction for the Y-series is

$$\frac{m_2(m_2^2-1)}{12}+\frac{m_3(m_3^2-1)}{12} = \frac{2(2^2-1)}{12}+\frac{2(2^2-1)}{12}=0.5+0.5=1.$$

Thus the rank correlation is:

$$R = 1-\frac{6\left[\sum D^2+\frac{m_1(m_1^2-1)}{12}+\frac{m_2(m_2^2-1)}{12}+\frac{m_3(m_3^2-1)}{12}\right]}{n(n^2-1)}$$

$$= 1-\frac{6[41+2+0.5+0.5]}{10(10^2-1)}=1-\frac{264}{990}=1-0.267=0.733.$$

Merits and Demerits of Rank Correlation Coefficient Method

Merits:

1. It is simple to calculate and easy to understand.
2. This method is very useful where the data are of a qualitative nature.
3. When we are given the ranks instead of actual data, this is the only method for finding out the degree of correlation.

Demerits:

1. This method cannot be used for calculating correlation in a grouped frequency distributions. It is applicable only to individual observations.
2. If the number of values is quite large, it becomes a difficult task to ascertain the ranks and their differences.
3. This method lacks precision as compared to Karl Pearson's method. It uses ranks only. Original values are not taken into account.

EXERCISE 15.7

1. The ranking of ten students is statistics and Mathematics are as follows:

Statistics	3	5	8	4	7	10	2	1	6	9
Mathematics	6	4	9	8	1	2	3	10	5	7

what is the coefficient of rank correlation. **[Ans.** – 0.297]

2. Calculate the coefficient of correlation from the following data by the method of rank correlation:

X	75	88	95	70	60	80	81	50
Y	120	134	150	115	110	140	142	100

[Ans. 0.9286]

3. Compute Rank correlation from the following table of Index Number of Supply (X) and price (Y):

X	115	134	120	130	124	128
Y	130	132	128	130	127	125

[Ans. 0.3286]

4. Ten competitors in a beauty contest are ranked by three judges in the following order:

First judge	1	5	4	8	9	6	10	7	3	2
Second judge	4	8	7	6	5	9	10	3	2	1
Third judge	6	7	8	1	5	10	9	2	3	4

Use the rank correlation to gauge which pair of judges have the nearest approach to common taste in beauty.

[**Ans.** $r_{12} = 0.5515$, $r_{23} = 0.7333$, $r_{31} = 0.0545$]

5. Following are the scores of ten students in a class and their I.Q.

Student	1	2	3	4	5	6	7	8	9	10
Score	35	40	25	55	85	90	65	55	45	50
I.Q.	100	100	110	140	150	130	100	120	140	110

Calculate the rank correlation coefficient between the score and I.Q. [**Ans.** 0.4909]

6. Calculate Spearman's rank correlation coefficient between advertisement cost and sales from the following data:

Adv. Cost (₹ 1000)	39	65	62	90	82	75	25	98	36	78
Sales (₹ 1000.000)	47	53	58	86	62	68	60	91	51	84

[**Ans.** 0.8182]

7. Obtain the Rank correlation from the following data:

X	68	64	75	50	64	80	75	40	55	64
Y	62	58	68	45	81	60	68	48	50	70

[**Ans.** 0.5636]

❑❑❑

CHAPTER 16

Basic Probability

INSIDE THIS CHAPTER

16.1. INTRODUCTION TO PROBABILITY

The words "chance", and "probability" are quite familiar to us. Statements like. "He or She may pass the examination with good marks", "Probability that it may rain today", convey the sense of uncertainty about the occurrence of some event. Ordinarily, it appears that there cannot be any exact measurement for these uncertainties, but in Mathematical Statistics, we do have methods for calculating the degree of certainty of events in numerical value, under certain conditions.

When, we perform experiments in science and engineering, repeatedly under identical conditions, we get almost the same result. There also exist experiments in which the outcome may be different even if the experiment is performed under identical conditions. In such experiments, the outcome of each experiment depends on chance.

Also a **random experiment** is defined as an experiment in which all the possible outcomes are known in advance and no personal bias is exercised. Throwing of an unbiased coin is a random experiments as out of two faces, any of the face *i.e.*, head or tail may come up. Similarly, throwing of an unbiased die is a random experiment as any of the six faces of the die may come up. In this experiment, there are six possibilities (1 or 2 or 3 or 4 or 6).

Some More Definitions

We define and explain certain terms which are used frequently.

(*a*) **Event and trial.** When an experiment is repeated under the same conditions, it results in any one of the several possible outcomes. Then, the possible outcomes are known as *events* or *cases* and the experiment is called a *trial.* In a throw of a coin the turning of head or tail is called an *event* and the throwing of coin is a called a *trial.*

(*b*) **Exhaustive events.** The outcomes of a random experiment are called **exhaustive**, if these cover all the possible outcomes of the experiment. In a throw of a coin, the possible outcomes are head and tail *i.e.,* these are two exhaustive cases. In the experiment of rolling a die, the outcomes 1, 2, 3, 4, 5, 6 (six cases) are exhaustive.

(*c*) **Favourable events.** The events which entail the required happening are said to be **favorable** events. For example, in a throw of two dice, the number of cases favourable to getting a sum 7 is 6, viz., (1.6); (6, 1); (2, 5) (5, 2); (3, 4); (4, 3).

(*d*) **Mutually exclusive events.** Two events are known as **mutually exclusive** when the occurrence of one of them excludes the occurrence of the other, *e.g.*, while tossing a coin we either get a head or tail but not both.

(*e*) **Independent events.** When the actual happening of one event does not influence in any why the probability of the happening of the other, the two events are called **independent events**. In throwing two coins at a time, the outcome of one is independent of the second. But in case a card is drawn from a pack of well shuffled cards and is not replaced, them, the second draw of the card is *dependent* on the first draw. The second draw is then a **dependent event**.

(*f*) **Equally likely events.** Events are said to be equally likely if there is no reason to expect any one in preference to any other. For example, in throw of a coin two cases *i.e.*, head and tail are equally likely to come.

(*g*) **Odds in favour of or against the trial.** If an experiment can succeed in m ways and fail in n ways, each of these ways being equally likely, then the odds are m to n in favour or n to m against the trial.

16.2. CLASSICAL DEFINITION OF PROBABILITY

Let an event A can happen in m ways, and fail in n ways where all ways are beings equally likely to occur, then the probability of the **happening of event** A is defined as

$$P(A) = \frac{(\text{Number of favourable cases})}{(\text{Total number of mutually exclusive and equally likely cases})}$$

$$= \frac{m}{m+n} = p \text{ (say);}$$

while that of its **failing** is defined as P (not A)

or $$P(\bar{A}) = \frac{n}{m+n} = q \text{ (say).}$$

Thus, $$P(A) + P(\bar{A}) = p + q = \frac{m}{m+n} + \frac{n}{m+n} = 1$$

From above, it may be noted that $P(A) = p$ is such that $0 \le p \le 1$. $P(\bar{A}) = q$ is called the complementary even. Also $0 \le q \le 1$.

Notes: (*i*) The probability of happening an event A, while another event B has already happened is called the *conditional probability* of A such that B has already happened. It is usually denoted by $P(A/B)$.This is the case when events A and B are not independent.

(*ii*) The probability of happening an event A or event B is usually denoted by $P(A + B)$.

(*iii*) The probability of happening of both events A and B is usually denoted by $P(AB)$.

SOLVED EXAMPLES

Example 16.1. Three coins are tossed, find the probability of getting at least 2 heads.

Solution. There are eight possibilities:

HHH, HHT, HTH, THH, THT, TTH, HTT, TTT.

Out of these possibilities, the required possibilties are four. Hence, the required probability is given by

$$p = \frac{\text{No. of favourable cases}}{\text{Total number of cases}} = \frac{4}{8} = \frac{1}{2}$$

Example 16.2. What is the probability that if a card is drawn at random from an ordinary pack of cards, it is: (*i*) a red card, (*ii*) a club, (*iii*) one of the count cards (jack or queen or king).

Solution. Number of exhaustive cases = 52

(*i*) There are 26 red cards and 26 black cards in an ordinary pack.

∴ Favourable cases = 26 (number of red cards)

$$\therefore \text{ Probability of getting a red card} = \frac{26}{52} = \frac{1}{2}$$

(*ii*) Number of clubs in a pack = 13

∴ Favourable cases = 13

$$\therefore \text{ Probability of getting a club} = \frac{13}{52} = \frac{1}{4}.$$

(*iii*) There are 4 × 3 = 12 court cards in a pack of cards.

∴ Number of favourable cases = 12

Number of exhaustive cases = 52

$$\therefore \text{ Probability of getting a count card} = \frac{12}{52} = \frac{3}{13}.$$

Example 16.3. A bag contains 4 white, 5 red and 7 black balls.

(*i*) What is the probability that three balls drawn at random are all red?

(*ii*) What is the probability that one is white, one is red and one is black?

Solution. (i) There are 16 balls in all. Three balls can be drawn in ${}^{16}C_3$ ways (exhaustive number of ways). Now, three red balls out 5 can be drawn in ${}^{5}C_3$ ways. Hence, the required probability is equal to

$${}^{5}C_3 \div {}^{16}C_3 = \frac{5\times4\times3}{1\times2\times3} \div \frac{16\times15\times14}{1\times2\times3} = \frac{1}{56}$$

(*ii*) Next, one white ball out of 4 can be drawn in ${}^{4}C_1$ ways, one red ball can be drawn in ${}^{5}C_1$ ways and one back ball in ${}^{7}C_1$ ways: what is more, each case of drawing a white ball can be associated with each case of drawing a read ball and each case of drawing a black ball.

Hence, the total number of favourable ways = ${}^{4}C_1 \times {}^{5}C_1 \times {}^{7}C_1 = 4 \times 5 \times 7 = 140$

$$\text{Hence, the required probability} = \frac{140}{{}^{16}C_3} = \frac{140}{560} = \frac{1}{4}.$$

Example 16.4. If 3 of 20 tyres in storage are defective and 4 of them are randomly chosen for inspection (*i.e.*, each tyre has the same chance of being selected), what is the probability that only one of the defective tyres will be included?

Solution. Given that defective tyres are 3.

Then non-defective tyres are 20 – 3 = 17

Selecting of 4 tyres from 20, which contains one defective is ${}^{3}C_1 \times {}^{17}C_3$ ways

Also the number of possible cases are ${}^{20}C_4$

$$\therefore \text{ The required probability} = \frac{{}^{3}C_1 \times {}^{17}C_3}{{}^{20}C_4} = \frac{8}{19}$$

(Here we are selecting 1 defective tyre from 3 tyres that can be done in ${}^3C_1$ ways and remaining 3 tyres can be taken from 17 non-defective tyres in ${}^{17}C_3$ ways.)

Example 16.5. What is the probability that a non-leap year contains 53 Sundays?

Solution. A non-leap year consists of 365 days.

Of these, there are 52 complete weeks and 1 extra day. That day may be any one of the 7 days.

So already we have 52 Sundays. For one more Sunday, the probability that getting a one more Sunday is 1/7.

Hence, the probability that a non-leap year contains 53 Sundays is 1/7.

16.3. THEOREMS ON PROBABILITY

There are two important theorems of probability, namely:

1. The addition theorem or the theorem on total probability.
2. The multiplication theorem or theorem on compound probability.

16.3.1. Addition Theorem or the Theorem on Total Probability

Statement: If A and B are two *mutually exclusive* and *equally likely* events of a random experiment, the probability of occurrence of the events "A or B" is the sum of the probabilities of the event A and B *i.e.*,

$$P(A \text{ or } B) = P(A) + P(B)$$

or

$$P(A + B) = P(A) + P(B)$$

Proof. Let there are n exhaustive, mutually exclusive and equally likely cases. Out of the total n cases let m_1 cases be favourable to event A and m_2 cases favourable to the event B. Therefore, total number of cases which are favourable to either A or B are $(m_1 + m_2)$.

Hence,
$$P(A \text{ or } B) = P(A+B) = \frac{m_1 + m_2}{n} = \frac{m_1}{n} + \frac{m_2}{n}$$
$$= P(A) + P(B)$$
$$= \text{Sum of the probabilities of the event } A \text{ and } B.$$

Notes: 1. The theorem can be extended to events. A, B, C, D ... i.e., $P(A + B + C + D + ...) = P(A) + P(B) + P(C) + P(D) + ...$

2. As a particular case, when the events A and B are **not mutually exclusive,** then there will be some outcomes, or cases which favour both A and B together and suppose this happens in m_k ways (this is included in both m_1 and m_2 favourable to A and B respectively). Thus, the total number of cases favouring either A or B both is $m_1 + m_2 - m_k$. Hence the probability of occurrence of A and B or both is given by

$$P(A+B) = \frac{m_1 + m_2 - m_k}{n} = \frac{m_1}{n} + \frac{m_2}{n} - \frac{m_k}{n} = P(A) + P(B) - P(AB)$$

where $P(AB)$ represents the probability of both A and B happening together. It may be noted that when A and B are mutually exclusive then $P(AB) = 0$.

3. $P(A+B+C) = P(A) + P(B) + P(C) - P(AB) - P(BC) - P(CA) + P(ACB)$

Here events A, B and C are not mutually exclusive. In case A, B and C are mutually exclusive, we have $P(AB) = P(BC) = P(CA) = P(ABC) = 0$

Thus,
$$P(A+B+C) = P(A) + P(B) + P(C)$$

16.3.2. Theorem on Multiplication of Probabilities or Compound Probability

Statement. The probability of the occurence of two *independent events* is the product of their separate probabilities *i.e.*,

$$P(AB) = P(A).\ P(B)$$

Proof. Let the two independent events be A and B. Let the events A succeed in m_1 ways and fail in n_1 ways, and the event B succeed in m_2 ways and fail in n_2 ways, all the ways in both the events A and B being equally likely. Now there are $(m_1 + n_1)$ ways in event A and $(m_2 + n_2)$ ways in event B. Each of the $(m_1 + n_1)$ ways can be associated with each of $(m_2 + n_2)$ ways. Thus in their simultaneous happening, the total number of ways are $(m_2 + n_1)\ (m_2 + n_2)$. Out of these total number of ways, we have $m_1\ m_2$ ways in which both the events succeed.

Hence, the probability that both the events succeed is

$$\frac{m_1 m_2}{(m_1 + n_1)\ (m_2 + n_2)} = \left(\frac{m_1}{m_1 \times n_1}\right) \times \left(\frac{m_2}{m_2 + n_2}\right)$$

$$= \text{Product of the probabilities of success of the two events.}$$

or $$P(AB) = P(A).\ P(B)$$

This is also true for any number of events say $A, B, C, D, \ldots\ldots\ldots\ldots$

Thus $$P(ABCD) = P(A).\ P(B).\ P(C).\ P(D) \ldots\ldots$$

Notes: 1. If $P(AA) = p_1$, $P(B) = p_2$, then

$P(AB) = p_1\, p_2$ *i.e.*, probability of happening or success of events A and B is equal to $p_1 \times p_2$.

2. If p_1 is the probability of A's success, then $q_1 = 1 - p_1$ is the probability of A's failure.
3. Probability that A succeeds and B fails $= p_1 \times (1 - p_2)$ is $p_1\, q_2$.
4. Probability that B succeeds and B fails $= p_2 \times (1 - p_1) = p_2\, q_1$.
5. Probability that both A and B fails $= (1 - p_1) \times (1 - p_2) = q_1\, q_2$
6. Probability that **atleast** one events succeeds = Probability that both the event A and B succeed + probability that A succeeds and B fails + probability that B succeeds and A fails.

$$= p_1 p_2 + p_1 q_2 + p_2 q_1 = p_1 p_2 + p_1 (1 - p_2) + p_2 (1 - p_1)$$
$$= 1 - (1 - p_1)(1 - p_2) = 1 - q_1\, q_2$$
$$= 1 - \text{the probability that } \textbf{both the events fail}$$

7. If events A and B are not independent, $P(AB) = P(A).\ P(B/A)$.

Repeated Trials

If the probability of the happening of an event in one trial is p, then the probability of its happening exactly 'r' times in 'n' trials is given by ${}^nC_r p^r q^{n-r}$, where $q\ (= 1 - p)$ is the probability of failure.

The chance of happening r times and failing $(n - r)$ times is given by $(p \times p \times \ldots\ldots)$ r times $\times (q \times q \times \ldots\ldots)$ $(n - r)$ times $= p^r \times q^{n-r}$.

Now the number of combinations of n-things into two groups of r and $(n - r)$ things

$$= \frac{n!}{r!(n-r)!} = {}^nC_r$$

Thus the chance of exactly 'r' successes and '$(n - r)$' failure in 'n' trials = ${}^nC_r p^r q^{n-r}$

or $$P(r) = {}^nC_r p^r q^{n-r}$$

Similarly, $P(n)$ = Probability of 'n' successes and no failure in trials = $^nC_np^n$

$P(0)$ = probability of no success and n failures = $^nC_0q^n$

and
$$P(1) = {}^nC_1pq^{n-1}$$
$$P(2) = {}^nC_2p^2q^{n-2}$$
$$P(3) = {}^nC_3p^2q^{n-3}$$
$$\ldots\ \ldots\ \ldots\ \ldots$$
$$P(n-1) = {}^nC_{n-1}p^{n-1}q$$

Thus, we see that in the expansion of $(q+p)^n$ first term gives the chance of no success second term gives the change of 1 success and $(n-1)$ failure, third term give 2 successes and $(n-2)$ failure, and the last term given n successes and no failure,

as
$$(q+p)^n = {}^nC_0q^n + {}^nC_1q^{n-1}\cdot p + {}^nC_2q^{n-2}p^2 + {}^nC_3q^{n-3}\cdot p^3 + \ldots + {}^nC_rq^{n-r}p_r + \ldots + {}^nC_np_n$$

SOLVED EXAMPLES

Example 16.6. Find the probability of drawing a card from a well shuffled pack such that the drawn card is either a king or queen.

Solution. The events that the card drawn is either a king or queen are mutually exclusive events. Let the event of the king card drawn be denoted by A and the queen card drawn be denoted by B. A pack containing 52 cards will have 4 kings and 4 queens.

Hence,
$$P(A) = \frac{4}{52},\quad P(B) = \frac{4}{52}$$
$$P(A+B) = P(A) + P(B) = \frac{4}{52} + \frac{4}{52} = \frac{2}{13} \text{ is the required result.}$$

Example 16.7. If the probability of the horse A winning the race is $\frac{1}{5}$ and the probability of the horse B winning the same race is $\frac{1}{6}$, what is the probability that one of the horses will win the race?

Solution. Probability of winning of the horse $A = \frac{1}{5}$

Probability of winning of the horse $B = \frac{1}{6}$

$$\therefore \quad P(A+B) = P(A) + P(B) = \frac{1}{5} + \frac{1}{6} = \frac{11}{30}$$

Example 16.8. A basket contains 20 apples and 10 oranges of which 5 apples and 3 oranges are defective. If a person takes out 2 at random, what is the probability that either both are apples or both are good?

Solution. Out of 30 items, two can be selected in $^{30}C_2$ ways.

So, exhaustive number of cases = $^{30}C_2$

Consider the following events:

$$A = \text{ getting two apples; } B = \text{ getting two good items}$$

$$\text{Required probability} = P(A \cup B) = P(A) + P(B) - P(A \cap B) \quad \ldots (i)$$

There are 20 apples, out of which 2 can be drawn in ${}^{20}C_2$ ways.

$$\therefore \quad P(A) = \frac{{}^{20}C_2}{{}^{30}C_2}$$

There are 8 defective pieces and the remaining 22 are good. Out of 22 good pieces, two can be selected in ${}^{22}C_2$ ways.

$$\therefore \quad P(B) = \frac{{}^{22}C_2}{{}^{30}C_2}$$

Since there are 15 pieces are good apples out of which 2 can be selected in ${}^{15}C_2$ ways, therefore

$$P(A \cap B) = \text{ Probaility of getting 2 peices which are good apples}$$

$$= \frac{{}^{15}C_2}{{}^{30}C_2}$$

From eqn. (*i*), required probability $= P(A) + P(B) - P(A \cap B)$

$$= \frac{{}^{20}C_2 +}{{}^{30}C_2} + \frac{{}^{22}C_2}{{}^{30}C_2} - \frac{{}^{15}C_2}{{}^{30}C_2} = \frac{316}{435}$$

Example 16.9. An urn *A* contains 2 white and 5 black balls, and an urn *B* contains 5 black and 6 white balls. Find the probability that at least one ball in white when one ball each is drawn at random from the urns *A* and *B*.

Solution. Probability of drawing a white ball

$$\text{from the urn } A = \frac{2}{7} = p_1 \text{ (say)}$$

Probability of not drawing a white ball

$$\text{from the urn } A = 1 - \frac{2}{7} = \frac{5}{7} = q_1 \text{(say)}$$

Similarly for the urn *B*, we have

$$p_2 = \frac{6}{11},\ q_2 = \frac{5}{11}$$

Therefore, the probability that at least one of the balls in white (the events being independent)

$$= 1 - q_1 q_2 = 1 - \frac{5}{7} \times \frac{5}{11} = \frac{52}{77}$$

Otherwise: The following are the possibilities and their respectively probabilities :

Possibility	**Probability**
(*i*) The first ʰall is white and second black	$\frac{2}{7} \times \frac{5}{11} = \frac{10}{77}$

(*ii*) The first ball is black and second white $\frac{5}{6}\times\frac{6}{11}=\frac{30}{77}$

(*iii*) The first ball is white and second also white $\frac{2}{7}\times\frac{6}{11}=\frac{12}{77}$

As (*i*), (*ii*), (*iii*) are mutually exclusive, and also exhaustive, therefore, the required probability

$=\frac{10}{77}+\frac{30}{77}+\frac{12}{77}=\frac{52}{77}$.

Example 16.10. Two aeroplanes bomb a target in succession. The probability of each correctly scoring a hit is 0.3 and 0.2 respectively. The second will bomb only if the first misses the target. Find the probability that :

(*i*) the traget is hit (*ii*) both fails to score hits.

Solution. Let A be the event that represents the 1st aeroplane and B represents the 2nd aeroplane.

Given that

$\therefore$ $P(A) = 0.3$ and $P(B) = 0.2$

which gives $P(\bar{A}) = 0.7$ and $P(\bar{B}) = 0.8$

Now

(*i*) The probability that the target is hit

$= P(A \text{ hits the target}) + P(A \text{ misses the target})\ P(B \text{ hits the target})$

$= 0.3+0.7\times 0.2 = 0.44$

(*ii*) The probability that both fail

$= P(\text{A misses the target})\ P(B \text{ misses the target})$

$= 0.7\times 0.8 = 0.56$

Example 16.11. A class has 10 boys and 5 girls. Three students are selected at random one after the other. Find the parbability that: (*i*) First two are boys and third is girl; (*ii*) First and third of same sex and second is of opposite sex.

Solution. Given that, class has 10 boys and 5 girls.

$\therefore$ Total number of students = 15

(*i*) The probility that the second student is boy $\frac{10}{15}$

And the probability that the second student is also boy $=\frac{9}{14}$

And the probability that the third student is girl $=\frac{5}{13}$

Hence the probability that the first two are boys and the third is girl $=\frac{10}{15}\frac{9}{14}\frac{5}{13}=\frac{15}{91}$

(*ii*) In this there are two cases:

Case 1: First and third are boys and second is a girl and

Case 2: First and third are girls and the second is a boy.

Hence, the required probability = (first boy) (second girl) (third boy) + (first girl) (second boy) (third girls) $= \frac{10}{15}\frac{5}{14}\frac{9}{13} + \frac{5}{15}\frac{10}{14}\frac{4}{13} = \frac{5}{21}$

Example 16.12. The probabilities of passing in subject *A*, *B*, *C* and *D* are $\frac{3}{4},\frac{2}{3},\frac{4}{5}$ and $\frac{1}{2}$ respectively, To qualify in the examination a student should pass in *A* and two subjects among the three, what is the probability of qualifying in that examination.

Solution. Given that the probabilities of passing in subject *A*, *B*, *C* and *D* are $\frac{3}{2},\frac{2}{3},\frac{4}{5}$ and $\frac{1}{2}$ respectively

i.e., $$P(A) = \frac{3}{4} \quad P(B) = \frac{2}{3} \quad P(C) = \frac{4}{5} \quad \text{and} \quad P(D) = \frac{1}{2}$$

Now there are four possibilities to qualify and examination:

1. Pass in subject *A*, *B*, *C* and fail in subject *D*
2. Pass in subject *A*, *C*, *D* and fail in subject *B*
3. Pass in all subject *A*, *B*, *D* and fail in subject B
4. Pass in subject *A*, *B*, *C* and *D*

The probabilities of not passing in subject *B*, *C* and *D* are $P(B^c) = \frac{1}{3}$, $P(C^c) = \frac{3}{4}$ and $P(D^c) = \frac{1}{2}$. Also all the four events are independent.

Now the probability of qualifying in that examination is the sum of the four cases

$$= P(A \cap B \cap C \cap D^c) + P(A \cap B \cap D \cap C^c) + P(A \cap C \cap D \cap B^c) + P(A \cap B \cap C \cap D)$$

$$= P(A)\,P(B)\,P(C)\,(D^c) + P(A)\,P(B)\,P(D)\,PC^c) + P(A)\,P(C)\,P(D)\,P(B^c) + P(A)\,P(B)\,P(C)P(D)$$

$$= \left\{\frac{3}{4}\times\frac{2}{3}\times\frac{4}{5}\times\frac{1}{2}\right\} + \left\{\frac{3}{4}\times\frac{2}{3}\times\frac{1}{2}\times\frac{1}{5}\right\} + \left\{\frac{3}{4}\times\frac{4}{5}\times\frac{1}{2}\times\frac{1}{3}\right\} + \left\{\frac{3}{4}\times\frac{2}{3}\times\frac{4}{5}\times\frac{1}{2}\right\} = \frac{11}{20}$$

Example 16.13. Players *A* and *B* throw a pair of dice. A wins if he throws 6 before *B* throws 7 and *B*, if he throws 7 before *A* throws 6. If *A* begins, show that his chance of winning is $\frac{30}{61}$.

Solution. When a pair of dice is thrown the total number of possible cases are 36.

The number of possible cases for getting a sum of are (1, 5), (2,4), (3, 3), (4, 2) and (5, 1) they are 5 in number.

$\therefore$ The probability of *A* getting the sum 6 $= \frac{5}{36}$

The probability of *A* not getting the sum 6 $= \frac{31}{36}$

$$\therefore \qquad 1 - \frac{5}{36} = \frac{31}{36}$$

Similarly, the number of possible cases for getting a sum of 7 are (1, 6), (2, 5), (3, 4), (4, 3) (5, 2) and (6, 1) which are 6 in number.

$\therefore$ The probability of B getting the sum 7 is $= \dfrac{6}{36}$

And the probability of B not getting the sum $7 = 1 - \dfrac{6}{36} = \dfrac{30}{36}$

If A starts the game: The probability of A wins the game = in 1st time A wins or the 3rd time A wins or 5th time A wins and so on.

$$= \frac{5}{26} + \frac{31}{36}\frac{30}{36}\frac{5}{36} + \frac{31}{36}\frac{30}{36}\frac{31}{36}\frac{5}{36} + \ldots = \frac{5}{36}\left[1 + \frac{31}{36}\frac{30}{36} + \frac{31}{36}\frac{30}{36}\frac{31}{36} + \ldots\right]$$

$$= \left(\frac{5}{36}\right)\left[1 + \frac{31}{36}\frac{30}{36} + \left(\frac{31}{36}\frac{30}{36}\right)^2 \ldots\right] = \left(\frac{5}{36}\right)\left[\frac{1}{1 - \left(\frac{31}{36}\right)\left(\frac{30}{36}\right)}\right] = \frac{30}{61}$$

Note: Suppose in the question given B starts the game. The probability of A wins the game 1st time B loses the game and A wins or B loses. A loses, B loses and wins and so on.

$$= \frac{30}{36}\frac{5}{36} + \frac{30}{36}\frac{31}{36}\frac{30}{36}\frac{5}{36} + \ldots = \frac{30}{36}\frac{5}{36}\left[1 + \frac{31}{36}\frac{30}{36} + \left(\frac{31}{36}\frac{30}{36}\right)^2 \ldots\right]$$

$$= \frac{30}{6}\left(\frac{5}{36}\right)\left[\frac{1}{1 - \left(\frac{31}{36}\right)\left(\frac{30}{36}\right)}\right]$$

Similarly, we can find the probability of B winning the game is the above two cases.

Example 16.14. ***A, B, C* are aiming to shoot a ballon. A will succeed 4 times out of 5 attempts. The chance of *B* to shoot the balloon is 3 out of 4 that of *C* is 2 out of 3. If the three aim the balloon simultaneously, then find the probability that at least two of them hit the balloon.**

Solution. From the given data,

$$P(A) = \frac{4}{5},\quad P(B) = \frac{3}{4} \text{ and } P(C) = \frac{2}{3}$$

which gives
$$P(A^C) = \frac{1}{5},\quad P(B^c) = \frac{1}{4} \text{ and } P(C^c) = \frac{1}{3}$$

Now the probability that at least two of them hit the balloon is

$$= P(A \cap B \cap C^c) + P(A \cap B^c \cap C) + (A^c \cap B \cap C) + P(A \cap B \cap C)$$

$$\therefore P(A)\,P(B)\,P(C^c) + P(A)\,P(B^c)\,P(C) + P(A^c)\,P(B)\,P(C) + P(A)\,P(B)\,P(C)$$

$$= \left\{\frac{4}{5} \times \frac{3}{4} \times \frac{1}{3}\right\} + \left\{\frac{4}{5} \times \frac{1}{4} \times \frac{2}{3}\right\} + \left\{\frac{1}{4} \times \frac{3}{4} \times \frac{2}{3}\right\} + \left\{\frac{4}{5} \times \frac{3}{4} \times \frac{2}{3}\right\}$$

$$= \frac{5}{6} = 0.83$$

EXERCISE 16.1

1. What is the probability of getting the king of hearts from a pack of cards at least once in 52 cards.

2. Problem in Mathematics is given to three students A, B, C whose chances of solving it are $\frac{1}{2}, \frac{1}{3}, \frac{1}{4}$ respectively. What is the probability that the problem will be solved? $\left[\textbf{Ans.} \frac{3}{4}\right]$

3. What is the change that a: (*i*) non-leap year, (*ii*) leap year should have fifty three Sundays. $\left[\textbf{Ans.} \frac{1}{7}, \frac{2}{7}\right]$

4. An urn contain 5 red and 10 black balls. Eight of them are placed in another urn. What is the change that the latter contains 2 red and 6 black balls? $\left[\textbf{Ans.} \frac{149}{429}\right]$

5. A committee consists of 9 students two of which are from 1st year, three from 2nd year and four from 3rd year. Three students are to be removed at random. What is the change that:
   (*i*) the three students belong to different classes.
   (*ii*) two belong to the same class and third to the different class, and
   (*iii*) the three belong to the same class? $\left[\textbf{Ans.} (i) \frac{2}{7}, (ii) \frac{55}{84}, (iii) \frac{11}{120}\right]$

6. Five men in a company of twenty are graduates. If 3 men are packed out of 20 at random, what is the probability that :
   (*i*) they are all graduates
   (*ii*) at least one is graduate? $\left[\textbf{Ans.}\ (i) \frac{1}{114}, (ii) \frac{137}{228}\right]$

7. Three cards are drawn at randow from an ordinary pack of 52 cards. Find the change that these are cards of clubs. $\left[\textbf{Ans.} \frac{11}{850}\right]$

8. Find the probability of throwing 7 with two dice in a single throw. Is this probability the same as that of getting 7 in two throws of single die? $\left[\textbf{Ans.} \frac{1}{6} \text{ yes}\right]$

9. From 20 tickets marked from 1 to 20, one ticket is drawn at random. Find the probability that it is marked with a multiple of 3 or 5. $\left[\textbf{Ans.} \frac{9}{20}\right]$

10. A bag contains 4 white, 5 green and 6 red balls. What is the probability that a ball drawn in either white or green? [**Ans.** 0.6]

11. From a pack of 52 cards, two cards are drawn. Find the probability that they are both kings if: (*i*) the first card is not replaced, (*ii*) it is replaced. $\left[\textbf{Ans.}\ (i) \frac{1}{221}, (ii) \frac{1}{169}\right]$

12. A student appears in an examination in four papers *i.e.,* first, second, third and fourth. His probabilities of passing the papers are respectively $\frac{2}{3}, \frac{4}{5}, \frac{5}{6}, \frac{3}{4}$. For qualifying the examination, he must pass in first paper and at least two other papers. Find the chances that he qualifies the examination. $\left[\textbf{Ans.} \frac{107}{180}\right]$

13. There are five letters and five addressed envelopes. Determine the change that all the letters are not dispatched in the right envelopes. $\left[\textbf{Ans.} \dfrac{119}{120}\right]$

14. The probability that a man A now aged 35 years will be live to 70 is 0.3 and that a man B now aged 45 years will live to 80 is 0.2, find the probability that at least one of these two persons will be alive for 35 years hence.

15. The probability that a 50 years old man will be alive at 60 is 0.83 and the probability that a 45 years and woman will be alive at 55 is 0.87. Find the probability that both will be alive 10 years hence.

[**Ans.** 0.72]

16. A and B throw alternately with a pair of dice. The one who throws 9 first wins. Show that the chances of their winning are 9 : 8.

❑❑❑

UNIT - VIII

OPERATION RESEARCH

17. Linear Programming : Problem Formulation and Graphical Solution

CHAPTER 17

Linear Programming : Problem Formulation and Graphical Solution

INSIDE THIS CHAPTER

17.1. INTRODUCTION

The central issue in many problems arising in business, industry and government is that various activities compete for limited resources. It is the responsibility of the decision maker to allocate the resources among the competing activities so as to achieve best results. Linear Programming problems deal with determining the best possible allocation of available resources among the competing activities so as to meet certain specifications. In particular, they may relate to situations where a number of sources, such as materials, machines, they may relate to situations where a number of resources, such as materials. The objective or goal can whatever the user wants so long it can be expressed in a linear manner. For example, the objective might be a maximize profits, to minimize costs or waste, or to minimize the number of employees. The constraints might reflect limited labour resources, limited raw materials, and limited demand for the products.

Linear programming is a mathematical technique for finding an optimum (maximum or minimum) value of a functions, called objective function of several independent variables, these variable being subject to various, restrictions (or constraints) expressed as equations or inequalities. The term 'Linear' indicates that the function to be maximized or minimized is linear is linear and that the corresponding constraints be represented by a system of linear inequalities or linear equations involving the variables.

A Linear Programming Problem has the following essential parts:

- **A set of non-negative restriction** to avoid getting negative solutions in which we are not restricted. For example, negative price or negative output do not mean anything.
- **A set of linear constraints** representing the limitations on the use of machineary, labour etc.
- **An objective function** to be maximized or minimized subject to the above constraints.

17.2. GENERAL MATHEMATICAL MODEL OF LINEAR PROGRAMMING PROBLEM

The term '*programming*' means planning and refers to a process of determining a particular program. The term '*linear*' means that all relationships involved in a particular program which can be solved by this method are linear. To make understandable the terminology of linear programming it is necessary to given the n\mathematical model of a general linear programming problem. Mathematically, the general linear programming problem (LPP) may be stated as:

$$\text{Optimize (Max. or Min.) } Z = c_1x_1 + c_2x_2 + ... + c_nx_n$$

subject to the linear constraints,

$$\left.\begin{array}{l} a_{11}x_1 + a_{12}x_2 + ... + a_{1n}x_n \quad (\le, \ =, \ \ge) b_1 \\ a_{12}x_1 + a_{22}x_2 + ... + a_{2n}x_n \quad (\le, \ =, \ \ge) b_2 \\ .... \\ .... \\ a_{11}x_1 + a_{12}x_2 + ... + a_{1n}x_n \quad (\le, \ =, \ \ge) b_1 \end{array}\right\} \quad ...(i)$$

and $\quad x_1, x_2 ...... x_n \ge 0 \quad ...(ii)$

where

(*i*) the linear function Z which is to be maximized or minimized is the objective function of the LPP.

(*ii*) $x_1, x_2, x_3, ... x_n$ are decision variables.

(*iii*) the equations/inequations (*i*) are the constraints of LPP.

(*iv*) in the set of the constraints (*i*) the expression means that each constraint may take any one of the three signs.

(*v*) the set of inequation (*ii*) is the set of non-negative restrictions of the general linear programming problem.

(*vi*) c_j (j = 1, 2, ..., n) represents per unit profit or cost to the jth variable.

(*vii*) b_i (i = 1, 2, ..., m) is the requirement or availability of the ith constraint.

(*viii*) a_{ij} (i = 1, 2, ..., m; 1, 2, ..., n) is refer to as the technological or substitution coefficients.

It is appropriate at this stage to give some definitions and concepts related to LPP.

1. **Solution:** A set of values of decision variables satisfying all the constraints of a linear programming problem is called a solution to that problem.
2. **Feasible Solution:** Any solution which also satisfies the non-negativity restrictions of the problem is called a feasible solution.
3. **Optimal Feasible Solution:** Any feasible solution which maximizes or minimizes the objective function is called an optimal feasible solution.
4. **Optimization Techniques:** The process of obtaining the optimal values are called optimization techniques.
5. **Feasible Region:** The common region determined by all the constraints and non-negativity restriction of an LPP is called a feasible region.
6. **Convex Region:** A region is said to be convex, if the line segment joining any two arbitrary points of the region lies entirely within the region. Feasible region of an LPP is always a convex region.

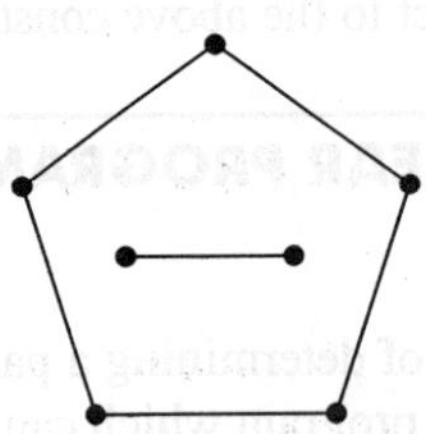

(i) Region is convex

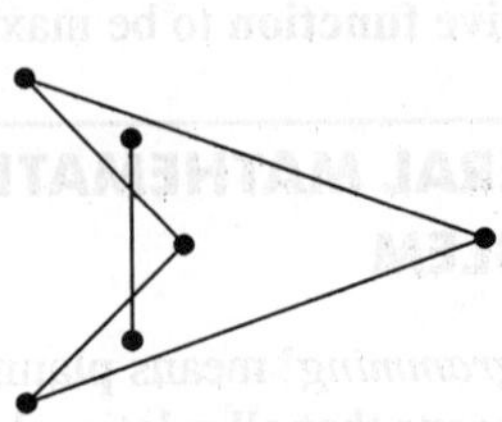

(ii) Region is not convex

Fig. 17.1

17.3. FORMULATION OF LINEAR PROGRAMMING PROBLEM

The formulation of linear programming problem as a mathematical model involves the following basic steps:

Step 1: Identify the decision variables and assign symbols $x_1, x_2, ...$ to them. These decision variables are those quantities whose values we wish to determine.

Step 2: Identify the set of constraints and express them as linear equations/inequations in terms of the decision variables. These constraints are the given conditions.

Step 3: Identify the objective function and express it as a linear function of decision variables. It might take the form of maximizing profit or production or minimizing cost.

Step 4: Add the non-negativity restrictions on the decision variables, as in the physical problems, negative values of decision variables have no valid interpretation.

To being with, a problem is to be presented in a linear programming form which requires defining the variables involves involved, establishing relationships between them and formulating the objective function and the constraints. We illustrate this through few examples, where in the stress will be on the analysis of the problem and formulation of the linear programming model.

SOLVED EXAMPLES

Example 17.1. (Production Allocation Problem) A furniture dealer deals in only two times, viz., tables and chairs. He has ₹ 5760 to invest and a space to store at most 20 pieces. A table costs him ₹ 360 and a chair ₹ 240. He can sell a table at a profit of ₹ 22 and a chair at a profit of ₹ 18. Assume that he can sell the items he buys. Formulate this problem as an LPP so that he can maximize the profit.

Solution. Let x_1 be the number of tables and x_2 the number of chairs. (x_1 and x_2 are decision variables)

It is obvious that the must have

$$x_1 \geq 0, \qquad x_2 \geq 0 \qquad \text{(Non-negative restrictions)}$$

(As the number of tables and chairs cannot be negative)

Profit on x_1 tables = ₹ $22x_1$

Profit on x_2 chairs = ₹ $22x_2$

Then the total profit that a furniture dealer gets after selling x_1 and tables and x_2 chairs is given by

$$Z = 22x_1 + 18x_2 \qquad \text{(Objective Function)}$$

The cost of x_1 tables = ₹ $360\, x_1$

The cost of x_2 chairs = ₹ $240\, x_1$

Now, in order to purchase x_1 tables and x_2 chairs which the dealer can invest is given by $360x_1 + 240x_2$ and the total space to store x_1 tables and x_2 chairs is given by $x_1 + x_2$.

Since the amount is not available for more than ₹ 5760 and the space to store tables and chairs is available only for 20 pieces, therefore

$$360x_1 + 240x_2 \leq 5760 \qquad \text{(Constraint)}$$

$$x_1 + x_2 \leq 20 \qquad \text{(Constraint)}$$

The mathematical formulation of LPP is as follows:

Maximum $Z = 22x_1 + 18x_2$

subject to the constraints

$$360x_1 + 240x_2 \leq 5760$$

$$x_1 + x_2 \leq 20 \quad \text{and} \quad x_1 \geq 0, \quad x_2 \geq 0.$$

Example 17.2. (Manufacture Problem) A manufacturer has three machines I, II and III installed in his factory. Machine I and II are capable of being operated for at the most 12 hours, whereas machine III must be operated at least 5 hours a day. He produces only two items, each requiring the use of the three machines.

The number of hours required for producing I unit of each of the item *A* and *B* on the three machines are given in the following table:

Item	*Number of hours required on the machines*		
	I	*II*	*III*
A	1	2	1
B	2	1	5/4

He makes a profit of ₹ 60 on item *A* and ₹ 40 on item *B*. Assuming that he can sell all that he produces, how many of each item should be produce so as to maximize the profit? Formulate the above problem as a linear programming problem.

Solution. Let x_1 be the number of items *A* and x_2 be the number of items *B* produced. The total profit on the production is $60x_1 + 40x_2$.

The objective of manufacturer is to minimize the profit. Then, the objective function to be maximized is

$$Z = 60x_1 + 40x_2$$

Time available on machines I, II and III is 12, 12 and 5 hours. So the capacity constraints are

$$x_1 + 2x_2 \leq 12$$
$$2x_1 + x_2 \leq 12$$
$$x_1 + \frac{5}{4}x_2 \geq 5 \quad i.e. \quad 4x_1 + 5x_2 \geq 20 \qquad \text{and} \qquad x_1, x_2 \geq 0$$

Therefore, the mathematical formulation of the problem is

$$\text{Maximize } Z = 60x_1 + 40x_2$$

subject to the constraints

$x_1 + 2x_2 \leq 12$ (Constraint on machine I)

$2x_1 + x_2 \leq 12$ (Constraint on machine II)

$4x_1 + 5x_2 \geq 20$ (Constraint on machine III)

and $x_1 \geq 0, \quad x_2 \geq 0$ (Non-negativity restrictions)

Example 17.3. (Diel Problem) A diet is to contain at least 400 units of carbohydrates, 500 units of fat and 300 units of protein. Two foods are available: F_1, which costs ₹ 2 per unit of F_2, which costs ₹ 2 per unit and F_2, which costs ₹ 4 per unit. A unit of food F_1 contains 10 units of carbohydrates, 20 units of fat and 15 units of protein; a unit of food F_2 contains 25 units of carbohydrate, 10 units of fat and 20 units of protein. Find the minimum cost for a diet that consists of a mixture of these two foods and also meet the minimum nutrition requirements. Formulate the problem as a linear programming problem.

Solution. Let the mixture contains x_1 kg of food F_1 and x_2 kg of food F_2.

We make the following table from the given data:

Resources / *Food*	*Carbohydrate*	*Fat*	*Protein* (₹/unit)	*Cost*
F_1	10	20	15	2
F_2	25	10	20	4
Requirement (in units)	400	500	300	

The minimum requirement of carbohydrate is 400 units.

Therefore, $10x_1 + 25x_2 \geq 400$

Similarly, the requirement of fat and protein is 500 and 300 units respectively.

Therefore, $20x_1 + 10x_2 \geq 500$ and $15x_1 + 20x_2 \geq 300$

Also $x_1 \geq 0, \quad x_2 \geq 0$

Cost of purchasing 1 unit of food F_1 is ₹ 2

Cost of purchasing 1 unit of food F_2 is ₹ 4

The total cost of purchasing x_1 units of food F_1 and x_2 units of food F_2 (in ₹) is

$$Z = 2x_1 + 4x_2$$

which is the objective function.

Therefore, the mathematical formulation of the problem is

$$\text{Minimum } Z = 2x_1 + 4x_2$$

subject to the constraints

$$10x_1 + 25x_2 \geq 400$$
$$20x_1 + 10x_2 \geq 500$$
$$15x_1 + 20x_2 \geq 300 \quad \text{and} \quad x_1 \geq 0, \quad x_2 \geq 0$$

Example 17.4. (Transportation Problem) There is a factory located at each of the two places *P* and *Q*. From these locations, a certain commodity is delivered to each of the three depots situated at *A*, *B*, *C*. The weekly requirements of the depots are, respectively, 5, 5 and 4 units of the commodity while the production capacity of the factories of *P* and *Q* are, respectively, 8 and 6 units. The cost of transposition per unit is given below:

To / From	Cost (₹ 1 unit)		
	A	*B*	*C*
P	16	10	15
Q	10	12	10

How many units should be transported from each factory to each depot in order that the transportation cost is minimum? Formulate the above linear programming problem mathematically.

Solution. Let x_1 units and x_2 units of the commodity be transported from the factory *P* to the depots *A* and *B* respectively. Since the factory *P* has the capacity of producing 8 units of commodity, therefore, $(8 - x_1 - x_2)$ units will be transported from the factory *P* to the depot *C*.

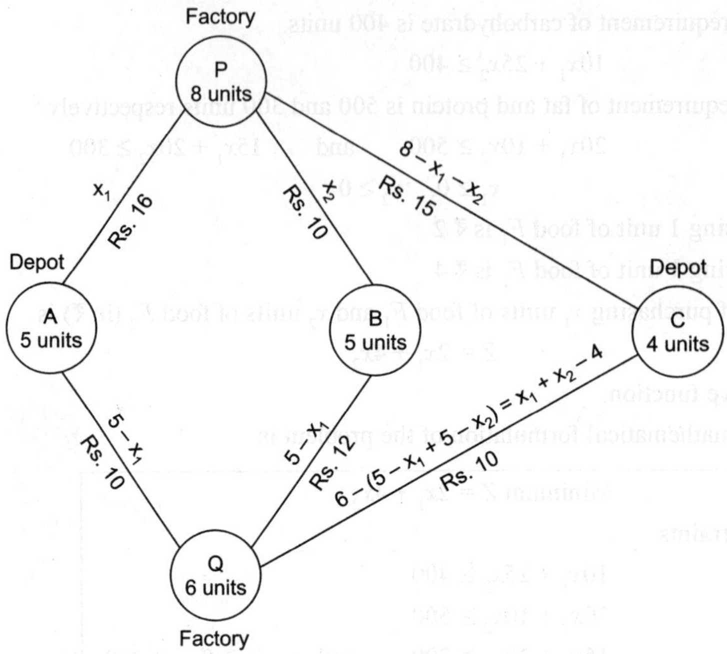

Fig. 17.2

Clearly, $x_1 \geq 0, x_2 \geq 0$ and $8 - x_1 - x_2 \geq 0$ *i.e.*, $x_1 + x_2 \leq 8$.

Now, the weekly requirement of the depot A is 5 units of the commodity and x_1 units are already transported from the factory P, therefore, the remaining $(5 - x_1)$ units are to be transported from the factory Q. Similarly $(5 - x_2)$ and $6 - (5 - x_1 + 5 - x_2) = x_1 + x_2 - 4$ units are to be transported from the factory Q to the depots B and respectively.

Clearly, $5 - x_1 \geq 0, \quad 5 - x_2 \geq 0$ and $x_1 + x_2 - 4 \geq 0$

i.e., $x_1 \leq 5, \quad x_2 \leq 5$ and $x_1 + x_2 \geq 4$.

The total transportation cost is given by

$$Z_1 = 16x_1 + 10x_2 + 15(8 - x_1 - x_2) + 10(5 - x_1) + 12(5 - x_2) + 10(x_1 + x_2 - 4)$$

or

$$Z_1 = x_1 - 7x_2 + 190.$$

The objective is to minimize $Z = x_1 - 7x_2$ since 190 is a constant and does not affect the optimal solution. Hence the above LPP can be stated mathematically as follows:

Minimize $Z = x_1 - 7x_2$

subject to the constraints

$$x_1 + x_2 \leq 8$$
$$x_1 \leq 5$$
$$x_2 \leq 5$$
$$x_1 + x_2 \leq 4$$
$$x_1 \geq 0, \quad x_2 \geq 0.$$

EXERCISE 17.1

1. What is meant by Linear Programming Problem? Given a brief description of the problem with illustrations.
2. What are the advantages of LPP? What are the limitations of Linear Programming?

3. A decorative items dealer deals in two items: Wall hangings and artificial plants. He has ₹ 15,000 to invest and a space to store at the most 800 pieces. A wall hanging cost him ₹ 300 and artificial plant ₹ 150. He can sell a wall-hanging at a profit of ₹ 50 and an artificial plant at a profit of ₹ 18. Assuming he can sell all the items that the buys; formulate a linear programming problem in order to maximize his profit.

[**Ans.** Max. $Z = 50x_1 + 18x_2$, s.t. $2x_1 + x_2 \le 100$; $x_1 + x_2 \le 80$; $x_1 \ge 0, x_2 \ge 0$]

4. A manufacturer produces two types of models M_1 and M_2. Each M_1 model require 4 hours of grinding and 2 hours of polishing; whereas each M_2 model requires 2 hours of grinding and 5 hours of polishing. The manufacturer has 2 grinders and 3 polishers. Each grinder work for 40 hours a week and each polisher work for 60 hours in a week. Profit on an M_1 model is ₹ 3 and on an M_2 model is ₹ 4. Whatever is produced in a week is sold in the market. How should the manufacturer allocate his production capacity to the two types of models so that he may take the maximum profit in a week.

[**Ans.** Minimize $Z = 3x_1 + 4x_2$
subject to the constraints
$4x_1 + 2x_2 \le 80$;
$2x_1 + 5x_2 \le 180$ and $x_1, x_2 \ge 0$]

5. The manager of an oil refinery must decide on the optimal mix of two possible blending processes of which the inputs and outputs per production run are as follows:

	Input (units)		*Output (units)*	
Process	***Crude A***	***Crude B***	***Gasoline X***	***Gasoline Y***
1	4	3	4	7
2	5	6	5	5

The maximum amounts available of crudes A and B are 225 units and 200 units respectively. Make demand shows that at least 150 units of gasoline X and 120 units of gasoline Y must be produced. The profit per production run from process 1 and process 2 are ₹ 200 and ₹ 300 respectively. Formulate the problem as a linear programming problem.

[**Ans.** Minimize $Z = 200x_1 + 300x_2$
subject to the constraints $4x_1 + 5x_2 \le 225$;
$3x_1 + 6x_2 \le 200$; $4x_1 + 5x_2 \ge 150$; $7x_1 + 5x_2 \ge 120$; $x_1, x_2 \ge 0$]

6. A diet is to contain at least 400 units of carbohydrate, 500 units of fat, and 300 units of protein. Two foods are available: F_1 which costs ₹ 2 per unit, and F_2, which costs ₹ 4 per unit. A unit of food F_1 contains 10 units of carbohydrate, 20 units of fat, and 15 units of protein, a unit of F_2 contains 25 units of carbohydrate, 10 units of fat, and 20 units of protein. Find the minimum cost for a diet that consists of a mixture of these two foods and also meets the minimum nutrition requirements. Formulate the problem as a LP problem.

[**Ans.** Minimize $Z = 2x_1 + 4x_2$
subject to the constraints $10x_1 + 25x_2 \ge 400$
$20x_1 + 10x_2 \le 500$; $15x_1 + 20x_2 \ge 300$; $x_1 \ge 0, x_2 \ge 0$]

7. An automobile industry is producing two models M_1 and M_2 of two wheelers. The profits of two two-wheelers are ₹ 1000 and ₹ 1200 respectively. Both the models require processing in three types of machine groups. The following table indicates the machine group hours per week and the time required on each machine group for each machine group for each number of two models *i.e.* M_1 and M_2. Formulate the problem in the linear programming form:

Machine-Group	*Hours Required by*		*Total Available Machine Group Hour Per Week*
	Model M_1	*Model M_2*	
Machine-Group-1	4	2	100
Machine-Group-2	6	8	240
Machine-Group-3	8	14	280

[**Ans.** Max. $Z = 1000x_1 + 1200x_2$, s.t. $4x_1 + 2x_2 \leq 100$; $6x_1 + 8x_2 \leq 240$; $8x_1 + 14x_2 \leq 280$; $x_1 \geq 0$, $x_2 \geq 0$]

8. Consider the following problem faced by a production planner in a soft drink plant. He has two bottling machines A and B. A is designed for 8-ounce bottle and B for 16 ounce bottles. However, each can be used on both types with some loss of efficiency. The following data is available.

Machine	*8-ounce bottles*	*16-ounce bottles*
A	100/minute	40/minute
B	60/minute	75/minute

The machine can be run 8-hour per day, 5 day week. Profit on 8 ounce bottle is 15 paise and on 16 ounce bottle is 25 paise. Weekly production of the drink cannot exceed 3,00,000 ounces and the market can absorb 25,000 eight-ounce bottles and 7,000 sixteen-ounce bottle per week. The planer wishes to maximize his profit subject, of course, to all production and marketing restrictions. Formulate this as linear programming problem.

[**Ans.** Maximize $Z = 0.15x_1 + 0.25x_2$

subject to the constraints $\left(\frac{x_1}{100}\right) + \left(\frac{x_2}{40}\right) \leq 2400$ (for machine A)

$\left(\frac{x_1}{60}\right) + \left(\frac{x_2}{75}\right) \leq 400$ (for machine B)

$8x_1 + 16x_2 \leq 3{,}00{,}000$; $0 \leq x_1$ 25000 and $0 \leq x_2 \leq 7000$]

9. Ajay wants to invest at most ₹ 12000 in bond A and B. According to the rules, he has to invest at least 2000 in bond A and at least ₹ 4000 in bond B. If the rate of interest on bond A is 8% per annum and the rate of interest on bond B is 10% per annum. Formulate LPP.

[**Ans.** Max. $Z = 0.08x_1 + 0.10x_2$, s.t. $x_1 + x_2 \leq 12000$; $x_1 \geq 2000$, $x_2 \geq 4000$, $x_1 \geq 0$, $x_2 \geq 0$]

10. An aeroplane can carry a maximum of 200 passengers. A profit of ` 400 is made on each first class ticket and a profit of ₹ 300 is made on each economy class ticket. The airline reserves atleast 20 seats for first class. However, atleast 4 time as many passenger prefer to travel by economy class than by the first class. How may tickets of each class must be sold in order to maximize profit for the airline? Formulate the problem as L.P. model. [**Ans.** Max. $Z = 400x_1 + 300x_2$, s.t. $x_1 + x_2 \leq 200$; $x_1 \geq 20$, $x_2 \geq 4x_1 \geq 0$, $x_2 \geq 0$]

17.4. THE GRAPH OF A LINEAR INEQUALITY

The graph of a linear inequality in two variables x_1 and x_2 is the set of all points (x_1, x_2) for which the inequality holds. The procedure for graphing a linear inequality is as follows:

Step I: Graph the corresponding linear equation, a line. This is done by finding any two points on the line and drawing a straight line through them. For convenience, we choose the two points so that each is on an axis.

Step II: Choose a point P not on the line. Normally, this point is taken as the origin as long as it does not lie on the line. Test P to see whether P satisfies the inequality.

Step III: If the coordinates of the chosen point P satisfy the inequality, then all points on the same side of the line as the point P satisfy the inequality. If the coordinates of the point P do not satisfy the inequality, then all points on the opposite side of the line from P satisfy the inequality.

SOLVED EXAMPLES

Example 17.5. Graph the system

$$3x_1 + 3x_2 \leq 36$$
$$5x_1 + 2x_2 \leq 50$$

Solution. We first graph the two lines:

$$3x_1 + 3x_2 \leq 36$$
$$5x_1 + 2x_2 \leq 50$$

The graphs are shown in Fig. 17.3(*a*). The graphs of the two linear inequalities are shown in Fig. 17.3(*b*), and the desired solution is the cross-hatched region.

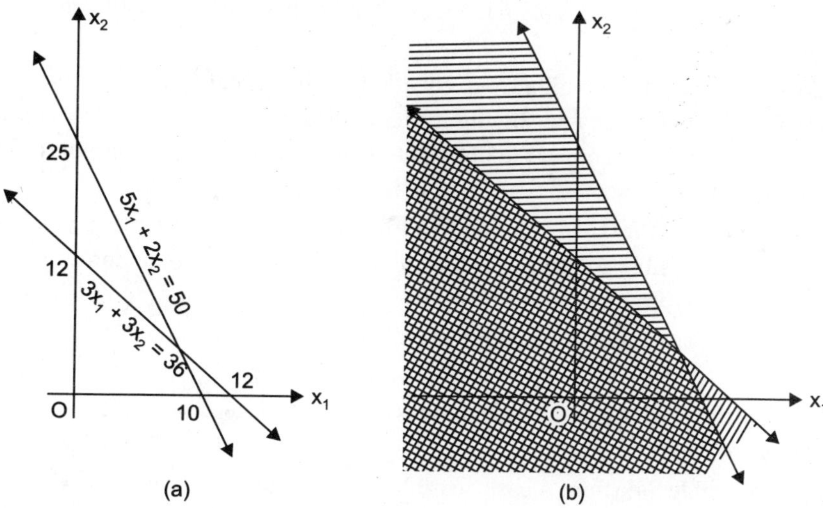

Fig. 17.3

Example 17.6. Graph the system

$$2x_1 + 4x_2 \leq 40 \quad ...(i)$$
$$5x_1 + 2x_2 \leq 50 \quad ...(ii)$$
$$x_1 \geq 0 \quad ...(iii)$$
$$x_2 \geq 0 \quad ...(iv)$$

Solution. We must shade in all points that satisfy each of the inequalities (*i*) to (*iv*). This can be done by first drawing the graphs of the four corresponding equalities. Once these lines have been drawn, we can mentally identify the four regions representing inequalities (*i*) to (*iv*) an shade the intersection of these regions. The desired region is shaded in Fig. 17.4. The inequalities in (*iii*) and (*iv*) are satisfied by the fact that we confined the graph to the first quadrant only.

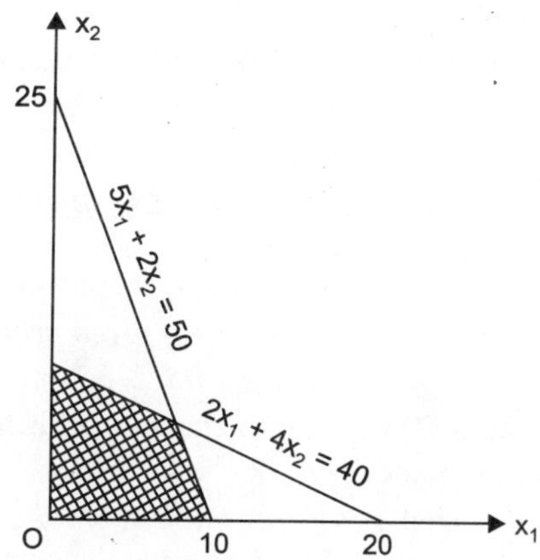

Fig. 17.4

17.5. GRAPHICAL SOLUTION METHOD

The graphical method for solving linear programming problems that involves only two variables is based on a well-defined set of logical steps. Following this systematic procedure, the given linear programming problem can be solved with minimum amount of computational effort. This method is based on a theorem, called extreme point theorem, which states as follows:

Extreme Point Theorem: An optimum solution to a linear programming problem. If it exists, occurs at one of the area of the feasible region.

Before summarizing the graphical solution algorithm, we must understand some definitions and concepts related to linear programming problem.

Feasible Solution: The set of values of decision variables $x_j (j = 1, 2, ..., n)$ which satisfy all the constraints and non-negativity conditions of an LPP simultaneously is said to constitute the feasible solution to that linear programming problem.

Infeasible Solution: The set of values of decision variables $x_j (j = 1, 2, ..., n)$ which do not satisfy all the constraints and non-negativity conditions of an LPP simultaneously is said to constituted the feasible solution to that linear programming problem.

Basic Solution: For a set of m simultaneous equations in n variables $(n > m)$, a solution obtained by setting $(n - m)$ variables equal to zero and solving for remaining m variables is called a basic solution. The $(n - m)$ variables whose value did not appear in this solution are called non-basic variables and the remaining m variables are called basic variables.

Basic Feasible Solution: A feasible solution to a linear programming problem which is also the basic solution is called the basic feasible solution. That is, all basic variables assume non-negative values. Basic feasible solution are of two types:

(*i*) *Degenerate*: A basic feasible solution is called degenerate if value of at least one basic variable is zero.

(*ii*) *Non-Degenerate*: A basic feasible solution is called non-degenerate if value of all m basic variables are non-zero and positive.

Optimum Basic Feasible Solution: A basic feasible solution which optimizes (maximizes or minimizes) the objective function value of the given linear programming problem is called optimum basic feasible solution.

Unbounded Solution: A solution which can increase or decrease the value of objective function of the linear programming problem indefinitely is called an unbounded solution.

17.6. GRAPHICAL SOLUTION PROCEDURE

The outlines of graphical procedure as follows:

Step 1: *Formulate the given problem as a linear programming problem.*

Step 2: *Construct the graph for the problem as follows:* 'Treat each inequality as though it were an equality and for each equation arbitrarily select two sets of points. Plot each set of points and connect with appropriate line'.

Step 3: *Identify the feasible region.* The feasible region (solution space) is that which satisfies all the constraints and for each equation simultaneously. For 'less than or equal to' and 'less than' constraints this is generally the region below these lines. For 'greater than or equal to' or 'greater than' constraint, this is generally the region which lies above the lines.

Step 4: *Locate the solution points.* This is accomplished by identifying the corners.

Step 5: *Evaluate the objective function* at each of the solution points.

Step 6: *Identify the optimal solution.* which corresponds to that corner point which gives the optimal value of the objective function. The procedure of the method may be understood by the following examples.

SOLVED EXAMPLES

Example 17.7. Use the graphical method to solve the following LP problem

$$\text{Maximize } Z = 60x_1 + 15x_2$$

subject to the constraints

$$x_1 + x_2 \leq 50$$
$$3x_1 + x_2 \leq 90$$
$$x_1 \geq 0 \quad x_2 \geq 0$$

Solution. Step 1: The given LP problem is already in mathematical form.

Step 2: We shall treat x_1 as the horizontal axis and x_2 as the vertical axis. Each constraint will be plotted on the graph by treating it as linear equations and then appropriate inequality conditions will be used to mark the area of feasible solutions.

Consider the first constraint $x_1 + x_2 \leq 50$. Treat it as equation $x_1 + x_2 = 50$. The easiest way to plot this line is to find any two points that satisfy the equation, then drawing a straight line through them. The two points are generally the points at which the line intersects the x_1 and x_2 axes. For example, when $x_1 = 0$ we get $x_2 = 50$. Similarly, when $x_2 = 0$ we get $x_1 = 50$.

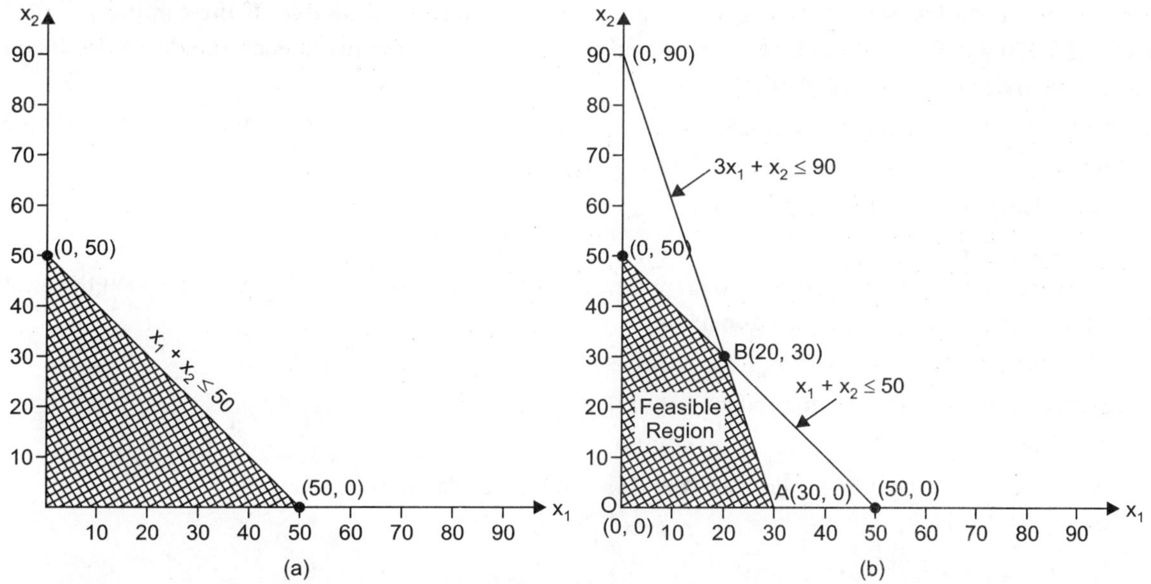

Fig. 17.5

These two points are then connected by a straight line as shown in Fig. 17.5(*a*). But the question is: where are the points satisfying $x_1 + x_2 \leq 50$. Any point above the constraint line violates the inequality condition. But any point below the line does not violate the constraint. Thus the inequality and non-negativity condition can only be satisfied by the shaded area (feasible region) as shown in Fig. 17.5(*b*). Similarly, the constraint $3x_1 + x_2 \leq 90$ is also plotted on the graph and is indicated by the shaded area.

Step 3: Since all the constraints have been graphed, the area which is bounded by all the constraints line including all the boundary points is called feasible region (or solution space). The feasible region is shown by the shaded area *OABCO*.

Step 4: Since the optimal value of the objective function occurs at one of the extreme points of the extreme points of the feasible region, it is necessary to determine the coordinates. The coordinates of extreme points of the feasible region are: $O(0, 0)$, $A(3, 0)$, $B(20, 30)$, $C(0, 50)$.

Step 5: Evaluate the objective function value at each extreme point of the feasible region as shown below:

Extreme Point	*Coordinate* (x_1, x_2)	*Objective Function* $Z = 60x_1 + 15x_2$
O	(0, 0)	60(0) + 15(0) = 0
A	(30, 0)	60(30) + 15(0) = 1800
B	(20, 30)	60(20) + 15(30) = 1650
C	(0, 50)	60(0) + 15(50) = 750

Step 6: Since we desire Z to be maximum, from step 5, we conclude that maximum value of $Z = 1800$ is achieved at the point $A(30, 0)$. Hence, the optimal solution to the given LP problem is $x_1 = 30, x_2 = 0$ and Max $Z = 1800$.

Example 17.8. A Toy Company manufacturers two types of doll; a basic version doll *A* and a deluxe version doll *B*. Each doll of type *B* takes twice as long to produce as one of type *A*, and the company would have time to make a maximum of 2000 per day if it produced only the basic version. The supply of plastic is sufficient to produce 1500 dolls per day (both *A* and *B* combined). The deluxe version requires a fancy dress of which there are only 600 per day available. If the company makes profit of ₹ 300 and ₹ 500 per doll respectively on doll *A* and *B*; how many of each should be produced per day in order to maximize profit?

Solution. Let x_1 dolls of type A and x_2 dolls of type B should be produced per day. The total profit for x_1 dolls of type A and x_2 dolls of type B is ₹ $3x_1 + 5x_2$

Let Z denote the maximum profit, then

$$Z = 3x_1 + 5x_2$$

Since total time per day consumed to prepare x_1 and x_2 dolls of type A and B is $x_1 + 2x_2$ and at the most 2000 dolls of type A may be produced in a body

Therefore, $x_1 + 2x_2 \le 2000$

Further plastic is available to produce only 1500 dolls.

Therefore, $x_1 + 2x_2 \le 1500$

Also fancy dress required for dolls of type B, available for 600 dolls per day only

So, $x_2 \le 1500$

Hence the given LPP is as follows:

$$\text{Maximize } Z = 3x_1 + 5x_2$$

subject to the constraints

$$x_1 + 2x_2 \le 2000$$
$$x_1 + x_2 \le 1500$$
$$x_2 \le 600$$

and $x_1 \ge 0, x_2 \ge 0$

Plot on a graph each constraint by first treating it as a linear equation in the same way as discussed earlier. Use the inequality condition of each constraint to mark the feasible region as shown in Fig. 17.6.

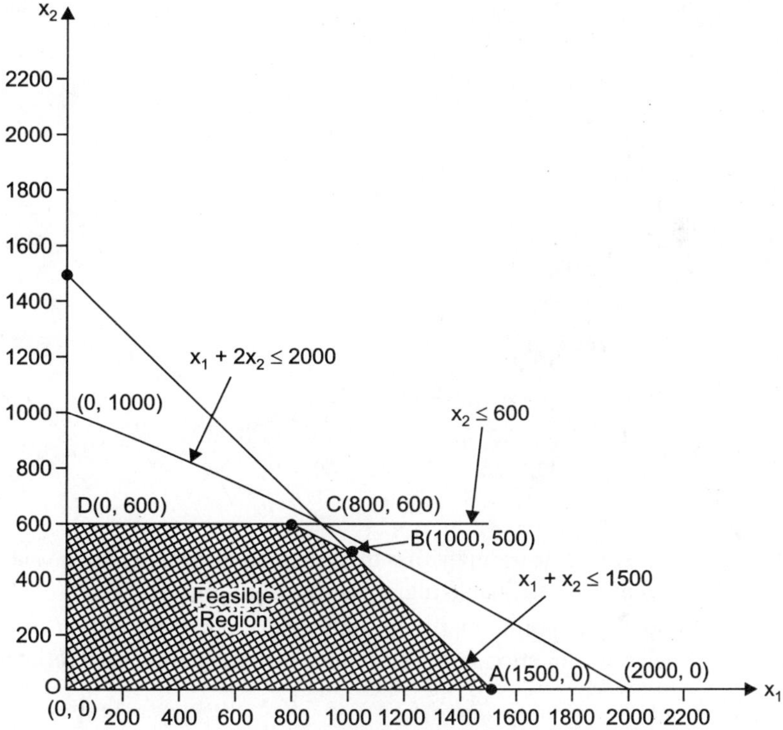

Fig. 17.6

The coordinates of extreme points of the feasible region are: $O(0, 0)$, $A(1500, 0)$, $B(1000, 500)$, $C(800, 600)$ and $D(0, 600)$. The value of the objective function at each of these extreme points is as follows:

Extreme Point	*Coordinate* (x_1, x_2)	*Objective Function* $Z = 3x_1 + 25x_2$
O	(0, 0)	3(0) + 5(0) = 0
A	(1500, 0)	3(1500) + 5(0) = 4500
B	(1000, 500)	3(1000) + 5(500) = 5500
C	(800, 600)	3(800) + 5(600) = 5400
D	(0, 600)	3(0) + 5(600) = 3000

The maximum value of the objective function $Z = 5500$ occurs at the extreme point $B(1000, 500)$. Hence the optimal solution to the given LP problem is $x_1 = 1000$, $x_2 = 500$ and Max. $Z = 5500$.

Example 17.9. Use the graphical method to solve the following LP problem.

$$\textbf{Minimize } Z = 20x_1 + 10x_2$$

subject to the constraints

$$x_1 + 2x_2 \leq 40$$
$$3x_1 + x_2 \geq 30$$
$$4x_1 + 3x_2 \geq 60$$

and $$x_1, x_2 \geq 0$$

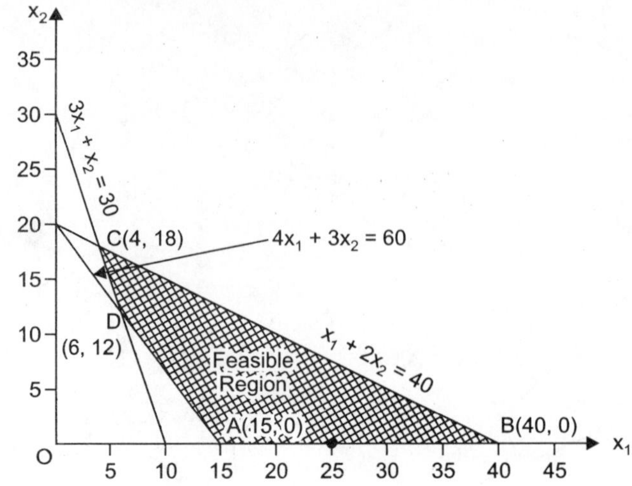

Fig. 17.7

Solution. Plot on a graph each constraint by first treating it as a linear equation. The use the inequality condition of each constraint to mark the feasible region as shown in Fig. 17.7.

The coordinates of the extreme points of the feasible region are: $A(15, 0)$, $B(40, 0)$, $C(4, 18)$ and $D(6, 12)$. The value of the objective function at each of these extreme points is as follows:

Extreme Point	*Coordinate* (x_1, x_2)	*Objective Function* $Z = 20x_1 + 10x_2$
A	(15, 0)	20(15) + 10(0) = 300
B	(40, 0)	20(40) + 10(0) = 800
C	(4, 18)	20(4) + 10(18) = 260
D	(6, 12)	20(6) + 10(12) = 240

The minimum value of the objective function $Z = 240$ occurs at the extreme point $D(6, 12)$.

Hence, the optimal solution to the give LP problem is: $x_1 = 6$, $x_2 = 12$ and Min $Z = 240$.

Example 17.10. A manufacturer has three machines I, II and III installed in his factory. Machines I and II are capable of being operated for at most 12 hours whereas machine III must be operated for at least 5 hours a day. She produces only two items *M* and *N* each requiring the use of all the three machines. The number of hours required for producing 1 unit of each of *M* and *N* on the three machines are given in the following table:

Items	*Number of hours required on machine*		
	I	*II*	*III*
M	1	2	1
N	2	1	1.25

She makes a profit of ₹ 600 and ₹ 400 on items *M* and *N* respectively. How many of each item should she produce so as to maximum her profit assuming that she can sell all the items that the produced? What will be the maximum profit?

Solution. Let the manufacturer produces x_1 and x_2 number of items of M and N respectively, then

$$x_1 \geq 0 \quad \text{and} \quad x_2 \geq 0$$

The total profit Z is given by

$$Z = 600x_1 + 400x_2$$

The total time required on machine I for producing x_1 and x_2 items of M and $N = x_1 + 2x_2$

But the time available on machine I for producing is at most 12 hour

$\therefore$ $$x_1 + 2x_2 \leq 12 \quad \text{(Constraint on machine I)}$$

Similarly, the total time required on machine II and machine III for producing x_1 and x_2 items of M and N are $2x_1 + x_2$ and $x_1 + 1.25x_2$, respectively

But the time available on machine II for producing is at most 12 hours and the operated time at least one machine III is 5 hours

$\therefore$ $$2x_1 + x_2 \leq 12 \quad \text{(Constraint on machine II)}$$

and $$x_1 + 1.2x_2 \geq 5 \quad \text{(Constraint on machine III)}$$

Thus, the mathematical formulation of the given problem is as follows:

Maximize $Z = 600x_1 + 400x_2$

subject to the constraint

$$x_1 + 2x_2 \leq 12 \quad \ldots(i)$$

$$2x_1 + x_2 \leq 12 \quad \ldots(ii)$$

$$x_1 + \frac{5}{4}x_2 \geq 4 \quad \ldots(iii)$$

and $$x_1 \leq 0, x_2 \leq 0 \quad \ldots(iv)$$

Plot on a graph each constraint by first treating it as a linear equation. The use the inequality condition of each constraint to mark the feasible region as shown in Fig. 17.8.

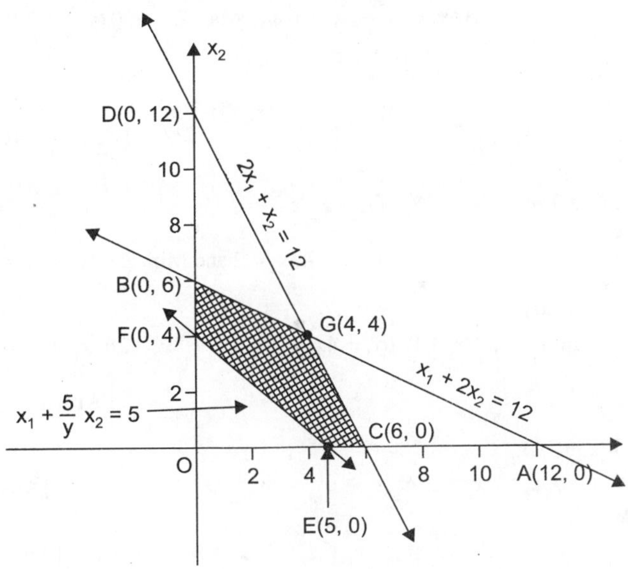

Fig. 17.8

Observe that the feasible region is bounded.

The coordinates of the corner points E, C, G, B and F of the feasible region $ECGBF$ are $E(5, 0)$, $C(6, 0)$, $B(0, 6)$ and $F(0, 4)$ respectively.

Note that the coordinates of G are obtained by solving the equation $x_1 + 2x_2 = 12$ and $2x_1 + x_2 = 12$. The value of the objective function at these points are given in the following table:

Extreme Point	*Coordinate* (x_1, x_2)	*Objective Function* $Z = 600x_1 + 400x_2$
E	(5, 0)	$600(5) + 400(0) = 3000$
C	(6, 0)	$600(6) + 400(0) = 3600$
G	(4, 4)	$600(4) + 400(4) = 4000$ ← Maximum
B	(0, 4)	$600(0) + 400(6) = 2400$
F	(0, 4)	$600(0) + 400(4) = 1600$

From the table, we find that Z is maximum at the $G(4, 4)$. Hence the manufacturer has to produce 4 units of each item M and N to get the maximum profit of ₹ 4,000.

EXERCISE 17.2

1. Explain the graphical method of solving a linear programming problem.
2. Explain how would you identify the cases of redundant constraint, no solution, multiple solution and unbounded solution from the graph of linear programming involving two variables. Give a rough sketch of each case.
3. Use graphical method to solve the following linear programming problems:

(*a*) Minimize $Z = 18x_1 + 10x_2$
subject to $4x_1 + x_2 \geq 20$, $2x_1 + 3x_2 \geq 30$, $x_1, x_2 \geq 0$ **[Ans.** $x_1 = 3$, $x_2 = 8$ min $Z = 134$]

(*b*) Maximize $Z = 15x_1 + 10x_2$
subject to $3x_1 + 2x_2 \leq 80$, $2x_1 + 3x_2 \leq 70$, $x_1, x_2 \geq 0$

$$\left[\textbf{Ans. } x_1 = \frac{80}{3}, x_2 = 0 \text{ and Max. } Z = 400 \text{ or } x_1 = 20, x_2 = 10 \text{ and Max. } Z = 400\right]$$

(*c*) Maximize $Z = 50x_1 + 30x_2$
subject to the constraints: $2x_1 + x_2 \geq 18$, $x_1 + x_2 \geq 12$, $3x_1 + 2x_2 \leq 34$, $x_1, x_2 \geq 0$
[Ans. $x_1 = 10$, $x_2 = 2$ and Max $Z = 560$]

(*d*) Minimize $Z = 2x_1 + 4x_2$
subject to $x_1 + 2x_2 \geq 8$, $x_1 + 4x_2 \geq 12$, $x_1 \geq 3$, $x_2 \geq 2$

$$\left[\textbf{Ans. } x_1 = 4, x_2 = 2 \text{ and Min. } Z = 16 \text{ or } x_1 = 3, x_2 = \frac{5}{2} \text{ and Min. } Z = 16\right]$$

(*e*) Minimize $Z = 30x_1 + 20x_2$
subject to the constraints: $x_1 + x_2 \leq 8$, $6x_1 + 4x_2 \geq 12$, $5x_1 + 8x_2 = 20$, $x_1, x_2 \geq 0$

$$\left[\textbf{Ans. } x_1 = \frac{4}{7}, x_2 = \frac{15}{7} \text{ and Min. } Z = 6\right]$$

(*f*) Maximize $Z = 10x_1 + 12x_2$
subject to $x_1 + x_2 \leq 60$, $x_1 - 2x_2 \geq 0$, $x_1, x_2 \geq 0$ **[Ans.** $x_1 = 2$, $x_2 = 0$ Min $Z = 2$]

(*g*) Minimize $Z = x_1 + 5x_2$
subject to the constraints: $x_1 + 4x_2 \leq 12$, $x_1 + x_2 \geq 2$, $x_1 \leq 8$, $x_1, x_2 \geq 0$ **[Ans.** $x_1 = 40$, $x_2 = 0$ Min $Z = 2$]

(*h*) Minimize $Z = 5x_1 + 3x_2$
subject to the constraints: $2x_1 + x_2 \geq 10$, $x_1 + 3x_2 \geq 15$, $x_1 \leq 10$, $x_2 \leq 8$, $x_1, x_2 \geq 0$
[Ans. $x_1 = 40$, $x_2 = 0$ Min $Z = 2$]

(*i*) Maximize $Z = 2x_1 + 3x_2$
subject to the constraints: $x_1 + x_2 \leq 30$, $x_2 \geq 3$, $0 \leq x_2 \leq 12$, $x_1 - x_2 \geq 0$ and $0 \leq x_1 \leq 20$
[Ans. $x_1 = 18$, $x_2 = 10$ Max. $Z = 72$]

(*j*) Maximize $Z = 40x_1 + 80x_2$
subject to the constraints: $2x_1 + 3x_2 \le 48, x_1 \le 15, 10x_1, x_1 \ge 0, x_2 \ge 0$
[**Ans.** $x_1 = 9, x_2 = 10$ Max. $Z = 1160$]

(*k*) Maximize $Z = 3x_1 + 4x_2$
subject to the constraints: $4x_1 + 2x_2 \le 80, 2x_1 + 5x_2 \le 180$ and $x_1, x_2 \ge 0$
$\left[\textbf{Ans. } x_1 = \frac{5}{2}, x_2 = 35 \text{ and Max. } Z = 147.5\right]$

(*l*) Minimize $Z = 4x_1 + 2x_2$
subject to the constraints: $x_1 + 2x_2 \ge 80, 2x_1 + 5x_2 \le 180$ and $x_1, x_2 \ge 0$
[**Ans.** $x_1 = 0.65, x_2 = 1.2$ and Min. $Z = 5$]

(*m*) Maximize $Z = 7x_1 + 3x_2$
subject to the constraints: $x_1 + 2x_2 \le 3, x_1 + x_2 \le 4, 0 \le x_1 \le \frac{5}{2}$ and $0 \le x_2 \le \frac{3}{2}$
[**Ans.** $x_1 = 2.5, x_2 = 1.5$ and Max. $Z = 22$]

(*n*) Minimize $Z = 3x_1 + 5x_2$
subject to the constraints: $x_1 + 2x_2 \le 2000, x_1 + x_2 \ge 1500, x_2 \le 600, x_1, x_2 \ge 0$
[**Ans.** $x_1 = 1000, x_2 = 500$ and Max. $Z = 5500$]

(*o*) Minimize $Z = 20x_1 + 30x_2$
subject to the constraints: $x_1 + 2x_2 \le 40, 3x_1 + x_2 \ge 30, 4x_1 + 3x_2 \ge 60, x_1, x_2 \ge 0$
[**Ans.** $x_1 = 15, x_2 = 0$ and Min. $Z = 300$]

4. Verify that the following problems have no feasible solution

(*a*) Maximize $Z = 15x_1 + 20x_2$
subject to the constraints: $x_1 + x_2 \ge 12, 6x_1 + 9x_2 \le 54, 15x_1 + 10x_2 \ge 90, x_1, x_2 \ge 0$

(*b*) Maximize $Z = 4x_1 + 3x_2$
subject to the constraints: $x_1 + x_2 \le 3, 2x_1 - x_2 \le 3, x_1 \ge 4, x_1, x_2 \ge 0$

5. Verify that the following problems have multiple solutions

(*a*) Maximize $Z = 6x_1 + 4x_2$
subject to the constraints: $x_1 + x_2 \le 5, 3x_1 + 2x_2 \le 12, x_1, x_2 \ge 0$

(*b*) Minimize $Z = 12x_1 + 9x_2$
subject to the constraints: $3x_1 + 6x_2 \ge 36, 4x_1 + 3x_2 \ge 24, x_1 + x_2 \le 15, x_1, x_2 \ge 0$

6. Verify that the following problem has an unbounded solution
Maximize $Z = 20x_1 + 30x_2$
subject to the constraints: $5x_1 + 2x_2 \ge 50, 2x_1 + 6x_2 \ge 20, 4x_1 + 3x_2 \ge 60, x_1, x_2 \ge 0$

7. Graph the feasible region of the following problem and identify a redundant constraint
Maximize $Z = 20x_1 + 30x_2$
subject to the constraints: $5x_1 + 2x_2 \ge 50, 2x_1 + 6x_2 \ge 20, 4x_1 + 3x_2 \ge 60, x_1, x_2 \ge 0$
[**Ans.** The constraint $4x_1 + 5x_2 \le 20$ is redundant]

8. An automobile manufacturer makes automobiles and trucks in a factory that is divided into two shops. Shop I, which perform the basic assembly operation, must work for 5 man-days on each truck but only 2 man-days on each automobile. Shop II which performs finishing operation must work 3 man-days for each automobile or truck that is produced. Because of man and machine limitations, shop I has 180 man-days per week available while shop II has 135 man-days per week. If the manufacturer makes a profit of ₹ 3000 on each truck and ₹ 2000 on each automobile, find graphically how many of each should be produced to maximize his profit.
[**Ans.** $x_1 = 15, x_2 = 30$ Max. $Z = 1,20,000$]

9. A soft drink firm has two bottling plants, one located at P and other at Q. Each plant produces three different soft drinks, *A, B* and *C* the capacities of two plants in number of bottles per day, are as follows:

Products	*Plants*	
	P	*Q*
A	3000	1000
B	1000	1000
C	2000	6000

10. A firm plants to purchase at least 200 quintals of scrap containing high quality metal *X* and low quality metal. *Y*. It decides that the scrap to be purchased must contain atleast 100 quintals of *X*-metal and no more than 35 quintals of *Y*-metal. The firm can purchase the scrap from two suppliers (*A* and *B*) in unlimited quantities. The percentage of X and Y metals in terms of weight in the scraps supplied by A and B is given below:

Metals	*Supplier A*	*Supplier B*
X	25%	75%
Y	10%	20%

The price of A's scrap is ₹ 200 per quintal and that of B's ₹ 400 per quintal. The firms wants to determine the quantities that it should buy from the two suppliers so that the total cost is minimized. Using linear programming techniques, solve it by graphic method.

[**Ans.** Minimize $Z = 200x_1 + 400x_2$ subject to: $x_1 + 3x_2 \geq 400$, $x_1 + x_2 \geq 200 + 2x_2 \leq 350$
$x_1, x_2 \leq 0$, $x_1 = 100$ and Min. $Z =$ ₹ 60,000]

❑❑❑

Index

J

L

M

N

O

P

Q

R

S

T

U

V

Z

❑❑❑

Win Prizes

ATTENTION READERS

Win Prizes

We request for your frank assessment regarding various aspects of the book as given below:

APPLIED MATHEMATICS–II

BY: Dr. Hari Arora; Dr. Anjali Naithani

Please Fill up the Sheet in Neat and Clean Writing

(i) What topics of your syllabus, important for your examination point of view, are not covered in this book ?

(ii) What are the chapters wherein the treatment of the subject matter is not systematic or organised or updated ?

(iii) Have you come across misprints/mistakes/factual inaccuracies in the book ? Please specify the chapters and the page numbers.

(iv) What other books on the same subject have you found/heard better than the present book ? Please specify in terms of price and quality.

(v) Further suggestions and comments for the improvement of this book.

(PS : If need be please attach a separate sheet to write your views)

OTHER DETAILS

(i) Who recommended you this book ? Please tick (✓) in the box near the option relevant to you.

☐ Teacher ☐ Friends ☐ Bookseller

....

....

....

....

(ii) Name of the recommending teacher, his designation and address:

....

....

....

....

(iii) Name and address of the bookseller you purchased the book from :

....

....

....

....

(iv) Name and address of your institution (Please mention the University or Board as the case may be)

....

....

....

....

(v) Your name and complete postal address:

....

....

....

....

(vi) Course you are enrolled in:

....

....

....

....

The best assessment will be awarded each month : The award will be in the form of our publications as decided by the Editorial Board.

Please mail the filled up coupon at your earliest to :

S.K. KATARIA & SONS®

4885/109, Prakash Mahal, Dr. Subhash Bhargav Lane,

Opposite Delhi Medical Association, Daryaganj, New Delhi-110002 (INDIA)

Phone: +91-11-23243489, 43551243; Mobile: +91-9871775858

e-mail: katariabook@gmail.com; katariabooks@yahoo.com

Website: www.skkatariaandsons.com